Naturally Occurring Acetylenes

Naturally Occurring Acetylenes

F. Bohlmann
T. Burkhardt
C. Zdero

Institute of Organic Chemistry
Technical University
Berlin, Germany

1973

ACADEMIC PRESS · London and New York

ACADEMIC PRESS INC. (LONDON) LTD.
24/28 Oval Road,
London NW1 7DX

United States Edition published by
ACADEMIC PRESS INC.
111 Fifth Avenue
New York, New York 10003

Library of Congress Catalog Card Number: 72-84348
ISBN: 0 12 111150 4

PRINTED IN GREAT BRITAIN BY
J. W. ARROWSMITH LTD., BRISTOL

Preface

Although several reviews of natural acetylenes have already been published, development in this field has been so rapid that an up to date and more detailed treatment of this large class of natural products now seems to be desirable.

While the majority of these compounds has been isolated from the family Compositae within the last ten years, the occurrence of acetylenes has been recognized in many other families, as well as in marine organisms. The number of known natural acetylenes has now increased to more than 650. Since the biogenesis has already been established for many compounds, the arrangement of this book is mainly along biogenetical lines. In addition the chemotaxonomical viewpoints are covered in detail in Chapters 6 and 7.

Technical University, Berlin
November, 1972

F. BOHLMANN
T. BURKHARDT
C. ZDERO

Contents

3

4

5

6

1

Introduction

1. Historical Background

By 1826[1], crystals had already been isolated from a Japanese essential oil of what we now know to be the polyacetylenic compound called dehydromatricaria ester (**1**). Then, in 1892[2], the first naturally occurring acetylene was discovered by A. Arnaud from the seed oil of *Picramnia tariri* DC. Ten years later, Arnaud[3] assigned the correct structure of the so-called tariric acid (**2**). His work was contemporary with the early work on carlina oxide (3)[4], his structure being elucidated only half a century later[5]. This compound was isolated from the essential oil of the Carlina thistle, *Carlina acaulis* L. F. Semmler[4] proposed two possible formulae, but he preferred the incorrect allenic structure (**4**). Influenced mainly by A. V. Bayer's work on some highly unstable acetylinic compounds, he argued: "Es ist von Hause aus sehr unwahrscheinlich, dass eine acetylenartige Verbindung vorliegt . . .".

Nearly 50 years later H. Gilman and A. Pfau and their coworkers[5] were able to establish the structure of the furane derivative as **3** by synthesis and spectral methods.

$$H_3C-(C\equiv C)_3-CH=CH-CO_2CH_3 \quad \mathbf{1}$$

$$H_3C-(CH_2)_{10}-C\equiv C-(CH_2)_4-CO_2H \quad \mathbf{2}$$

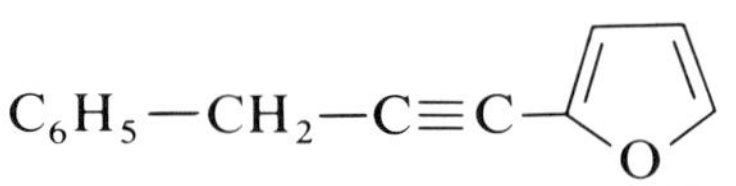

3

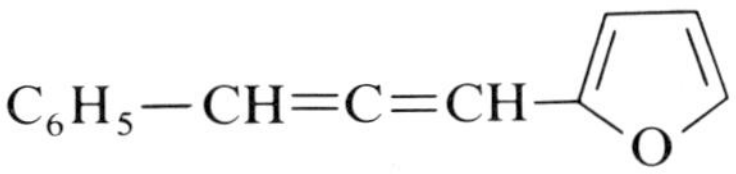

4

At the same time Russian workers[6] characterized the first natural diacetylene, the lachnophyllum ester (**5**), isolated from the essential oil of *Lachnophyllum gossypinum* Bge.—a member of Aserbeidschan Compositae. During their investigations they tried to distill the ester, which caused a violent explosion. In order to ascertain the reason for this, they elucidated the structure by classical methods.

Some years later N. A. Sörensen and his co-workers[7] isolated a further diacetylenic ester from the scentless mayweed—*Matricaria inodora* L.—the widespread matricaria ester (**6**). It must be credited to the Norwegian school at Trondheim that we now know that acetylenic compounds are by no means rare in common garden and wild plants of the large Compositae ("daisy") family. At Trondheim, they were able to show that these compounds, and some others, could be isolated from different members of this family.

$$H_3C{-}CH_2{-}CH_2{-}(C{\equiv}C)_2CH{=}CH{-}CO_2CH_3 \quad \mathbf{5}$$

cis

lachnophyllum ester

$$H_3C{-}CH{=}CH{-}(C{\equiv}C)_2{-}CH{=}CH{-}CO_2CH_3 \quad \mathbf{6}$$

cis *cis*

matricaria ester

However, only when new methods of structure determination became available did the natural acetylenes give rise to a new large group of natural products. Around 1950, E. R. H. Jones and his coworkers at Manchester and F. Bohlmann and his coworkers at Brunswick started their work on synthesis and spectroscopy of polyacetylenes.

Knowing the spectral regularities, the number of known new acetylenes increased rapidly in the following years. With the use of n.m.r. and mass spectroscopy the number of natural acetylenes is now more than 650, isolated from about 15 different families of higher plants as well as from algae and micro-organisms.

Several reviews on this subject have already been written[8–13], but in the last five years our knowledge of natural acetylenes has produced a detailed summary of the many different structures and of the biogenesis in this class of natural products.

It seems to be reasonable to follow biogenic considerations in the arrangement of this book.

2. Isolation

Most of the acetylenes, especially polyacetylenes, are thermally unstable. This instability is most marked in condensed phases—especially in the crystalline state where we have the typical phenomenon of photodecomposition. Many polacetylenes exposed to daylight develop quite striking colours. The resulting products, which may be red or deep blue, are insoluble solids with graphite-like structures. If bulky terminal groups are present, polyacetylenes are much more stable. For example, the entire series of di-tert. butylpolyynes compounds, with up to 12 conjugated triple bonds, has been prepared[14,15]. Thus, in the condensed phase the stability depends on the distance of parallel orientated polyacetylene chains. Knowing these properties it follows that only careful methods of isolation can be used.

In the beginning, the first natural acetylenes were isolated mainly by steam distillation using the methods of essential oil production; but this method is clearly unsuitable for the more unstable as well as the polar compounds. Generally, careful solvent extraction at room temperature is preferred. In higher plants especially, the acetylenes are normally concentrated in "oil channels". Therefore, extractions with ether–petroleum ether mixtures are sufficient even for highly polar compounds. It is only fungal materials that sometimes require extraction by more polar solvents. Often fungal metabolites are accumulated in the aqueous culture media and are extractable by different solvents, depending on the polarity of the special compounds.

Normally, the extracts are first separated by bulk chromatography to obtain fractions with different polarity (e.g. hydrocarbons, carbonyl compounds, alcohols). Usually aluminia or silica preparations are used in column chromatography as well as in thin-layer chromatography. (TLC). Repeated column chromatography and finally TLC are usually successful in separation and isolation of the acetylenes. But sometimes partial chemical reactions (manganese dioxide oxidation, sodium borohydride reduction) and partial saponification are necessary to separate highly-complex mixtures of acetylenes, especially if they are co-occurring with other compounds. Highly-polar acetylenes have been most successfully separated and purified by counter-current distribution, often as their methyl esters.

Fractional crystallisation, even at low temperature, is of very limited use since it may lead to considerable losses due to the previously mentioned instability, especially in the crystalline state.

Gas chromatography (GLC) is not very useful. Only the more simple unpolar samples can be identified by this method. Even then, isolation from crude extracts are unsuccessful as the concentrations are normally too low.

3. Structure Determination

3.1. Spectroscopic Methods

Structure elucidation methods in the field of natural acetylenes are mostly spectroscopic. Most important are the electronic spectra. Simple polyacetylenes and also polyynenes give highly characteristic u.v. spectra with very sharp vibrational fine structure. The band spacing is characteristically about 2200 cm as the ground state stretching frequency of the triple bond, and therefore gives a wider spacing than polyene chromophors (1500–1600 cm).

Normally we can observe two sets of bands, the longer-wavelength-series of lower intensity and the shorter wavelength-series with very high intensity (ε up to 10^6). All combinations of conjugated triple and double bonds give special types of u.v. spectra. In Table I and in Figs. 1 to 31 typical spectra are combined.

TABLE I

U.V. MAXIMA OF POLYYN-ENES

$R-[CH=CH]_n-[C\equiv C]_{n'}-[CH=CH]_{n''}-R$ in nm ($\varepsilon \times 10^{-3}$)

n	n′	n″	Longer wavelengths group				Lower wavelengths group		
2	1	1	306(28)	290(31)					
2	1	2	338(55)	316(52)					
—	2	—	255(0·2)	240(0·4)	230(0·3)				
1	2	—	280(15)	264(18)	251(13)	238(6)	210(42)		
1	2	1	311(17)	292(20)	276(14)	261(8)	244(23)	235(30)	230(31)
2	2	—	309(28)	294(33)			236(40)	226(24)	
2	2	1	335(26)	313(36)	294(26)	277(14)	264(26)	249(29)	
3	2	—	332(63)	315(61)			250(22)	242(12)	
3	2	1	356(40)	330(51)	313(35)		274(24)	263(28)	
—	3	—	307(0·1)	286(0·2)	269(0·2)	255(0·2)	210(150)		
1	3	—	328(14)	307(20)	288(15)	271(8)	242(110)	231(70)	
1	3	1	353(18)	328(25)	308(19)	289(10)	267(57)	254(70)	245(65)
2	3	—	347(33)	325(41)	305(26)	288(14)	269(115)	259(58)	
2	3	1	369(29)	342(39)	320(30)		289(81)	273(96)	242(66)
3	3	—	356(47)	333(56)	312(31)		283(92)	271(47)	
—	4	—	355(0·1)	330(0·2)	308(0·2)	287(0·2)	233(275)	223(200)	211(90)
1	4	—	375(7)	348(11)	324(9)	303(6)	272(155)	258(108)	239(81)
1	4	1	391(9)	362(15)	337(12)	315(8)	287(86)	271(106)	258(81)
2	4	—	382	356	331	311	290	273	
—	5	—	392(0·1)	362(0·2)	338(0·2)	312(0·2)	262(350)	249(250)	236(110)
1	5	—	410	378	349	327	285	265	

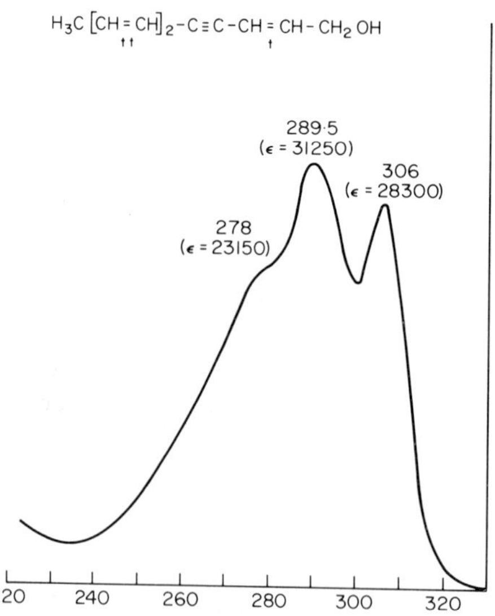

FIG. 1.

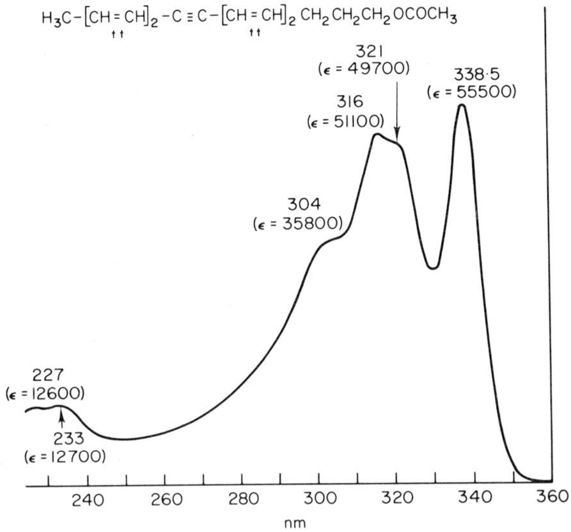
$H_3C-[CH=CH]_2-C\equiv C-[CH=CH]_2CH_2CH_2CH_2OCOCH_3$
321
(ε = 49700)
338·5
(ε = 55500)
316
(ε = 51100)
304
(ε = 35800)
227
(ε = 12600)
233
(ε = 12700)
240
260
280
300
320
340
360
nm

FIG. 2.

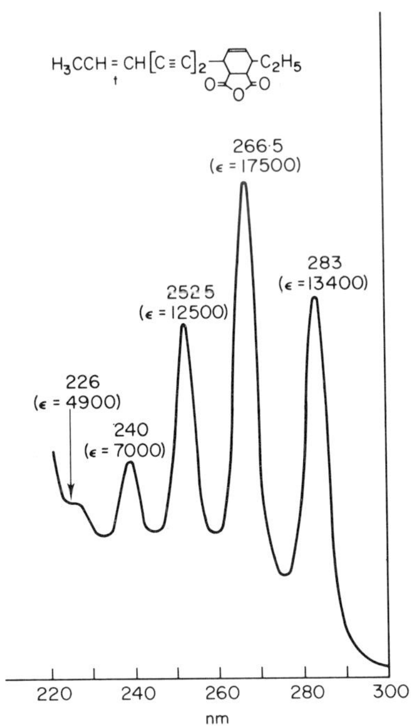
$H_3CCH=CH[C\equiv C]_2$ C_2H_5
266·5
(ε = 17500)
283
(ε = 13400)
252·5
(ε = 12500)
226
(ε = 4900)
240
(ε = 7000)
220
240
260
280
300
nm

FIG. 3.

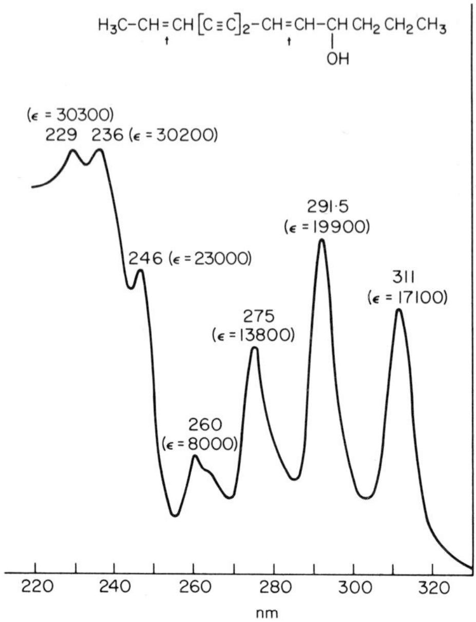

H3C-CH=CH[C≡C]2-CH=CH-CH CH2 CH2 CH3
OH
(ε = 30300)
229 236 (ε = 30200)
246 (ε = 23000)
260 (ε = 8000)
275 (ε = 13800)
291·5 (ε = 19900)
311 (ε = 17100)
220 240 260 280 300 320
nm

FIG. 4.

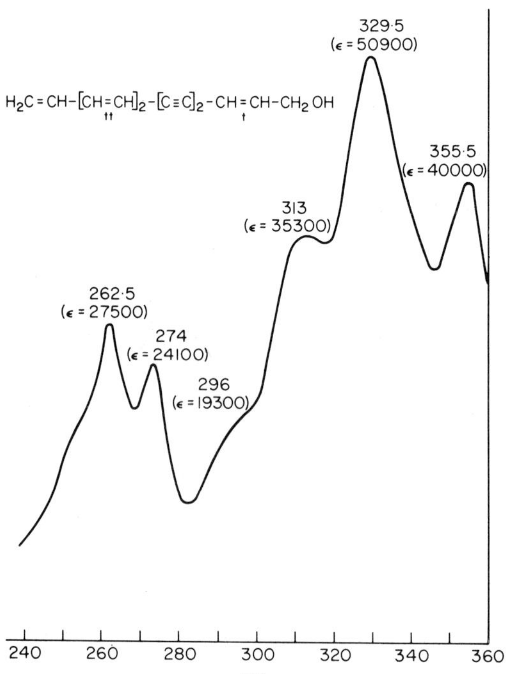

H2C=CH-[CH=CH]2-[C≡C]2-CH=CH-CH2 OH
262·5 (ε = 27500)
274 (ε = 24100)
296 (ε = 19300)
313 (ε = 35300)
329·5 (ε = 50900)
355·5 (ε = 40000)
240 260 280 300 320 340 360
nm

FIG. 5.

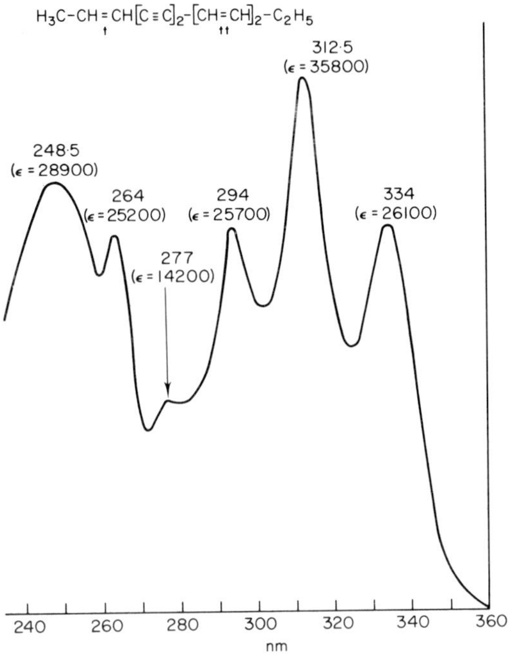
H3C-CH=CH[C≡C]2-[CH=CH]2-C2H5
312·5
(ε = 35800)
248·5
(ε = 28900)
264
(ε = 25200)
294
(ε = 25700)
334
(ε = 26100)
277
(ε = 14200)
240
260
280
300
320
340
360
nm

FIG. 6.

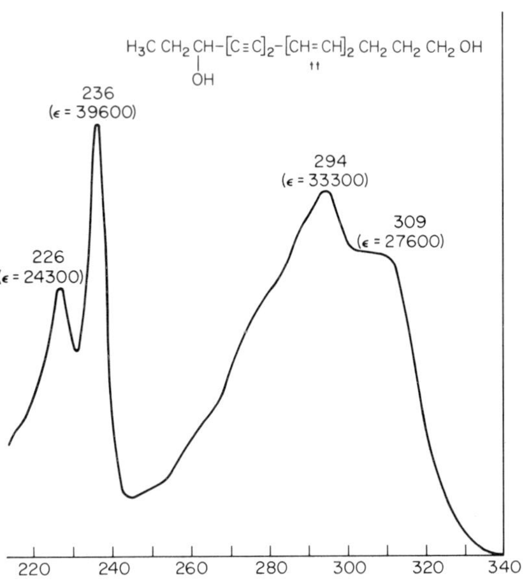
H3C CH2 CH-[C≡C]2-[CH=CH]2 CH2 CH2 CH2 OH
OH
236
(ε = 39600)
294
(ε = 33300)
309
(ε = 27600)
226
(ε = 24300)
220
240
260
280
300
320
340
nm

FIG. 7.

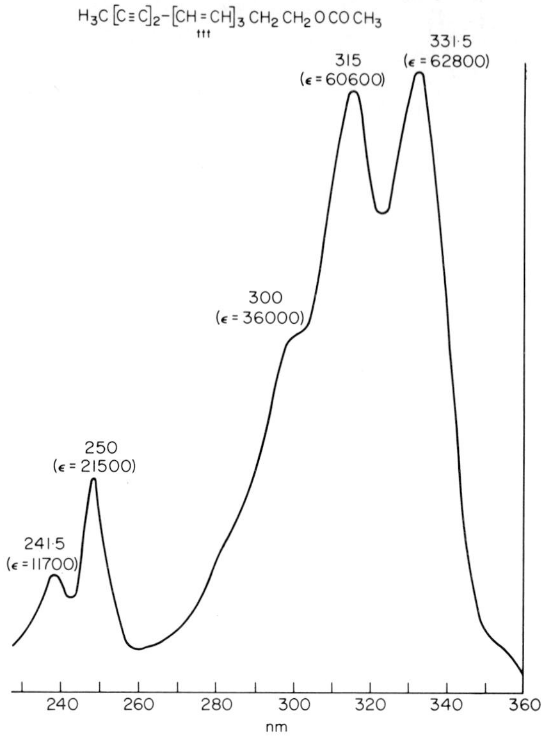

H3C [C≡C]2–[CH=CH]3 CH2 CH2 O CO CH3
ttt
331·5
(ε = 62800)
315
(ε = 60600)
300
(ε = 36000)
250
(ε = 21500)
241·5
(ε = 11700)
240
260
280
300
320
340
360
nm

FIG. 8.

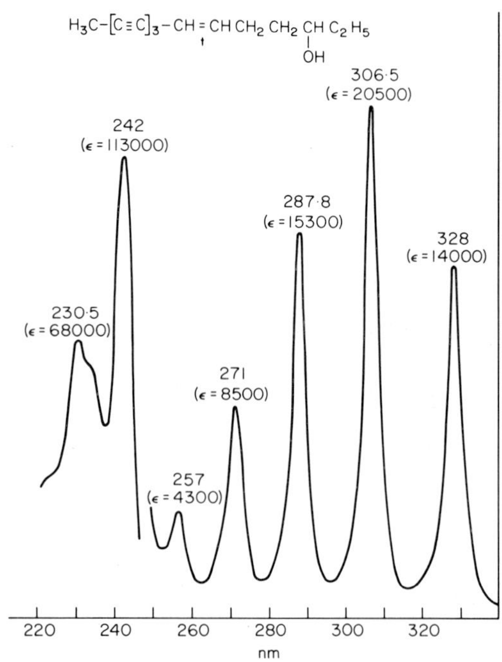

H3C–[C≡C]3–CH=CH CH2 CH2 CH C2 H5
t
OH
306·5
(ε = 20500)
242
(ε = 113000)
287·8
(ε = 15300)
328
(ε = 14000)
230·5
(ε = 68000)
271
(ε = 8500)
257
(ε = 4300)
220
240
260
280
300
320
nm

FIG. 9.

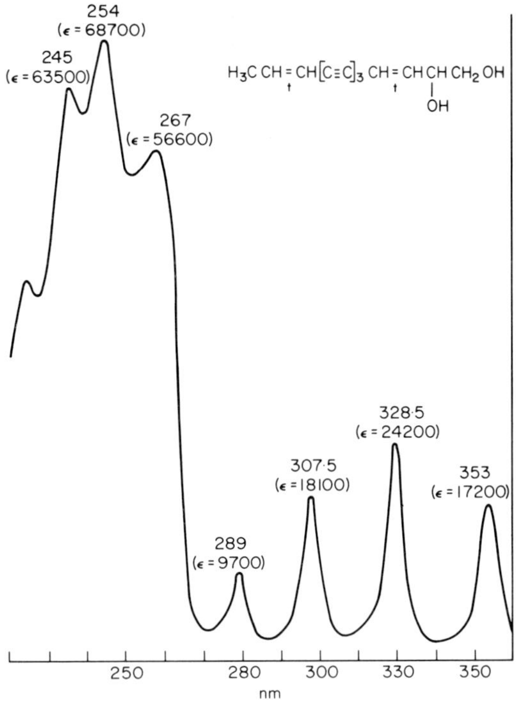
254
(ε = 68700)
245
(ε = 63500)
267
(ε = 56600)
H3C CH = CH[C≡C]3 CH = CH CH CH2 OH
OH
328·5
(ε = 24200)
307·5
(ε = 18100)
353
(ε = 17200)
289
(ε = 9700)
250
280
300
330
350
nm

FIG. 10.

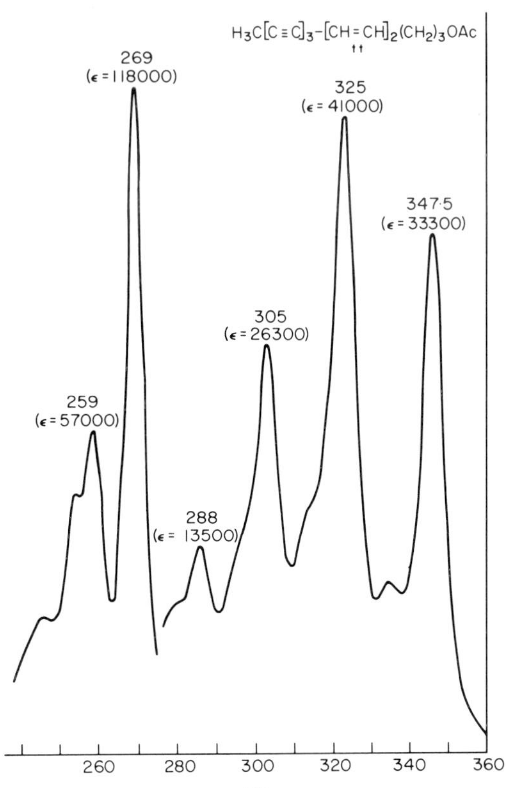
H3C[C≡C]3–[CH = CH]2(CH2)3OAc
269
(ε = 118000)
325
(ε = 41000)
347·5
(ε = 33300)
305
(ε = 26300)
259
(ε = 57000)
288
(ε = 13500)
260
280
300
320
340
360
nm

FIG. 11.

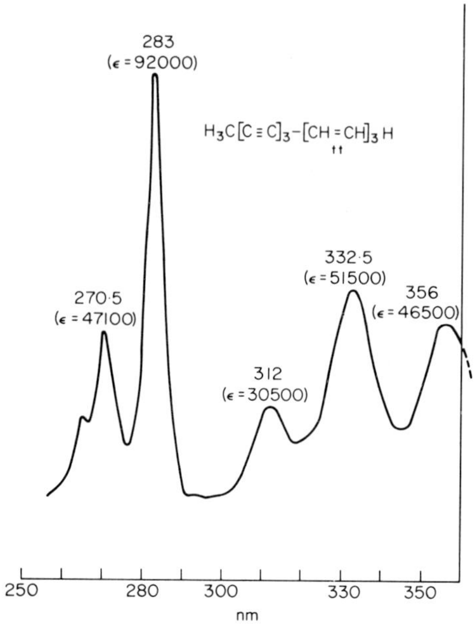

FIG. 12.

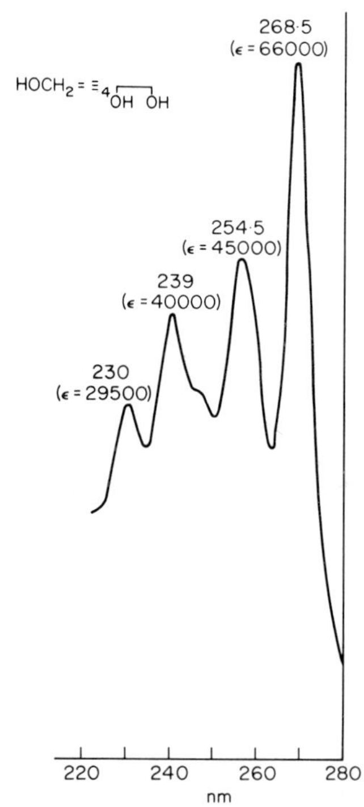

FIG. 13A.

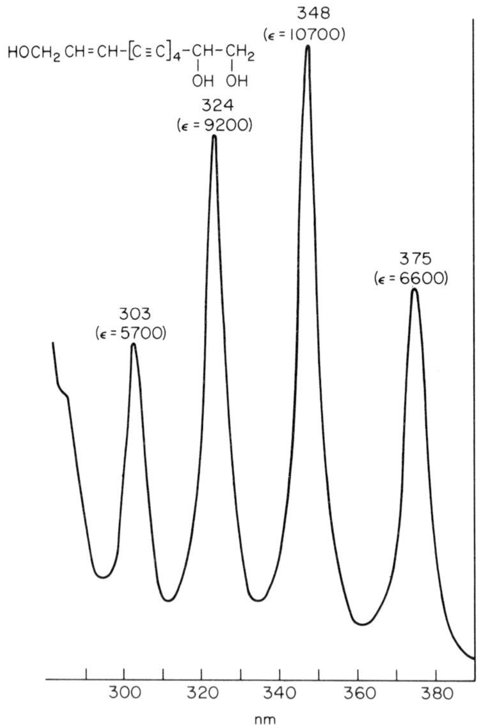

FIG. 13B.

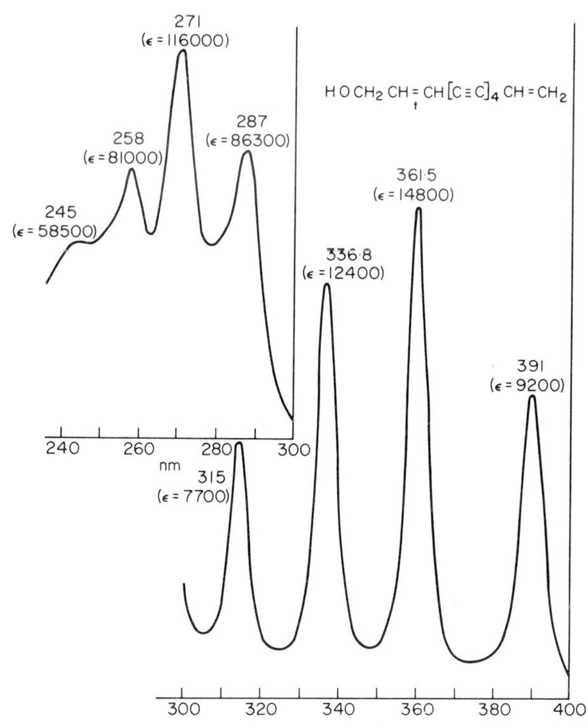

FIG. 14.

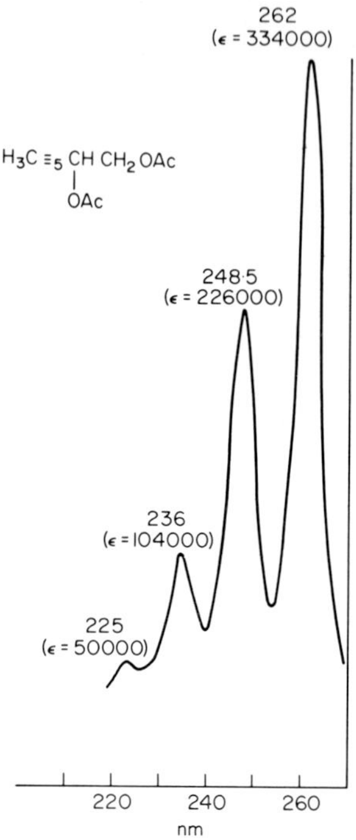

FIG. 15A.

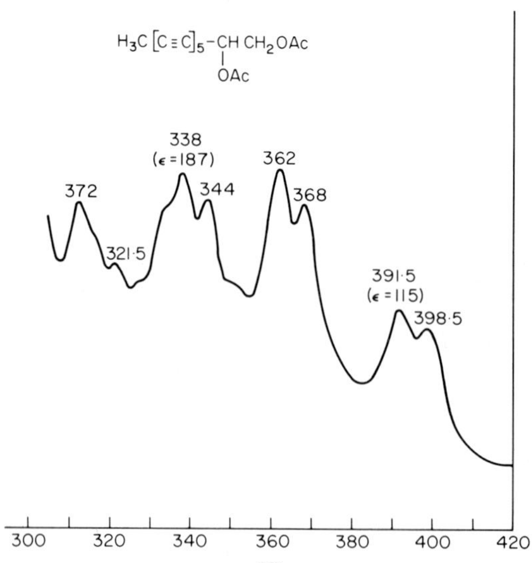

FIG. 15B.

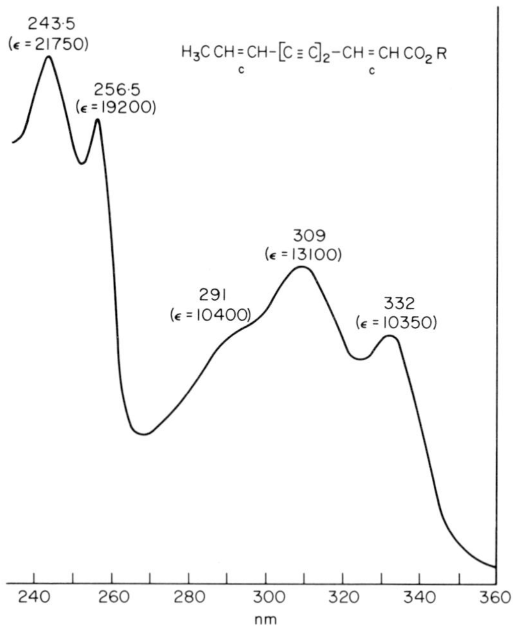
243·5
(ε = 21750)
H3C CH = CH-[C ≡ C]2-CH = CH CO2 R
c
c
256·5
(ε = 19200)
309
(ε = 13100)
291
(ε = 10400)
332
(ε = 10350)
240
260
280
300
320
340
360
nm

Fig. 16.

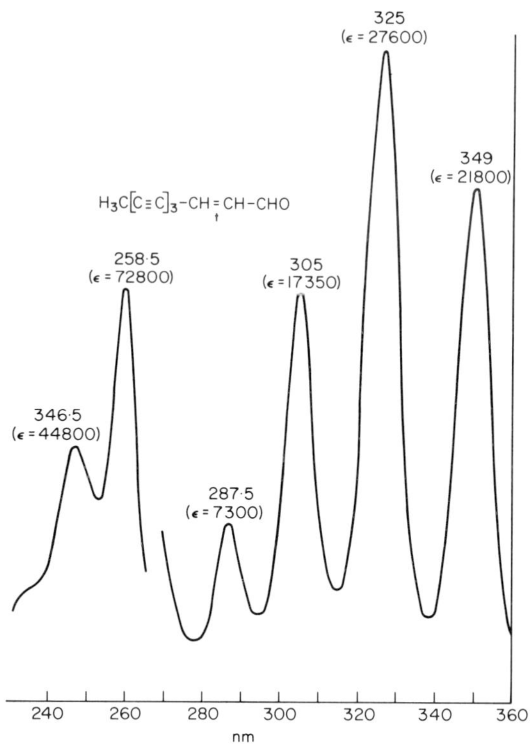
325
(ε = 27600)
349
(ε = 21800)
H3C[C≡C]3-CH = CH-CHO
t
258·5
(ε = 72800)
305
(ε = 17350)
346·5
(ε = 44800)
287·5
(ε = 7300)
240
260
280
300
320
340
360
nm

Fig. 17.

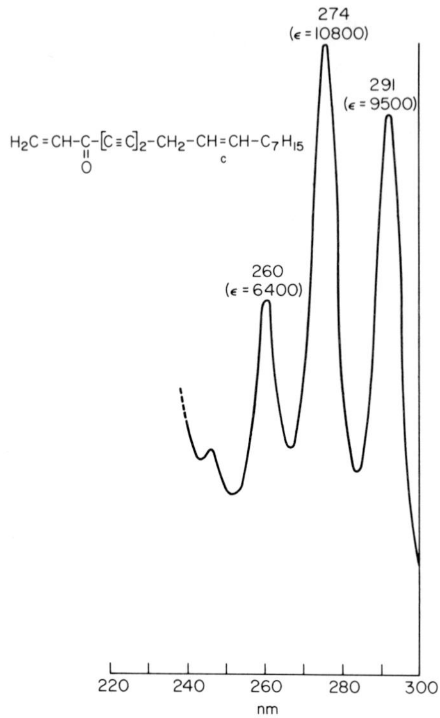

274
(ε = 10800)
291
(ε = 9500)
H2C = CH-C-[C ≡ C]2-CH2-CH = CH-C7H15
O
c
260
(ε = 6400)
220
240
260
280
300
nm

FIG. 18.

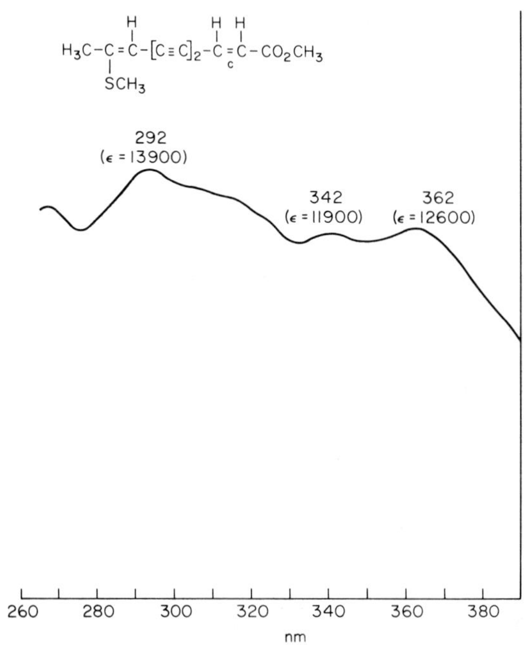

H
H H
H3C-C = C-[C ≡ C]2-C = C-CO2CH3
c
SCH3
292
(ε = 13900)
342
(ε = 11900)
362
(ε = 12600)
260
280
300
320
340
360
380
nm

FIG. 19.

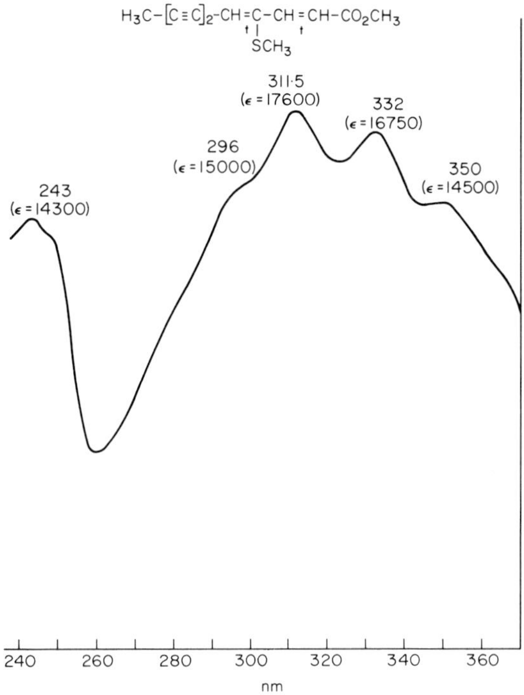

H3C–[C≡C]2–CH=C–CH=CH–CO2CH3
SCH3
311·5
(ε = 17600)
332
(ε = 16750)
296
(ε = 15000)
350
(ε = 14500)
243
(ε = 14300)
240
260
280
300
320
340
360
nm

FIG. 20.

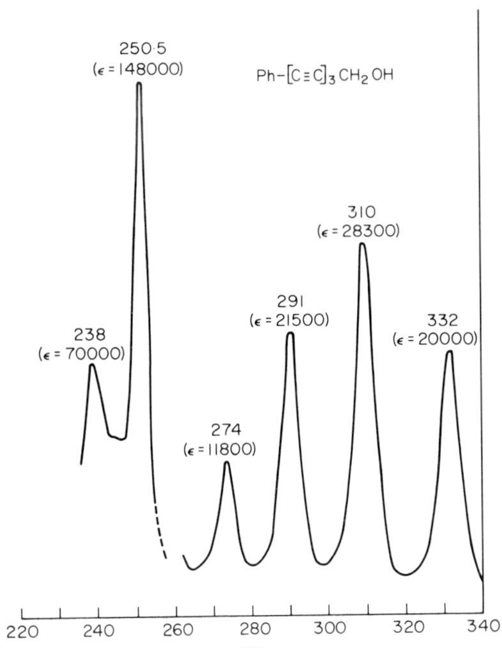

250·5
(ε = 148000)
Ph–[C≡C]3CH2OH
310
(ε = 28300)
291
(ε = 21500)
332
(ε = 20000)
238
(ε = 70000)
274
(ε = 11800)
220
240
260
280
300
320
340
nm

FIG. 21.

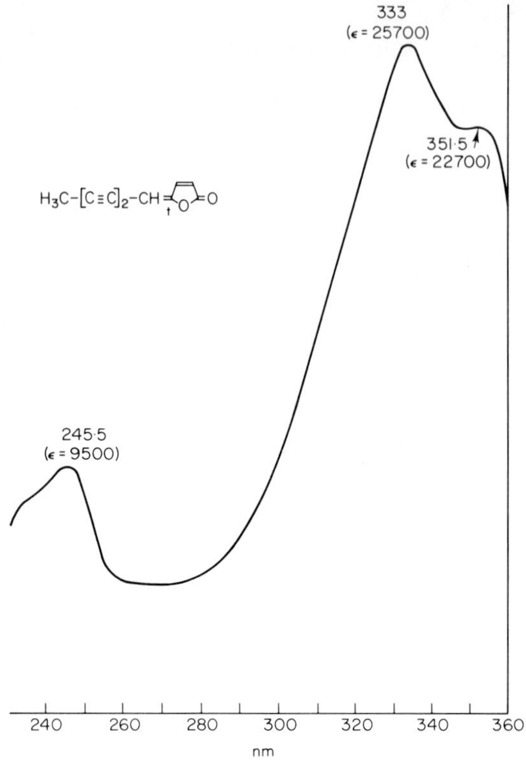
333
(ε = 25700)
351·5
(ε = 22700)
$H_3C-[C\equiv C]_2-CH=$
245·5
(ε = 9500)
240
260
280
300
320
340
360
nm

FIG. 22.

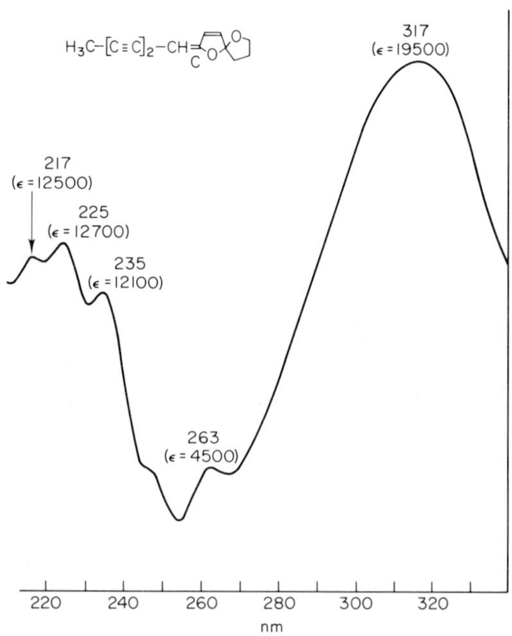
$H_3C-[C\equiv C]_2-CH=$
317
(ε = 19500)
217
(ε = 12500)
225
(ε = 12700)
235
(ε = 12100)
263
(ε = 4500)
220
240
260
280
300
320
nm

FIG. 23.

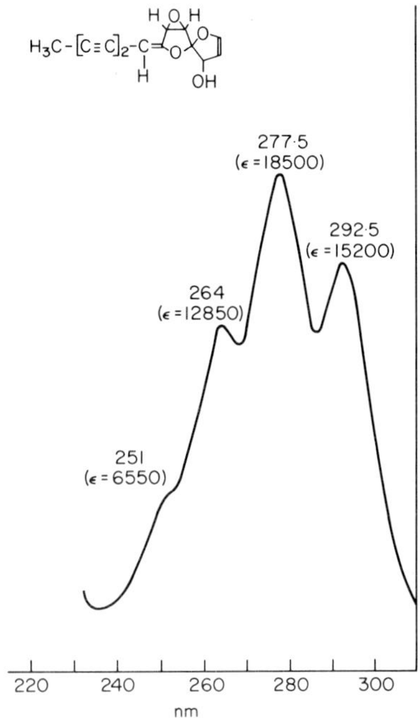
277·5
(ε = 18500)
292·5
(ε = 15200)
264
(ε = 12850)
251
(ε = 6550)
220
240
260
280
300
nm

FIG. 24.

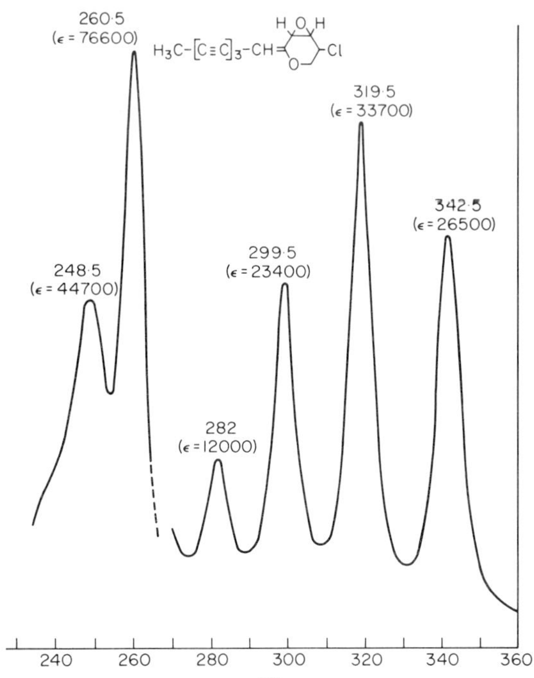
260·5
(ε = 76600)
319·5
(ε = 33700)
342·5
(ε = 26500)
299·5
(ε = 23400)
248·5
(ε = 44700)
282
(ε = 12000)
240
260
280
300
320
340
360
nm

FIG. 25.

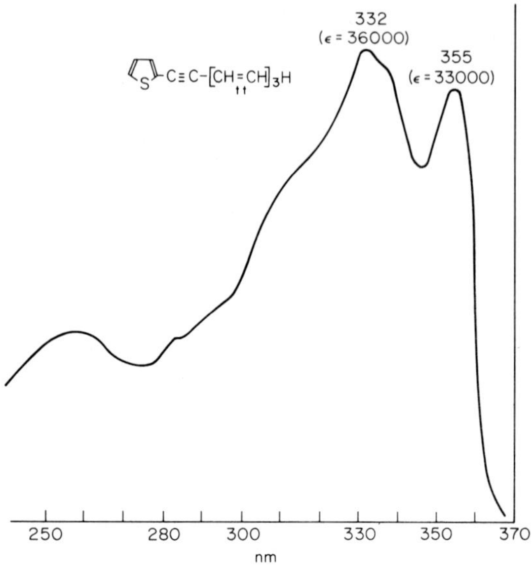
332
(ε = 36000)
355
(ε = 33000)
250
280
300
330
350
370
nm

FIG. 26.

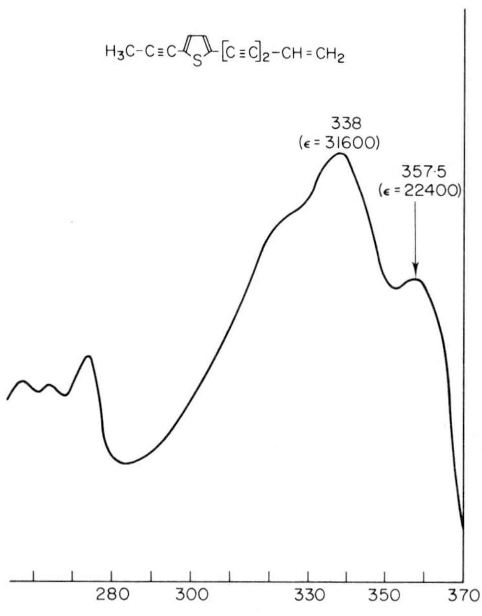
H3C-C≡C-⟨S⟩-[C≡C]2-CH=CH2
338
(ε = 31600)
357·5
(ε = 22400)
280
300
330
350
370
nm

FIG. 27.

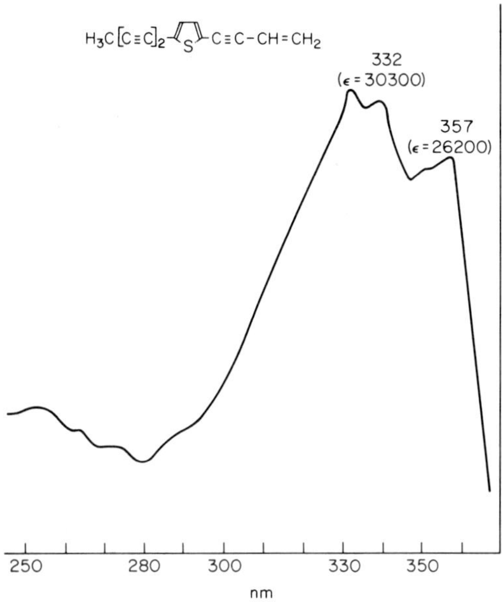
H3C[C≡C]2-⟨S⟩-C≡C-CH=CH2
332
(ε = 30300)
357
(ε = 26200)
250
280
300
330
350
nm

FIG. 28.

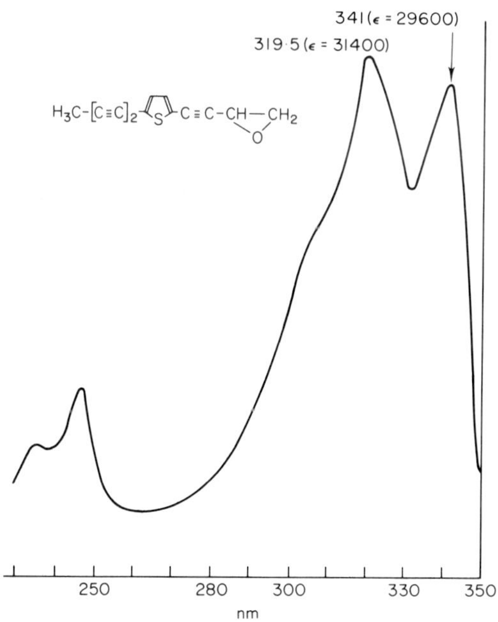
341 (ε = 29600)
319·5 (ε = 31400)
H3C-[C≡C]2-⟨S⟩-C≡C-CH—CH2
O
250
280
300
330
350
nm

FIG. 29.

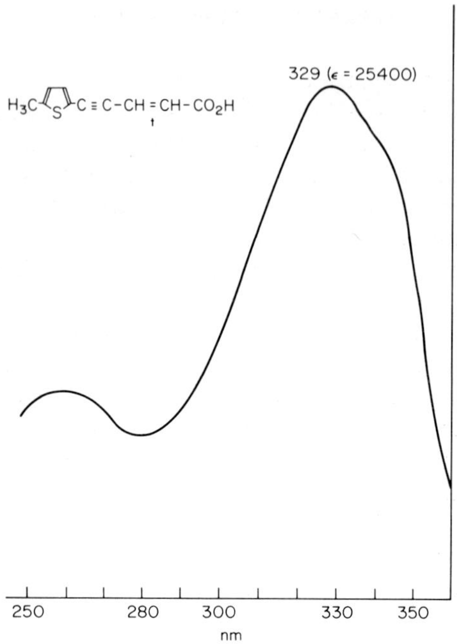
329 (ε = 25400)
H3C–C≡C–CH=CH–CO2H
250
280
300
330
350
nm

FIG. 30.

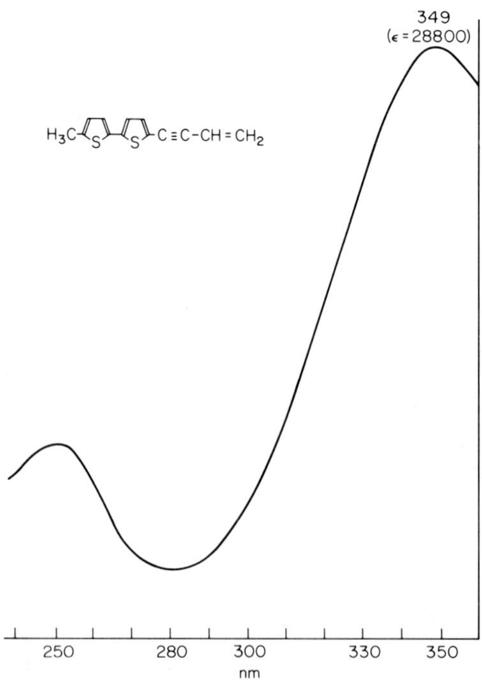
349
(ε = 28800)
H3C–C≡C–CH=CH2
250
280
300
330
350
nm

FIG. 31.

Not only are the wavelength of the observed maxima important, but the relative intensities of the two band series are also useful diagnostic features. Figures 1 to 15 show the u.v. spectra of simple polyynenes with different numbers and arrangements of triple and double bonds, while Figs 16 to 18 show the influence of carbonyl conjugation. Figures 19 and 20 are examples of sulphur conjugation effects, while Figs 21 to 31 represent spectra indicating the influence of conjugation with ring systems. The longer wavelength series is increased if double bonds are in conjugation with the polyyne chain.

Ethylenic groups have a larger effect than carbonyl groups, though both types of unsaturation give very similar position of the maxima. The type of spectrum is also influenced by the special grouping of triple and double bonds. If we compare the u.v. spectra of an ene-triyn-ene and of a triyne-diene, the differences are very clear, as shown in Figs 10 and 11.

More complicated endgroups in conjugation with the polyyne chain unfortunately lead to less characteristic u.v. spectra. For example, the widespread spiroketalenolether of type **7** shows a u.v. spectrum with maximum at 318 nm without fine structure (see Fig. 23). The same effect can be observed, if we have a thiophene ring in the chromophoric system. Compound **8** only has a broad maximum without any typical separation of the bands (see Fig. 28):

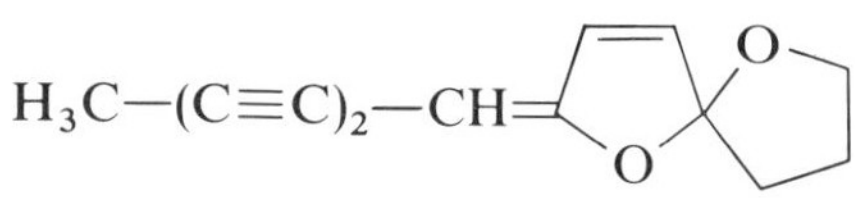

7

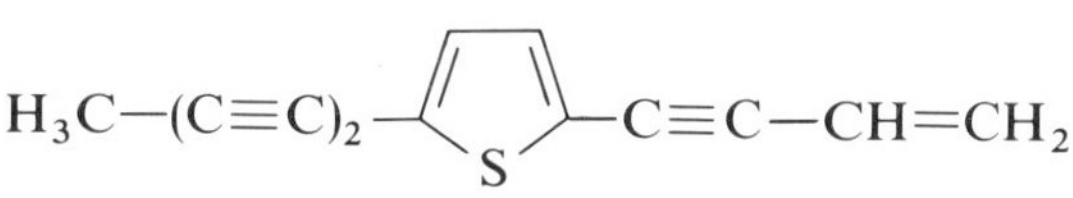

8

Several other combinations give the same effect (see, e.g. Figs 19, 20, 22 and 26 to 31). Nevertheless, u.v. spectroscopy is a very important and sensitive method in the detection of polyacetylenes, even in crude extracts. Only simple acetylenes, di- and tri-acetylenes without further conjugation, are difficult to detect by the method. Even the intense maximum of a triacetylene is often hidden by other chromophores because of its short wavelength position (*ca.* 210 nm).

U.v. spectra are also important in quantitative analysis to estimate concentrations as normal weighing procedures are impossible due to the instability of many of those compounds.

The use of infrared spectroscopy (i.r.) in the field of natural acetylenes is not as important as u.v. Clearly, the usual detection of functional groups (e.g. hydroxyl-, different kind of carbonyl-, nitrile, amide-, and enolethers) is also useful. The acetylenic stretching frequency is usually weak, but if carbonylgroups are in conjugation with the acetylene linkage, this band, which occurs around 220 cm, is of high intensity, as shown in Fig. 32. The intensity is sometimes an important tool in establishing the position of the triple bond in polyacetylenic carbonyl compounds. The nearer the triple bond is located to the carbonyl group, the stronger is the —C≡C— stretching band.

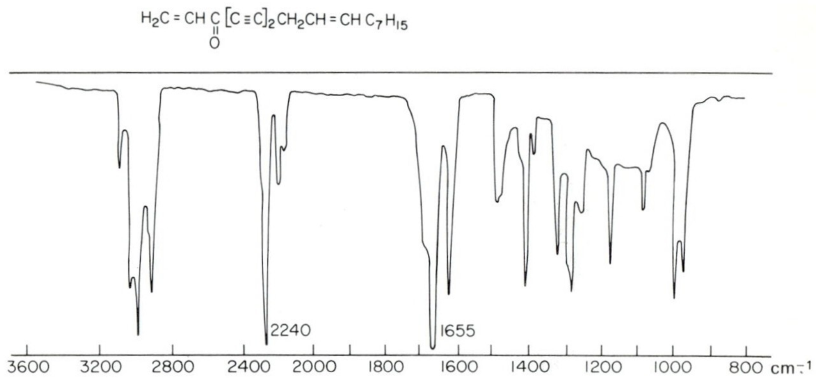

FIG. 32.

Terminal ethynyl groups can be detected by characteristic bands at 3300 and 2050/cm. With allenes we have the same situation as with acetylenes. The allene stretching frequency at about 1950/cm also is only intense, if we have polar groups in conjugation. Here and also in case of acetylenes the position of the bands depends from the special situation and can be changed in a range of about 100/cm.

The assignment of configuration of disubstituted double bonds is based on the i.r. absorption around 950 cm in *trans*-compounds and about 720 cm in *cis*-compounds. Vinyl end groups give rise to characteristic bands near 920 cm, but this position can be shifted by conjugation or polar groups to higher wave numbers. *Trans–trans* dienes give an absorption band near 990 cm.

In the last ten years n.m.r. spectroscopy has become the most important method in structure elucidation of new acetylenes. Before this method became available, the result of the u.v. and i.r. spectra normally were insufficient to determine the complete structure of new acetylinic compounds. Therefore, chemical methods such as hydrogenation and degradation were necessary, but which often turn out to be very difficult if only a few milligrams were available. The application of n.m.r. in combination with u.v. and i.r. spectra often allows elucidation of the complete structure even with very small amounts of material.

the fragmentation pattern of different deuterium labelled compounds, the most likely scheme is that given below:

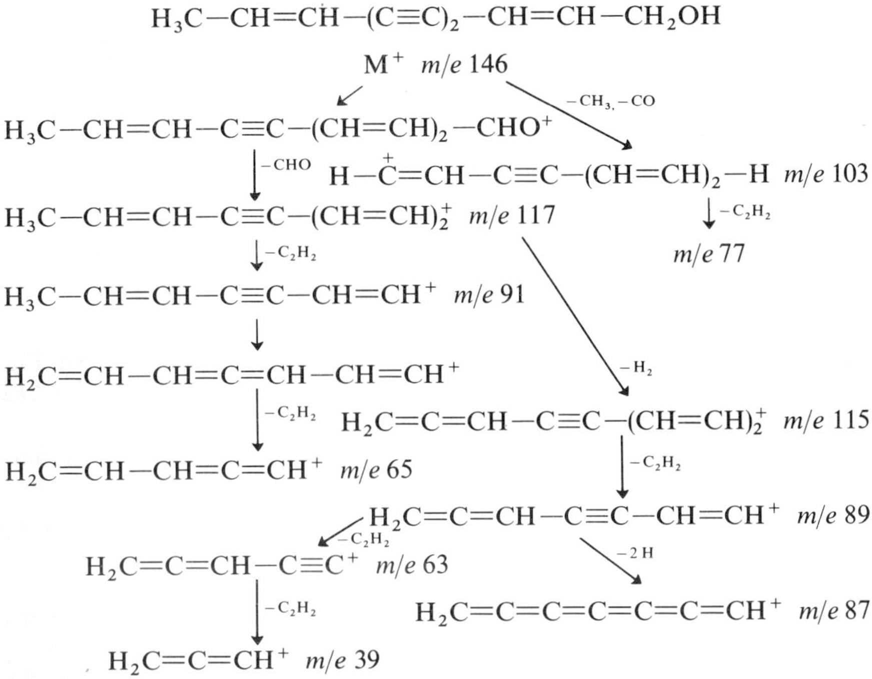

Moreover, the high tendency of McLafferty fragmentation is typical in polyacetylenes especially at low voltage as shown in the following scheme[17]:

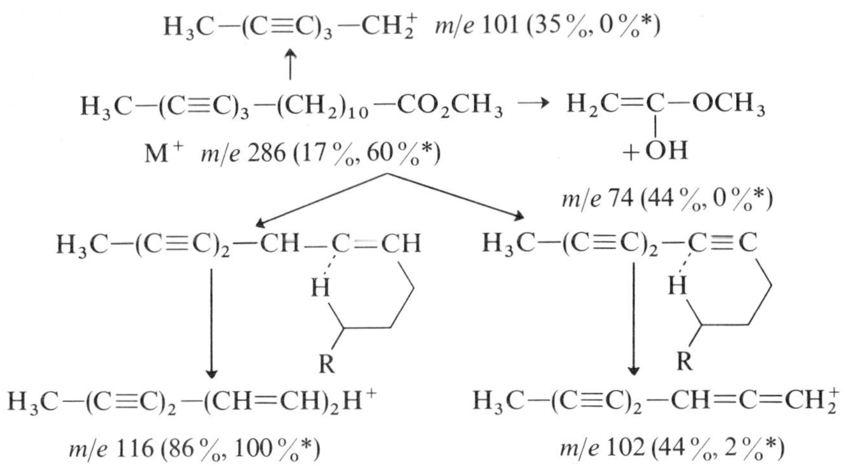

* 10 eV.

Beside these more special fragmentations, the natural acetylenes show the usual ones with the exception that, very often, fragmentation starting with ionisation on hetero-atoms is less pronounced because of the facile ionisation of the polyyn-ene system.

3.2. Chemical Methods

Having all the physical data, only a few structures are normally still possible and usually, only a few chemical reactions are necessary for the final structure elucidation. Sometimes, compounds showing no molecular peak in the mass spectrum have to be hydrogenated. By identifying the hydrogenation products by gas chromatography or mass spectroscopy, useful statements about the original structure are obtainable.

But we have to bear in mind that allylic hydroxyls can sometimes be removed by hydrogenation. Also, polyynene-epoxids can give rise to isomerisation during hydrogenation and that alcohols as well as ketones are the products. In some cases partial hydrogenation with Lindlar's catalyst is also of interest, as the resulting polyenes at once show the original number of conjugated unsaturated bonds.

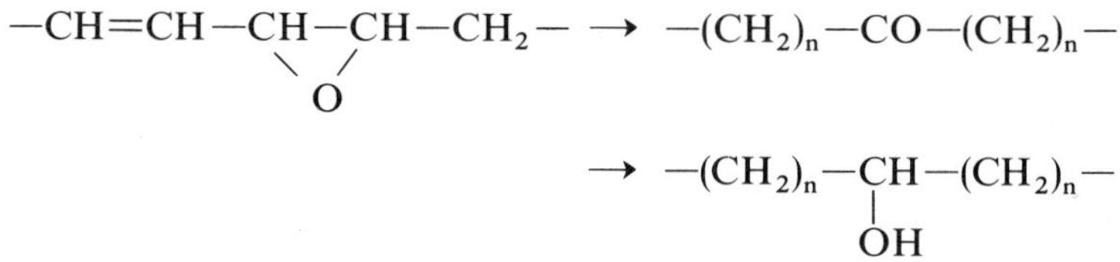

Oxidative degradation of polyynenes are possible either by ozonisation or epoxydation followed by hydrolysis to the diol and periodate cleavage.

The identification of the resulting products can be achieved by the usual methods such as TLC (GLC) or mass spectroscopic measurements. As double bonds are more reactive than triple bonds, partial oxidations are possible.

Oxidation with permanganate is only of limited interest. Only in the presence of large saturated or phenyl groups is this method useful.

If, in polyynenes, a diene group is present, the formation of a maleic anhydride adduct can be used to obtain crystalline derivatives or to recognise the chromophoric system:

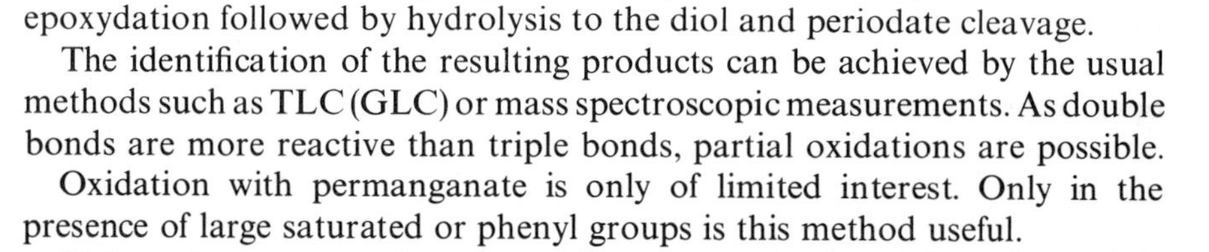

This method was used to establish the chromophoric system of an ene-triyn-diene. The u.v. spectra of this type of chromophore was not known at that time, but the maleic anhydride adduct shows the typical spectrum of an en-triyne.

In cases where the adducts were difficult to purify, the adducts with malein-*N*-phenyl-imides are to be preferred, as these adducts easily can be purified by chromatography.

Sometimes isomerisation with proton acceptors may also be useful. Compounds having an allenic group are often isomerised very rapidly to the corresponding acetylenic compounds. Also, nucleophilic additions can give some further conclusions on the structure as, e.g. the resulting enolethers give typical u.v. spectra. The velocity of this nucleophilic additions depends on the number of conjugated triple bonds[18]. Therefore, in the case of tetraynes and pentaynes care has to be taken to prevent the polyyne from undesigned reaction in alkaline media. Quite general polyynes become more and more reactive against nucleophilic reactants with increasing numbers of conjugated triple bonds, while electrophilic reactions become more and more difficult.

In some cases, partial oxidations with manganese dioxide are very useful. The velocity of allylic hydroxyl oxidations depends on the special environment. Thus, secondary hydroxyls react faster than primary ones, bis-allylic faster than allylic and these faster than propargylic hydroxyls. In this way complex mixtures of different alcohols can be separated.

Also sodium borohydride reduction is sometimes helpful in elucidation of complicated chromophoric systems or again in separation of complex mixtures, e.g. esters and aldehydes or ketones. In such a case also mild saponification may be useful to remove the easily reacting —O-acetates from the other carbonyl derivatives.

Sometimes two or three possibilities for the final structure still may be open. Then syntheses of the presumed structures are necessary.

4. General Consideration of the Biogenesis

The first statement on the biogenesis of natural acetylenes was the incorporation of acetyl-CoA and malonyl-CoA. Thus in *Polyporus anthracophilus* the ester **6** was uniformly labelled into alternate carbon atoms of the chain by feeding [1-^{14}C]-acetate[19] :

$$H_3C-\overset{\times}{C}H=CH-\overset{\times}{C}\equiv C-\overset{\times}{C}\equiv C-\overset{\times}{C}H=CH-\overset{\times}{C}O_2CH_3 \quad \mathbf{6}$$

Similarly, using [2-^{14}C]-malonate it was shown that ^{14}C in the alcohol **10** from *Tricholoma grammopodium* was incorporated specifically into the

C_1–C_8- part of the chain and not in the terminal C_2-unit ($C_9 + C_{10}$), showing that acetate acts as the starting molecule:

$$\underset{10}{H_3C}-C{\equiv}\overset{\times}{C}-C{\equiv}\overset{\times}{C}-C{\equiv}\overset{\times}{C}-CH{=}\overset{\times}{C}H-\underset{1}{C}H_2OH \quad \mathbf{10}$$

The same results were obtained by feeding labelled acetate to higher plants. These results and the knowledge that nearly all natural acetylenes have unbranched carbon chains led to the supposition that the acetylenes are formed from fatty acids.

Looking at the structures of many acetylenes having an isolated *cis*-double bond it turns out, that the position of this double bond normally is the same as in oleic acid, if we count the carbon atoms beginning with the terminal methyl group of oleic acid (**11**). The following examples may show that:

$$C_{18}: H_3C-(CH_2)_7-\underset{cis}{CH{=}CH}-(CH_2)_7-CO_2H \quad \mathbf{11}$$

$$C_{17}: H_3C-CH{=}CH-C{\equiv}C-C{\equiv}C-CH_2-\underset{cis}{CH{=}CH}-(CH_2)_5-CH{=}CH_2 \quad \mathbf{12}$$

$$C_{14}: H_3C-C{\equiv}C-C{\equiv}C-C{\equiv}C-CH_2-\underset{cis}{CH{=}CH}-(CH_2)_4-OR \quad \mathbf{13}$$

In order to establish the proposed relationship between oleic acid and other possible precursors of the natural acetylenes feeding experiments with ^{14}C- and ^{3}H-labelled compounds were necessary. In the case of higher plants, it turns out that emulsions of the different lipophilic precursors are easily incorporated by the aerial parts or even by the whole plant. Maximum conversion is normally observed after about 48 hours. The freshly cut plants then have to be extracted and the corresponding acetylenes can be isolated in the usual manner.

Rigorous purification is extremely important and transformations to other derivatives are sometimes necessary in order to be sure that the observed activity is constant. Special degradations are then necessary to determine the exact position of the label. Very often, double labelling is helpful. The results of many such experiments have produced a good review of the different biogenetical pathways which exist in natural acetylenes.

With labelled oleic acid, it could be shown that the natural acetylenes, both from micro-organism[20] and higher plants[13], are formed from oleic acid. Linoleic, crepenynic and dehydrocrepenynic acids are also incorporated and followed by the corresponding diyne and triyne esters. Therefore the following scheme is most likely, e.g. for the biogenesis of the important precursor **13**.

$$H_3C{-}(CH_2)_7{-}CH{=}CH{-}(CH_2)_7{-}CO_2R \quad \textbf{11}$$
cis

$$\downarrow -[H]$$

$$H_3C{-}(CH_2)_4{-}CH{=}CH{-}CH_2{-}CH{=}CH{-}(CH_2)_7{-}CO_2R \quad \textbf{14}$$
cis

$$\downarrow -[H]$$

$$H_3C{-}(CH_2)_4{-}C{\equiv}C{-}CH_2{-}CH{=}CH{-}(CH_2)_7{-}CO_2R \quad \textbf{15}$$
cis

$$\downarrow -[H]$$

$$H_7C_3{-}CH{=}CH{-}C{\equiv}C{-}CH_2{-}CH{=}CH{-}(CH_2)_7{-}CO_2R \quad \textbf{16}$$

$$\downarrow -[H]$$

$$H_7C_3{-}C{\equiv}C{-}C{\equiv}C{-}CH_2{-}CH{=}CH{-}(CH_2)_7{-}CO_2R \quad \textbf{17}$$

$$\downarrow -[H]$$

$$H_3C{-}C{\equiv}C{-}C{\equiv}C{-}C{\equiv}C{-}CH_2{-}CH{=}CH{-}(CH_2)_7{-}CO_2R \quad \textbf{18}$$

$$\downarrow [O]$$

$$H_3C{-}C{\equiv}C{-}C{\equiv}C{-}C{\equiv}C{-}CH_2{-}CH{=}CH{-}$$
cis

$$CH_2{-}CH_2{-}CH_2{-}CH_2{-}OR \quad \textbf{13}$$

By feeding the labelled homologous oleic acid esters with varying numbers of methylene groups between the double bond and ester group, it turns out that only oleic acid itself is a precursor and not the acids already shortened; in the case of the corresponding triyne acid series, the incorporation increases from C_{18} to C_{16} and C_{14} or, if we have odd-numbered carbon chains in the resulting acetylenes, from C_{18} to C_{17} and C_{15}. The other chain lengths of C_{17} and C_{15}, or in the second case C_{16} and C_{14}, are not incorporated.

Therefore, the scheme previously discussed can be specified in the following way:

$$H_3C{-}C{\equiv}C{-}C{\equiv}C{-}C{\equiv}C{-}CH_2{-}CH{=}CH{-}(CH_2)_7{-}CO_2R \quad \textbf{18}$$
cis

2 × β-oxidation ↙   ↘ 1 × α-oxidation, 2 × β-oxidation

$$H_3C(C{\equiv}C)_3{-}CH_2{-}CH{=}CH{-}(CH_2)_3{-}CO_2R \quad \textbf{19}$$
cis

$$\downarrow [H]$$

$$H_3C(C{\equiv}C)_3{-}CH_2{-}CH{=}CH{-}(CH_2)_4{-}OR \quad \textbf{13}$$

$$H_3C{-}(C{\equiv}C)_3{-}CH_2{-}CH{=}CH{-}(CH_2)_2{-}CO_2R \quad \textbf{20}$$

$$\downarrow [O]$$

$$H_3C{-}(C{\equiv}C)_3{-}CH_2{-}CH{=}CH{-}(CH_2)_3{-}OR \quad \textbf{21}$$

19, **20**, **13** and **21** are all important precursors for a large number of natural acetylenes which are formed by more specific biochemical reactions as shown in detail in Chapter 2.

However, some other schemes are necessary to explain the large variety of existing acetylenes. Crepenynic acid (**15**) can be transformed also to **22** and **23** and the diyne **17** to **24**, **25** and probably to **26**:

$$\mathbf{15} \rightarrow H_3C{-}CH_2{-}CH_2{-}CH{=}CH{-}C{\equiv}C{-}CH_2{-}\underset{cis}{CH{=}CH}{-}R \quad \mathbf{22}$$

$$\downarrow$$

$$H_3C{-}CH{=}CH{-}CH{=}CH{-}C{\equiv}C{-}CH_2{-}\underset{cis}{CH{=}CH}{-}R \quad \mathbf{23}$$

$$\mathbf{17} \rightarrow H_3C{-}CH{=}CH{-}C{\equiv}C{-}C{\equiv}C{-}CH_2{-}\underset{cis}{CH{=}CH}{-}R \quad \mathbf{24}$$

$$\rightarrow H_3C{-}CH_2{-}\underset{OH}{CH}{-}C{\equiv}C{-}C{\equiv}C{-}CH_2{-}\underset{cis}{CH{=}CH}{-}R \quad \mathbf{25}$$

$$\rightarrow H_2C{=}CH{-}\underset{OH}{CH}{-}C{\equiv}C{-}C{\equiv}C{-}CH_2{-}\underset{cis}{CH{=}CH}{-}R \quad \mathbf{26}$$

The sequences are, in part, still not established. **25** may be in some cases, the precursor of **24**, but perhaps **26** may arise from **22**, which possibly can also be a precursor of **24**.

Up to now it seems to be possible that different routes are involved in the biogenesis of acetylenes having such endgroups. These problems will be discussed in more detail in Chapters 2 and 3.

Obviously it is not always established at what stage the necessary biochemical reactions really occur. The nature of the group R— in **22**–**26** must therefore perhaps vary from case to case.

Besides the chain-shortening to C_{14} and C_{13} mentioned earlier, some further pathways are still necessary to explain the biogenesis of all the known acetylenes. The results of feeding experiments with labelled compounds and the co-occurrence of some acetylenes lead to the conclusion that the following routes also operate in special cases:

(1) C_{17}-compounds

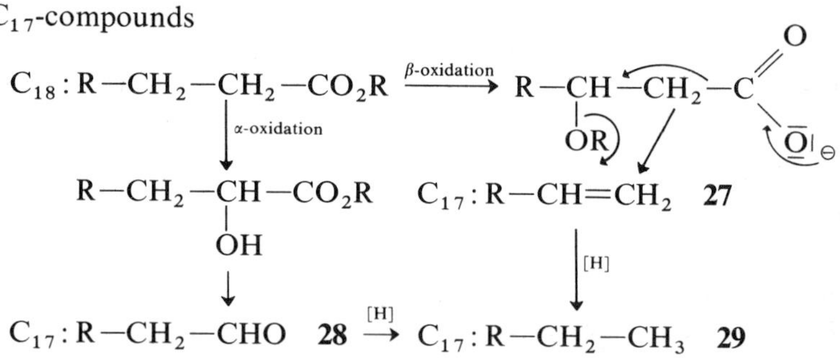

(2) C_{10}-compounds

$$C_{18}\text{-acids (e.g. } \mathbf{17, 18, 22, 23, 24}) \xrightarrow{4 \times \beta\text{-oxidation}} C_{10}\text{-compounds}$$

or

$$\mathbf{18} \xrightarrow{[O]} H_3C{-}(C{\equiv}C)_3{-}CH{=}CH{-}CH(OH){-}(CH_2)_7{-}CO_2R \quad \mathbf{30}$$

$$\downarrow [O]$$

$$H_3C{-}(C{\equiv}C)_3{-}CH{=}CH{-}CO{-}(CH_2)_7{-}CO_2R \quad \mathbf{31}$$

$$\downarrow [O]$$

$$H_3C{-}(C{\equiv}C)_3{-}CH{=}CH{-}COO{-}(CH_2)_7{-}CO_2R \quad \mathbf{32}$$

$$\downarrow$$

$$H_3C{-}(C{\equiv}C)_3{-}CH{=}CH{-}CO_2CH_3 \quad \mathbf{1}$$

It could be shown that **31** is indeed a very good precursor for the dehydromatricariaester (**1**). Bayer–Villiger-like oxidation of **31** (first proposed by Sir E. R. H. Jones)[22] is more likely since a triglyceride is isolated with linoleic and linolenic acid and the acid **35**[21] probably formed by the following reaction sequence[22]:

$$H_3C{-}(CH_2)_4{-}\underset{cis}{CH{=}CH}{-}CH_2{-}\underset{cis}{CH{=}CH}{-}CH_2{-}CH{=}C{=}CH{-}(CH_2)_3{-}CO_2R \quad \mathbf{33}$$

$$\downarrow [O]$$

$$H_3C{-}(CH_2)_4{-}\underset{cis}{CH{=}CH}{-}\underset{trans}{CH{=}CH}{-}CH(OH){-}CH_2{-}CH{=}C{=}CH{-}(CH_2)_3{-}CO_2R \quad \mathbf{34}$$

$$\downarrow [O]$$

$$H_{11}C_5{-}CH{=}CH{-}CH{=}CH{-}CO{-}O{-}CH_2{-}CH{=}C{=}CH{-}(CH_2)_3{-}CO_2R \quad \mathbf{35}$$

The biogenesis of C_{17}-compounds of type **27** is established by feeding experiments, while the second route (**28** → **29**, **27** → **29**) is not really proved.

However, there is a striking difference in the compounds derived from Compositae and Umbelliferae. The first family always has compounds of type **27** while the second one has always those of type **29**.

Further oxydative degration of the C_{18}-acids also leads to the relatively small number of C_{16}-compounds and in the same way, from compounds of type **28** to the rare C_{15}-acetylenes.

Also, in some case only α-oxidation seems to be a biochemical route, as in *Centaurea* species. Often the whole series of different chain lengths is found, e.g.:

$$H_3C{-}CH{=}CH{-}(C{\equiv}C)_2{-}(CH{=}CH)_2{-}(CH_2)_n{-}CHO \quad \mathbf{36}$$

$$n = 2-6$$

Degration at the other end of the acetylenic precursors is only of limited importance. Relatively often the elimination of one carbon atom can be observed. The mechanism of this reaction sequence is probably the following:

$$H_3C{-}C{\equiv}C{-}R \rightarrow HO{-}CH_2{-}C{\equiv}C{-}R \rightarrow OCH{-}C{\equiv}C{-}R$$

$$\swarrow \qquad \searrow$$

$$HO_2C{-}C{\equiv}C{-}R \rightarrow H{-}C{\equiv}C{-}R$$

Only in two cases is the elimination of two carbons stated (see Chapter 3, Sections 3.2 and 5.2):

$$H_3C{-}CH_2{-}CH_2{-}(C{\equiv}C)_2{-}R \rightarrow H_3C{-}CO{\mid}CH_2{-}(C{\equiv}C)_2{-}R$$

$$\downarrow$$

$$H_3C{-}(C{\equiv}C)_2{-}R$$

All the other biochemical transformations necessary to get the large variety of the natural acetylenes are more or less common ones.

Allylic oxidation is very important and is accompanied by allylic rearrangement.

A: $$R{-}CH_2{-}C{\equiv}C{-} \rightarrow R{-}\underset{\displaystyle OH}{\underset{\mid}{C}H}{-}C{\equiv}C{-}$$

B: $$R{-}CH_2{-}CH{=}CH{-} \rightarrow R{-}\underset{\displaystyle OH}{\underset{\mid}{C}H}{-}CH{=}CH{-}$$

C: $$R{-}C{\equiv}C{-}CH_2{-}CH{=}CH{-} \rightarrow R{-}C{\equiv}C{-}\underset{\displaystyle OH}{\underset{\mid}{C}H}{-}CH{=}CH{-}$$

D: $$R{-}C{\equiv}C{-}CH_2{-}CH{=}CH{-} \rightarrow R{-}C{\equiv}C{-}CH{=}CH{-}\underset{\displaystyle OH}{\underset{\mid}{C}H}{-}$$

The mechanism of pathway D might be best represented by the following scheme.

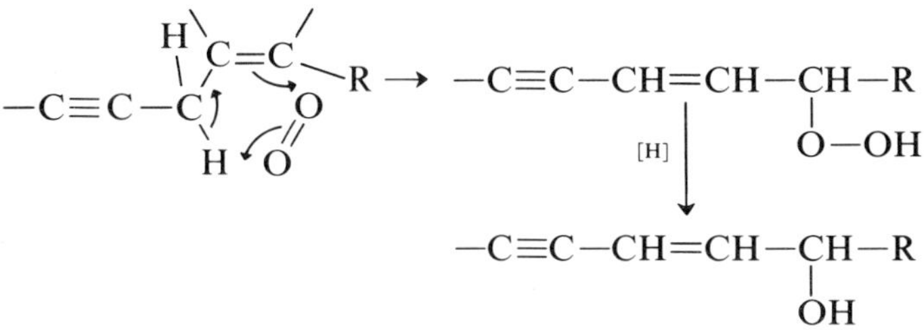

So far no case is known where oxygen functions are introduced when not in an allylic position or by β-oxidation. In compounds having an oxygen function in other positions, it turns out that a double bond has first been introduced which has then been hydrogenated:

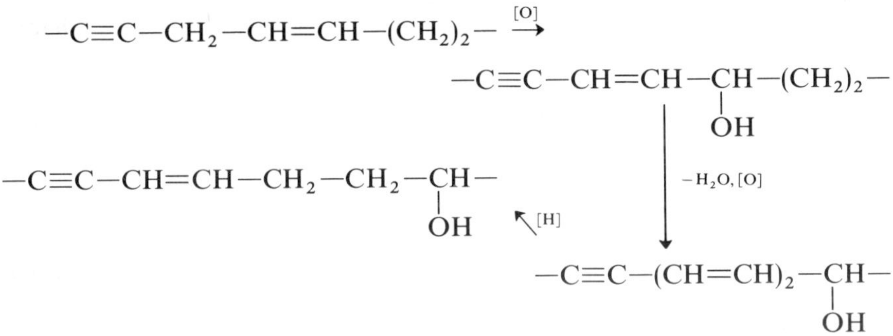

A second usual oxidative reaction of polyynenes is the epoxidation, which often is followed by secondary transformations:

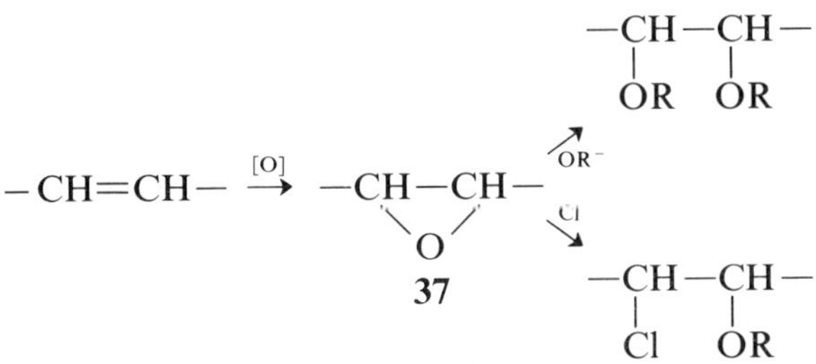

37 38 39

Compounds of type **38** and **39** can be precursors of special ring compounds (see Chapter 2, Section 2.7.1.).

Formation of sulphur compounds is also a very common reaction in the field of natural acetylenes. The formal addition of H_2S leads to the large group of acetylenic thiophenes:

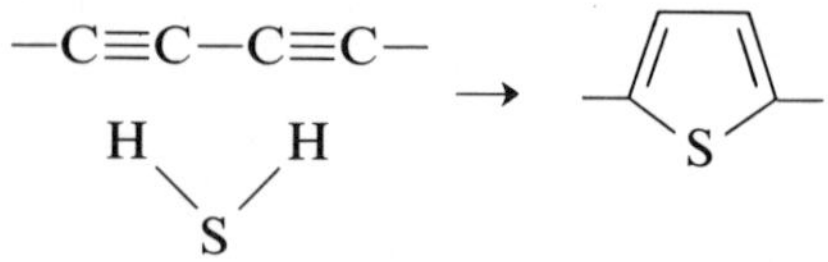

40

Beside this, the formation of methyl thio-ethers can very often be observed:

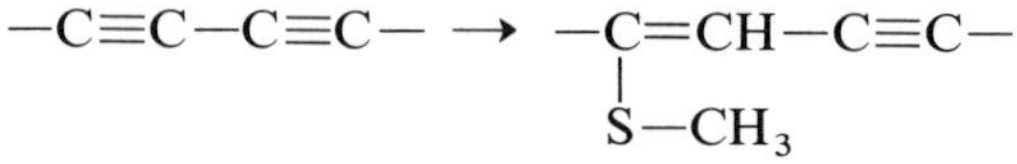

41

Here it could be shown that this must be a two-step reaction and therefore the formation of thiophenes and methyl thio-ethers probably starts from a common intermediate, while double addition leads to the disulphides of type **42**:

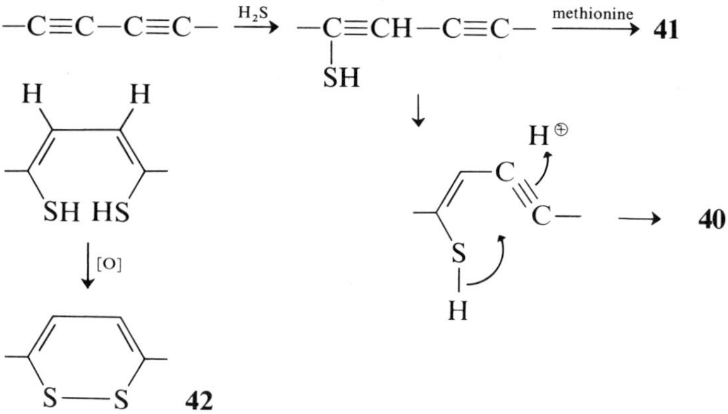

Characteristic of natural acetylenes is the further formation of *O*-heterocycles and aromatic compounds as shown in the following few examples of the unequivocally established cases:

$$H_3C-(C\equiv C)_3-CH_2-CH=CH-(CH_2)_3-OH \quad \mathbf{21}$$

cis

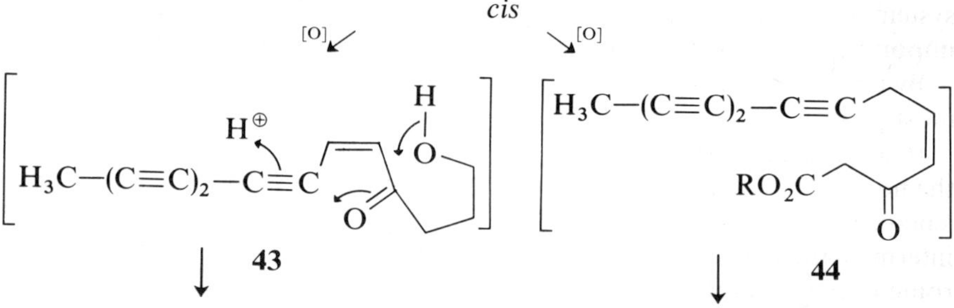

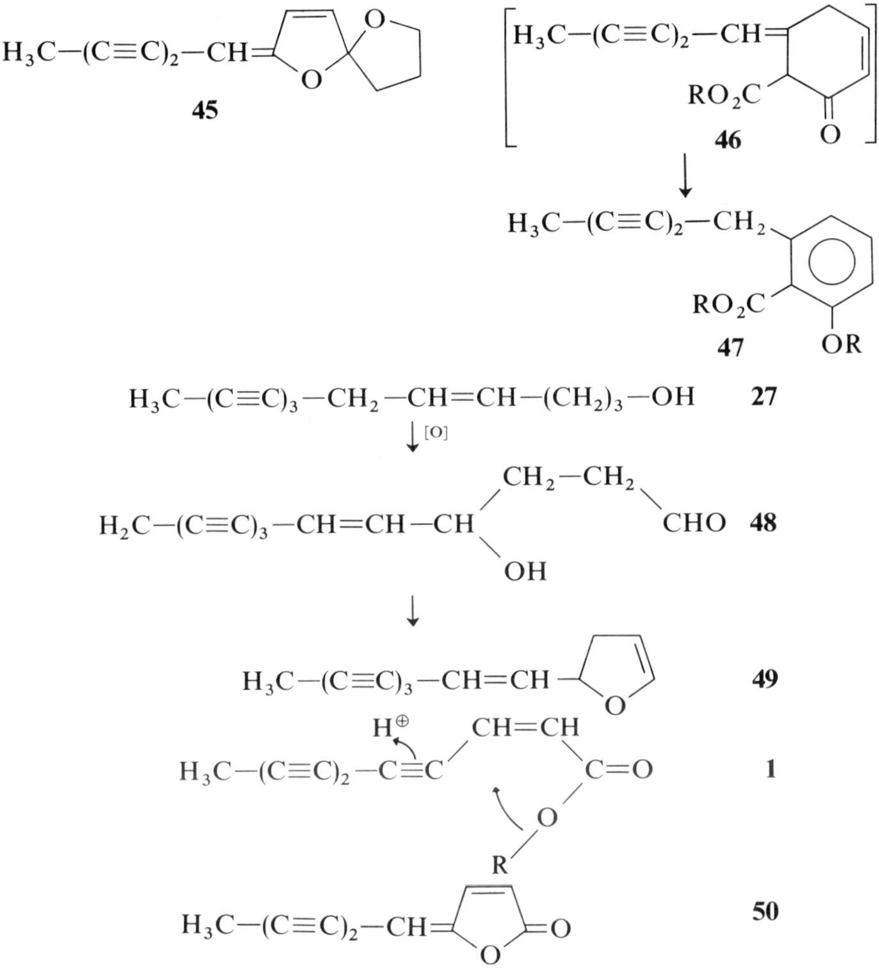

45 46 47 27 48 49 1 50

Further reaction pathways will be shown in Chapters 2 and 3. Still an open question is the enzymatic way of triple bond formation. Up to now this enzyme system has not been isolated and therefore a more defined mechanism of this important reaction cannot be given.

By using cell-free preparations, it could be shown that the chloroplasts are most probably necessary for the triple bond formation[23].

It seems to be possible that we have here some kind of cooperation with the normal fatty acid biogenesis. If an enzyme system is present in the acetylene-producing organism, which can transfer hydrogens to the unsaturated intermediate precursors of fatty acid synthesis, this would be an attractive route of triple bond formation.

2

Compounds with Methyltriyne Endgroups and Related Compounds

As already mentioned in Section 4 of Chapter 1, the C_{18}-triyne acid **5** must be formed by subsequent dehydrogenation of oleic and crepenynic acids, respectively (**1**). Also dehydrocrepenynic acid (**2**)[24] as well as the diyne acid **3** are precursors. Therefore the following sequence is the most likely natural pathway:

$$H_3C{-}(CH_2)_4{-}C{\equiv}C{-}CH_2{-}\underset{cis}{CH{=}CH}{-}(CH_2)_7{-}CO_2R \quad \mathbf{1}$$

$$\downarrow -[H]$$

$$H_3C{-}(CH_2)_2{-}CH{=}CH{-}C{\equiv}C{-}CH_2{-}\underset{cis}{CH{=}CH}{-}(CH_2)_7{-}CO_2R \quad \mathbf{2}$$

$$\downarrow -[H]$$

$$[H_3C{-}(CH_2)_2{-}C{\equiv}C{-}C{\equiv}C{-}CH_2{-}\underset{cis}{CH{=}CH}{-}(CH_2)_7{-}CO_2R] \quad \mathbf{3}$$

$$\downarrow -[H]$$

$$[H_3C{-}CH{=}CH{-}C{\equiv}C{-}C{\equiv}C{-}CH_2{-}\underset{cis}{CH{=}CH}{-}(CH_2)_7{-}CO_2R] \quad \mathbf{4}$$

$$\downarrow -[H]$$

$$[H_3C{-}C{\equiv}C{-}C{\equiv}C{-}C{\equiv}C{-}CH_2{-}\underset{cis}{CH{=}CH}{-}(CH_2)_7{-}CO_2R] \quad \mathbf{5}$$

By synthesis of the series of the labelled acid esters **6–11**[25] and feeding experiments with all these compounds it was possible to decide, at what stage the dehydrogenation steps are starting. It turns out that only oleic acid ester is an useful precursor[25]. All compounds having a methyl-triyne endgroup and those which are related to such a group are formed starting with **5**, which then will be degradated to shorter chain-lengths[25]:

$$H_3C{-}(CH_2)_7{-}\underset{cis}{CH{=}CH}{-}(CH_2)_n{-}CO_2CH_3 \quad n = 7: \mathbf{6}$$

n = 6: **7**
n = 5: **8**
n = 4: **9**
n = 3: **10**
n = 2: **11**

In the next Section the relatively few acetylenes having a C_{17}-, C_{16}- or C_{15}-carbon chain will be discussed.

1. C_{17}-, C_{16}- and C_{15}-compounds

The most widespread C_{17}-acetylene is an established hydrocarbon, the so-called "centaur X_3", first recognised by Löfgren[26,27] from *Centaurea cyanus* L., the common cornflower. Some years later, this hydrocarbon was isolated from *Centaurea cyanus* L. and from the common mugwort—*Artemisia vulgaris* L.[28]. The low melting compound (m.p. 18°) shows the typical u.v. spectrum of a triyne-diene and therefore gives an adduct with maleic anhydride. The i.r. spectrum shows the *trans–trans* configuration of the diene grouping and furthermore shows an isolated vinyl group. Oxidation with permanganate leads to adipic acid and hydrogenation gives *n*-heptadecane. Therefore, the structure must be **14**.

The hydrocarbon **15** with the *trans–cis* configuration is also present in the root extract of *Artemisia vulgaris* L. The structures have been established by synthesis[29] starting with the aldehyde **12**, which was obtained by oxidative coupling of pentadiyne with pent-4-yn-2-en-1-ol and subsequent manganese dioxide oxidation:

$$H_3C{-}C{\equiv}C{-}C{\equiv}C{-}H + H{-}C{\equiv}C{-}CH{=}CH{-}CH_2OH$$

$$\downarrow$$

$$H_3C{-}(C{\equiv}C)_3{-}CH{=}CH{-}CH_2{-}OH$$

$$\downarrow MnO_2$$

$$H_3C{-}(C{\equiv}C)_3{-}CH{=}CH{-}CHO \quad \mathbf{12}$$

Wittig reaction with the ylene **13** gives the *trans–trans*, as well as the *trans–cis* hydrocarbon. The ylene **13** was prepared in the following way:

$$\mathrm{Cl{-}(CH_2)_4{-}Br} \xrightarrow[\text{(2) NaI; (3) CN}^-]{\text{(1) Na}-\text{C}\equiv\text{C}-\text{H}} \mathrm{N{\equiv}C{-}(CH_2)_4{-}C{\equiv}C{-}H} \xrightarrow{\text{CH}_3\text{OH/H}^+}$$

$$\mathrm{RO_2C{-}(CH_2)_4{-}C{\equiv}C{-}H} \xrightarrow[\text{Lindlar}]{\text{H}_2} \mathrm{RO_2C{-}(CH_2)_4{-}CH{=}CH_2} \xrightarrow[\text{(2) PBr}_3]{\text{(1) LiAlH}_4}$$

$$\mathrm{Br{-}(CH_2)_5{-}CH{=}CH_2} \xrightarrow[\text{(2) LiBu}]{\text{(1) Ph}_3\text{P}} \mathrm{Ph_3P{=}CH{-}(CH_2)_4{-}CH{=}CH_2} \quad \mathbf{13}$$

$$\mathbf{12} + \mathbf{13} \rightarrow \mathrm{H_3C{-}(C{\equiv}C)_3{-}CH{=}CH{-}CH{=}CH{-}(CH_2)_4{-}CH{=}CH_2}$$

trans *trans* **14** centaur X_3

trans *cis* **15**

The other known C_{17}-compounds are closely related to **14**. Although **16** has not been isolated up to now, it is quite probable that the following biogenetical pathway exists in this group:

$$\mathbf{6} \xrightarrow{[\text{O}]} \left[\mathrm{H_3C{-}(CH_2)_7{-}\underset{cis}{CH{=}CH}{-}(CH_2)_5{-}\underset{\displaystyle OR}{\underset{|}{CH}}{-}CH_2{-}C({=}O)O^{\ominus}}\right]$$

$$\downarrow$$

$$[\mathrm{H_3C{-}(CH_2)_7{-}\underset{cis}{CH{=}CH}{-}(CH_2)_5{-}CH{=}CH_2}]$$

$$\downarrow -[\text{H}]$$

$$[\mathrm{H_3C{-}(C{\equiv}C)_3{-}CH_2{-}\underset{cis}{CH{=}CH}{-}(CH_2)_5{-}CH{=}CH_2}] \quad \mathbf{16}$$

$$\downarrow [\text{O}]$$

$$\mathrm{H_3C{-}(C{\equiv}C)_3{-}CH{=}CH{-}\underset{\displaystyle OH}{\underset{|}{CH}}{-}(CH_2)_5{-}CH{=}CH_2} \quad \mathbf{17}$$

$$\downarrow -[\text{H}_2\text{O}]$$

14, 15

$$\mathrm{H_3C{-}(C{\equiv}C)_3{-}(CH{=}CH)_2{-}(CH_2)_4{-}CH{=}CH_2} \quad \mathbf{14}$$

$$\downarrow [\text{O}]$$

$$\mathrm{H_3C{-}(C{\equiv}C)_3{-}(CH{=}CH)_2{-}\underset{\displaystyle OR}{\underset{|}{CH}}{-}(CH_2)_3{-}CH{=}CH_2}$$

18: R = H

19: R = Ac

$$\downarrow [\text{O}]$$

$$\mathrm{H_3C{-}(C{\equiv}C)_3{-}(CH{=}CH)_2{-}CO{-}CH_2}$$

$$\mathrm{CH_2CH_2CH{=}CH_2} \quad \mathbf{20}$$

The formation of **14** from oleic acid as well as from β-hydroxy-oleic acid[30] has been established using labelled materials, while the incorporation of **5** is very low[31]. Perhaps **14** owes its existence only to a side reaction and it is formed slowly, while oleic acid is more rapidly transformed to **5** and to the different isolated acetylenes formed from this precursor. **17**, obviously the precursor of **14**, has also been found in *Artemisia selengensis* Auct.[32]. **18**[32] and **19**[33], oxidation products of **14**, are present too in the same *Artemisia* species, respectively in *Dahlia* tubers. **20** is a minor acetylene from the roots of *Artemisia vulgaris* L.[34].

In the structure elucidation of **17** and **18** manganese dioxide oxidation leading to the corresponding ketones was the key step together with the observed u.v. spectra typical for triyne-diene chromophores.

The only members with 16 and 15 carbon atoms in this series are the aldehydes **21** and **22**, the analogous alcohols **23** and **24**[35] and the C_{16}-hydrocarbon **25**:

$$H_3C{-}(C{\equiv}C)_3{-}(CH{=}CH)_2{-}(CH_2)_n{-}CHO \qquad \textbf{21}: n = 4 \quad \textbf{22}: n = 3$$

$$H_3C{-}(C{\equiv}C)_3{-}(CH{=}CH)_2{-}(CH_2)_n{-}CH_2OH \qquad \textbf{23}: n = 4 \quad \textbf{24}: n = 3$$

$$H_3C{-}(C{\equiv}C)_3{-}(CH{=}CH)_2{-}(CH_2)_3{-}CH{=}CH_2 \qquad \textbf{25}$$

21 and **22** occur together with aldehydes having other chromophoric systems (see Chapter 3, Sections 1.1 and 1.2.7) in *Centaurea* species[36]. They must be formed by α-oxidation. **25** has been isolated from different *Chrysanthemum* species[37] and probably is built up by the same pathway as **14**, starting with the corresponding C_{17}-acid. **25** is only isolated from species where C_{13}-acetylenes occur and which are biosynthesised, starting with **5** by α-oxidation.

2. Acetylenes Derived from the C_{14}-triyn-(8,10,12)-en-(5)-ol-(1)

2.1. Simple Triynes

As already discussed in Chapter 1, Section 4, the C_{14}-alcohol **27** is a very important precursor for many acetylenes.

Double β-oxidation of the C_{18}-triyne acid would lead to the C_{14}-acid, which then has to be reduced to give **27**. This alcohol has been isolated in very minute amounts from *Centaurea moschata* L.[38].

The corresponding acetate **28** can be synthesised in the following way[38]:

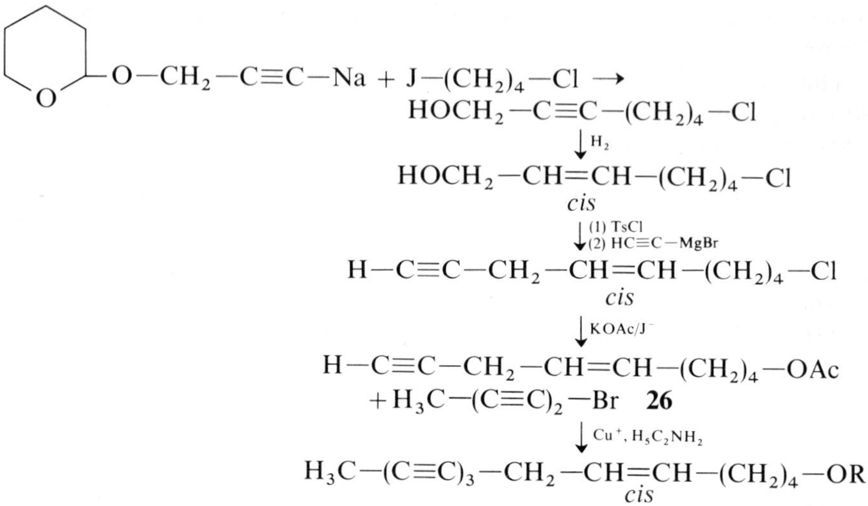

27: R = H **28**: R = Ac

The coupling reaction of bromoacetylenes of type **26** with ethynyl derivatives is a very important synthetic method introduced by Cadiot and Chodkiewicz[39]. This synthesis creates the possibility of introducing the label at different stages, either by involving tritium in the hydrogenation step or with 5-^{14}C—**26** in the coupling reaction. The resulting labelled compounds have been very important in establishing the role of **27** as a very common precursor. The acetate **28** is also a natural product and it has been found in small amounts in *Chrysanthemum serotinum* L.[40].

Closely related to **27** are the diol **33** and the monoacetate **34** both isolated from *Centaurea muricata* L.[41]. **33**, directly formed from **27**, is a further important key intermediate. It has been synthesised in the following way[42]:

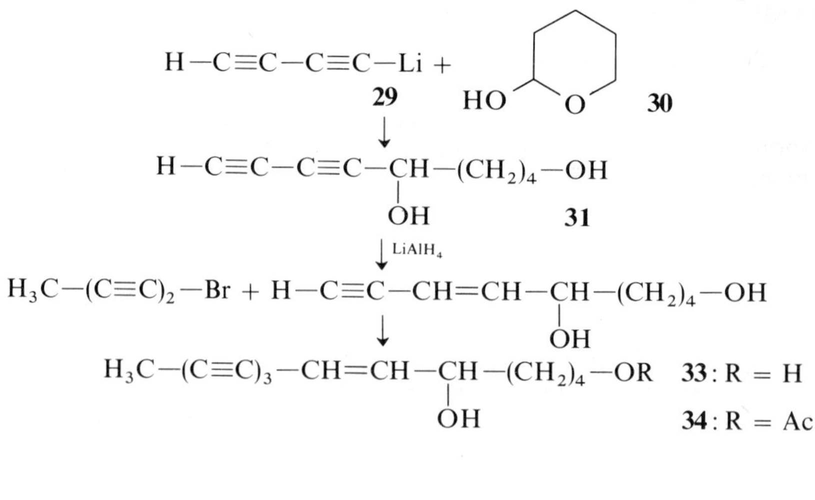

While **27** and **33** seem to occur only in very small amounts, the products of water elimination of **33** or **34** are widespread in species of the family Compositae.

The acetate **35** was first isolated from *Tanacetum vulgare* L.[43] and the structure was established by synthesis[44]. The alcohol **36** is present in *Chrysanthemum atratum* Jacq.[45] and in many other Compositae:

$$\underset{\qquad\qquad\qquad\qquad t,t}{H_3C-(C\equiv C)_3-(CH=CH)_2-(CH_2)_3-OR} \qquad \begin{array}{l}\mathbf{35}: R = Ac\\ \mathbf{36}: R = H\end{array}$$

As shown later, **36** is the precursor of the widespread pentaynene **94** (see Chapter 2, Section 2.6). It is further transformed by oxidative reactions to the triynes **37**–**39**, isolated from different numbers of the tribus Heliantheae and Anthemideae:

$$\mathbf{36} \xrightarrow{[O]} H_3C-(C\equiv C)_3-(CH=CH)_2-\underset{\displaystyle OR}{\underset{|}{CH}}-(CH_2)_2-OR'$$

37: R = R′ = H[46] **38**: R = R′ = Ac[47] **39**: R = Ac, R′ = H[48]

37 has been synthesised by the following way:[49]

$$H_3C-(C\equiv C)_3-CH=CH-CHO + Ph_3P=CH-CO-CH_2-CO_2R$$

$$\downarrow$$

$$H_3C-(C\equiv C)_3-(CH=CH)_2-CO-CH_2-CO_2R$$

$$\downarrow \text{(1) } NaBH_4 \text{ (2) } LiAlH_4$$

$$\mathbf{37}$$

37 is the precursor of the tridecatriyne-triene (see Section 2.7.2) and of "artemisia ketone" (see next Section). Epoxidation of **36** leads to the alcohol **40**[40], which is the precursor of a tetrahydropyrane (see Section 2.3) isolated together with **40** from *Chrysanthemum serotinum* L.[40]:

$$\mathbf{36} \xrightarrow{[O]} H_3C-(C\equiv C)_3-CH=CH-\underbrace{CH-CH}_{O}-(CH_2)_3-OH \qquad \mathbf{40}$$

Another very common transformation of **36** is shown by the existence of the thiophenes **42** and **43**, isolated from *Xeranthemum cylindraceum* Sibth.[50]:

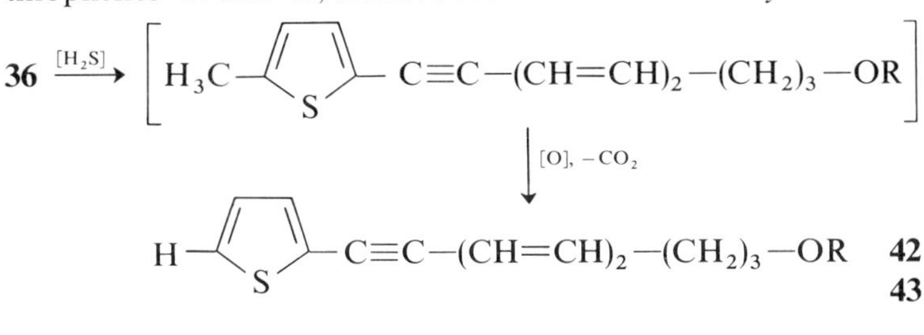

41

42: R = H
43: R = Ac

The elimination of the terminal methyl group is also very common (see Chapter 1, Section 4). The *in vitro* addition of H_2S to polyynes is easily possible as shown by K. E. Schulte and coworkers[51]. Several thiophenes have been prepared using this method:

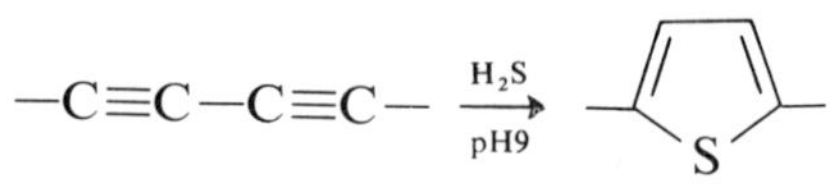

2.2. Compounds Related to the Tetradecatriyn-(8,10,12)-en-(6)-on-(3)

From the roots of the common mugwort, *Artemisia vulgaris* L., Sörensen and his coworkers[28] have isolated a ketone with a triyn-ene chromophore. Perhydrogenation of the ketone gave *n*-tetradecanone-(3). Therefore, the most probable structure is **44**, established by synthesis[52]:

$$\mathrm{H{-}C{\equiv}C{-}CH{=}CH{-}CH_2{-}Br + H_2\underset{\displaystyle CO_2R}{\underset{|}{C}}{-}CO{-}C_2H_5}$$

$$\xrightarrow{\mathrm{NaOC_2H_5,\ KOH,\ -CO_2}}$$

$$\mathrm{H_3C{-}C{\equiv}C{-}C{\equiv}C{-}H + H{-}C{\equiv}C{-}CH{=}CH{-}CH_2{-}CH_2{-}CO{-}C_2H_5}$$

$$\downarrow \mathrm{Cu_2Cl_2, O_2}$$

$$\mathrm{H_3C{-}(C{\equiv}C)_3{-}CH{=}CH{-}CH_2{-}CH_2{-}CO{-}CH_2{-}CH_3} \quad \mathbf{44}$$

"artemisia ketone"

The name "artemisia ketone" unfortunately was first used for a rare terpene ketone.

44 is also very widespread, but only in plants belonging to the tribus Anthemideae, a very large group in the family Compositae.

The elucidation of the biogenesis of this ketone first was a problem, as the oxygen function was located in an unusual position. Feeding experiments however show that **44** is formed from **27**[53]. To learn more about the detailed pathway, further labelled compounds have been synthesised. It turns out that we have to assume the following reaction sequence since **27**, **33**, **37** and the corresponding alcohol of the ketone **45** have been incorporated with increased yields[54]:

$$27 \xrightarrow{[O]} 33 \xrightarrow[{[O]}]{-[H_2O]} H_3C-(C{\equiv}C)_3-(CH{=}CH)_2-\underset{\displaystyle OH}{\underset{|}{CH}}-CH_2-CH_2-OH \quad \mathbf{37}$$

$$\downarrow [O]$$

$$[H_3C-(C{\equiv}C)_3-\underset{t,t}{(CH{=}CH)_2}-CO-CH_2-CH_2-OH]$$

$$\downarrow -H_2O$$

$$H_3C-(C{\equiv}C)_3-\underset{t,t}{(CH{=}CH)_2}-CO-CH{=}CH_2 \quad \mathbf{45}$$

$$\downarrow [H]$$

$$\mathbf{44}$$

The biogenesis of **44** is a typical case, showing that in this field subsequent hydrogenation of double bonds is the only way to introduce oxygen function formal in a not allylic position.

The corresponding alcohol **46** is also present in several species of the tribus Anthemideae[45]:

$$H_3C-(C{\equiv}C)_3-CH{=}CH-CH_2-CH_2-\underset{\displaystyle OH}{\underset{|}{CH}}-CH_2-CH_3 \quad \mathbf{46}$$

The corresponding dehydro product of **44** has not been isolated so far in nature, but the thiophene **48** is obviously formed from **47**:

$$\mathbf{45} \xrightarrow{[H]} H_3C-(C{\equiv}C)_3-(CH{=}CH)_2-CO-C_2H_5 \quad \mathbf{47}$$

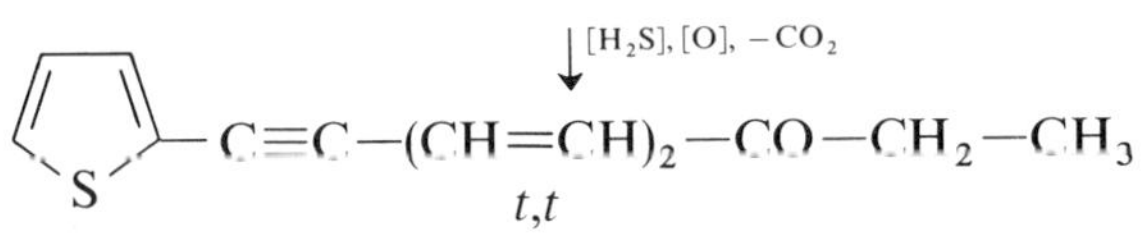

48

48 has been isolated from the roots of scentless mayweed[9,55] together with the thiophene **49**, closely related to **44**:

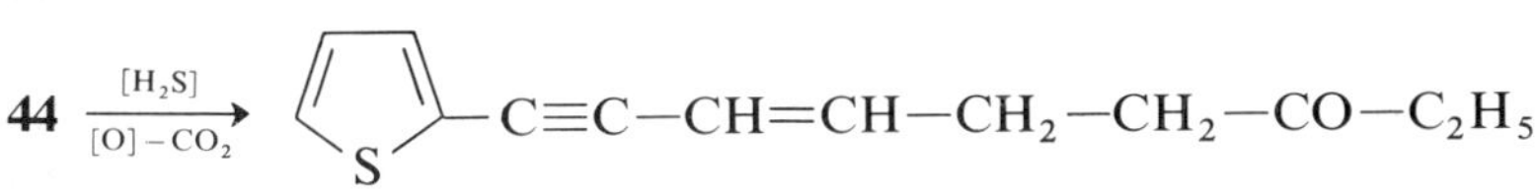

49

49 is relatively widespread in the tribus Anthemideae; it has been synthesised by the following route[56]:

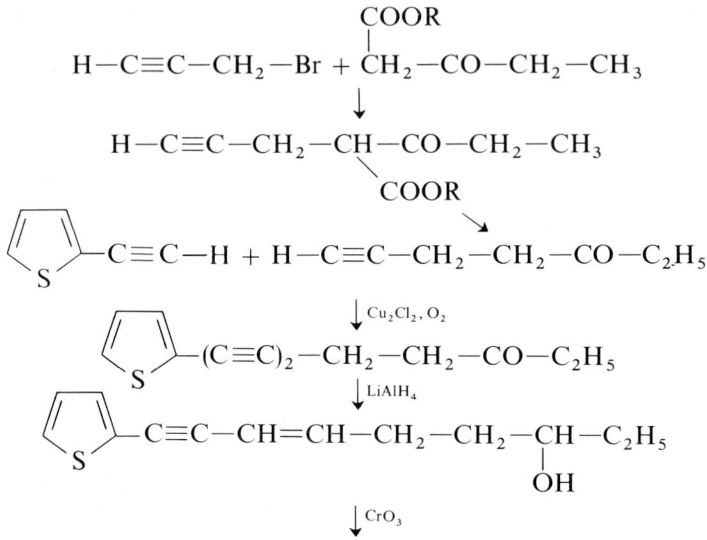

49

Many other thiophenes closely related to **48** and **49** have been isolated, e.g. from *Anthemis saguramica* Sosn.[57]:

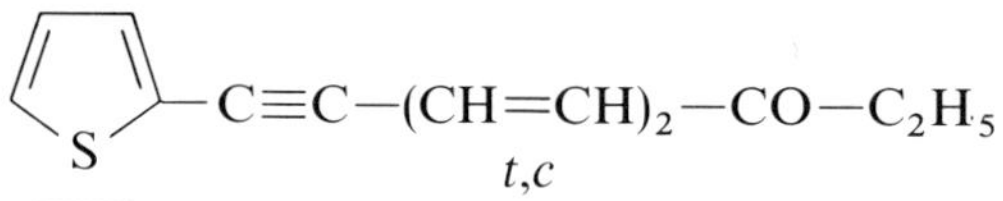
50

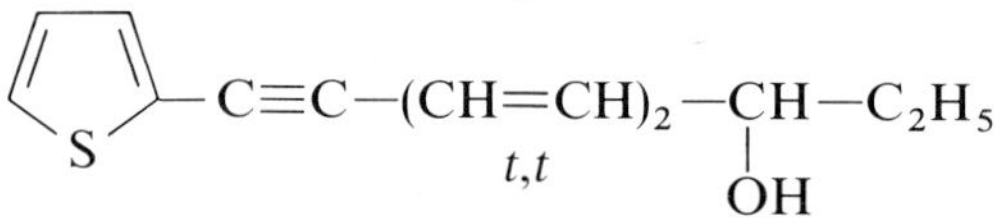
51

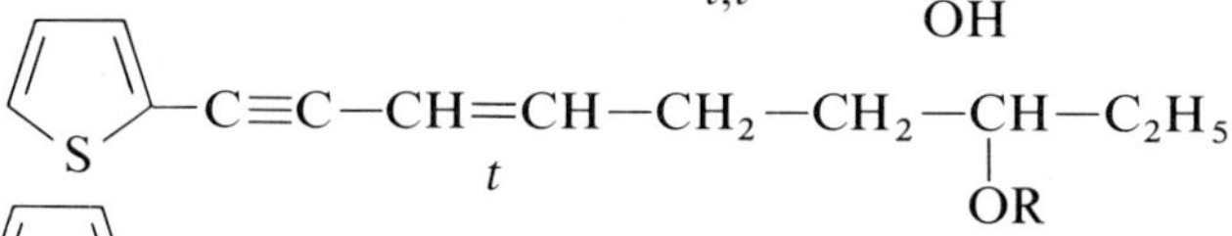
52: R = H

53: R = Ac

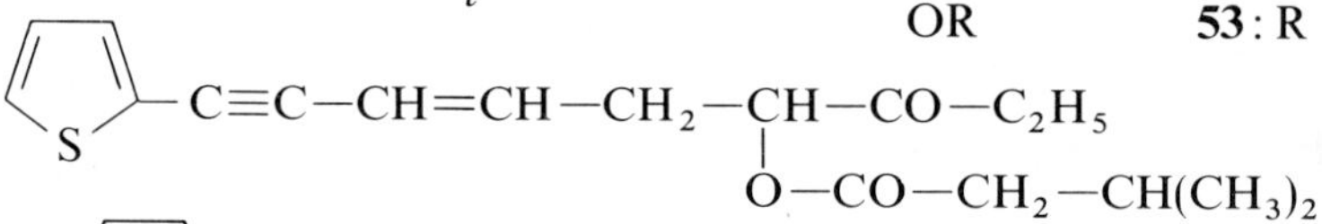
54

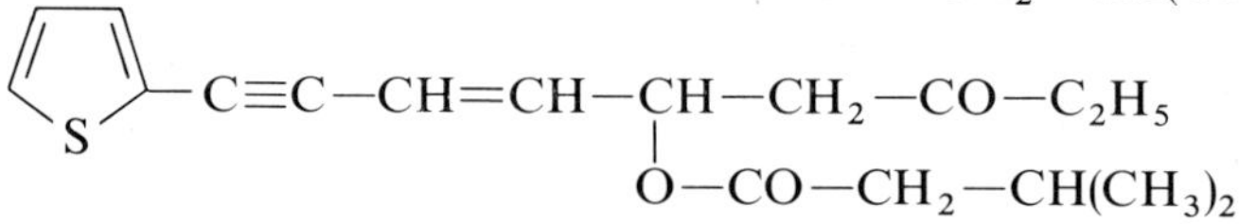
55

In the case of the ester ketones **54** and **55**, the n.m.r. spectra turn out to be very useful for structure determination. The following signals have been observed for the protons of **55**:

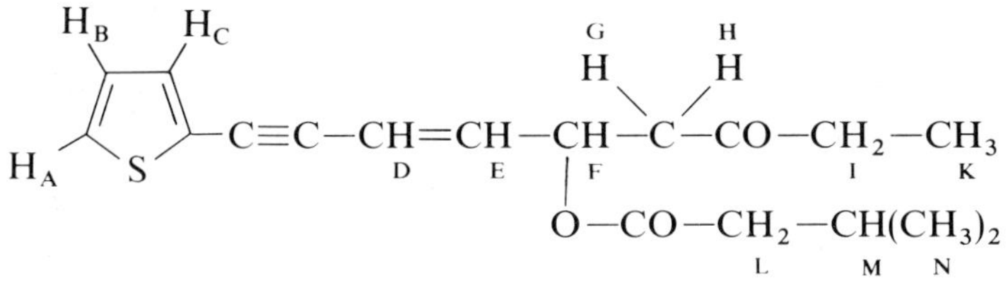

55

H_A *dd* 2·82	H_D *d* 4·15	H_I *q* 7·64 (J = 7 Hz)
H_B *dd* 3·10	H_E *dd* 3·90	H_K *t* 8·98 (J = 7 Hz)
H_C *dd* 2.90	H_F *ddd* 4·36	H_L *m* 7·84
	H_G *dd* 7·25	H_M *m* 8·4
	H_H *dd* 7·43	H_N *d* 9·07 (J = 6 Hz)

J_{AB} = 5·0 Hz; J_{AC} = 1 Hz; J_{BC} = 3·7 Hz; J_{DE} = 15·5 Hz;

J_{EF} = 5·5 Hz; J_{FG} = 7 Hz; J_{FH} = 6·5 Hz; J_{GH} = 15 Hz.

Position and coupling constants in the spectrum of **55** are clearly only in agreement with the proposed structure. The biogenetic relationships are obvious, but the occurrence of **55** may point out that perhaps **48** can also be formed by oxidation of **49** to the corresponding alcohol, which may be transformed to **48** by elimination of water. All the discussed thiophenes show broad, uncharacteristic u.v. spectra.

Another very rare ketone is found together with **44** in the roots of *Anacyclus radiatus* Lois.[58]. This ketone (**58**) may be formed in the following way:

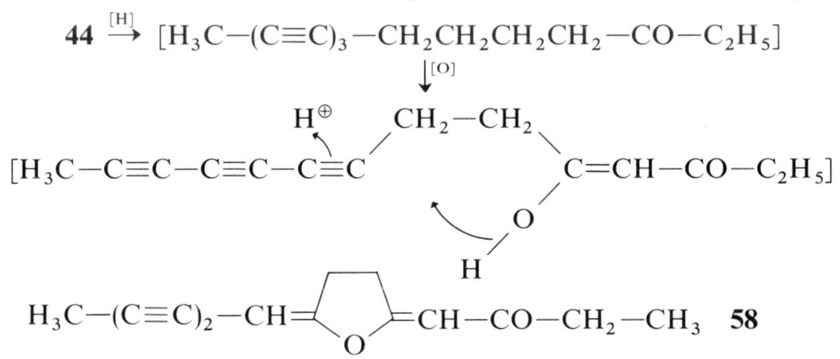

58

The n.m.r. data of **58** unequivocally establish the structure. The acetylenic methyl group signal is typically split by long range coupling to a doublet at 8·0τ, while the presence of the ethyl-keto group is indicated by a quartet at 7·60 and a triplet at 8·97τ. The methylene groups give rise to a complex A_2B_2-multiplet centred at 7·02τ. A triplet at 3·99 (J = 1·5 Hz) and a multiplet at 5·32τ represents the two olefinic protons.

The u.v. spectrum of **58** is of some interest. It shows that in this case resonance across the ring oxygen has to be assumed to be the u.v. maximum at 292 nm which is distinctly different from that resulting from simple superimposition of the spectra of a diyne enolether and a vinylogue ester. Even in the corresponding alcohol the normal diyne enolether chromophore does not appear.

The pronounced resonance can also be seen from the extremely shifted carbonyl band in the i.r. of **58** (1625/cm) which may be overlapped by the enolether band.

2.3. Tetrahydropyranes

Beside other *O*-heterocycles, some tetrahydropyranes have also been isolated from different Compositae species. The first compound of this type was found while looking for a substance responsible for the fish poison property of a South American Compositae plant—*Ichthyothere terminalis*[59]. It turns out that this fish poison was a polyacetylene. The same compound has been isolated from *Dahlia coccinea* Cav. var. coccinea[60], from *Chrysanthemum serotinum* L.[40] and from *Penttia globifera* N.[35].

Its u.v. spectrum is typical for a triyn-ene and the i.r. spectrum shows the presence of an *O*-acetate and a *trans*-double bond. Saponification of the optically active acetate gives an alcohol, which is also present in the plants. Mass-spectrometric determination the molecular formula of the alcohol showed this to be $C_{14}H_{14}O_2$. The n.m.r. spectrum of the acetate together with the already mentioned other data are only in agreement with the structure **59**. The biogenesis of **59** is more or less clear[38], the precursor is **40**, co-occurring with **59** in *Chrysanthemum serotinum* L.[40], itself formed from the very common precursor **5** (see Chapter 2, Section 2.1) as shown by feeding experiments:

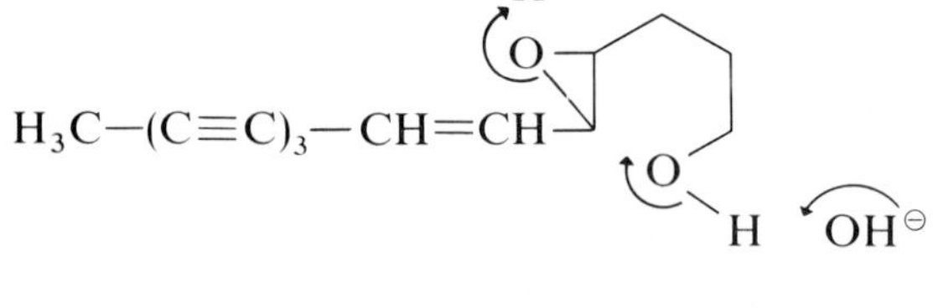

40

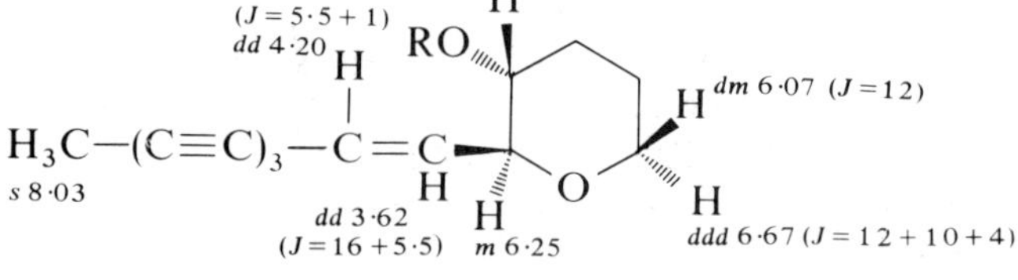

59: R = Ac

60: R = H

59 →

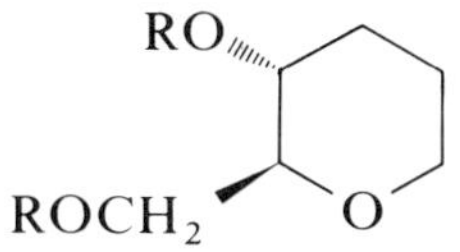

61

The absolute configuration given in **59** is established by degradation to the diol derivative **61** also synthesised from D-glucose[60] giving a compound with opposite configuration.

Similar compounds with an ene-diyn-ene chromophore are known too (see Chapter 3, Section 3.2).

Two further epimeric tetrahydropyranes have been isolated from *Centaurea muricata* L. Again the u.v. spectra show the presence of a triyn-ene system. While the two epimeric optically active alcohols are not to separate, the methyl ethers obtained with methanolic sulphuric acid can be separated by TLC. Lithium alanate reduction of the alcohols gives the diol **33** (see Chapter 2, Section 2.1) so together with the n.m.r. data, the structures must be **62** and **64**:

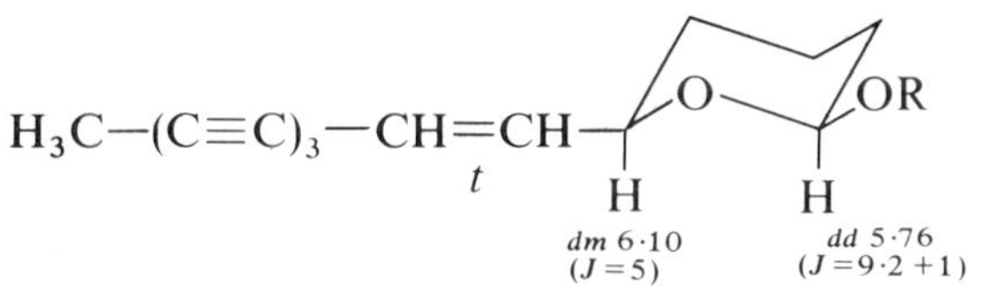

62: R = H
63: R = CH_3

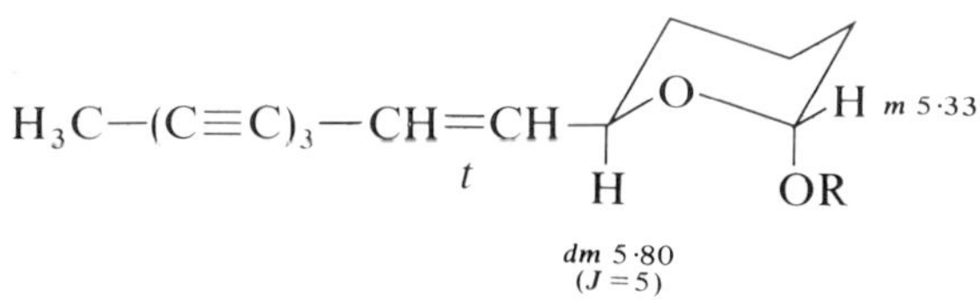

64: R = H
65: R = CH_3

The inseparable epimeric alcohols crystallise in different modifications. One, melting at 106°, is very photosensitive, while the other, melting at 109°, is much more stable. Atomic models show that the reason for this somewhat surprising difference is perhaps due to two arrangements in the crystalline state (see [41]). Only in one of these two possibilities are the polyyne chains parallel orientated, a necessary supposition to give the polymeric coloured products produced by irradiation.

The epimeric semi-acetales **62** and **64** also have been synthesised[61]. Grignard reaction of dehydrometricarianal (**67**) with the ketale **66** after hydrolysis and glycol cleavage leads to the desired polyynes:

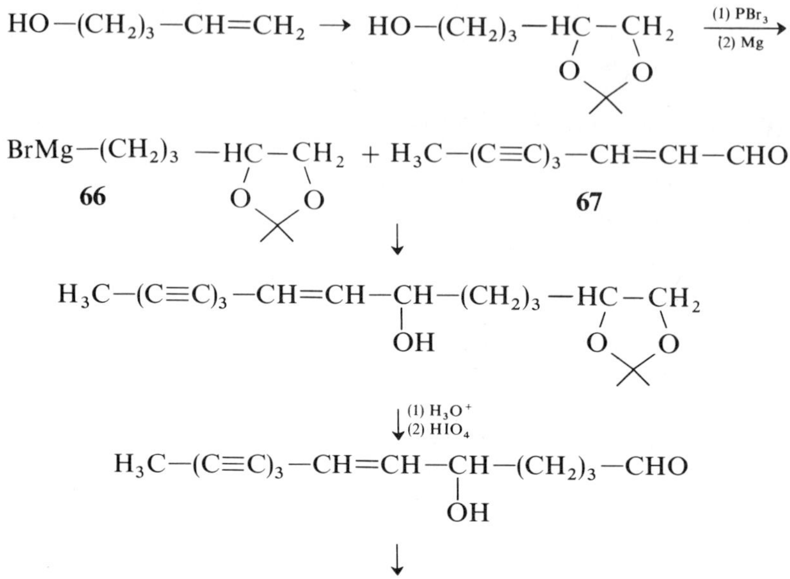

62 and **64**

33 clearly is also the biogenetical precursor of **62** and **64**.

2.4. Spiroketalenolethers

Beside other polyacetylenes obtained from the roots of *Tanacetum vulgare* L.[62], an optically active crystalline compound has been isolated with the molecular formula $C_{14}H_{14}O_2$. The u.v. spectrum is not very characteristic and shows only weakly-defined maxima at 320 and 310 nm, while from the i.r. spectrum the presence of an enolether double bond could be recognised. Ozone cleavage leads to hexa-diyn-al, while hydrogenation in the presence of small amounts of water gives a diketoalcohol its structure being elucidated as **70** by further chemical reactions. These results, together with the n.m.r. spectrum, are only in agreement with structure **68**:

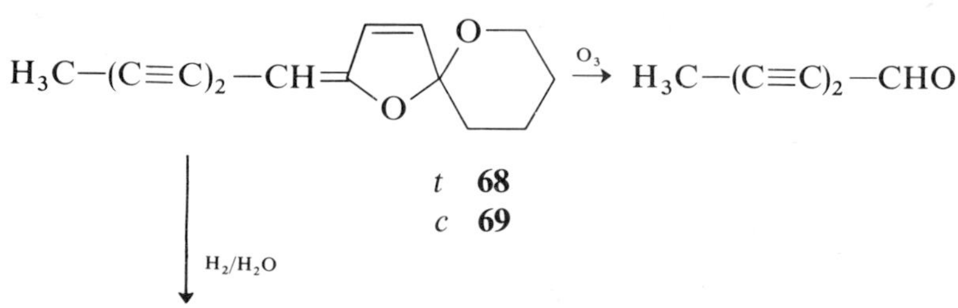

t **68**
c **69**

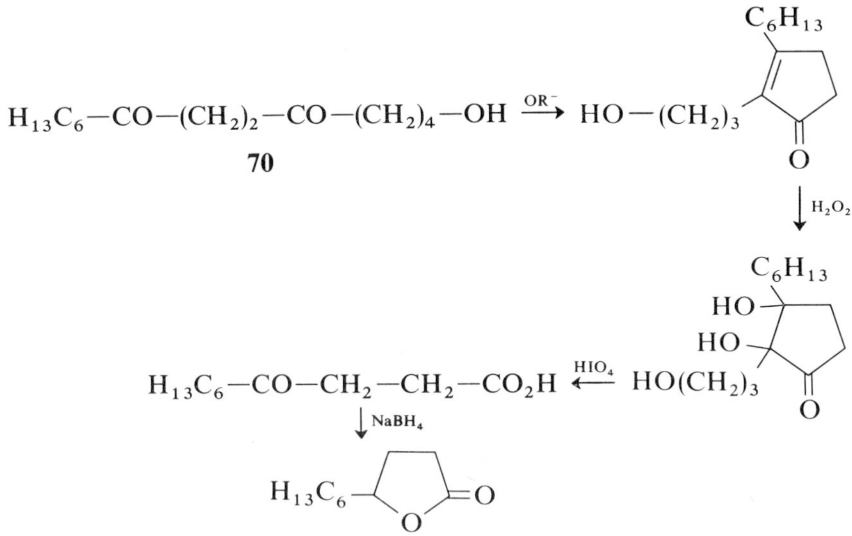

In the meantime, this enolether has been isolated from many species belonging to the tribus Anthemideae together with the corresponding *cis*-isomer **69**[63]. The structures are further established by synthesis[64], following the same routes as in the synthesis of the corresponding enolethers with the five membered ring (see Chapter 2, Section 3.2).

The biogenesis of **68** and **69** is well established by tracer studies with different labelled precursors. Again the important alcohol **27** is the starting molecule, which is transformed by oxidation to the ketoalcohol **71**, which then cyclises to the enolethers **68** and **69**[42]:

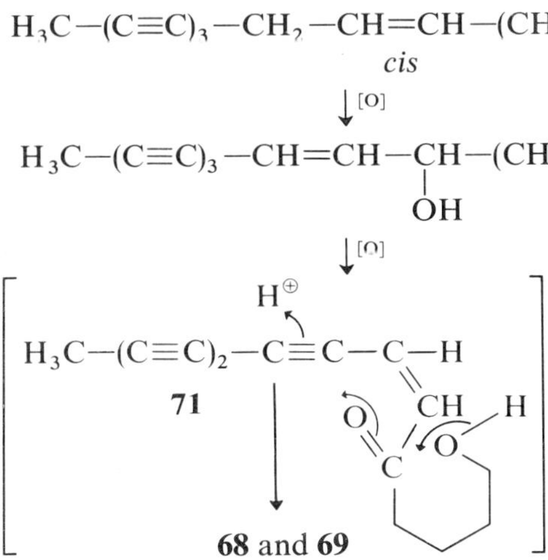

The reaction of **71** is also possible by *in vitro* treatment of the *trans*-configurated ketone **71** with tert.-butylate during irridation[42].

As in many other cases **68** and **69** also are transformed by oxidation to a number of further natural acetylenes. Epoxidation leads to the *cis–trans* isomeric epoxides **72** and **73**:

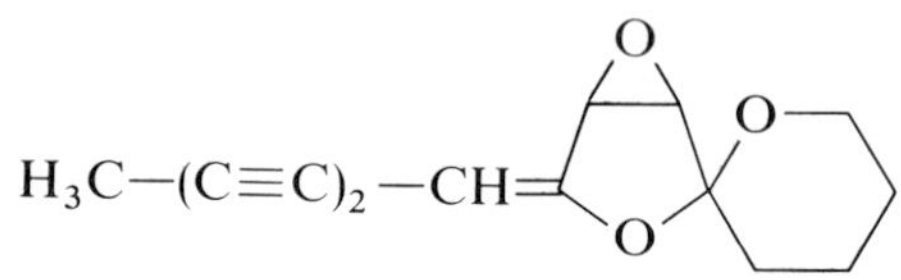

cis **72**[65]
trans **73**[35]

A second oxidation step is the introduction of oxygen in the six-membered ring. From different species again only from the tribus Anthemideae two pairs of *cis–trans* isomeric esters (**74**–**77**) have been isolated:

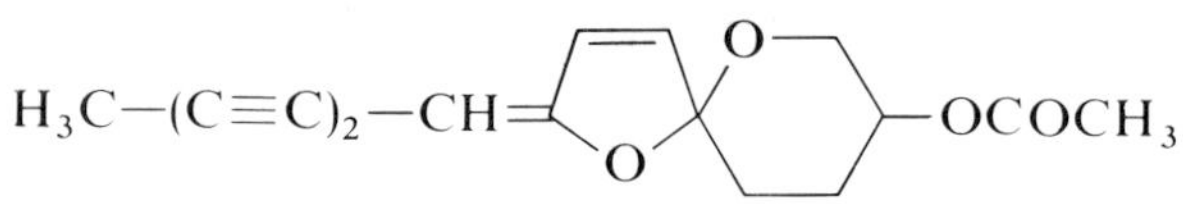

cis **74**[48]
trans **75**[48]

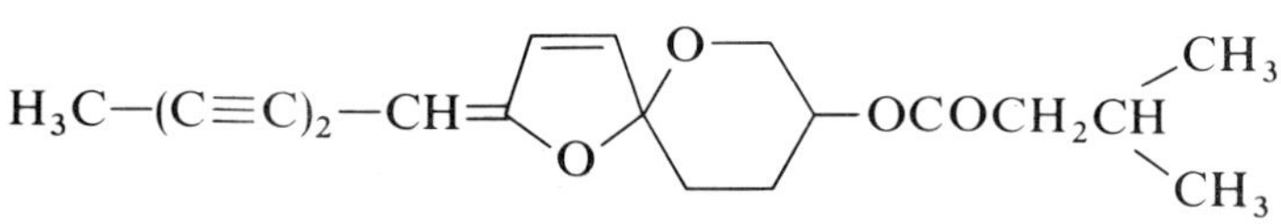

cis **76**[52]
trans **77**[52]

Again the n.m.r. spectra give the most important conclusions in structure determination of these esters (see the end of this section).

A further typical structure variation has also been observed in the field of the spiroketalenolethers. From *Chrysanthemum carinatum* Schousb., an enolether with an acetylene-H stretching band in the i.r. has been isolated in very small amounts[66]. All the other spectral data are very similar to those of **69**. Therefore, the structure must be **79**. As in other cases we have to assume that **79** is formed by oxidation of the methyl group in **69** followed by subsequent decarboxylation:

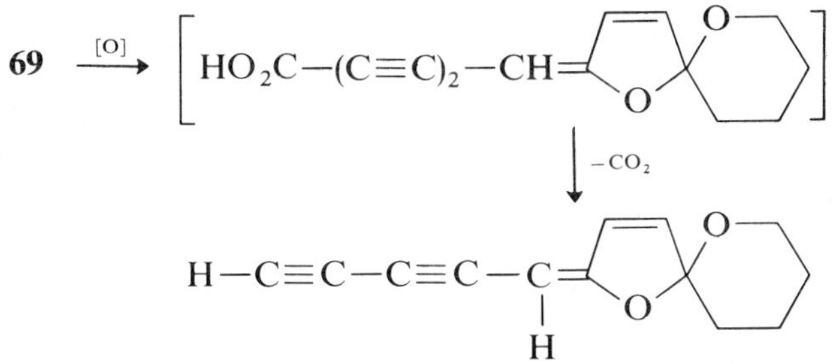

78

79

79 must be the precursor of a number of sulphur compounds also isolated from the same *Chrysanthemum* species and several others. The relatively uncharacteristic u.v. spectra (with maxima around 335 nm) of the two non-polar compounds give no direct statements on the possible structures. But the n.m.r. spectra show that we must deal with methyl-thioethers of six-membered spiroketalenolethers. A careful analysis of the spectra using decoupling techniques finally gives the structures **80** and **81** as the only possible ones. These compounds are clearly formed from the isomeric enolethers **79** by a formal addition of methyl mercaptan:

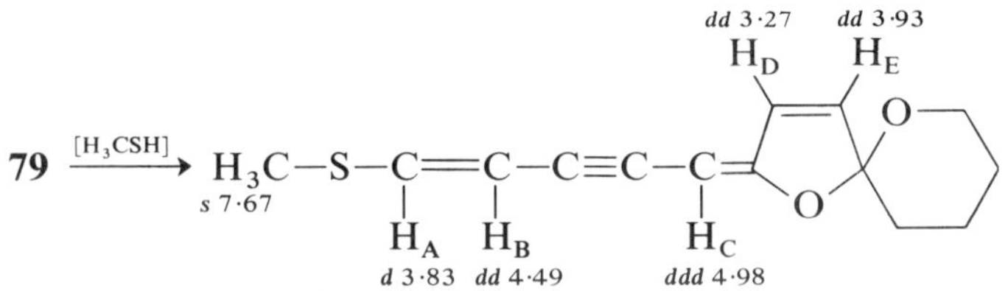

80

(J_{AB} = 10 Hz; J_{BC} = 2·8 Hz; J_{CD} = 0·6 Hz; J_{CE} = 1·9 Hz; J_{DE} = 5·5 Hz)

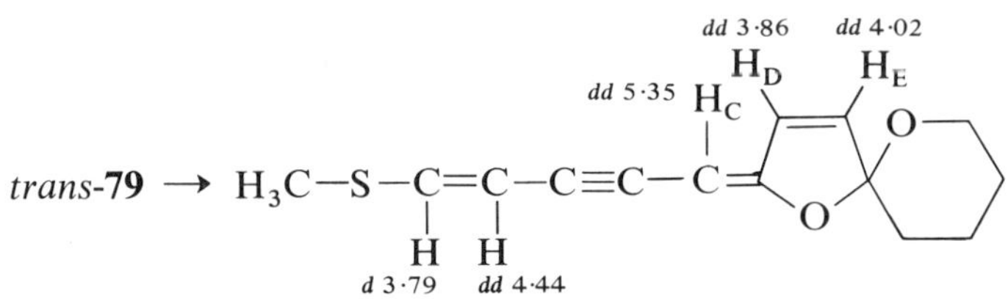

81

(J_{CE} = 0·7 Hz; J_{DE} = 5·5 IIz)

Clearly oxidation products of **80** and **81** are the much more polar compounds isolated from the same *Chrysanthemum* species[67]. The n.m.r. spectra are only in agreement with the structures **82–84**:

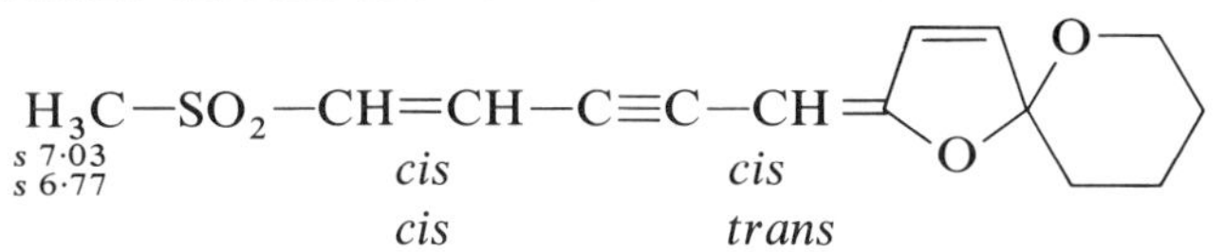

82

83

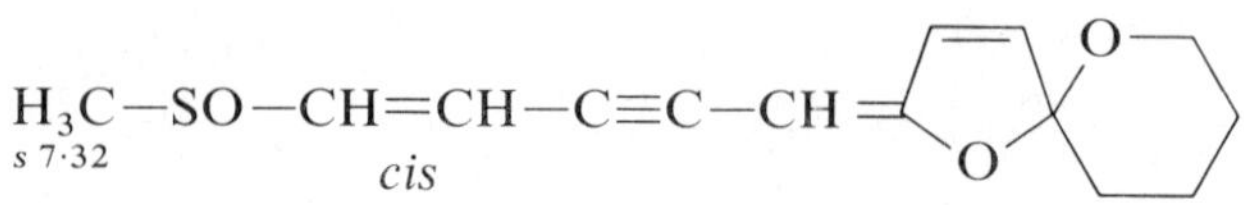

84

Similar sulphur compounds are also known with a five-membered ring enolether (see Section 3.2).

Finally a spiroketal-enolether should be mentioned having no acetylene group but surely is formed also from **79**. This compound first was isolated from *Chrysanthemum balsamita* L.[68] and later from several other members of the tribus Anthemideae. It shows a broad u.v. maximum at 327 nm. The molecular formula determined as $C_{13}H_{14}O_2S$ and an i.r. band at 860/cm lead to the conclusion that a thiophene compound is present. Oxidation gives thiophene carboxylic acid and after hydrogenation the lactone **86**. All data are only consistent with the structure **85**, the compound itself is probably formed by a formal H_2S-addition to **79**:

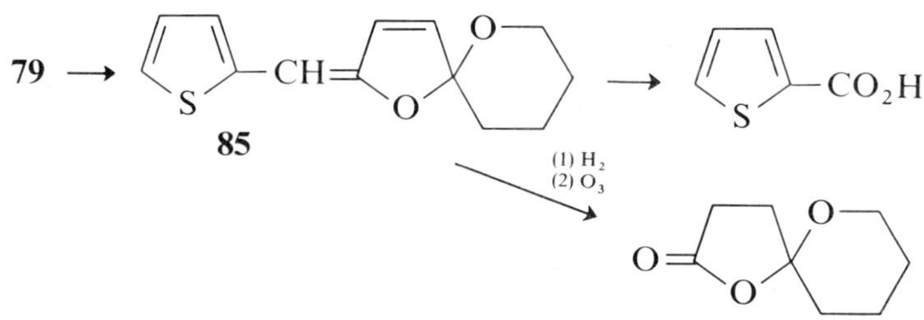

The structure has been further established by synthesis in the following way[56]:

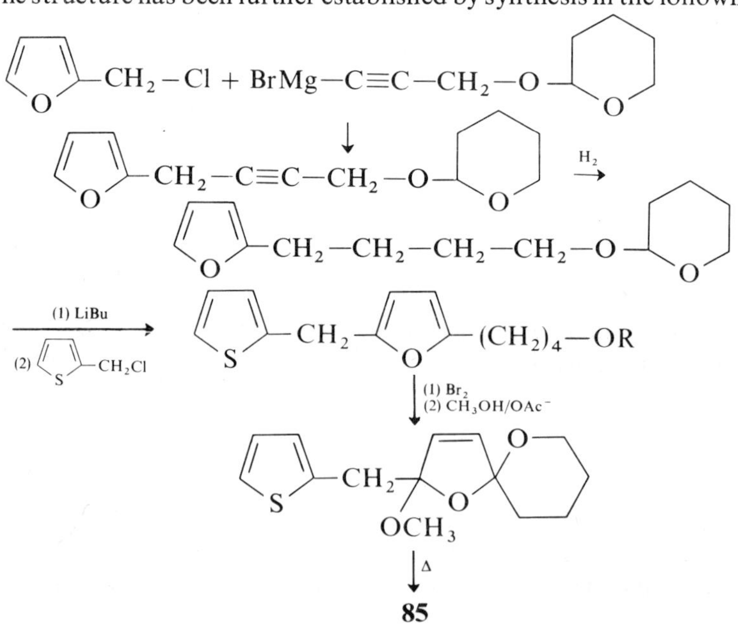

2.4.1. N.m.r. data of the spiroketalenolethers

The n.m.r. spectra of the enolethers are very characteristic, especially in connection with the elucidation of the configuration of the enolether double bond as shown by comparison of the data of **68** and **69**[63]. The *trans*-enolether proton always is shifted to higher fields, while in the case of a *cis*-proton the coupling H_{AC} is larger and the position of the signals of H_B and H_C are more differentiated.

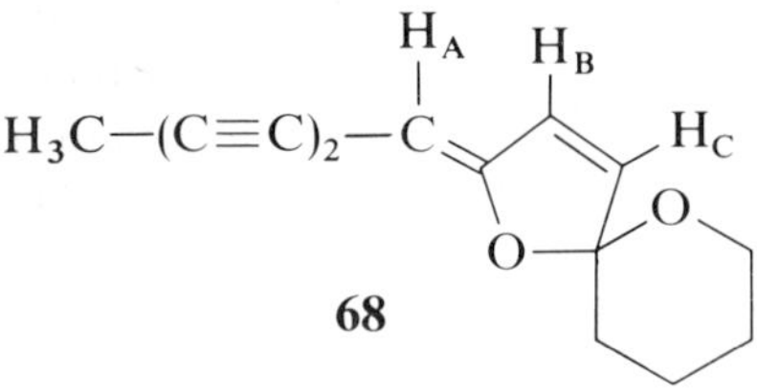

68

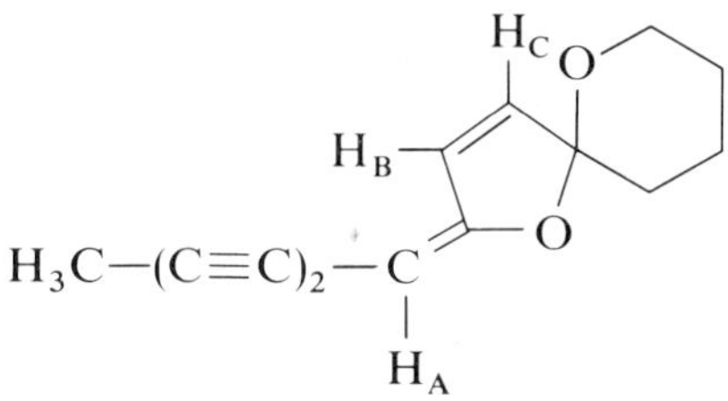

69

H_A *m* 5·46
H_B, H_C } AB*q* 3·89

H_A m 5·13
H_B *d* 3·38 (J = 5·5 Hz)
H_C *dd* 3·88 (J = 5·5 + 1·6 Hz)

A further example of the usefulness of n.m.r. spectroscopy in this field is the analysis of the spectrum of **74**[66]:

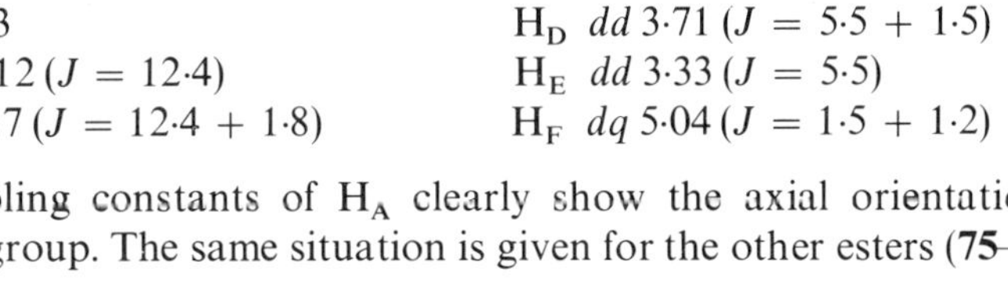

74

H_A *m* 5·13
H_B *dm* 6·12 (J = 12·4)
H_C *dd* 5·87 (J = 12·4 + 1·8)

H_D *dd* 3·71 (J = 5·5 + 1·5)
H_E *dd* 3·33 (J = 5·5)
H_F *dq* 5·04 (J = 1·5 + 1·2)

The coupling constants of H_A clearly show the axial orientation of the *O*-acetyl group. The same situation is given for the other esters (**75**–**77**)[66].

2.5. Aromatic Acetylenes

In 1958 N. A. Sörensen and J. S. Sörensen[69] had already isolated a phenyl-acetylene with a very characteristic u.v. spectrum. In comparison with a triyn-ene system, the observed maxima are all shifted to longer wavelengths because of the phenyl group in conjugation with a triyne. The structure was elucidated as **87** from the u.v. data and following hydrogenation to *n*-heptyl benzene:

87

The interesting question of the biogenesis of this compound has been established by several feeding experiments. Firstly, it appears that **87** is composed of acetate units which show the relationship with the other natural acetylenes[38]. **87** is then formed by feeding different labelled **27**[70]. As a result of these experiments, the following proposal is the most likely for the biogenesis of **87**:

$$H_3C{-}(C{\equiv}C)_3{-}CH_2{-}CH{=}CH{-}(CH_2)_4{-}OH \quad \mathbf{27}$$

cis

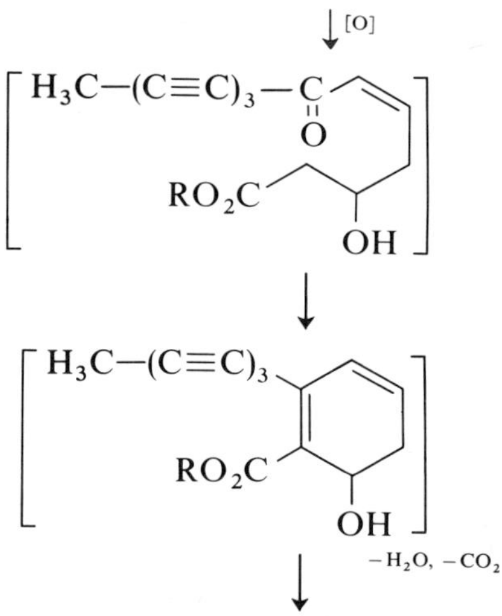

87

The postulated intermediates are very probably those indicated by feeding experiments with [14-^{14}C-2·3-^{3}H]-**27** since half of the tritium at C_3 is incorporated[70].

Simple oxidation products of **87** are the alcohol **88** and the corresponding acetate **89** isolated from different *Bidens* and *Coreopsis* species[71]:

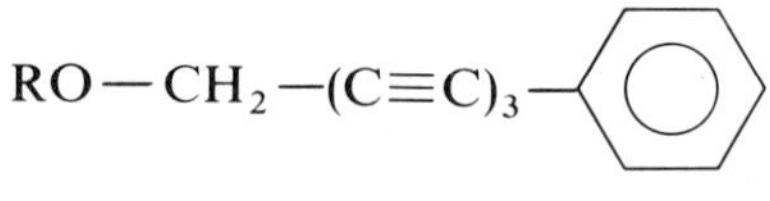

88: R = H
89: R = Ac

The synthesis of the two compounds has been realized by oxydative coupling of phenyl acetylene with penta-2,4-diyn-1-ol[71]:

$$HO-CH_2-C{\equiv}C-C{\equiv}C-H + H-C{\equiv}C-PH \rightarrow \mathbf{88} \xrightarrow{Ac_2O} \mathbf{89}$$

A further transformation product of **87** is the thiophene **90** isolated also from *Coreopsis* species[72], clearly formed by formal H_2S-addition to **87**:

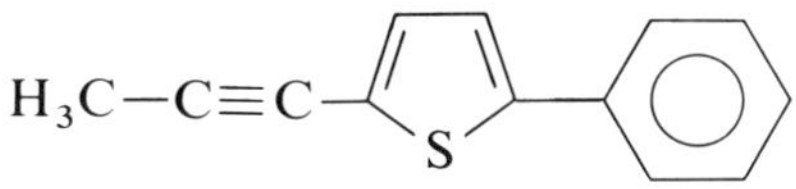

90

An acetate has also been isolated from a *Coreopsis* specy and which turns out to be **91** and not **92**[73]. These two compounds can be distinguished most easily by mass spectroscopy; as in **92** the fragmentation with loss of *O*-acetyl is much more favoured due to the stability of the resulting cation[73]:

91

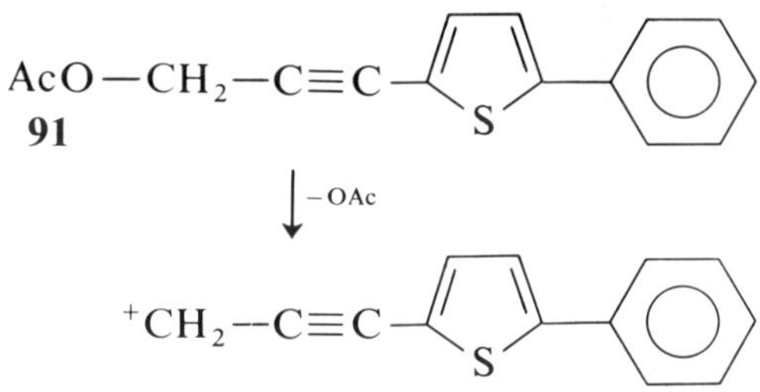

92

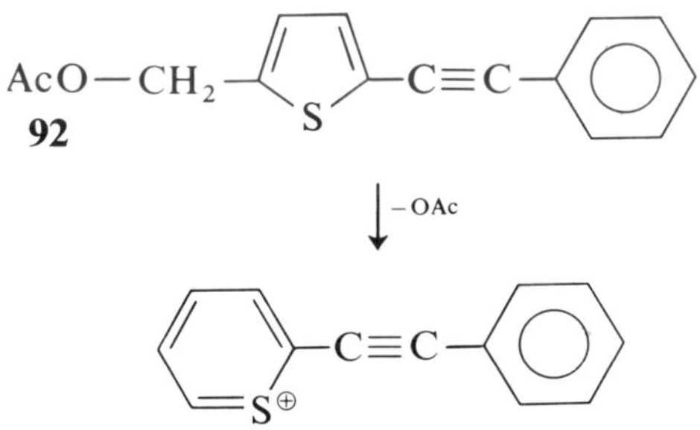

A thiophene of type **92** also has been found in a *Bidens* species[74]. The yellow crystals with a broad u.v. maximum, on sodium borohydride reduction, yield an alcohol identical with the saponification product of **92**. Therefore, the structure of this thiophene must be **93**:

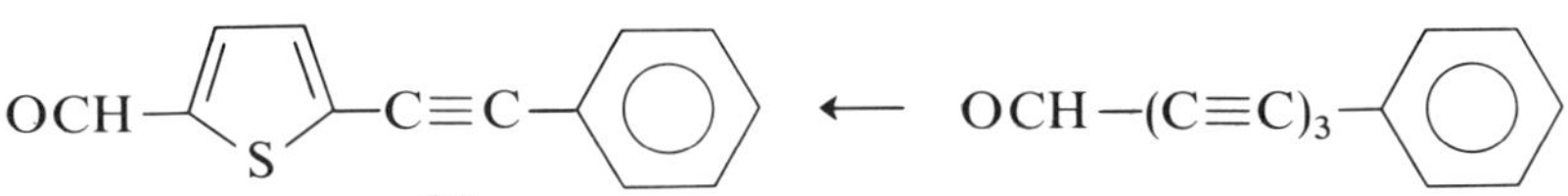

93

It seems to be probable that **91** is formed as mentioned via **90** and **87**, and **93** by formal addition of H_2S to the corresponding phenyltriyne aldehyde[73].

In vitro H_2S-addition under alkaline conditions to **87** as well as to **88** only gives thiophenes of type **92**[73].

2.6. Compounds Related to the Tridecapentaynenes

In 1953 Sörensen and his coworkers published the proposed structure of a widespread hydrocarbon isolated in very small amounts from different representatives of the family Compositae. This compound crystallises as yellow needles at low temperatures from petroleum ether solutions, but at room temperature these crystals immediately give a black polymer. The u.v. maxima and the isolation of *n*-tridecane by hydrogenation lead to the structure **94**[75].

Some years later, this structure has been established by synthesis[76]. Starting with hexadiynal, this hydrocarbon has been built up in the following way:

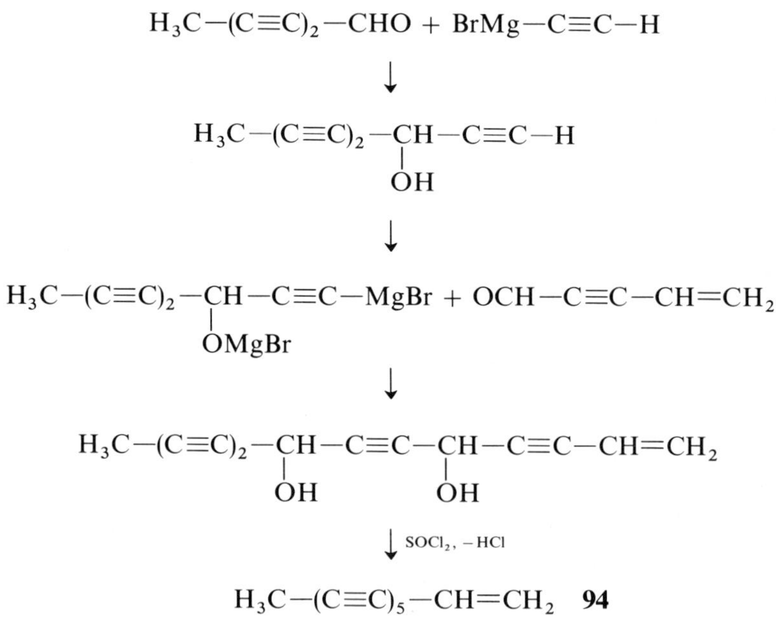

94

The biogenesis of this important polyyne is represented by the following pathway in analogy to that one established for the corresponding tetrayndiene (see Section 2.7.1.).

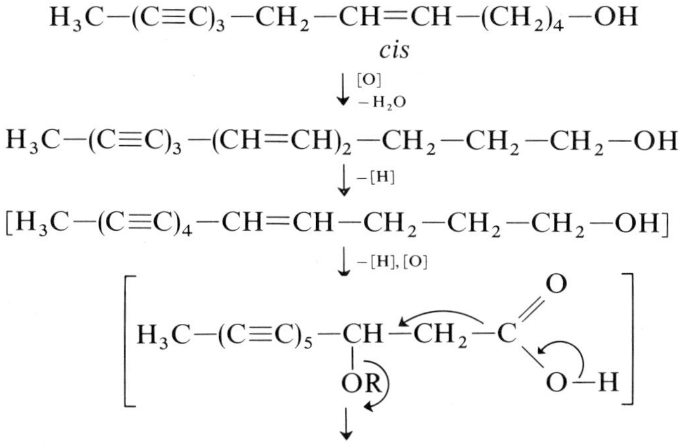

27

36

94

This widespread polyyne is transformed to a large number of other acetylenes, which will be discussed in the following Sections.

2.6.1. Oxidation products

So far, only two simple oxidation products of **94** have been isolated. In addition to many other polyynes, a polar crystalline compound has been isolated, sparingly soluble in unpolar solvents, from the roots of *Bidens leucanthus* L. The extremely unstable polyyne has therefore been acetylated yielding well-defined crystallised needles of a more soluble compound. The very characteristic n.m.r. spectrum and the typical pentayne u.v. spectrum are only in agreement with structure **96**[77]; therefore the natural polyyne must be **95**, being the first example of a natural simple pentayne:

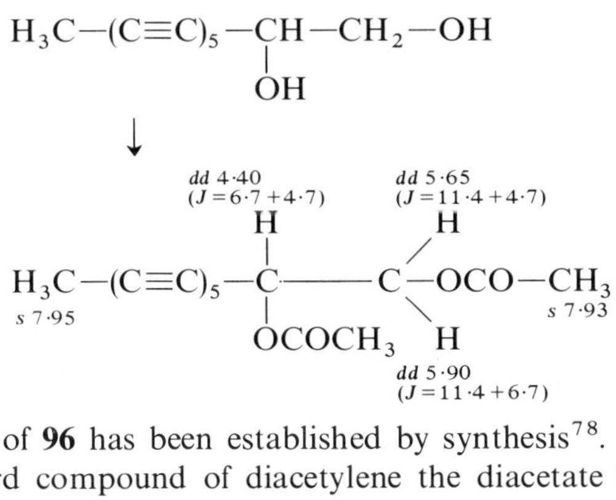

95

96

The structure of **96** has been established by synthesis[78]. Starting with the mono-Grignard compound of diacetylene the diacetate **97** has first been synthesised; on coupling with heptatriyne this yields the diacetate **96**:

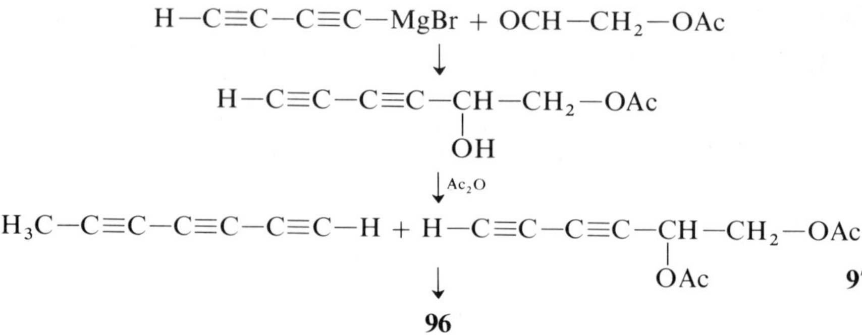

97

96

The corresponding epoxide **98** is present in very small amounts in the roots of *Centaurea deusta* Ten.[79]. It has been identified by hydrolysis to the diol **95**, which gives on periodate oxidation the extremely unstable aldehyde **99** with u.v. maxima at 419, 388, 360, 337, 289, 278, 260 and 247 nm.

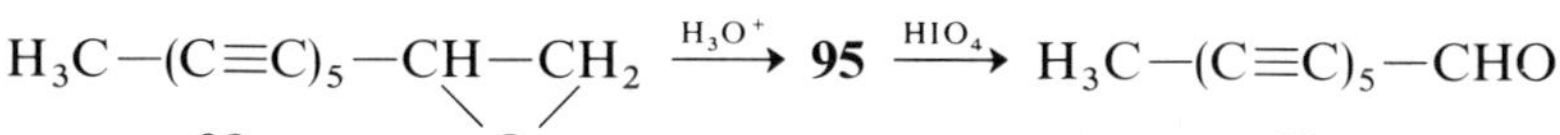

98 99

The u.v. data do not agree with those reported for a pentayne-ketone and so its structure is still not established[80].

2.6.2. Addition products

A number of different addition products of **94** are also formed in many Compositae. The first compounds of this type have been isolated from *Flaveria repanda* Lag. Beside **94**, the root extract yields two relatively non-polar compounds showing uncharacteristic u.v. maxima only at 360 nm. The separation of these two crystalline polyynes is very difficult. Both compounds have the molecular formula $C_{14}H_{10}S$. The i.r. spectra show the presence of a conjugated vinyl group, while the n.m.r. spectra are best interpreted by the *cis–trans*-isomeric thioethers **100** and **101**[81]:

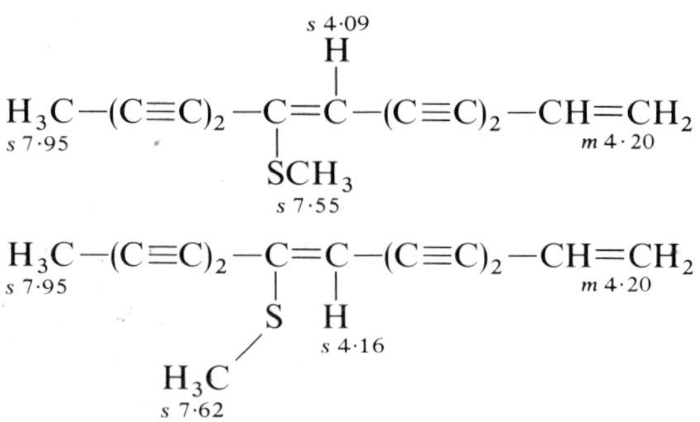

100

101

The structures have been established by synthesis[82]:

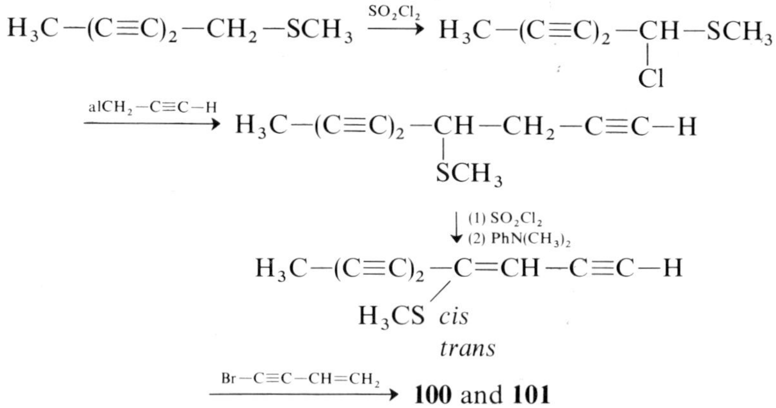

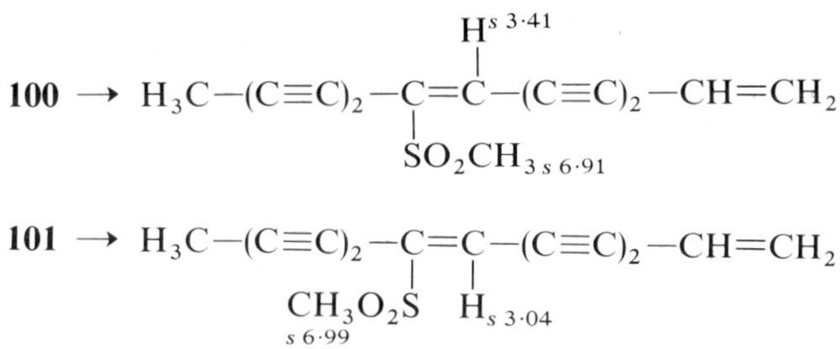

102

103

The n.m.r. signals of the corresponding sulphones **102** and **103** clearly establish the configurations, as the deshielding effect of the sulphone group leads to a downfield shift of the *cis*-olefinic proton.

The corresponding optically active *trans*-sulphoxide **104** has been isolated from the roots of *Flaveria repanda* Lag.[81], while the roots of *Helenium hoopesii* A. Gray contain the *cis*-isomer **105**[83]:

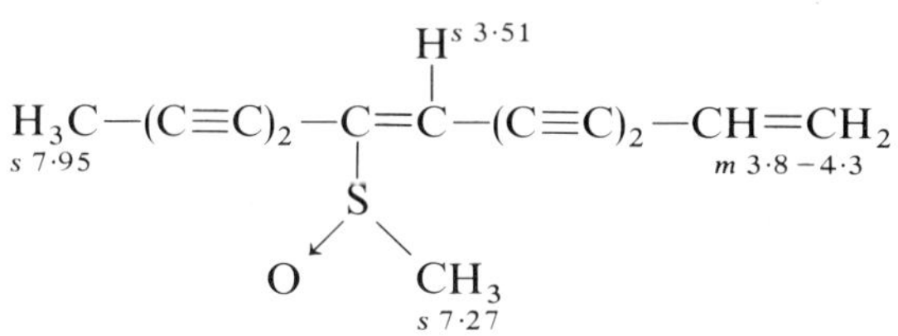

104

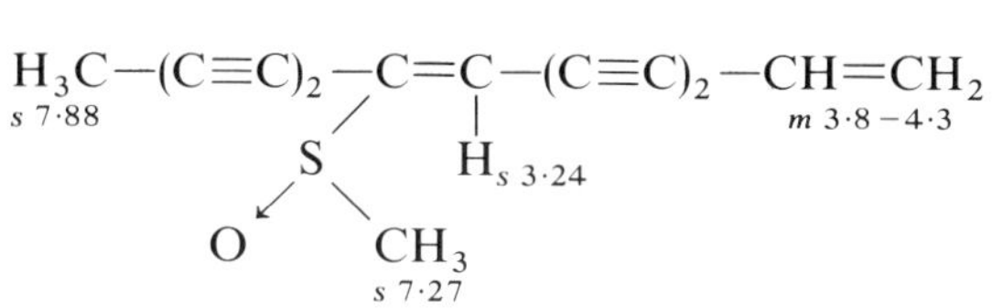

105

Surprisingly, the roots of *Helenium tenuifolium* Nutt. contain the stereoisomeric sulphone **106** as is clearly shown by the n.m.r. spectrum[83]:

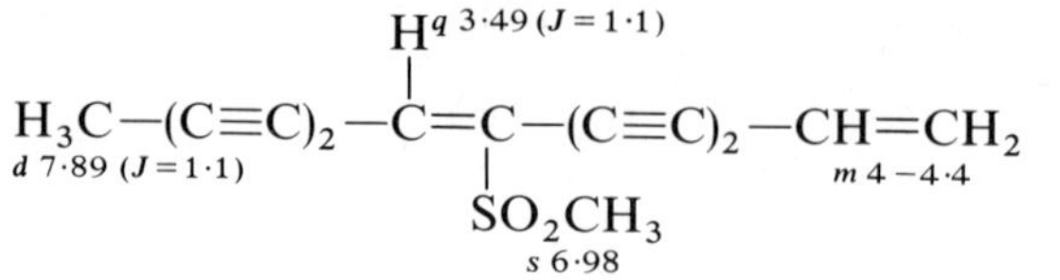

106

From *Gaillardia aristata* Pursh. and also from *Helenium hoopesii* A. Gray, two crystalline sulphones have been isolated[83]. The less polar one shows a U.V. maxima typical for a diynene shifted by about 10 nm to longer wavelengths (λ max: 293, 276, 261, 218 nm). The n.m.r. spectrum and the molecular formula of the optically active acetylene is only in agreement with the structure **107**:

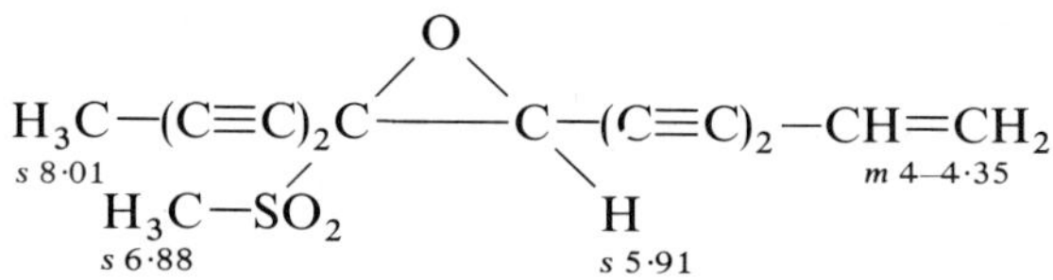

107

The position of the signal of the epoxy-proton is best represented with a *cis*-configuration, especially if we compare the observed τ-value with that of the second epoxy-proton signal in the di-epoxide **108**, which turns out to be the more polar compound:

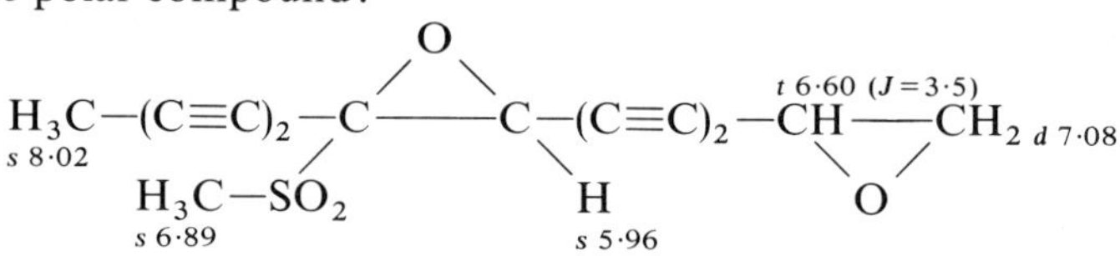

108

107 and **108** are the first naturally occurring epoxy-sulphones.

The biogenesis of these sulphur compounds is established by feeding labelled **94**[84], which shows that **100** and **101** are formed from the pentayne **94**. Surely the other compounds are biochemically oxidation products of **100** and **101**.

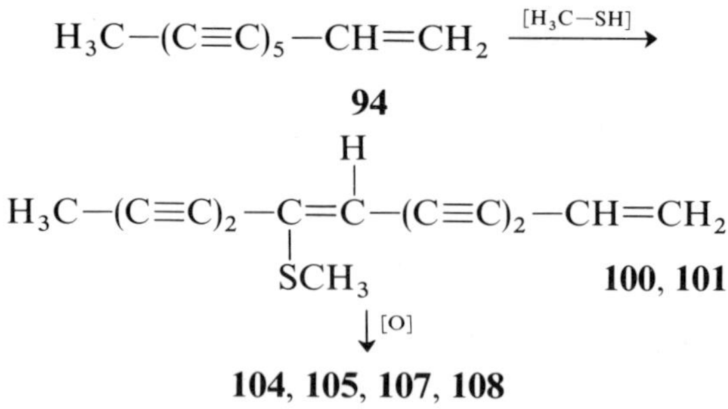

The biogenetic pathway to **106** is still not clear. Perhaps the addition of a sulphur compound has taken place in an earlier stage, perhaps already the triyne **36** is a precursor. Subsequent transformation in the discussed manner (see Section 2.6) then could give **106**[85]:

$$H_3C-(C\equiv C)_3-(CH=CH)_2-(CH_2)_3-OH \qquad \mathbf{36}$$

$$\downarrow CH_3SH$$

$$\left[H_3C-(C\equiv C)_2-CH=\underset{\displaystyle SCH_3}{\underset{|}{C}}-(CH=CH)_2-(CH_2)_3-OH \right] \qquad \mathbf{109}$$

$$\downarrow$$

$$\mathbf{106}$$

Different directions of such additions have been observed also in other cases (see Section 5.2).

The nature of the sulphur compound formed first in all these cases is not quite clear. Probably, as in the formation of the enolthioethers derived from dehydromatricariaester (see Section 5.2), we have a two-step reaction:

$$-C\equiv C-C\equiv C- \longrightarrow -\underset{\displaystyle SH}{\underset{|}{C}}=CH-C\equiv C- \longrightarrow -\underset{\displaystyle SCH_3}{\underset{|}{C}}=CH-C\equiv C-$$

$$\mathbf{110}$$

Again the intermediate of type **110** seems to be the precursor for the large number of thiophenes, which will be discussed in the next Section.

2.6.3. Thiophenes

Thiophene acetylenes which are all derived from the pentayneene **94** have been isolated from a large number of *Compositae* species. As previously mentioned, the most probable thiophene formation is the cyclisation of a precursor of type **110**:

110 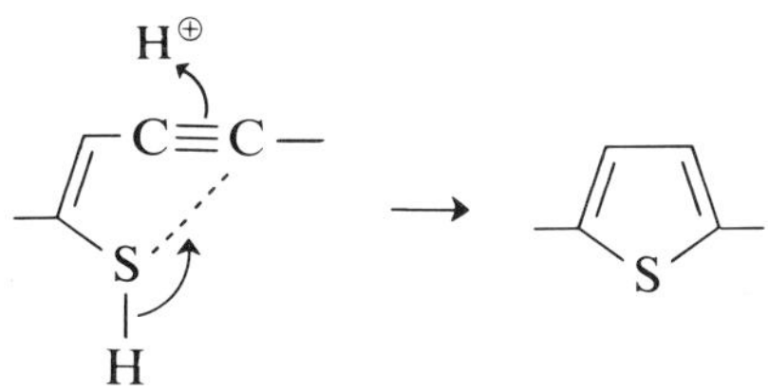

By feeding experiments it has been shown that several of these thiophenes are formed from **94**[84].

From the four possible thiophenes directly related to **94**, only two are known; the diynes **111**[50] and **112**[46]:

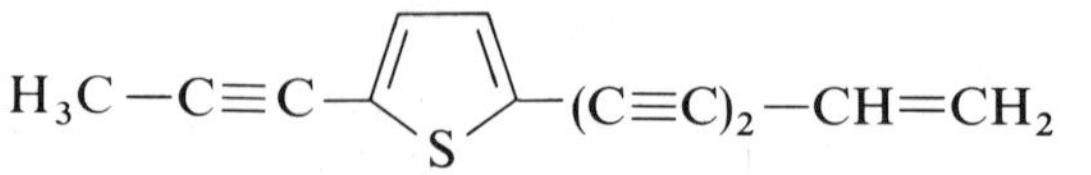

111

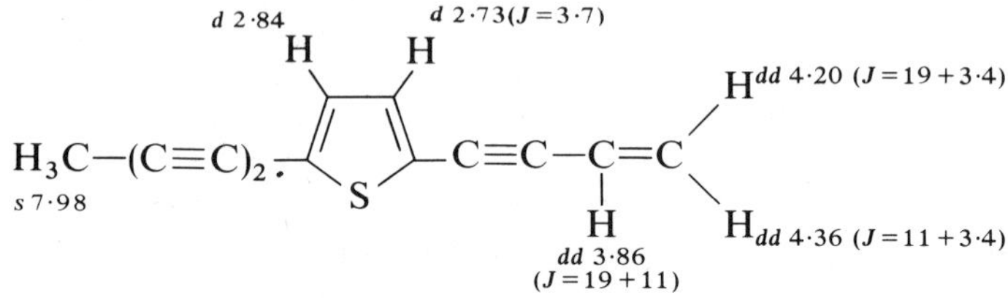

112

The spectral data for **111** and **112** are very similar. The corresponding hydrogenation products have therefore been produced. Comparison with the synthetic dialkylthiophenes allows the correct assignments.

As well as **111**, a polar thiophene has been isolated from the roots of *Rudbeckia triloba* L. The spectral data at once show that its structure is **113**[74], an oxidation product of **111**:

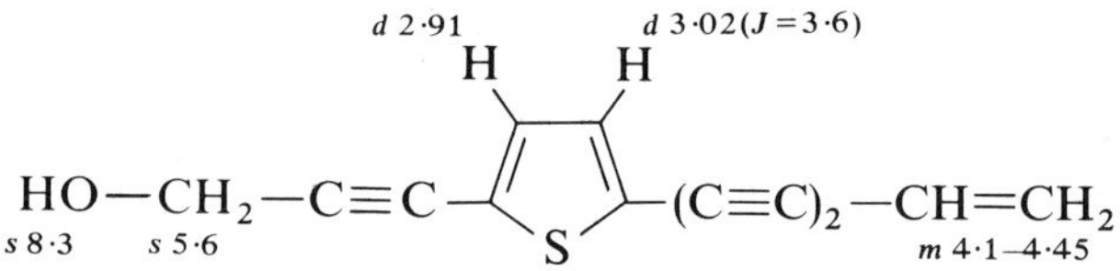

113

All the other known monothiophenes related to **94** are transformation products of **112**. From *Echinops* species several compounds with very similar u.v. spectra have been isolated[46]. The most non-polar one shows characteristic n.m.r. signals which are only in agreement with structure **114**, clearly derived from **112** by epoxidation:

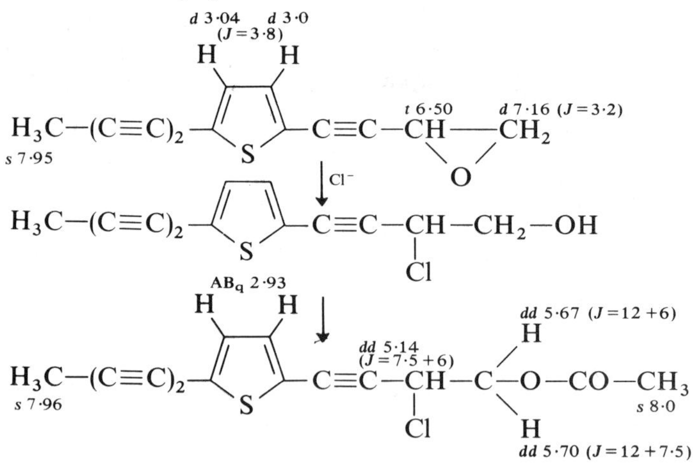

114

115[46]

116[46]

114 → 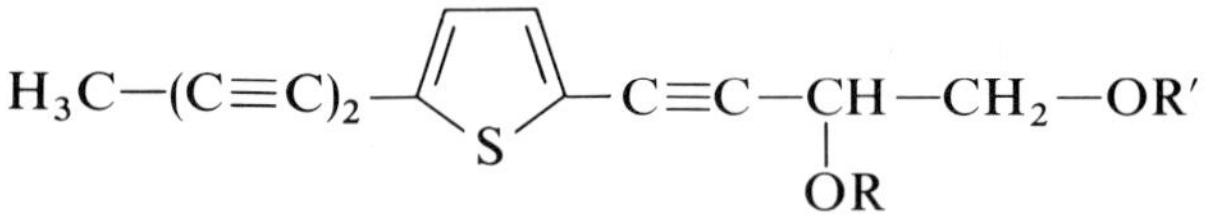

117[46] : R = R′ = H
118[46] : R = R′ = Ac
119[46] : R = H, R′ = Ac

In the same plant, a number of thiophenes have been found and which are all derived from **114**. Ring opening by chloride ions leads to the chlorohydrine **115** and the corresponding acetate **116**, while hydrolysis gives the diol **117** and then the diacetate **118** and the monoacetate **119**, which has been isolated from *Eclipta erecta* L.[86].

The structures are well established by correlation of the different thiophenes. So **115** and **116** yield **114** on treatment with diluted alkali, while **115** could be obtained by acidic methanolysis of **116**. Hydrolysis of **114** gives the diol **117**, which, on acetylation, forms the diacetate **118**. Periodate oxidation of **117** leads to the aldehyde **120**[46], while the position of the free hydroxyl in **119** is established by oxidation with manganese dioxide, yielding the ketone **121**[86]:

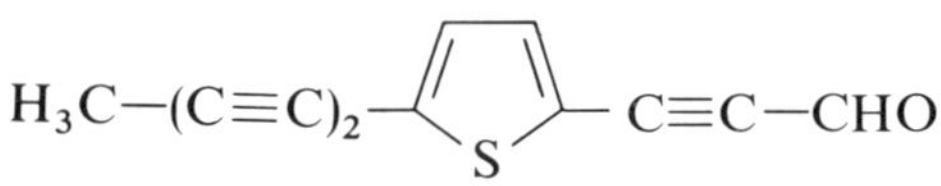

120

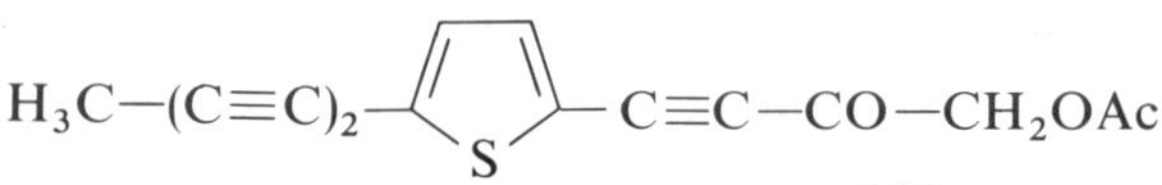

121

The roots of *Eclipta erecta* L. also contain two thiophenes showing the same molecular formula as **115** and **116**, respectively. The spectral data, as well as the transformation to **114** by alkaline treatment, are only in agreement with the structures **122** and **123**[86]:

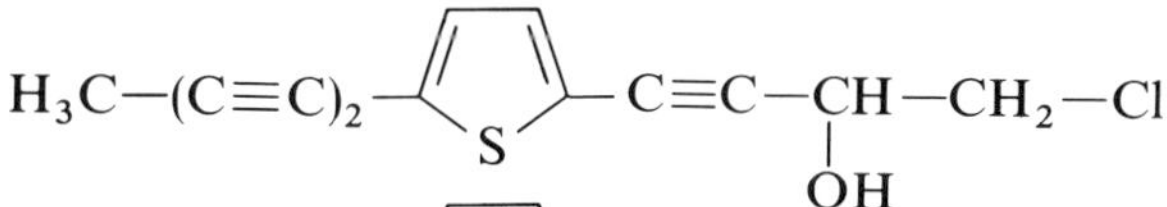 **122**

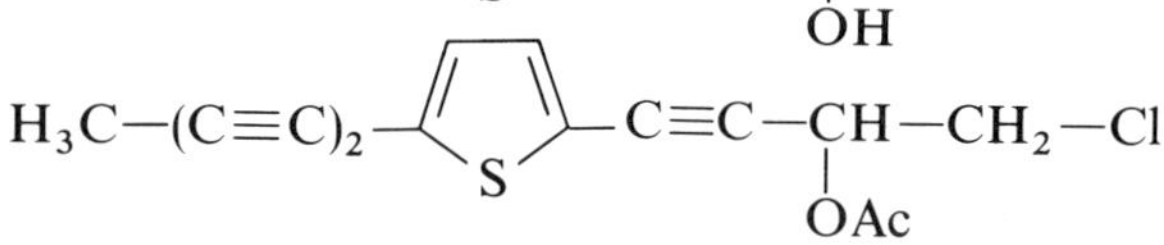 **123**

Also from *Eclipta erecta* L., two further thiophenes have been isolated showing the presence of a —C≡C—H group in the i.r. spectrum. The spectral data of the less polar compound are only in agreement with structure **124**, while those of the polar thiophene are best represented by **125**[86]:

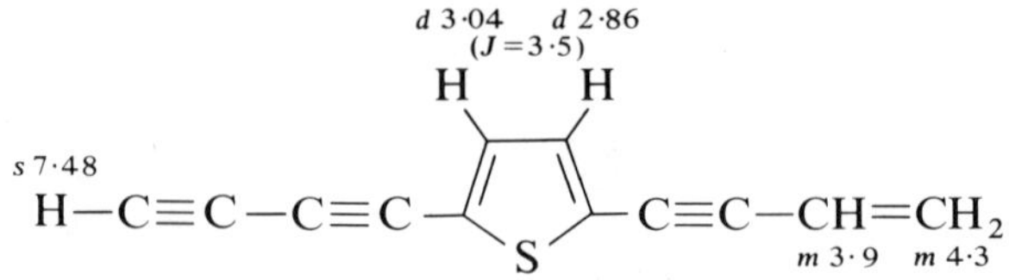

124

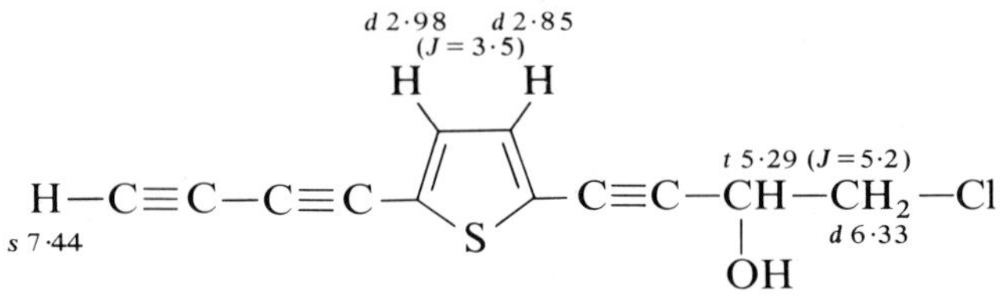

125

An alcohol and the corresponding acetate are also present in the roots of *Echinops* species. Molecular formula and the spectral data clearly establish the structures of the two thiophenes **126** and **127**[46].

The biogenesis of these two acetylenes is not really established, but it seems probable that they are formed both from **112**—perhaps by reduction of **114**:

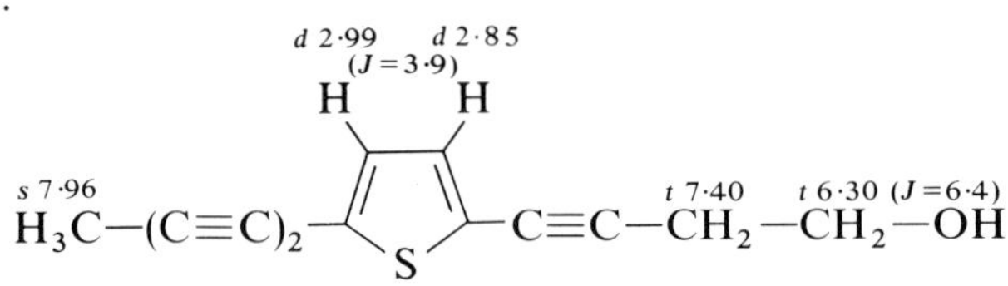

126

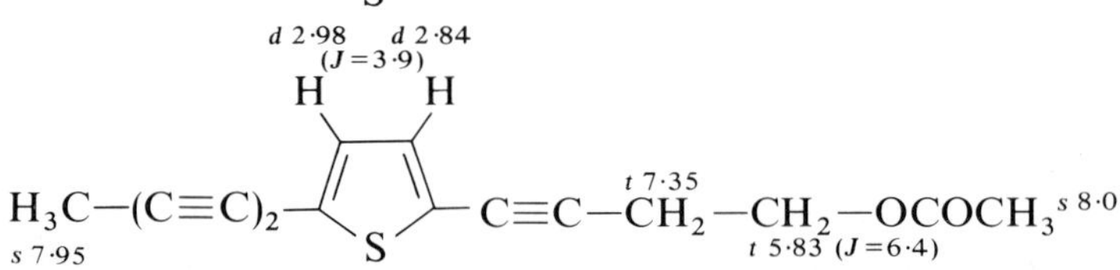

127

The structures of **126** and **127**, as well as many others, have been established further by synthesis. A convenient route is the method of Stevens and Castro[87]. Starting with 2,5-di-iodothiophene the reaction with the cuprous salt of butyn-(1)-ol-(4) gives a thiophene, which on subsequent reaction with the cuprous salt of pentadiyne yields the alcohol **126**[88]:

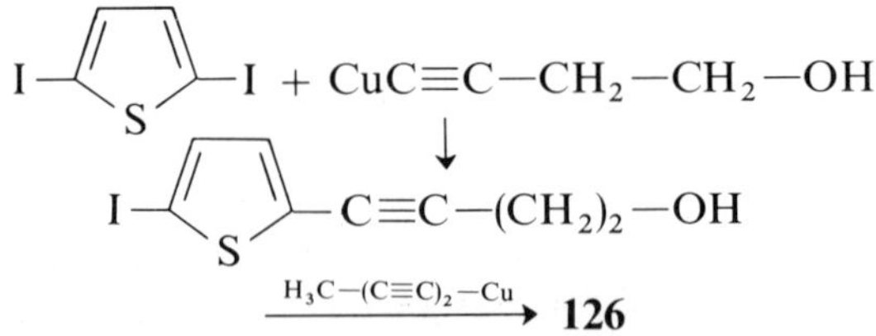

Similarly diiodothiophene on reaction with the cuprous salt of pentadiyne leads to the corresponding monoiodo-thiophene, which gives (with the cuprous salt of butenyne) the thiophene **112**[88]:

Similarly, the following acetate has been synthesised and can be transformed by the usual methods to the thiophenes **114–119**[88]:

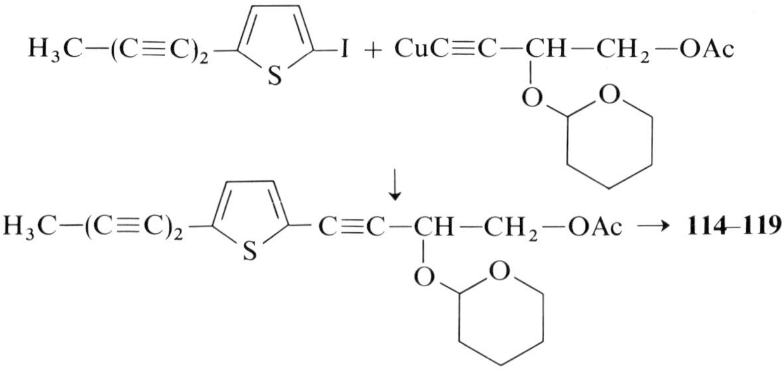

111 has been synthesised by the reaction of di-iodothiophene, first with the cuprous salt of propyne, and then with that one of hexadiyneene[88]:

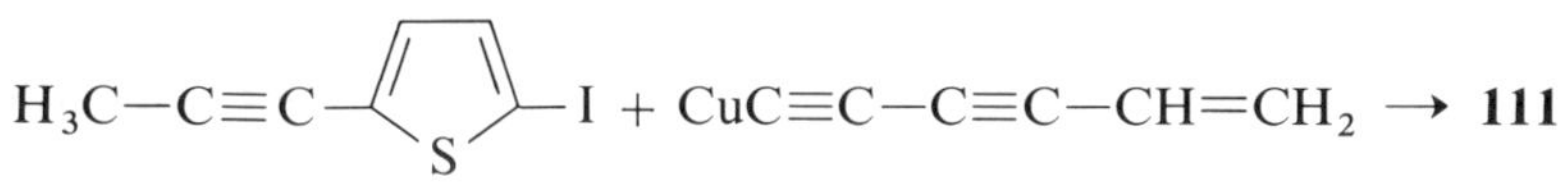

Besides thiophenes, a few dithio compounds have been isolated from different Compositae. Together with **111** and **112**, these red coloured compounds have often been found difficult to separate from the thiophenes. The u.v. spectra show beside maxima in the region of 340–350 nm weak absorption at 496 and 499 nm, respectively. The molecular formula for both compounds have been estimated as $C_{13}H_8S_2$. TLC of these red oils gives the thiophenes **111** and **112**. Also heating *in vacuo* at 80° leads to the transformation to **111** and **112** and elementary sulphur. Therefore, together with the spectral data, the only possible structures for these new acetylenes are **128** and **129**[89]:

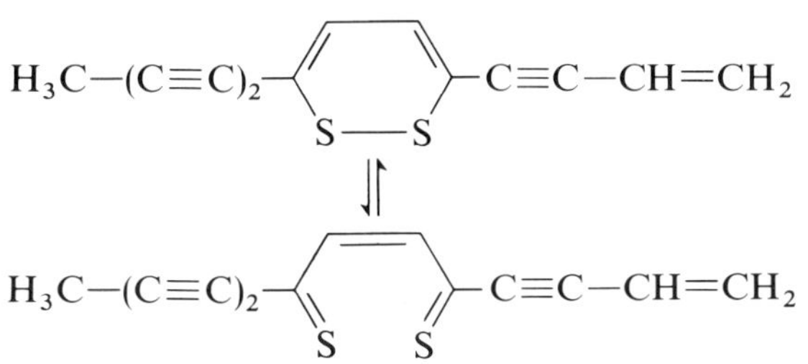

128

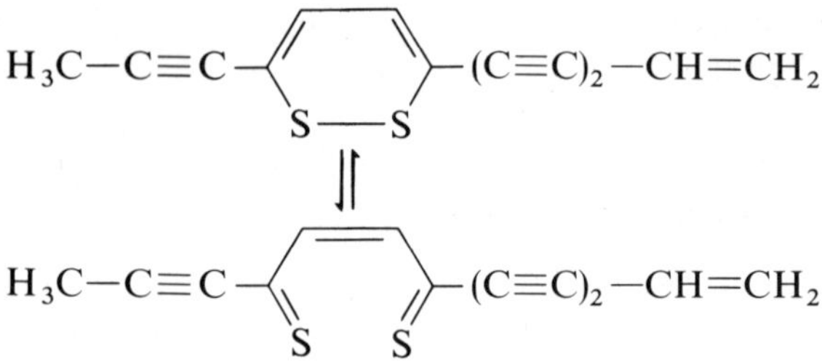

129

Perhaps the possible valence tautomerism is of importance in connection with the observed colour.

The very smooth transformation of the dithio compounds to thiophenes suggests that these compounds could be important in the biogenesis of the thiophenes, if we assume the addition of hydrogen disulphide in the following way:

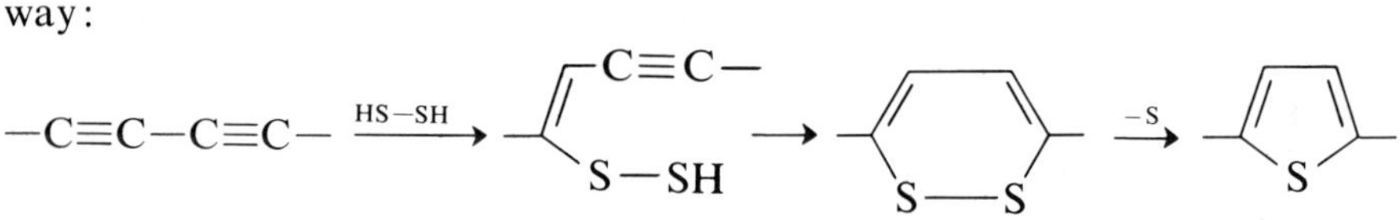

Chemical reactions of polyynes with sodium disulphide, however, leads to the corresponding thiophenes and five membered disulphides[90]:

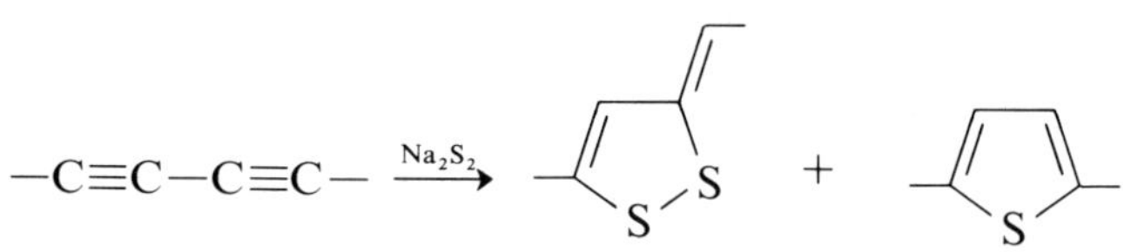

Therefore, this assumption is not very likely.

In addition to the monothiophenes, a considerable number of dithiophenes related to **94** have also been found in many *Compositae* species.

From *Buphthalmum salicifolium* L., a crystalline compound has been isolated showing a molecular formula $C_{13}H_{10}S_2$, formerly known to be a diaddition product of H_2S to **94**. The spectral data clearly show this to be a dithiophene of structure **130**[91], as the methyl singlet located at 7·59 is only in agreement with a thiophene methyl and not with an acetylenic one.

Together with **130**, the corresponding alcohol **131**[85] and the acetate **132**[81] are also present in the same plant:

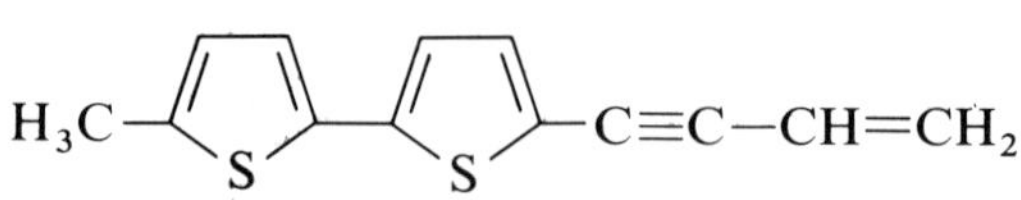

130

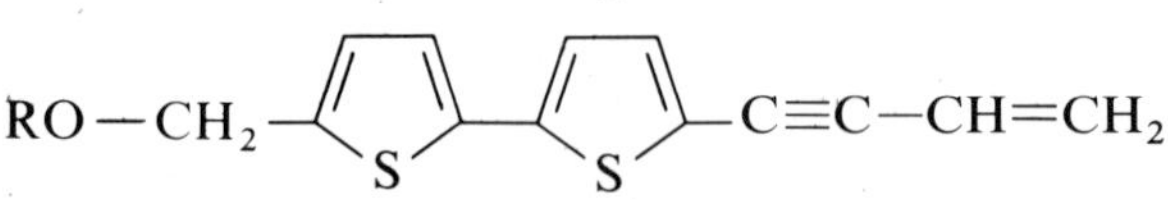

131: R = H
132: R = Ac

131 could be synthesised by formylation of the dithienyl derivative **137** and subsequent reduction of the resulting aldehyde with sodium borohydride[81].

A complex mixture of further esters have been found in the roots of *Eclipta erecta* L. The mixture is hard to separate but finally gives the esters **133–136**[86]:

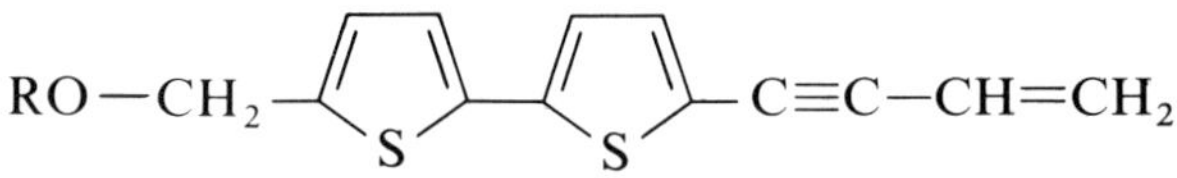

133: R = 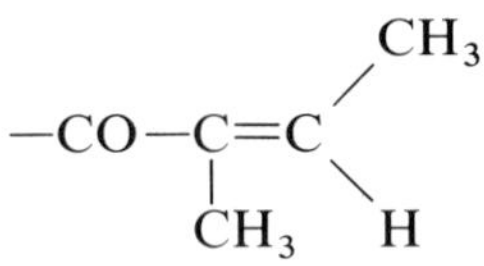

135: R = $-CO-CH(CH_3)_2$

134: R = 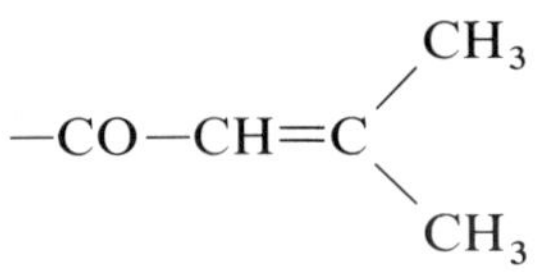

136: R = 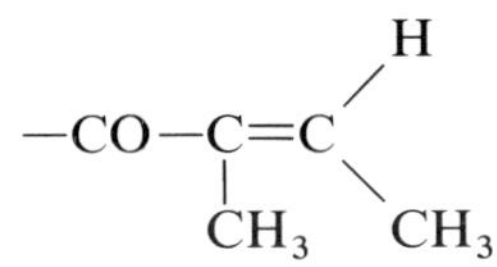

The nature of the ester residue is elucidated by mass and n.m.r. spectroscopy and synthesis[86].

Already well known is a further dithienyl derivative first isolated from the roots of *Tagetes erectus* L.[92]. This oily compound has the structure **137** and biogenetically is clearly formed by oxidation of **131** and subsequent decarboxylation[78]:

131 → 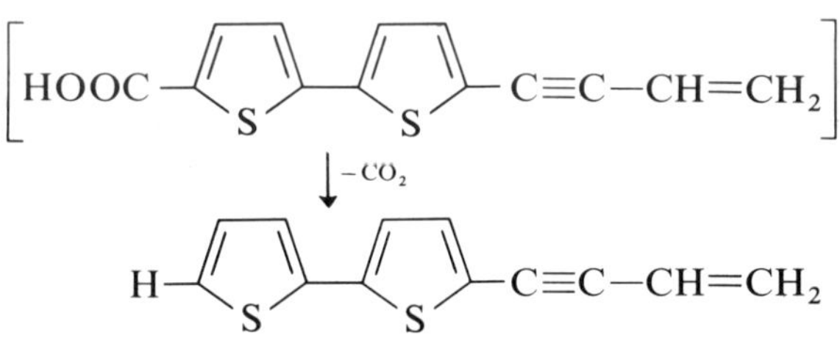

137

The structure of **137** has been established by synthesis starting with dithienyl acetylene in the following way[93]:

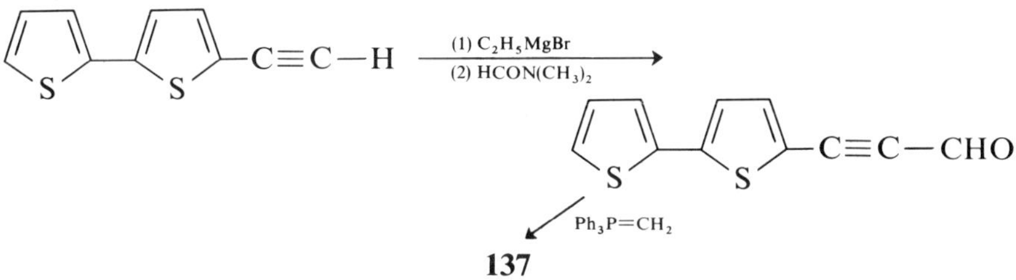

137 itself is a precursor of several other dithienyl derivatives most of them first isolated from the roots of *Echinops sphaerocephalus* L.[46].

As well as the diol **138**, the diacetate **139**, both mono-acetates **140** and **141**, and the alcohol **142** have been identified[46]:

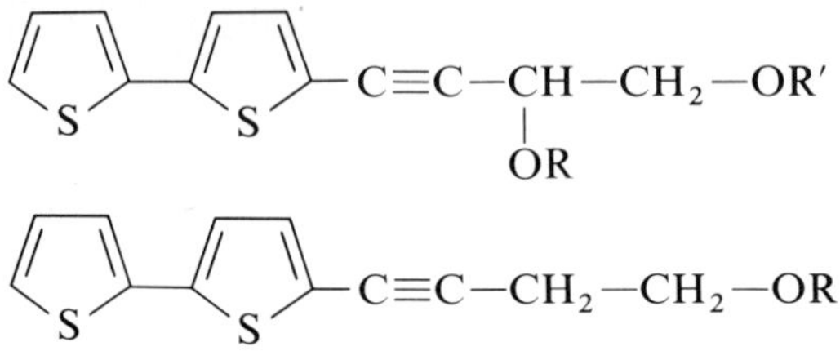

138: R = R′ = H
139: R = R′ = Ac
140: R = H, R′ = Ac
141: R = Ac, R′ = H
142: R = H
143: R = Ac[36]

The acetate **143** is present in *Tagetes* roots[93].

The structures of **138**–**143** have been elucidated by spectral methods as well as by some chemical reactions. The diol **138** easily is transformed on periodate oxidation to the aldehyde **144**, while the position of the free hydroxyl in **140** and **141** is estimated by the reaction with manganese dioxide; only **140** yields a ketone (**145**):

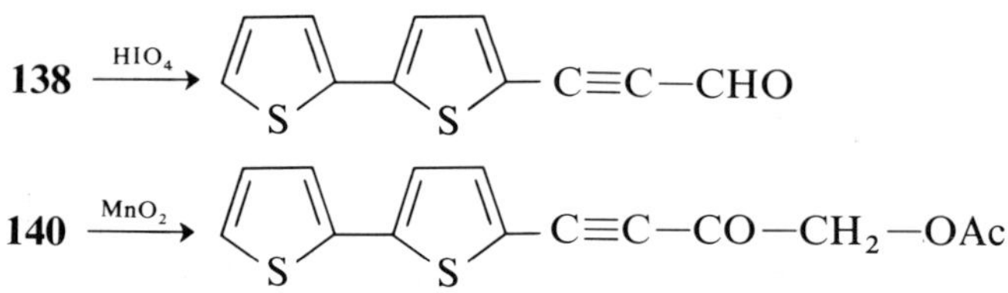

144

145

A further dithienyl compound isolated from *Tagetes minutus* L. has the structure **147**[94]. It is therefore probable that again all these compounds are biogenetically derived from the epoxide **146**, which has not yet been isolated from natural sources:

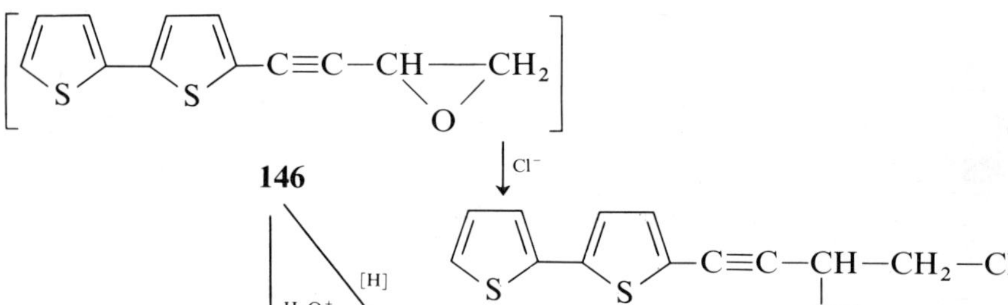

146 **138** **142** **147**

From *Berkheya adlamii* Hook., a compound has been isolated which shows the presence of an enolester group in its i.r. spectra (1782/cm)[95]. The spectral data are only in agreement with the structure **148**, while the configuration of the double bond could only be estimated by comparison of the n.m.r.

data of both isomers obtained synthetically:

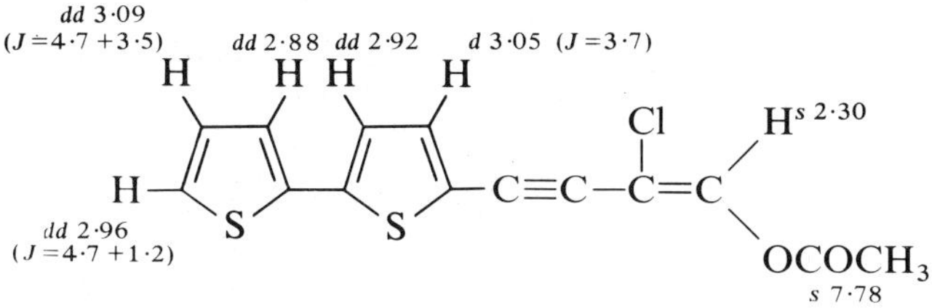

148

In the fragmentation pattern of **148**, some special pathways are observed:

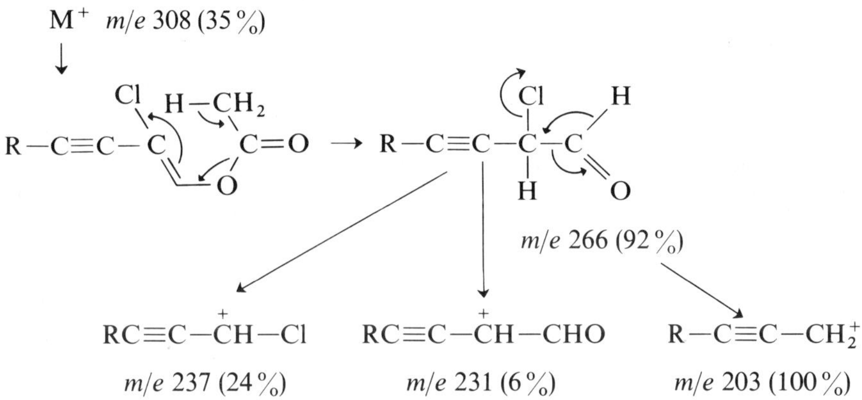

148 is synthesised in the following way[95]:

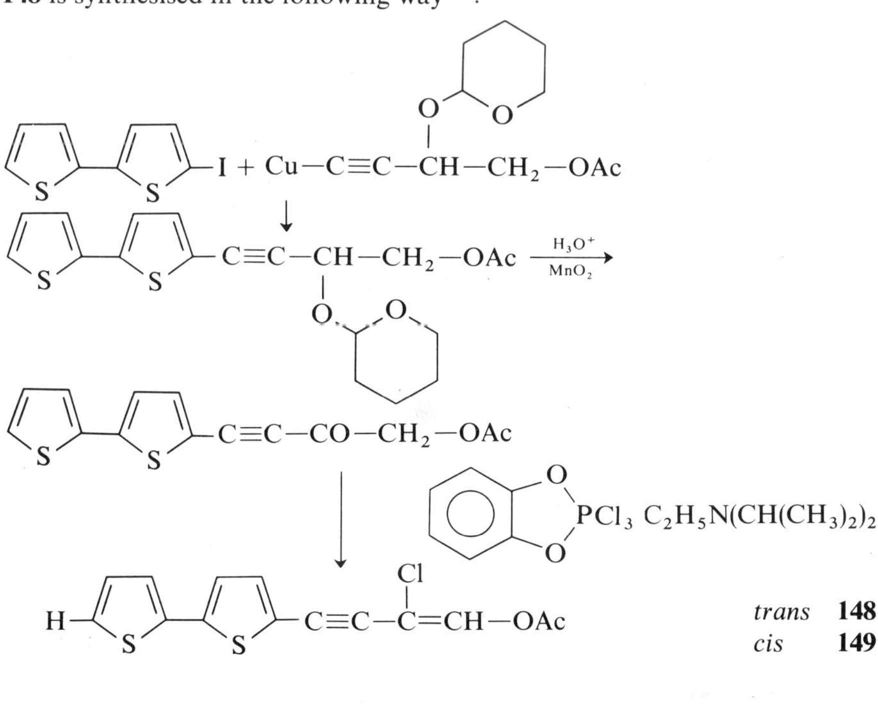

trans **148**
cis **149**

Biogenetically **148** is probably also formed via **146** and **150** in the following sequence:

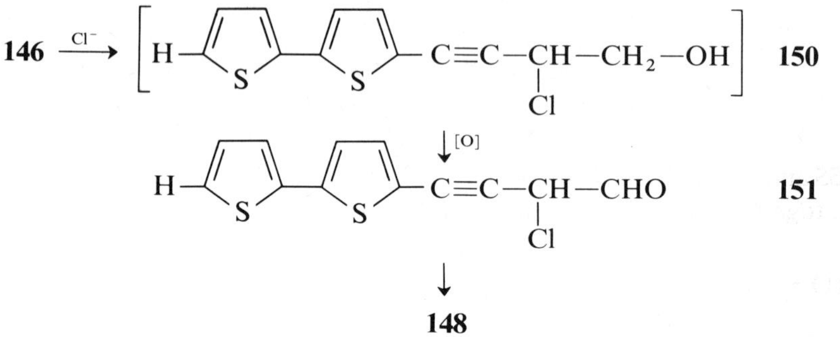

Closely related to **130** are a number of terthienyls isolated from *Tagetes* and *Eclipta* species. Zechmeister[96] had already shown in 1947 that a crystalline compound from the roots of Marigold *Tagetes erectus* L. has the structure **157**. The relation to **137** was first proposed[84], feeding with labelled **137** shows that this is not a precursor of **157**[97].

Later, several substituted terthienyls were isolated and the correct biogenetical sequence now seems to be clear[86]:

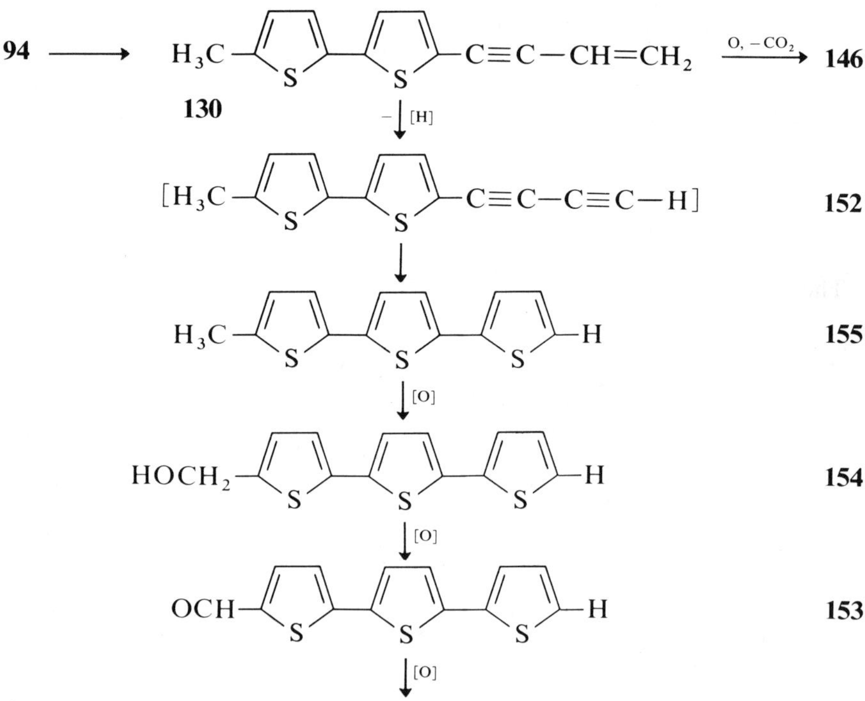

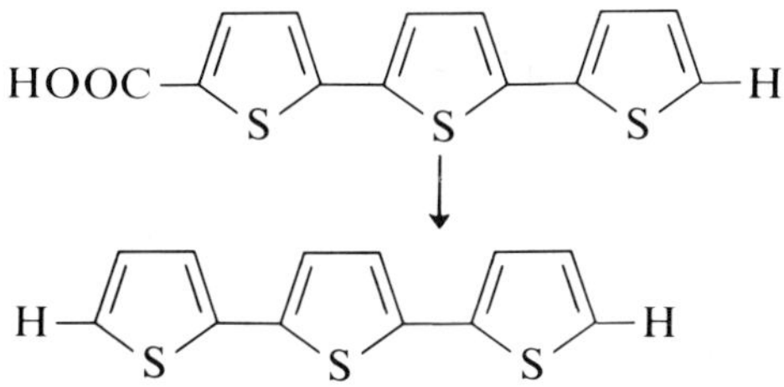

156

157

155 and also **153** and **154** have been found in the root extract of *Eclipta erecta* L. together with some esters of **154**[86]:

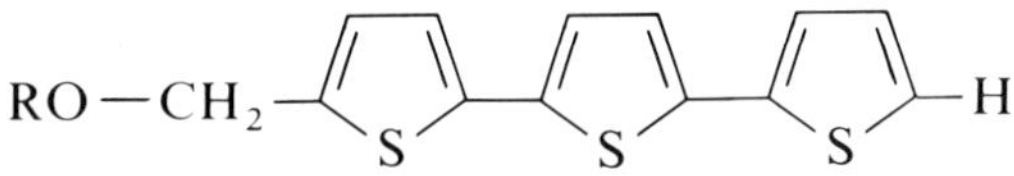

154: R = H

158: R = $-COCH_3$

159: R = $-CO-CH(CH_3)_2$

160: R = $-CO-CH{=}C(CH)(CH)$

161: R = $-CO-C(CH_3){=}C(CH_3)(H)$

162: R = $-CO-C(CH_3){=}C(H)(CH_3)$

The only dithienyls derived from **111** are the compounds **163**[74] and **164**[74]:

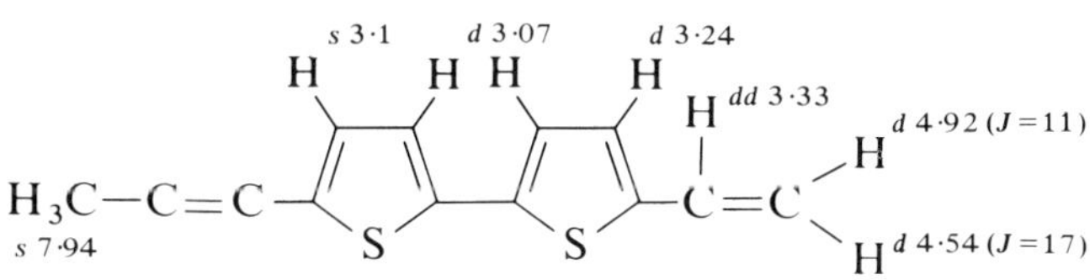

163

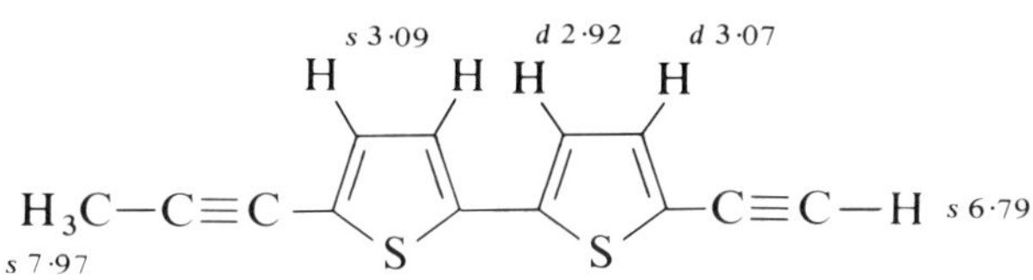

164

164 isolated from *Tagetes erectus* L. flowers is the dehydrogenation product of **163**. This is of interest in connection with the biogenesis of terthienyl (**157**), where the isomeric compound **152** is proposed as an intermediate.

From the roots of several Berkheya species, some further dithienyl-derivatives have been isolated[35]. All these compounds are very probably formed as **164** from the following dehydro-compound, itself formed by dehydrogenation of **111**:

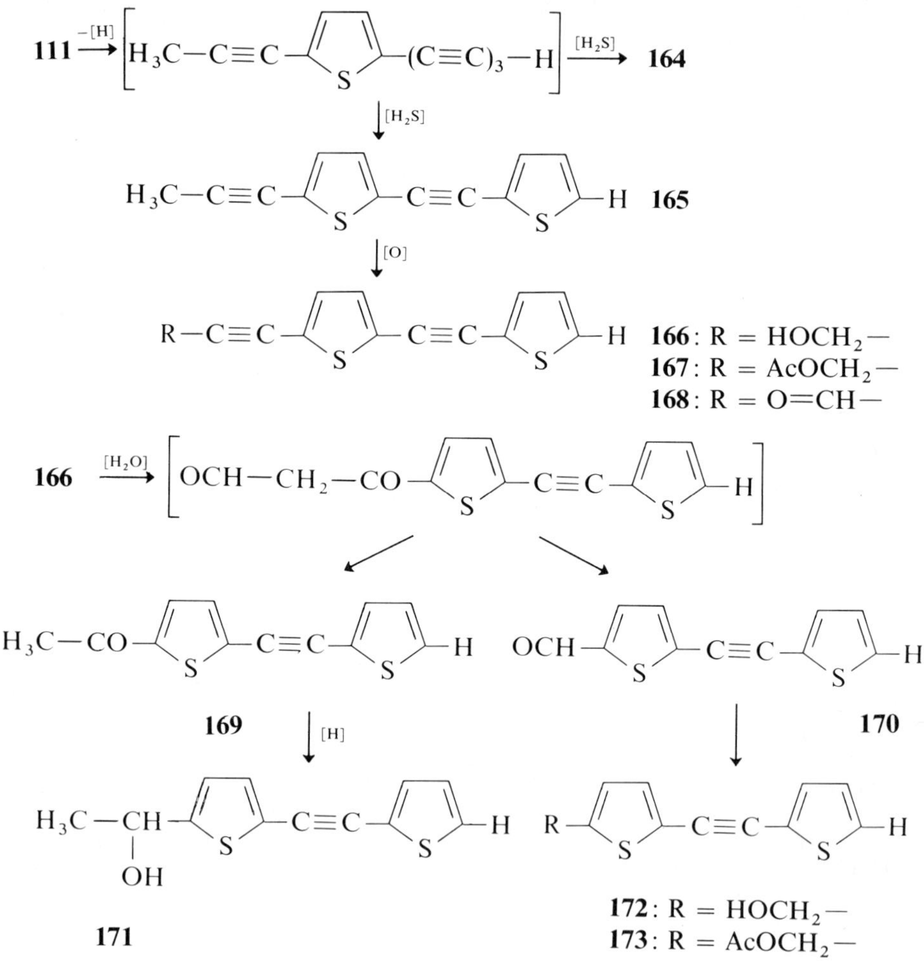

164, **165**

166: R = $HOCH_2-$
167: R = $AcOCH_2-$
168: R = O=CH−

166, **169**, **170**, **171**

172: R = $HOCH_2-$
173: R = $AcOCH_2-$

The degradation of **166** to **169** and **170** is hypothetical and perhaps other pathways can also be proposed.

If we summarise the formation of thiophenes starting with the pentayn-ene **94** the following scheme shows the main pathways:

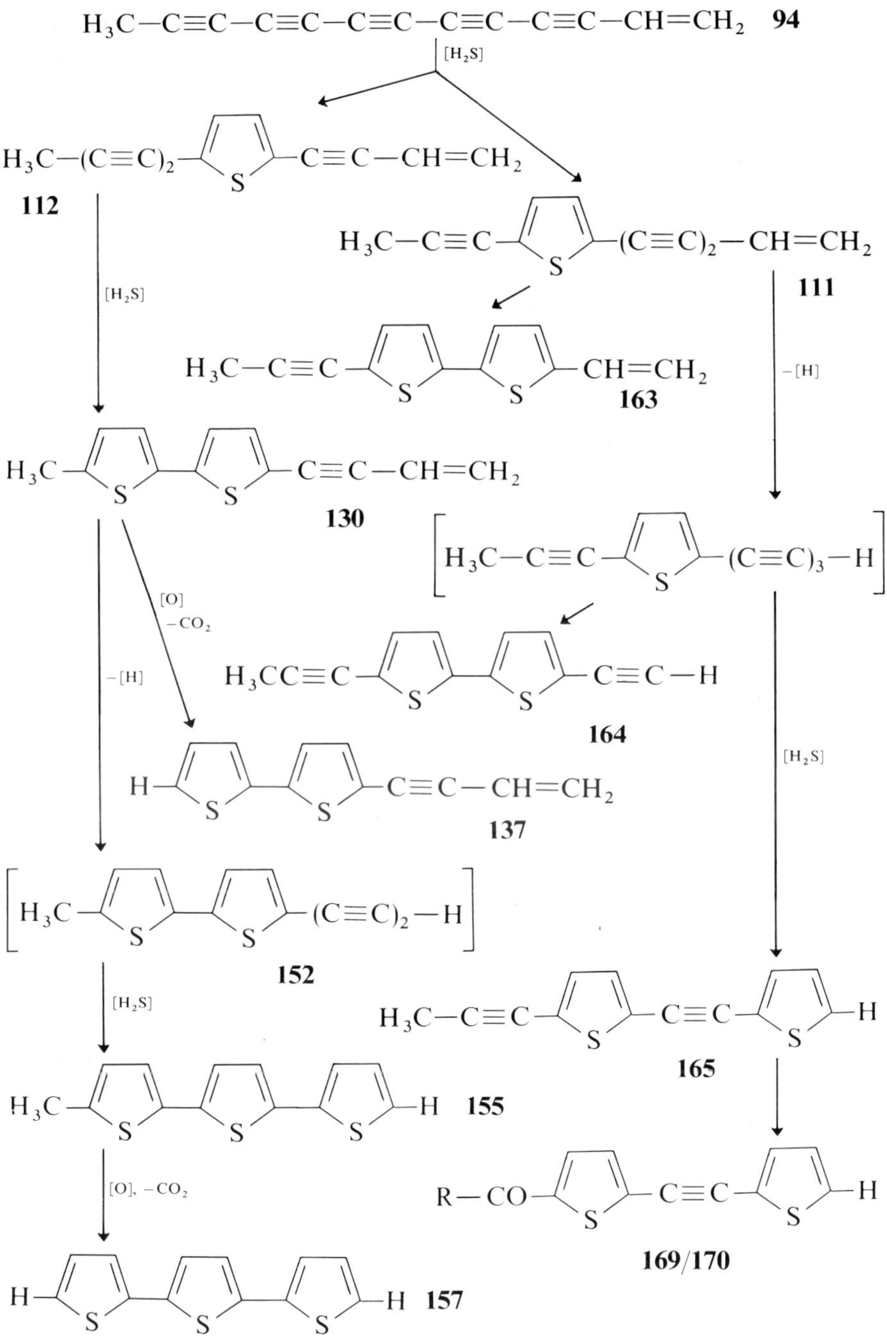

Especially **112**, **130**, **137**, **155** and **165** are further transformed to several already known derivatives by different routes, mainly by oxidation steps.

2.7. Compounds Related to Other Hydrocarbons

Beside the pentayn-ene **94** some other hydrocarbons have been isolated from different Compositae being again precursors for further acetylenes.

2.7.1. Tridecatetrayne-diene and related acetylenes

The biogenetic transformation sequence of **27** leading to the formation of the two *cis–trans*-isomeric tetrayne-dienes is shown by tracer studies[98]:

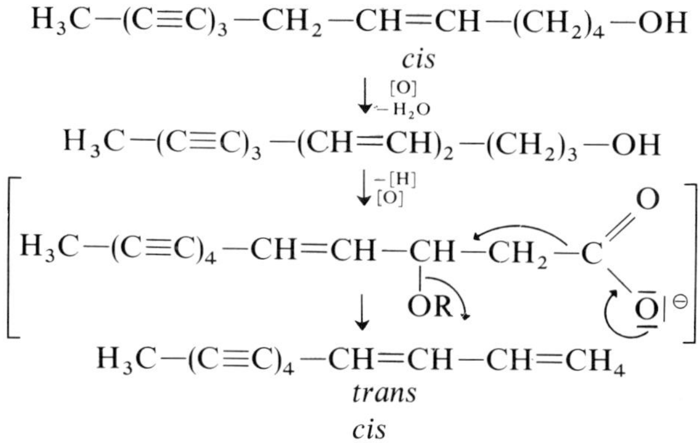

27

36

174

175 (*trans*)

176 (*cis*)

The *trans*-isomer **175** has been isolated first from *Rudbeckia* species[50]. The structure follows from the characteristic u.v. spectrum and the identity with synthetic material[99], obtained by the following reaction sequence with ^{14}C-methyliodide also as 1-^{14}C-**175**:

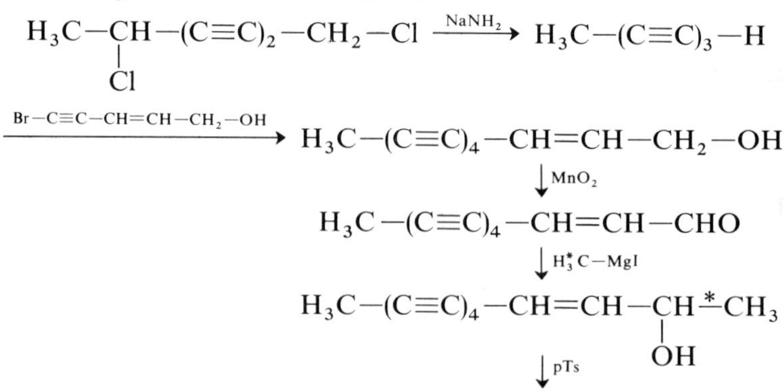

175

The *cis*-isomer **176** is present in the tubers of *Dahlia* species[33].

Some oxidation products of the hydrocarbons have also been isolated. A very sensitive crystalline compound has been separated from the complex mixture of the polyynes present in the aerial parts of *Centaurea ruthenica* Lam.[100]. The u.v. spectrum clearly shows the tetrayn-ene chromophore, while mild hydrolysis leads to a diol (later isolated from *Bidens* species[35]) which on subsequent reaction with periodate gives the highly unstable aldehyde **179**. Therefore, the structure of the first tetrayne from *C. ruthenica* has to be **177**:

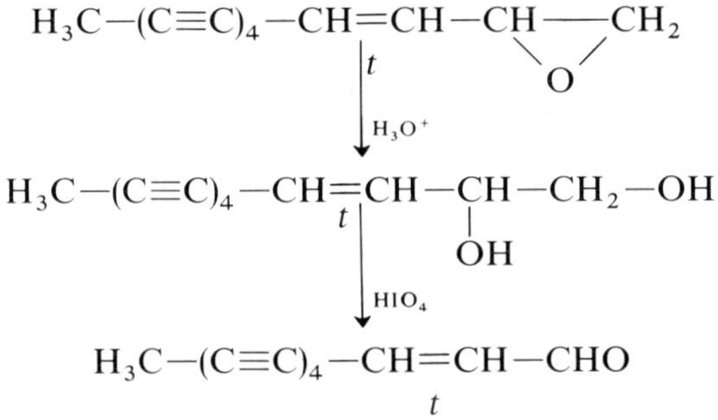

177

178

179

Two further tetrayn-enes have been isolated from the same plant[100], both giving the epoxide **177** on treatment with diluted alkali together with the elimination product **182**. Thus, only the structures **180** and **181** are in agreement with these properties and spectral data:

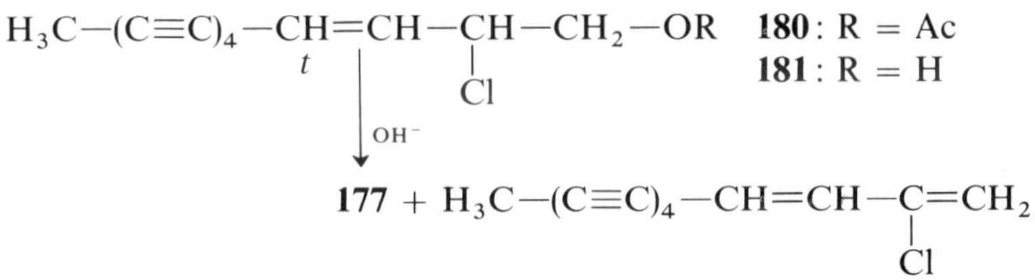

180: R = Ac
181: R = H

182

The structure **180** has also been confirmed by synthesis[101]:

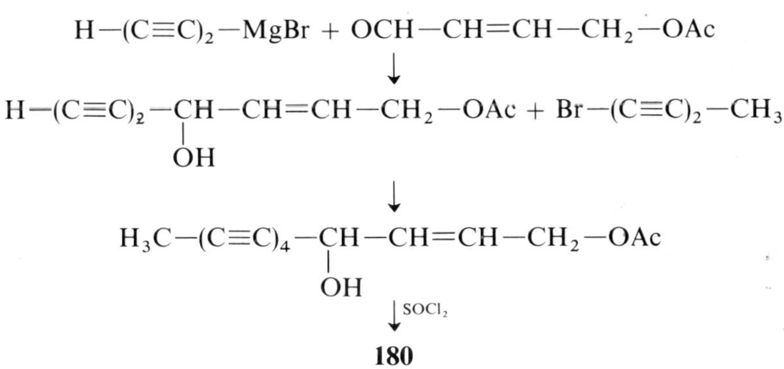

Several polyynes which turn out to be closely related to **177** have been isolated from *Anaphalis* and *Gnaphalium* species. The roots of *Anaphalis margaritacea* B. et H. contain, beside the pentayne **94**, an unstable crystalline compound showing a broad uncharacteristic u.v. maximum at 341 nm. The highly complex n.m.r. spectrum however is only in agreement with the structure **183**[102]:

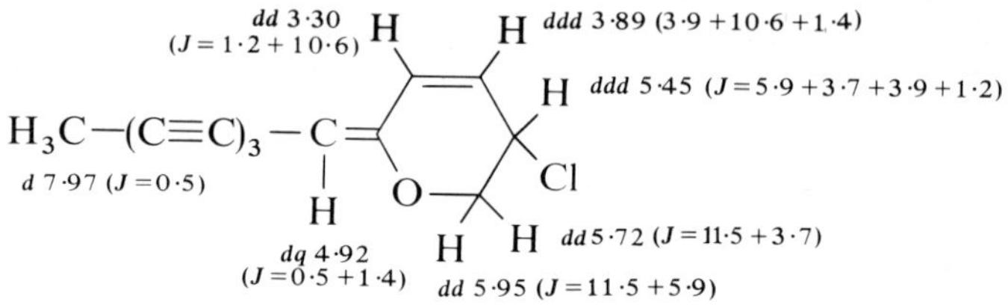

183

Together with **183**, the *trans*-isomer has been found in another *Anaphalis* species[103]. The n.m.r. spectrum of **184** is quite similar to that of **183**, but the signals of the olefinic protons are different. These differences are typical for *cis–trans*-isomeric enolethers of the following type:

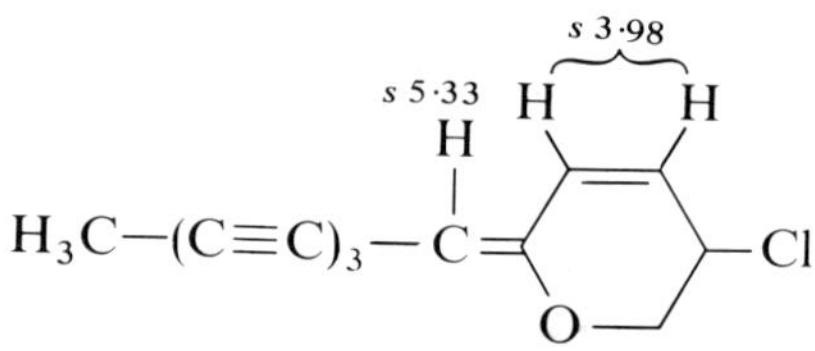

184

Somewhat more polar is a second compound showing u.v. maxima of a triyne-enolether (λ max: 343, 320, 300, 282, 261, 249 nm). The n.m.r. spectrum shows that this compound is the epoxide of **183**[103]:

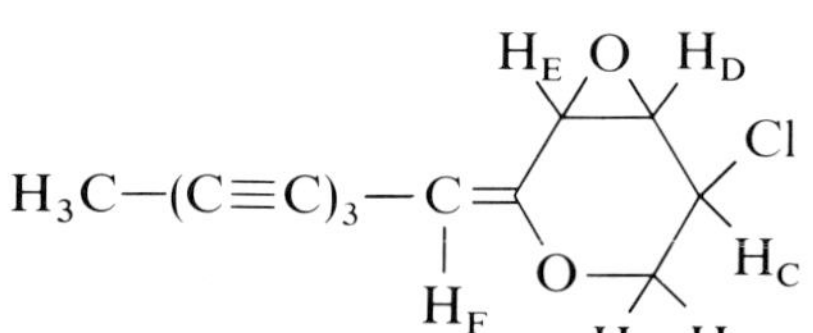

185

H_A	*ddd* 5·86	H_D	*m* 6·54
H_B	*dd* 5·53	H_E	*d* 5·65
H_C	*m* 6·37	H_F	*s* 5·15

$$J_{AB} = 12\text{ Hz},\ J_{AC} = 2{\cdot}5\text{ Hz},\ J_{BC} = 2\text{ Hz},\ J_{AD} = 1\text{ Hz},\ J_{DE} = 4\text{ Hz}$$

The alcohol **186** has also been found in very small amounts in the polar fractions of the *Anaphalis* root extracts[35]:

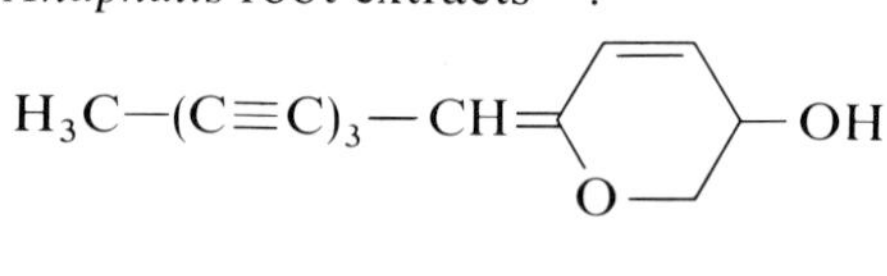

186

Two somewhat more polar acetylenes than **183** have been isolated in *Gnaphalium* species and have spectral data very close to those of **183** and **184**. Again, these compounds are *cis–trans*-isomers as clearly can be recognised from the n.m.r. data[103]:

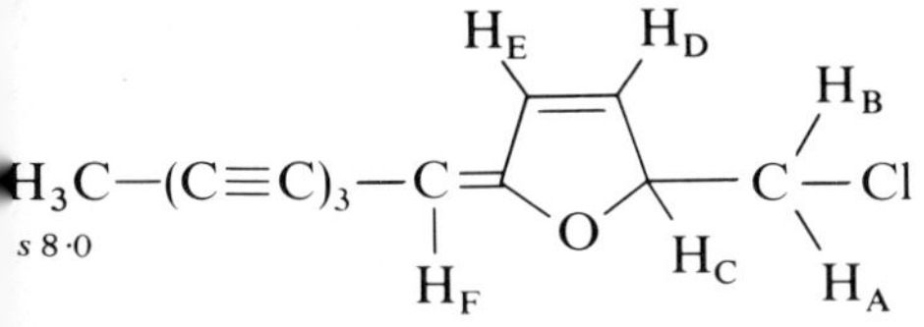

187

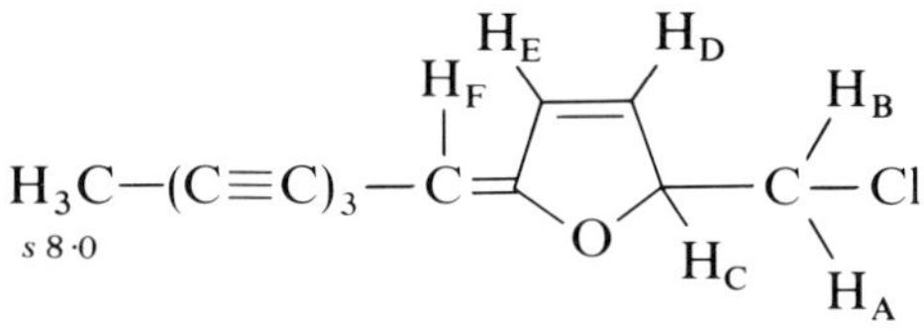

188

	187	188
H_A	*dd* 6·58	*dd* 6·56
H_B	*dd* 6·33	*dd* 6·22
H_C	*dddd* 4·67	*dddd* 4·64
H_D	*ddd* 3·42	*m* 3·52
H_E	*dd* 3·12	*dd* 3·42
H_F	*m* 5·10	*m* 5·45
	$J_{AB} = 10$, $J_{AC} = 6{\cdot}5$, $J_{BC} = 4{\cdot}5$	$J_{AB} = 10$, $J_{AC} = 7$, $J_{BC} = 4{\cdot}5$
	$J_{CD} = 2$, $J_{CE} = 1{\cdot}5$, $J_{DE} = 5{\cdot}5$	$J_{CD} = 2$, $J_{CE} = 2$, $J_{DE} = 5{\cdot}5$
	$J_{DF} = 1{\cdot}5$	

The structures **187** and **188** are confirmed by a synthesis commencing with the tetrayne already used in the synthesis of **180**[104]:

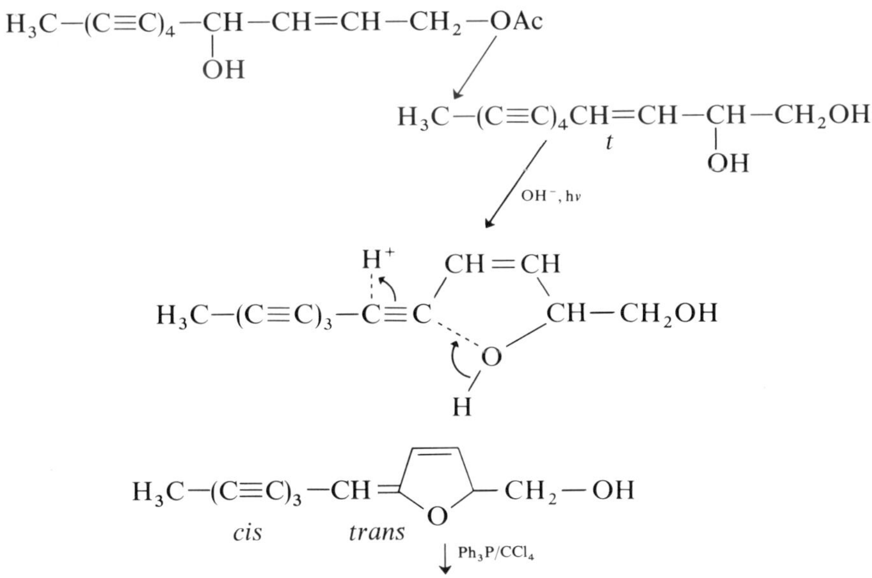

187 + 188

All these compounds very probably are formed from *cis*-**177**, while **177** biogenetically is derived from **175**[99], which itself is formed from **27**[98]. The following reaction sequence is most likely:

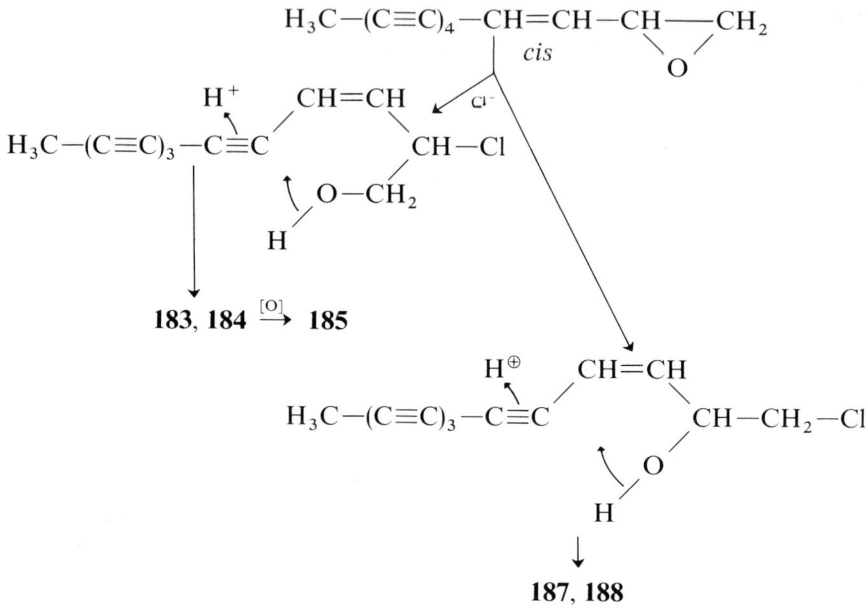

This would mean, that ring-opening in the case of the *trans*-epoxide **177** gives the chlorohydrine derivatives **180** and **181**, while the corresponding *cis*-isomer leads to the *cis*-alcohols, which at once cyclisise to enolethers of type **183** or **187**. Perhaps direct transformation of *cis*-**177** by nucleophilic attack is also possible:

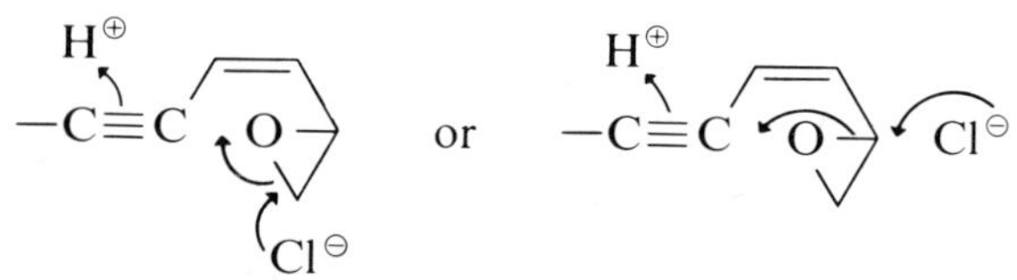

As in the case of the pentayne **94**, the precursor **175** of a few thiophenes is also isolated from different Compositae. The roots of *Rudbeckia amplexicaulis* Vahl. contain **175** together with the thiophene **189**[50]:

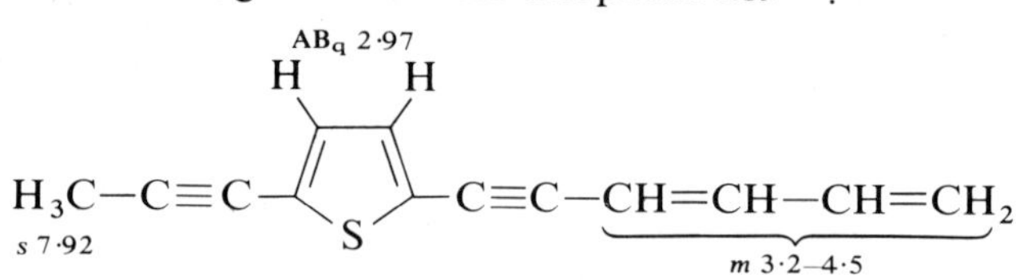 **189**

The complex n.m.r. spectrum makes exact structure elucidation impossible, but hydrogenation again gives 2-propyl-5-hexyl-thiophene which is identical with authentic material. The *trans*-configuration of the double bond is evident from the i.r. spectrum.

The corresponding dithio compound has also been found in *Rudbeckia* species[89]:

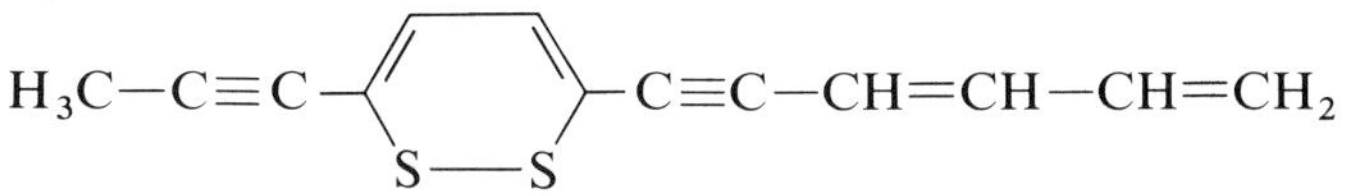

190

The dithienyl derivative **191** has also been isolated[74] from a *Rudbeckia* species; this is clearly also formed from the tetrayne **175**:

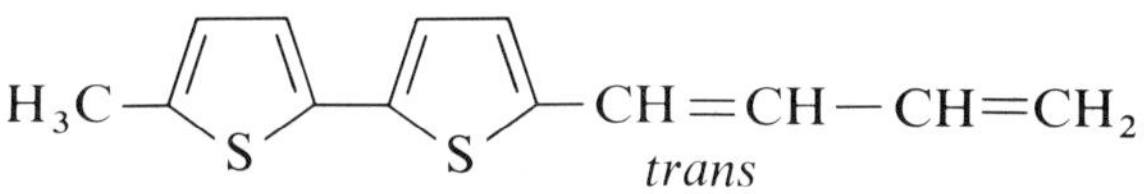

191

2.7.2. Tridecatriyne-triene and related acetylenes

Biogenetically the triyne **192** is formed from **27** too[38]:

$$H_3C{-}(C{\equiv}C)_3{-}CH_2{-}CH{=}CH{-}(CH_2)_4{-}OH \quad \textbf{27}$$

cis

$\downarrow$ [O], $-H_2O$ [O]

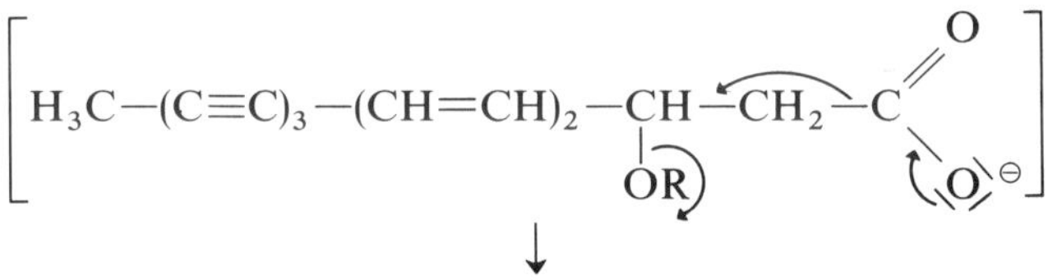

$\downarrow$

$$H_3C{-}(C{\equiv}C)_3{-}CH{=}CH{-}CH{=}CH{-}CH{=}CH_2$$

trans	*trans*	**192**
trans	*cis*	**193**

The crystalline hydrocarbon **192** was first isolated from *Achillea* species[105], but it has subsequently been found in several others belonging to the tribus Anthemideae. The structure is established by hydrogenation leading to *n*-tridecane and by the reaction with maleic anhydride, which gives an adduct showing a triyn-ene chromophore. Finally the structure could be established by synthesis using the Wittig reaction[105]:

$$H_3C{-}(C{\equiv}C)_3{-}CH{=}CH{-}CHO + Ph_3P{=}CH{-}CH{=}CH_2$$

$$\downarrow$$

192

The *cis*-isomer **193** is present in *Dahlia* tubers[60]. Only a few oxidation products related to **192** are known. The most widespread one is the epoxide **194**, isolated from several members of the tribus Anthemideae. The structure has been elucidated mainly by chemical methods[43].

Only mild hydrolysis leads to the corresponding diol **195**, which is also a natural product isolated from *Pentzia pilulifera*[106]:

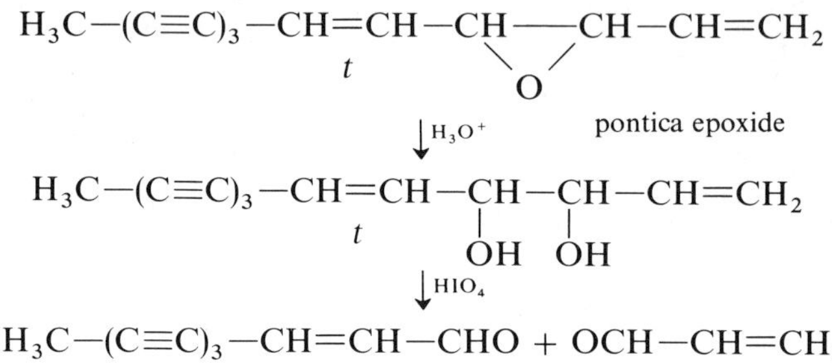

194

195

Periodate oxidation gives dehydromatricarianal and acroleine, clearly establishing the structure. Epoxidation of **192** also gives the epoxide **194** as well as the isomeric compound[43].

The absolute configuration has been shown by oxidative degradation and is shown to be **194a**[107]:

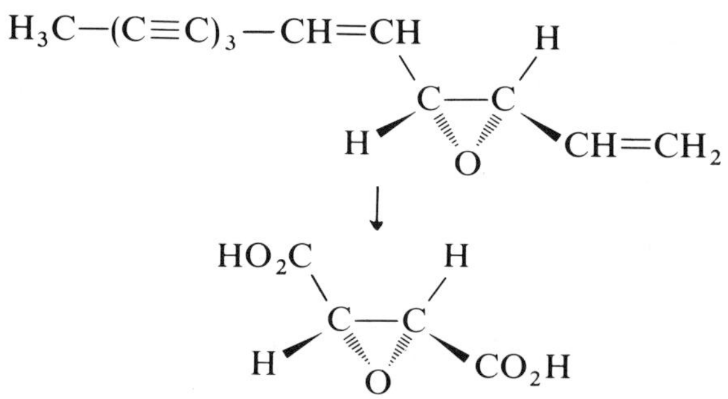

194a

Only from *Anthemis cota* L. has an isomeric epoxide been isolated. The not very clear u.v. spectrum is formed by overlapping a triyne and a diene spectrum with some kind of interaction. Hydrogenation leads to *n*-tridecanon-(5), *n*-tridecanol-(5) and the hydrogenated epoxide, which on hydrolysis and periodate cleavage leads to *n*-valeraldehyde and *n*-octylaldehyde. Therefore, the structure of the epoxide has to be **196**[105]:

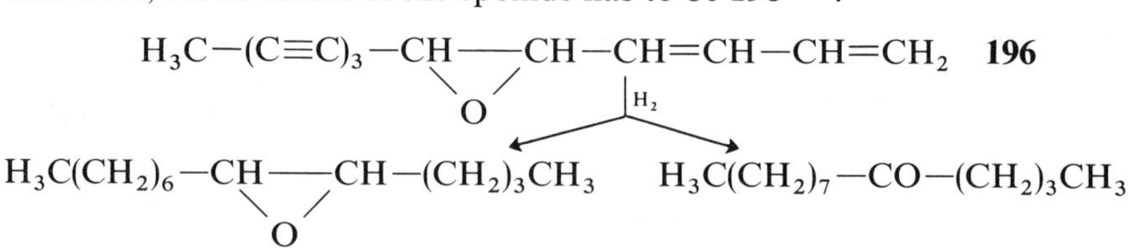

196

From *Buphthalmum salicifolium* L., a compound which is difficult to purify has been isolated and shows a u.v. maxima at 365 and 346 nm[108]. The spectral data show that this acetylene is most probably formed by formal addition of methyl mercaptane to **192**, but the position of the —S—CH_3 group cannot be estimated exactly from the n.m.r. spectra. The mass spectrum of the perhydro compound, however, clearly establishes the position of the —SCH_3 group, as the main fragment is $C_8H_{17}S^+$. Therefore, the structure of the thioether has to be **197**:

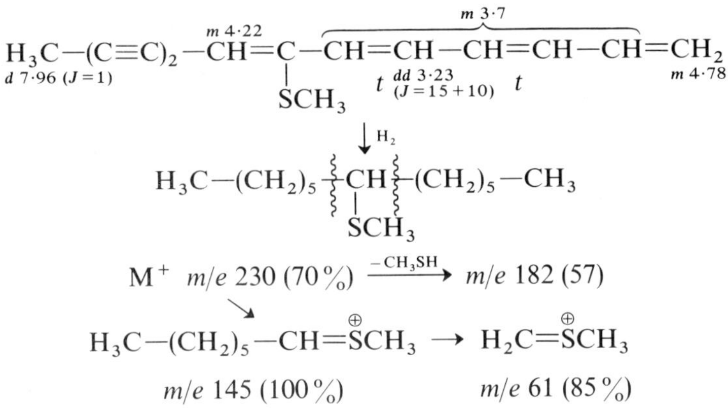

197

In the biogenesis of **197**, the direction of the sulphur addition is interesting, since it is normally observed in the opposite direction. (See however **106**, Section 2.6.2 and **415–418**, Section 5.2).

By formal addition of H_2S followed by oxidative elimination of the methyl-endgroup **192** is transformed to the thiophenes **198** and **199**, isolated from *Xeranthemum* species[50]:

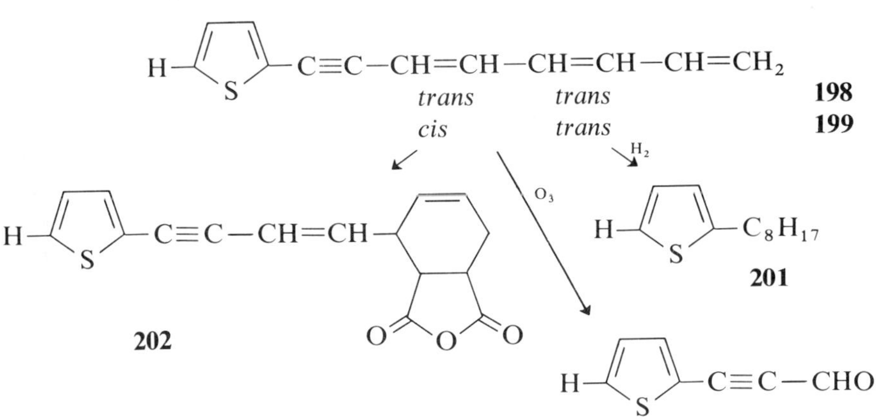

198 trans trans
199 cis trans
201
202
200

The structures have been established by chemical reactions. **199** can be isomerised by irradiation to **198**, which on ozonisation leads to the aldehyde **200** and by hydrogenation to *n*-octyl-thiophene (**201**). With maleic anhydride the adduct **202** has been obtained.

198 has been synthesised by the following reaction sequence[36]:

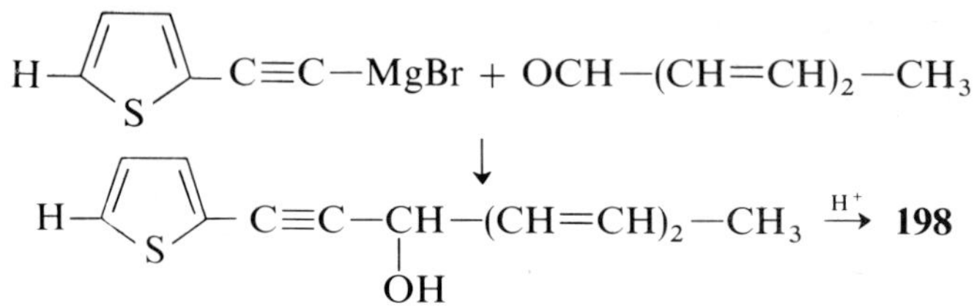

2.7.3. Other acetylenes

Up to now only two open chain compounds have been isolated from higher plants having a C_{11}-carbon chain.

From *Psilothonna tagetes* belonging to the genus *Gamolepis* (tribus Senecioneae) two hydrocarbons difficult to separate by TLC have been isolated, whose structures have been elucidated by spectral methods as well as by synthesis[109]:

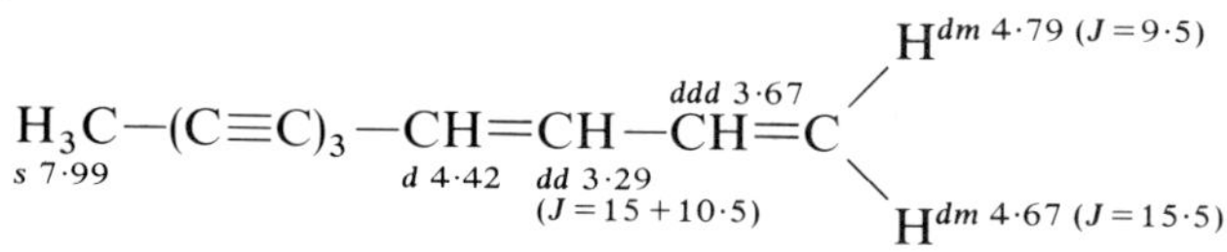

203

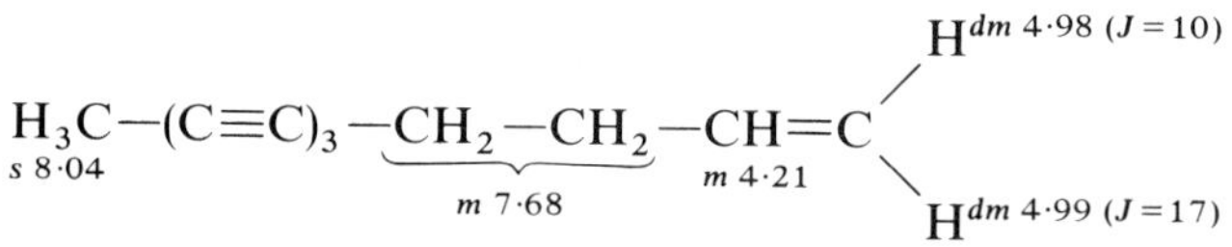

204

203 shows the typical triyne-diene U.V. spectrum, however, all maxima are shifted by about 4 nm to shorter wavelengths due to the missing substituent.

The structures of **203** and **204** have been established by synthesis[109]:

$$H_3C{-}(C{\equiv}C)_3{-}CH{=}CH{-}CHO + CH_3MgBr$$

$$\searrow H_3C{-}(C{\equiv}C)_3{-}CH{=}CH{-}CH(OH){-}CH_3 \xrightarrow{-H_2O} \mathbf{203}$$

$$H_3C{-}(C{\equiv}C)_3{-}CH_2{-}OH \xrightarrow{SOCl_2} H_3C{-}(C{\equiv}C)_3{-}CH_2{-}Cl$$

$$\swarrow +BrMg{-}CH_2{-}CH{=}CH_2$$

204

Though the biogenesis of these hydrocarbons is not established, the following route seems to be very likely. By feeding labelled **205**, good incorporation into **203** and **204** can be observed[35]:

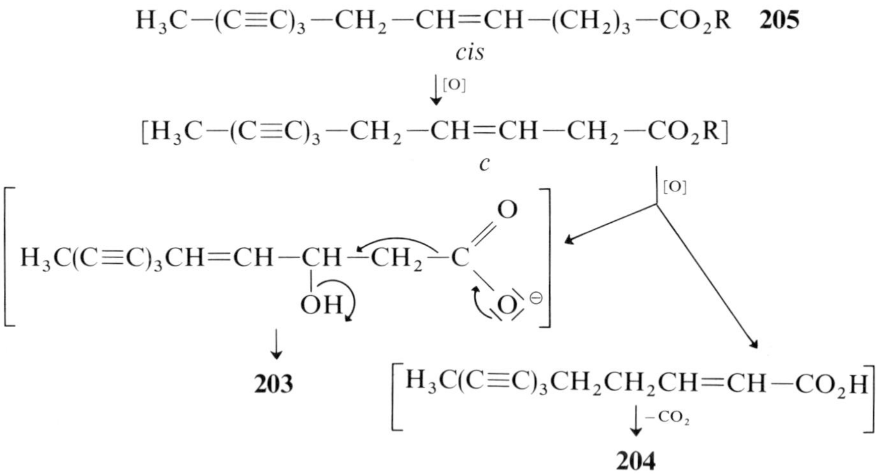

3. Compounds Related to the C_{13}-triyn-(7,9,11)-en-(4)-ol-(1)

3.1. Simple Acetylenes

As already mentioned in Chapter 1, Section 4 the C_{18}-triyn-ene acid **5** is the common precursor for **27** as well as for the C_{13}-triyne **207**[25] as shown by several feeding experiments using labelled precursors. Furthermore, it turns out that in the latter case the first step in the transformation of the C_{18}-chain is α-oxidation leading to the C_{17}-triyne, which then is further degradated by twice β-oxidation to the C_{13}-precursor as shown in the following sequence[25]:

$$C_{18}: \quad [H_3C-(C\equiv C)_3-CH_2-CH\overset{cis}{=}CH-(CH_2)_7-CO_2R] \qquad \mathbf{5}$$

$$\downarrow \alpha\text{-oxidation}$$

$$C_{17}: \quad [H_3C-(C\equiv C)_3-CH_2-CH\overset{cis}{=}CH-(CH_2)_6-CO_2R]$$

$$\downarrow 2 \times \beta\text{-oxidation}$$

$$C_{13}: \quad [H_3C-(C\equiv C)_3-CH_2-CH\overset{cis}{=}CH-(CH_2)_2-CO_2R] \qquad \mathbf{206}$$

$$\downarrow [H]$$

$$C_{13}: \quad [H_3C-(C\equiv C)_3-CH_2-CH\overset{cis}{=}CH-(CH_2)_3-OR] \qquad \mathbf{207}$$

As the corresponding C_{14}-triyne **27** also the acetate **207** has been isolated from natural sources[35]. Both, the acetate **207** and the alcohol **206** are important precursors and therefore have been synthesised with ^{14}C- and ^{3}H-labelling using the following routes[38,110]:

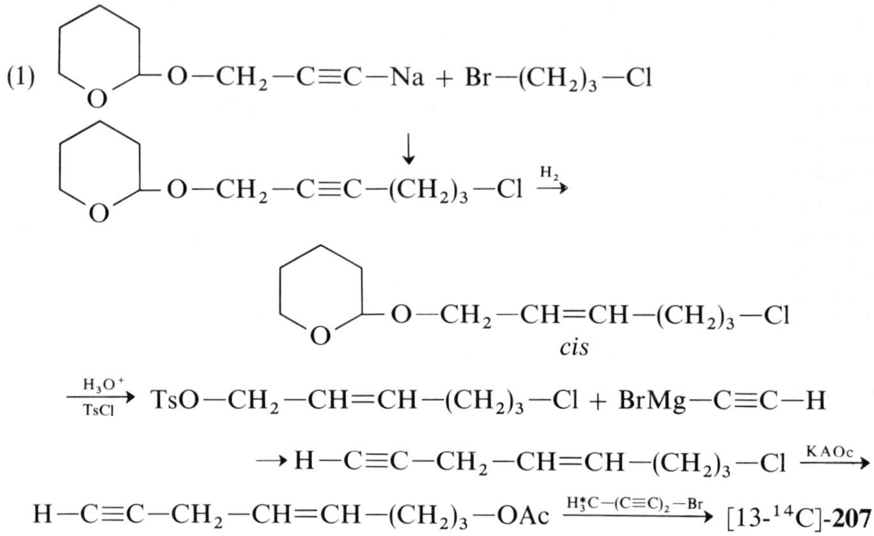

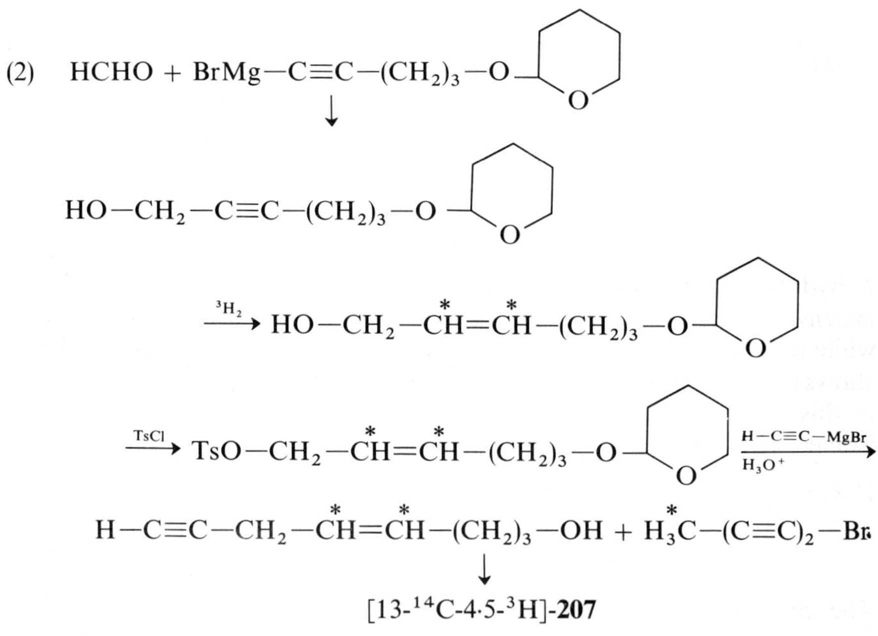

$$\text{(3)}\quad H{-}C{\equiv}C{-}CH_2{-}\underset{c}{CH{=}CH}{-}(CH_2)_3{-}OAc \xrightarrow[\text{(2) } CrO_3 \text{ (3) } CH_2N_2]{\text{(1) } CH_3OH/H^+}$$

$$H{-}C{\equiv}C{-}CH_2{-}\underset{c}{CH{=}CH}{-}(CH_2)_2{-}CO_2CH_3 + H_3\overset{*}{C}{-}(C{\equiv}C)_2{-}Br$$

$$[13\text{-}^{14}C]\text{-}\ \mathbf{206}$$

The acetate **207** is present in the aerial parts of *Chrysanthemum croaticum* Horw. together with the acetate **208**[35]. From different *Chrysanthemum* species an alcohol and an acetate have been isolated showing the typical U.V. spectra of triyne-dienes. Hydrogenation of the acetate gives *n*-tridecanol acetate. Therefore, when considered together with the spectral data, only structure **208** (R = Ac) agrees with these results[62]. Saponification of **208** (R = Ac) leads to the corresponding alcohol, identical with the natural compound[37]. The structure of **208** could be established by synthesis[62]:

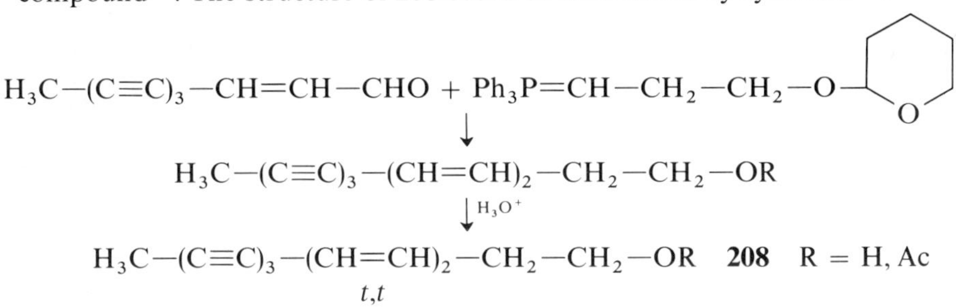

208 R = H, Ac

Biogenetically **208** is formed from **207** by the following route[38]:

$$\mathbf{207} \xrightarrow{[O]} H_3C{-}(C{\equiv}C)_3{-}CH{=}CH{-}\underset{OH}{CH}{-}(CH_2)_3{-}OH \xrightarrow{-H_2O} \mathbf{208}$$

A hydrogenation product of **208** has been isolated first from *Matricaria matricarioides* L. Hydrolysis and subsequent hydrogenation gives *n*-tridecanol, while ozonisation leads to hexadiynoic acid. The maleic anhydride adduct shows the characteristic U.V. spectrum of an ene-diyne. Therefore, the structure of this acetate isolated as the "all-*trans*"- as well as the *trans–cis–trans*-isomer has to be **209**[111] and **209a**[112]:

$$H_3C{-}(C{\equiv}C)_2{-}CH{=}CH{-}CH{=}CH{-}CH{=}CH{-}CH_2{-}CH_2{-}OAc$$

trans	*trans*	*trans*	**209**
trans	*cis*	*trans*	**209a**

The concentration of **209a** seems to vary depending on the season[112].

The structure of **209** has been established by synthesis[111]:

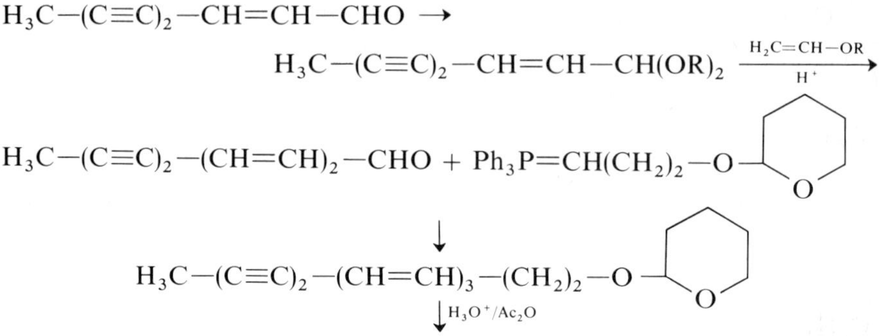

209

The biogenesis of **209** could also be elucidated by feeding labelled **207**[54]. The following reaction sequence seems to be most likely:

$$\mathbf{207} \xrightarrow[-H_2O]{[O]} \mathbf{208} \xrightarrow{[H]} \mathbf{209}$$

Together with further acetylenes, two alcohols have been isolated from the roots of *Chrysanthemum leptophyllum* DC., the structures being elucidated by spectral methods as **210** and **211**[106]:

$$H_3C{-}(C{\equiv}C)_2{-}(CH{=}CH)_3{-}(CH_2)_2{-}OH \quad \mathbf{210}$$

$$t,t,t$$

$$\downarrow [O], -CO_2$$

$$H{-}(C{\equiv}C)_2{-}(CH{=}CH)_3{-}(CH_2)_2{-}OH \quad \mathbf{211}$$

211 represents a further case, where the methyl endgroup is probably eliminated by oxidation and subsequent decarboxylation.

3.2. Spiroketalenolethers

As in the case of **27**, **207** is also the precursor of a considerable number of spiroketalenolethers.

From the roots of *Matricaria matricarioides* L., a pair of *cis–trans*-isomeric acetylenes has been isolated and the structures elucidated by chemical and spectroscopic methods[62]. The u.v. spectra of the isomers are not very characteristic and show only broad, slightly different maxima. The well-defined i.r. spectra show the presence of acetylenic as well as enolether bonds. Hydrogenation of the optically-active compounds with molecular formula $C_{13}H_{12}O_2$ in the presence of small amount of water leads to a diketo alcohol, its structure being estimated by further reactions as shown in the following sequences in analogy to **68** (see Section 2.4):

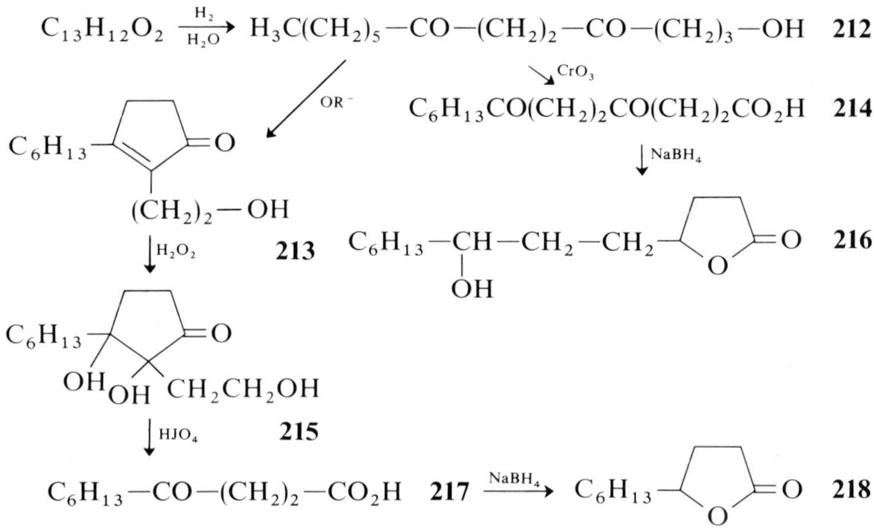

The structures of **214** and **217** have been established by comparison with authentic materials.

The two acetylenes can be isomerised by u.v. irradiation and also by acid treatment. Ozonisation leads to hexadiynal and osmium tetroxide yields a diol showing a diyn-ene-enolether u.v. spectrum. The n.m.r. spectra, together with all these results, are only in agreement with the structures **219** and **220**[62]:

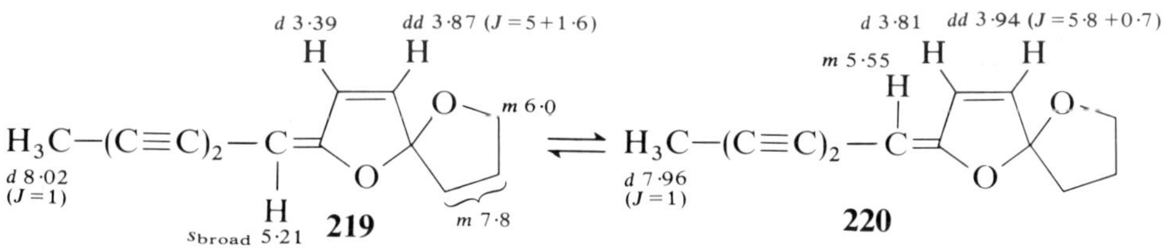

The somewhat surprising results of the hydrogenation only can be explained in the following manner:

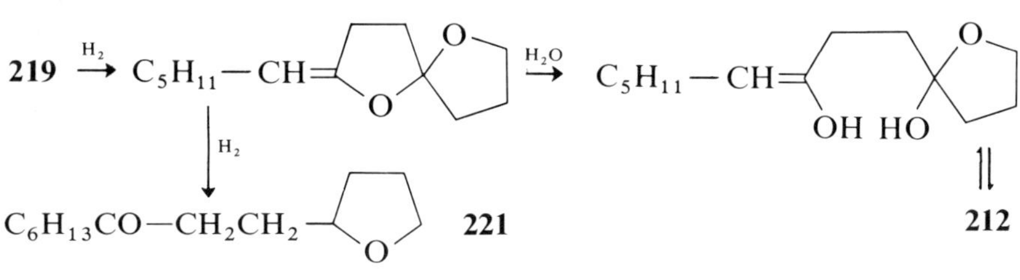

Beside **212**, **221**[23] has also been isolated from the hydrogenation probably formed by hydrogenolysis of the ketal oxygen bond. The *cis–trans*-isomerisation using acidic methanol leads to racemic **220**. Therefore, the following reaction mechanism is most likely:

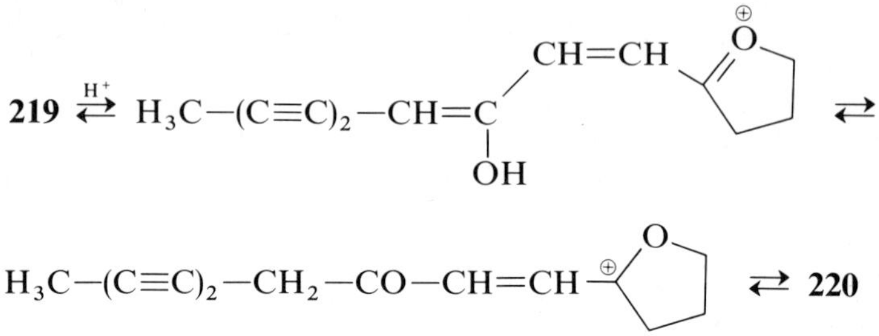

219 and **220** have been synthesised by two different methods. The first commences with the furane derivative **222**[56]:

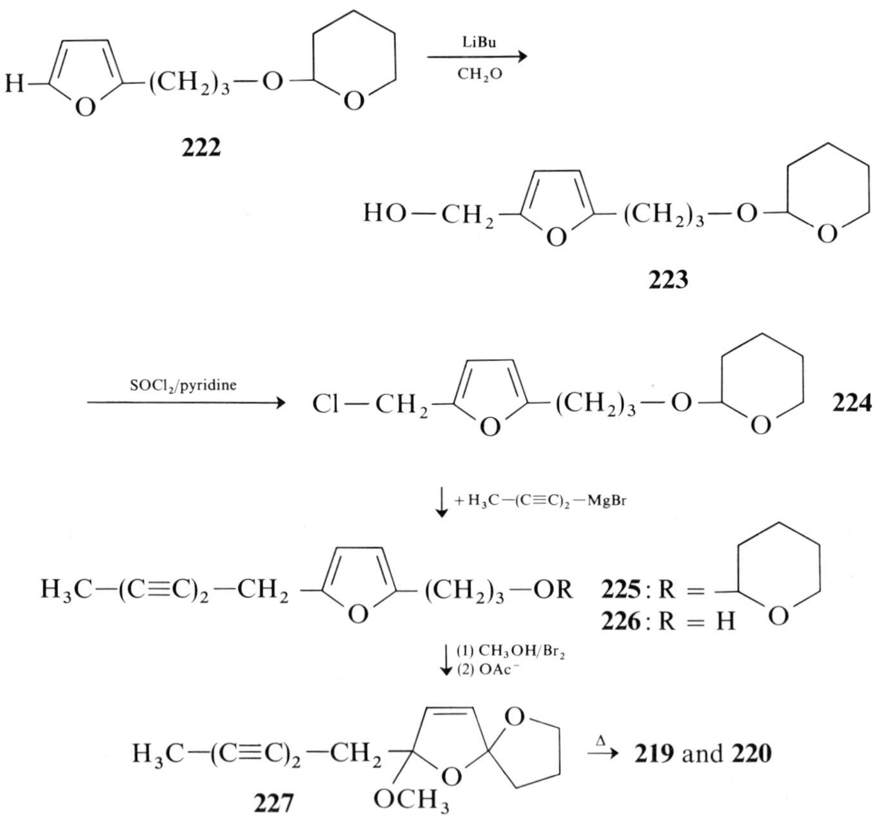

The second route is closely related to the biogenesis[42]:

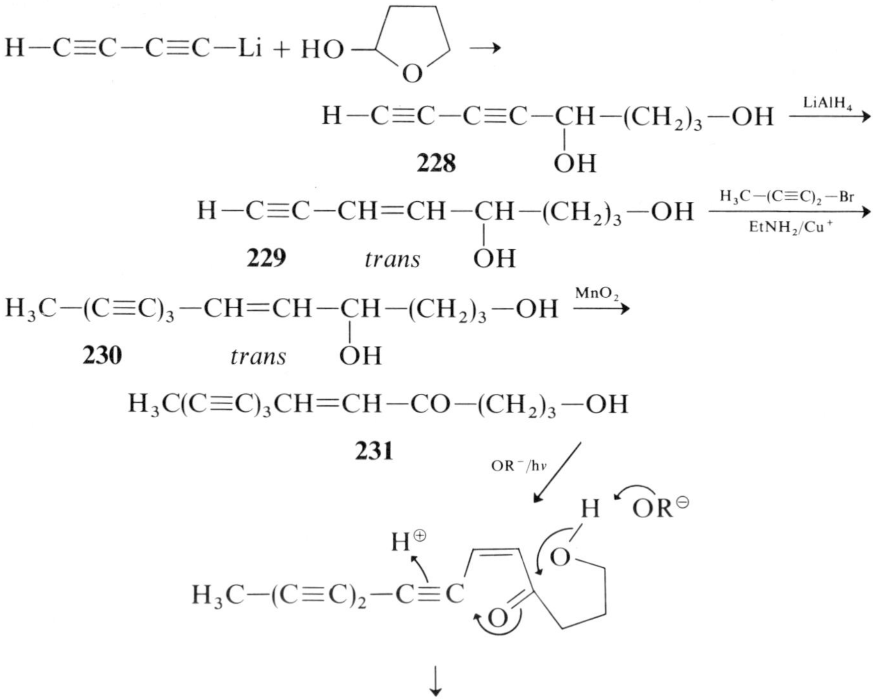

219 + 220

Several feeding experiments with different labelled precursors clearly establish the biogenetic pathway to **219** and **220** as shown in the following sequence[38]:

$$H_3C{-}(C{\equiv}C)_3{-}CH_2{-}CH{=}CH{-}(CH_2)_3{-}OH \quad \mathbf{207}$$

cis

$$\downarrow [O]$$

$$H_3C{-}(C{\equiv}C)_3{-}CH{=}CH{-}CH(OH){-}(CH_2)_3{-}OH$$

$$\downarrow [O]$$

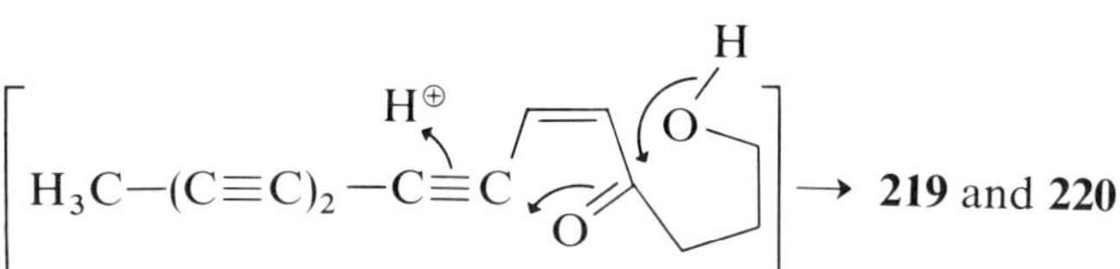

232

Using double labelled **207** (13-^{14}C, 4·5-^{3}H) this sequence could be further proved by localisation of the tritium using the different hydrogenation products. The described pathway requires the loss of one tritium isotope at C_4. This could be shown by isolation of **212** and **221** followed by exchanging the movable hydrogens[38]:

$$\mathbf{219} \xrightarrow{H_2} C_6H_{13}{-}CO{-}CH_2{-}\overset{*}{C}H_2{-}CO{-}(CH_2)_3{-}OH[5\text{-}^3H]\text{-} \quad \mathbf{212}$$

$\rightarrow C_6H_{13}{-}CO{-}CH_2{-}\overset{*}{C}H_2$ 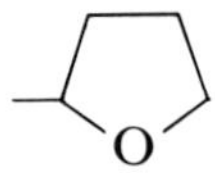 [5-^{3}H]- **221**

As expected, only in **212** could the tritium be exchanged, thus clearly establishing the proposed mechanism. **219** and **220** are very common in the tribus Anthemideae. From *Chrysanthemum leptophyllum* D.C., an acetylene has been isolated showing very similar spectral data to those of **219**, with the exception of the methyl signal in the n.m.r. spectrum. However both the n.m.r. and i.r. spectra show the presence of an acetylenic proton. Therefore, the only possible structure is **233**[106]:

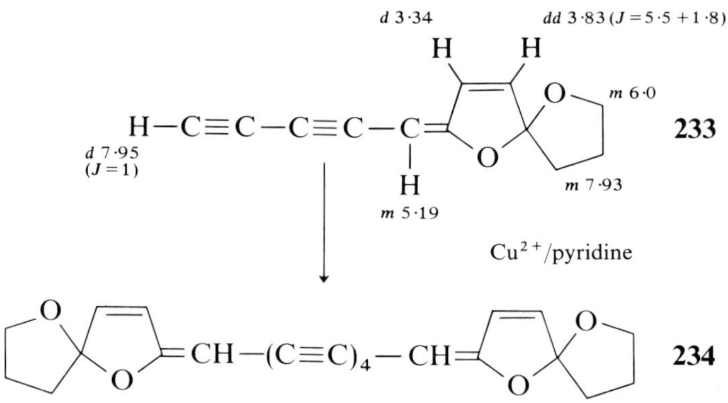

233

234

Dimerisation using copper acetate in pyridine leads to the highly unstable tetrayne **234** with u.v. maxima at 440, 405, 376 and 345 nm[106].

233 is certainly formed by oxidative degradation of **219** and is the precursor of some sulphur compounds isolated from *Chrysanthemum* species[65].

Such a mixture of *cis–trans* isomeric thioethers is present in the roots of *Chrysanthemum coronarium* L. together with other enolethers. It is therefore difficult to purify.

While one of the isomers could be crystalline, the other one only crystallises as the sulphone derivative obtained by treatment with peracid.

The spectral data of the optically-active thioethers are only in agreement with structures **235** and **236**[65]:

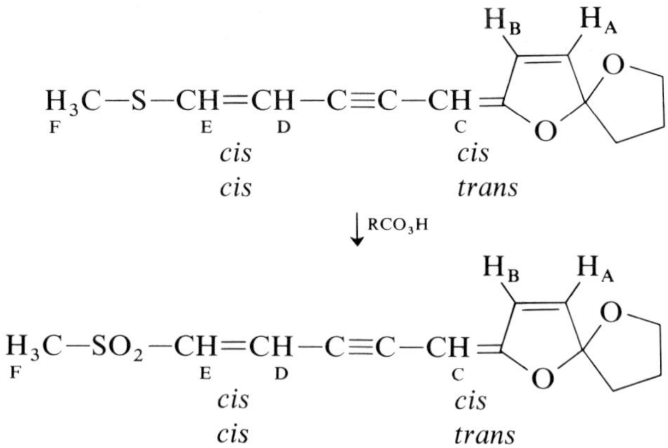

235
236
237
238

N.m.r. signals of **235–238**:

	H_A	H_B	H_C	H_D	H_E	H_F
235:	*dd* 3·91	*dd* 3·22	*ddd* 5·03	*dd* 4·45	*d* 3·77	*s* 7·68
236:	*d* 3·79	*dd* 3·75	*dd* 5·35	*dd* 4·40	*d* 3·72	*s* 7·66
237:	*dd* 3·64	*dd* 2·89	*ddd* 4·88	*m* 3·52		*s* 6·95
238:	*m* 3·62		*d* 5·20	*m* 3·42		*s* 6·76

Coupling constants of

235: $J_{AB} = 5{\cdot}5 \quad J_{AC} = 1{\cdot}9 \quad J_{BC} = 0{\cdot}6 \quad J_{CD} = 2{\cdot}7 \quad J_{DE} = 9{\cdot}8$ Hz
236: $J_{AB} = 5{\cdot}6 \quad J_{AC} = 0{\cdot}7 \quad J_{CD} = 2{\cdot}6 \quad J_{DE} = 9{\cdot}9$ Hz

The structures are established by synthesis following the route for preparing **219** and **220**[49]:

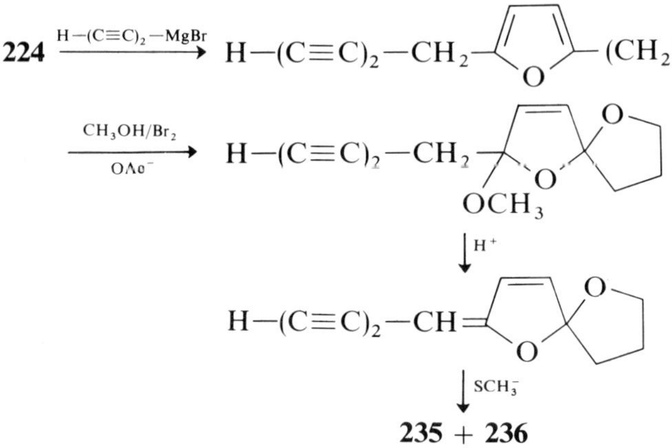

cis **233**
trans **239**

Small amounts of the corresponding *trans–trans*- and *trans–cis*-isomers formed by *cis*-addition of methyl mercaptane have been isolated as well as **235** and **236**. This synthesis also establishes the structure of **233**.

From *Chrysanthemum coronarium* L. also the oxidation products of **235** and **236**, the sulphoxides **240** and **241** have been isolated[65]. Their structures are elucidated clearly by n.m.r. spectroscopy:

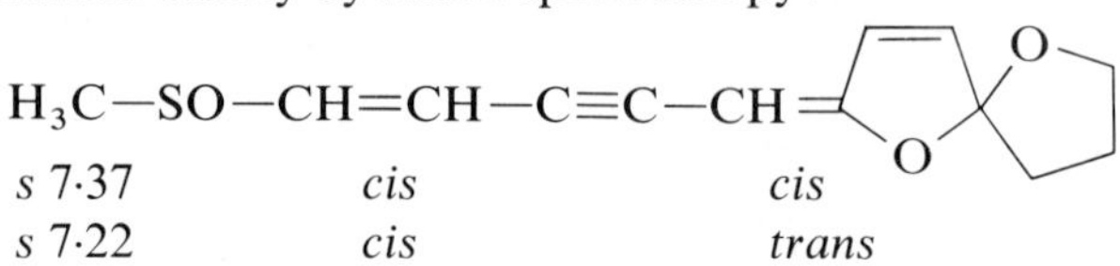

240
241

235 and **236** are biogenetically related to **219** and **220** as well as to **233** and **239**. With *Chrysanthemum carinatum* Schousb., it has been shown that methionine is incorporated, probably again in two steps since double-labelled methionine gives different incorporation of the $^{3}H_3C$-group and the ^{35}S[113]:

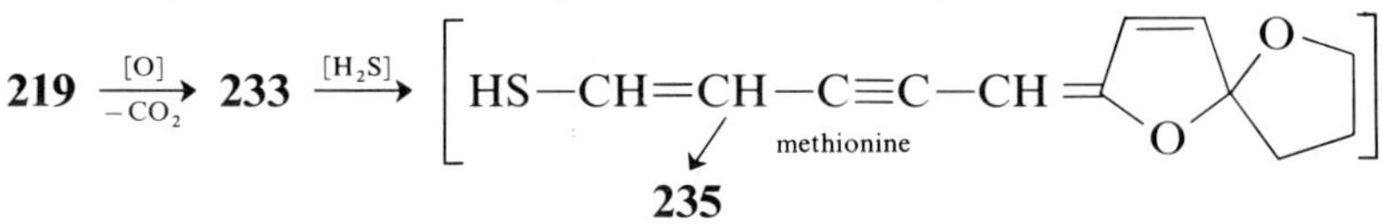

240 and **241** are clearly oxidation products of **235** and **236**.

Direct oxidation products of **219** and **220** are also very common. From *Chrysanthemum parthenifolium* Pers., an acetylene has been isolated showing the typical u.v. maxima of a diyn-enolether. The molecular formula was estimated as $C_{13}H_{12}O_3$ and the n.m.r. spectrum established structure **242** for this enolether[37]:

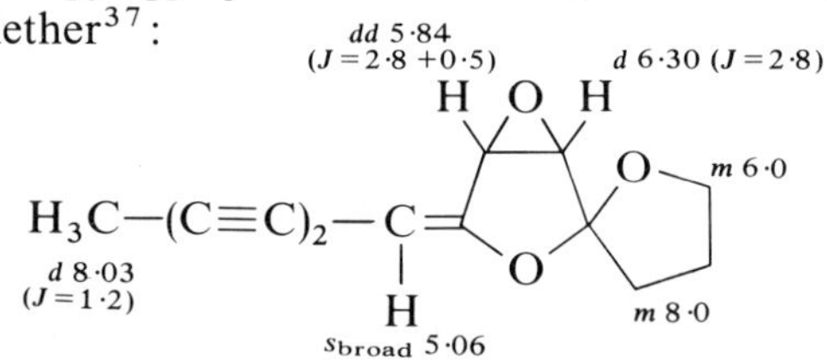

242

The aerial parts of *Chrysanthemum mawii* Hook. contain an enolether[37] being the *cis*-isomer of a second enolether isolated from *Chrysanthemum parthenicum* Bernh.[45]. The structures are elucidated from the spectral data and have been established by synthesis[37]:

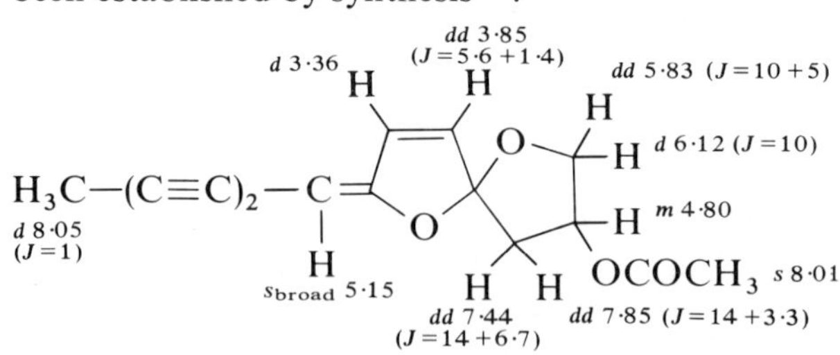

243

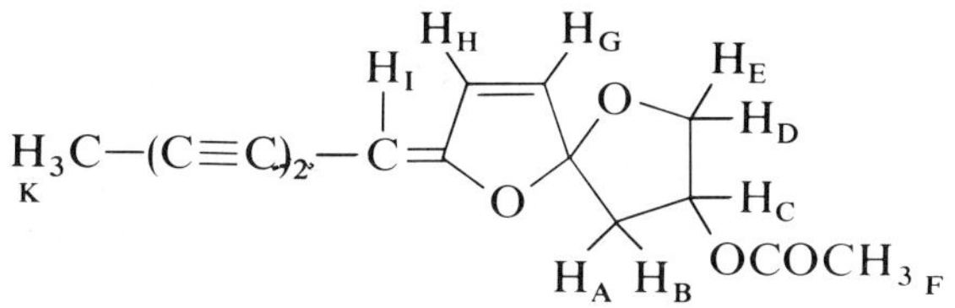

244

H_A	*dd* 7·82	H_F	*s* 7·97
H_B	*dd* 7·32	H_G	*dd* 4·00
H_C	*dddd* 4·68	H_H	*d* 3·89
H_D	*dd* 5·76	H_I	*dq* 5·46
H_E	*dd* 6·14	H_K	*d* 7·99

The coupling constants are:

$$J_{AB} = 14 \quad J_{AC} = 3 \quad J_{BC} = 6{\cdot}6 \quad J_{CD} = 4{\cdot}8 \quad J_{CE} = 2 \quad J_{DE} = 10{\cdot}1$$
$$J_{GH} = 5{\cdot}5 \quad J_{GI} = 1 \quad J_{IK} = 1{\cdot}2\ \text{Hz}$$

The synthesis of **243** and **244** is achieved by the following route[37]:

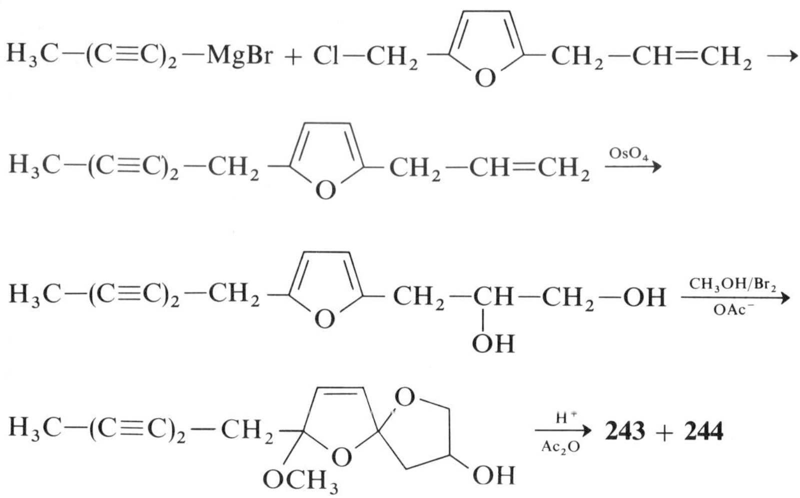

In connection with the configuration of **243**, a second synthesis is of interest[114]. Hydroboration of the natural enolether **248** leads to a mixture of the epimeric alcohols **246** and **247**. Only **247** shows a hydrogen bond in the i.r. spectrum. Model considerations let suppose, that we deal with the following configurations:

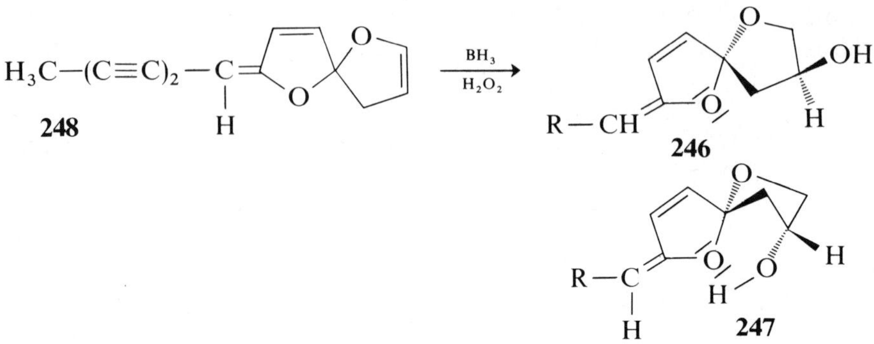

248 246 247

Only the n.m.r. spectrum of **246** is very similar to that of **243**. Therefore, the configuration of the natural acetates are most probably of type **246**.

From the roots of *Chrysanthemum leucanthemum* L. beside many other acetylenes again a pair of *cis–trans*-isomeric optically active enolethers have been isolated. The molecular formula is $C_{13}H_{10}O_2$, while the u.v. spectra are very similar to those of **219** and **220** respectively. The i.r. spectra, again very well-defined, show a second enolether stretching band and the hydrogenation gives the diketo alcohol **212**. Therefore, it is most likely that we have here the enolethers **248** and **249**[62]. The n.m.r. data clearly establish this though the interpretation is somewhat complicated due to the large number of coupling partners:

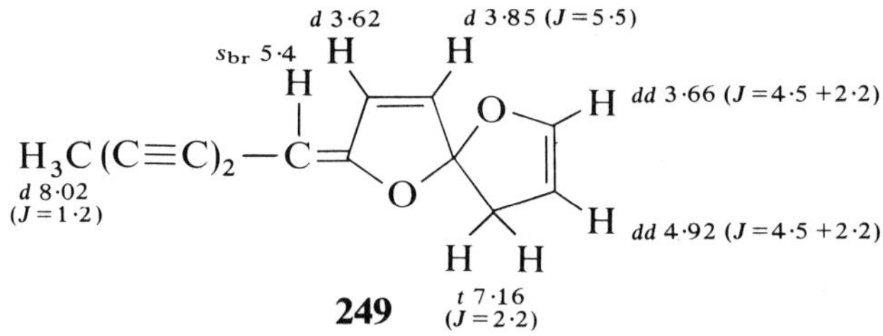

249

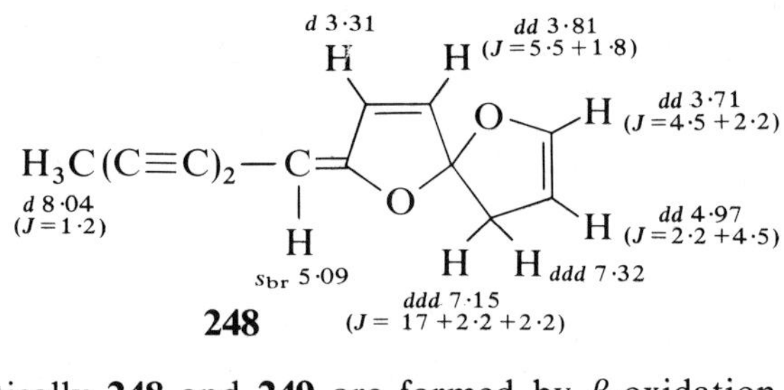

248

Biogenetically **248** and **249** are formed by β-oxidation of **219** and **220** respectively, followed by stereospecific elimination as shown by special feeding experiments[38,110]. Double labelled **250** (13-^{14}C-1-^{3}H) is incorporated into **248** by loss of the tritium, while [13-^{14}C-4·5-^{3}H]-**207** is incorporated in the expected manner. This only can be explained in the following manner:

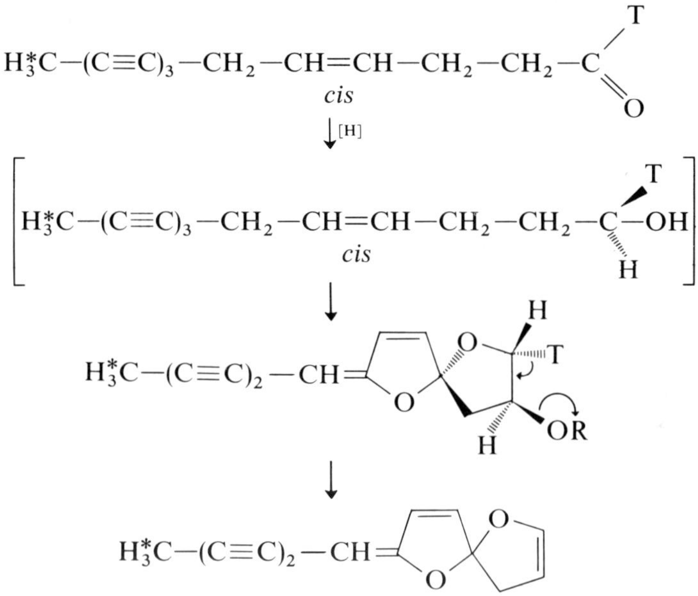

250

207

248

This would mean, that **250** is first reduced stereospecifically and then is transformed to **219**, which stereospecifically again is oxidized to **246**. Finally *trans*-elimination would lead to **248** with loss of the tritium at C_1.

The degradation to the enolether **251** with —C≡C—H endgroup is also known[106] in the series of **248**, which again are biochemically transformed to the methylthioethers **252** and **253**[67] isolated from *Chrysanthemum carinatum* Schousb.:

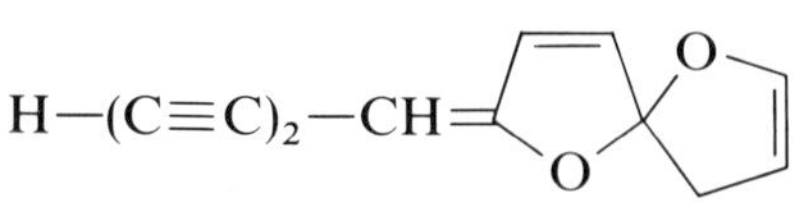

251

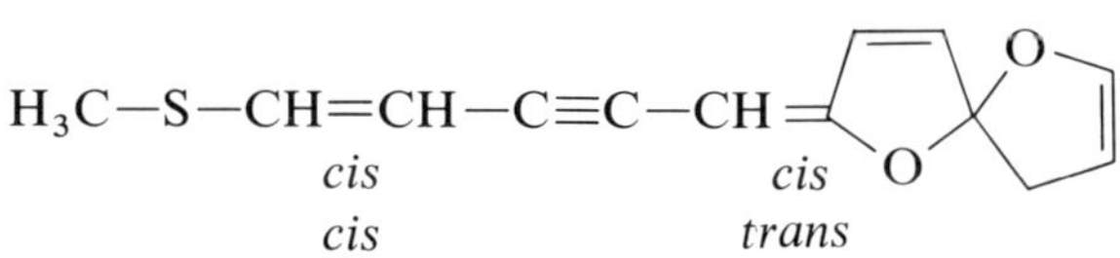

252
253

Oxidation products of **248** and **249** are also very common, especially in *Chrysanthemum* species. From the roots of *Chrysanthemum leucanthemum* L. two optically active crystalline enolethers have been isolated showing molecular formula $C_{15}H_{12}O_4$. The i.r. spectra show the presence of an acetate group, which can be saponified to give an alcohol with the molecular formula $C_{13}H_{10}O_3$. These properties indicate that the enolethers, which are interconvertible by u.v. irridation, have only an additional oxygen function. Catalytic hydrogenation again gives an anomalous crystalline hydrogenation product, which on saponification and subsequent periodate oxidation yields the keto-acid **217**. Direct periodate oxidation, under acidic conditions, of the alcohol obtained from the natural acetate gives the known lactone **257**. These results are in agreement with the structures **254** and **255**. The n.m.r. spectra also agree with these proposals[62]:

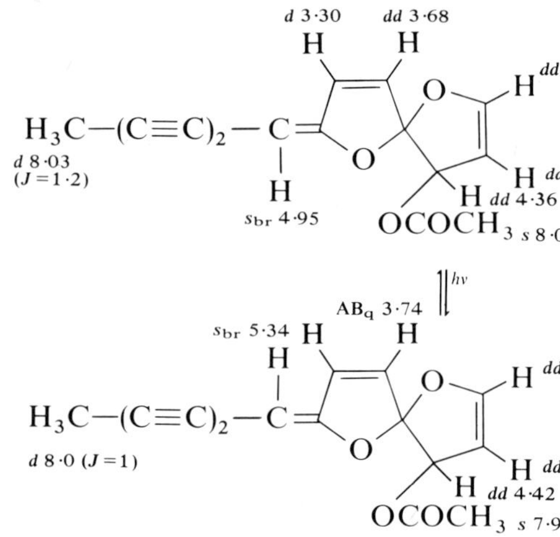

254

255

254 $\xrightarrow{H_2}$ $C_6H_{13}-CO-CH_2-CH_2-CO-CH(OAc)-CH_2-CH_2-OH$ **256**

(1) OH^- (2) HIO_4

$C_6H_{13}-CO-CH_2-CH_2-COOH$ **217**

H_3O^+

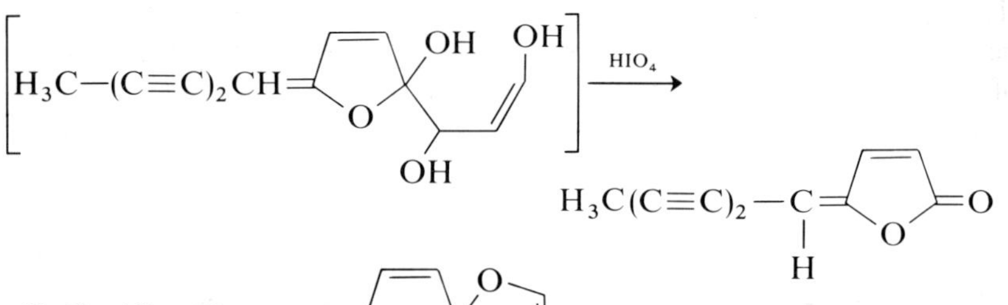

257

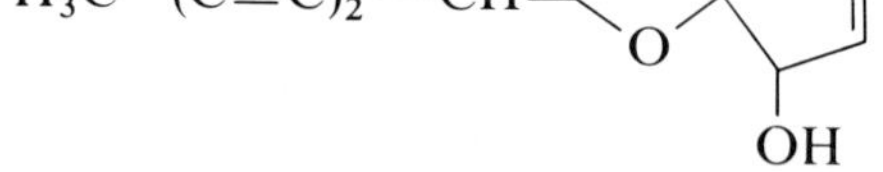

cis: **258**
trans: **259**

The corresponding alcohols **258** and **259** also have been isolated from *Chrysanthemum* species[65].

Again, the degradation of **254** with elimination of the methyl group has been observed[106]:

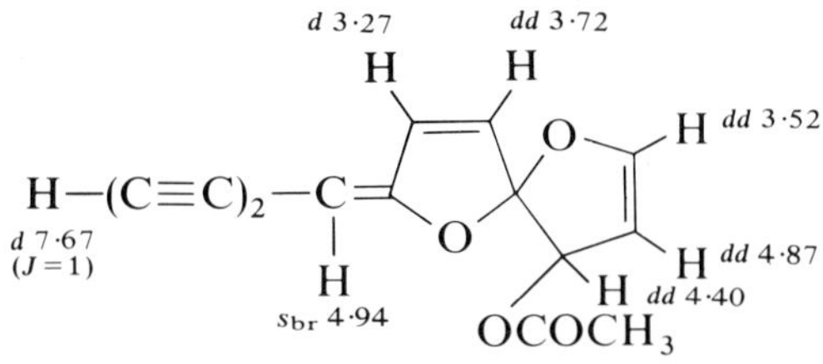

260

As in this case compounds such as **260** are usually found together with the corresponding methyl derivative. The separation is normally achieved by silver salt formation of the terminal acetylene-H compound, which then can be separated and purified after cyanide hydrolysis of the isolated silver salt.

An isovalerate has been also isolated[106] from the roots of *Chrysanthemum* monspeliense beside the acetates **254** and **260**. Again the n.m.r. data are very useful for structure elucidation of **261**:

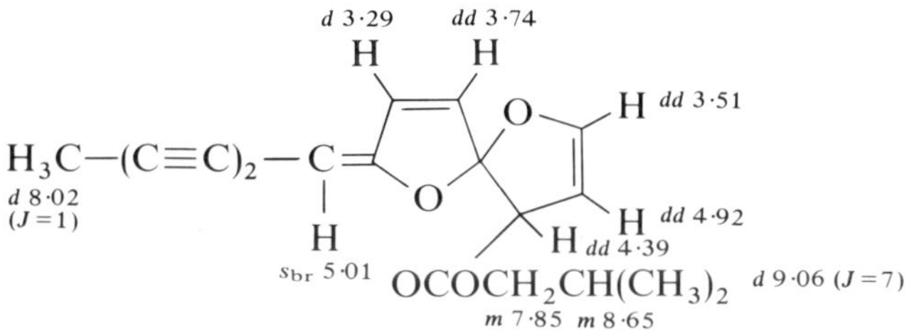

261

Further oxidation of enolethers of type **254** leads to a number of epoxides also very common especially in *Chrysanthemum* species. As already mentioned, the roots of *Chrysanthemum leucanthemum* L. contain a number of acetylenes; among these there is also an acetate with molecular formula $C_{15}H_{12}O_5$. The u.v. spectrum, typical for a diyn-enolether, indicate that we are dealing again with an epoxide. The n.m.r. spectrum and other properties of the compound agree with structure **262**[115]:

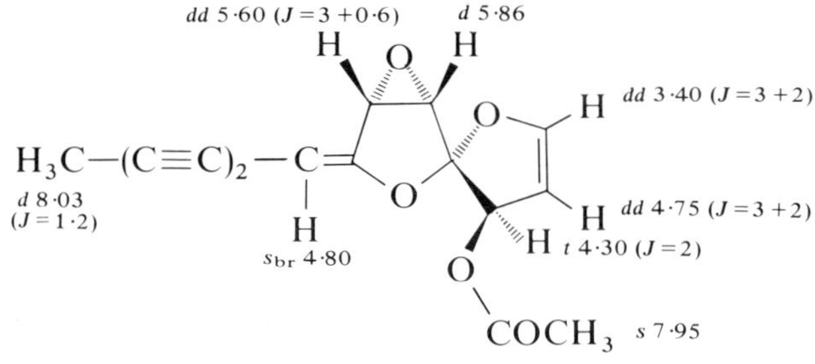

262

The corresponding alcohol **263**, obtained by very mild saponification of **262**, has also been isolated from *Chrysanthemum* species[37]:

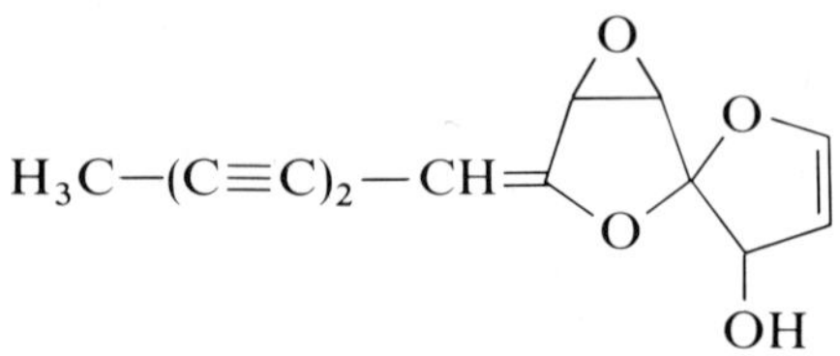

263

Further esters of type **262** have been found in *Chrysanthemum* species. A very rich source of such compounds is *Chrysanthemum monspeliense*, containing as much as four different esters of the alcohol **263**. The mixture of these esters is difficult to separate. The structures have been elucidated mainly by n.m.r. and mass spectroscopy. As well as **262** it turns out that this plant contains the esters **264–266**[106]:

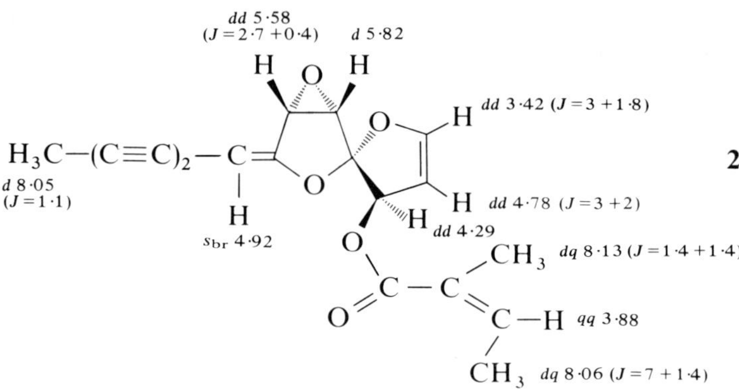

264

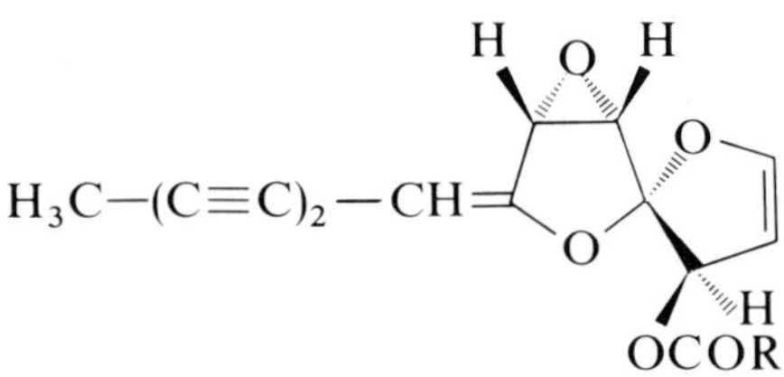

265: R = $-CH_2CH(CH_3)_2$

266: R = $-CH{=}C(CH_3)_2$

The configuration of all the epoxide esters is not established, but model considerations together with the n.m.r. spectra indicate that the given configuration is most likely since a typical shift of the epoxide proton at C_5 is always observed in an unsaturated ester. This seems only to be possible if the particular hydrogen atom is very close to the ester carbonyl; this is only true in the given configuration or in the enantiomer. The absolute configuration of all spiroketalenolethers is completely unknown. Chemical investigations of this problem seems to be extremely difficult due to the easy racemisation of the assymetric ketal centre.

3.3. Aromatic Compounds

Again, in the tribus Anthemideae, a considerable number of aromatic acetylenes have been identified. A typical example is the occurrence of a number of diacetylenic compounds isolated from the roots of *Chrysanthemum frutescens* L. The main compound showing only an uncharacteristic u.v. maximum at 280 nm ($\varepsilon = 3000$) is a crystalline ester with weak acetylene bands only in the i.r. spectrum. Hydrogenation gives an oily ester with the molecular formula $C_{15}H_{22}O_3$, while that of the natural product is $C_{15}H_{14}O_3$. Therefore, we can assume that only two triple bonds are hydrogenated. It is further evident from the i.r. spectrum that an aromatic ring is present. Methoxyl group analysis shows the presence of two $-OCH_3$ groups. Therefore, most probably this acetylene is a derivative of a methoxybenzoic ester. The u.v. spectrum indicates an *o*-methoxybenzoic ester. This could be confirmed by permanganate oxidation yielding *o*-methoxyphthalic acid. These results are only in agreement with structure **267**[116], which has been confirmed by synthesis[117]:

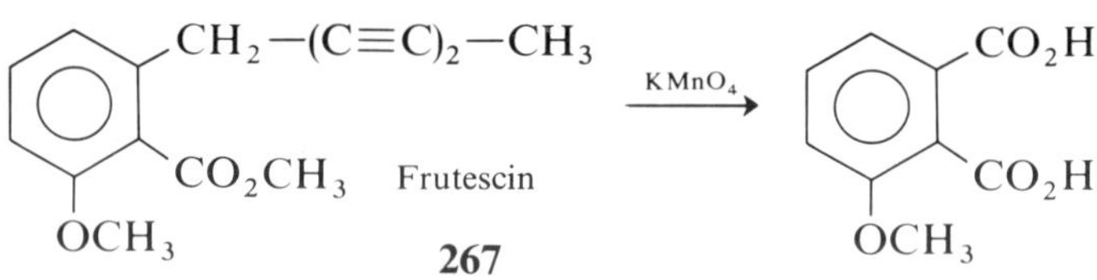

267

Starting with the aldehyde **268** the ester **267** has been prepared using the following reaction path:

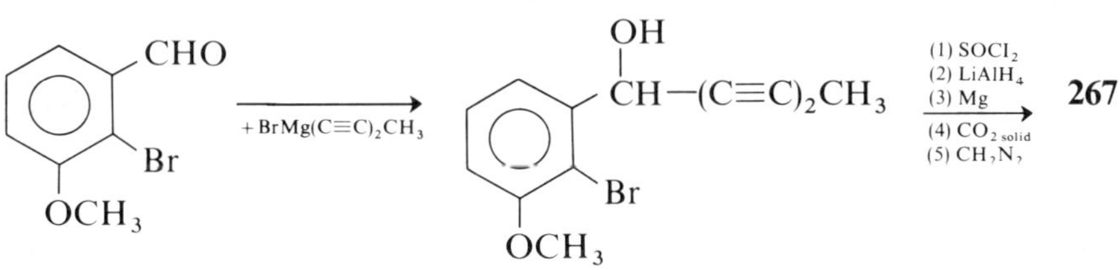

268 **269**

Together with **267**, a second ester is present in the same fraction of the root extract. This one shows a strong $-C{\equiv}C-H$ band in the i.r. spectrum and has the molecular formula $C_{14}H_{12}O_3$. All other data are very similar. Therefore, the structure can only be **270**, and is further established by oxidative coupling to the crystalline tetrayne **271** and also by oxidation to methoxyphthalic acid[116]:

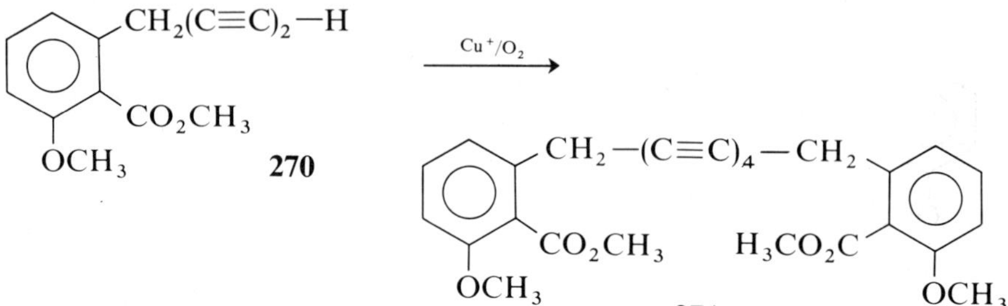

270 **271**

A somewhat more polar ester also found in the root extract of *Chrysanthemum frutescens* L. is optically active and has the molecular formula $C_{17}H_{16}O_5$. The result of acidic *trans*-esterification as well as the stronger i.r. band at 1750/cm let suppose, that this compound might be the *O*-acetate of **267**. This compound most likely has structure **272**[116]:

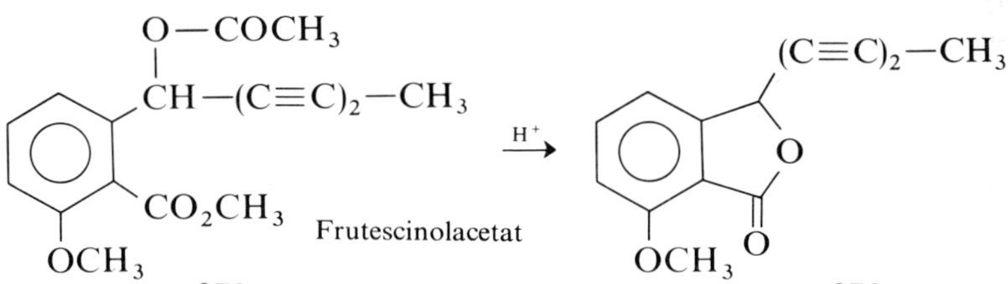

272 **273**

This assumption has been confirmed by synthesis of **273**[118]:

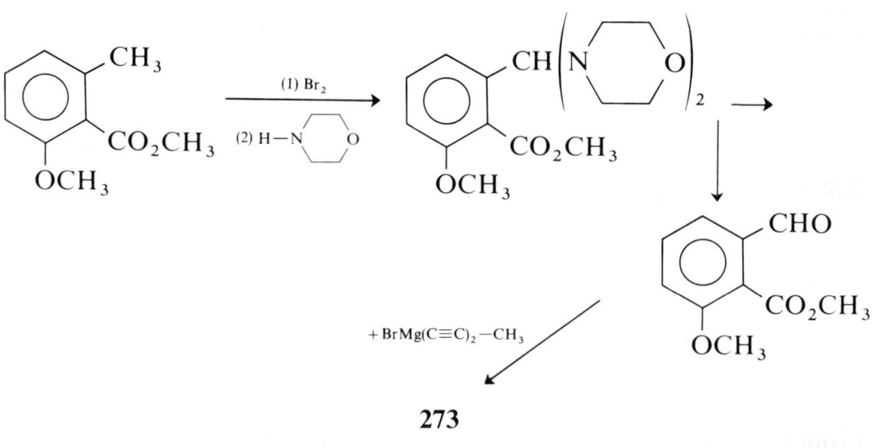

273

The desmethyl compound **274** has also been isolated together with **272**. By converting **274** to its silver salt, it could be separated easily from **272**.

Again acidic *trans*-esterification leads to a γ-lactone[116]:

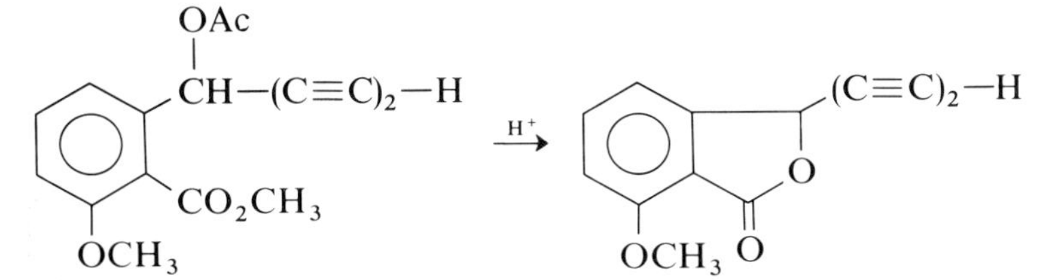

274 275

A further very similar crystalline compound has also been isolated from *Chrysanthemum frutescens* L. It shows a more typical u.v. spectrum with maxima at 330, 292, 276, 265 and 215 nm[116]. The strong acetylene stretching band and the molecular formula estimated as $C_{15}H_{12}O_4$ show it to be the keto-ester **276** as a further derivative of **267**.

This was confirmed by hydrogenation and borohydride reduction, which finally yields the lactone **277** also obtained from **272** by hydrogenation and transesterification:

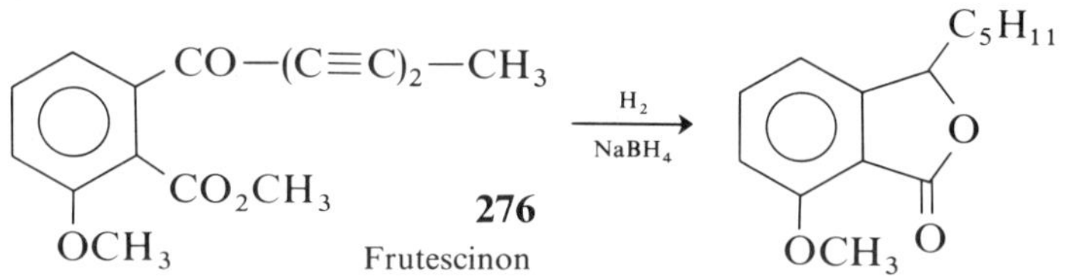

276 Frutescinon **277**

The structure has since been established by synthesis[117]. Manganese dioxide oxidation of **269** followed by ketalisation leads to the ketal **279**, which could be converted to **276** as follows:

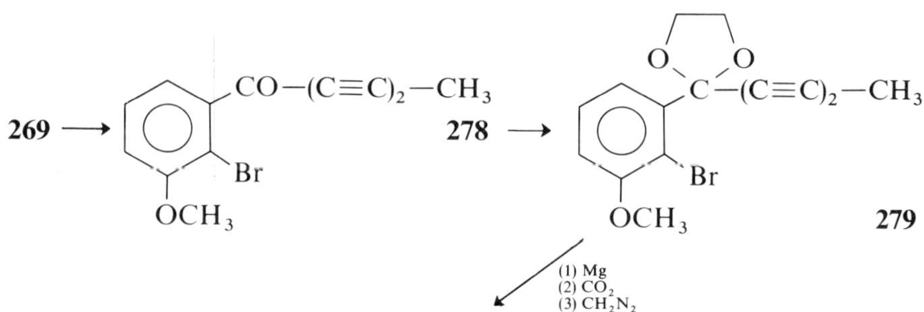

278 **279** **276**

From the roots of *Artemisia capillaris*[119], and also from several other species belonging to the tribus Anthemideae[120], a crystalline compound has been isolated showing the typical u.v. spectrum of an isocoumarin. Partial ozonisation leads to phthalic acid and pentynoic-(3)-acid. Therefore, the structure **280** has been proposed for this isocoumarine:

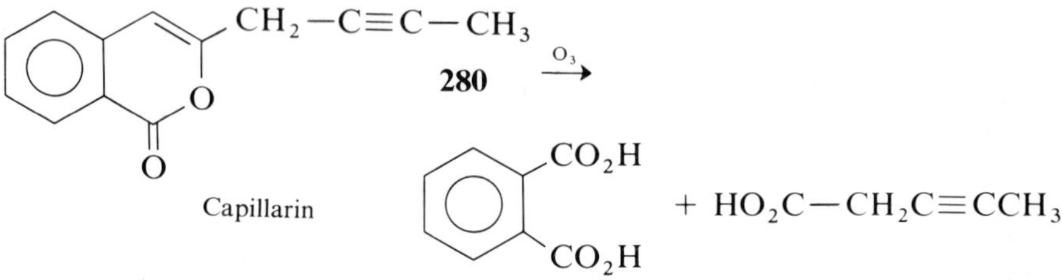

280

Capillarin

Two further isocoumarines have been isolated from the roots of *Anthemis fuscata* Brot., with structures elucidated as **281** and **282**. As in the same plant, **280** is also present but it seems likely that **281** and **282** are biogenetically related to **280**[121]:

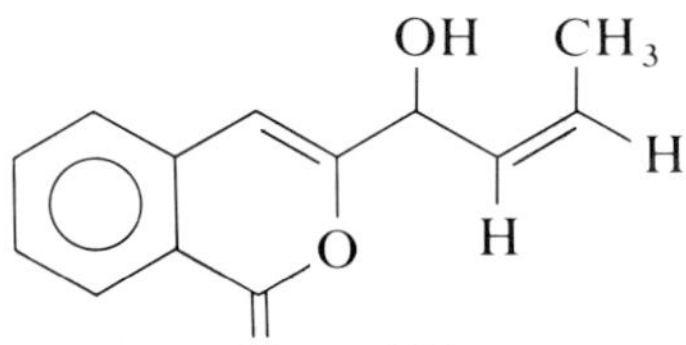

281

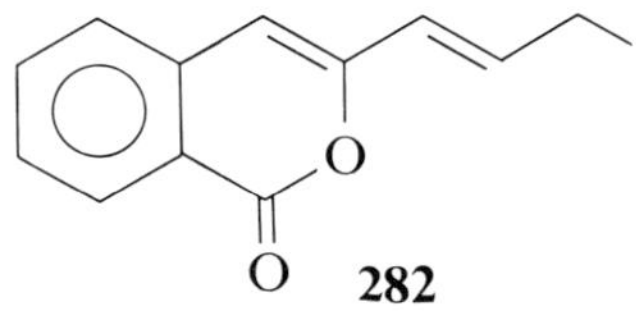

282

281 could be formed by hydrogenation and oxidation, while **282** might be formed by hydrogenation and subsequent isomerisation.

The biogenesis of all these aromatic acetylenes is more or less established. Feeding experiments using labelled **206** show that **267** and the other aromatics most likely are formed by the following pathway[38,110,122]:

$$H_3C-(C\equiv C)_3-CH_2-CH=CH-CH_2-CH_2-CO_2R \quad \textbf{206}$$

cis

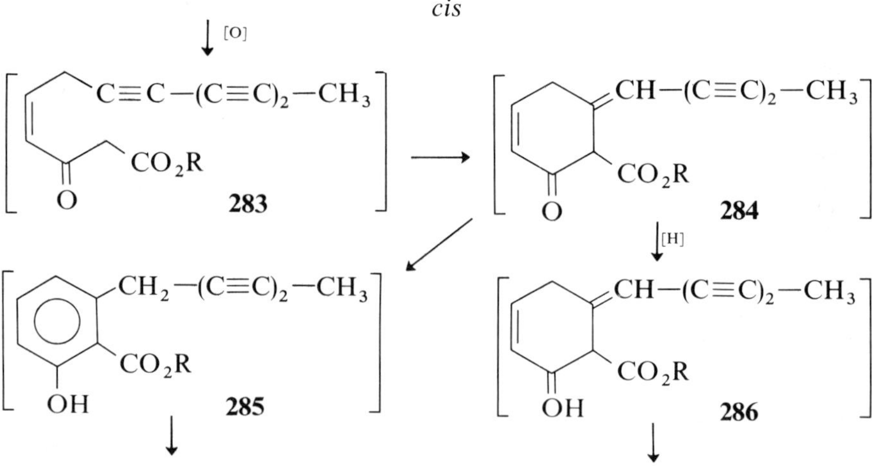

283 **284** **285** **286**

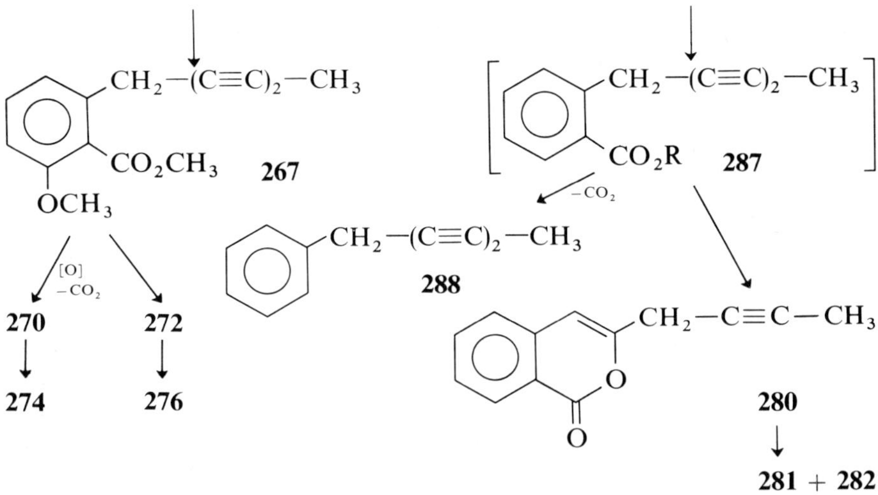

The most important step would be β-oxidation of **206** followed by a "Michael addition" leading to the ketoester **284**, which can be converted into a number of compounds isolated from the above species.

267 has been synthesised by a biogenetic-like pathway as follows[123]:

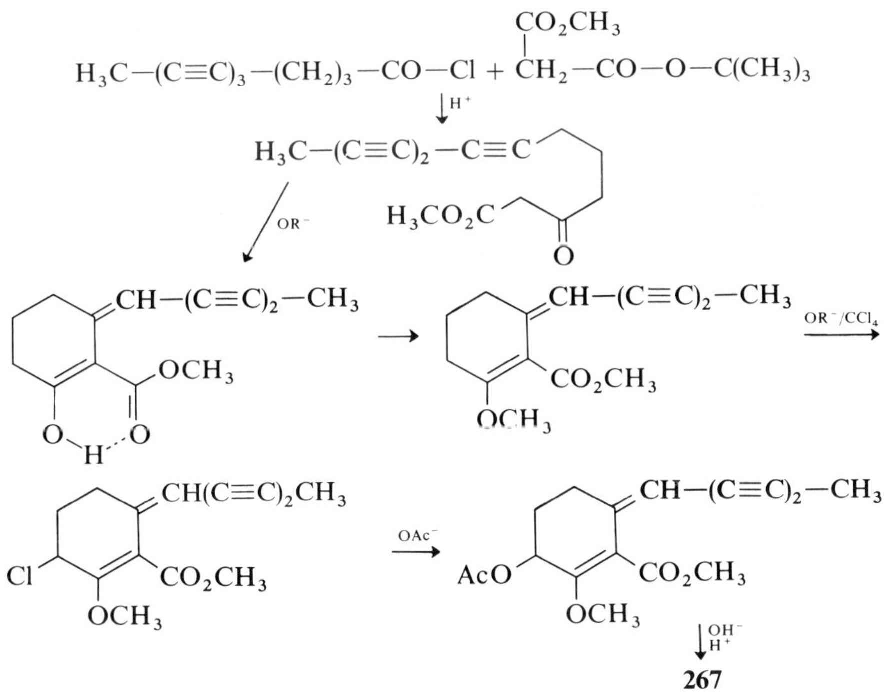

The crucial reaction step again is a "Michael addition". The introduction of the last double bond being first a real problem has been achieved by chlorination of the anion using carbon tetrachloride, nucleophilic substitution of the allylic chloride by acetate and alkaline elimination finally leads to frutescin (**267**).

288 and a number of derivatives are also found in different species of the tribus Anthemideae. **288** has been isolated first from *Artemisia capillaris*[124]. Together with **288**, a second hydrocarbon is also present in the roots of *Artemisia dracunculus* L. and it shows a very similar u.v. spectrum, but from the i.r. spectrum it is evident that this compound has a terminal ethylenic group. Therefore, this must be the phenylderivative **289**[48], which can be dimerised to the tetrayne:

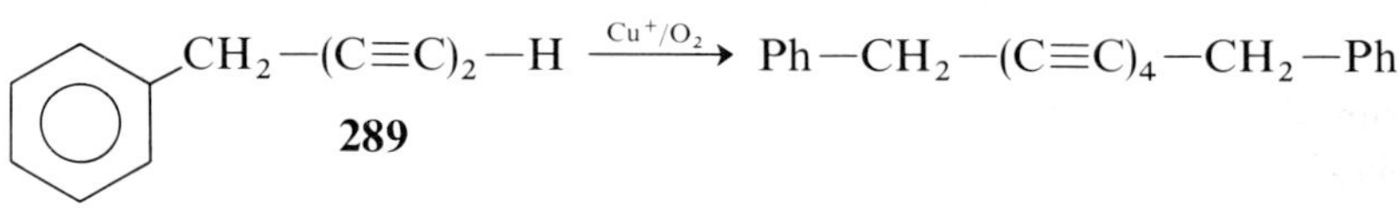

289

Structure **289** is confirmed also by synthesis[120].

The roots of *Chrysanthemum frutescens* L. contain an optically active acetate, which on extremely mild saponification leads to a crystalline alcohol; normal conditions yield a ketone. Molecular formula and spectral data, as well as manganese dioxide oxidation leading to a ketone, are only in agreement with structure **291** for the alcohol—and consequently **290** for the natural acetate[120]:

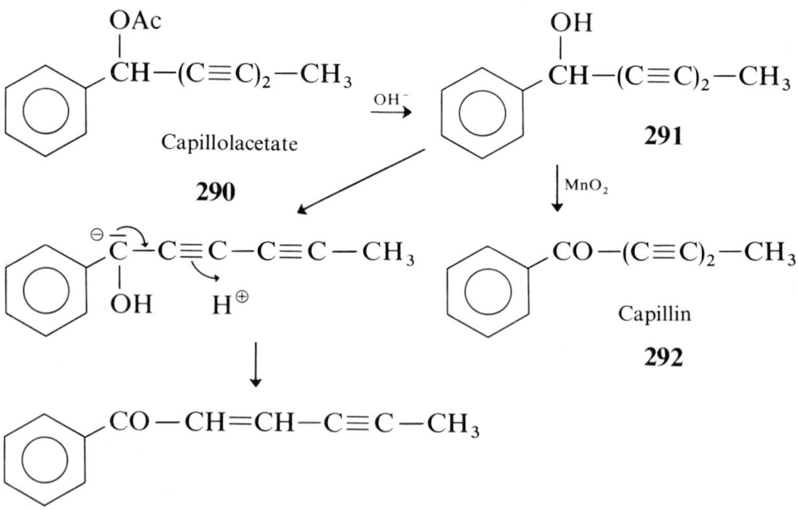

Capillolacetate **290**

291

Capillin **292**

The mechanism of ketone formation is probably characterised by isomerisation via the anion.

292 is also present in the root of *Chrysanthemum frutescens* L. and many other species. It was first identified in the roots of *Artemisia capillaris*[125]. Capillin (**292**) is also converted by elimination of the methyl endgroup. The structure of the resulting ketone **293**[63], isolated from *Chrysanthemum segetum* L., can be easily confirmed by synthesis[63]:

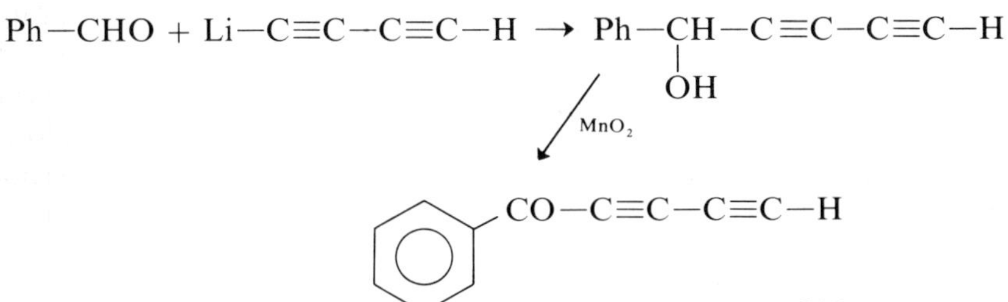

293

From the same plant several sulphur-containing compounds have been isolated. The less polar compound with molecular formula $C_{12}H_{12}S$ and a u.v. maximum at 281 nm shows a strong acetylenic band in the i.r. Partial hydrogenation leads to a thioether, which on hydrolysis gives 5-phenyl-penten-(2)-al. Hydrogenation with three moles of hydrogen leads to the mercaptoether **296**. Considered therefore, together with the n.m.r. spectrum, the only structure possible for the natural thioether[63] is **294**:

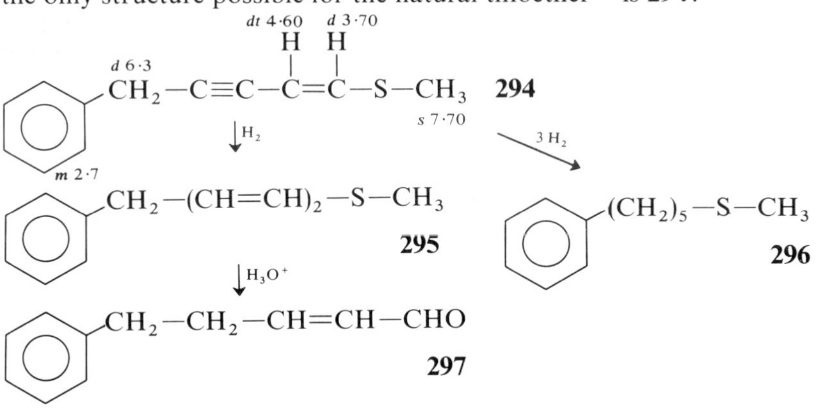

294 can be prepared by addition of methylmercaptide to **289**[126]:

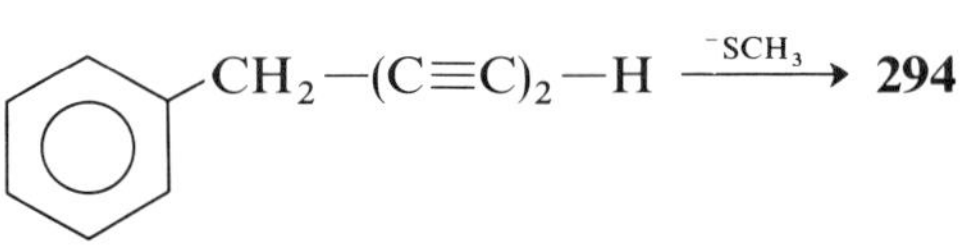

Also from *Chrysanthemum segetum* L. a pair of *cis–trans*-isomeric sulphur compounds have been isolated[63]. The structures have been confirmed by borohydride reduction and subsequent partial hydrogenation followed by hydrolysis yielding phenylpentadienal. By u.v. irradiation the two compounds are interconvertible and thus, together with the n.m.r. data, only structures **298** and **299** are possible:

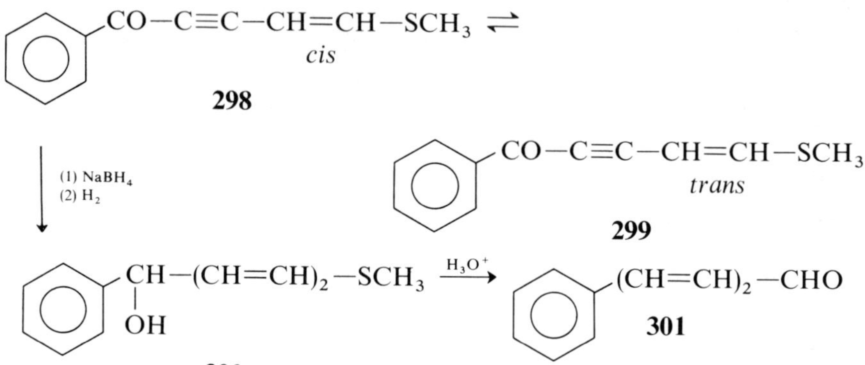

298 **299** **300** **301**

Both isomers can easily be prepared by addition of methyl mercaptide to **293**[126]:

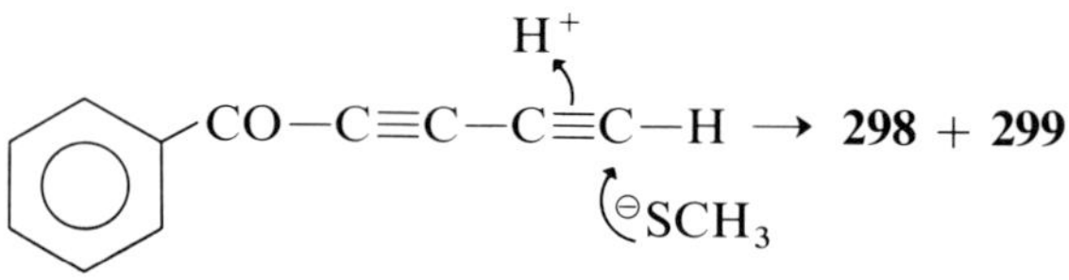

In this case both isomers are formed directly, while addition to **289** only gives the *cis*-isomer.

By feeding labelled **293** it has been shown, that also in the plants this pathway is possible[99]:

$$\text{Ph}\overset{*}{-}\text{CO}-(\text{C}{\equiv}\text{C})_2-\text{H} \rightarrow \text{Ph}\overset{*}{-}\text{CO}-\text{C}{\equiv}\text{C}-\text{CH}{=}\text{CH}-\text{SCH}_3 \rightarrow$$
$$\text{Ph}\overset{*}{-}\text{CO}-\text{C}{\equiv}\text{C}-\text{CH}{=}\text{CH}-\text{SO}_2\text{CH}_3 \xrightarrow{\text{KMnO}_4} \text{Ph}\overset{*}{-}\text{CO}_2\text{H}$$

A hydrogenation product of **292** has been described also as a constituent of *Artemisia capillaris*[127]. The structure has been confirmed by synthesis[128].

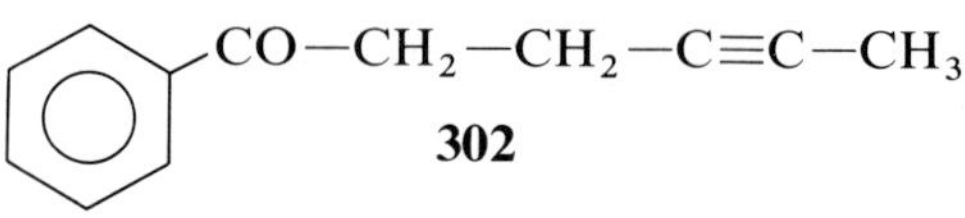

302

Oxidation products of **292** are also known. From the aerial parts of *Lonas inodora*, two ketones have been isolated with u.v. spectra very close to that of **292**. By i.r., the less polar compound clearly shows the presence of a further carbonyl group, while from the n.m.r. spectrum it is evident that only structure **303** is possible[129]. The more polar compound is an alcohol (i.r.: 3615/cm). The n.m.r. spectrum shows it to be the keto-alcohol **304**[129]. The structures of **303** and **304** have been established by synthesis[129]:

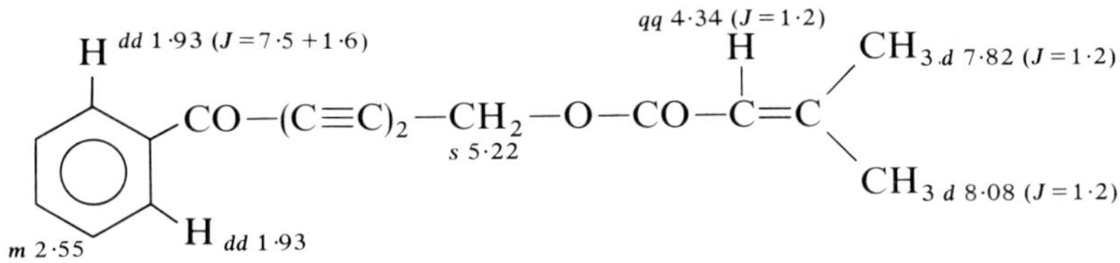

303

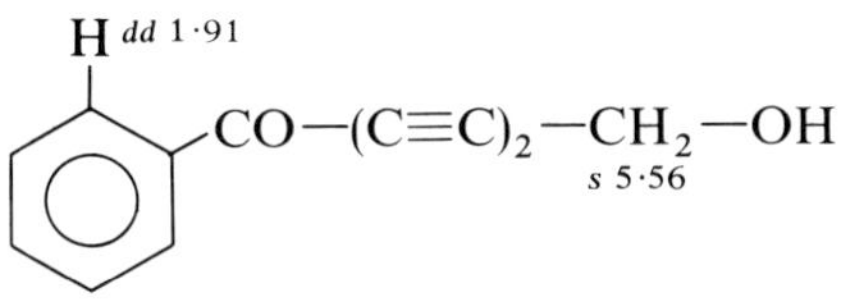

304

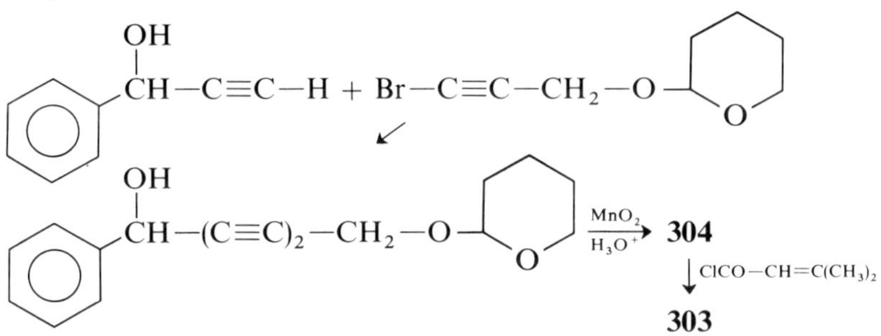

Two more drastically transformed aromatic acetylenes have been isolated from the aerial parts of *Anthemis fuscata* Brot.[130]. In addition to **280–282**, already isolated from the roots, two phenols have been found. The more polar phenol can be oxidized using manganese dioxide yielding a ketone with a characteristic u.v. spectrum (λ max: 359, 300, 284, 271, 243, 213 nm). The n.m.r. spectrum is not well resolved, but that of the unpolar compound together with the molecular formula $C_{12}H_{10}O_2$ only agrees with the structure **305**. Therefore, the second compound with molecular formula $C_{12}H_{10}O_2$ should be **306**[130]:

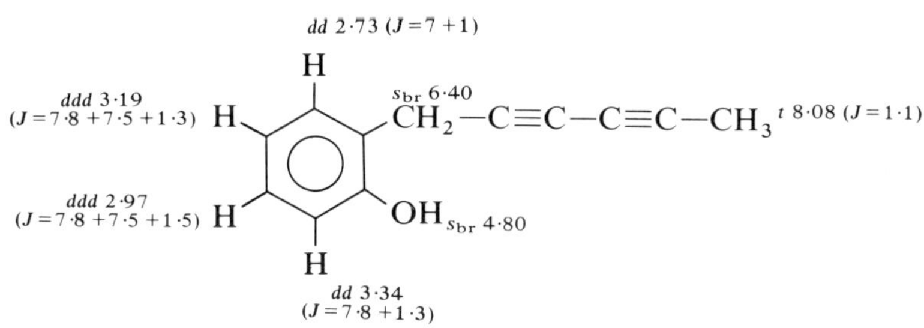

305

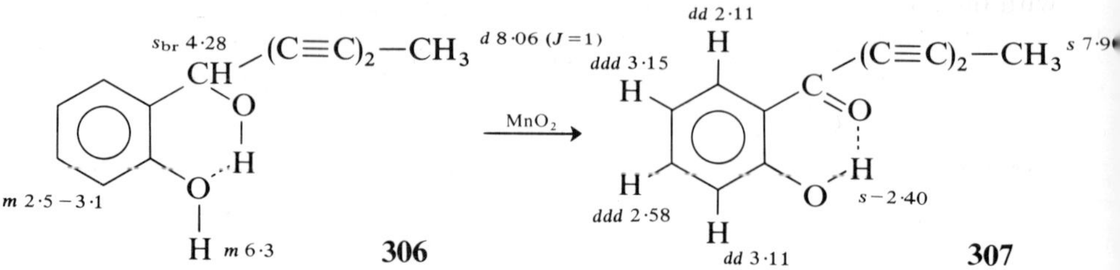

306 307

The structures **306** and **307** have been confirmed by synthesis[130]:

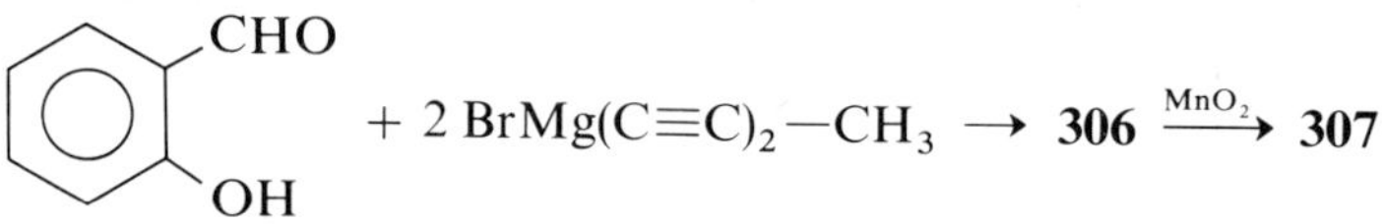

The biogenesis of **305** and **306** is not really established, but it seems certain that these phenolic compounds are closely related to the other aromatic acetylenes. Most probably the intermediate **287** (not yet isolated) is the common precursor. Biochemical oxidation might be a future method of elimination of the carboxyl group leading to the phenols **305** and **306**:

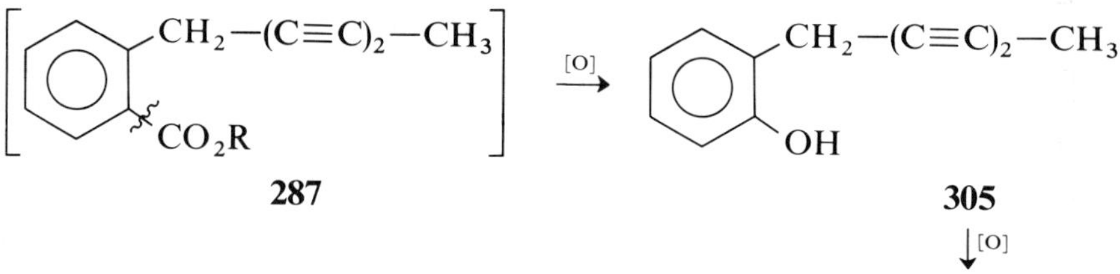

287 305 306

The root extract of *Chrysanthemum carinatum* Schousb. contains besides many other acetylenes, the thioether **308**[67]. The structure has been elucidated by chemical and spectral methods. The crystalline sulphur compound with molecular formula $C_{13}H_{12}O_2S$ can be converted with peracids to a crystalline sulphone showing u.v. maxima at 313, 273, 261 and 231 nm. The i.r. spectrum exhibits acetylenic bands (2200, 2180/cm) and a strong band at 1635/cm, which is due to a carbonyl group as borohydride reduction leads to an alcohol, with an allenic band. Manganese dioxide oxidation of the alcohol does not give the original ketone but a compound with u.v. maximum at 250 nm and a CO- stretching band at 1675/cm. These results clearly show, that a strongly polarised carbonyl group is present. Therefore, together with the n.m.r. spectrum, the following interpretation is most likely:

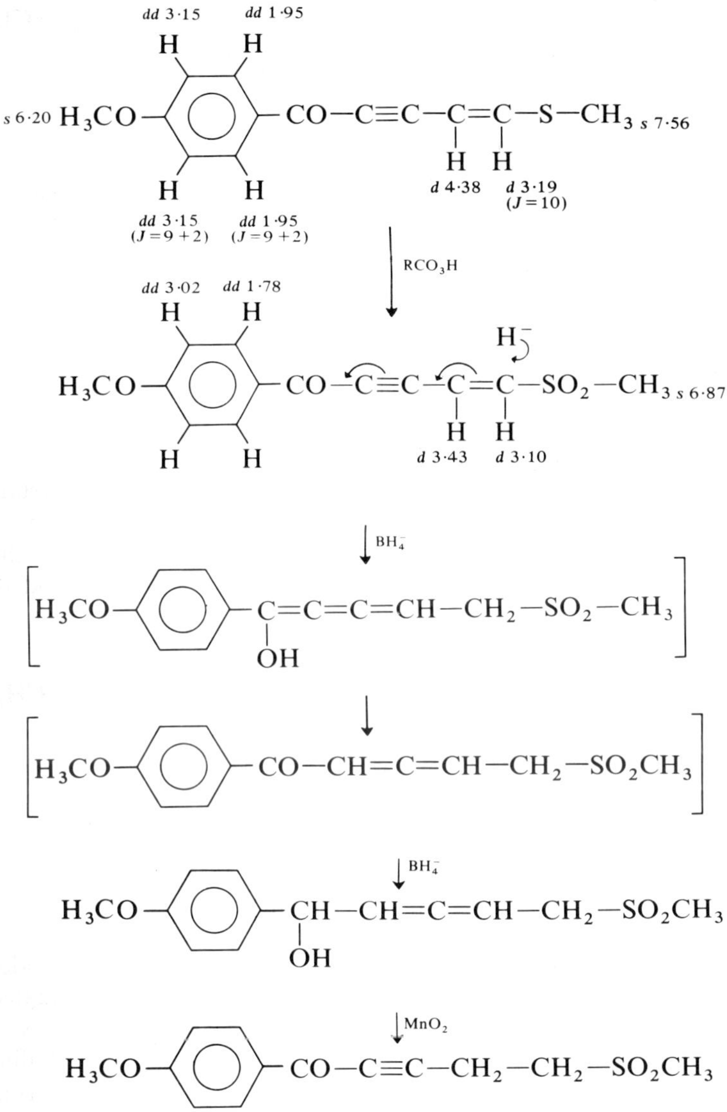

308

309

310

311

The highly polarisised sulphone **309** is probably attacked by borohydride in the γ-position of the keto group leading to an enolic intermediate, which then gives the allenic ketone and finally the alcohol **310**, which is isomerised during manganese dioxide oxidation leading to **311**.

The structure of **308** has been confirmed by synthesis starting with anisaldehyde[67]:

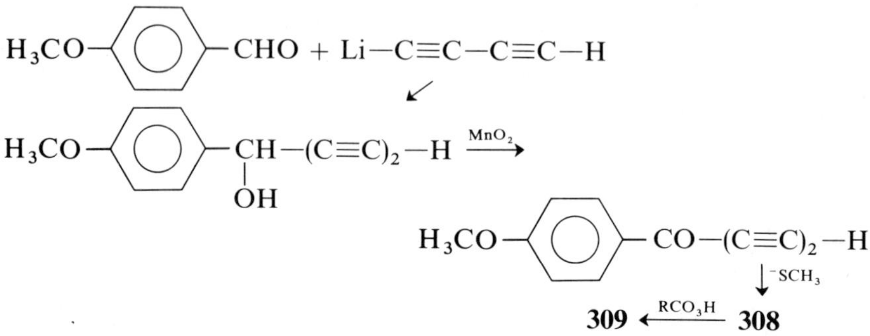

Nothing is known of the biogenesis of **308**. Perhaps it is only an oxidation product of **298** that is finally methylated:

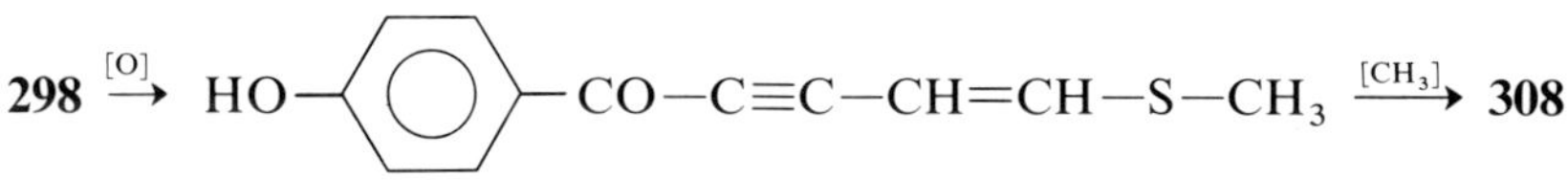

3.4. Other Ring Compounds

Several acetylenes have been isolated from different species belonging to the tribus. Anthemideae having a furan ring. The first member of this group has been found during the investigation of the root extract of *Chrysanthemum leucanthemum* L.[45,62].

A crystalline optically active compound showing the typical u.v. maxima of a triyn-ene system has been isolated from the petrolether eluates. The molecular formula estimated as $C_{13}H_{10}O$, the spectral data and the results of some reactions are only in agreement with the structure **312**:

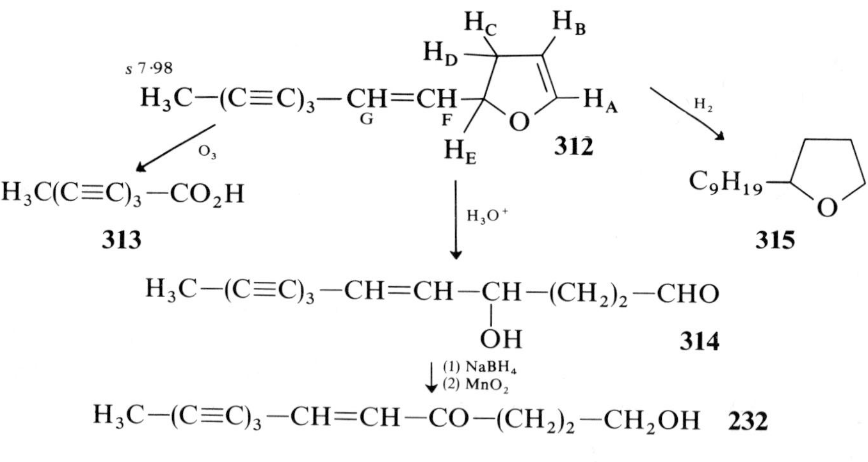

The degradation products **315** and **232** are identical with synthetic compounds. The *cis* configuration of the double bond is evident from the interpretation of the very complicated n.m.r. spectrum[45]:

H_A *ddd* 3·57 H_B *ddd* 4·98 H_C *dddd* 4·42 H_D *dddd* 7·02
H_E *dddd* 4·42 H_F *dd* 3·50 H_G *dd* 4·26

$J_{AB} = J_{AC} = J_{AD} = J_{BC} = J_{BD} = 2{\cdot}5$ Hz $J_{CD} = 16$ $J_{CE} = 8$ $J_{DE} = 11$
$J_{EF} = 8$ $J_{EG} = 1{\cdot}5$ $J_{FG} = 11$ Hz

A second unpolar compound, closely related to **312**, has been found in *Chrysanthemum leucanthemum* L. and also in other species[45]. This compound, with molecular formula $C_{13}H_8O$, shows u.v. maxima at 361 and 332·5 nm very similar to those of a triyne-triene-chromophore however with a shift to longer wavelengths of about 5 nm. This together with the n.m.r. spectrum only agrees with structure **316**:

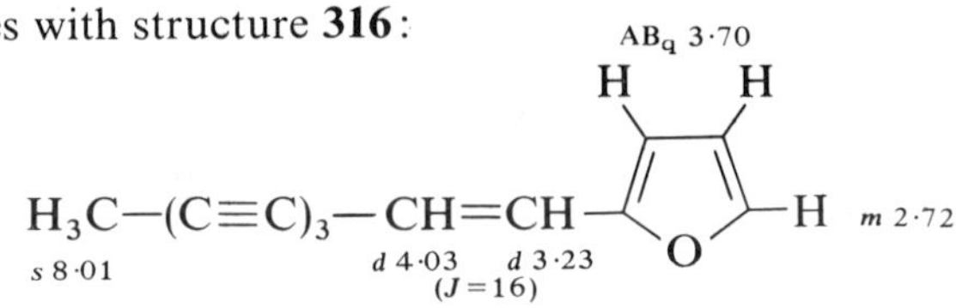

316

316 has been synthesised by Wittig reaction[45]:

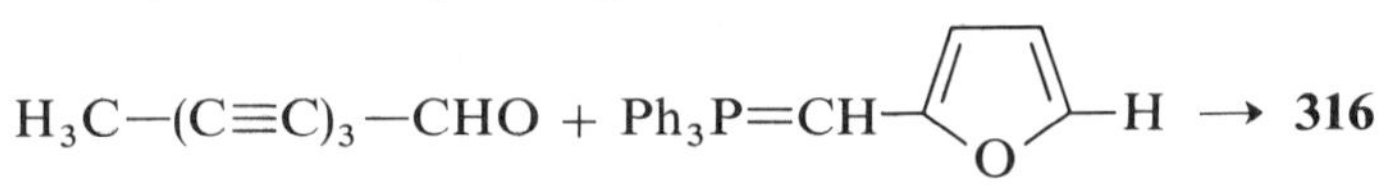

Biogenetically, **312** and **316** are formed also from **207**, as is shown by feeding experiments[38]:

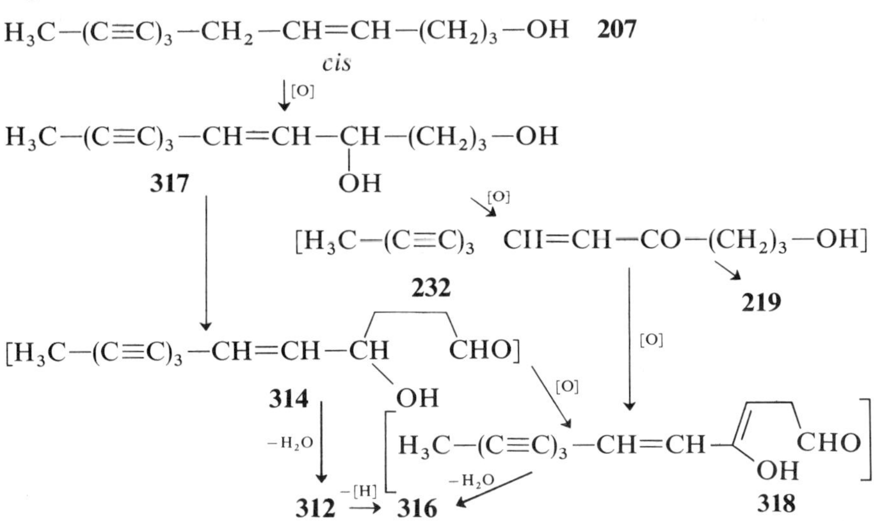

This scheme clearly shows that **317** is the key intermediate, which can lead to the spiroketalenolethers, first by oxidation to **232**, or by oxidation to **314**, and subsequent transformation to **312**. The pathway to **318** is not yet clear since dehydrogenation of **312**, as well as elimination of water from the ketoaldehyde, **318** seem both to be likely possibilities.

A further possible pathway, starting with **318**, is the formation of two aldehydes isolated from the roots of *Chrysanthemum silvaticum* Hoffmgg. L.K. together with spiroketalenolethers. The structure of the less polar compounds has been established by spectral methods as **319**[37]:

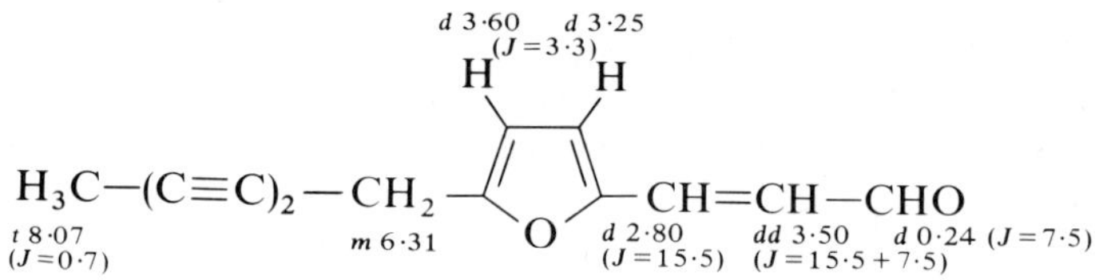

319

The structure has been confirmed by synthesis[37]. Starting with chloromethyl furfurolacetal, the reaction with pentadiyne magnesium bromide leads to the furan derivative **310**, which is transformed by vinylether condensation to **319**:

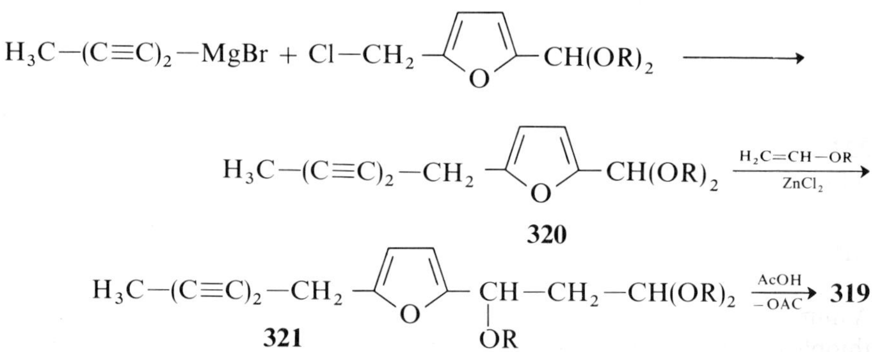

Biogenetically, **319** is most probably formed in the following way:

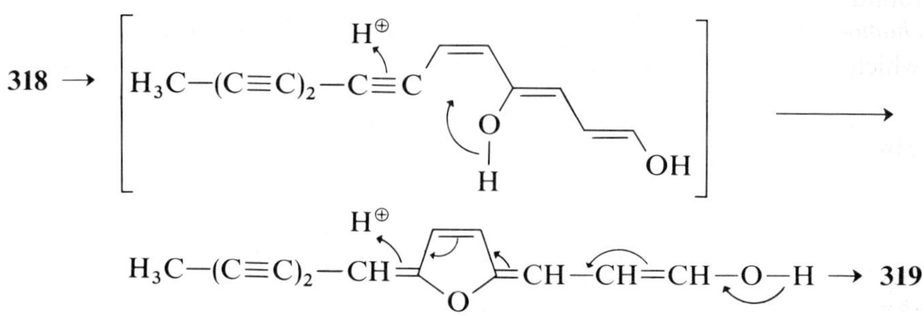

the thioether **323** found in *Chrysanthemum coronarium* L. together with several others[65] is closely related to **319**:

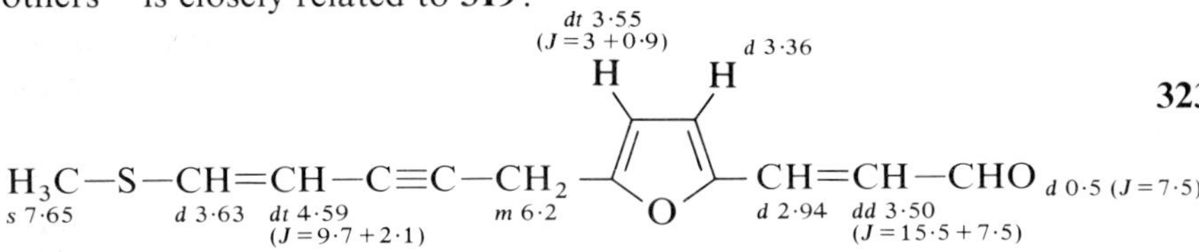

323

323 could only be separated from other compounds by boro-hydride reduction. Manganese dioxide oxidation then leads to the original aldehyde, which can be synthesised by the following sequence[131]:

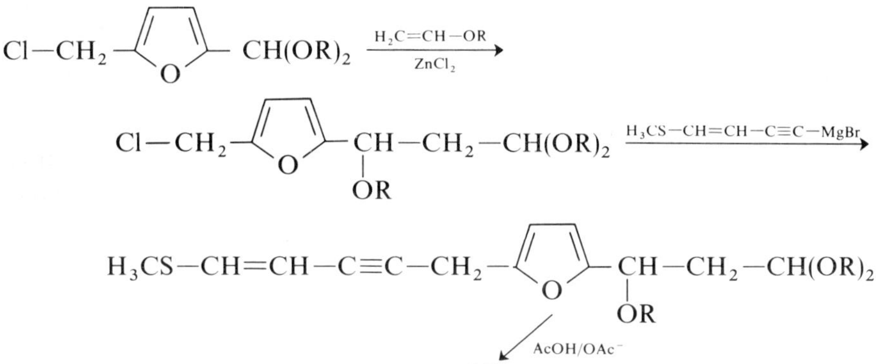

323

323 surely is formed from **319** by elimination of the methyl endgroup and subsequent formal addition of methylmercaptan:

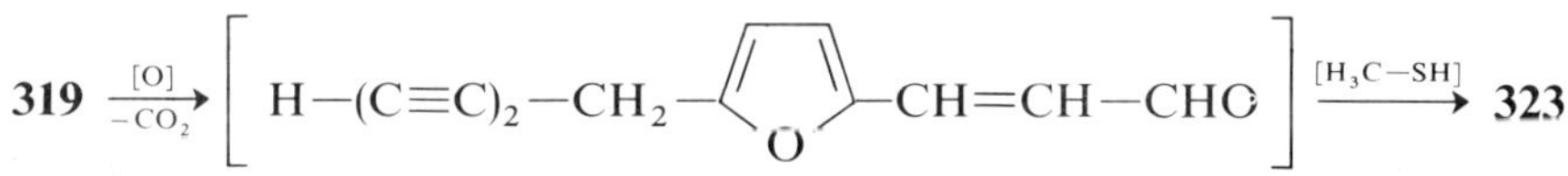

A number of further transformation products, starting with **316**, are found as thiophene compounds. Although the thiophene **324** formed by formal addition of H_2S to **316** is not known, several degradation products have been found in different members of the tribus Anthemideae. The roots of *Santolina chamaecyparissus* L. contain, together with related compounds, an acetate which turns out to be the thiophene furane acetylene **325**[132]:

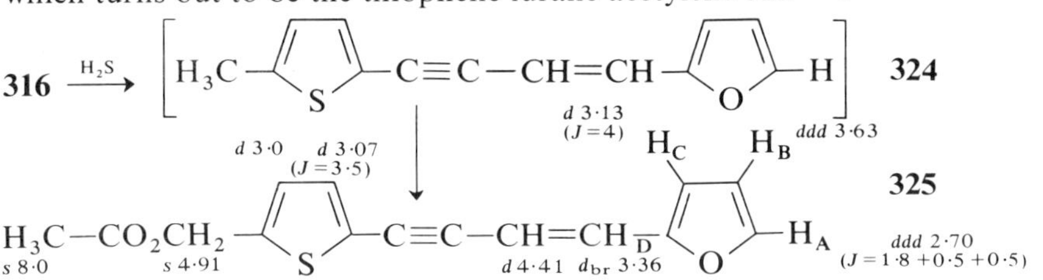

324

325

Long range coupling between H_A and H_D, H_B and H_D, respectively ($J = 0{\cdot}5$ Hz) has been observed. The downfield shift of H_C is due to the deshielding effect of the triple bond only possible with a *cis*-configurated double bond.

As well as the *cis–trans*-isomeric alcohols the corresponding *trans*-acetate **326** are also present in the roots of the *Santolina* species[133]. The isomeric aldehydes have also been isolated:

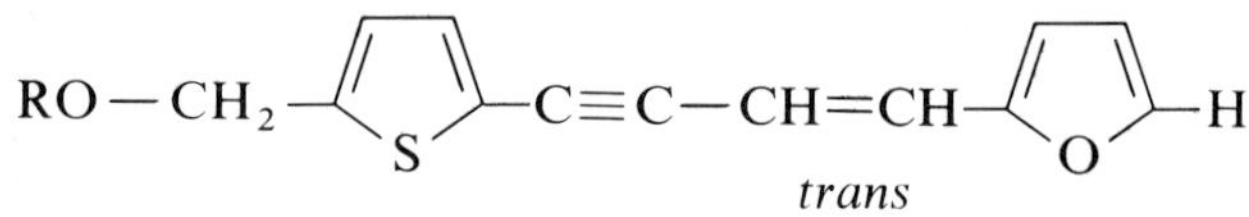

326: R = Ac
327: R = H
328: R = H

cis

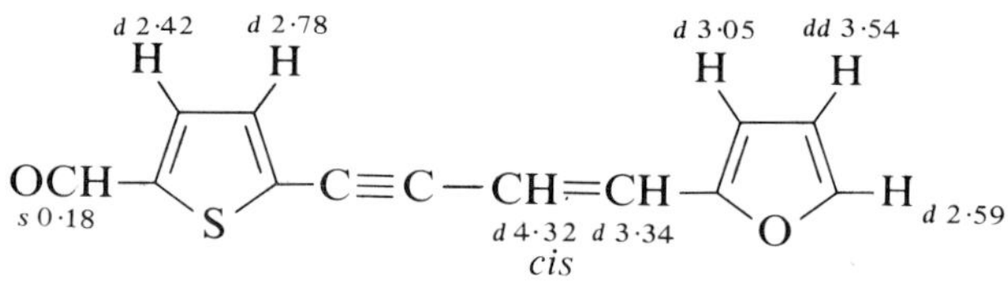

329

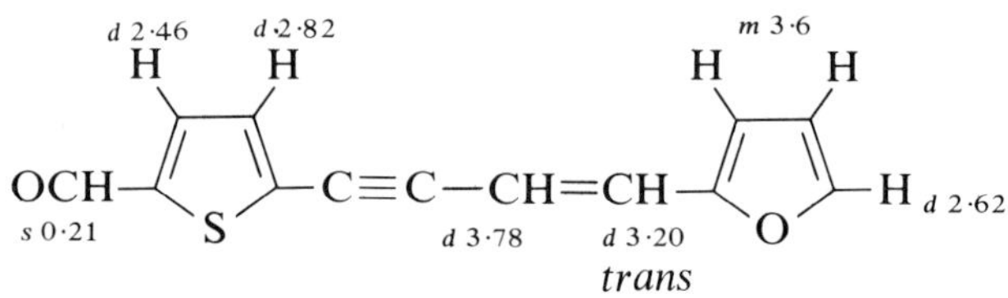

330

Again, the different chemical shifts of the β-furan protons (H_C) in the *cis–trans*-isomeric aldehydes are very pronounced.

Much more widespread are the *cis–trans*-isomeric thiophenes **331** and **332** first isolated from the roots of *Santolina pinnata* Vir.[68]. The structure of these compounds has been elucidated by degradation and synthesis:

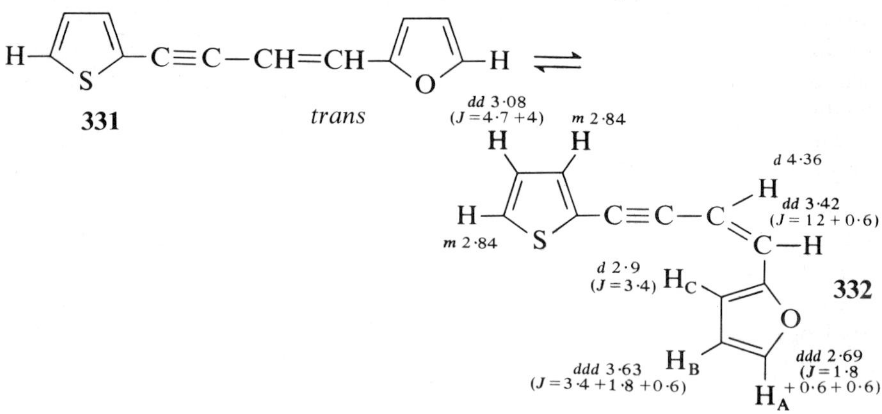

331 **332**

In the n.m.r. spectrum of **332**, the long range couplings and the deshielding of H_C are again very pronounced.

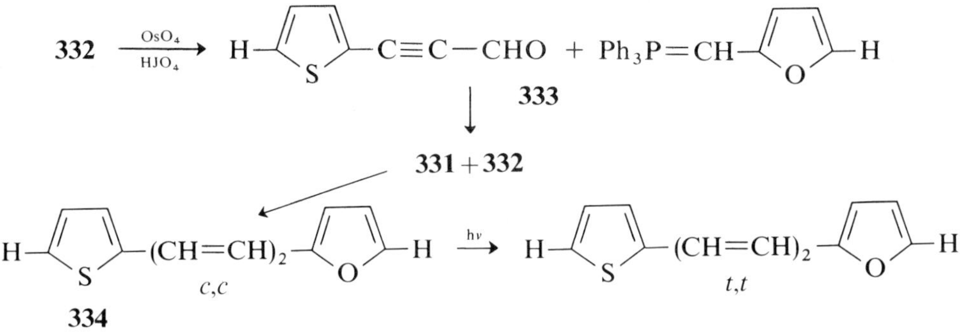

Remarkable is the u.v. absorption shift on hydrogenation of **332** (about 10 nm)[68].

4. Special Acetylenes from Micro-organisms with more than Ten Carbon Atoms

Many micro-organisms have been investigated in the search for new antibiotics. In this connection, compounds have been found which turn out to be polyynes. Later, Jones and his school especially, investigated a large number of different fungi (mainly Basidiomycetes) looking for further acetylenes. Acetylenes are fairly widespread, though the variation in structures are not as large as in the case of the Compositae acetylenes. In general, the compounds isolated from micro-organisms are normally more polar because of the presence of many polar functional groups. However, by feeding experiments with labelled oleic acid, it has been shown that this acid also seems to be the common precursor. As in the case of acetylenes from the higher plants, we have also in the field of microorganisms to distinguish between compounds having ten (or less) carbon atoms and those with more carbon atoms. This section will deal first with the second group. The remaining compounds will be discussed in Section 5.4.

An acid most closely related to oleic acid isolated from the culture medium of *Poria sinuosa* Fr.[134] is the C_{14}-triyne **335**. Purification only could be achieved by counter current distribution. The observed u.v. spectrum is typical for a triyne-carbonyl system, while a *cis* double bond is evident from the i.r. spectrum. The i.r. spectrum of the corresponding diester shows the presence of two different conjugated and an unconjugated ester groups. Hydrogenation yields *n*-tetradecane-dioic acid, while alkaline permanganate gives glutaric acid and alkaline isomerisation leads to a diacid **336** with a real triyneene-chromophore. These results only agree with structure **335**:

$$HO_2C{-}(C{\equiv}C)_3{-}CH_2{-}\underset{cis}{CH{=}CH}{-}(CH_2)_3{-}CO_2H \quad \textbf{335}$$

$$\xrightarrow{H_2} HO_2C(CH_2)_{12}CO_2H \qquad \xrightarrow{KMnO_4} HO_2C{-}(CH_2)_3{-}CO_2H$$

$$\downarrow OH^-$$

$$HO_2C{-}CH_2{-}(C{\equiv}C)_3{-}CH{=}CH{-}(CH_2)_3{-}CO_2H \quad \textbf{336}$$

Biogenetically, **335** is formed in the following way:

oleic acid

$$\downarrow -[H]$$

$$[H_3C{-}(C{\equiv}C)_3{-}CH_2{-}\underset{cis}{CH{=}CH}{-}(CH_2)_7{-}CO_2R]$$

$$\downarrow \text{(twice)}\beta\text{-oxidation}$$

$$[H_3C{-}(C{\equiv}C)_3{-}CH_2{-}\underset{cis}{CH{=}CH}{-}(CH_2)_3{-}CO_2R]$$

$$\downarrow [O]$$

335

Mycomycin is a C_{13}-compound, probably closely related to **335** and isolated from the *Actinomycete Nocardia* acidophylus. The structure of this highly unstable, optically active compound has been established mainly by chemical method. Hydrogenation leads to *n*-tridecanoic acid, while mild alkali treatment gives Isomycomycin showing the characteristic u.v. maxima of a triyne-diene-chromophore. The i.r. spectrum shows the presence of an allenic and an acetylene H-group. These results only agree with the structures **336** and **337**[135]:

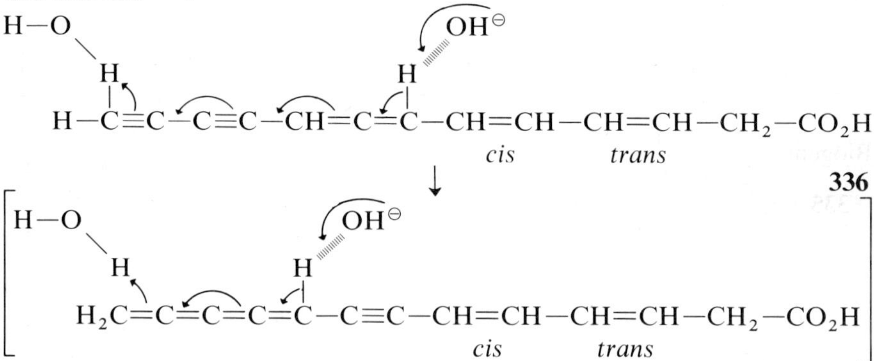

336

$$\downarrow$$

$$H_3C{-}(C{\equiv}C)_3{-}\underset{t,t}{(CH{=}CH)_2}{-}CH_2{-}CO_2R$$

337: R = H
338: R = CH_3

By synthesis the structure of **338** has been confirmed[136]:

$$H_3C{-}(C{\equiv}C)_2{-}H + H{-}C{\equiv}C{-}(CH{=}CH)_2{-}CH_2{-}CN$$
$$\downarrow$$
$$H_3C{-}(C{\equiv}C)_3{-}(CH{=}CH)_2{-}CH_2{-}CN$$
$$\downarrow\; CH_3OH/H_3O^+$$
$$\mathbf{338}$$

The *trans–trans*-isomer of **336** itself also can be prepared[137]:

$$H(C{\equiv}C)_2Li + OCH{-}(CH{=}CH)_2{-}N(CH_3){-}Ph$$
$$\downarrow$$
$$H(C{\equiv}C)_2{-}(CH{=}CH)_2{-}CHO$$
$$\downarrow\; BH_4^-$$
$$H{-}(C{\equiv}C)_2{-}(CH{=}CH)_2{-}CH_2OH + Br{-}C{\equiv}C{-}CO_2H$$
$$\downarrow$$
$$HO_2C{-}(C{\equiv}C)_3{-}(CH{=}CH)_2{-}CH_2OH$$
$$Cu^{2+} \downarrow \Delta$$
$$H(C{\equiv}C)_3{-}(CH{=}CH)_2{-}CHO$$
$$\downarrow\; Ph_3P{=}CH{-}CO_2CH_3$$
$$\overset{H^{\oplus}}{} \quad H{-}(C{\equiv}C)_2{-}C{\equiv}C{-}CH{=}CH{-}CH{=}CH{-}CH{=}CH{-}CO_2CH_3 \quad \underset{H^{\ominus}}{}$$
$$\downarrow$$
$$H{-}(C{\equiv}C)_2{-}CH{=}C{=}CH{-}\underset{t,t}{(CH{=}CH)_2}{-}CH_2{-}CO_2CH_3$$

Biogenetically **336** probably is formed in the following way:

$$\mathbf{335} \xrightarrow[\text{isomerisation}]{-CO_2} [H{-}(C{\equiv}C)_2{-}CH{=}C{=}CH{-}\underset{cis}{CH{=}CH}{-}(CH_2)_3{-}CO_2H]$$
$$\downarrow\; [O]$$
$$[H{-}(C{\equiv}C)_2{-}CH{=}C{=}CH{-}CH{=}CH{-}\underset{OH}{CH}{-}CH_2{-}CH_2{-}CO_2H]$$
$$\downarrow\; -H_2O$$
$$\mathbf{336}$$

The only further C_{13}-acetylenes from micro-organisms related to the triyne acids are two tetraynes isolated from *Fistula hepatica* (Huds.) Fr. The unpolar part of the extracts gives a hydrocarbon, which turns out to be **339**[138], as hydrogenation yields *n*-tridecane and oxidation *n*-valeric acid. The structure has been further confirmed by synthesis[138]:

$$H{-}C{\equiv}C{-}C{\equiv}C{-}Na + Br{-}(CH_2)_3{-}CH_3 \rightarrow H{-}(C{\equiv}C)_2{-}C_4H_9$$

$$\swarrow Br{-}(C{\equiv}C)_2{-}CH_3$$

$$H_3C{-}(C{\equiv}C)_4{-}CH_2{-}CH_2{-}CH_2{-}CH_3 \quad \mathbf{339}$$

The mixture in the polar parts could only be separated by counter current distribution. The crystalline main compound obtained again shows the presence of a tetrayne-chromophore, while the n.m.r. spectrum exhibits a $H_3C{-}C{\equiv}C$-signal at 7·9τ. Periodate oxidation using special conditions was a powerful tool in further structural information. Beside formaldehyde the highly unstable tetrayne aldehyde **341** could be obtained, which on borohydride reduction yields a tetrayne alcohol. Subsequent hydrogenation finally leads to *n*-decanole. Diluted base gives a triyne-enolether, which on ozonolysis leads to L-(−)-lyxonic acid-γ-lactone (**343**). These results are only in agreement with structure **340** and the absolute configuration is also established:

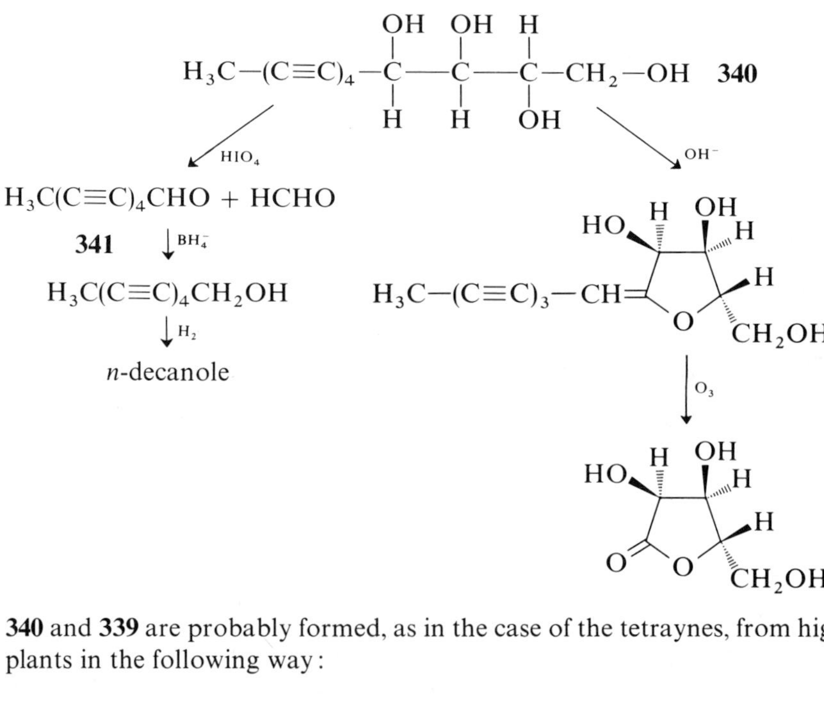

342

343

340 and **339** are probably formed, as in the case of the tetraynes, from higher plants in the following way:

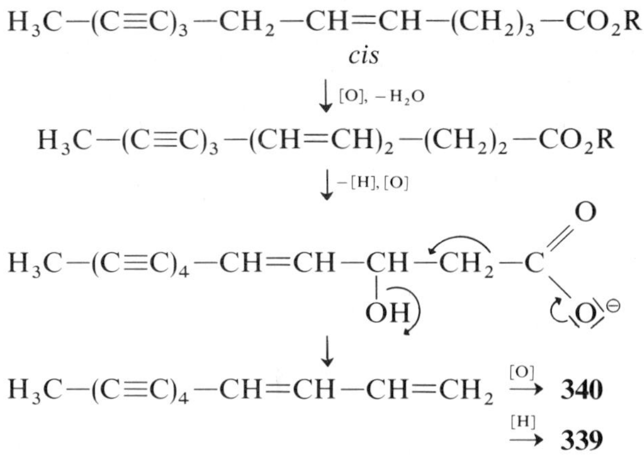

Up to now only feeding of labelled acetate has shown that **340** is built up by acetate units in the usual manner. Following **335** and **336** are Odyssic acid and its transformation products[139]. Although the biogenetic pathways of these compounds are not really established, this acid (**345**) is very likely formed in the following way:

$$H_3C-(C\equiv C)_3-CH_2-CH=CH-(CH_2)_3-CO_2R$$

$$\downarrow [O], \text{ isomerisation}$$

$$H_3C-(C\equiv C)_2-CH=C=CH-CH=CH-CH_2-CO_2R \quad \mathbf{344}$$

$$\downarrow [H_2O]$$

$$H_3C-(C\equiv C)_2-CH=C=CH-\underset{\displaystyle OH}{\underset{|}{C}H}-CH_2-CH_2-CO_2H \quad \mathbf{345}$$

Odyssic acid

The corresponding lactone and the degradated C_{11}-compounds have been isolated[140] together with **345**. This kind of methyl endgroup elimination is still more common with acetylenes from micro-organisms rather than those from higher plants. It could be shown with a cell-free medium that acetylene carboxylic acids are easily decarboxylated[141].

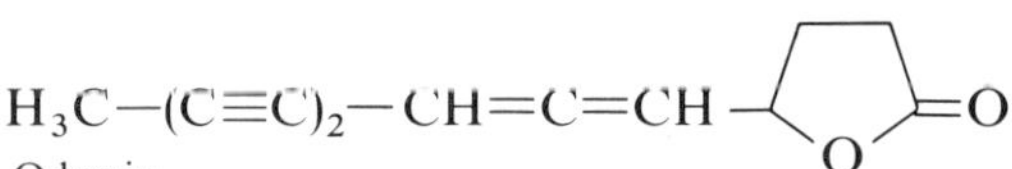 **346**

Odyssin

$$H-(C\equiv C)_2-CH=C=CH-\underset{\displaystyle OH}{\underset{|}{C}H}-CH_2-CH_2-CO_2H \quad \mathbf{347}$$

Nemotinic acid

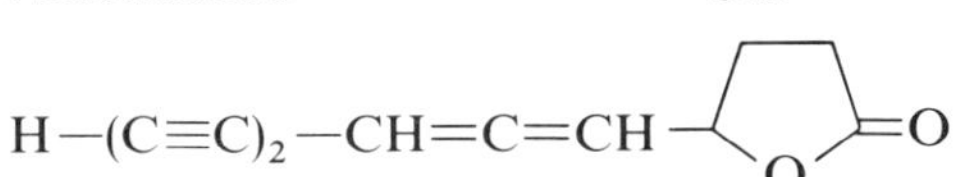 **348**

Nemotin

Again **345–348** are easily isomerised by alkali giving the iso-compounds with a triyne-chromophore. In the case of **346** and **348** also elimination of water occurs[142]:

$$\mathbf{345} \xrightarrow{OH^-} H_3C{-}(C{\equiv}C)_3{-}CH_2{-}\underset{\displaystyle OH}{\underset{|}{CH}}{-}(CH_2)_2{-}CO_2H \quad \mathbf{349}$$

$$\mathbf{346} \xrightarrow{OH^-} H_3C{-}(C{\equiv}C)_3{-}CH{=}CH{-}(CH_2)_2{-}CO_2H \quad \mathbf{350}$$

$$\mathbf{347} \xrightarrow{OH^-} H{-}(C{\equiv}C)_3{-}CH_2{-}\underset{\displaystyle OH}{\underset{|}{CH}}{-}(CH_2)_2{-}CO_2H \quad \mathbf{351}$$

$$\downarrow H^+$$

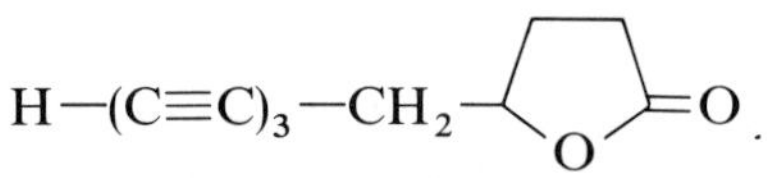

352

Isonemotinic acid lactone

$$\mathbf{348} \xrightarrow{OH^-} H{-}(C{\equiv}C)_3{-}CH{=}CH{-}(CH_2)_2{-}CO_2H \quad \mathbf{353}$$

Nemotin A

All these isomers have been prepared by synthesis[143,144], e.g. **352** can be prepared using the following route[144]:

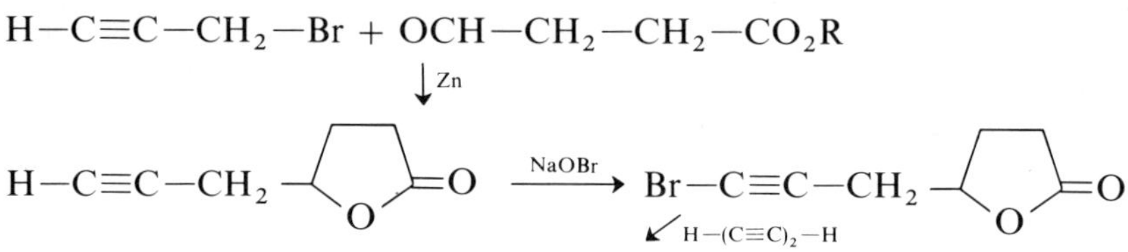

352

Derivatives of **345**, respectively, **351** are the glycosides **354** and **355** as well as the methyl ester **356**[145]:

$$H_3C{-}(C{\equiv}C)_2{-}CH{=}C{=}CH{-}\underset{\displaystyle O\text{-xylose (D)}}{\underset{|}{CH}}{-}(CH_2)_2{-}CO_2H \quad \mathbf{354}$$

$$H{-}(C{\equiv}C)_2{-}CH{=}C{=}CH{-}\underset{\displaystyle O\text{-xylose (D)}}{\underset{|}{CH}}{-}(CH_2)_2{-}CO_2H \quad \mathbf{355}$$

$$H{-}(C{\equiv}C)_2{-}CH{=}C{=}CH{-}\underset{\displaystyle OH}{\underset{|}{CH}}{-}(CH_2)_2{-}CO_2CH_3 \quad \mathbf{356}$$

A number of further allenic compounds have been isolated from different Basidiomycetes.

The neutral fractions from *Odontia bicolor* contain four metabolites with the ene-diyne-chromophore, two of them polar and two less polar. Both non-polar components have an i.r. spectrum showing the presence of free ethynyl, allene and hydroxyl groups. Hydrogenation of the less polar metabolite gives *n*-undecanol, while the polar leads to *n*-decanol. These results are only in agreement with the structures **357** and **361**[146]. The more polar of the two polar compounds, on treatment with acids in acetone, forms a ketal, which could be purified by chromatography. Again it has the u.v. absorption characteristic of an ene-diyne and from the i.r. the presence of free ethynyl and allene is evident. Hydrogenation leads to 1,3-*O*-isopropylidene-decane-1,3-diol and *n*-decanol. Therefore, the parent metabolite has to be formulated as **362**[146]. The less polar of the two polar acetylenes shows very similar spectral data. Hydrogenolysis gives *n*-undecanol, which on hydrogenation of its alkali isomerisation product followed by methylation gives 1,4-dimethoxy-undecane. Therefore, the only possible structure for this metabolite is **358**. *Flammula sapinea* Fr. also contains **357** and **361** as well as **362**[146]. However, a further polar metabolite with very similar spectral data could not be hydrogenolysed. It forms an isopropylidene derivative, which on hydrogenation leads to the ketal of *n*-undecane-diol-(1,3). As the parent diol does not react with periodate, the only possible structure is **360**:

C_{11}: $H{-}(C{\equiv}C)_2{-}CH{=}C{=}CH{-}(CH_2)_4{-}OH$ **357**

$H{-}(C{\equiv}C)_2{-}CH{=}C{=}CH{-}CH(OH){-}(CH_2)_3{-}OH$ **358**

$\downarrow OH^-$

$H{-}(C{\equiv}C)_3{-}CH_2{-}CH(OH){-}(CH_2)_3{-}OH$ **359**

$H{-}(C{\equiv}C)_2{-}CH{=}C{=}CH{-}CH_2{-}CH(OH){-}CH_2{-}CH_2{-}OH$ **360**

C_{10}: $H{-}(C{\equiv}C)_2{-}CH{=}C{=}CH{-}(CH_2)_3{-}OH$ **361**

$H{-}(C{\equiv}C)_2{-}CH{=}C{=}CH{-}CH(OH){-}CH_2{-}CH_2{-}OH$ **362**

All the isolated allenes are optically active. However, the absolute configuration seems to be different. Most probably **357**, **358** and **360**–**362** have the R-configuration as they are all laevorotatory[146]:

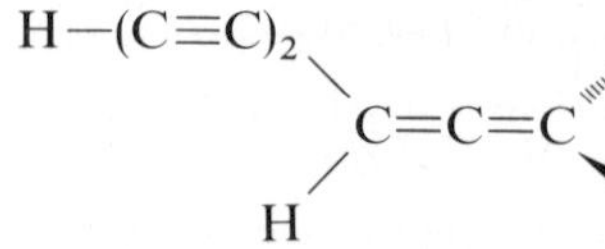 **363**

Preliminary biosynthetic studies have indicated that the diyne–allene system is formed by enzymatic isomerisation of a conjugated triyne precursor and not by isomerisation of an interrupted triyne grouping[146].

Several interesting metabolites have been found by extensive counter-current distribution of the extracts from *Drosophila subatrata*[147]. The first acetylene shows an u.v.-spectrum of an ene-diyne system, which, on hydrogenation, gives *n*-undecanoic acid. From the i.r. spectrum of the ester it is evident that the so-called Drosophilin C has a terminal ethynyl and an unconjugated carboxyl group. Therefore only the structures **364** and **365** are possible:

$$H{-}C{\equiv}C{-}CH_2{-}(C{\equiv}C)_2{-}\underset{cis}{CH{=}CH}{-}CH_2{-}CO_2H \quad \mathbf{364}$$

$$H{-}C{\equiv}C{-}CH_2{-}\underset{cis}{CH{=}CH}{-}(C{\equiv}C)_2{-}CH_2{-}CO_2H \quad \mathbf{365}$$

To decide between these possibilities alkaline isomerisation is very useful. Mild conditions lead to an allene diyn–ene system, while somewhat stronger conditions give a triyn–ene chromophore still with an unconjugated carboxyl group as shown by i.r. spectroscopy. These facts only agree with **364**[147]:

$$\mathbf{364} \rightarrow H_2C{=}C{=}CH{-}(C{\equiv}C)_2{-}CH{=}CH{-}CH_2{-}CO_2H \quad \mathbf{366}$$

$$\downarrow$$

$$H_3C{-}(C{\equiv}C)_3{-}CH{=}CH{-}CH_2{-}CO_2H \quad \mathbf{367}$$

366 is present in the original extracts too. The isolation of **364** from natural sources is of considerable interest, as the connection with the general biogenesis of related compounds is still uncertain. In higher plants, acetylenes with an isolated vinyl endgroup have often been found:

$$H_2C{=}CH{-}\underset{|}{C}H{-}(C{\equiv}C)_2{-} \xrightarrow[?]{-[H]} H{-}C{\equiv}C{-}CH_2{-}(C{\equiv}C)_2{-}$$

Perhaps in the case of micro-organisms, dehydrogenation to a terminal isolated ethynyl group is possible. The other possibility would be that **367** is the precursor and that isomerisation via **366** would lead to **364**.

By feeding labelled crepenynic acid it has been shown that **364** and **366** are formed from this acid using the carbon atoms C_8–C_{18}[22]. Therefore, the proposed pathway seems to be very likely.

All the other acetylenes from microorganism will be discussed in Section 5.4 and in Chapter 3, Section 8.

5. Compounds related to Dehydromatricariaester

5.1. Simple Derivatives

As already mentioned in Chapter 1, Section 4, several acetylenic compounds are known which are biogenetically related to the widespread C_{10}-ester named "dehydromatricariaester". This was first isolated from the roots of the common mugwort—*Artemisia vulgaris* L.—most probably in 1826[1]. In 1955 Stavholt and Sörensen[28] again investigated the essential oil of the *Artemisia* specy. They were able to get a highly-unsaturated crystalline ester, which turns out to have molecular formula $C_{11}H_8O_2$. Hydrogenation leads to methyl caprate and therefore most likely this ester should have structure **368** or **369**:

$$H_3C{-}(C{\equiv}C)_3{-}\underset{cis}{CH{=}CH}{-}CO_2CH_3 \quad \mathbf{368}$$

$$H_3C{-}CH{=}CH{-}(C{\equiv}C)_3{-}CO_2CH_3 \quad \mathbf{369}$$

At that time it was difficult to decide between these two possibilities. Spectral studies, however[148], led to the conclusion that only structure **368** is possible. This has been further confirmed by synthesis of the *trans*-isomer **370**, which is also very common in *Compositae* species, especially in those belonging to the tribus Anthemideae[149]. Some years later the *cis*-isomer **368** was also prepared by synthesis. The *trans*-compound is easily obtained by oxidative coupling of pentadiyne with methylpenten-(4)-ynoate[150]:

$$H_3C{-}C{\equiv}C{-}C{\equiv}C{-}H + H{-}C{\equiv}C{-}\underset{t}{CH{=}CH}{-}CO_2CH_3$$

$$\downarrow$$

$$H_3C{-}C{\equiv}C{-}C{\equiv}C{-}C{\equiv}C{-}\underset{t}{CH{=}CH}{-}CO_2CH_3 \qquad \mathbf{370}$$

The *cis*-isomer has been obtained via the following route[151]:

$$H_3C{-}(C{\equiv}C)_2{-}H + H{-}C{\equiv}C{-}\underset{cis}{CH{=}CH}{-}CH_2OH$$

$$\downarrow$$

$$H_3C{-}C{\equiv}C{-}C{\equiv}C{-}C{\equiv}C{-}\underset{c}{CH{=}CH}{-}CH_2OH$$

$$\downarrow MnO_2$$

$$H_3C{-}C{\equiv}C{-}C{\equiv}C{-}C{\equiv}C{-}\underset{c}{CH{=}CH}{-}CHO$$

$$\downarrow (1)\ CrO_3,\ (2)\ CH_2N_2$$

$$\mathbf{368}$$

The problem here is to avoid isomerisation during the oxidation of the alcohol and in the cyclisation of the *cis*-acid, which easily gives the butenolide **436** (see Section 5.3).

Biogenetically, **368** and **370** are formed from oleic acid and from C_{18}-triyne acid **5**[35] most probably (as already mentioned in Chapter 1, Section 4) in the following way:

$$H_3C-(C\equiv C)_3-CH_2-CH=CH-(CH_2)_7-CO_2R \qquad \mathbf{5}$$

$$\downarrow [O]\ C$$

$$H_3C-(C\equiv C)_3-CH=CH-CO-(CH_2)_7-CO_2R \qquad \mathbf{371}$$

$$\downarrow [O]$$

$$H_3C-(C\equiv C)_3-CH=CH-CO-O-(CH_2)_7-CO_2R \qquad \mathbf{372}$$

$$\downarrow$$

$$\mathbf{360} + \mathbf{370}$$

$$+HO-(CH_2)_7-CO_2R \qquad \mathbf{373}$$

Labelled **371** is readily incorporated in **368**[35]. Therefore, this previously outlined pathway (Chapter 1, Section 4) is very likely, though the second fragment **373** has not yet been isolated and other triynes are also converted within the plants to **368**[35].

A number of simple derivatives of **368** have been isolated from micro-organisms (see Section 5.4) but also some from higher plants. Especially rich are some representatives of the genus Lactuca (subfamily: Liguliflorae of Compositae). The roots of *Lactuca plumieri* Gren. and Godr.[152] contain the previously synthesised triynes **374**, **375** and **376**. **374** and **375** also have been isolated from micro-organisms[153], while **376** also has been found in *Tridax trilobata*[154]:

$$H_3C-(C\equiv C)_3-CH=CH-CHO \qquad \mathbf{374}$$

trans

$$H_3C-(C\equiv C)_3-CH=CH-CH_2OH \qquad \mathbf{375}$$

trans

$$H_3C-(C\equiv C)_3-CH_2-CH_2-CH_2OH \qquad \mathbf{376}$$

The main compound however obviously is a diol, which could be oxidised by manganese dioxide to a triynone, while alkaline treatment yields a diyn-enolether, both present in the root extract too. A further compound with u.v. maxima at 337, 315, 296, 279 and 248 nm on sodium borohydride reduction yields an alcohol, which, on allylic rearrangement, leads to **375**; the peracid oxidation of the original acetylene gives an epoxide identical with a further acetylene from the *Lactuca* species. The following scheme shows the elaborate relationships:

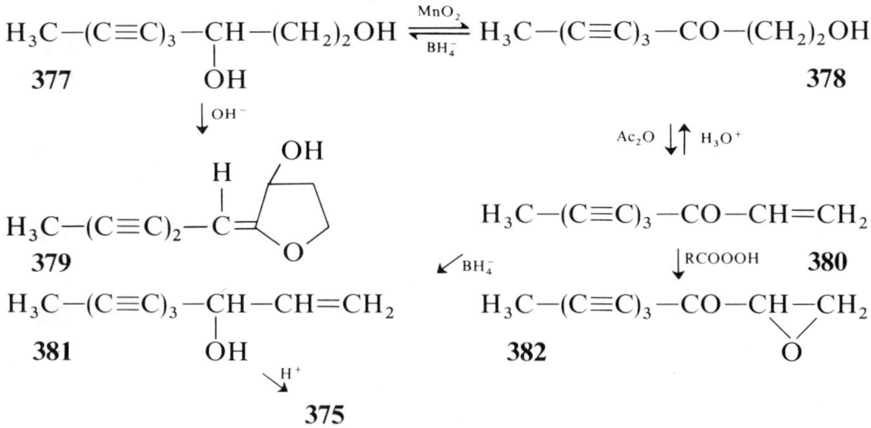

379 could be oxidised using manganese dioxide to give the ketone **383**, while, on ozonolysis, **379** gives (R)-(+)-2-hydroxybutyrolactone (**384**). Thus, **377** and **379** have (R)-configuration:

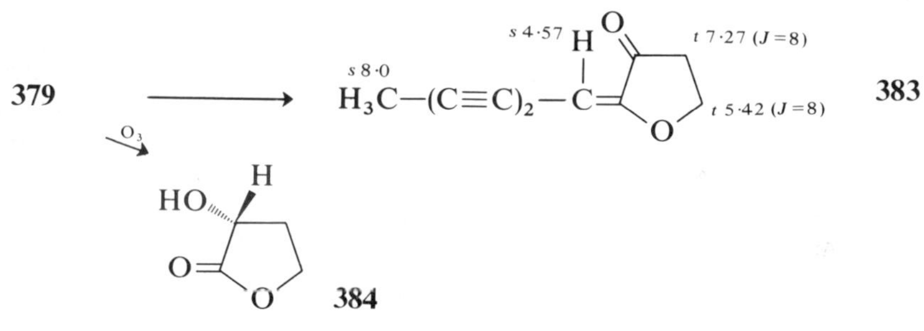

381, the only product not isolated from the *Lactuca* species, is the key intermediate and has been synthesised by coupling pentadiyne with 1-bromopenten-(4)-yn-(1)-ol-(3):

$$H_3C-(C\equiv C)_2-H + Br-C\equiv C-\underset{\displaystyle OH}{\underset{|}{C}H}-CH=CH_2 \rightarrow \mathbf{381}$$

All these compounds (**374**–**382**) are certainly closely related biogenetically. However, it is not quite so apparent what the particular biogenetic pathway is. Perhaps **375**, probably a reduction product of **370**, is the common precursor. But this would mean that **375** might be converted within the plant by allylic rearrangement to **381**. Another possibility would be that **375** is first reduced to **376**, which then is oxidised to **377** and further to **378**, which, with elimination of water, would lead to **380**. The last possibility seems to be more likely as **377** is the main constituent. The following biogenetic pathways would explain the formation of the whole series:

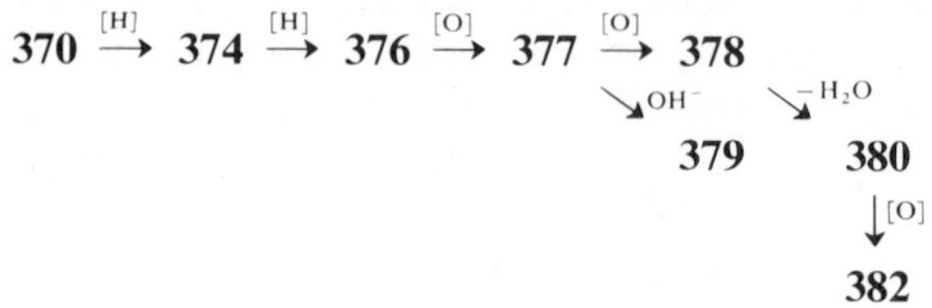

Feeding experiments are necessary in order to confirm these proposals.

A further derivative of **370** has been isolated from different *Achillea* species[107]. From the i.r. spectrum it is evident that we are dealing with a secondary amide having a *trans* double bond. On hydrogenation, an amide is obtained which on alanate reduction gives *n*-decyl-isobutyl-amine. Therefore, the structure of the amide has to be **385**, which could be established by synthesis[107]:

$$H_3C{-}(C{\equiv}C)_3{-}\underset{trans}{CH{=}CH}{-}CO{-}NH{-}CH_2{-}CH(CH_3)_2 \quad \mathbf{385}$$

Biogenetically, **385** is probably formed by reaction of **370** with valine:

$$\mathbf{370} \longrightarrow \left[H_3C{-}(C{\equiv}C)_3{-}CH{=}CH{-}CONH{-}\underset{\displaystyle CO_2H}{\underset{|}{CH}}{-}CH(CH_3)_2\right] \quad \mathbf{386} \xrightarrow{-CO_2} \mathbf{385}$$

A compound closely related to **386** has been isolated from micro-organisms (see Section 5.4).

5.2. Addition Products

A considerable number of acetylenes being formal methyl mercaptane addition products of **368** or **370** are found in the genus *Anthemis*.

The roots of *Anthemis tinctoria* L. contain a crystalline compound showing a broad and not very characteristic u.v. spectrum. The molecular formula, estimated as $C_{12}H_{12}O_2S$, indicates that we have a methyl mercaptane addition product of **368**[155]. The i.r. spectrum shows strong bands for acetylenic linkage, a conjugated carbonyl group and double bonds, while the n.m.r. spectrum shows three methyl group signals [*s* 6·25; *s* 7·6; *d* 7·78 (J = 0·7)] and two olefinic signals [*s*(br) 4·82 (1H) and ABq 3·87 (2H)].

The interpretation of the n.m.r. spectrum is not easy due to the missing coupling constants. The methyl group signals must be explained as $-CO_2CH_3$, $-S-CH_3$ and $=\overset{|}{C}-CH_3$, while the olefinic situation is unclear. The assumed presence of a methylthioether could be confirmed by the conversion to a sulphone (*s* 7·07) showing the typical u.v. spectrum of an en-diyne-ene-ester. Therefore, the structure of the thioether must be **387**:

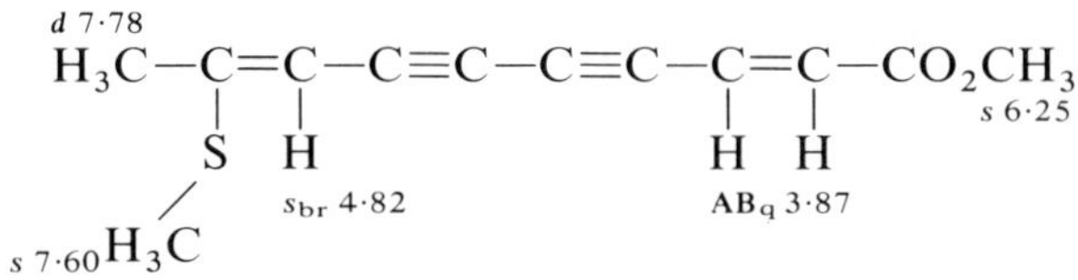

387

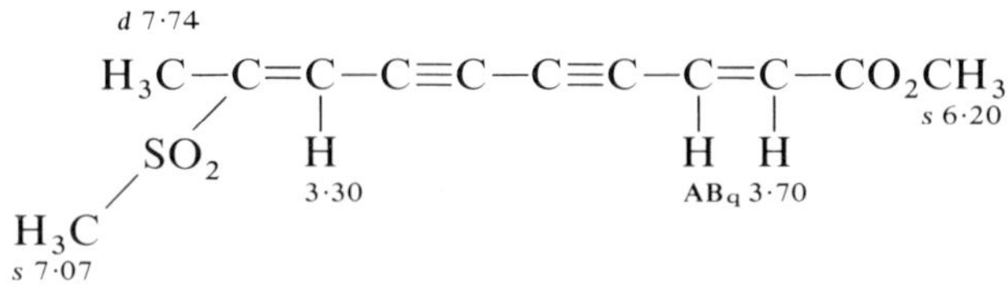

388

Due to the *cis*-configuration of the Δ^2-double bond in the n.m.r. spectrum, only an AB-quartet can be observed for the two olefinic protons. This is typical of such a group (A) since two effects are operating here; the influence of the carbonyl group and the magnetic anisotropy of the triple bonds:

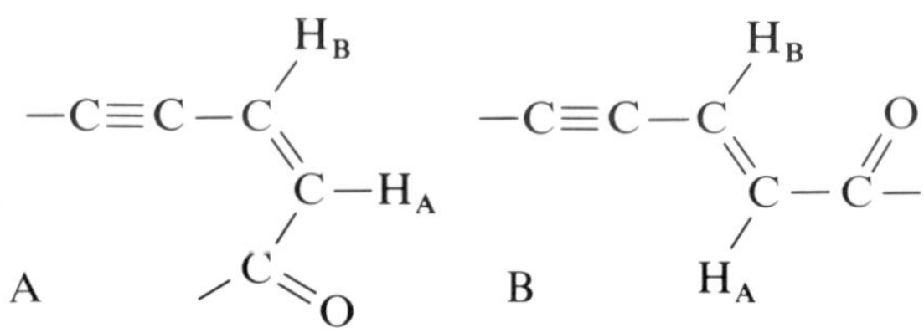

The corresponding *trans*-situation (B) always gives the expected doublets as H_B is now shifted by the anisotropy effect of the carbonyl group to lower fields.

An isomeric thioether has also been isolated from the roots of *Anthemis tinctoria* L., which turns out to have the configuration **389**[156]:

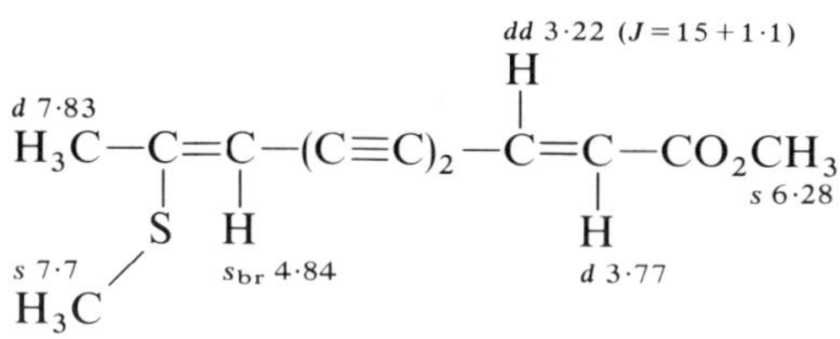

389

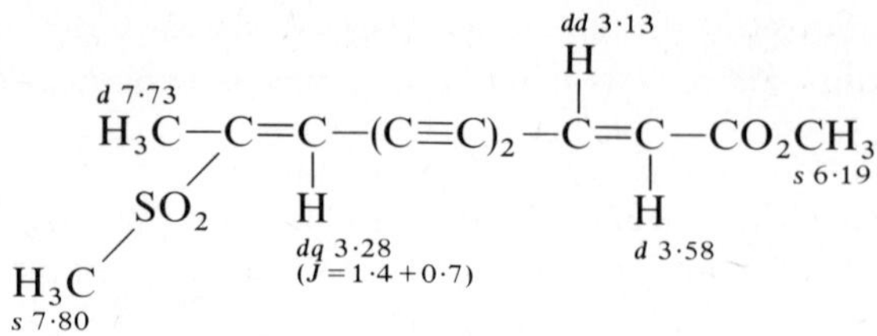

390

The configuration of the Δ^8-double bond only could be clearly established by synthesis of all possible isomers[157], which could be achieved by the following route:

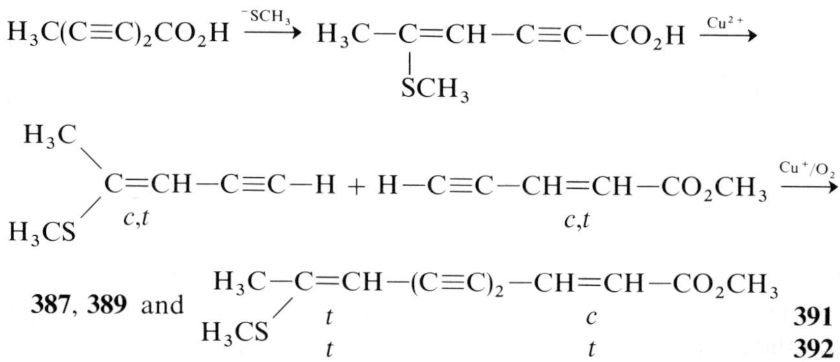

391
392

By preparing also the sulphones from **391** and **392** and by measuring all the n.m.r. spectra, sufficient data are available to estimate all configurations exactly. Due to the magnetic anisotropy of the sulphone group the proton at C_8 is always shifted to lower fields in the *cis*-cases.

By feeding C_1-labelled **370**, it has been shown that **387** is formed by formal addition of methyl mercaptan. **387** isolated as the sulphone **388** had activity only at C_1 as shown by degradation[157]:

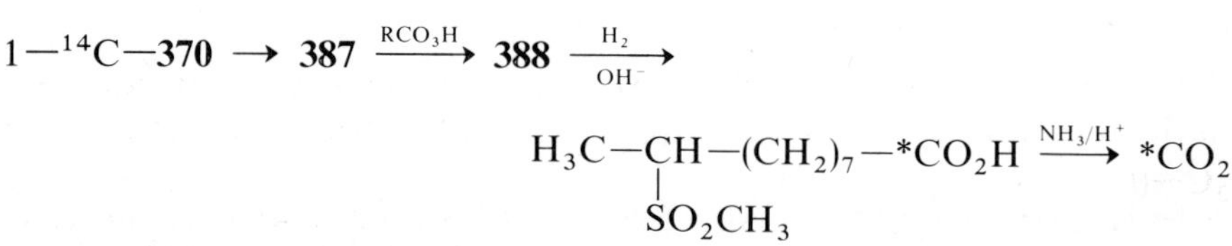

Most probably all the other thioethers derived from **368** or **370** are formed in the same manner.

The roots of *Anthemis carpatica* Willd. contain a further thioether also with molecular formula $C_{12}H_{12}O_2S$ but giving a very different u.v. spectrum with a maximum at 347 nm. The n.m.r. spectrum together with the other data only agrees with structure **393**[155]. The configuration of the thioenol ether can again be estimated by preparing the corresponding sulphone together with that of a second naturally occurring thioether, which is the 4-*trans*-isomer of **393**[85]:

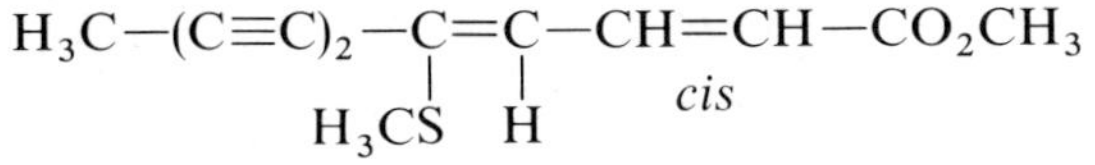

393

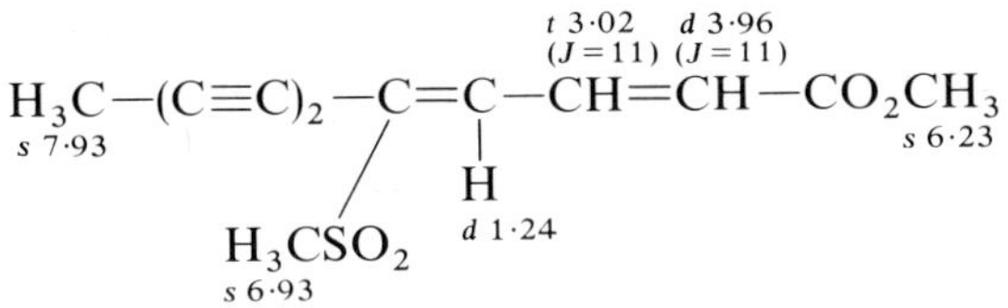

394

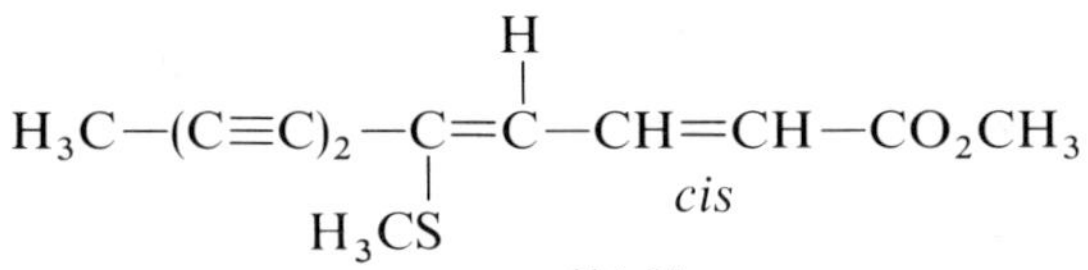

395

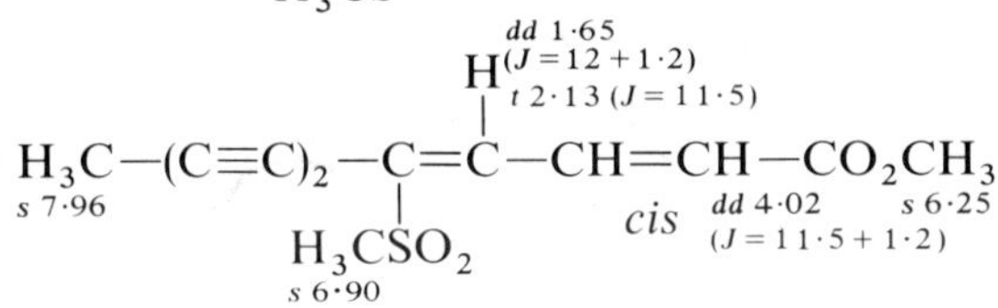

396

Furthermore, the structures and configurations have been confirmed by synthesis of all four possible isomers, using the following route[158]:

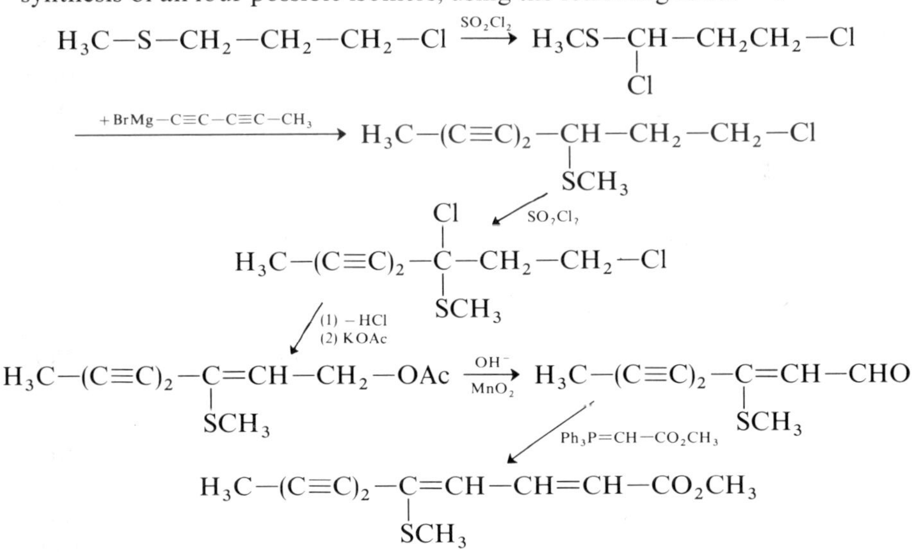

c	*c*	**393**
t	*c*	**395**
c	*t*	**397**
t	*t*	**398**

From the roots of *Anthemis cota* L. a crystalline thioether has been isolated showing an u.v. maximum at 356 nm and i.r. bands in agreement with the presence of triple bonds, estercarbonyl and two different double bonds. The n.m.r. spectrum clearly shows that only structure **399**[155] is possible, while the configuration follows from the n.m.r. spectrum of the sulphone **400**:

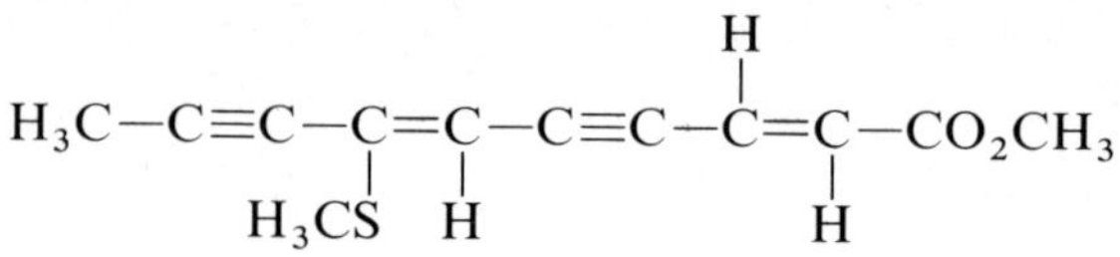
399

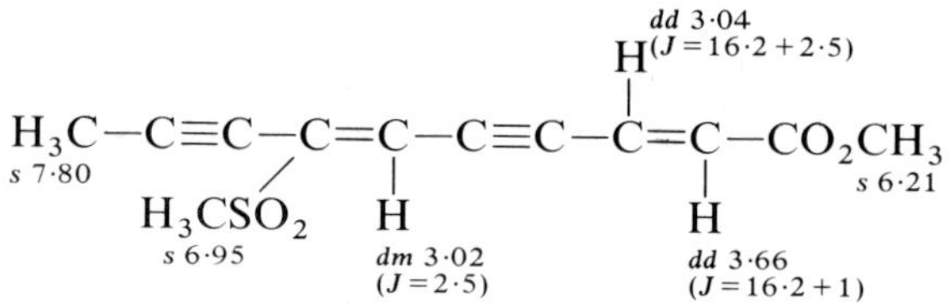
400

The other *cis–trans*-isomers of **399** are also found in *Anthemis* species[155,156]. Again the n.m.r. spectra of the sulphones have been very helpful in structure determination:

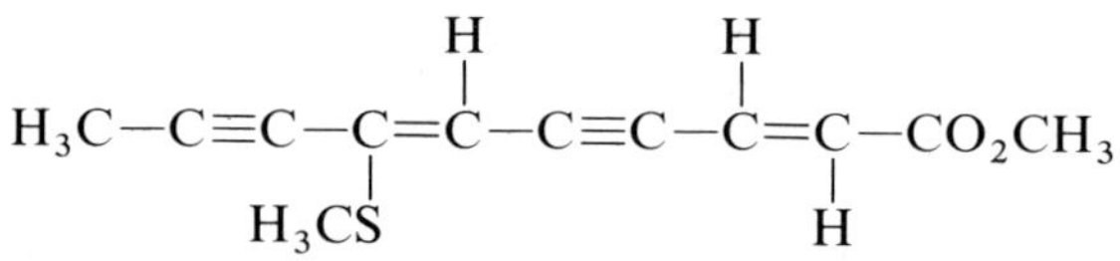
401

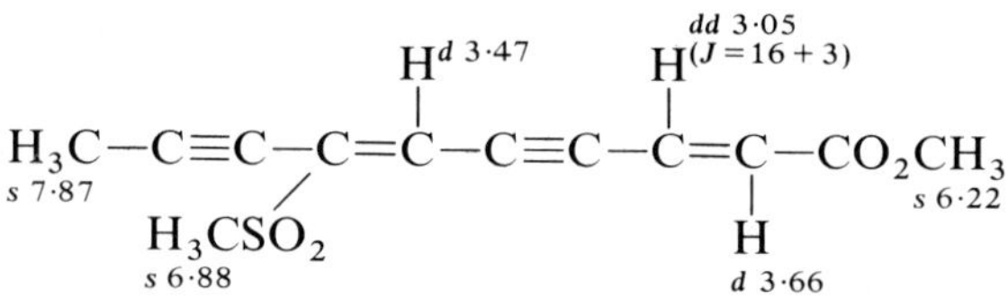
402

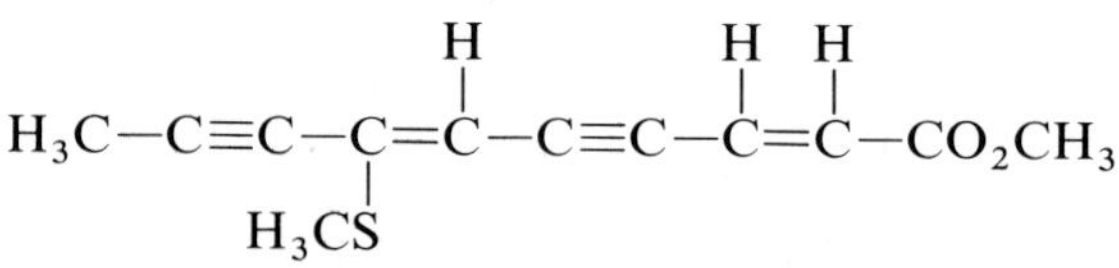
403

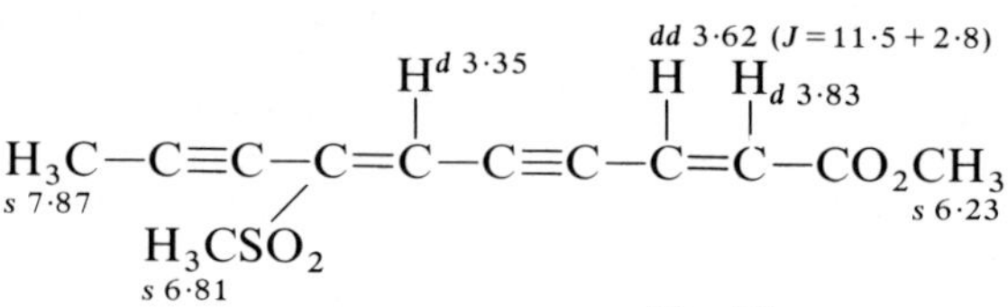
404

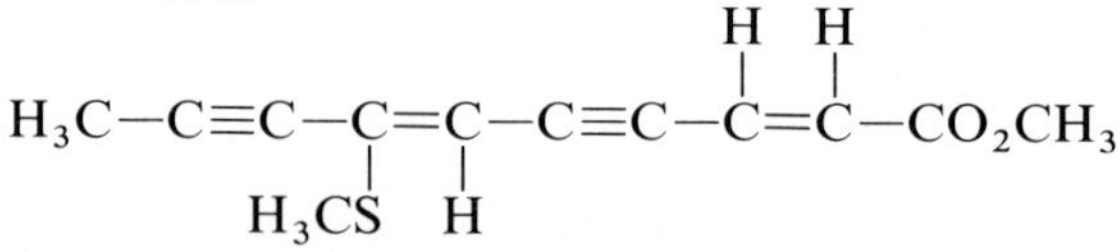
405

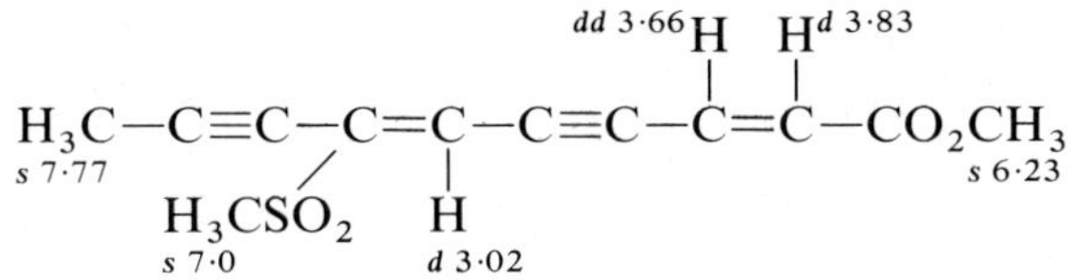

406

The structures of all isomers have been established by synthesis via the following route[159]:

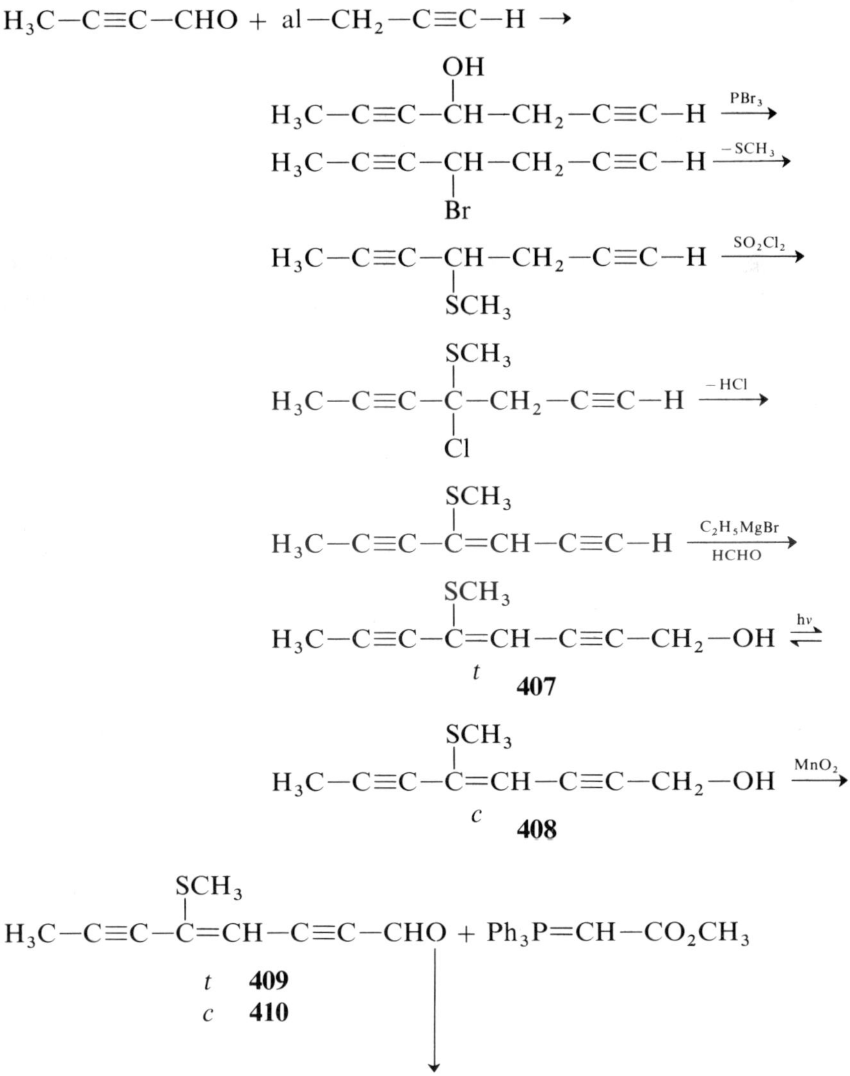

399, 401, 403, 405

As **409** and **410** could be separated and the configurations of their double bonds are clear, the following Wittig reaction always yields the Δ^2-*cis–trans*-isomeric pairs and thus the configurations of all isomers are confirmed.

A hydrogenation product of **393** has also been found in the root extracts of different *Anthemis* species[155]. The spectral data of the oily ester are only in agreement with the structure **411**; again the n.m.r. spectrum of the crystalline sulphone was very helpful in elucidation of the configuration of the double bonds:

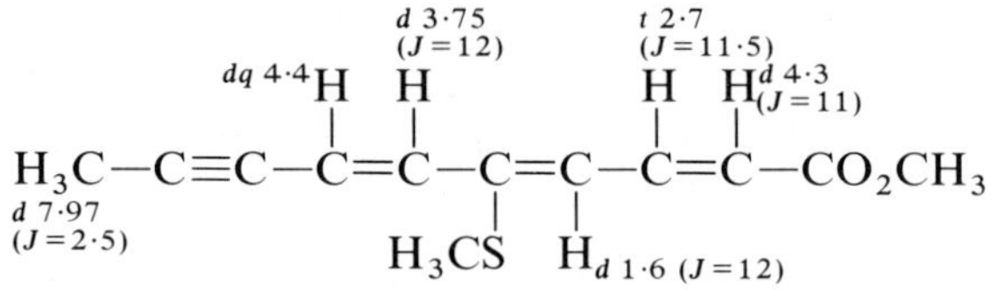

411

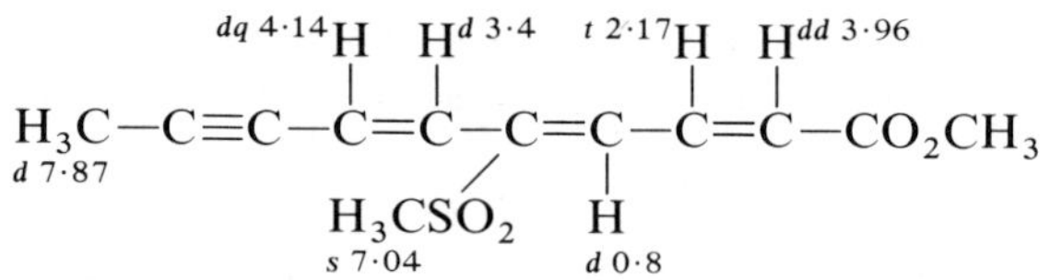

412

The extremely low field position of the hydrogenes at C_3 and C_4 in **412** are only understandable excepting the following configuration:

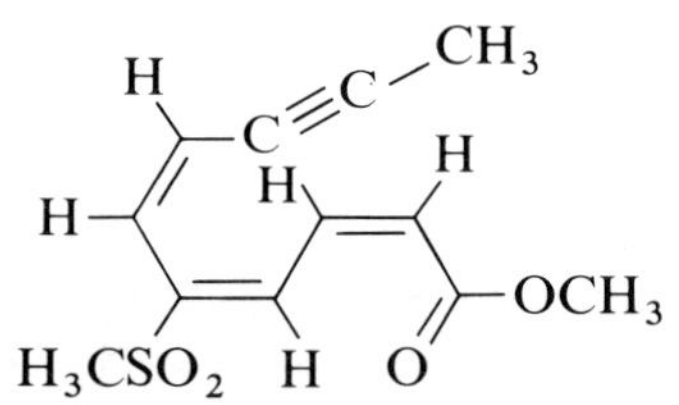

412

The roots of *Anthemis austriaca* Jacq. contain together with many other acetylenes a complex mixture of thioether-esters. One of these esters has a very long wavelength u.v.-absorption at 390 nm and in the infrared a strong cumulene band at 2050/cm together with an —C≡C— stretching band. This compound, very difficult to separate from other thioethers, turns out to be cumulene **414**[160] with a *cis*-double bond.

Most probably the configuration of the cumulene bond is *trans*, but on TLC, an equilibrium with the corresponding *cis*-isomer appears:

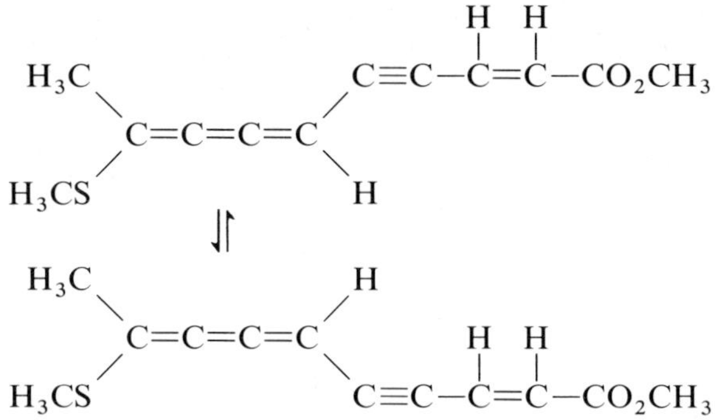

413

414

414 clearly is also formed by formal 1,4-addition of methylmercaptan on *cis*-dehydromatricariaester (**368**).

A number of thioethers with unusual position of the *S*-methyl-group have been found together with the normal ones (**387**, **405** and **411**) in the root extract of a variety of *Anthemis tinctoria* L.[85]. Using spectral methods, especially n.m.r., the structures as well as the configurations of these three thioethers could be elucidated. Again the prepared sulphones turns out to be very useful. The *trans–trans*-isomer **418** not isolated from the natural source could be obtained by u.v. irridation of the *cis–trans*-isomer **415**:

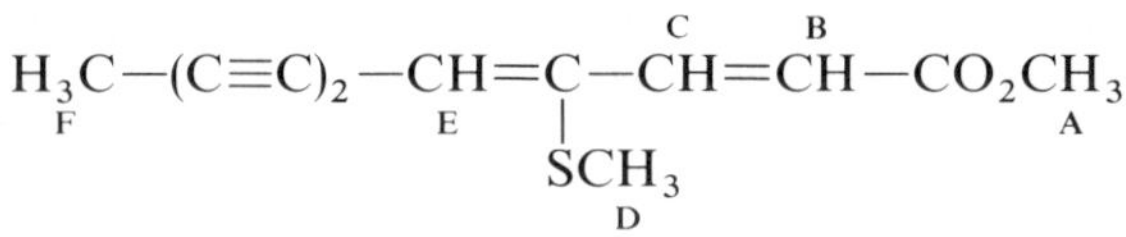

t	*c*	**415**
c	*c*	**416**
c	*t*	**417**
t	*t*	**418**

The observed n.m.r. signals are summarized in the following table:

	A	B	C	D	E	F
415:	*s* 6·32	*d* 4·13 $J = 11{\cdot}7$	*dd* 3·57 $J_{BC} = 11{\cdot}7$ $J_{CE} = 1{\cdot}5$	*s* 7·72	*m* 4·29	*d* 7·98 $J_{EF} = 1{\cdot}2$
416:	*s* 6·33	*d* 4·07 $J = 11{\cdot}5$	*dd* 3·51 $J = 11{\cdot}5 + 1{\cdot}4$	*s* 7·73	*m* 4·61	*d* 8·01 $J = 1$
417:	*s* 6·27	*d* 3·80 $J = 16$	*s* 2·25 $J = 16$	*s* 7·67	*m* 4·49	*d* 7·96 $J = 1$
418:	*s* 6·30	*d* 3·72 $J = 15{\cdot}5$	*d* 2·80 $J = 15{\cdot}5$	*s* 7·53	*m* 3·91	*d* 7·94 $J = 1$

$$\underset{F}{H_3C}-(C{\equiv}C)_2-\underset{E}{CH}{=}\overset{C}{\underset{\underset{D}{SO_2CH_3}}{C}}-\overset{B}{CH}{=}CH-CO_2\underset{A}{CH_3}$$

t	*c*	**419**
c	*c*	**420**
c	*t*	**421**
t	*t*	**422**

The n.m.r. signals of the sulphones are summarised in the following table:

	A	B	C	D	E	F
419:	*s* 6·28	*d* 3·89	*dd* 3·20	*s* 6·85	*m* 3·12	*d* 7·94
		J = 12	*J* = 12·5 + 1·5			*J* = 1·4
420:	*s* 6·23	*d* 3·71	*dd* 3·35	*s* 7·04	*m* 3·30	*d* 7·98
		J = 12	*J* = 12 + 1			*J* = 1
421:	*s* 6·19	*d* 3·16	*d* 2·30	*s* 7·03	*m* 3·08	*d* 7·91
		J = 16·5	*J* = 16·5			*J* = 1
422:	*s* 6·21	*d* 3·17	*d* 2·23	*s* 7·04	*m* 3·09	*d* 7·92
		J = 16	*J* = 16			*J* = 1

Once again the configuration of the Δ^4-double bond follows from the comparison of the n.m.r. signals in the thioethers and the corresponding sulphones. In the *cis*-configuration, a shift of about 1·4τ is observed, while in the *trans* case, only a shift of about 0·8τ is usual. However, in the latter case, the situation is much more complicated due to the de-shielding effects of both sulphone and carbonyl groups.

By synthesis, the structures and configurations have been confirmed[161]. Starting with the triyne **423**, addition of methyl mercaptide leads to the *trans*-thioether **424**, while thermal addition yields the *cis*-isomer **425**. Manganese dioxide oxidation gives the two aldehydes **426** and **427** its n.m.r. spectra, clearly establishing the configurations. Finally, Wittig reaction allows preparation of all four isomers with definite configurations:

$$\underset{\mathbf{423}}{H_3C-(C{\equiv}C)_3-CH_2OH} \xrightarrow{CH_3-SH} H_3C-(C{\equiv}C)_2-CH{=}\underset{\underset{SCH_3}{|}}{C}-CH_2OH$$

t	**424**
c	**425**

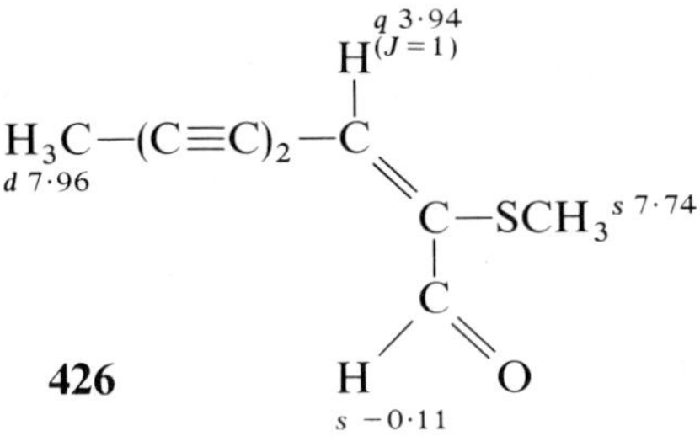

426

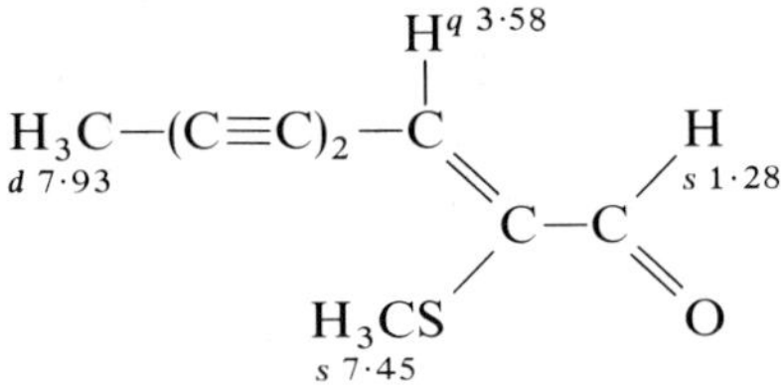

427

$$\mathbf{426} \text{ or } \mathbf{427} + Ph_3P{=}CH{-}CO_2CH_3 \longrightarrow \mathbf{415\text{–}418}$$

The biogenetical formation of **415–417** probably requires a route different from that in the case of **393**, clearly formed from **368** following the polarisation of **368**. Perhaps **415–417** are formed by methyl mercaptan addition in an earlier stage:

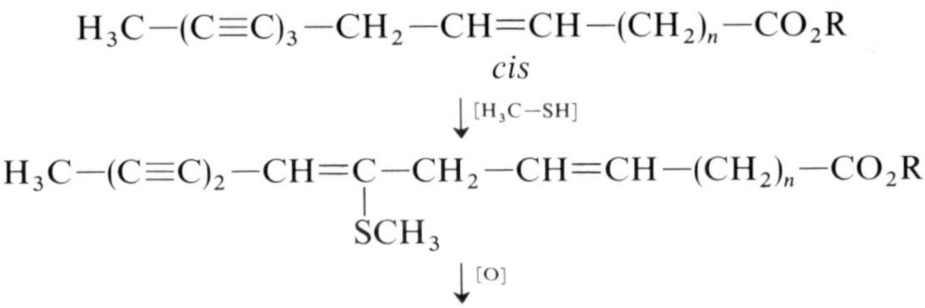

428

415–417

428 then could be transformed to **415–417**, as in the case of **368** (see Section 5.1), which together with **370** is surely the precursor of all the above mentioned thioethers (except **415–417**).

The general question of thioether formation has been studied using double-labelled methionine with the assumption that perhaps these thioethers are formed in the following manner[99]:

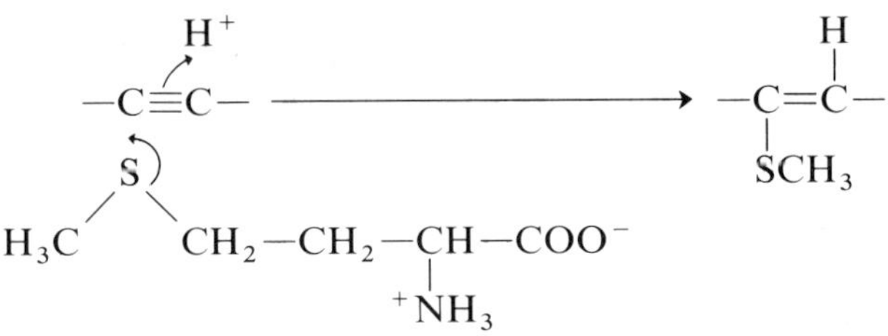

But as already discussed in Chapter 1, Section 4 it turns out that this is a two-step reaction as the label of the $[^3H{-}CH_3]{-}^{35}S$-methionine is incorporated in a different amount. By feeding experiments with *Anthemis tinctoria* L., the thioethers **401** and **411** isolated as sulphones show very different ^3H and ^{35}S activity. This result is only in agreement with the assumption that methionine is mostly used as a methyl donor, while first most probably by using in part also methionine as a sulphur source a HS-compound is formed, which then is methylated or transformed to a thiophene.

5.3. Ring Compounds

Ring compounds derived from dehydromatricariaester are also common.

The only aromatic compound related to **368** has been isolated from *Anthemis tinctoria* L., together with the dihydrothioether **411**. The structure of this crystalline thioether, showing u.v. maxima at 305 and 229 nm, has been established from the spectral data[155] and by synthesis[162] starting with the acetophenone derivative **429**:

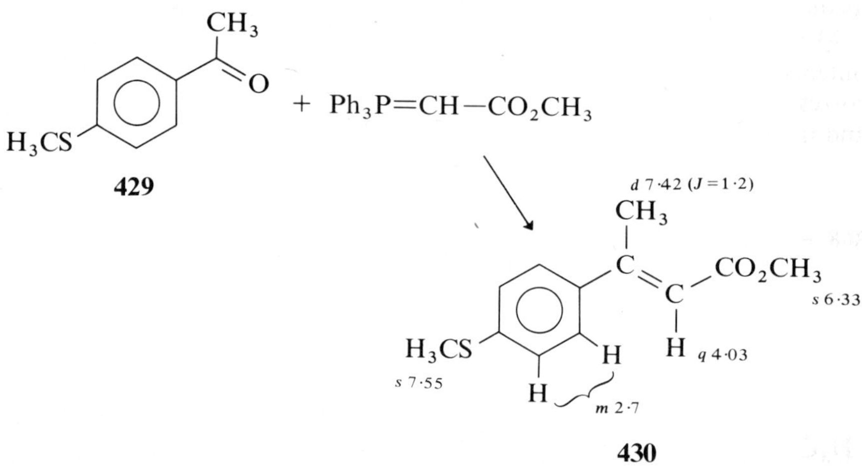

As **430** has been found together with **411**, it seems likely that biogenetic relationships exist. By feeding labelled **370**[162], **430** is indeed formed using **370** and **401** as precursors[159]. Therefore, the following route might be the biogenetic pathway:

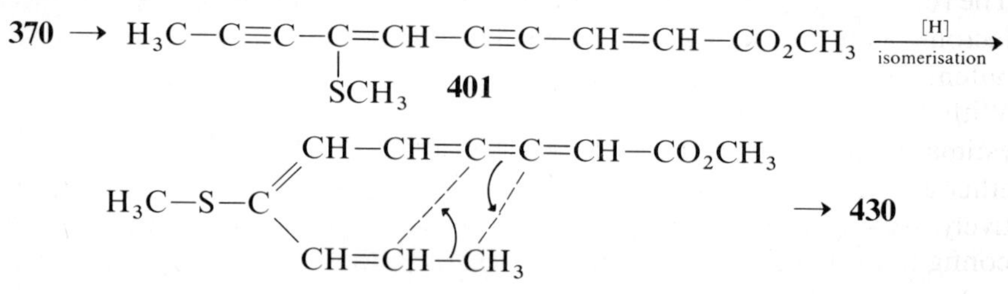

By degradation of the sulphone of **430**, it could be established that the methyl group in the side chain has its origin in the methyl group of **370**[163]:

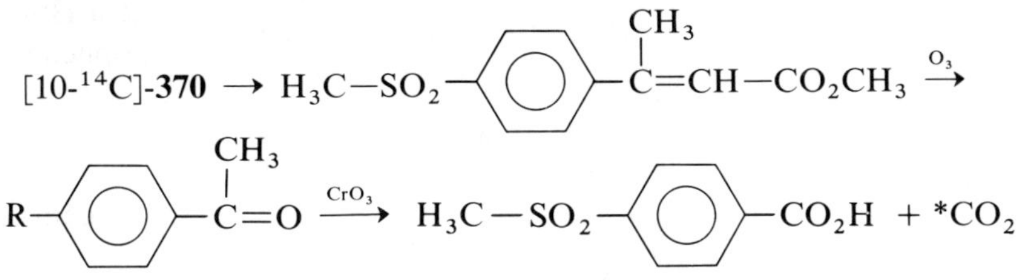

434

All the detected activity occurred in the isolated carbondioxide, while **434** was inactive. The real mechanism of the formation of **430** is still difficult to understand. The conversion of **401** by partial hydrogenation and isomerisation to **431** seems to be the most likely formation of the necessary intermediate, though such a methyl migration is as yet unknown.

More widespread in the tribus Anthemidae are the *cis–trans*-isomeric butenolides **436**[156] and **437**[62], clearly formed by cyclisation of the acid corresponding to **368**. Also the acid **435** is easily converted *in vitro* into **436** and traces of **437**:

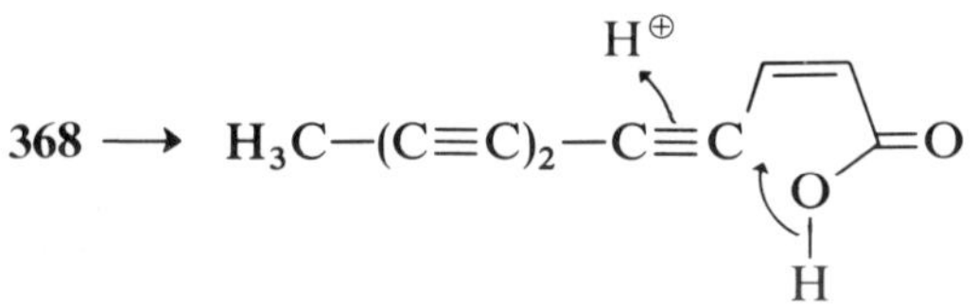

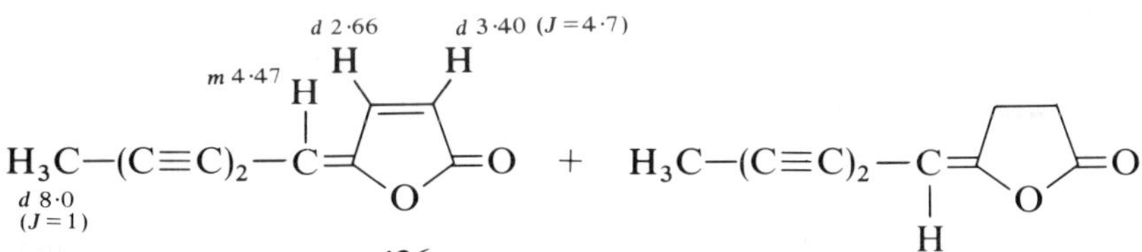

436 **437**

436 is the precursor of three thioethers isolated from *Anthemis* species[156]. The roots of *A. rigescens* Willd. contain beside a number of other acetylenes a lactone with u.v. maximum at 386 nm and infrared bands typical for a butenolide (1780, 1720, 1105, 1065, 945, 880/cm) and a triple bond (2240/cm). With peracid a sulfone could be obtained, while the molecular formula, estimated as $C_{11}H_{10}O_4S$ for the sulfone, and n.m.r. data both for the thioether and the sulfone clearly establish the structures as **438** and **439**, respectively. Again the proton shift in the spectrum of the sulphone indicates the configuration of the thioenolether double bond:

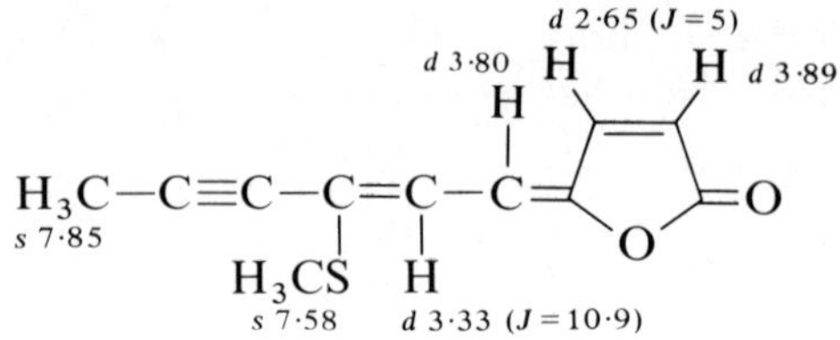

438

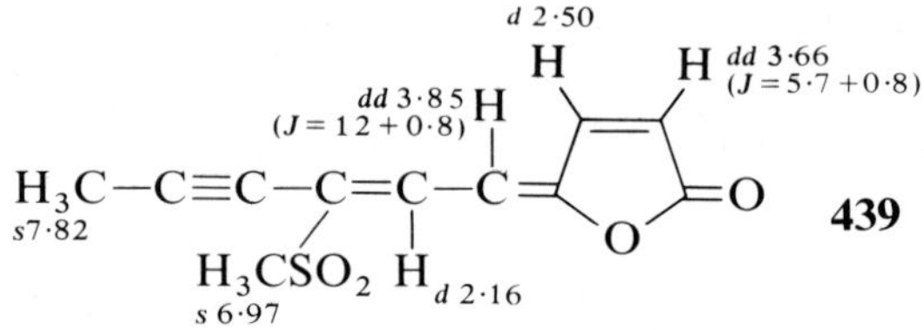

439

The isomeric thioether **440** is present in the roots of *Anthemis kelwayi*[156]:

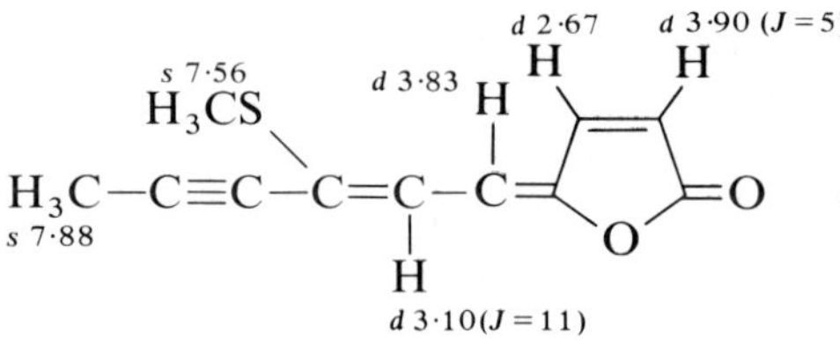

440

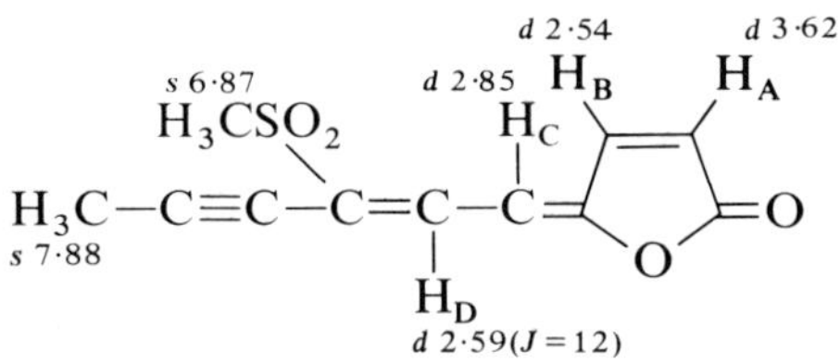

441

Again the configuration of the thioether double bond follows from the proton shift of H_D. The configuration of the second double bond in **438** and **440** follows from the n.m.r. signals of H_C. Only in the spectrum of **439** can a small coupling with H_A can be seen, while in *cis* cases, a larger one can be expected.

Also in the root extract of *Anthemis rigescens* Willd., a third isomer is present in very small amounts and is only isolated as the sulphone. The spectral data however clearly show the presence of **442** and **443**, respectively[156]:

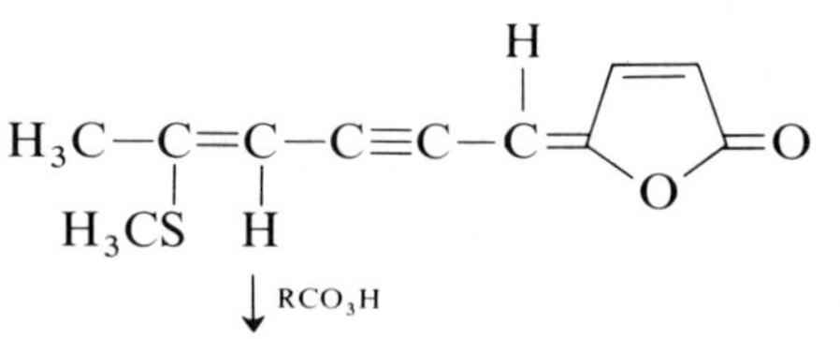

442

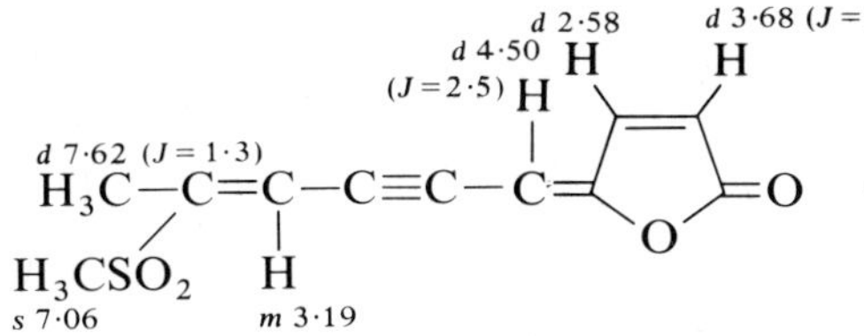

443

The structure of **442** has been confirmed by synthesis[164], which is best achieved by u.v. isomerisation and ring closure in the following way:

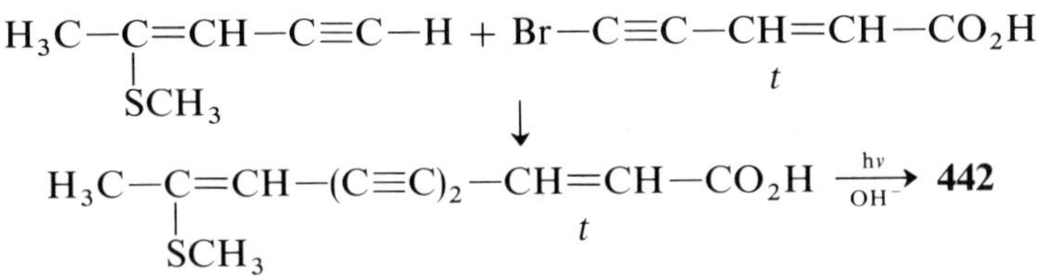

The roots of *Anthemis austriaca* Jacq.[165] contain further lactones, two cumulenes and one acetylene. The structure of the last one, with u.v. maximum at 362 nm, follows clearly from the observed signals in the n.m.r. spectrum of the original thioether and its sulphone:

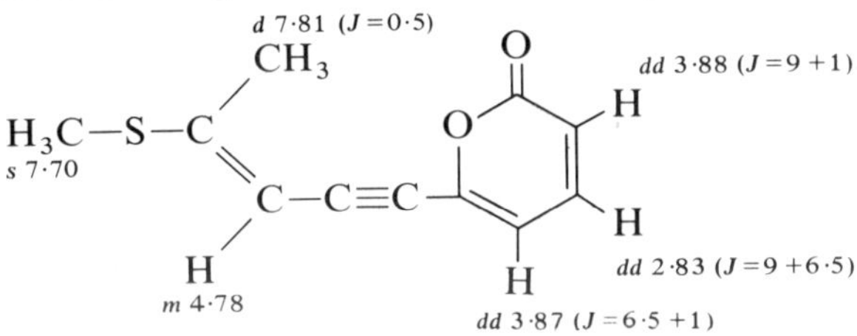

444

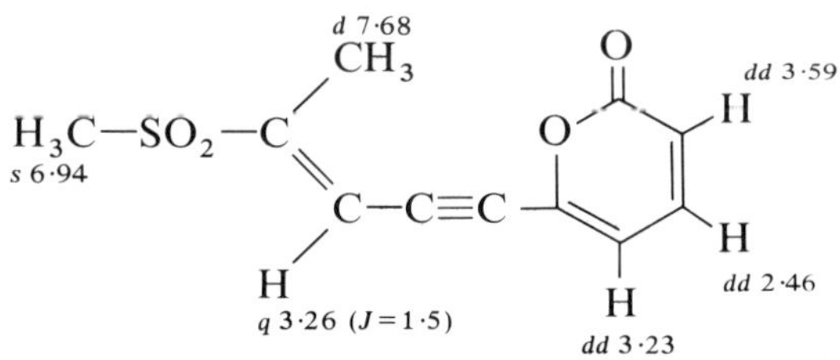

445

The structure of **444** further has been confirmed by synthesis[166]:

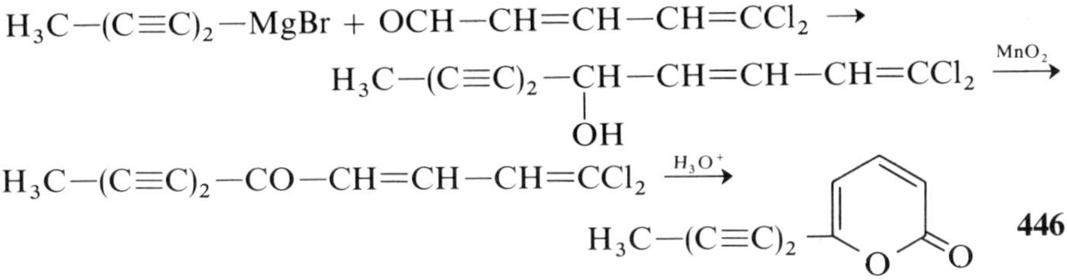

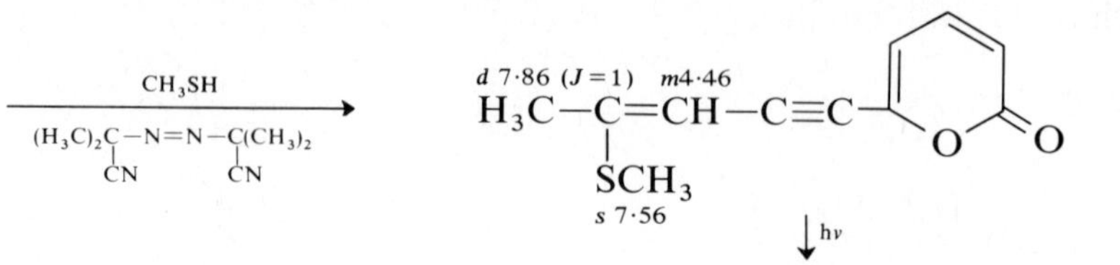

447

444

The data of synthetic **444** and **447** clearly establish the proposed configuration of the first naturally occurring acetylenic pyrone.

The structures of the *cis–trans*-isomeric cumulenes also isolated from *Anthemis austriaca* Jacq.[165] clearly follow from the spectral data, while the configuration of the cumulene bond can only be estimated by partial hydrogenation and conversion of the resulting dihydrocompounds to the sulphones. The cumulenes themselves are too sensitive for reaction with peracids, therefore the corresponding sulphones could not be prepared:

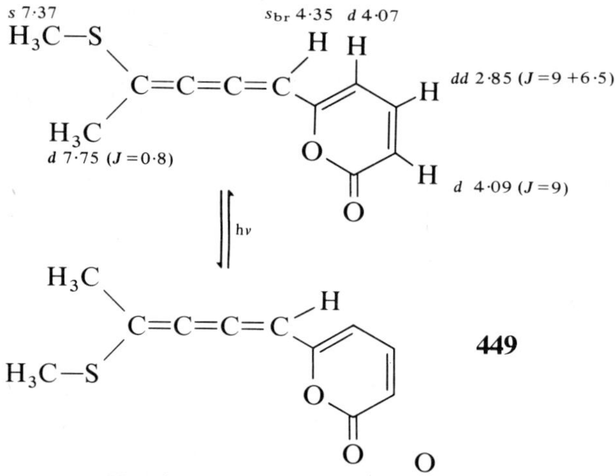

448

449

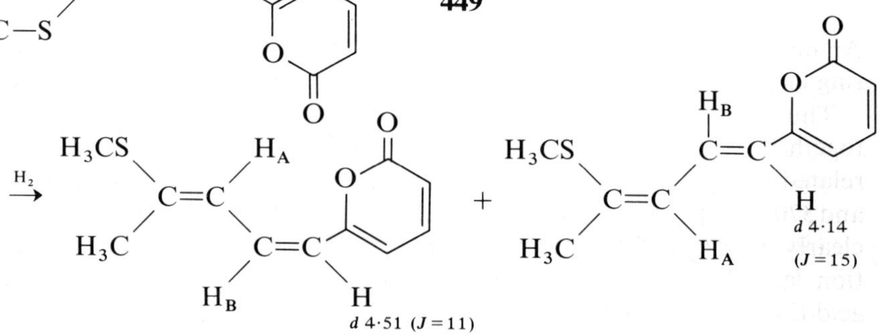

448 $\xrightarrow{H_2}$ **450** + **451**

The sulphone of **450** shows a strong downfield shift (2·88 → 1·88) in the n.m.r. spectrum for H_A, while that for H_B is little influenced (3·55 → 3·45). This is only possible if we deal with configuration **450** and not with **452**, which clearly also establishes the configuration of **448**:

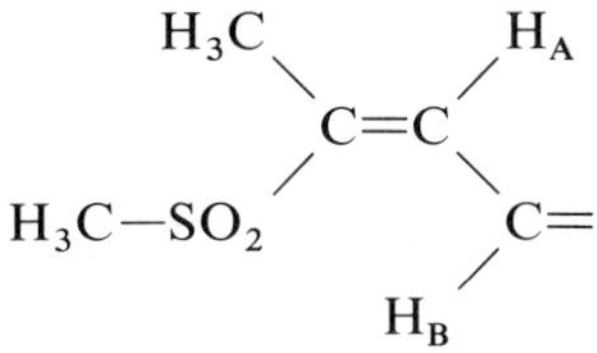

452

The isomeric cumulene **449** could not be obtained in a pure state. It isomerises very rapidly, even during TLC. Biogenetically, the pyrones **444**, **448** and **449** are formed from **370**, as is shown by feeding experiments with both labelled **370** and **5**[166]. The thioether **387** is the most likely an intermediate for **444**:

370 → 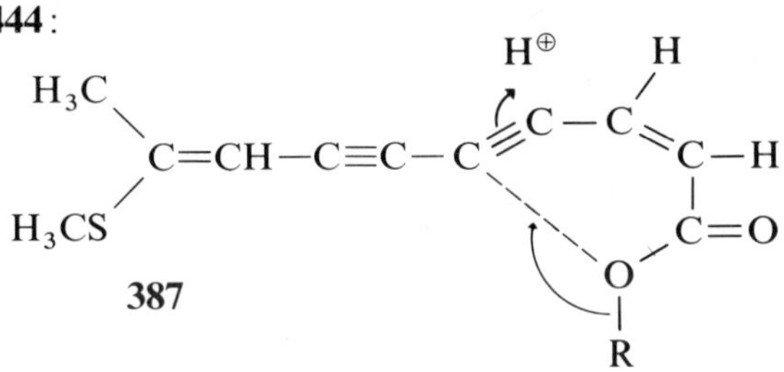 → **444**

387

In the same manner **448** and **449** should be formed via **413**.

The thioether butenolides are very probably formed from the lactone **436**, following the normal polarisation of the triple bonds:

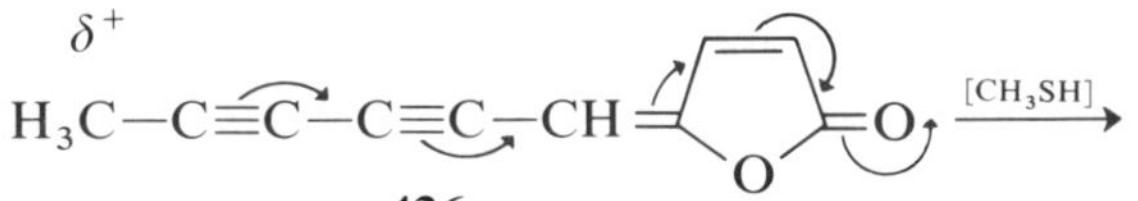

436

An introduced *S*-methyl group (as in **387**) possibly changes the situation and ring closure to a pyrone then becomes possible.

Thiophenes have also been found in the tribus Anthemideae very often together with **368** and **370** and their addition products; all are closely related to **368** and **370**. The roots of *Anthemis nobilis* L. contain besides **368** and **370** two *cis–trans*-isomeric thiophenes with molecular formula $C_{11}H_{10}O_2$ clearly indicating formal H_2S-addition products of **368**. Partial hydrogenation leads to an ester, which, on oxidation, gives 5-thiophene-carboxylic acid-(2). Therefore, the structures of these two esters must be **453** and **454**[68]:

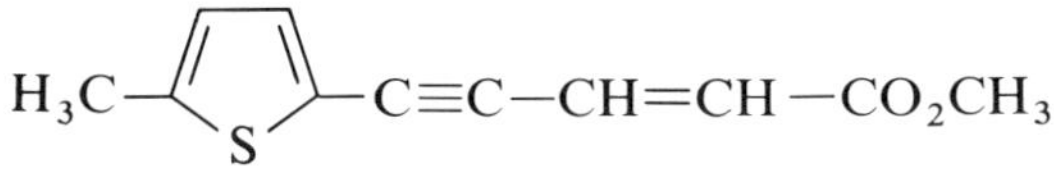

t **453**
c **454**

By synthesis, these structures could be confirmed using the following route[126]:

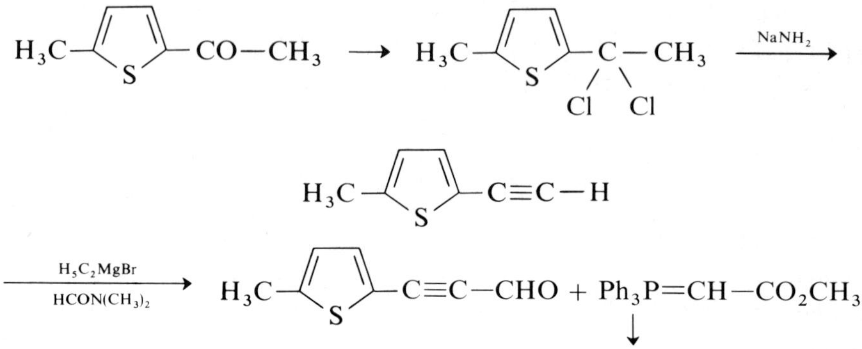

453 + 454

The same plant also contains a butenolide (i.r.: 1795, 1765, 1645, 1100, 1065, 885/cm) with a u.v. maximum at 371 nm and of molecular formula $C_{10}H_8O_2S$. The n.m.r. spectrum clearly shows the presence of a 2,5-disubstituted thiophene ring, one of the substituents being methyl [*dq* 3·31τ (J = 3·5 + 0·9) and *d* 2·86τ (J = 3·5)]. All data only agree with structure **455**[167], which could be established by synthesis[167]:

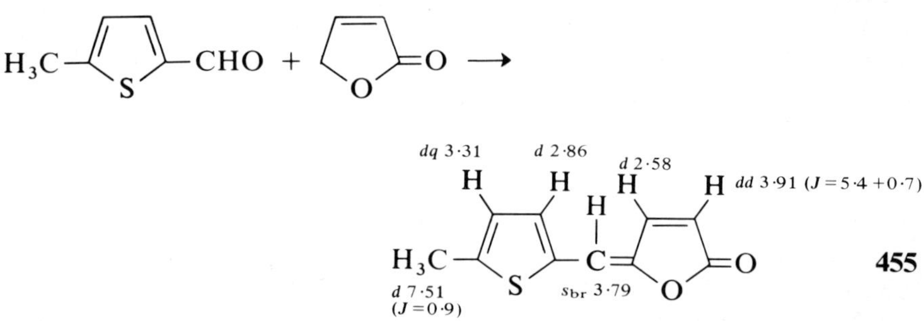

455

455 is very probably formed by cyclisation of the acid corresponding to **454**:

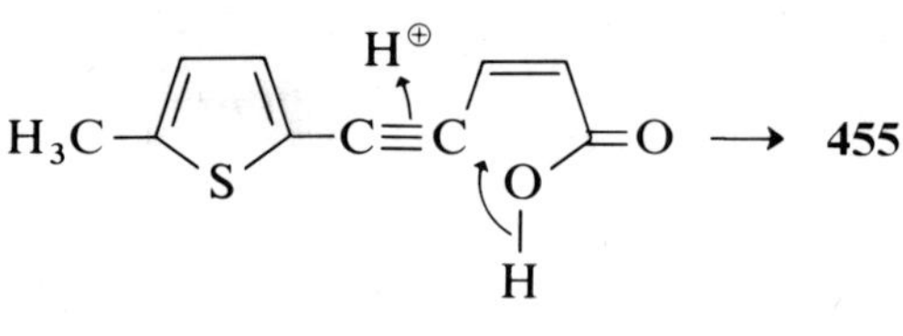

A degradation product of **453** has been found in the roots of *Anthemis fuscata* Brot.[45]. Th u.v. maximum of this ester is subject to a shift of about

10 nm to shorter wavelengths. The structure is established by synthesis[68] as **456**:

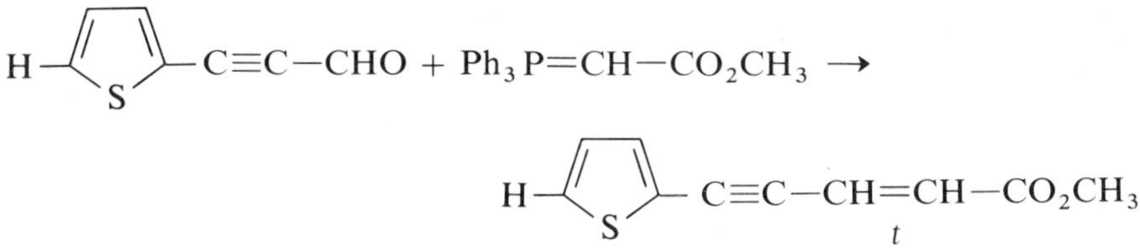

456

Probably a biochemical reduction product is the aldehyde **457** isolated from *Anthemis saguramica* Sosn.[57]. The n.m.r. spectrum clearly demonstrates the presence of a monosubstituted thiophene:

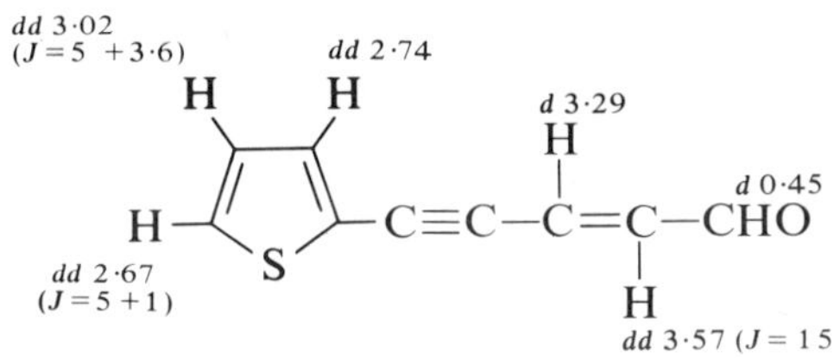

457

From the roots of the common tansy *Tanacetum vulgare* L., a further thiophene ester, with a u.v. spectrum nearly the same as that of **454** has been isolated[168]. Hydrogenation leads to 5-(*n*-propyl)-2-thienyl-propionate. Therefore, this ester must have structure **458**. The *trans*-isomer is also present in the tansy roots:

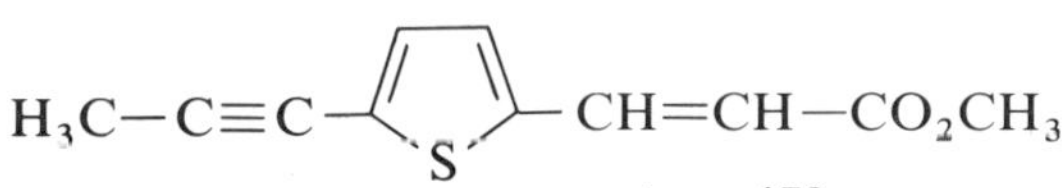

cis **458**
trans **459**

Synthesis of **459** has been achieved by Wittig reaction[169]:

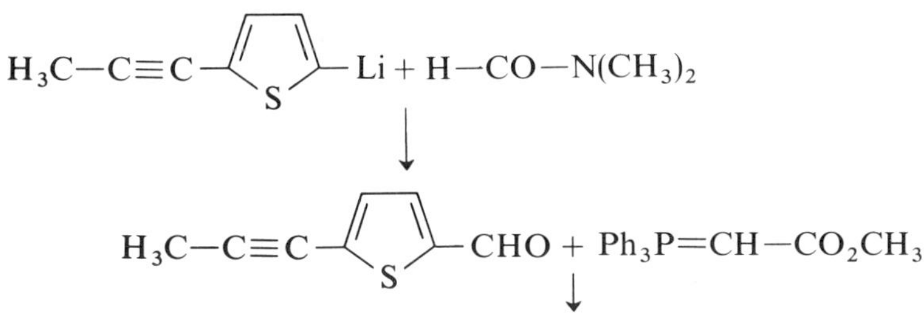

459

Two thiophene ketones have been isolated from the roots of *Artemisia arborescens* L.[170]. The somewhat unusual structures are elucidated by spectral methods and by chemical reactions as **460** and **461**:

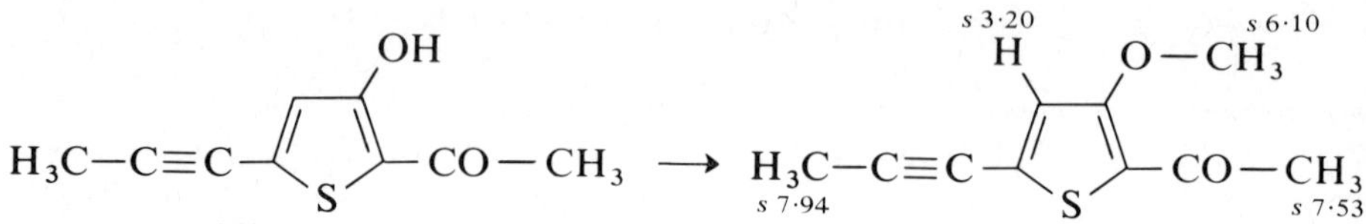

460 **461**

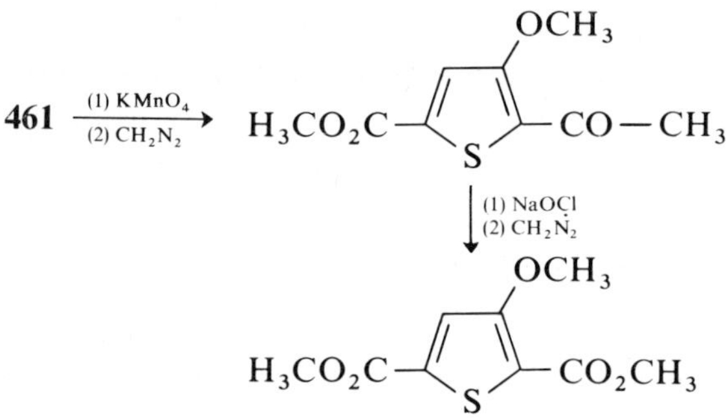

463

462

The diester **462** was identical with an authentic sample and the structures have been further established by synthesis[126]:

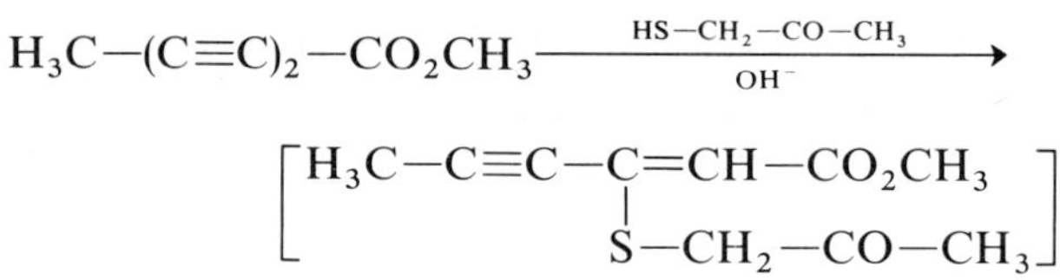

→ **460**

The biogenesis of **460** still is not clear. Perhaps it is formed in the following way:

$[H_3C-C\equiv C-C\equiv C-CH_2-CO-CH_2-CO-CH_3]$ **464**

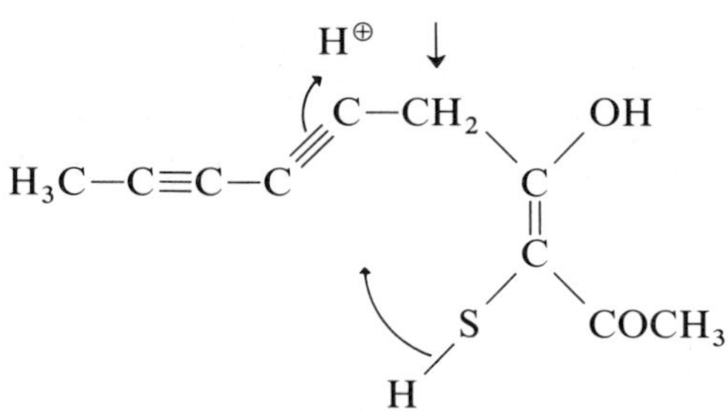

→ **460**

So far, no feeding experiments have been accomplished. Perhaps other pathways should also be discussed.

The co-occurrence of thioenolethers and thiophenes derived from dehydromatricariaester makes it most likely that the following scheme are operating in the tribus Anthemideae:

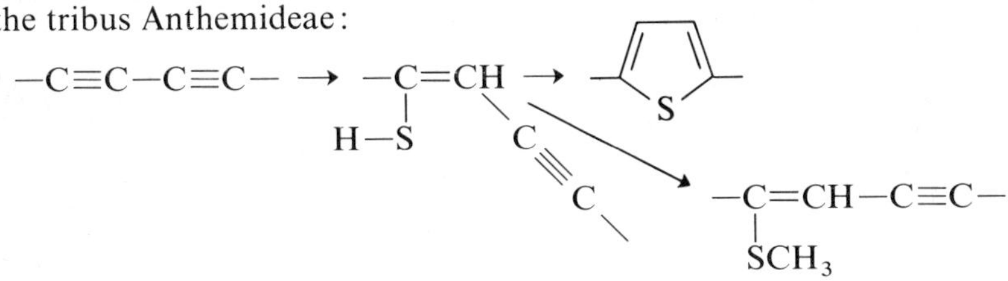

5.4. Acetylenes from Micro-organisms derived from C_{10}-compounds

A considerable number of acetylenes have been isolated from micro-organisms most of them probably also derived from dehydromatricaria-ester (**370**). By feeding labelled **370** this assumption could be established for several members of this series of natural acetylenes[171].

From *Pleurotus ulmarius* among other compounds dehydromatricaria acid (**465**)[153] has been isolated, purified as its methylester. Together with **465** also the aldehyde **466** and the alcohol **467** are present[153]. The *cis*-isomer of **467** has been found in the culture medium of *Fistulina hepatica*[138]:

$$\underset{t}{H_3C{-}(C{\equiv}C)_3{-}CH{=}CH{-}CO_2R} \qquad \mathbf{465}: R = H; \quad \mathbf{370}: R = CH_3$$

$$\underset{t}{H_3C{-}(C{\equiv}C)_3{-}CH{=}CH{-}CHO} \qquad \mathbf{466}$$

$$\underset{t\ /\ c}{H_3C{-}(C{\equiv}C)_3{-}CH{=}CH{-}CH_2{-}OR} \qquad \mathbf{467}: (R = H, CH_3); \quad \mathbf{468}: R = H$$

The dehydromatricariaester **370** has been isolated from a *Clitocybe* species. In the same genus the methyl ether **467** (R = CH_3) is also present[172].

An oxidation product of **465** is also present in *Pleurotus ulmarius*. The structure has been established by hydrogenation of the ester with subsequent oxidation and hydrolysis yielding sebacic acid; spectral data clearly indicate the presence of a triyn-enoic acid with a further hydroxy group. Therefore, only structure **469** is possible[138]:

$$HO{-}CH_2{-}(C{\equiv}C)_3{-}CH{=}CH{-}CO_2R \qquad \mathbf{469}: R = H; \quad \mathbf{470}: R = CH_3$$

$$\big\downarrow \text{(1) } H_2 \text{ (2) } CrO_3 \text{ (3) } OH^-$$

$$HO_2C{-}(CH_2)_8{-}CO_2H$$

By synthesis the structure of **469** has been established[153]:

$$\underset{}{HO{-}CH_2{-}(C{\equiv}C)_2{-}H} + Br{-}C{\equiv}C{-}\underset{t}{CH{=}CH}{-}CO_2H \rightarrow \mathbf{469}$$

A further oxidation product has been found in the culture medium of *Poria sinuosa* Fr. isolated as the crystalline diester. Hydrogenation leads to dimethyl sebacate, while lithium aluminium hydride reduction gives the diol **473**, identical with authentic material. Therefore, the natural compound must be **471**[134]:

$$RO_2C{-}(C{\equiv}C)_3{-}\underset{t}{CH{=}CH}{-}CO_2R \qquad \mathbf{371}: R = H \quad \mathbf{472}: R = CH_3$$

$$\downarrow LiAlH_4$$

$$HO{-}CH_2{-}(C{\equiv}C)_3{-}\underset{t}{CH{=}CH}{-}CH_2{-}OH \quad \mathbf{473}$$

473 itself is also a metabolite obtained from micro-orgamisms[22].

Hydrogenated compounds have also been isolated as well as oxidation products. The culture medium of *Merulius lacrymans* contains the di-acid **474** probably formed by hydrogenation of **471**[153]:

$$HO_2C{-}(C{\equiv}C)_3{-}CH_2{-}CH_2{-}CO_2H \quad \mathbf{474}$$

Further products of this type are present in the culture medium of *Poria sinuosa* Fr. The optically active acid, purified as its ester shows a u.v. maxima typical of a triyne, while the i.r. spectrum indicates that a diol is present. Periodate does not affect the ester. Hydrogenation followed by alanate reduction, however, gives a triol consuming one mole of periodate. The structure must be **475**[134]:

$$HO{-}CH_2{-}(C{\equiv}C)_3{-}CH_2{-}\underset{OH}{\underset{|}{CH}}{-}CO_2R \quad \mathbf{475}$$

$$\downarrow (1)\ H_2;\ (2)\ LiAlH_4$$

$$HO{-}CH_2{-}(CH_2)_7{-}\underset{OH}{\underset{|}{CH}}{-}CH_2OH \quad \mathbf{476}$$

$$\downarrow HIO_4$$

$$HCHO$$

A second ester turns out to be **477**, a hydrogenation product of **475**[134]:

$$HO{-}CH_2{-}(C{\equiv}C)_2{-}\underset{t}{CH{=}CH}{-}CH_2{-}\underset{OH}{\underset{|}{CH}}{-}CO_2H \quad \mathbf{477}$$

From *Daedalea juniperina* Murr. a dihydro compound has been isolated together with **467**. The i.r. spectrum shows the presence of an allene, while the u.v. spectrum is typical for an ene-diyne. Therefore the structure of this compound yielding *n*-decanol on hydrogenation is **478**[146]:

$$H_3C{-}(C{\equiv}C)_2{-}CH{=}C{=}CH{-}CH_2{-}CH_2{-}OH \quad \mathbf{478}$$

An oxidation product of **467** is present in the culture medium of *Coprinus quadrifidus* together with **473**[173]. The structure follows from the result of the comparison with the manganese dioxide oxidation product of **473**, which was not identical with the natural hydroxy aldehyde. Therefore, this must be **479**:

$$OCH{-}(C{\equiv}C)_3{-}\underset{t}{CH{=}CH}{-}CH_2{-}OH \quad \mathbf{479}$$

The oxidation of the terminal methyl group is an important pathway in the formation of a number of C_9-acetylenes very common in micro-organisms. As shown with cell-free extracts of *Coprinus quadrifidus* acetylenic acids are easily transformed to the corresponding ethynyl compounds[141]:

$$HO_2C{-}C{\equiv}C{-}C{\equiv}C{-}R \rightarrow H{-}C{\equiv}C{-}C{\equiv}C{-}R$$

The same organism contains the triynes **480** and **481**, their structures easily established by spectral methods[173]:

$$H{-}(C{\equiv}C)_3{-}CH{=}CH{-}CH_2{-}OH \quad \mathbf{480}$$

$$H{-}(C{\equiv}C)_3{-}CH{=}CH{-}CHO \quad \mathbf{481}$$

The structures have been confirmed by synthesis using the modified Cadiot–Chodkiewicz condition starting with diacetylene[144]:

$$H{-}C{\equiv}C{-}C{\equiv}C{-}H + Br{-}C{\equiv}C{-}CH{=}CH{-}CH_2{-}OH \rightarrow \mathbf{480} \xrightarrow{MnO_2} \mathbf{481}$$

The same fungus further produces some polar compounds, which can be separated by counter-current distribution. The major compound is a crystalline triol which is very photo sensitive. Therefore, the presence of a free triyne system seems to be most likely. While i.r. spectral measurement was impossible due to the insolubility, the formation of a silver salt indicates the presence of a free ethynyl group. Hydrogenation leads to nona-triol-(1,2,3), identified by periodic cleavage and yielding *n*-heptanal. Therefore, the structure has to be **482**[173]. Under alkaline conditions, **482** is converted into the enolether **483** showing the characteristic u.v. maxima of a diyn-enolether.

Ozonolysis gives L-erythro-lactone, and the absolute configuration of **482** and **483** is therefore established:

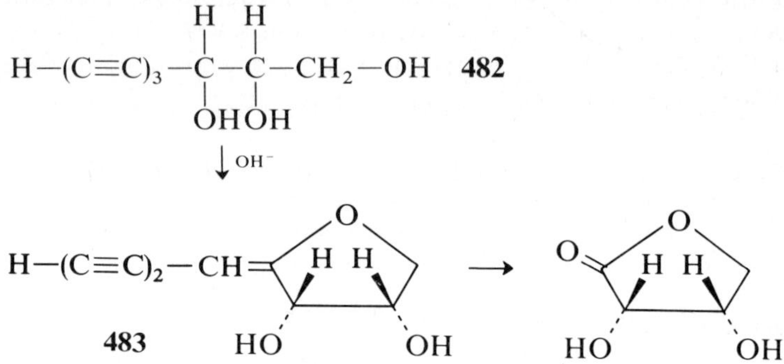

482

483

Together with **482**, the epoxide **484** also occurs[173] and this is very probably the precursor of **482**, itself formed by oxidation of **467**:

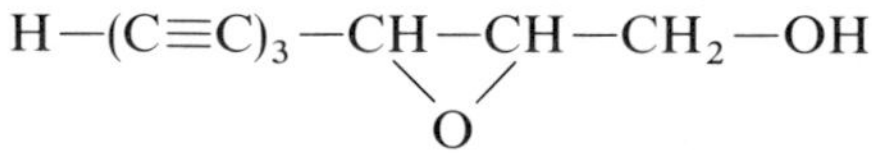

484

The structure of **484** has been established by synthesis[174]:

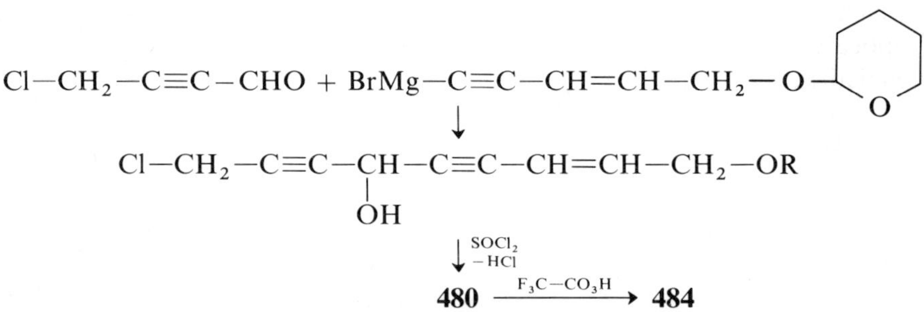

An acid closely related to **480** and purified as its methyl ester has been obtained from *Poria sinuosa*. The structure follows from the u.v. spectrum, is very similar to that of **370** and, from the i.r. spectrum, which shows the presence of a free ethynyl group (3310/cm) and a conjugated ester group (1730, 947/cm). These are only in agreement with **485** for the original acid[134]:

$$H{-}C{\equiv}C{-}C{\equiv}C{-}C{\equiv}C{-}\underset{t}{CH{=}CH}{-}CO_2R \qquad \textbf{485}: R = H \qquad \textbf{486}: R = CH_3$$

The ester **486** has also been obtained[22] from other micro-organisms. Again from *Polyporus anthracophilus*, a very polar compound has been isolated[134] which, after counter-current distribution, could be purified as its methyl ester. The u.v. spectrum identical with that of **486** and the i.r. spectrum showing the presence of free ethynyl-, ester- and amide-linkage agree with the structure **488**, indicating that the natural substance must be **487**. This has been finally confirmed by hydrolysis yielding L-valine and hydrogenation of **488**, which gives nonanoyl-L-valine ester.

$$\mathrm{H{-}(C{\equiv}C)_3{-}\underset{t}{CH{=}CH}{-}CO{-}NH{-}\underset{\displaystyle CO_2R}{\underset{|}{CH}}{-}CH_2(CH_3)_2}$$

487: R = H

488: R = CH_3

The isolation of **487** is very interesting, as extracts of higher plants contain several isobutylamides, very probably formed by decarboxylation of compounds of type **487** (see Chapter 3, Section 6).

Several formal hydrogenation products of **485** and **480**, respectively, also have been found in culture media of fungi. *Aleurodiscus roseus* contains several allenes[175]. Among others, a C_9-alcohol, its spectral data being identical with those of Marasin, has been isolated from *Marismius ramealis*[176]. The alcohol with the u.v. spectrum of a diyn-ene and i.r. bands typical for free ethynyl- and allene groups together with the result of hydrogenation, which leads to *n*-nonanol, are only in agreement with structure **489** for this optically active allene. Under alkaline conditions, **489** is transformed to an enolether. **489** can be prepared by alanate reduction of **480**[144]:

$$\mathbf{480} \xrightarrow{\mathrm{LiAlH_4}} \mathrm{H{-}(C{\equiv}C)_2{-}CH{=}C{=}CH{-}CH_2{-}CH_2OH} \quad \mathbf{489}$$

$$\downarrow \mathrm{OH^-}$$

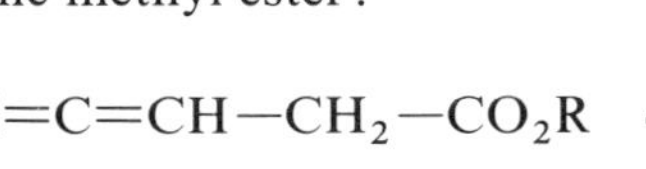

Whereas (+)-**489** has been obtained from *Marismius ramealis*, the *Aleurodiscus* species gives (−)-**489**. This organism also contains the corresponding acid **490**, together with the methyl ester:

$$\mathrm{H{-}(C{\equiv}C)_2{-}CH{=}C{=}CH{-}CH_2{-}CO_2R} \quad \mathbf{490}\mathrm{: R = H, CH_3}$$

Two very polar compounds were also isolated from the same fungus. Both, on hydrogenation, form nonane-1,2,3-triol, but unlike **482** the configuration of the 2,3-hydroxyls are *threo*. The spectral data of one of the compounds are identical with those of **482**, but the data of the other compound clearly indicate a diyn-ene; its structure is therefore **492**, which also occurs as its 3-epimeric triol:

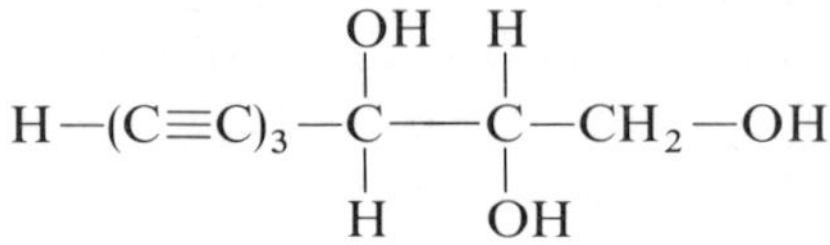

491

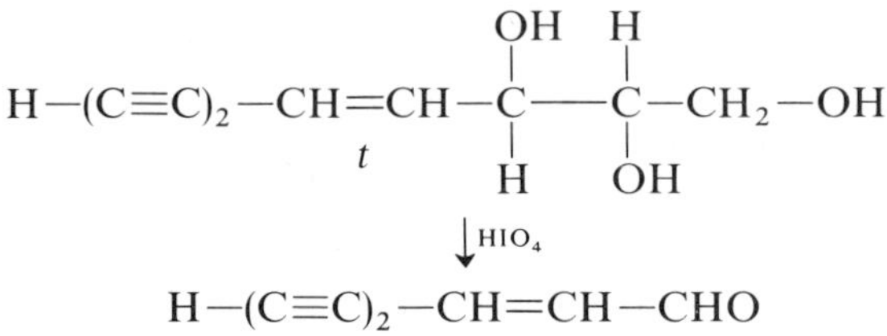

492

493

Drosophila subatrata contains, among several other polynes, a very unstable crystalline acid[147], which on hydrogenation and reduction yields *n*-nonanol and therefore must be **494** as the u.v. spectrum indicates a monosubstituted ene-diyne. The i.r. spectrum shows the presence of free ethynyl and *cis*-double bond, together with an unconjugated carboxyl group:

$$\mathrm{H{-}C{\equiv}C{-}C{\equiv}C{-}\underset{cis}{CH{=}CH}{-}CH_2{-}CH_2{-}CO_2H} \qquad \textbf{494}$$

$$\mathrm{H{-}C{\equiv}C{-}C{\equiv}C{-}\underset{trans}{CH{=}CH}{-}CH_2{-}CH_2{-}CH_2{-}OH} \qquad \textbf{495}$$

The corresponding alcohol **495** is also present in this organism.

In some cases, acetylenes have also been isolated from fungi, which have probably been formed by further degradation of the precursor originally present. Together with **489**, the corresponding C_8-compound giving on hydrogenation *n*-octanol has been found in *Cortinellus berkeleyanus*[146]. The spectral data are only in agreement with structure **496**:

$$\mathrm{H{-}C{\equiv}C{-}C{\equiv}C{-}CH{=}C{=}CH{-}CH_2{-}OH} \qquad \textbf{496}$$

Several of the previously mentioned acetylenic allenes have now been synthesised by S. R. Landor[177] and his co-workers. **496** has been obtained in the following way:

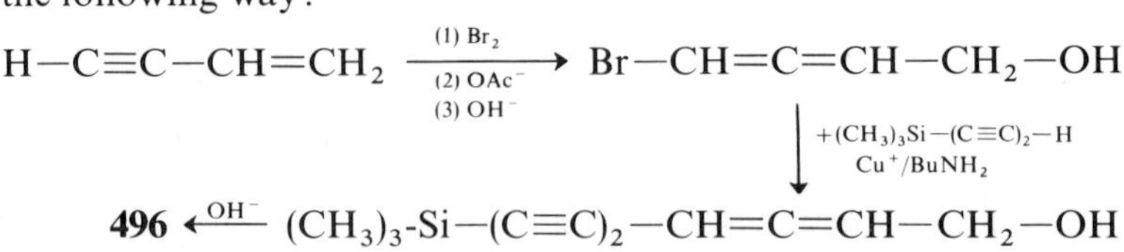

Compound **397**, which is very similar, could also be synthesised:

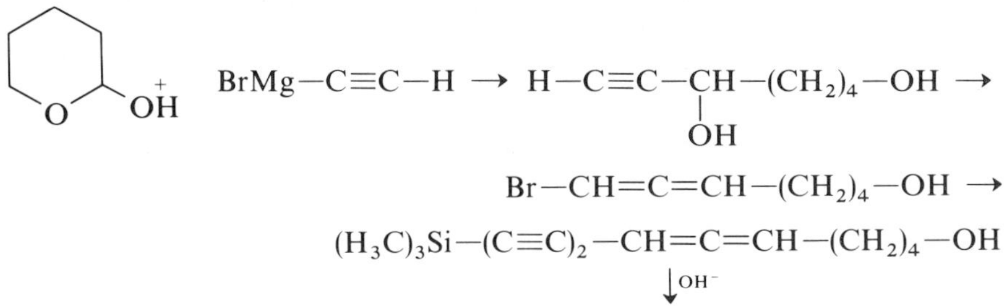

397

In both cases, the key step is the possibility of Cadiot–Chodkiewicz coupling of bromo allenes with protected diacetylene. The alanate reduction of ene-ynes leading to allenes has been improved by use of partially-blocked alanate. By using a sugar derivative as blocking group, optically active allenes could be obtained. Marasmin (**489**) and **478** have been synthesised by these methods:

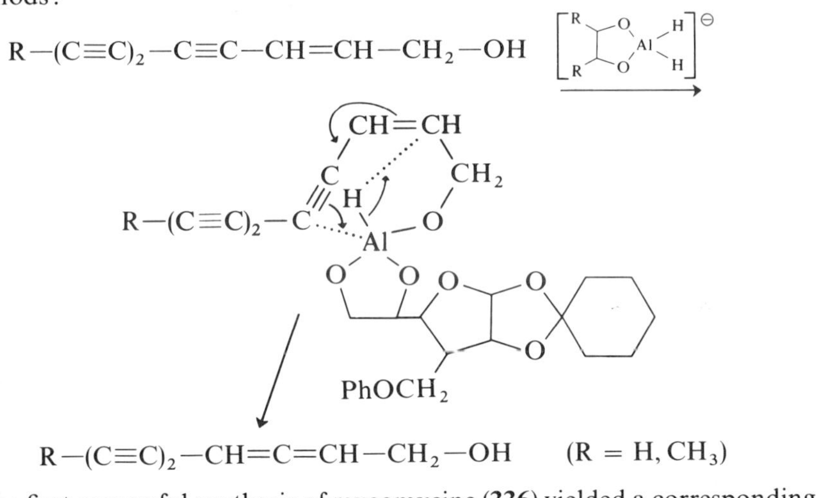

The first successful synthesis of mycomycine (**336**) yielded a corresponding alcohol:

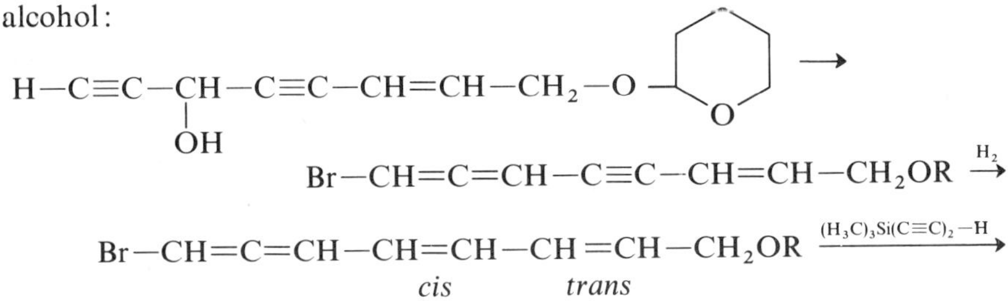

$$(H_3C)_3Si{-}(C{\equiv}C)_2{-}CH{=}C{=}CH{-}\underset{cis}{CH{=}CH}{-}\underset{trans}{CH{=}CH}{-}CH_2OH \xrightarrow{OH^-}$$

$$H{-}(C{\equiv}C)_2{-}CH{=}C{=}CH{-}\underset{cis}{CH{=}CH}{-}\underset{trans}{CH{=}CH}{-}CH_2{-}OH$$

Perhaps it will be possible to introduce the missing carboxyl group.

The octatriynoic-acid **497** is probably formed by oxidation of **465**. It is isolated from micro-organisms[22]:

$$H_3C{-}(C{\equiv}C)_3{-}CO_2H \quad \mathbf{497}$$

497 is clearly the precursor of the hydroxy-acid **498**[178] and of Agrocybin (**499**), which is isolated from the culture medium of *Agrocybe dura*[179]:

$$HO{-}CH_2{-}(C{\equiv}C)_3{-}CO_2H \qquad \mathbf{498}$$

$$HO{-}CH_2{-}(C{\equiv}C)_3{-}CONH_2 \quad \mathbf{499}$$

These structures have been confirmed by synthesis[179]:

$$HOCH_2{-}C{\equiv}C{-}C{\equiv}C{-}H + H{-}C{\equiv}C{-}CO_2CH_3 \rightarrow$$

$$HOCH_2{-}(C{\equiv}C)_3{-}CO_2CH_3 \xrightarrow{NH_3} \mathbf{499}$$

As free triacetylenes have also been isolated from a micro-organism[180], the oxidation product of **498**, the diacid, might possibly be transformed by decarboxylation to hexatriyne:

$$HO_2C{-}(C{\equiv}C)_3{-}CO_2H \xrightarrow{-CO_2} H{-}(C{\equiv}C)_3{-}H$$

Up to the present time, only a few thiophene acetylenes have been found in micro-organisms. *Daedalea juniperina* contains the aldehyde **500**, its structure being established by degradation and synthesis, probably by the following biogenetic pathway[181,182]:

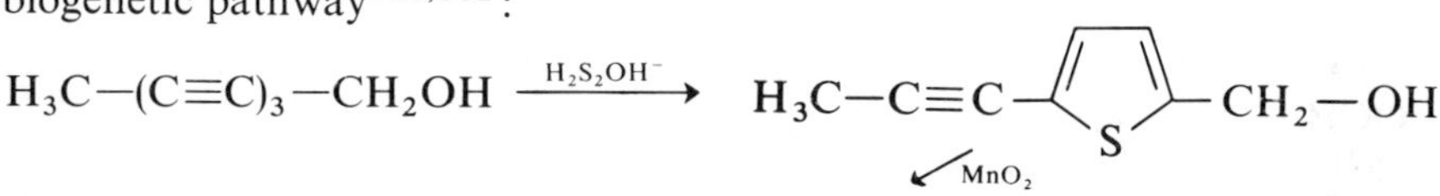

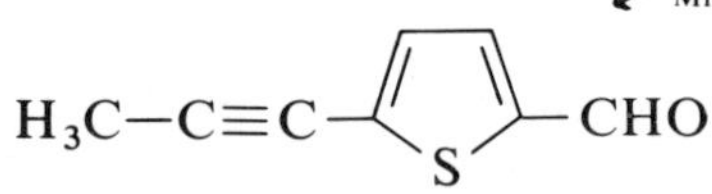
500

From the same organism the thiophenes **501** and **502** have been obtained[183]:

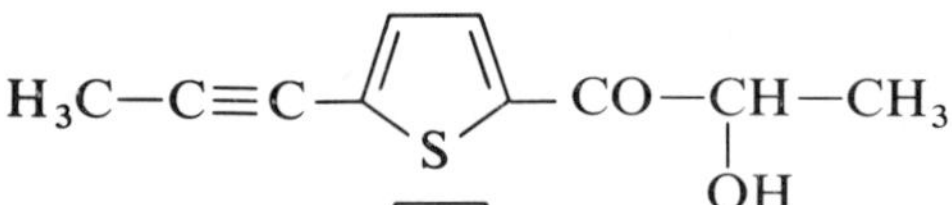
501

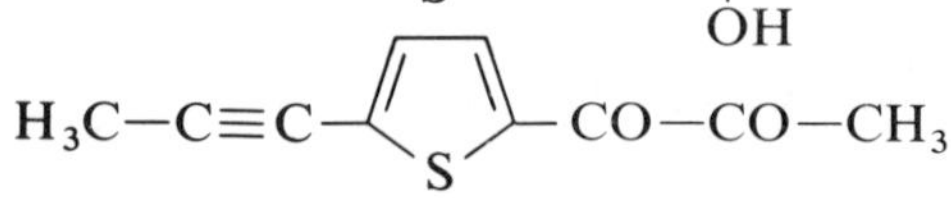
502

The isolation of these thiophenes indicates that **500** might also be formed by oxidative degradation of **501**. It is more probably formed by degradation of the corresponding diol **504**, which is perhaps formed by epoxidation of **503**. This latter compound could be the reduction product of **459** which is present in higher plants (see Section 5.3):

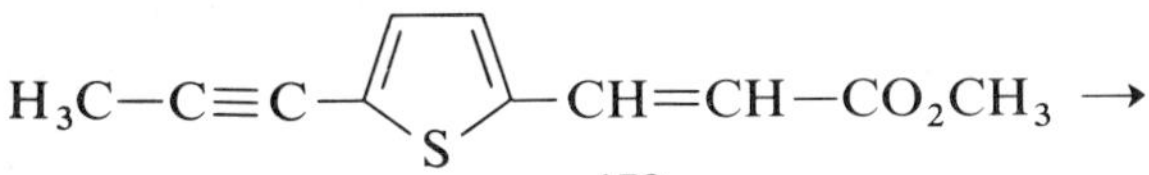

459

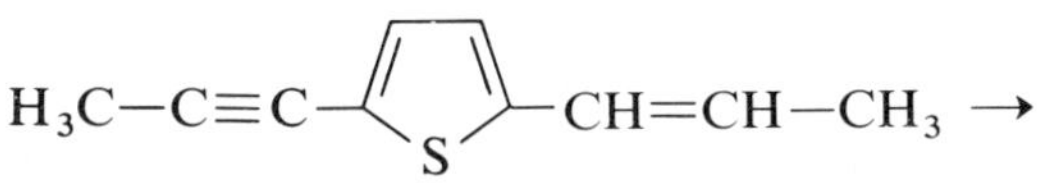

503

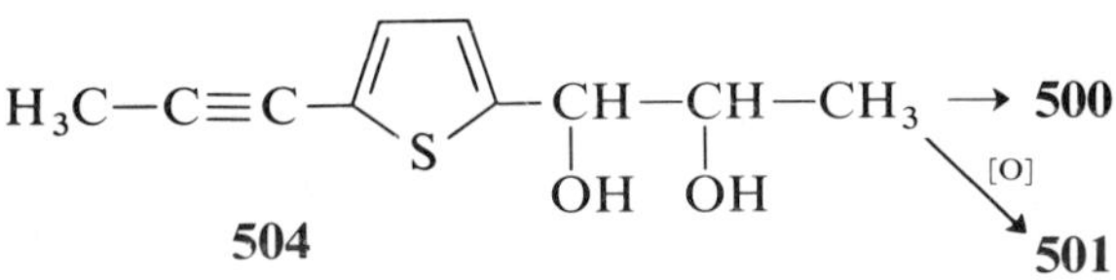

504

As shown in this chapter nearly all the C_8—C_{10}-acetylenes from fungi are more or less closely related to dehydromatricariaester (**370**), which is an important precursor in the higher plants too, especially in the tribus Anthmideae. If we put together the necessary biochemical pathways starting with **370** respectively the acid **465**, we come to the following scheme:

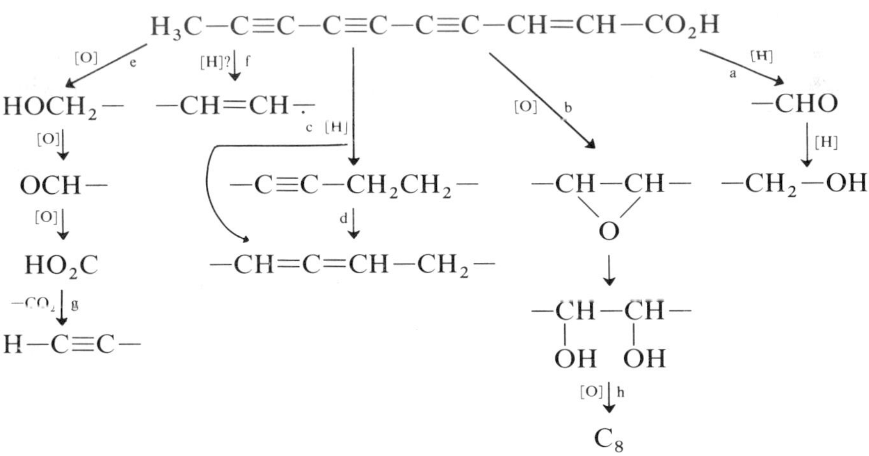

Further degradations e.g. route (h), a chain shortening subsequent to (b) and (d), also seem to be very probable.

Only some of these pathways in this scheme really have been established[171]. By feeding [1-^{14}C]-**370** it could be shown, that with *Tricholoma grammopodium* **370** is transformed in a yield of 46% to **467**, clearly establishing the importance of route (a). By feeding [1-^{14}C]-**466**, which also is converted in high yields to **467** by the same organism, a second example can be given. This fungus also shows the importance of route (b) and (e) as **370** is transformed here in about 5% yield to **471** and with 11% incorporation to **474**, while further conversion using routes (e), (g), (a) and (c) leads to **482** with 4% of incorporation in *Coprinus quadrifidus*:

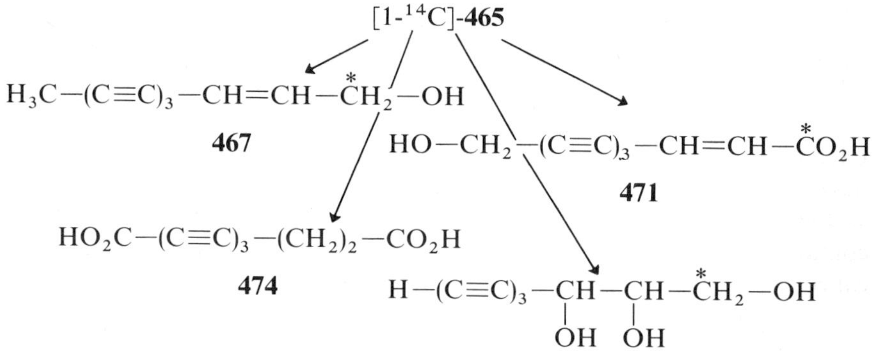

467 **471** **474** **482**

Using the culture medium of *Papulospora polyspora*[184] it could be shown, that **467** also can be oxydised back to **466** and **465** while the mycel seems to reduce **465** and **466** to **467**. Thus it seems probable that in this fungus, **467**, **466** and **465** are continually being interconverted by an oxidation-reduction cycle. A number of further alcohols have been treated with the culture fluid clearly establishing, that the enzymes present are specific for the alcohol **467** and the aldehyde **466**, as the other compounds are converted only with very small rates into the corresponding acids.

3

Compounds without Methyltriyne Endgroups and Related Derivatives

1. C_{18}- and C_{17}-compounds

1.1. C_{18}-acetylenes

Except those discussed in Chapter 4, only a few C_{18}-acetylenic compounds have been found in plants. In 1964, Mikolajczak and co-workers[185] investigated the seed oil of *Crepis foetida* L., family Compositae. The seed oil contained a fatty acid shown to be *cis*-9-octadecen-12-ynoic acid, conveniently called crepenynic acid **1**.

$$H_3C(CH_2)_4{-}C{\equiv}C{-}CH_2{-}\underset{c}{CH{=}CH}{-}(CH_2)_7CO_2H \quad \mathbf{1}$$

It was the first C_{18}-acetylene to be isolated from a compositae species with an interrupted conjugation and *cis*-ethylenic bond, analogous to linoleic acid in containing methylene-interrupted unsaturation. The purification and establishment of the structure had presented difficulties due to the presence of other long chain fatty acids and to the susceptibility of **1** to oxidation. The structure of the acid was elucidated by hydrogenation and degradation reactions. **1** was hydrogenated to stearic acid by the absorption of three moles of hydrogen. The orientation of the unsaturated system could be determined by the following oxidation experiments:

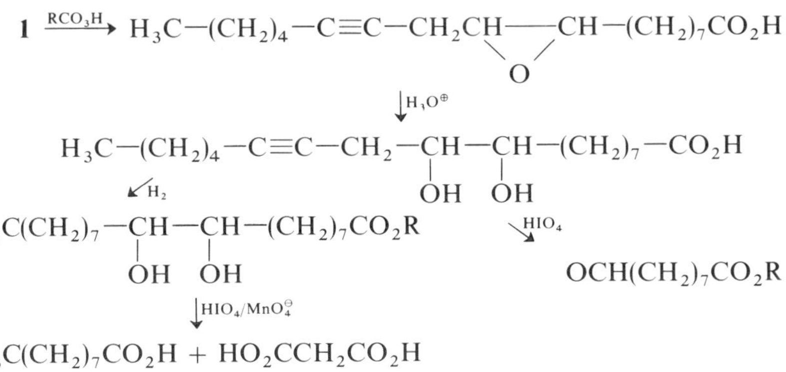

The double bond in **1** was epoxidised with peracetic acid and the resulting epoxide was converted into the *threo* acetylenic diol. Periodate cleavage of the diol yielded a steam-volatile acetylenic aldehyde and azela aldehydic acid. Furthermore, **1** was oxidised with periodate/permanganate and the crude mixture was converted, using diazomethane, into the corresponding methyl esters, which were analysed by GLC. These measurements showed the presence of a nonane dioic acid, *n*-hexanoic acid and malonic acid. These results place the unsaturation at C_9 and C_{12}.

As **1** is recognised to be a key intermediate in the biogenesis of polyacetylenes from fatty acids, it seemed desirable to develop a synthesis for preparing labelled crepenynic acid.[186] Heptyne-(1) was condensed with ethylene oxide to give non-3-yn-1-ol, which was brominated with triphenylphosphinedibromide to 1-bromo-non-3-yne. The latter was transformed to non-3-yn-1-yl-triphenylphosphoniumbromide. Finally, the ylide was condensed with methyl-8-formyloctanoate (obtained by ozonolysis of methyl oleate):

$$H_3C(CH_2)_4{-}C{\equiv}C{-}H \xrightarrow{\triangledown_{O}} H_3C(CH_2)_4{-}C{\equiv}C{-}CH_2CH_2OH \xrightarrow{Ph_3PBr_2}$$

$$H_3C(CH_2)_4{-}C{\equiv}C{-}CH_2CH_2Br \rightarrow$$

$$H_3C(CH_2)_4C{\equiv}C{-}CH_2CH_2\overset{\oplus}{P}Ph_3 \xrightarrow[\substack{(2)\ OCH(CH_2)_7CO_2CH_3 \\ (3)\ OH^{\ominus}}]{(1)\ BuLi} \mathbf{1}$$

The natural and synthetic methyl ester had identical i.r., n.m.r., and mass spectra. No evidence of the presence of the *trans*-isomer was found.

The i.r. spectrum of the autoxidised methyl crepenynate indicates hydroxyl absorption and conjugated *trans*-enynic absorption. The u.v. spectrum shows conjugated unsaturation, λ_{max} 234 nm. GLC analysis indicates a number of short chain materials, the probable results from cleavage of hydroperoxides formed by autoxidation. The autoxidation products are very interesting since there is a second fatty acid having both double and triple bonds, but in conjugation. This enynolic acid **2** was found as one of several unusual triglyceride substituents in *Helichrysum bracteatum* seed oil:

$$H_3C(CH_2)_4{-}C{\equiv}C{-}\underset{t}{CH{=}CH}{-}\underset{\substack{| \\ OH}}{CH}{-}(CH_2)_7CO_2H \quad \mathbf{2}$$

The structure of **2** has been elucidated by oxidative degradation and spectral properties.[187] The u.v. spectrum of the methyl ester indicates a conjugated trans-enyne chromophore with an absorption maximum at 228 nm. The ester is also indicated to be a monohydroxy ester by an i.r. maximum in the OH region. The saturated hydroxyester was obtained by hydrogenation of the unsaturated compound, its structure being established by oxidative cleavage with chromium trioxide in acetic acid. The products obtained, nonanoic and decanoic acid, together with half esters of octanedioic and nonanedioic acid, placed the hydroxyl at C_9 on a normal C_{18}-skeleton. The n.m.r. spectrum of the methyl ester is in accordance with the structure assigned on chemical grounds. The signals of the protons on C_{10} and C_{11} would be expected to split each other into doublets. The signal due to the C_{10}-proton is further split into a pair of relatively wide spaced doublets (τ 3·7–4·1; J = 6 Hz) by coupling with the proton at C_9. Apparently there is long-range coupling between the proton at C_{11} and the two protons at C_{14} as evidenced by the pair of closely spaced triplets at τ 4·1 and 4·5 (J = 1·8 Hz).

Starting at crepenynic acid, helenynolic acid **2** is certainly formed by oxidation of the activated methylene group by allylic rearrangement, as outlined in the scheme below:

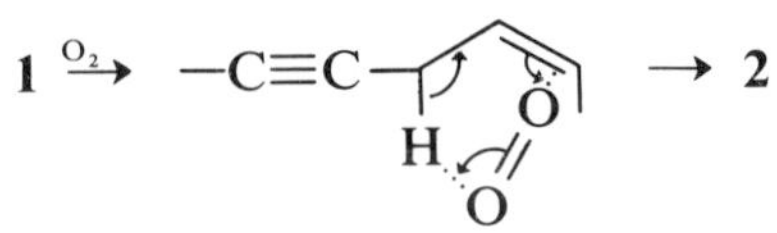

Certain long-chain alcohols—cicutoxin (**183**) and oenanthetoxin (**172**) (see Section 1.2.6)—have an analogous grouping, but with a more extended system of conjugation. It seems very plausible that those C_{18}-acetylenic acids are precursors in the biogenetic pathway leading to more unsaturated acetylenes. The finding that dehydro-crepenynic acid **3** is a precursor of phenyl-heptatriyne and other polyen-ynes[24] supports this assumption and the hypothesis of Bu'Lock about the origin of the naturally occurring acetylenes.[20]

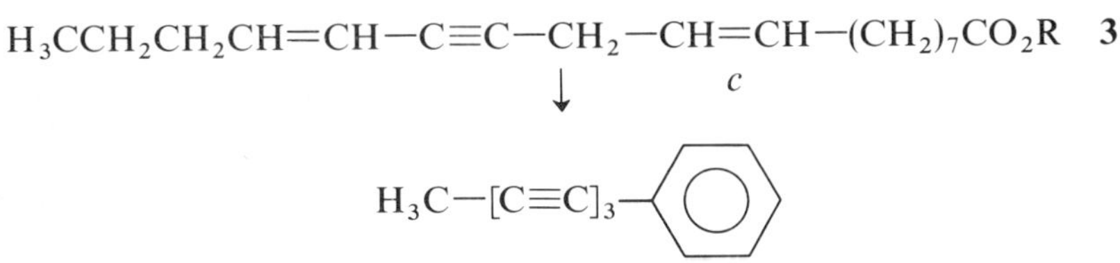

As there are many natural acetylenes having obviously structural and functional relationships to common fatty acids, it seems very probable that the more highly unsaturated compounds are derived from less unsaturated acids by dehydrogenation steps. In Bu'Lock's hypothesis, the whole range of acetylenes in micro-organisms and in higher plants of the Compositae and Umbelliferae can be derived from linoleic acid **5** via its acetylenic dehydro derivative, crepenynic acid **1**. Incorporation of labelled oleic acid **4** in several polyenyne compounds also support this hypothesis[25,30]. A probable sequence from oleic acid **4** to the various acetylenes shows the predominant role of **1**, **2** and **3** (see also Chapter 2)

$$\underset{}{H_3C(CH_2)_7-\underset{c}{CH=CH}-(CH_2)_7-CO_2R} \qquad \mathbf{4}$$

$$\downarrow -[H]$$

$$H_3C(CH_2)_4-\underset{c}{CH=CH}-CH_2-\underset{c}{CH=CH}-(CH_2)_7-CO_2R \qquad \mathbf{5}$$

$$\downarrow -[H]$$

$$H_3C(CH_2)_4-C\equiv C-CH_2-\underset{c}{CH=CH}-(CH_2)_7-CO_2R \qquad \mathbf{1}$$

$$\swarrow -[H] \qquad\qquad \searrow [O]$$

$$\mathbf{3} \qquad\qquad\qquad \mathbf{2}$$

$$\downarrow -[H]\ \alpha\text{- and } \beta\text{-oxidation}$$

natural acetylenes

In some cases, structural features of **1** or **2** remain recognisable throughout the series, whilst in others they are obscured. For example, in the C_{17}-compound falcarinone (**114**) most of the molecule of **1** is recognisable. As previously noted, the grouping of **2** is still apparent in the C_{17}-acetylenic diol oenanthetoxin (**172**). Otherwise, there are many compounds with no superficial relationship to **1** or **2** as, for example, the C_{10}-triynene compound dehydromatricarianol. It has been shown, however[188], that labelled oleic acid **4** is converted by the fungus *Tricholoma grammopodium* into **1** and dehydromatricarianol:

$$\mathbf{4} \rightarrow \mathbf{1} \rightarrow H_3C-(C\equiv C)_3CH\underset{t}{=}CHCH_2OH$$

dehydromatricarianol

From *Tricholoma grammopodium* to which (10–^{14}C)oleic acid had been administered, labelled linoleic **5** and crepenynic acid **1** were isolated. Degradation of **1** and **5** demonstrated that the C_{18}-chain of both acids was derived, intact, from oleic acid. Thus, it is established that, as shown in the scheme above, oleic acid undergoes dehydrogenation to the diene, linoleic acid, and to the enyne, crepenynic acid. The stages at which chain shortening reactions succeed and the nature of such reactions are known already in part. Others need to be established.

Starting from C_{18}-acids, there are some C_{17}-compounds which may well arise by α-oxidation, whilst for the C_{14}-series, two steps of β-oxidation are the most probable mechanism (see Chapter 2, Sections 1.2, 1.3, 1.4 and 3).

For the common C_{10}-series, further β-oxidation or direct oxidation are the possibilities.

An important connecting link in the biogenesis of natural acetylenes is the ketoacetate **6** isolated from *Cosmos sulphureus* Cav.[189]

$$H_3C(CH{=}CH)_2{-}C{\equiv}C{-}(CH{=}CH)_2(CH_2)_4{-}CO{-}CH_2CH_2OAc \quad \mathbf{6}$$

The structure of **6** has been determined by spectroscopic data and finally established by synthesis. Its i.r. spectrum reveals the presence of an acetate- and a keto-group as well as *trans*-double bonds (1750, 1250, 1730, 3030, 1640, and 980/cm). A weak signal at 2180/cm indicates an acetylenic group. The u.v. spectrum shows the characteristic absorption of an diene-yn-diene chromophore (338, 317, 304, 235, 229 nm). The n.m.r. spectrum of **6** includes a complex multiplet in the olefinic region and a doublet at τ 8·21, assigned to a methylgroup attached to a diene-grouping. Two triplets centred at τ 5·7 and 7·38 (J = 6·5 Hz) indicate the endgroup $-COCH_2CH_2OAc$. A multiplet at τ 7·5–8·0 affords two methylene groups standing beside unsaturated carbon atoms and a multiplet at τ 8·5 corresponds to two further $-CH_2$-groups. The mass spectrum of the hydrogenated product, with a molecular ion peak at *m/e* 326 and major peaks at *m/e* 266 ($M - CH_3CO_2H$) and *m/e* 239 ($M - CH_2CH_2OAc$) furnishes valuable confirmation of the proposed structure. Peaks at *m/e* 130 and *m/e* 70 correspond to the cleavage shown below:

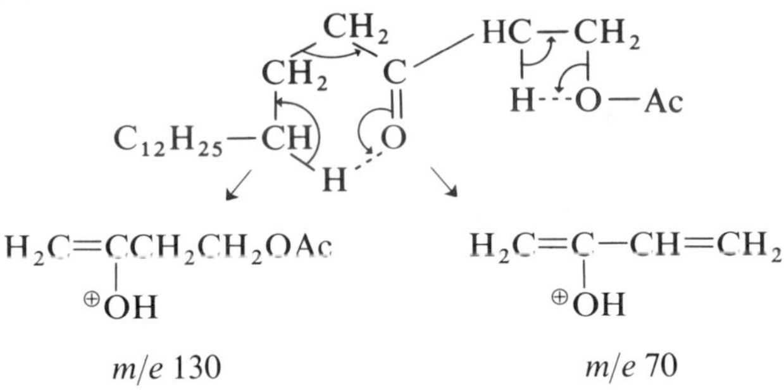

Final proof of the structure was provided by synthesis[189].

Jones and his co-workers[190] have isolated from an Umbellifer (*Pastinaca sativa* L.) a very closely related C_{18}-analogue of falcarinone (**114**)—a C_{17}-polyacetylenic compound (see Section 1.6)—the additional atom of the C_{18}-compound **7** being present as an aldehyde group:

$$H_2C{=}CH{-}CO{-}(C{\equiv}C)_2CH_2\underset{cis}{CH{=}CH}{-}(CH_2)_7{-}CHO \quad \mathbf{7}$$

The purification of **7** was difficult, as the acetylenic compound closely resembled falcarinone (**114**) not only in its u.v. absorption but also in its great instability. The main difference between the i.r. spectra of **114** and **7** is the presence of bands (2730 and 1730/cm) attributable to an aldehyde group. These data and the extensive identity of the n.m.r. spectra of falcarinone and the keto-aldehyde **7** strongly suggested the structure. The mass spectrum with a molecular ion peak at *m/e* 270 and peaks at *m/e* 157 and *m/e* 55, corresponding to the expected cleavage of the $C_7—C_8$ and $C_{15}—C_{16}$ bonds, respectively, gives confirmation of the proposed structure **7** and of the structural relationship between **7** and crepenynic acid **1**, both possessing an interrupted conjugation and the *cis*-double bond of oleic acid **4**. This supports our assumption, that the related C_{17}-compounds like falcarinone derive naturally by loss of a carbon atom from C_{18}-fatty acids and that the loss occurs from the more saturated end of the molecule. The aldehyde group of **7** is precisely in the expected position.

The occurrence of the long chain aldehydes **8** and **9** is very interesting in regard to the importance of acetylenic compounds with interrupted conjugation and *cis*-ethylenic bonds as biogenetic precursors. The aldehydes **8** and **9** were isolated from *Centaurea deusta* Ten.[79]. The determination of the chain-length was rather difficult, as the plant extract contains many other aldehydes. The aldehyde mixture was reduced with sodium borohydride and, after hydrogenation, the saturated alcohols were separated and determinated by GLC.

$$\underset{cis}{H_3CCH{=}CH}{-}(C{\equiv}C)_2CH_2\underset{cis}{CH{=}CH}{-}(CH_2)_n{-}CHO \qquad \mathbf{8}: n = 7 \quad \mathbf{9}: n = 6$$

In higher plants, especially in *Centaurea* species, there are a few polyenes, all of them containing a tetraene grouping but having different chain length.

$$H_3CCH_2(CH{=}CH)_4{-}(CH_2)_n{-}CHO$$

10: n = 7[192]
11: n = 6[192]
12: n = 5[192,36]
13: n = 4[36]

$$H_3CCH_2(CH{=}CH)_4{-}(CH_2)_n{-}CO{-}CH_2CH_2OR$$

14: R = H, n = 5[193] **15**: R = Ac, n = 5[193]
16: R = H, n = 4[193] **17**: R = Ac, n = 4[193]
18: R = H, n = 3[193] **19**: R = Ac, n = 3[193]

10, **11**, and **12** were isolated from *Zoegea baldshuanica* C. Winkl. together with other polyenynes as a complex mixture. After reducing the aldehyde group with sodium borohydride, an extensive separation was possible, but there were still homologous series, being assigned by mass spectroscopic measurements.

From the leaves of *Centaurea lippii*, the tetraen-aldehyde **13** was isolated and converted by reducing with sodium borohydride to the crystallising alcohol **20**. Hydrogenation of **20** yielded *n*-pentadecanol.

$$\mathbf{13} \xrightarrow{NaBH_4} H_3CCH_2{-}(CH{=}CH)_4(CH_2)_4CH_2OH \quad \mathbf{20}$$

Oxidative degradation yielded propanal, which was characterised as 2,4-dinitrophenylhydrazone.

14–20 have been found in *Saussurea pectinata* Bunge. The non-polar fraction of the leaf extract shows bands in the i.r. spectrum at 1750 and 1730/cm attributable to a keto-acetate. These data and the data of the n.m.r. spectrum strongly suggest the partial structure $-CH_2-CO-CH_2CH_2OAc$ (triplet 7·63τ (J = 6 Hz); triplet 5·79τ (J = 6·5 Hz), and singlet 8·06τ). Furthermore the presence of an end-standing ethyl-group is recognisable by a triplet at τ 9·0 and a multiplet at τ 7·85. Eight olefinic protons give rise to a complex multiplet centred at τ 3·4. As there is a multiplet at τ 8·7, which is difficult to interpret, very little information is available on the chain length of the compounds. The presence of a mixture of homologous keto-acetates is confirmed by the mass spectrum. All fragments occur threefold, respectively, for **13**, **17** and **19**. (See scheme for the keto-acetate **19**.)

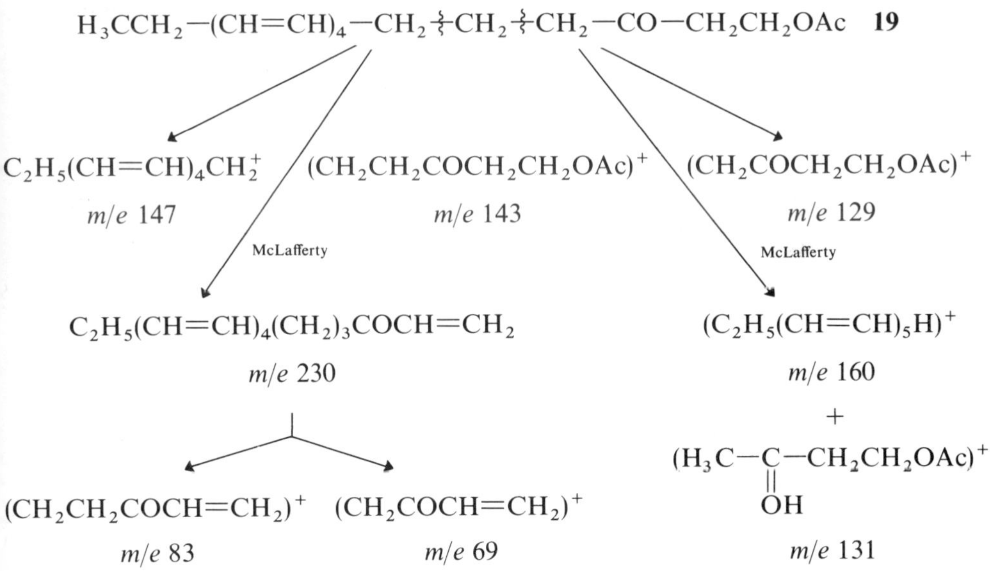

A similar fragmentation is observed for **15** and **17**. Peaks at *m/e* 188 and *m/e* 174 are smaller than the corresponding peak *m/e* 160. The intense peak at *m/e* 160 is caused by the favoured McLafferty rearrangement, followed by the formation of a conjugated pentaene.

The polar fraction contains the corresponding keto-alcohols **14**, **16** and **18**. Their fragmentation follows the same pattern. For instance compound **18** shows the following fragments:

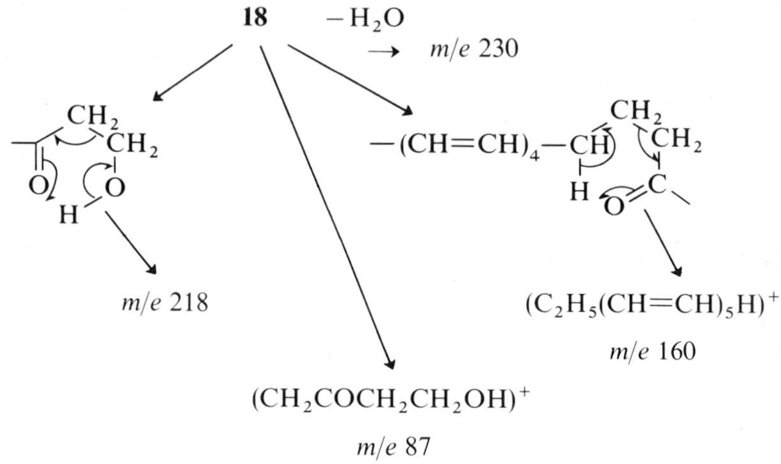

At first sight, the tetraene grouping seems to be somewhat unusual. For their formation an allene isomerisation seems to be very likely. The biogenesis of these compounds probably follows the sequence shown below:

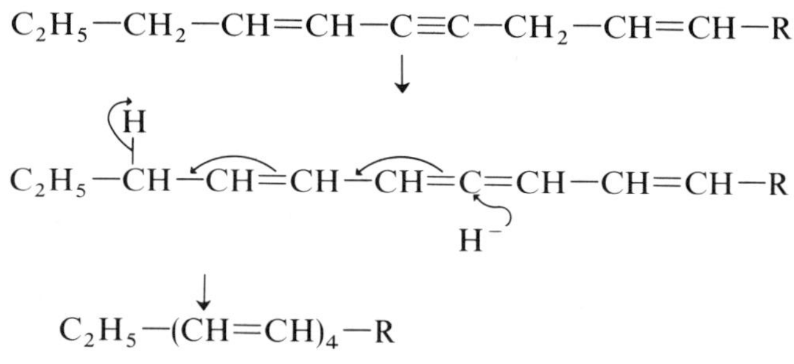

Obviously, the tetraenes are by-products in the oxidative degradation of C_{18}- or C_{17}-compounds.

1.2. C_{17}-acetylenes

1.2.1. General considerations of the biogenesis of the C_{17}-polyacetylenes

The multiplicity of the C_{17}-polyacetylenic compounds can be distributed into special groups basing on a distinct chromophoric system in the "distant" part of the molecule (i.e. the part further from the carboxyl group). The main groups are:

(A) $C_3H_7—CH=CH—C≡C—R$

(B) $H_3C—(CH=CH)_2—C≡C—R$

(C) $C_3H_7—(C≡C)_2—R$

(D) $H_3C—CH=CH—(C≡C)_2—R$

The entire range of the C_{17}-polyacetylenes can be derived from oleic acid (**4**) via linoleic acid (**5**) and its acetylenic 12,13-dehydro derivative, crepenynic acid (**1**), first described in the seed oil of *Crepis foetida* L., as mentioned in Section 1.1.

The necessary transformations involve further dehydrogenation steps, as shown in the scheme below:

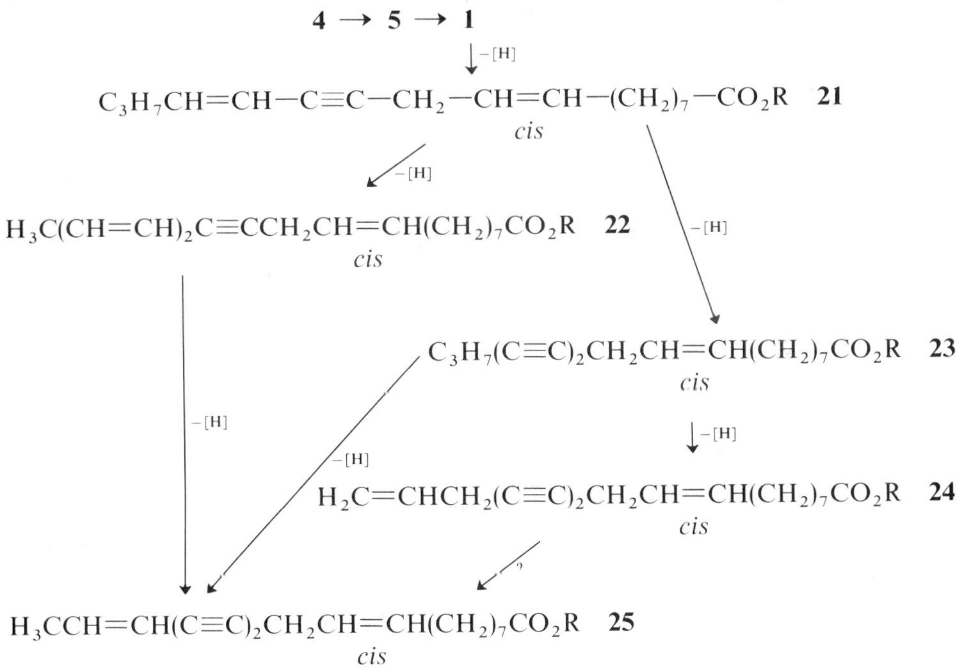

Another classification is given by the character of the endgroup in the "proximal" part of the molecules. Only in Compositae there are C_{17}-compounds with a terminal isolated ethylenic bond, while in Umbelliferae these endgroups are saturated.

It is proposed that chain shortening intervenes by β-oxidation of the C_{18}-acid and loss of CO_2, as shown in the scheme below:

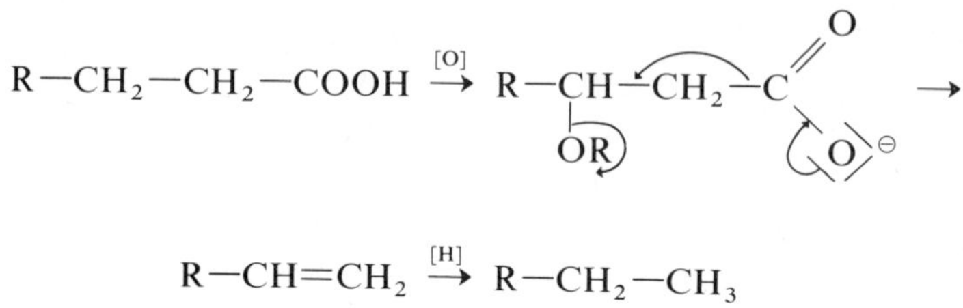

Typical endgroup for compounds isolated from Compositae

Typical endgroup for compounds isolated from Umbelliferae

Support for this contention comes from the results of labelling techniques. It has been shown, that β-hydroxy-oleic acid is incorporated into dehydro-falcarinone (**52**) (see Section 1.2.5). However, more evidence about the sequence must be sought and, in particular, the stage at which the chain-shortening reaction intervenes needs to be established. As outlined in the scheme above, the saturated endgroup may well arise by hydrogenation of the terminal vinyl group.

1.2.2. Compounds related to dehydro-crepenynic acid

Dehydro-crepenynic acid (**3**) itself has not been isolated from higher plants until now. As mentioned in Section 1.1, **3** has been found to be a precursor for some special polyacetylenes like phenylheptatriyne and a spiroketal-enol-ether. In the biogenesis of the poly-en-ynes, **3** seems to be rapidly converted into further intermediates, but there are some C_{17}-polyacetylenes which might arise from further oxidation and dehydrogenation, respectively, from **3**.

A complex mixture of polyacetylenes with an en-yne-diene-chromophore has been isolated from *Aster schreberi* Nees[194]. The purification of the mixture presented difficulties. In the least polar fraction, beside terpene-hydrocarbons, a mixture of two en-yne-diene-hydrocarbons was isolated. The i.r. spectra display bands for a *trans-trans*-diene-grouping (990, 957, and 910/cm). The mass spectrum of the mixture exhibits molecular ions due to $C_{17}H_{24}$ and $C_{17}H_{22}$. The n.m.r. spectrum consists of complex multiplets in the olefinic region, which are difficult to interpret. In view of these spectral results, it is only possible that the unknown main compound has the structure **26**. This structural proposal was confirmed by the reaction with maleic anhydride. Ozone cleavage of the adduct yields butyraldehyde:

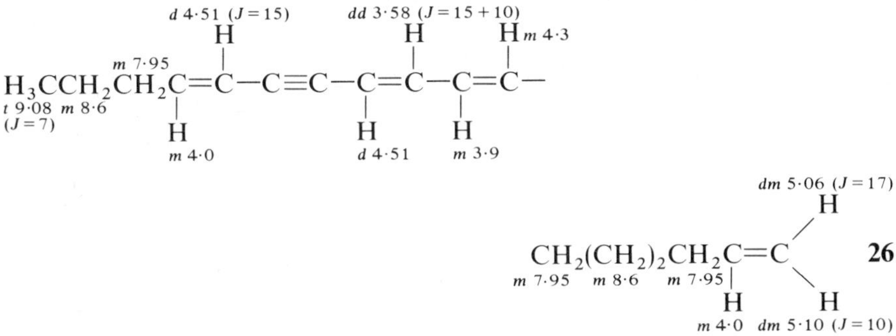

26

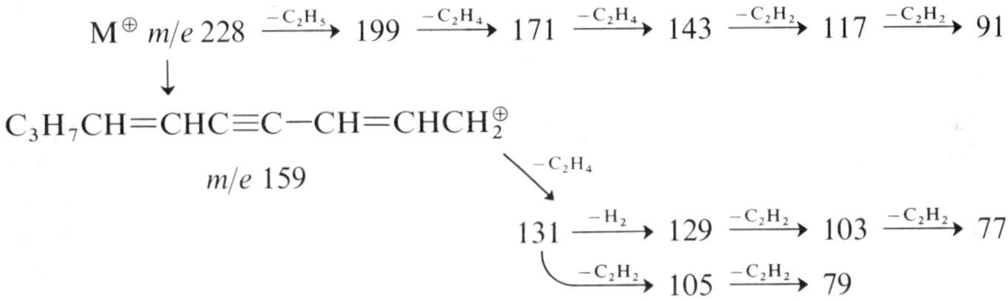

The structural assignment for the second hydrocarbon (**27**), containing an additional double bond, is derived from its mass spectrum. The metastable peak at *m/e* 111·9 indicates that *m/e* 159 arises from *m/e* 226. This fragmentation pattern is favoured in compounds like **27**. The intense peak at *m/e* 67 is caused by the cation $H_2C{=}CH{-}CH{=}CH{-}CH_2^+$[194]. The isolation of the corresponding alcohol (**35**) (see below) furthermore confirms the proposed structure of **27**.

$$\underset{t}{C_3H_7{-}CH{=}CH}{-}C{\equiv}C{-}\underset{t,t}{(CH{=}CH)_2}{-}CH_2{-}CH_2{-}(CH{=}CH)_2H \quad \mathbf{27}$$

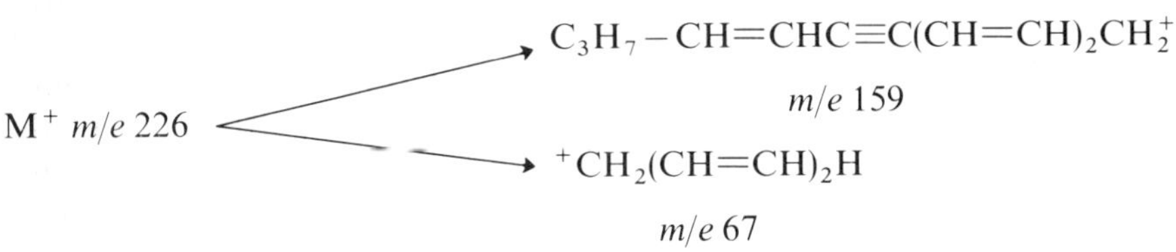

The more polar fraction of the plant extract contained a mixture of some alcohols, which could be separated only incompletely from each other. Only from the most polar fraction, purified by repeated TLC, one compound was obtained in a pure state. Oxidation with manganese dioxide yielded an aldehyde (2760, 1700, and 1690/cm). The substance was thus deduced to be a long chain allylic alcohol. The i.r. spectra of the alcohol and the aldehyde agree with the presence of a *trans-trans*-diene-grouping and an additional trans-configurated double bond. The n.m.r. spectrum of the aldehyde exhibits a *cis*-double bond conjugated to the aldehyde function: $-CH_2-CH{=}CH-CHO$ (*dtd* 7·38τ (J = 7·5, 7, 1·5 Hz); *dt* 3·53τ (J = 10, 7·5 Hz); *m* 4·0τ; *d* 0·02τ (J = 7·5 Hz)). The alcohol was converted into the acetate. Ozone-cleavage of the acetate yielded butyre aldehyde. These data and reaction results strongly indicate that the aldehyde is **29** and accordingly the original alcohol is **28**. The mass spectra of both confirm the structures:

$$\underset{t}{H_3CCH_2CH_2CH{=}CH}-C{\equiv}C-\underset{t,t}{(CH{=}CH)_2}CH_2CH_2CH_2\underset{c}{CH{=}CH}-CH_2OH \quad \mathbf{28}$$

$$M^+\ m/e\ 244 \xrightarrow{-CH_2CH=CHCHO} 173 \xrightarrow{-C_2H_2} 145 \xrightarrow{-H_2} 143 \xrightarrow{-C_2H_2} 117$$

$$M^+ \searrow C_3H_7CH{=}CHC{\equiv}C(CH{=}CH)_2CH_2^+\ (m/e\ 159) \xrightarrow{-H_2} 157 \xrightarrow{-C_2H_2} 131 \xrightarrow{-H_2} 129 \xrightarrow{-C_2H_2} 103 \xrightarrow{-C_2H_2} 77$$

$$M^+ \xrightarrow{\text{McLafferty}} C_3H_7CH{=}CHC{\equiv}C(CH{=}CH)_3H^+\ (m/e\ 172)$$

$$H_3CCH_2CH_2CH{=}CH-C{\equiv}C-(CH{=}CH)_2CH_2CH_2CH_2\underset{c}{CH{=}CH}-CHO \quad \mathbf{29}$$

$$M^+\ m/e\ 242 \rightarrow 172 \rightarrow 143 \rightarrow 117 \rightarrow 91$$

$$M^+\ m/e\ 242 \rightarrow 159 \rightarrow 131 \rightarrow 129 \rightarrow 128$$

The less polar alcohol fraction yielded a mixture of ketones after oxidation with manganese dioxide; this was separated only in part. The more polar oxidation product is deduced by spectral methods to be the ketone **30** and accordingly the original alcohol is **31**. Sodium borohydride reduction converts **30** into **31**.

$$C_3H_7-\underset{t}{CH{=}CH}-C{\equiv}C-\underset{t,t}{(CH{=}CH)_2}CH_2CH_2CH_2CO-CH{=}CH_2 \quad \mathbf{30}$$

$$C_3H_7-\underset{t}{CH{=}CH}-C{\equiv}C-\underset{t,t}{(CH{=}CH)_2}CH_2CH_2CH_2-\underset{\displaystyle OH}{CH}-CH{=}CH_2 \quad \mathbf{31}$$

It seemed desirable to prove the structures of **28** and **31** by synthesis. The route chosen[195] involved preparation of the aldehydes **A** and **B**, which were obtained by the following routes:

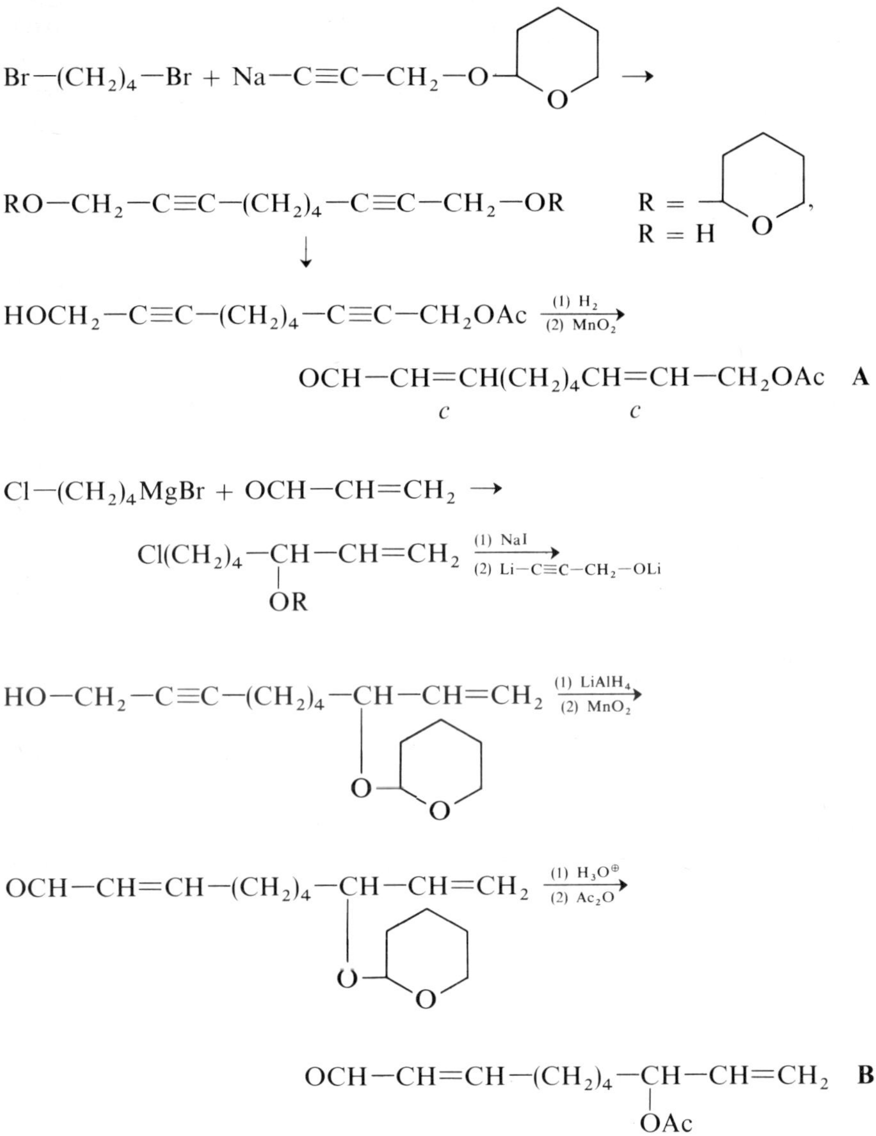

The magnesium bromide derivative of hept-3-en-1-yne was condensed with the aldehydes **A** and **B** to give the corresponding carbinols after allylic rearrangement. The carbinols were converted into the corresponding bromides and dehydrohalogenated with 1,5-diaza-bicyclo-(5,4,0)-undec-5-ene to the acetates of **28** and **31**. Saponification of the acetates yielded the isomeric alcohols **28** and **31**, identical with the natural compounds.

$$\mathbf{A} + C_3H_7CH{=}CH{-}C{\equiv}C{-}MgBr \rightarrow$$

$$C_3H_7CH{=}CH{-}C{\equiv}C{-}\underset{\displaystyle OH}{\underset{|}{CH}}{-}CH{=}CH{-}$$

$$(CH_2)_4CH{=}CH{-}CH_2OAc \longrightarrow$$

$$C_3H_7{-}CH{=}CH{-}C{\equiv}C{-}(CH{=}CH)_2{-}(CH_2)_3{-}\underset{cis}{CH{=}CH}{-}CH_2OAc$$

$$\downarrow OH^{\ominus}$$

$$\mathbf{28}$$

$$\mathbf{B} + C_3H_7CH{=}CH{-}C{\equiv}C{-}MgBr \longrightarrow$$

$$C_3H_7CH{=}CH{-}C{\equiv}C{-}\underset{\displaystyle OH}{\underset{|}{CH}}{-}(CH_2)_4{-}\underset{\displaystyle OAc}{\underset{|}{CH}}{-}CH{=}CH_2 \xrightarrow[\substack{(2)\ -HBr\\(3)\ OH^-}]{(1)\ PBr_3} \mathbf{31}$$

The less polar oxidation product of the alcohol fraction consists in a mixture of two ketones, which could not be separated. The mass spectrum, however, exhibits molecular ions due to $C_{17}H_{22}O$ and $C_{17}H_{20}O$, respectively. The structures of **32** and **34** were assigned on the basis of i.r., n.m.r., and mass spectra. The original compounds were received by reduction with sodium borohydride, therefore the structures of the original alcohols had to be **33** and **35**:

$$H_3CCH_2{-}\underset{\displaystyle OH}{\underset{|}{CH}}{-}\underset{t}{CH{=}CH}{-}C{\equiv}C{-}(CH{=}CH)_2{-}(CH_2)_4{-}CH{=}CH_2 \quad \mathbf{33}$$

$$MnO_2 \downarrow\uparrow NaBH_4$$

$$H_3CCH_2{-}CO{-}CH{=}CH{-}C{\equiv}C{-}(CH{=}CH)_2{-}(CH_2)_4{-}CH{=}CH_2$$

$$\mathbf{32}$$

$$H_3CCH_2{-}\underset{\displaystyle OH}{\underset{|}{CH}}{-}CH{=}CH{-}C{\equiv}C{-}$$

$$(CH{=}CH)_2{-}CH_2CH_2{-}(CH{=}CH)_2H \quad \mathbf{35}$$

$$\downarrow MnO_2$$

$$H_3CCH_2{-}CO{-}CH{=}CH{-}C{\equiv}C{-}$$

$$(CH{=}CH)_2{-}CH_2CH_2{-}(CH{=}CH)_2H \quad \mathbf{34}$$

From the aerial parts of *Aster schreberi* Nees, besides **26**, **27**, **28**, **31**, **33** and **35**, very small amounts of two polyacetylenes with a dien-yne chromophore (λ max 278, 267 nm) were isolated. The i.r. spectrum displays bands for a saturated ketone function at 1730/cm, a *trans-trans*-diene-grouping at 990/cm and a vinyl group at 910/cm. The n.m.r. spectrum confirms the presence of the following groupings: $CH_3CH_2{-}CO{-}CH_2CH_2{-}$: *t* 8·97τ ($J$ = 7 Hz); *q* 7·65τ (J = 7 Hz); *m* 7·48τ (4H); $={C}{-}(CH_2)_2{-}C=$: *m* 7·84τ and $=C{-}(CH_2)_4{-}CH{=}CH_2$: *m* 7·9τ (4H); *m* 8·5τ (4H); *m* 4·0τ (1H); *dd* 5·08τ (1H) (J = 10, 1·5); *dd* 4·98τ (1H) (J = 16, 1·5).

These data and, in particular, the mass spectrometric fragmentation pattern, strongly indicate that the ketones have the structures **36** and **37**:

$$H_3CCH_2{-}CO{-}CH_2CH_2{-}C{\equiv}C{-}\underset{t,t}{(CH{=}CH)_2}{-}CH_2CH_2{-}(CH{=}CH)_2H \quad \mathbf{36}$$

$$H_3CCH_2{-}CO{-}CH_2CH_2{-}C{\equiv}C{-}(CH{=}CH)_2{-}(CH_2)_4{-}CH{=}CH_2 \quad \mathbf{37}$$

Further evidence for these structures was obtained by converting **36** and **37** to the corresponding alcohols.

The spectrum of **26**, **27**, **28**, **31**, **33**, **35**, **36** and **37** permits the following interpretation of the biogenetical route. The precursor for all these substances is oleic acid (**4**); via **5**, **1** and **21**, they are formed by several oxidation and dehydrogenation steps, as shown in the scheme below (see also Section 1.2.1):

4 → **5** → **1**

↓

$$C_3H_7{-}CH{=}CH{-}C{\equiv}C{-}CH_2{-}CH{=}CH{-}(CH_2)_7{-}CO_2R \quad \mathbf{21}$$

↓ [O]

$$C_3H_7CH{=}CH{-}C{\equiv}C{-}CH{=}CH{-}\underset{OH}{CH}{-}(CH_2)_7{-}CO_2R \quad \mathbf{38}$$

↓ −[H₂O]

$$C_3H_7CH{=}CH{-}C{\equiv}C{-}(CH{=}CH)_2{-}(CH_2)_6{-}CO_2R \quad \mathbf{39}$$

↓ [O]

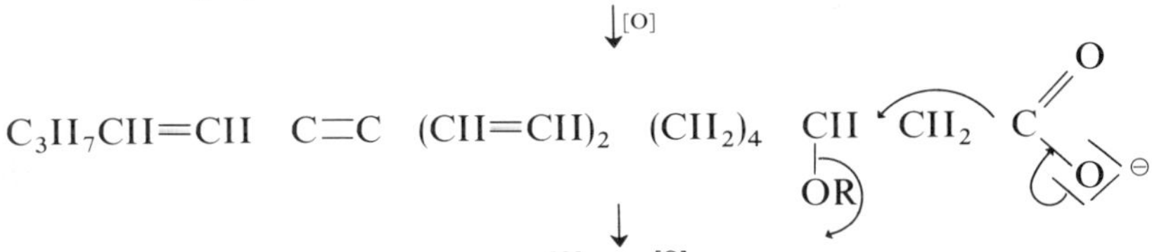

40

↓

33 ←[O]— **26** —[O]→ **31** **36**

33 —[O], [H]↓ **37**; **26** —[O]↓ **28**; **31** —−[H₂O]↓ **27**; **35** —[O], [H]↑ **36**

37 **28** **27** —[O]→ **35**

The stage at which chain shortening occurs is not yet clear.

The *cis*-configuration of the 2,3-double bond in **28** is remarkable. As it is proposed that **28** is formed from **26** or **31**, a *trans*-configuration double bond would have been expected, as observed in many other cases.

The occurrence of the saturated ketones **36** and **37** is very interesting too, as it seems very likely that they are formed from the corresponding conjugated ketones by biological hydrogenation. This is shown in similar examples by labelling experiments:

$$\mathbf{33} \xrightarrow{[O]} C_2H_5{-}CO{-}CH{=}CH{-}C{\equiv}C{-}R \xrightarrow{[H]} \mathbf{37}$$

1.2.3. Compounds containing a dien-yne-diene chromophore

In the preceding section (3.1.2.2) compounds with a propyl-en-yne-grouping were described. Further dehydrogenation leads to a methyl dien-yne-grouping (see Section 3.1.2.1). There are only a few poly-en-ynes, which may be derived from a compound related to **22**.

From the leaves of *Serratula gmelini* Ldb.[196], some homologous aldehydes (**41**, **42** and **43**) were obtained. The structures were elucidated by spectral data; for instance, the i.r. shows the presence of an aldehyde group and the all-*trans*-configuration of the double bonds (995/cm). The mass spectrum exhibits peaks due to a homologous series of aldehydes with the chain lengths C_{16}, C_{17} and C_{18}, where at the C_{17}-aldehyde **42** is the main product. Reduction with sodium borohydride yielded the corresponding alcohols **44**, **45** and **46**, which also occur in small amounts in nature. The C_{16}-alcohol **46** is known from another plant, *Dahlia merckii* Lehm.[197]

$$H_3C(\underset{t,t}{CH{=}CH})_2{-}C{\equiv}C{-}(\underset{t,t}{CH{=}CH})_2{-}(CH_2)_n{-}CHO$$

41: n = 6
42: n = 5
43: n = 4

42 M$^+$ *m/e* 242

↓ and →

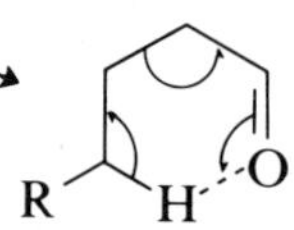

$$H_3C(CH{=}CH)_2C{\equiv}C(CH{=}CH)_2CH_2^+$$ *m/e* 157

↓ $-CH_3$

m/e 142

$$H_2C{=}CHOH^+$$ *m/e* 44

$$H_3C(CH{=}CH)_2{-}C{\equiv}C{-}(CH{=}CH)_2{-}(CH_2)_n{-}CH_2OH$$

44: n = 6
45: n = 5
46: n = 4

The existence of the C_{18}-keto-acetate **6** with the same chromophoric system (see Section 1.1) permits the interpretation that all these compounds arise from C_{18}-acetylenes by α-oxidation:

$$R{-}CH_2CH_2CO_2H \xrightarrow{[O]} R{-}CH_2{-}\underset{\displaystyle OH}{\underset{|}{CH}}{-}CO_2H \rightarrow$$

$$R{-}CH_2CHO \xrightarrow{[H]} RCH_2CH_2OH$$

Considering the structural relationship, the following biogenetic pathway seems to be very likely (see also Section 1.2.1):

$$\mathbf{4} \rightarrow \mathbf{5} \rightarrow \mathbf{1} \rightarrow \mathbf{21} \rightarrow \mathbf{22} \rightarrow \underset{\displaystyle \mathbf{44}}{\underset{\downarrow}{\mathbf{41}}} \rightarrow \underset{\displaystyle \mathbf{45}}{\underset{\downarrow}{\mathbf{42}}} \rightarrow \underset{\displaystyle \mathbf{46}}{\underset{\downarrow}{\mathbf{43}}}$$

The leaf extract of *Tridax trilobata* Hemsl.[154] contains, *inter alia*, a complex mixture of different carbonyl compounds, whose separation and purification had presented difficulties. Very small amounts of a ketone were isolated, but it could not be obtained in a pure state, although it had been subjected to repeated TLC. The mass spectrum leads to the molecular formula $C_{17}H_{20}O$. The ketone was reduced with sodium borohydride and the alcohol obtained showed the characteristic u.v. absorption bands of an en-yne-diene chromophore. Manganese dioxide oxidation, however, did not yield the original ketone, but a dihydro-ketone which showed a dien-yn-enone chromophore in the u.v. In view of these spectral results, the unknown compound must have structure **47**:

$$H_2C{=}CH{-}CO{-}CH{=}CH{-}C{\equiv}C{-}(CH{=}CH)_2{-}(CH_2)_4{-}CH{=}CH_2$$

47

$$\mathbf{47} \xrightarrow{NaBH_4} H_3CCH_2{-}\underset{\displaystyle OH}{\underset{|}{CH}}{-}\underset{t}{CH{=}CHC}{\equiv}\underset{t,t}{C(CH{=}CH)_2}(CH_2)_4CH{=}CH_2 \quad \mathbf{33}$$

$$\downarrow MnO_2$$

$$\underset{\substack{t\ 8{\cdot}81\\(J=7\ Hz)}}{H_3C}{-}\underset{q\ 7{\cdot}51}{CH_2}{-}CO{-}\underset{m\ 3{\cdot}3\text{–}4{\cdot}3}{CH{=}CH}{-}C{\equiv}C{-}$$

$$\underset{m\ 3{\cdot}3\text{–}4{\cdot}3}{(CH{=}CH)_2}{-}\underset{m\ 7{\cdot}93}{CH_2}{-}\underset{m\ 8{\cdot}6}{(CH_2)_2}{-}\underset{m\ 7{\cdot}93}{CH_2}{-}CH{=}C\begin{matrix}\diagup H^{dm\ 5{\cdot}1}\\ \diagdown H_{dm\ 5{\cdot}08}\end{matrix} \quad \mathbf{32}$$

Obviously, sodium borohydride has also attacked the reactive double bond in **47**.

Furthermore the carbonyl fraction contains small amounts of two *cis, trans* isomeric aldehydes, yielding alcohols when reduced with sodium borohydride. The alcohols give the typical u.v. spectrum of a dien-yne-diene-chromophore. The structures of the aldehydes were assigned doubtlessly on the basis of n.m.r. and mass spectra, but there was no evidence about the configuration. An answer was obtained by examinations of model substances. According to that, the aldehydes were deduced to be **48** and **49**.

$$\underset{d\,0{\cdot}53}{\mathrm{OCH}}-\underset{t,t}{(\mathrm{CH{=}CH})_2}-\mathrm{C{\equiv}C}-(\mathrm{CH{=}CH})_2-(\mathrm{CH_2})_4-\mathrm{CH{=}CH_2} \qquad \mathbf{48}$$

$$\underset{d\,0{\cdot}66}{\mathrm{OCH}}-\underset{c,t}{(\mathrm{CH{=}CH})_2}-\mathrm{C{\equiv}C}-\mathrm{R} \qquad \mathbf{49}$$

The presumed biogenesis of **48** and **49** was confirmed by tracer studies. (^{3}H)-labelled **22** has been transformed into **48**[35]. This experimental result agrees with the assumption about the biogenetic route, as described in Section 1.2.1.

$$\mathbf{4} \rightarrow \mathbf{5} \rightarrow \mathbf{1} \rightarrow \mathbf{21} \rightarrow \mathrm{H_3C(CH{=}CH)_2C{\equiv}CCH_2CH{=}CH(CH_2)_7CO_2R} \quad \mathbf{22}$$

$$\mathbf{21} \xrightarrow{?} \mathbf{47} \qquad \mathbf{22} \xrightarrow{[\mathrm{O}],\ -[\mathrm{CO_2}],\ [\mathrm{H_2O}]} \mathbf{48}, \mathbf{49}$$

It may be presumed that **47** too is derived naturally from **22** or **21**, although there are no experimental supports until now. In particular, it is not yet clear whether the following transformations are possible:

$$\mathrm{H_3C{-}CH{=}CH{-}CH{=}CH{-}C{\equiv}C{-}R}\ \mathbf{21} \xrightarrow{\mathrm{B}} \mathrm{H_2C{=}CH{-}CH_2{-}CH{=}CH{-}C{\equiv}C{-}R} \xrightarrow{[\mathrm{O}]} \mathrm{H_2C{=}CH{-}CO{-}CH{=}CH{-}C{\equiv}C{-}R}$$

$$\mathbf{21} \xrightarrow{\mathrm{A}} \mathrm{HOCH_2{-}(CH{=}CH)_2{-}C{\equiv}C{-}R} \xrightarrow{[\mathrm{O}]} \mathrm{H_2C{=}CH{-}CO{-}CH{=}CH{-}C{\equiv}C{-}R}$$

Two mechanistic routes could be postulated, which would lead to **47**. The first of these (A) would involve an allylic rearrangement. This problem needs to be investigated by tracer studies.

1.2.4. Compounds containing a diyne-CH_2-ene grouping

As mentioned in Section 1.2.1, octadeca-(12,14)-diyn-(9)-ene-ynoic acid (**23**) probably arise by dehydrogenation from dehydrocrepenynic acid (**21**):

$$\mathrm{H_3CCH_2CH_2CH{=}CH{-}C{\equiv}C{-}R} \xrightarrow{-[\mathrm{H}]} \mathrm{H_3CCH_2CH_2{-}(C{\equiv}C)_2{-}R}$$

The conversion of the double bond into a triple bond could be achieved by dehydrogenation by structurally and stereochemically specific enzymes.

23 plays an important part in the reaction sequence leading to natural acetylenes. By means of ^{14}C- and ^{3}H-labelling techniques, it has been shown that some natural acetylenes arise from **23**. There are some C_{17}-compounds still containing the structural feature of **23**.

From the roots of *Chrysanthemum frutescens* L. **50** was isolated[198].

$$H_3C-CH_2-CH_2-(C{\equiv}C)_2-CH_2-\underset{cis}{CH{=}CH}-(CH_2)_5-CH{=}CH_2 \qquad \mathbf{50}$$

The structure was elucidated by spectral methods and confirmed by synthesis.

A corresponding 15-hydroxy-compound was isolated from *Coreopsis gigantea*[199].

$$H_3C-CH_2-\underset{\displaystyle OH}{\underset{|}{CH}}-(C{\equiv}C)_2-CH_2-\underset{cis}{CH{=}CH}-(CH_2)_5-CH{=}CH_2 \qquad \mathbf{51}$$

51 was oxidised with manganese dioxide to the corresponding ketone. The spectral characteristics of the ketone consist with the structure proposed for **51**.

Extracts of *Erodiophyllum elderi* F. Müller[200] contain small amounts of a non-polar acetylenic substance, the separation and purification of which was difficult. The substance closely resembled dehydrofalcarinone (**52**)—see Section 1.2.5—in its u.v. and i.r. absorption. The n.m.r. spectrum, however, strongly suggests a structure different from dehydrofalcarinone. There is indeed an isolated vinyl group: $-CH_2-CH{=}CH_2$: *m* 7·88τ (2H); *ddt* 4·26τ (1H)(J = 17·5, 10, 6 Hz); *m* 5·07τ (1H)(J = 17·5 Hz); *dm* 5·13τ (1H)(J = 10 Hz), but there are no signals for a conjugated terminal ethylenic bond.

$$H_2C{=}CH-CO-(C{\equiv}C)_2-CH_2-\underset{cis}{CH{=}CH}-(CH_2)_5-CH{=}CH_2 \qquad \mathbf{52}$$

dehydrofalcarinone

Instead of this, the n.m.r. spectrum confirms the presence of a propyl-endgroup: *t* 8·97τ (3H) (J = 7 Hz), *tq* 8·42τ (2H) (J = 7, 6·7 Hz), and *t* 7·66τ (2H) (J = 6·7 Hz). The last triplet at τ 7·66 could be ascribed to a methylene group adjacent to a diyne-grouping. Narrow multiplets at τ 3·88 and τ 7·38 indicate, together with the u.v. and i.r. data, the following grouping:

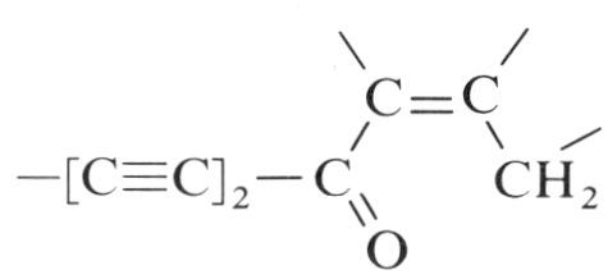

Besides a molecular ion peak at m/e 242, the mass spectrum of the compound exhibits an ion due to α-cleavage of the ketone at m/e 119 and furthermore loss of CO to m/e 91. The foregoing data indicate that the ketone is **53**:

$$H_3C{-}CH_2CH_2{-}(C{\equiv}C)_2{-}CO{-}\underset{cis}{CH{=}CH}{-}(CH_2)_5{-}CH{=}CH_2 \quad \mathbf{53}$$

Hydrogenation of the ketone **53** yields the corresponding *n*-heptadecanone-(8) (**54**), which was characterised by mass spectroscopy:

$$\mathbf{53} \xrightarrow{H_2} H_3C{-}(CH_2)_6{-}CO{-}(CH_2)_8{-}CH_3 \quad \mathbf{54}$$

$$\underset{m/e\ 127}{C_7H_{15}CO^+} \leftarrow M^+\ m/e\ 254 \rightarrow \underset{m/e\ 155}{C_9H_{19}CO^+}$$

$$\downarrow$$

$$\underset{m/e\ 142}{C_7H_{14}{-}\underset{^+OH}{C}{=}CH_2} \leftarrow$$

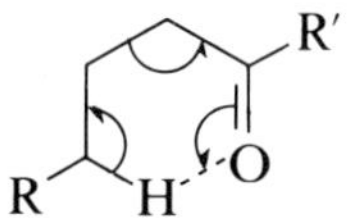

$$\rightarrow \underset{m/e\ 170}{C_9H_{19}{-}\underset{^+OH}{C}{=}CH_2}$$

The more polar fractions of *Erodiophyllum elderi* consist of a mixture of alcohols. Partial oxidation yielded only one ketone, identical with **53**; accordingly the original alcohol was **55**. Reduction of **53** formed an alcohol; the n.m.r. of this compound confirms the structure of **55**:

$$\underset{\substack{t\ 8{\cdot}98\\(J=7)}}{H_3C}{-}\underset{tq\ 8{\cdot}43}{CH_2}{-}\underset{t\ 7{\cdot}75}{CH_2}{-}(C{\equiv}C)_2{-}$$

$$\overset{H^{m\ 4{\cdot}94}}{\underset{\underset{s\ 8{\cdot}1}{OH}}{C}}{-}\underset{\substack{cis\\m\ 4{\cdot}55}}{CH{=}CH}{-}\underset{m\ 7{\cdot}95}{CH_2}{-}\underset{m\ 8{\cdot}65}{(CH_2)_3}{-}\underset{m\ 7{\cdot}95}{CH_2}{-}\underset{\substack{H\\ddt\ 4{\cdot}25\\(J=17+10{\cdot}6)}}{C}{=}C\begin{matrix}H^{dm\ 5{\cdot}08\ (J=17)}\\H_{dm\ 5{\cdot}12\ (J=10)}\end{matrix} \quad \mathbf{55}$$

The alcohols, which were not very rapidly oxidised by manganese dioxide, were separated by repeated TLC. The less polar substance displayed no distinct u.v. absorption. However, it showed an intense peak at 2260/cm in the i.r. Furthermore, the i.r. spectrum displayed bands for a hydroxyl group (3620/cm) and a terminal vinyl group (3080, 1645, 920/cm). The mass spectrum with a molecular ion peak at *m/e* 246, together with the n.m.r. data, strongly suggested the structure **56** for the alcohol:

$$H_3CCH_2CH_2{-}(C{\equiv}C)_2{-}\underset{\displaystyle OH}{\overset{\displaystyle H^{t\ 5\cdot61\ (J=6)}}{C}}{-}(CH_2)_7{-}CH{=}CH_2 \quad \mathbf{56}$$

Further confirmation was obtained by hydrogenation of **56** to the corresponding *n*-heptadecanol-(8), which was identified by mass spectroscopy:

$$\mathbf{56} \xrightarrow{H_2} C_7H_{15}{-}\underset{\displaystyle OH}{CH}{-}C_9H_{19} \quad \mathbf{57}$$

$$M^+\ m/e\ 256$$

$$C_7H_{15}CHOH^+\ (m/e\ 129) \leftarrow \quad \xrightarrow{-H_2O} m/e\ 238 \quad \rightarrow C_9H_{19}CHOH^+\ (m/e\ 157)$$

The third alcohol showed the typical u.v. absorption of an ene-diyne-chromophore. Oxidation with manganese dioxide yielded a ketone with an u.v. spectrum of a diyn-en-one (λ max: 310, 292 nm). The mass spectrum with a molecular ion peak at *m/e* 242 and major peaks at *m/e* 160 and 145—corresponding to the expected McLafferty rearrangement and α-cleavage of the ketone—furnished valuable confirmation of the structure **59**, accordingly the original alcohol is **58**:

$$H_3CCH_2CH_2{-}(C{\equiv}C)_2{-}\underset{t}{CH{=}CH}{-}\underset{\displaystyle OH}{CH}{-}(CH_2)_5{-}CH{=}CH_2 \quad \mathbf{58}$$

$$\downarrow MnO_2$$

$$C_3H_7{-}(C{\equiv}C)_2{-}CH{=}CH{-}CO{-}(CH_2)_5{-}CH{=}CH_2 \quad \mathbf{59}$$

$$M^+\ m/e\ 242$$

$$\xleftarrow{\text{McLafferty}} C_3H_7(C{\equiv}C)_2CH{=}CHCOCH_3]^{\oplus}\ (m/e\ 160) \qquad \rightarrow R{-}CO^+\ (m/e\ 145) \xrightarrow{-CO} m/e\ 117$$

The simultaneous occurrence of **55** and **58** in *Erodiophyllum elderi* F. Müller provides circumstantial evidence in support of the following sequence of cellular reactions:

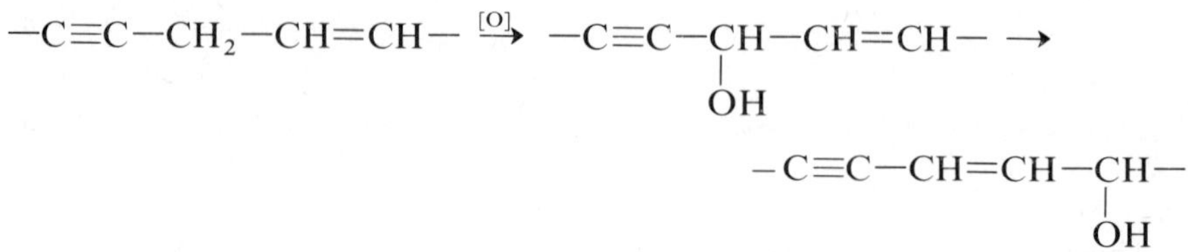

A hydrocarbon, which could well arise by dehydratisation from **58** was obtained from the aerial parts of *Coreopsis gigantea*[199]:

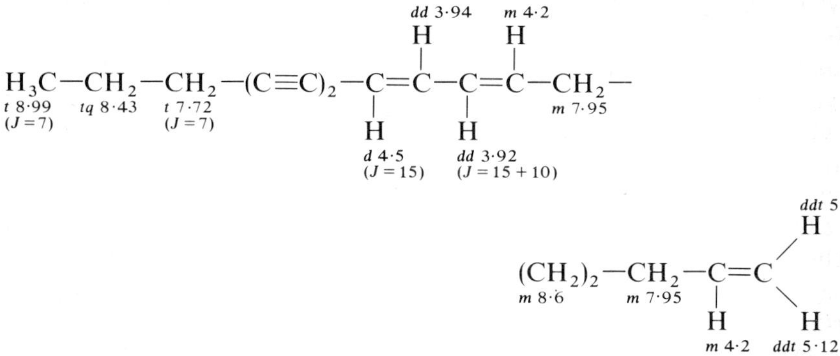

60

The structure of **60** was established by spectral methods. The corresponding 15-hydroxy-compound **61** was also isolated from the roots of *Coreopsis gigantea.*

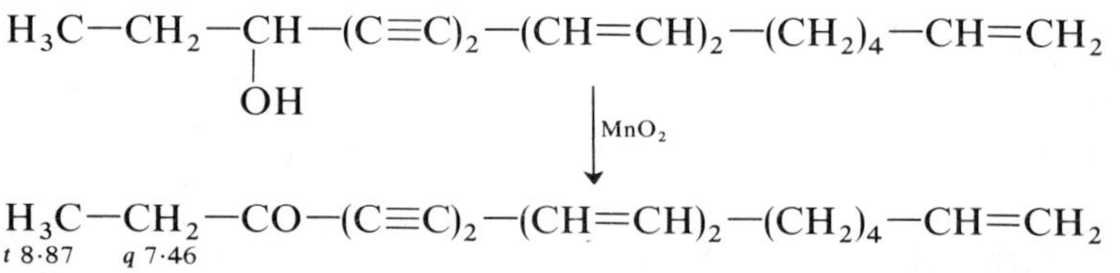

61

62

The structure of **61** was proved by n.m.r. structural analysis as well as mass spectroscopic measurements. Evidence for this structure was obtained by converting **61** to the corresponding ketone **62** by manganese dioxide oxidation.

Two polyacetylenic glycosides are derivable from **58**. From the roots of *Serratula gmelini* Ldb. some glycosides were isolated[196]. L-rhamnose and the monoacetate of L-rhamnose has been found to be linked as mono units to the aglycon **58**:

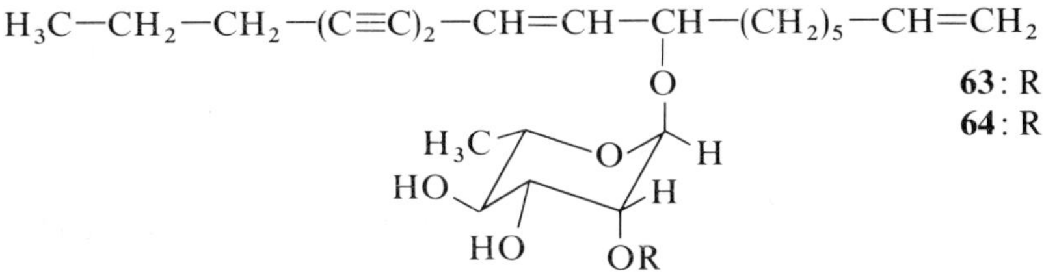

63: R = H
64: R = Ac

The purification and structure determination of the glycoside mixture had presented difficulties. The u.v. spectrum reveals a diyne-ene-chromophore. The polar properties and an intense hydroxyl band in the i.r. suggest that the compounds are polyols. Furthermore the i.r. displays bands for a terminal vinyl bond (1850, 1645, 925/cm) and a *trans*-disubstituted double bond (960/cm). Periodate cleavage yielded no product and manganese dioxide oxidation was negative too. The n.m.r. spectrum was not very helpful. The range of 5·5–7τ is a complex multiplet and does not allow any conclusion about the oxygen functional groups. Therefore the mixture was transformed by treatment with acetone and toluene *p*-sulphonic acid into the acetonides. From these and the triacetates, respectively, distinguishable n.m.r. spectra were obtained.

The acetonides could be separated by repeated TLC. High resolution mass spectra yielded the exact molecular formulae $C_{26}H_{38}O_5$ and $C_{28}H_{40}O_6$. Furthermore, a characteristic fragmentation pattern is recognisable, which must belong to a glycoside residue with a molecular constitution of $C_9H_{15}O_5$ and $C_{11}H_{17}O_6$, respectively, of the acetonides. These spectral evidences indicate that we deal with a deoxy-sugar and its corresponding mono-acetate. Experiments of glycoside cleavage under usual conditions gave no reactions. But alcoholysis with traces of toluene *p*-sulphonic acid yielded a mixture of a polyacetylenic alcohol, a polyacetylenic ethyl ether, and an ethyl glycoside. The structure of the latter was elucidated after transformation into the corresponding triacetate. The n.m.r. spectrum of the ethyl glycoside is only consistent with the ethylglycoside of L-rhamnose **65**.

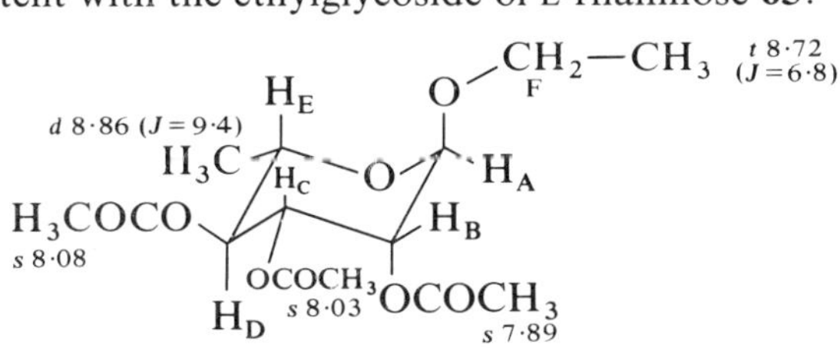

H_A *d* 5·39	H_E *dq* 6·15	J_{AB} = 1·7 Hz
H_B *dd* 4·96	H_{F_1} *dq* 6·49	J_{BC} = 3·3 Hz
H_C *dd* 4·85	H_{F_2} *dq* 6·27	J_{CD} = 9·5 Hz
H_D *dd* 5·15		J_{DE} = 9·5 Hz

Structure and configuration of the sugar residue were confirmed and proved by synthesis of **65**[196].

The n.m.r. spectrum, and in particular the mass spectrometric fragmentation pattern of the obtained ethyl ether, strongly indicates that the ether has the structure **66**:

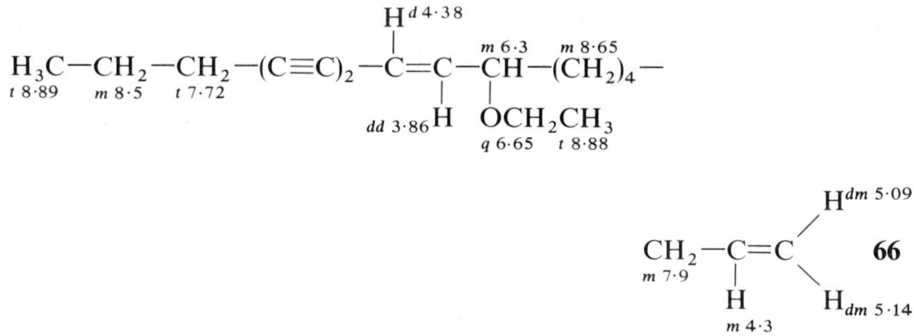

66

Base–peak in the mass spectrum is *m/e* 175, confirming the position of the oxygen functional group. The fragment *m/e* 131, occurring in the glycoside mass spectrum too, may be derived by hydride shift and simultaneous cleavage from *m/e* 243:

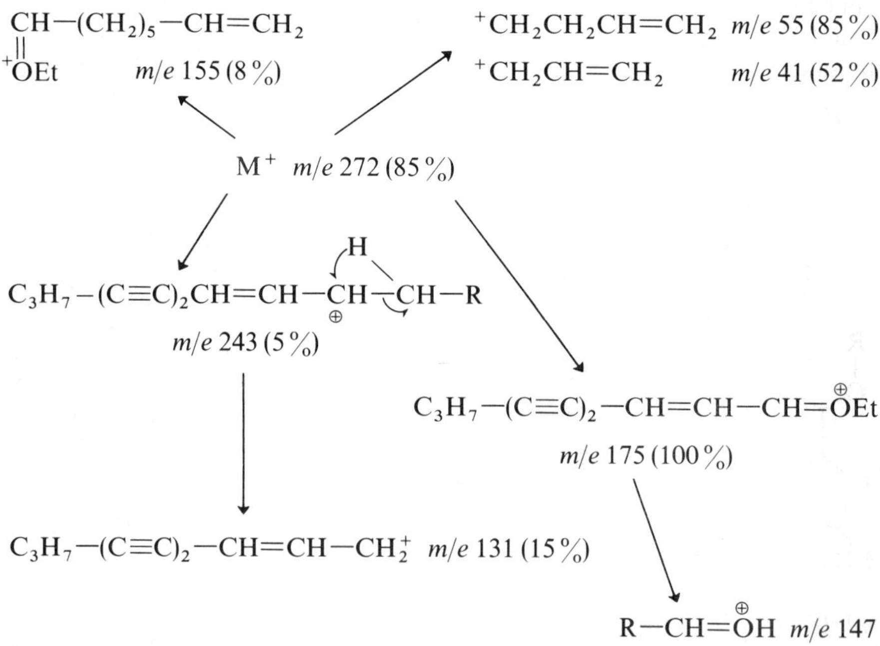

With the determination of the polyenyne chain and the sugar residue, the elucidation of the rhamnoside **63** was completed. The mass spectrum of **63** shows the following fragmentation pattern:

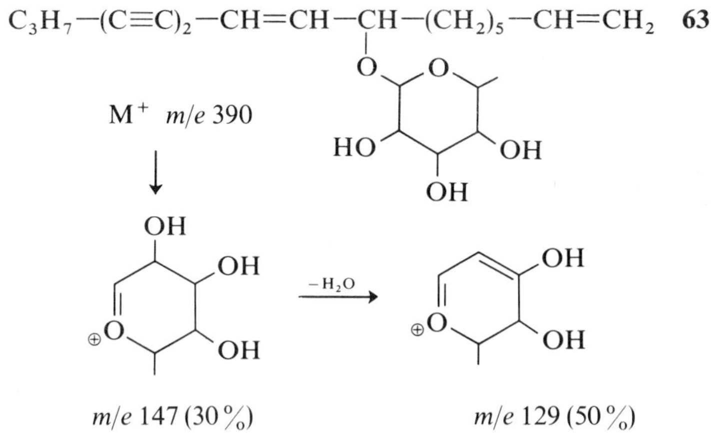

The acetonide **67** was the only derivative which was obtained from **63** in a pure state by repeated TLC. The n.m.r. spectrum of **67** is very similar to that of **63**; the only conspicuous difference is that the acetal proton is showing a singlet at τ 5·14. Obviously, a bond angle deformation is caused by the condensed five-membered ring, which leads to a detraction of the vicinal coupling constant. The mass spectrum of **67** shows the following characteristic fragments:

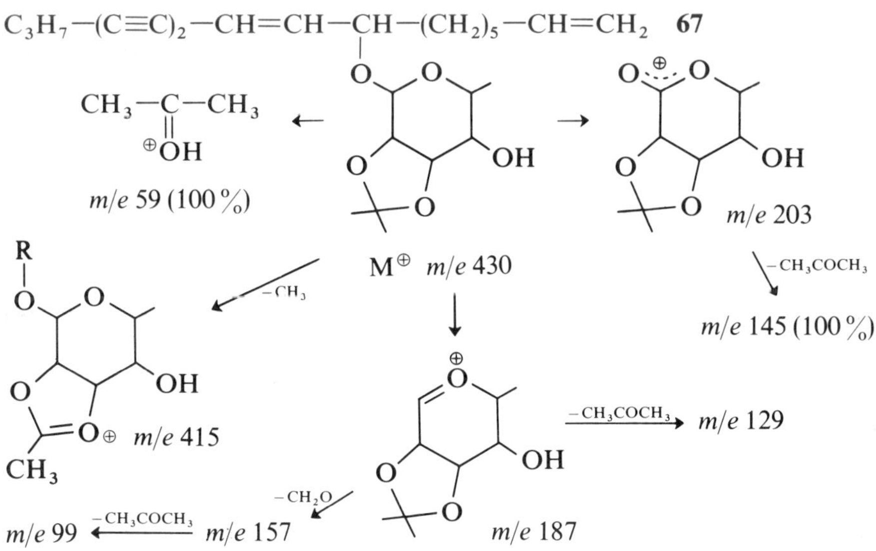

Together with the triol **63** the monoacetate **64** was isolated:

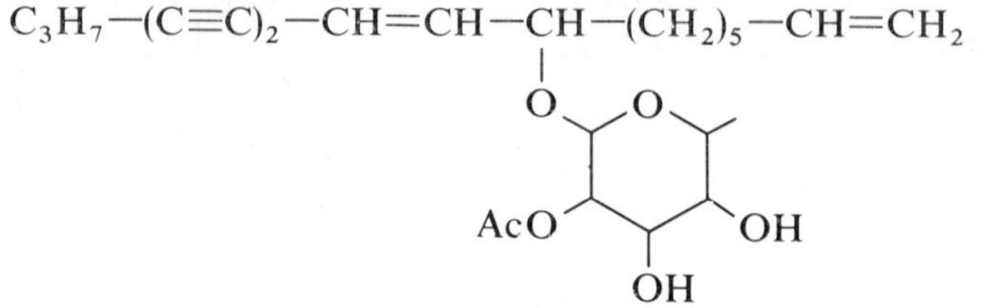

64

The position of the *O*-acetate grouping was determined by n.m.r. spectroscopy of the hydrogenated product. Acetylation of **64** yielded a triacetate, which was identical with that obtained from **63**.

1.2.5. Compounds related to dehydrofalcarinone

Examination of plants belonging to the family of Compositae—the largest and most widely dispersed family of flowering plants—reveals that many of them contain the C_{17}-ketone dehydrofalcarinone **52**:

$$H_2C{=}CH{-}CO{-}(C{\equiv}C)_2{-}CH_2{-}CH{=}CH{-}(CH_2)_5{-}CH{=}CH_2 \quad \mathbf{52}$$

The highly unstable polyacetylenic ketone has been isolated first from the roots of *Artemisia campestris* L.[105]. Purification of **52** is very difficult. In order to obtain **52** in a pure state, it was converted, using methanol and sulphuric acid, to the corresponding methanol adduct **69** and subsequently subjected to column chromatography on silica gel

$$\mathbf{52} \xrightarrow{CH_3OH/H_3O^+}$$

$$H_3CO{-}CH_2{-}CH_2{-}CO{-}(C{\equiv}C)_2CH_2CH{=}CH(CH_2)_5CH{=}CH_2 \quad \mathbf{69}$$

The structure of **69** was elucidated by u.v. and i.r. spectra and their relationship to dihydro-methoxyfalcarinone, as **69** gave spectra identical to those of the latter, up to an additional vinyl band in the i.r. at 915/cm (see Section 1.2.6).

When **69** was treated with toluene-*p*-sulphonic acid in benzene, **52** was formed and characterised by spectral methods. The compound shows absorption in the i.r. due to a terminal vinyl group (915/cm), acetylenic bonds (2240, 2150/cm), and an unsaturated ketone (1670/cm). Light absorption in diethyl ether: λ max 290·5, 274, 259, 245·5, and 208·5 nm; (ε = 9060, 10,400, 6240, 3840, and 17,600). The structure was finally established by synthesis[191]: Hept-(6)-en-ynoic acid **70** is reduced with lithium aluminium hydride to the corresponding alcohol **71**. This is converted into the bromo derivative **72**, which furnishes the acetylenic alcohol **74**; this is hydrogenated with Lindlar's catalyst to **75**. The toluene-*p*-sulphonate **76** is converted into the hydrocarbon **77** by a Grignard reaction. Cadiot–Chodkiewicz coupling[39] of **77** with 5-bromo-pent-(1)-en-(4)-yn-(3)-ol gives **78**, which is oxidised with manganese dioxide to **52**. The spectra and physical properties of synthesised **52** were identical with the natural product:

$$HO_2C(CH_2)_4CH{=}CH_2 \rightarrow R{-}(CH_2)_5{-}CH{=}CH_2 \rightarrow$$

70 **71**: R = OH; **72**: R = Br

$$RO{-}CH_2{-}C{\equiv}C{-}(CH_2)_5{-}CH{=}CH_2 \rightarrow$$

73: R = 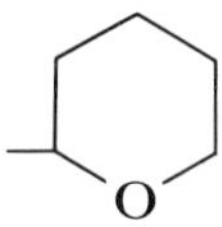 ; **74**: R = H

$$RO{-}CH_2{-}\overset{H}{C}{=}\overset{H}{C}{-}(CH_2)_5{-}CH{=}CH_2$$

75: R = H; **76**: R = p–Ts

$$H{-}C{\equiv}C{-}CH_2{-}CH{=}CH{-}(CH_2)_5{-}CH{=}CH_2$$

77

$$+\ H_2C{=}CH{-}\underset{OH}{CH}{-}C{\equiv}C{-}Br$$

$$H_2C{=}CH{-}\underset{OH}{CH}{-}(C{\equiv}C)_2{-}CH_2{-}CH{=}CH{-}(CH_2)_5{-}CH{=}CH_2 \xrightarrow{MnO_2} \mathbf{52}$$

78

Besides **52** and other polyacetylenes, **79** was isolated from *Artemisia scoparia* W. *et* K.[32].

$$\mathbf{52} \xrightarrow{[H_2O]} HOCH_2{-}CH_2{-}CO{-}(C{\equiv}C)_2CH_2CH{=}CH(CH_2)_5CH{=}CH_2 \quad \mathbf{79}$$

Using the spectral data, the compound was deduced to be a C_{17}-hydroxy-ketone. It is proposed, that **79** arises from **52** by hydratisation of the reactive $-CO{-}CH{=}CH_2$-grouping.

From the roots of a Compositae species, *Artemisia atrata* Lam., besides **52** the corresponding alcohol **78** has been isolated[201]:

$$H_2C{=}CH{-}\underset{OH}{CH}{-}(C{\equiv}C)_2{-}CH_2{-}CH{=}CH{-}(CH_2)_5{-}CH{=}CH_2 \quad \mathbf{78}$$

The structure of this oily compound, named dehydrofalcarinol, was elucidated by n.m.r. spectroscopy and confirmed by oxidation with manganese dioxide yielding **52**.

From the aerial parts of *Artemisia crithmifolia* L. a highly unstable acetylenic compound was isolated in traces. The compound showed a diyn-on-ene chromophore (292, 277, 262 nm) and must be a hydroxy-ketone of the dehydrofalcarinolone type[201]:

$$H_2C{=}CH{-}CO{-}(C{\equiv}C)_2{-}\underset{\displaystyle OH}{\underset{|}{CH}}{-}CH{=}CH{-}(CH_2)_5{-}CH{=}CH_2 \qquad \mathbf{80}$$

80 gave a typical i.r. spectrum (3625, 2240, 2150, 1655, and 916/cm) and manganese dioxide oxidation yielded a falcarindione-like chromophore:

$$\mathbf{80} \xrightarrow{MnO_2} H_2C{=}CH{-}CO{-}(C{\equiv}C)_2{-}CO{-}\overset{m\ 3{\cdot}5{-}4{\cdot}0}{CH{=}CH}{-}\ (cis)$$

$$(CH_2)_5{-}CH{=}CH_2 \qquad \mathbf{81}$$

The n.m.r. spectrum of **81** only agrees with the proposed structure of a diketone. The ethylenic protons give rise to complex signals in the olefinic region (τ 3·5–5). The keto-group deshields the protons of both the *cis*-double bond ($\tau < 4$) and the allylic methylene group (τ 7·4) and thus determine the position of the hydroxy group in **80**. The extremely unstable diketone **81** was also isolated from *Artemisia campestris* L. in traces, but without success in purification.

It seems most likely that there is a close relationship between the C_{17}-polyacetylenes **52**, **78**, **80**, and **81** in the biogenesis (see Section 1.2.1):

$$H_3C{-}(CH_2)_7{-}\underset{cis}{CH{=}CH}{-}(CH_2)_7{-}CO_2R \qquad \mathbf{4}$$

$$\downarrow [O]$$

$$H_3C{-}(CH_2)_7{-}CH{=}CH{-}(CH_2)_5{-}\underset{\displaystyle OH}{\underset{|}{CH}}{-}CH_2{-}CO_2R \qquad \mathbf{82}$$

$$\downarrow -[H]$$

$$C_3H_7{-}(C{\equiv}C)_2{-}CH_2{-}CH{=}CH{-}(CH_2)_5{-}\underset{\displaystyle OH}{\underset{|}{CH}}{-}CH_2{-}C({=}O)O^{\ominus} \qquad \mathbf{83}$$

$$\downarrow -[H], [O]$$

$$H_2C{=}CH{-}\underset{\displaystyle OH}{\underset{|}{CH}}{-}(C{\equiv}C)_2{-}CH_2{-}CH{=}CH{-}(CH_2)_5{-}CH{=}CH_2 \qquad \mathbf{78}$$

$$\downarrow [O]$$

$$\mathbf{52} \xrightarrow{[O]} \mathbf{80} \xrightarrow{[O]} \mathbf{81}$$

Experiments with labelled oleic acid **4** and β-hydroxy-oleic acid (**82**) provided evidence of the reality of β-oxidation to the terminal vinyl group. Both **4** and **82** were incorporated into **52**, but more evidence about the sequence of the steps and the specificity of the enzyme systems must also be sought.

A C_{17}-polyacetylenic hydrocarbon was isolated from the aerial parts of *Chrysanthemum maximum* Ramond and identified as **84**[202]:

$$\underset{cis}{H_3C{-}CH{=}CH}{-}(C{\equiv}C)_2{-}CH_2{-}\underset{cis}{CH{=}CH}{-}(CH_2)_5{-}CH{=}CH_2 \quad \mathbf{84}$$

The structure of **84** was determined by spectral data, using u.v., i.r., and n.m.r. spectroscopy. Light absorption gives typical absorption bands for an diyne-ene chromophore (281, 265, 238, and 226 nm), and the i.r. spectrum reveals the presence of a terminal vinyl group (915, 995, 1640/cm) and acetylenic bonds (2250/cm). As there is no signal in the range of 950/cm which is attributable to a *trans* double bond, **84** must contain a *cis* ene-diyne-chromophore. The n.m.r. spectrum of **84** supports this assignment. A double doublet at τ 8·12 (J = 1·5 Hz) is attributable to a $CH_3{-}CH{=}CH{-}$ grouping. A broad doublet at τ 6·01 (J = 5 Hz) indicates a methylene group between triple- and double bonds. A multiplet corresponding to four protons at τ 8·0 is most likely due to two methylene groups close to double bonds. A multiplet at τ 8·7 can be ascribed to three normal methylene groups. The olefinic region (τ: 4–5), however, is very complex and can be interpreted only incompletely. There is a multiplet at τ 5–5·25, showing the typical splitting of a terminal vinyl group. A doublet quartet at τ 4·0 (J = 11+7 Hz) can be assigned to an olefinic proton of the *cis*-propylene grouping. Catalytic hydrogenation of **84** yielded *n*-heptadecane, which was identified by GLC. The structure of **84** was finally confirmed synthetically by way of Cadiot–Chodkiewicz coupling of 1-bromo-pent-(1)-yn-(3)-ene with dodec-(1)-yne-(4,11)-diene:

$$H_3C{-}CH{=}CH{-}C{\equiv}C{-}Br$$

$$+\ H{-}C{\equiv}C{-}CH_2{-}CH{=}CH{-}(CH_2)_5{-}CH{=}CH_2 \rightarrow \mathbf{84}$$

The corresponding *trans*-compound **85** has been isolated from *Coreopsis gigantea*[199]:

$$\underset{trans}{H_3C{-}CH{=}CH}{-}(C{\equiv}C)_2{-}CH_2{-}\underset{cis}{CH{=}CH}{-}(CH_2)_5{-}CH{=}CH_2 \quad \mathbf{85}$$

In the root extract of *Solidago virgaurea* two C_{17}-polyacetylenes with ene-diyne chromophore were detected[203]. One of them is identical with **84**, the n.m.r. spectrum of the other (not quite pure) indicates epoxy-protons, vinyl-protons, and *cis*-double bond protons as well as a double doublet due to a *cis*-propenyl endgroup. The foregoing data lead to structure **86**:

$$H_3C{-}\underset{c}{CH{=}CH}{-}(C{\equiv}C)_2{-}CH_2{-}\underbrace{CH{-}CH}_{O}{-}(CH_2)_5{-}CH{=}CH_2 \quad \mathbf{86}$$

Mass spectra of both **84** and **86** were obtained by combined GLC–mass spectroscopy. The molecular peaks *m/e* 226 and 242, respectively, were in good agreement with the formula given.

The alcohol **87** can be isolated from an *Anthemis tinctoria* hybrid (Botanical garden, Bucharest)[85]:

$$R{-}\underset{c}{CH{=}CH}{-}(C{\equiv}C)_2{-}CH_2{-}\underset{c}{CH{=}CH}{-}(CH_2)_5{-}CH{=}CH_2$$

87: R = HO—CH$_2$—
88: R = OCH—

87 gives an i.r. spectrum consistent with the presence of a hydroxyl group and a vinyl group (3630, 925/cm). An ene-diyne-chromophore is established by the u.v. spectrum (λ max: 282·5, 267, 251·5 nm). As there was no success in purification **87** was converted, using manganese dioxide, to the corresponding aldehyde (**88**). The n.m.r. spectrum of **88** confirmed the proposed structure of **87**:

$$O{=}\underset{\substack{dd\ 0{\cdot}46\\(J=7+1{\cdot}2)}}{C(H)}{-}\underset{dd\ 3{\cdot}7}{CH}{=}\underset{\substack{dd\ 3{\cdot}44\\(J=11+1{\cdot}2)}}{CH}{-}(C{\equiv}C)_2{-}\underset{\substack{d\ 6{\cdot}91\\(J=5{\cdot}6)}}{CH_2}{-}\underset{m\ 4{\cdot}5}{CH{=}CH}{-}\underset{m\ 7{\cdot}9}{(CH_2)}{-}$$

$$\underset{m\ 8{\cdot}6}{(CH_2)_3}{-}\underset{m\ 7{\cdot}9}{CH_2}{-}\underset{m\ 4{\cdot}3{-}5{\cdot}2}{CH{=}CH_2} \quad \mathbf{88}$$

Catalytic hydrogenation and acylation of **87** yielded *n*-heptadecanol-(1)-acetate, as shown by GLC. The alcohol **89**, closely related to **87**, was isolated from *Silybum marianum* Gaertn. (tribus Cynareae, subtribus Carduinea)[204]:

$$H_3C{-}\underset{c}{CH{=}CH}{-}(C{\equiv}C)_2{-}\underset{OH}{CH}{-}\underset{c}{CH{=}CH}{-}(CH_2)_5{-}CH{=}CH_2 \quad \mathbf{89}$$

The structure was elucidated from its i.r. spectrum, which shows the presence of a terminal vinyl group (920, 1645/cm), and the *cis*-configuration of the double bonds (730/cm). Furthermore, there are bands attributable to a hydroxy group (3625/cm) and triple bonds (2240/cm). The ene-diyne chromophore is responsible for the u.v. absorption (λ max: 284, 268·5, 254, 246·5 nm).

Manganese dioxide oxidation, followed by hydrogenation, yields *n*-heptadecanone-(8). The very complex n.m.r. spectrum exhibits signals for the $CH_3{-}CH{=}CH{-}C{\equiv}$-grouping: *dd* τ 8·06 ($J = 7 + 1{\cdot}5$ Hz).

The occurrence of **89** and **84** in the same plant permits the interpretation that **84** is a precursor of **89**, which surely arises by allylic oxidation of the activated methylene group.

An acetylenic alcohol (**90**) with an ene-diyne chromophore (λ max: 282, 267, 253 nm) occurs in *Anthemis rudolfiana* hort.[85]:

$$H_3C{-}CH{=}CH{-}(C{\equiv}C)_2{-}\underset{OH}{CH}{-}\overset{O}{\widehat{CH{-}CH}}{-}(CH_2)_5{-}CH{=}CH_2 \quad \textbf{90}$$

The structure of **90** was elucidated by n.m.r. spectroscopy and confirmed by degradation reactions. The n.m.r. spectrum exhibits the following characteristic signals: a doublet corresponding to one proton at τ 5·72 is most likely due to a proton situated between an epoxide group and an acetylenic bond, and a multiplet at τ 7·02 contains the epoxide protons. The OH-group gives a signal at τ 7·53. Spin-decoupling of the proton between the epoxide group and the acetylenic bond results in a doublet with a coupling constant of 4·2 Hz for the epoxide protons. The *cis*-coupling constant of a *cis*-epoxide is of that order, while a *trans*-epoxide shows a coupling constant of about 2 Hz[203]. The double bond protons were also shown to be *cis* in agreement with the i.r. spectrum, as there is no evidence for a *trans* configurated double bond in the i.r. Oxidation with manganese dioxide yields a substance with an extended chromophore (λ max: 322, 304, 271, 241, 233 nm), and a n.m.r. spectrum, where the signals at τ 5·72 and τ 7·53 have disappeared. The oxidation product appears to be **91**. Treatment of **90** with sulphuric acid yielded the triol **92**, which was converted by periodate cleavage to the corresponding aldehydes. Reduction with sodium borohydride yields the alcohols **93** and **94**. Both alcohols yield *n*-octyl alcohol after catalytic hydrogenation.

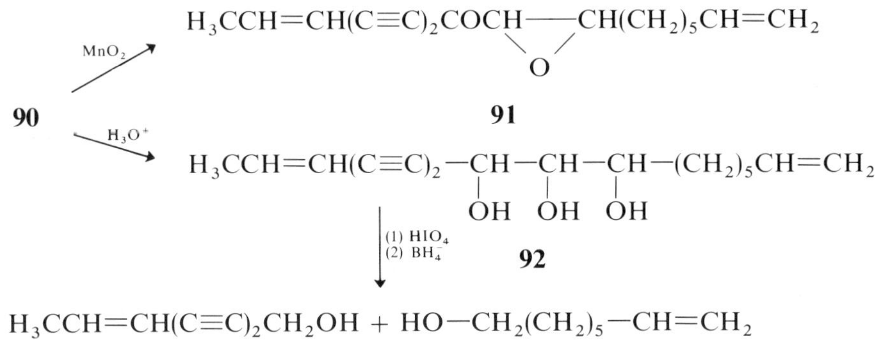

93 **94**

The alcohol **90** is very interesting, as it is proposed that **84** is also a precursor of **90** in the biogenetic pathway:

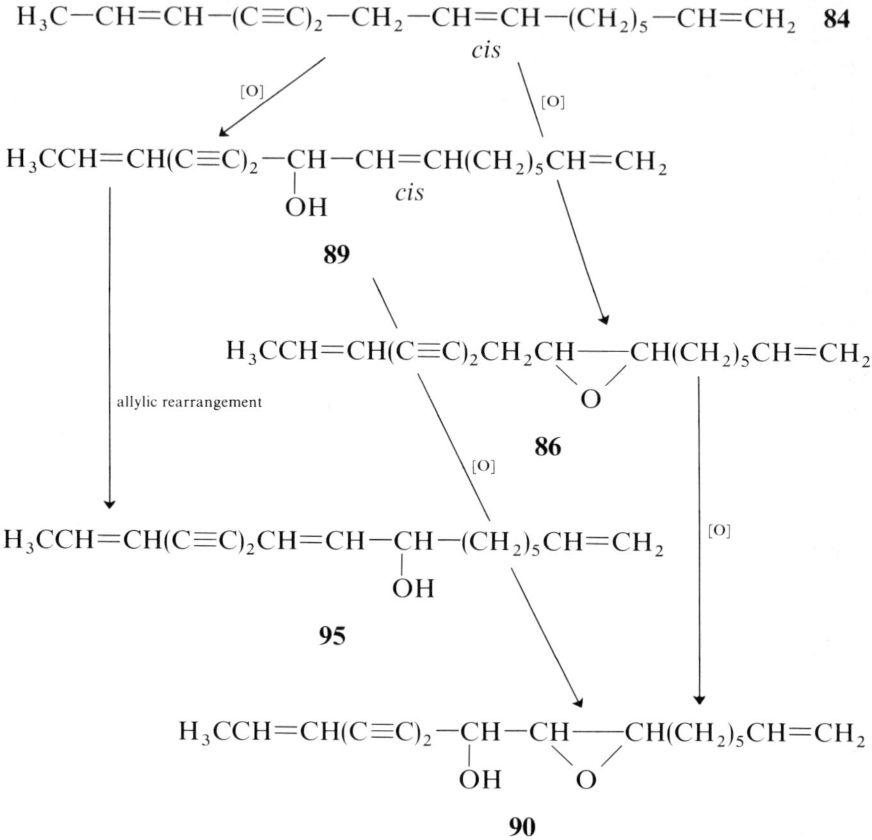

Obviously the formation of the epoxide prevents the allylic rearrangement to **95**, which is usually observed.

95 was isolated from the roots of *Cousinia hystrix* C. A. May[205]. Many attempts were made to get the compound, which could almost certainly be identified as an acetylenic alcohol with an ene-diyn-ene-chromophore, in a pure state; but there have been still some unknown impurities. The substance, purified by repeated column chromatography and TLC, gave an i.r. spectrum consistent with the presence of a hydroxyl group (3640/cm), triple bonds (2240/cm), and a monosubstituted vinyl group. There were no signs for *trans*-configurated double bonds. The n.m.r. exhibits resonances due to a $CH_3{-}CH{=}CH{-}$(*cis*)-grouping, the vinyl group and methylene groups. Using manganese dioxide, **95** was converted to the corresponding ketone **96**

and subsequently subjected to chromatography and crystallisation. Melting at 33°, the ketone shows an ene-diyne-ene-one chromophore (λ max: 335·5, 314·5, 307, 260, 248·5, 233 nm), whilst the i.r. spectrum shows an intense band at 2200/cm, according to acetylenic bonds and a splitted carbonyl band at 1698 and 1674/cm. An intense band at 1584/cm could be ascribed to a double bond, which is conjugated to the carbonyl group. The terminal vinyl group gives rise to characteristic bands at 1642, 1000 and 919/cm. The n.m.r. spectrum confirms the presence of a $\equiv$C—CH=CH—CO—(*cis*)-grouping (AB quartet τ 3·79), a propylene group (*dd* 8·03 τ (J = 6·6 + 1·5 Hz); *dq* 3·7τ) (J = 11 + 6·6), and a vinyl group (*m* 4·5τ (1H) and *m* 4·9–5·2τ (2H)). Furthermore there are signals, which were attributable to methylene adjacent to a double bond (τ 8·0) and normal methylenes (τ 8·5). These data strongly indicate that the ketone has the proposed structure **96**:

H_3C—CH=CH—$(C\equiv C)_2$—CH=CH—CO—CH_2
dd 8·03 *dq* 3·7 *m* 4·3 ABq 3·79 *t* 7·27
t *t*

$(CH_2)_3$—CH_2—CH=CH_2 **96**
m 8·5 *m* 8·0 *m* 4·5 *m* 4·9–5·2

The corresponding *cis-trans* alcohol **97** was obtained from the root extract of *Jurinea mollis*[196]:

H_3C—CH=CH—$(C\equiv C)_2$—CH=CH—CH(OH)—$(CH_2)_5$—CH=CH_2 **97**
cis *t*

97 occurs as aglycon in some polyyne glycosides (**98**, **99** and **100**), all obtained from root extracts of *Serratula gmelini* Ldb.[196] (see also Section 1.2.4):

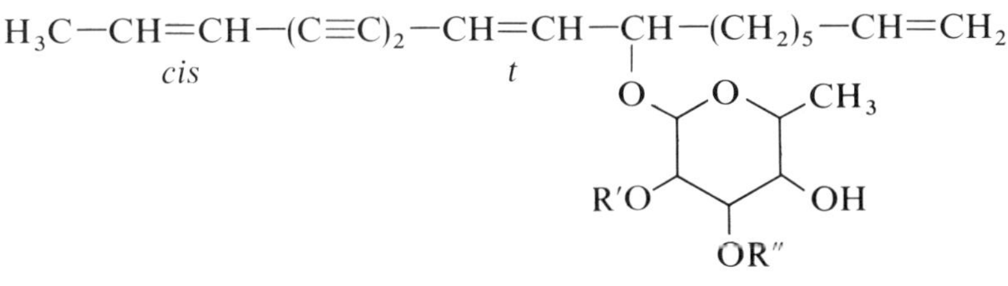

98: R′ = R″ = H
99: R′ = $COCH_3$
R″ = H
100: R′ = R″ = $COCH_3$

The structures of the three glycosides were elucidated in the same manner as described in Section 1.2.4 (**63**, **64**).

The so-called "centaur X_4" (**101**), a hydrocarbon with a diene-diyn-ene chromophore, was first identified spectroscopically by Löfgren[27]. It was, however, isolated first in 1958 from the roots of *Centaurea ruthenica* Lam.[206]:

$$\underset{}{H_3C{-}}\underset{t}{CH{=}CH}{-}(C{\equiv}C)_2{-}\underset{t}{CH{=}CH}{-}\underset{t}{CH{=}CH}{-}(CH_2)_4{-}CH{=}CH_2$$

101

In the i.r. spectrum, the compound shows absorption due to a terminal vinyl group (912, 990/cm) and to two conjugated double bonds (1625, 945/cm). Treatment with maleic anhydride gave an adduct, which was still showing the characteristic u.v. spectrum of an ene-diyne chromophore. Ozone cleavage of the adduct did not yield adipic acid, which was obtained by the same reaction from the polyyne. These experimental results were only consistent with the structure **101**. Finally, the proposed structure **101** was confirmed by synthesis[207]. Matricarianal (**513**) (see Chapter 3.7.1) was coupled with the ylen **102** to the hydrocarbon **101** and the product purified by crystallisation. The spectral data of the synthesised "all-*trans*"-compound were identical with those of the natural product:

$$\underset{\mathbf{513}}{H_3CCH{=}CH(C{\equiv}C)_2CH{=}CH{-}CHO} + \underset{\mathbf{102}}{Ph_3P{=}CH(CH_2)_4CH{=}CH_2}$$

$$\downarrow$$

$$\mathbf{101}$$

The corresponding *cis-trans-trans* isomer was obtained from the roots of *Aster bellidiastrum* Scop.[208]:

$$H_3C{-}\underset{c}{CH{=}CH}{-}(C{\equiv}C)_2{-}\underset{t}{CH{=}CH}{-}\underset{t}{CH{=}CH}{-}(CH_2)_4{-}CH{=}CH_2$$

103

The structure **103** was assigned on the basis of i.r. and n.m.r. spectra and mass spectroscopic analysis.

Besides dehydrofalcarinone (**52**), some C_{17}-polyynes related to **101** and **103** were obtained by extracting the leaves and flowering tops of *Tridax trilobata* Hemsl.[209]. The less polar fractions contain besides **52** two polyacetylen-compounds, the separation of which was difficult. The i.r. spectra of the compounds show aldehyde proton peaks at 2750 cm. The u.v. absorption bands indicate a dien-diyn-en-al chromophore. One of the two aldehydes, melting at 28·5°C, shows absorbance in the i.r. due to a *trans-trans*-diene, a *trans*-disubstituted double bond and a terminal vinyl group (1655, 995, 964, 922/cm). The n.m.r. spectrum consists of a double doublet at τ 0·49, indicating the aldehyde proton adjacent to a double bond; furthermore, a complex multiplet of 6 olefinic protons in the range of τ 3·1–4·6, three vinyl protons (τ 4·9–5·1), two methylene groups adjacent to double bonds and, in addition, two further normal methylene groups. In view of these spectral results, it seemed probable that the unknown aldehyde is structurally related to the all-*trans*-hydrocarbon **101** and consequently the structure has to be **104**:

$$\underset{\substack{H \\ dd\ 0{\cdot}49}}{O{=}}C-\underset{t}{CH=CH}-(C\equiv C)_2-\underset{t}{CH=CH}-\underset{t}{CH=CH}-(CH_2)_4-CH=CH_2$$

104

Reduction with sodium borohydride yielded the corresponding "all-*trans*" alcohol **105**, which was isolated from the roots of *Isotigma peucedanifolium* Less. too[71]:

$$\mathbf{104} \underset{MnO_2}{\overset{NaBH_4}{\rightleftharpoons}} HO-CH_2-CH=CH-(C\equiv C)_2-(CH=CH)_2-(CH_2)_4-CH=CH_2 \quad \mathbf{105}$$

The structure of **105** was proved by synthesis[191].

Wittig reaction of acetoxy-matricarianal (**106**) with the ylen **107** yielded the acetate of **105**. Saponification of the latter formed the alcohol **105**, the spectral properties of which were found to be identical with those of the natural compound:

$$H_3C-CO-O-CH_2-CH=CH(C\equiv C)_2CH=CH-CHO$$

106

$$+ Ph_3P=CH(CH_2)_4CH=CH_2$$

107

$$\downarrow$$

$$RO-CH_2-CH=CH-(C\equiv C)_2-(CH=CH)_2-(CH_2)_4-CH=CH_2$$

$R = COCH_3$

105: $R = H$

The second aldehyde differs from **104** in the i.r. The i.r. spectrum shows bands similar to those of **104**, but lacked the band of the *trans*-disubstituted double bond at 964/cm of the latter. In comparison with **104**, the u.v. spectrum shows only little red shift of the long wave absorption bands. The n.m.r.

spectrum of the unknown aldehyde differs essentially in the position of the aldehyde proton (τ 0·12), caused by the special stereochemistry of the adjacent double bond, which must be *cis*-configurated. In such a configuration, the aldehyde proton is shifted to lower fields, caused by the anisotropic effect of the triple bond:

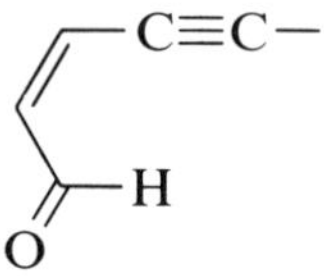

The above evidences indicate that the aldehyde is the corresponding *cis*-isomer **108**:

$$\underset{\substack{dd\ 0{\cdot}12 \\ (J=7+3{\cdot}3\ \text{Hz})}}{\text{H}}\!\!-\!\overset{\text{O}}{\overset{\|}{\text{C}}}\!-\underset{cis}{\text{CH=CH}}-(\text{C}\equiv\text{C})_2-\underset{t,t}{(\text{CH=CH})_2}-(\text{CH}_2)_4-\text{CH=CH}_2 \quad \mathbf{108}$$

Treatment of **108** with sodium borohydride gave an alcohol **109**, which was isolated too from the more polar fractions of the plant extract:

$$\mathbf{108} \rightarrow \text{HOCH}_2-\underset{cis}{\text{CH=CH}}-(\text{C}\equiv\text{C})_2-\underset{t,t}{(\text{CH=CH})_2}-(\text{CH}_2)_4-\text{CH=CH}_2$$

109

109 differs from **105** by the missing trans-double bond absorption at 960/cm in the i.r.

The occurrence of **85**, **89**, **95** and **101** in nature suggests, but does not probe, that the conjugated double bonds may be derived from **85** via **89**, **95**, etc. as outlined in the following scheme including allylic oxidation followed by dehydration:

$$-\text{C}\equiv\text{C}-\text{CH}_2-\text{CH=CH}-\text{CH}_2- \xrightarrow{[\text{O}]} -\text{C}\equiv\text{C}-\underset{\text{OH}}{\underset{|}{\text{CH}}}-\text{CH=CH}-\text{CH}_2-$$

$$-\text{C}\equiv\text{C}-\text{CH=CH}-\underset{\text{OH}}{\underset{|}{\text{CH}}}-\text{CH}_2- \xrightarrow{-[\text{H}_2\text{O}]}$$

$$-\text{C}\equiv\text{C}-\text{CH=CH}-\text{CH=CH}- \xrightarrow{[\text{O}]}$$

$$-\text{C}\equiv\text{C}-\text{CH=CH}-\text{CH=CH}-\underset{\text{OH}}{\underset{|}{\text{CH}}}-$$

An alcohol, closely related to the latter in the scheme above, was isolated from *Coreopsis gigantea*[199]:

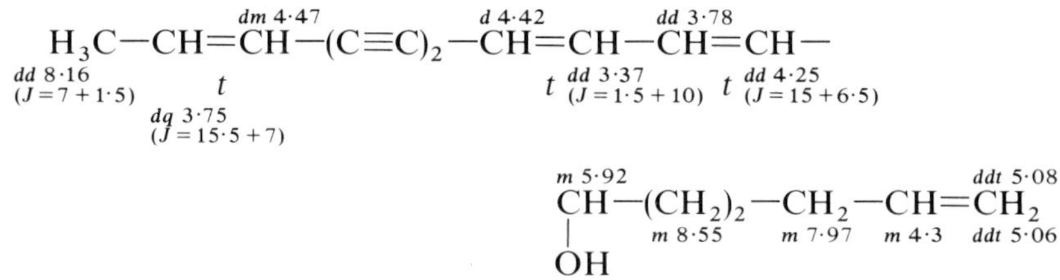

110

110 was converted into the corresponding ketone by manganese dioxide oxidation:

$$\mathbf{110} \xrightarrow{MnO_2} H_3CCH{=}CH(C{\equiv}C)_2(CH{=}CH)_2{-}CO{-}(CH_2)_3CH{=}CH_2$$

111

In order to confirm the structure, **111** was synthesised. Vinyl ether condensation of matricarianal-acetal yielded the aldehyde **112**, which was further transformed with pentenyl-magnesiumbromide to the racemic alcohol. Treatment of the alcohol with manganese dioxide afforded **111**, the properties of which were identical to those obtained from natural material:

$$H_3C{-}CH{=}CH{-}(C{\equiv}C)_2{-}CH{=}CH{-}CH(OR)_2 \xrightarrow[(2)\ H_3O^+]{(1)\ H_2C{=}CH{-}OR/ZnCl_2}$$

$$H_3C{-}CH{=}CH{-}(C{\equiv}C)_2{-}(CH{=}CH)_2{-}CHO$$

$$+ BrMg(CH_2)_3{-}CH{=}CH_2 \rightarrow \mathbf{110} \xrightarrow{MnO_2} \mathbf{111}$$

112

The "all-*trans*"-6-acetoxy-heptadeca-(1,7,9,15)-tetraene-(11,13)-diyne (**113**), the acetate related to **110**, was isolated from *Dahlia coccinea* Cav. var. coccinea by Jones and his co-workers[33]:

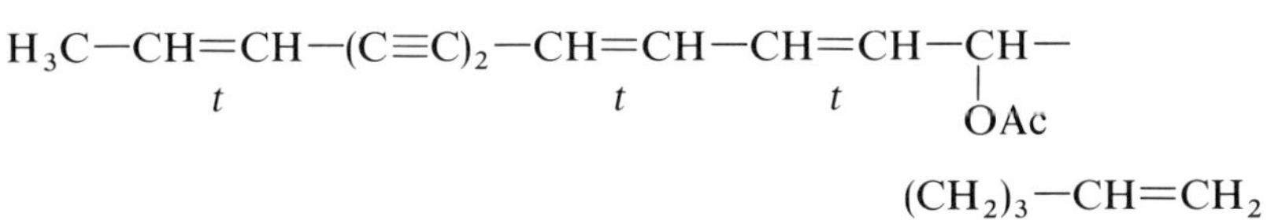

113

The acetate was available only in very small quantities. Its i.r. spectrum showed the presence of the acetate grouping (1730, 1225/cm), a terminal vinyl group and the all-*trans* configuration of the double bonds. The corresponding alcohol gave a ketone with manganese dioxide, proving the allylic nature of the hydroxyl group. The structure was finally established by synthesis.

1.2.6. Compounds related to falcarinone

In the preceding chapters, C_{17}-compounds possessing a terminal vinyl group were described. Their formation by a β-oxidation of C_{18}-acids, followed by loss of CO_2, seems to be very likely:

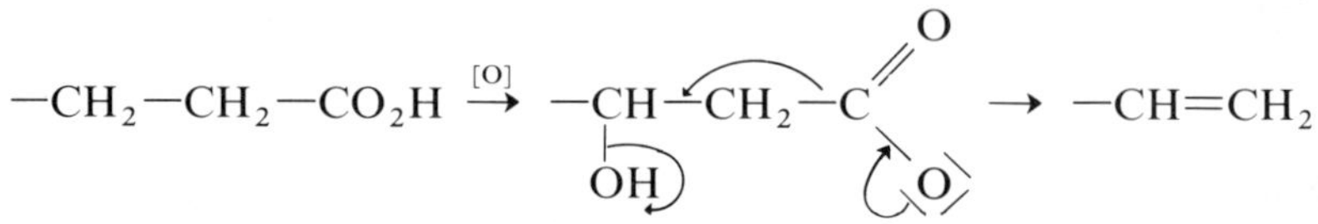

These compounds are known only from Compositae; corresponding C_{17}-polyacetylenes possessing a saturated endgroup were obtained from Umbelliferae. This fact suggests that the endgroup may be derived from a vinyl group by hydrogenation:

$$-CH{=}CH_2 \xrightarrow{[H]} -CH_2-CH_3$$

On the other hand, Jones has obtained the C_{18}-ketoaldehyde **7** (see Section 1.1) closely related to the C_{17}-ketone falcarinone (**114**):

$$H_2C{=}CH-CO-(C{\equiv}C)_2-CH_2-CH{=}CH-(CH_2)_6-CH_3 \qquad \mathbf{114}$$

$$H_2C{=}CH-CO-(C{\equiv}C)_2-CH_2-CH{=}CH-(CH_2)_6-CH_2-CHO \qquad \mathbf{7}$$

This finding permits the interpretation that **114** is derived naturally by loss of a carbon atom from C_{18}-fatty acids. No results are as yet available concerning the exact biosynthetic pathway in forming the saturated endgroup of polyacetylenes in the umbelliferous plants.

Falcarinone (**114**) has been first described as a constituent of *Falcaria vulgaris* Bernh.[210]. The highly unstable ketone **114** could be obtained pure by repeated column chromatography on silica gel (**114** had been decomposed on an alumina column). The oily compound obtained absorbs in the i.r. region at 1652/cm and this is due to a conjugated ketone. The intensity of the bands, attributable to triple bonds (2131, 2160, 2086/cm) is consistent with the position of the acetylenic bonds adjacent to the carbonyl group. The u.v. spectrum suggests the presence of an en-one-diyne chromophore (λ max: 290·7, 274·3, 260, 246, 208 nm). In order to obtain a stable derivative, **114** was treated with methanol and sulphuric acid. This gave a product, which was homogeneous by TLC. The methanol addition product showed an u.v. spectrum shifted to shorter wavelengths (λ max: 282, 266·5, 253, 240 nm), comparable to a diyne-one chromophore. This suggested that there exists a carbonyl group adjacent to triple bonds as well as a terminal vinyl group.

This special structure explains the reactivity of the terminal ethylenic bond to nucleophilic additions:

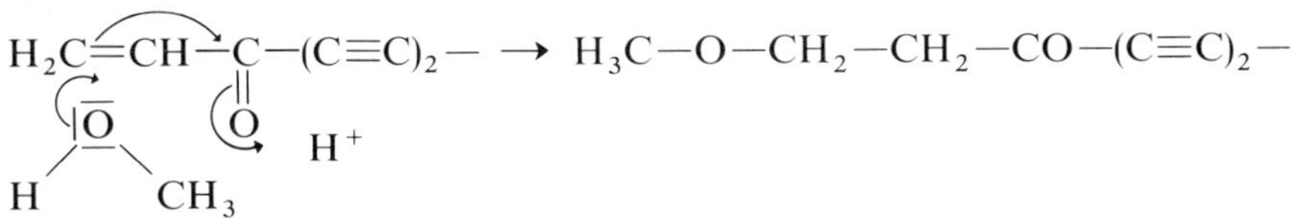

Dehydrofalcarinone (**52**) shows an analogous reactivity, as mentioned in Section 1.2.5.

In order to elucidate the carbon skeleton, **114** was hydrogenated to *n*-heptadecane-(3)-one. The position of the carbonyl-group was established by converting **114** into the corresponding polyyne carbinol, which was further hydrogenated and dehydrated to an olefin mixture. Oxidation with chromic acid gave a mixture of acids, which was esterified with diazo-methane. The esterification yielded methyl tetradecanoate and methyl pentadecanoate. Therefore, the oxygen function has to be located at the C-atom 3. Cleavage of the double bonds in **114** by ozonolysis resulted in the following products: *n*-octylaldehyde and formaldehyde, identified as 2,4-dinitrophenylhydrazones. These experimental results agree with the proposed structure **114**.

The influence of the isolated double bond on the u.v. absorption is very interesting. A synthesised model compound **115** showed absorption maxima shifted to shorter wavelengths (4 nm), although possessing an additional methyl substituent:

$$H_2C{=}CH{-}CO{-}(C{\equiv}C)_2{-}CH_2{-}CH{=}CH{-}(CH_2)_6{-}CH_3 \quad \mathbf{114}$$

$$H_3C{-}CH{=}CH{-}CO{-}(C{\equiv}C)_2{-}CH_3 \quad \mathbf{115}$$

The roots of *Carum carvi* L.—the ordinary caraway—contain two highly unstable C_{17}-polyacetylenes, closely related to falcarinone (**114**). The polyynic mixture was separated by column chromatography. A fraction that travelled with the solvent front through the silica gel column used to separate the components, was shown to be a diketone and was identified as falcarindione (**116**). The more polar compound was shown to be falcarinolone (**117**) on the basis of chemical and spectroscopic evidence[201]:

$$H_2C{=}CH{-}CO{-}(C{\equiv}C)_2{-}CO{-}\underset{cis}{CH{=}CH}{-}(CH_2)_6{-}CH_3 \quad \mathbf{116}$$

$$H_2C{=}CH{-}CO{-}(C{\equiv}C)_2{-}\underset{OH}{\underset{|}{CH}}{-}\underset{cis}{CH{=}CH}{-}(CH_2)_6{-}CH_3 \quad \mathbf{117}$$

Both compounds formed the more stable methanol adducts comparable to falcarinone. The dihydromethoxy derivatives lacked a *trans* double bond absorption in the i.r. The stereochemistry about the isolated double bond could be assigned on the basis of the observed band at 675/cm as *cis* configurated.

116 and **117** differ from **114** by the additional oxygen function at C_8.

Falcarinone (**114**) is rather widespread in nature and has been isolated from about eighty other genera of the family Umbelliferae.

Another closely related diacetylene, falcarinol (**118**) has been isolated from *Falcaria vulgaris* Bernh.[201] and many other Umbelliferae. The structure of falcarinol (**118**) was established by spectral properties and confirmed by synthesis:

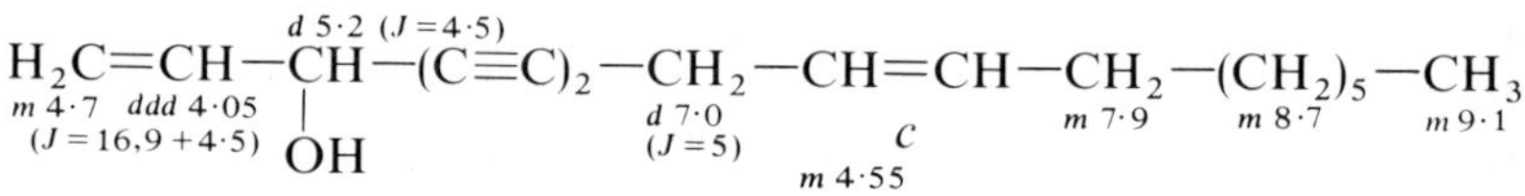

118

Panaxynol, isolated from the Araliaceae *Panax Ginseng* C. A. Meyer turned out to be identical with falcarinol, as shown by Takahashi[211].

In 1966 the rather improbable structure **119** was proposed for a toxic acetylenic compound isolated from the common carrot, *Daucus carota* L.[212]:

$$H_2C{=}CH{-}\underset{\displaystyle OH}{\underset{|}{C}H}{-}CH_2{-}(C{\equiv}C)_2{-}CH_2{-}CH{=}CH{-}C_6H_{13} \qquad \mathbf{119}$$

This structure (**119**) was assigned on the basis of the spectra of the compound itself, the u.v. spectrum of a "dehydration" product of the alcohol **119** with toluene-*p*-sulphonic acid and the chain lengths of the fatty acids obtained by alkaline permanganate oxidation.

Re-investigation of the plant yielded the presence of a C_{17}-alcohol with an interrupted chromophore. On the basis of the spectra and some of the chemical data quoted for carotatoxin[212], the latter was identical with falcarinol. This was proved by repeating the isolation procedure given for carotatoxin and by oxidising the alcohol with manganese dioxide to give falcarinone (**114**)[213].

The reaction of falcarinol (**118**) with toluene-*p*-sulphonic acid in benzene is a very interesting one. Treatment with the acid gives three products, all of which exhibit ene-diyne absorption. The most polar of these (slightly more polar than the original alcohol) was the expected primary alcohol **120**. Its structure was deduced from the spectral data and by oxidation with manganese dioxide to an aldehyde[214]:

$$\mathbf{118} \xrightarrow{H^+} RO{-}CH_2{-}CH{=}CH{-}(C{\equiv}C)_2{-}CH_2{-}\underset{cis}{CH{=}CH}{-}C_7H_{15}$$

120: R = H
121: R = $SO_2C_6H_4CH_3$

The compound of intermediate polarity, readily hydrolysed to the primary alcohol, proved to be the tosylate **121**. The least polar product contains neither hydroxyl nor tosyl groups. The base peak in its mass spectrum was the tropylium ion (*m/e* 91). This peak often occurs in the spectra of polyacetylenes, but is rarely so strong. Other intense peaks are *m/e* 304, apparently the molecular ion, and *m/e* 219. Both, the i.r. and the n.m.r. spectra (τ 2·77 (5H)), suggest the presence of a phenyl group. Thus, the benzenoid product was the hydrocarbon **122**, formed by electrophilic substitution of the solvent:

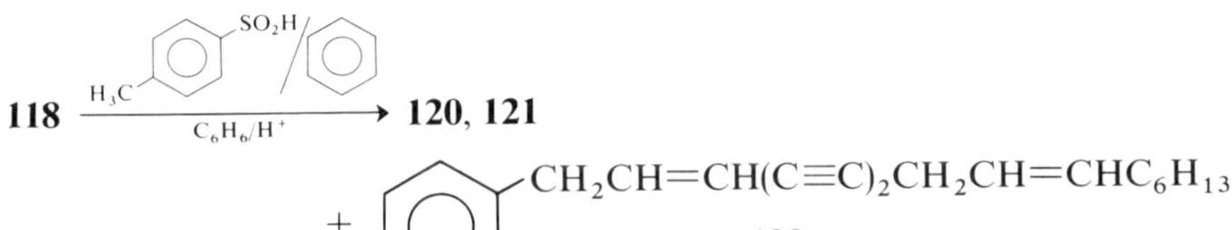

122

From the most polar fractions of *Falcaria vulgaris* Bernh. a diol was obtained[201], which was converted by manganese dioxide oxidation to a diketone, identical on its spectral properties with falcarindione (**116**). These observations are consistent with structure **123**.

$$H_2C{=}CH{-}\underset{OR}{\underset{|}{C}H}{-}(C{\equiv}C)_2{-}\underset{OR}{\underset{|}{C}H}{-}CH{=}CH{-}(CH_2)_6{-}CH_3 \quad \xrightarrow{MnO_2} \mathbf{116}$$

123: R = H
124: R = Ac

Chromatography of an extract of fresh carrots gave besides **118** and **117** falcarindiol (**123**) too[214]. Falcarindiol gave with acetic anhydride and pyridine the diacetate **124**.

A monoacetate of falcarindiol has been isolated from the roots of *Daucus carota* L.[214]. Its spectra reveal the presence of an acetoxy and a hydroxyl group as well as double bonds. The ethylenic protons give rise to complex signals in the range τ 4–5.

$$\underset{\text{OAc}}{H_2C{=}CH{-}CH}{-}(C{\equiv}C)_2{-}\underset{\text{OH}}{CH}{-}\underset{cis}{CH{=}CH}{-}(CH_2)_6{-}CH_3 \quad \textbf{125}$$

125 gave a diacetate and was oxidised to the ketoacetate **126**.

$$\textbf{125} \xrightarrow{MnO_2} \underbrace{H_2C{=}CH}_{m\ 4{\cdot}0\text{–}4{\cdot}8}{-}\underset{\underset{s\ 7{\cdot}89}{\text{OAc}}}{CH}(C{\equiv}C)_2CO\underset{m\ 3{\cdot}6-4{\cdot}0}{CH{=}CH}{-}\underset{m\ 7{\cdot}35}{CH_2}{-}\underset{m\ 8{\cdot}7}{(CH_2)_5}{-}\underset{t\ 9{\cdot}1}{CH_3} \quad \textbf{126}$$

The keto group of **126** deshields the protons of both, the *cis* double bond ($\tau < 4$) and the allylic methylene group (τ 7·35). These shifts are consistent with those observed for the corresponding protons in the diketone **116**[210] and thus determine the position of the hydroxy group in the natural material.

The roots of *Falcaria vulgaris* Bernh. have been found to contain a further alcohol shown to be **127** on the basis of chemical and spectroscopic evidence:

$$HO{-}\underset{\substack{t\ 6{\cdot}22\\(J=6)}}{CH_2}{-}\underset{\substack{t\ 7{\cdot}26\\(J=6)}}{CH_2}{-}CO{-}(C{\equiv}C)_2{-}\underset{\substack{d\ 6{\cdot}92\\(J=6)}}{CH_2}{-}\underset{\substack{c\\m\ 4{\cdot}58}}{CH{=}CH}{-}(CH_2)_6{-}CH_3 \quad \textbf{127}$$

Treatment of **127** with toluene-*p*-sulphonic acid yielded falcarinone (**114**).

A diol, closely related to **127**, was obtained in very small amounts from the roots of *Aegopodium podagraria* L.[35]. The compound shows a diyn-one-like chromophore in the u.v. (λ max: 284, 268, 254, 241 nm). Its i.r. spectrum displays bands for a ketone adjacent to triple bonds (1685/cm), hydroxy-groups (broad shoulder at 3500/cm), and triple bonds (2240, 2150/cm). The structure therefore should be **128**.

$$HO{-}CH_2{-}CH_2{-}CO{-}(C{\equiv}C)_2{-}\underset{\text{OH}}{CH}{-}CH{=}CH{-}C_7H_{15} \quad \textbf{128}$$

The 1,9-diene-4,6-diyne unsaturation is a common feature of the polyacetylenes mentioned above. They show in their mass spectra the typical allylic fission, which results in a strong peak at $M - C_6H_{13}$; this property can be used conveniently for the characterisation of this type of polyacetylenes.

The investigation of the constituents of *Falcaria vulgaris* Bernh.[201] gave very small amounts of a ketone, showing an ene-diyn-one chromophore (λ max: 310, 291, 275 nm). Reduction with sodium borohydride yielded an ene-diyne (λ max: 283, 267, 257 nm). It is very probable that the unknown ketone is **129**:

$$HO{-}CH_2{-}CH_2{-}CO{-}(C{\equiv}C)_2{-}CH{=}CH{-}C_8H_{17} \quad \textbf{129}$$

The investigation of *Falcaria vulgaris* Bernh.[201] have further resulted in the discovery of a ketone (i.r.: 1660/cm) with an uncharacteristic u.v. absorption at 342 nm. The strong intensity of a band at 2180/cm in the i.r. shows the presence of a conjugated acetylenic ketone. The presence of bands at 1008/cm in the i.r. exhibits some conjugated *trans*-disubstituted double bonds. Borohydride reduction yielded an alcohol with the characteristic absorption of a triene-yne (λ max: 311, 297 nm). The compound was converted by hydrogenation to *n*-heptadecanone-(3). The above evidences indicate that the ketone is **130**:

$$H_2C{=}CH{-}CO{-}C{\equiv}C{-}(CH{=}CH)_3{-}(CH_2)_5{-}CH_3 \quad \mathbf{130}$$

It seemed desirable to prove the structure of **130** by synthesis. The route chosen involved preparation of 1-bromo-non-(2)-ene by treatment of *n*-hexyl magnesium bromide with acrolein and bromination of the obtained alcohol with phosphoric tribromide. Wittig reaction of the corresponding ylene with pent-(2)-en-(4)-yn-(1)-al yielded tetradeca-(3,5,7)-trien-(1)-yne (**131**), which was converted into the corresponding Grignard-compound **132**. Reaction of **132** with acrolein gave the alcohol **133**, which was further oxidised with manganese dioxide to give **130**:

$$H_3C(CH_2)_5{-}MgBr + OCH{-}CH{=}CH_2 \rightarrow$$

$$H_3C(CH_2)_5{-}\underset{\displaystyle OH}{\underset{|}{CH}}{-}CH{=}CH_2 \xrightarrow{PBr_3}$$

$$H_3C(CH_2)_5{-}CH{=}CH{-}CH_2Br \rightarrow$$

$$H_3C(CH_2)_5CH{=}CHCH{=}P(Ph)_3 \xrightarrow{+OCH-CH=CH-C\equiv C-H}$$

$$\underset{\mathbf{131}}{H_3C(CH_2)_5(CH{=}CH)_3{-}C{\equiv}C{-}H} \rightarrow$$

$$\underset{\mathbf{132}}{H_3C(CH_2)_5(CH{=}CH)_3{-}C{\equiv}C{-}MgBr} + OCH{-}CH{=}CH_2 \rightarrow$$

$$\underset{\mathbf{133}}{H_3C(CH_2)_5(CH{=}CH)_3{-}C{\equiv}C{-}\underset{\displaystyle OH}{\underset{|}{CH}}{-}CH{=}CH_2} \xrightarrow{MnO_2} \mathbf{130}$$

134 is shown to occur in the leaves of *Opopanax chironium* Kch.[215]:

$$H_2C{=}CH{-}\underset{\displaystyle OH}{\underset{|}{CH}}{-}(C{\equiv}C)_2{-}(CH{=}CH)_2{-}(CH_2)_5{-}CH_3 \quad \mathbf{134}$$

This structure was established by u.v., and n.m.r. structural analysis as well as by mass spectroscopic measurements.

From the roots of *Oenanthe crocata* L. (Hemlock water dropwort) a hydrocarbon was isolated from the least-polar chromatographic fractions showing an u.v. absorption due to an ene-diyne chromophore and with a molecular formula $C_{17}H_{24}$[216]. The n.m.r. spectrum shows signals attributable to a $CH_3{-}CH{=}CH{-}(C{\equiv}C)_2{-}CH_2{-}CH{=}CH{-}$ grouping:

t *c*

dd 8·19τ (3H) (J = 6·5, 1·6 Hz); *dq* 3·75τ (1H) (J = 15, 6·5 Hz); *d* 4·5τ (1H) (J = 15 Hz); *d* 7·0τ (2H) (J = 5 Hz); *m* 4·6τ (2H). Taking these facts into consideration, structure **135** is indicated.

$$H_3C{-}CH{=}CH{-}(C{\equiv}C)_2{-}CH_2{-}CH{=}CH{-}(CH_2)_6{-}CH_3 \quad \mathbf{135}$$

trans *cis*

The stereochemistry of the isolated double bond could be assigned on the basis of the observed band in the i.r.

From the roots of *Heracleum sphondylium* L. besides furocoumarins as main constituents, two acetylenic alcohols were obtained, which could not be separated completely from each other[217]. The u.v. spectrum reveals an ene-diyne-like chromophore (λ max: 284, 269, 254 nm). Both alcohols were oxidised with manganese dioxide. One of the obtained ketones shows the typical u.v. absorption of falcarinone. The other one shows an u.v. absorption attributable to an ene-diyn-on-ene chromophore. The n.m.r. and mass spectra of the alcohols and ketones, respectively, were consistent with the proposed structures **136**, **137**, **138**, and **139**:

$$C_3H_7{-}(C{\equiv}C)_2{-}CH(OH){-}CH{=}CH{-}(CH_2)_6{-}CH_3 \quad \mathbf{136}$$

cis

$$\downarrow MnO_2$$

$$C_3H_7{-}(C{\equiv}C)_2{-}CO{-}CH{=}CH{-}(CH_2)_6{-}CH_3 \quad \mathbf{137}$$

cis

$$H_3C{-}CH{=}CH{-}(C{\equiv}C)_2{-}CH(OH){-}CH{=}CH{-}(CH_2)_6{-}CH_3 \quad \mathbf{138}$$

cis *cis*

$$\downarrow MnO_2$$

$$H_3C{-}CH{=}CH{-}(C{\equiv}C)_2{-}CO{-}CH{=}CH{-}(CH_2)_6{-}CH_3 \quad \mathbf{139}$$

cis *cis*

The ketones **137** and **139** give as the main fragment *m/e* 160 and 158, respectively:

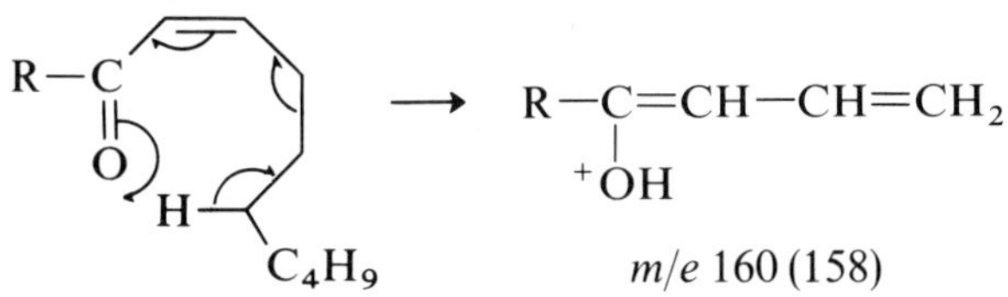

m/e 160 (158)

This fragmentation pattern confirms the position of the oxygen function and the *cis*-configuration of the isolated double bond. Both double bonds in **139** must be *cis*-configurated, as there is no band at 950/cm to be seen in the i.r.

A hydrocarbon was obtained from the roots of *Oenanthe crocata* L., which arise by oxidation of the activated methylene group of **135** to **138**, followed by allylic rearrangement and dehydration. A reasonable biosynthetic pathway for these reaction sequence is outlined already in Section 1.2.5:

$$\mathbf{135} \xrightarrow{[O]} \mathbf{138} \xrightarrow{-[H_2O]} \mathbf{140}$$

High-resolution mass spectroscopic measurements of the compound obtained from *Oenanthe crocata*[216] reveals the molecular formula $C_{17}H_{22}$. The structural assignments for the hydrocarbon are derived from the n.m.r., the i.r. shows a *trans-trans* double bond (987/cm), and the u.v. exhibits an ene-diyne-diene chromophore (λ max: 337, 315, 296, 266, 250 nm). These observations are only consistent with structure **140**:

$$H_3C{-}CH{=}CH{-}(C{\equiv}C)_2{-}CH{=}CH{-}CH{=}CH{-}CH_2{-}(CH_2)_4{-}CH_3 \qquad \mathbf{140}$$

Extracting the leaves of *Cicuta virosa* L., a ketone was obtained from the least polar fractions[218]. In order to obtain it in a pure state conversion by sodium borohydride to the corresponding alcohol was necessary and oxidisation into the original ketone. Although the reaction sequence was repeated, the compound could not be obtained pure. The u.v. spectrum of the alcohol reveals an ene-diyne-triene-chromophore (λ max: 356, 333 nm). Taking this into consideration, the structure must be **141**:

$$H_3C{-}CH{=}CH{-}(C{\equiv}C)_2{-}CH{=}CH{-}CH{=}CH{-}CH{=}CH{-}CO{-}C_3H_7 \qquad \mathbf{141}$$

To prove the proposed "all-*trans*" structure **141** was synthesised (as outlined in the scheme below) and gave a product which was proved to be homogeneous by TLC, and which seemed to be identical with the natural product when compared by TLC with the original ketone. However, the u.v. spectrum of the crude, natural ketone does not agree completely with that of the synthetic product; perhaps due to *cis-trans*-isomerism.

The synthesis route chosen involved preparation of the carbinol **142**, to be followed by partial hydrogenation to **143**. Treatment with mineral acid yielded the aldehyde **144**, which was converted into the carbinol **145** by Grignard reaction with ethinyl magnesium bromide; **145** was further transformed into the corresponding alcohol **146** by Cadiot coupling with 1-bromo-pent-(1)-yn-(3)-ene. Finally, manganese dioxide oxidation yielded the desired ketone **141**:

$$H_3C{-}O{-}CH{=}CH{-}C{\equiv}C{-}MgBr + OCH{-}CH{=}CHC_3H_7 \rightarrow$$

$$H_3C{-}O{-}CH{=}CH{-}C{\equiv}C{-}\underset{\displaystyle OH}{\underset{|}{CH}}{-}CH{=}CH{-}C_3H_7 \xrightarrow{H_2}$$

142

$$H_3C{-}O{-}(CH{=}CH)_2{-}\underset{\displaystyle OH}{\underset{|}{CH}}{-}CH{=}CHC_3H_7 \xrightarrow{H_3O^+}$$

143

$$OCH{-}(CH{=}CH)_3C_3H_7 + H{-}C{\equiv}C{-}MgBr \rightarrow$$

144

$$H{-}C{\equiv}C{-}\underset{\displaystyle OH}{\underset{|}{CH}}{-}(CH{=}CH)_3C_3H_7 + Br{-}C{\equiv}C{-}CH{=}CH{-}C_3H_7 \rightarrow$$

145

$$H_3CCH{=}CH(C{\equiv}C)_2{-}\underset{\displaystyle OH}{\underset{|}{CH}}{-}(CH{=}CH)_3C_3H_7 \xrightarrow{H^+}$$

146

$$H_3CCH{=}CH(C{\equiv}C)_2(CH{=}CH)_3\underset{\displaystyle OH}{\underset{|}{C}}HC_3H_7$$

$$\downarrow MnO_2$$

141

Oenanthetone, a C_{17}-polyacetylenic ketone, was first obtained from the roots of *Oenanthe crocata*[219]. The nature of the oxygen function is indicated by i.r. spectroscopic evidence. It shows a very strong absorption band at 1704/cm. Hydrogenation yielded *n*-heptadecanone-(4); 7 moles of hydrogen being absorbed in the presence of palladium. The compound shows bands at about 2200 and 2120/cm, furthermore at 1640 and 1600/cm (conjugated C=C-stretching), and 990/cm. The last band is due to the out of plane C—H vibrations of a *trans*-diene grouping. Absorption bands characteristic of *cis* double bonds are absent. The chromophore contains two acetylenic links. This was shown by partial hydrogenation in the presence of Lindlar's catalyst, which is selective even for acetylenes with many conjugated double bonds. After two moles of hydrogen had been absorbed the short wavelengths bands in the u.v. had disappeared; the long wavelength band system had been replaced by a new system, which was typically polyene in type and both more intense and of a longer wavelength than the original system. In view of these spectroscopic results it was proposed that oenanthetone has the structure **147**:

$$H_3C{-}\underset{t}{CH{=}CH}{-}(C{\equiv}C)_2{-}\underset{t}{CH{=}CH}{-}\underset{t}{CH{=}CH}{-}CH_2{-}CH_2{-}CO{-}C_3H_7 \qquad \textbf{147}$$

oenanthetone

This structure has been confirmed by synthesis[220].

Two further ketones were obtained from *Oenanthe crocata* L. besides oenanthetone (147)[216]. The u.v. spectra show the presence of an ene-diyn-ene and a diyne-diene chromophore. As they were isolated in only very small amounts it has been impossible to establish the structure distinctly. The ketones were assigned structures **148** and **149**, largely on the basis of mass spectroscopic measurements:

$$H_3C{-}CH{=}CH{-}(C{\equiv}C)_2{-}CH{=}CH{-}(CH_2)_4{-}CO{-}C_3H_7 \qquad \textbf{148}$$

$$H_3C{-}CH_2{-}CH_2{-}(C{\equiv}C)_2{-}CH{=}CH{-}CH{=}CH{-}(CH_2)_2{-}CO{-}C_3H_7 \qquad \textbf{149}$$

From the alcohol fraction of *Oenanthe crocata*, a carbinol was isolated which gave oenanthetone **147** when oxidised with the Jones reagent; accordingly the original alcohol is **150**:

$$H_3C{-}CH{=}CH{-}(C{\equiv}C)_2{-}CH{=}CH{-}CH{=}CH{-}(CH_2)_2{-}\underset{OH}{\underset{|}{CH}}{-}C_3H_7 \qquad \textbf{150}$$

$$\textbf{150} \xrightarrow{CrO_3} \textbf{147}$$

Evidence for this was obtained by converting oenanthetone (**147**) to the corresponding alcohol, which proved to be identical to the natural compound.

An alcohol with an interrupted conjugation was isolated in very small amounts from the leaves of *Opopanax chironium* Kch.[215]. The presence of a hydroxy group was shown by the i.r. (3620/cm). Furthermore, the i.r. spectrum displayed bands for triple bonds at 2230 and 2150/cm and a *trans*-disubstituted double bond at 960/cm. In addition the u.v. spectrum revealed an ene-diyne chromophore. The n.m.r. spectrum shows a two-proton doublet at 6·95τ ($J = 5$ Hz), indicating an isolated methylene group adjacent to triple bonds and a *cis* double bond, respectively. In view of these spectral results, it is clear that the unknown alcohol **151** is structurally related to the former described hydrocarbon **135**:

$$\mathrm{HO{-}CH_2{-}\underset{t}{CH{=}CH}{-}(C{\equiv}C)_2{-}CH_2{-}\underset{c}{CH{=}CH}{-}(CH_2)_6{-}CH_3} \qquad \mathbf{151}$$

From the roots of *Oenanthe crocata* L.[216] and *Falcaria vulgaris* Bernh.[201], an alcohol closely related to **151** was obtained. It differs from **151** by the hydrogenated *trans* double bond. The structure of the alcohol was elucidated by n.m.r. analysis:

$$\mathrm{HO{-}\underset{t\,6\cdot33}{CH_2}{-}\underset{m\,8\cdot3}{CH_2}{-}\underset{t\,7\cdot66}{CH_2}{-}(C{\equiv}C)_2{-}\underset{\substack{d\,7\cdot05\\(J=5\ Hz)}}{CH_2}{-}}$$

$$\mathrm{\underset{\substack{cis\\m\,4\cdot61}}{CH{=}CH}{-}\underset{m\,7\cdot95}{CH_2}{-}\underset{m\,8\cdot72}{(CH_2)_5}{-}\underset{t\,9\cdot11}{CH_3}} \qquad \mathbf{152}$$

The investigation of the constituents of *Azorella trifurcata* (Gaertn.) Pers.[217] gives some ene-diynes, which has been identified to be hydroxy acetates (1750, 1240/cm). The structures were assigned on the basis of i.r. and n.m.r. spectra and the molecular formula obtained from the mass spectrum. Taking these facts into consideration only structures **153** and **154** can be suggested. The more polar fractions of the plant extract contain the diol **155**:

$$\mathrm{\underset{s\,7\cdot98}{H_3C}{-}CO{-}O{-}\underset{\substack{dd\,5\cdot24\\(J=6\cdot5+1\cdot5)}}{CH_2}{-}\underset{cis}{\overset{dt\,3\cdot9}{CH}{=}\overset{d\,4\cdot39}{CH}}{-}(C{\equiv}C)_2{-}}$$

$$\mathrm{\overset{dm\,4\cdot9}{\underset{\substack{|\\OH}}{CH}}{-}\underset{cis}{\overset{m\,4\cdot55}{CH}{=}CH}{-}\underset{m\,7\cdot85}{CH_2}{-}\underset{m\,8\cdot7}{(CH_2)_n}{-}\underset{tm\,9\cdot1}{CH_3}} \qquad \mathbf{153}\text{: n = 3} \quad \mathbf{154}\text{: n = 5}$$

$$\underset{dd\ 5{\cdot}74}{HO{-}CH_2}{-}\overset{dt\ 3{\cdot}79}{CH}{=}\underset{\substack{cis\\ d\ 4{\cdot}50}}{CH}{-}(C{\equiv}C)_2{-}$$

$$\underset{|\atop OH}{\overset{dm\ 4{\cdot}88}{CH}}{-}\overset{m\ 4{\cdot}55}{CH}{=}\underset{cis}{CH}{-}\underset{m\ 7{\cdot}92}{CH_2}{-}\underset{m\ 8{\cdot}7}{(CH_2)_5}{-}\underset{t\ 9{\cdot}12}{CH_3} \qquad \mathbf{155}$$

The position of the secondary hydroxyl group is indicated by the n.m.r. data and the manganese dioxide oxidation products **156**, **157** and **158**:

$$\left.\begin{matrix}\mathbf{153}\\ \mathbf{154}\end{matrix}\right\rangle \xrightarrow{MnO_2} \underset{dd\ 5{\cdot}22}{AcO{-}CH_2}{-}CH{=}\underset{\substack{cis\\ m\ 3{\cdot}5{-}4}}{CH}{-}(C{\equiv}C)_2{-}$$

$$CO{-}CH{=}\underset{cis}{CH}{-}(CH_2)_n{-}CH_3 \qquad \mathbf{156}: n = 4 \quad \mathbf{157}: n = 6$$

$$\mathbf{155} \xrightarrow{MnO_2} \underset{\substack{d\ -0{\cdot}27\\ (J=7)}}{OCH}{-}\underset{m\ 3{\cdot}3-3{\cdot}9}{CH{=}CH}{-}(C{\equiv}C)_2{-}CO{-}CH{=}CH{-}(CH_2)_6{-}CH_3$$

158

Two diols were obtained from the most polar fractions of *Opopanax chironium* Kch.[215]. The u.v. spectra show the presence of an ene-diyn-ene chromophore and a diyn-ene chromophore, respectively. The ene-diyn-ene contains two allylic hydroxy-groups and when oxidised with manganese dioxide, yields a compound with a u.v. spectrum similar to deca-(2,8)-diene-(4,6)-dial. The structure of the oxidised product therefore is **160**. The n.m.r. spectrum of the alcohol exhibits the following characteristic signals: *trans*-$HO{-}CH_2{-}CH{=}CH{-}C{\equiv}$: *s* 7·97τ (1H), *dd* 5·78τ (2H) (J = 5, 1·5 Hz), *dt* 3·61τ (1H) (J = 16·5 Hz), *ddt* 4·18τ (1H) (J = 16, 1, 1·5 Hz).

The additional coupling constant of 1 Hz in the olefinic signal of the β-proton indicates a long range coupling over two acetylenic bonds with a further olefinic proton. The n.m.r. signals of the $\equiv C{-}CH{=}CH{-}CH(OH){-}CH_2{-}$ grouping consists in a *ddd* 4·23τ (1H) (J = 16, 1, 1 Hz), a *dd* 3·7τ (1H) (J = 16, 5·5 Hz), a *dt* 5·82τ (1H) (J = 6, 5·5 Hz) and a *s* 7·97τ (1H). In the mass spectrum the molecular ion peak at m/e 260, agreeing with $C_{17}H_{24}O_2$, is observed. Taking all these facts into consideration, structure **159** is suggested:

$$HO{-}CH_2{-}CH{=}CH{-}(C{\equiv}C)_2{-}CH{=}CH{-}\underset{|\atop OH}{CH}{-}(CH_2)_6{-}CH_3 \qquad \mathbf{159}$$

$$\big\downarrow MnO_2$$

$$OCH{-}CH{=}CH{-}(C{\equiv}C)_2{-}CH{=}CH{-}CO{-}(CH_2)_6{-}CH_3 \qquad \mathbf{160}$$

The second diol with ene-diyne absorption could not be obtained in a pure state. Mass spectroscopic measurements give a molecular ion peak at *m/e* 262, due to $C_{17}H_{26}O_2$. Therefore, structure **161** seems to be very likely:

$$HO{-}CH_2{-}CH_2{-}CH_2{-}(C{\equiv}C)_2{-}CH{=}CH{-}\underset{\displaystyle OH}{\underset{|}{CH}}{-}(CH_2)_6{-}CH_3 \qquad \mathbf{161}$$

The hydroxy aldehyde **162**, closely related to **159**, was obtained from *Oenanthe crocata* L.[216]:

$$OCH{-}CH{=}CH{-}(C{\equiv}C)_2{-}CH{=}CH{-}\underset{\displaystyle OH}{\underset{|}{CH}}{-}(CH_2)_6{-}CH_3 \qquad \mathbf{162}$$

(OCH: *dd* 0·51 ($J = 5 + 2$); CH=CH: *t*, *s* 3·43; CH=CH: *dd* 3·61, *t*, *dd* 4·22; CH(OH): *m* 5·86; $(CH_2)_6$: *m* 8·7; CH_3: *t* 9·11)

Manganese dioxide oxidation of **162** yielded the corresponding keto-aldehyde **163**, the spectral properties of which are identical to those of the oxidation product of **159**:

$$\mathbf{159},\ \mathbf{162} \xrightarrow{MnO_2} OCH{-}CH{=}CH{-}(C{\equiv}C)_2CH{=}CH{-}CO{-}CH_2{-}(CH_2)_5CH_3 \qquad \mathbf{163}$$

(OCH: *dd* 0·42 ($J = 4{\cdot}5 + 3{\cdot}2$); CH=CH: *m* 3·4; CH_2: *t* 7·52)

The monoacetate, corresponding to **161** also was isolated from *Oenanthe crocata* L.[216]. As it could not be obtained in a pure state, the acetate **164** was converted by mild saponification to the alcohol **161**, known from *Opopanax chironium* Kch. The structure of the latter was elucidated by n.m.r. and mass spectroscopic measurements:

$$AcO{-}CH_2{-}CH_2{-}CH_2{-}(C{\equiv}C)_2{-}CH{=}CH{-}\underset{\displaystyle OH}{\underset{|}{CH}}{-}(CH_2)_6{-}CH_3 \qquad \mathbf{164}$$

$$HO{-}CH_2{-}CH_2{-}CH_2{-}(C{\equiv}C)_2{-}CH{=}CH{-}\underset{\displaystyle OH}{\underset{|}{CH}}{-}(CH_2)_6{-}CH_3 \qquad \mathbf{161}$$

(HO: *s* 8·4; CH_2: *t* 6·23; CH_2: *tt* 8·22; CH_2: *t* 7·55; CH=CH: *d* 4·28, *dd* 3·71; CH(OH): *dt* 5·82; OH: *s* 8·4; $(CH_2)_6$: *m* 8·7; CH_3: *t* 9·12)

M^+ *m/e* 262

$\xrightarrow{-C_7H_{15}}$ *m/e* 163 $\xrightarrow{-2H_2O}$ *m/e* 127 (100%)

$\xrightarrow{-CH_2CH_2OH}$ *m/e* 217

The extracts from the leaves and stems of *Cotula plumosa* (Hook f.) Hook f.[221] were dominated in the u.v. by broad banded chromophores. The 280–290 nm band especially is rippled by a fine structure spectrum indicative of poly-enynes. High-vacuum distillation and chromatography furnished a polar compound with an ene-diyne-diene chromophore. A band at 3300/cm in the i.r. indicated an alcohol; another band at 1020/cm indicated that the alcohol function might be primary and allylic. An acetylenic peak is found at 2190/cm and absorption bands appearing at 1638 and 985/cm indicate a conjugated diene. There is no band in the 950–960/cm region, where the third double bond should be found if it occurred *trans*-configurated adjacent to a triple bond. Instead, there occurs a band at 718/cm, indicating *cis*-configuration of the third double bond. *n*-Heptadecanol was obtained by catalytic hydrogenation of the compound. Oxidation with manganese dioxide yielded an aldehyde, which showed a u.v. spectrum similar to synthetic pentadeca-(2,8,10,14)-tetraene-(4,6)-diyne-(1)-al. In view of these results the partial constitution

$$\underset{}{HOCH_2{-}}\underset{c}{CH{=}CH}{-}(C{\equiv}C)_2{-}\underset{t}{CH{=}CH}{-}\underset{t}{CH{=}CH}{-}R$$

seems to be very likely. The residue R— has been deduced from degradation. Ozonisation followed by esterification of the acid fraction with diazomethane led to methyl-*n*-heptanoate and the structure of the alcohol then must be **165**:

$$HOCH_2{-}\underset{c}{CH{=}CH}{-}(C{\equiv}C)_2{-}\underset{t,t}{(CH{=}CH)_2}{-}(CH_2)_5{-}CH_3 \quad \mathbf{165}$$

165 is the 2-*cis* isomer of oenanthetol (**166**), one of the non-toxic components isolated from the umbelliferous plant *Oenanthe crocata* L.

Oenanthetol (**166**) is accompanying oenanthetoxin (**172**) in the roots of *Oenanthe crocata* L.[219]:

$$HO{-}CH_2{-}\underset{t}{CH{=}CH}{-}(C{\equiv}C)_2{-}\underset{t,t}{(CH{=}CH)_2}{-}(CH_2)_5{-}CH_3 \quad \mathbf{166}$$

oenanthetol

The i.r. spectrum indicates the presence of disubstituted acetylenic and *trans* double bonds, and the absence of *cis* double bonds. A strong band near 3600/cm indicates the hydroxyl group, whilst hydrogenation yielded *n*-heptadecanol-(1). Use of Lindlar's catalyst produced a compound containing a conjugated pentaene chromophore; consequently, the two triple bonds are present in conjugation with three double bonds. The unsaturated system extends from $C_{(2)}$ to $C_{(11)}$ inclusive, shown by the isolation of *n*-heptanoic acid after complete oxidation with potassium permanganate. Confirmation of the assigned structure was obtained by synthesis of **166**[220], prepared by condensing the extremely useful ylide compound **167** with propargyl aldehyde. The resulting dien-yne **168** was cross-coupled with pentenynol and from the mixture oenanthetol was isolated completely identical with the material from *Oenanthe crocata*:

$$\mathrm{H{-}C{\equiv}C{-}CHO + Ph_3P{=}CH{-}CH{=}CH{-}C_6H_{13}} \qquad \textbf{167}$$

$$\mathrm{HOCH_2{-}CH{=}CH{-}C{\equiv}C{-}H + H{-}C{\equiv}C{-}(CH{=}CH)_2{-}C_6H_{13}} \qquad \textbf{168}$$

$$\downarrow \mathrm{Cu_2Cl_2/O_2}$$

$$\textbf{166}$$

Oenanthetol has also been obtained by Lythgoe *et al.*[222], who coupled pentenynol to give deca-2,8-diene-4,6-diyne-1,10-diol, the monobenzoate of which was oxidised to the aldehyde **169** with manganese dioxide. **169** was condensed with heptyl magnesiumbromide and the secondary alcohol dehydrated with toluene-*p*-sulphonic acid in boiling toluene to give all-*trans*-heptadeca-2,8,10-triene-4,6-diyne-1-ol, identical with natural oenanthetol:

$$\mathrm{BzO{-}CH_2CH{=}CH{-}(C{\equiv}C)_2CH{=}CH{-}CHO + BrMgCH_2{-}C_6H_{13}}$$

$$\downarrow \textbf{169}$$

$$\mathrm{BzOCH_2CH{=}CH{-}(C{\equiv}C)_2{-}CH{=}CH{-}\underset{\displaystyle OH}{\underset{|}{C}H}{-}CH_2{-}C_6H_{13}}$$

$$\downarrow \text{toluene-}p\text{-sulphonic acid}$$

$$\textbf{166}$$

From the aerial parts of *Opopanax chironium* Kch.[215] a complex mixture of several acetylenic compounds was obtained, containing, *inter alia*, oenanthetone **147** and the acetate **170** corresponding to oenanthetol **166**. Careful hydrolysis of **170** yielded oenanthetol, which was oxidised to the corresponding aldehyde **171**:

$$\mathrm{AcO{-}CH_2{-}CH{=}CH{-}(C{\equiv}C)_2{-}(CH{=}CH)_2{-}C_6H_{13}} \qquad \textbf{170}$$

$$\downarrow$$

$$\textbf{166}$$

$$\downarrow$$

$$\mathrm{OCH{-}CH{=}CH{-}(C{\equiv}C)_2{-}(CH{=}CH)_2{-}C_6H_{13}} \qquad \textbf{171}$$

The i.r. spectrum of **171** is characterised by a band at 2200/cm, due to acetylenic bonds; the aldehyde function is indicated by bands at 2750 and 1700/cm in the i.r. spectrum. Double bond bands appear at 995/cm (conjugated diene) and 965 cm attributable to a *trans*-disubstituted double bond.

The poisonous character of *Oenanthe crocata* has been known for many centuries. It is said to have been used for both medicinal and criminal purposes. The most poisonous parts of *Oenanthe crocata* are the roots. They resemble a bunch of small white carrots giving the plant the name "five finger death", and exude considerable quantities of a yellow oil when broken. By extraction of these roots with ether and chromatography of the extract, the toxic material oenanthetoxin (**172**) was isolated as an extremely unstable colourless crystalline diol with a composition agreeing to $C_{17}H_{22}O_2$[219]. *Oenanthe crocata* is responsible for much poisoning of stock and especially children in cases of poisoning among humans[223].

$$\underset{}{HO-CH_2-}\underset{t}{CH=CH}-(C\equiv C)_2-\underset{t,t}{(CH=CH)_2}-CH_2CH_2-\underset{\displaystyle OH}{\underset{|}{CH}}-C_3H_7 \qquad \mathbf{172}$$

oenanthetoxin

The toxin exhibits an absorption spectrum typical of a polyacetylene with an ene-diyne-diene chromophore. Accompanying the toxin in the plant is the relatively non-toxic alcohol oenanthetol (**166**), the ketone oenanthetone (**147**), and many polyacetylenes closely related to the formers[216]. **172** is an optically active laevorotatory carbinol which was hydrogenated to heptadecane-(1,14)-diol by absorption of seven moles of hydrogen. While using Lindlar's catalyst a substance containing a conjugated pentaene chromophore was produced. Consequently the triple bonds (indicated in the i.r. spectrum as bands at 2200 and 2120/cm) are present in conjugation with three double bonds. Oenanthetoxin dimethyl ether forms a maleic anhydride adduct with the u.v. spectrum of an ene-diyne, so oenanthetoxin itself must be either **173** or **174**:

$$R-CH=CH-(C\equiv C)_2-(CH=CH)_2-R \qquad R-(C\equiv C)_2-(CH=CH)_3-R$$

173 **174**

Synthesis of a model compound containing the ene-diyne-diene system **173** and the complete identity of the spectra of the synthetic and natural materials afforded convincing proof of the presence of the ene-diyne-diene system. The Grignard reagent from hept-2-ene-4,6-diyne **175** was condensed with hexa-2,4-dienal **176** and the resulting carbinol **177** rearranged by treatment with acid to the model chromophore compound **178**:

$$H_3C{-}CH{=}CH{-}(C{\equiv}C)_2{-}MgBr + OCH{-}(CH{=}CH)_2{-}CH_3$$

175 **176**

$$\downarrow$$

$$H_3C{-}CH{=}CH{-}(C{\equiv}C)_2{-}\underset{\displaystyle OH}{\underset{|}{CH}}{-}(CH{=}CH)_2{-}CH_3 \quad \mathbf{177}$$

$$\downarrow$$

$$H_3C{-}CH{=}CH{-}(C{\equiv}C)_2{-}(CH{=}CH)_2{-}\underset{\displaystyle OH}{\underset{|}{CH}}{-}CH_3 \quad \mathbf{178}$$

The orientation of this unsaturated system in the oenanthetoxin molecule was determined by oxidation of the maleic anhydride adduct of oenanthetoxin dimethyl ether. The production of methoxyacetic acid shows oenanthetoxin to be all-*trans* heptadeca-2,8,10-triene-4,6-diyne-1,14-diol.

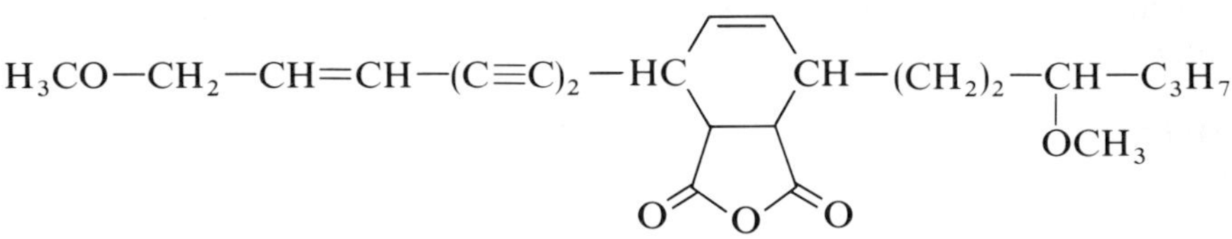

$$\downarrow MnO_4^-$$

$$H_3CO{-}CH_2{-}CO_2H$$

The assigned structure of oenanthetoxin was finally confirmed by synthesis[220]. Oenanthetoxin was obtained in the racemic form by coupling of pent-2-ene-4-yne-1-ol with the ketone **179**. The hydroxyketone from the cross-coupling was then reduced with lithium aluminium hydride to give (±)-oenanthetoxin **172**, melting 20° below the natural (+)-form but with practically identical i.r. and u.v. spectra:

$$HOCH_2{-}CH{=}CH{-}C{\equiv}C{-}H$$

$$+ H{-}C{\equiv}C{-}(CH{=}CH)_2(CH_2)_2{-}CO{-}C_3H_7 \quad \mathbf{179}$$

$$\downarrow Cu_2Cl_2/O_2$$

$$HO{-}CH_2{-}CH{=}CH{-}(C{\equiv}C)_2{-}(CH{=}CH)_2{-}CH_2CH_2{-}CO{-}C_3H_7$$

180

$$\downarrow LiAlH_4$$

(±)-oenanthetoxin **172**

Careful reinvestigation of *Oenanthe crocata* showed that the roots contain the monoacetate **181** of oenanthetoxin too[216]:

$$AcO-CH_2-CH=CH-(C\equiv C)_2-(CH=CH)_2-CH_2CH_2-\underset{\displaystyle OH}{\underset{|}{CH}}-C_3H_7 \qquad \textbf{181}$$

The acetate bands appear in the i.r. spectrum at 1750 and 1250/cm. Saponification of the acetate furnished an alcohol which, after crystallisation, melted at 87°C. It gave no melting point depression with **172**. Their u.v. and i.r. spectra were identical, the latter also with regard to the intensity relations of the acetylenic bands at 2200 and 2120/cm, the second has about double the intensity of the first. The comparison of i.r. curves of a number of synthetic and naturally occurring acetylenic compounds reveal that a 2120/cm peak intensity higher than that of the 2200/cm peak is characteristic of some trans-isomers, the *cis*-isomers exhibit the normal intensity relations.

The ketone **180** was also obtained from *Oenanthe crocata* L.[216]. Manganese dioxide oxidation yielded the keto-aldehyde **182**, the structure of which was elucidated by spectral methods, especially n.m.r. and mass spectrometric analysis:

$$HO-CH_2-CH=CH-(C\equiv C)_2-CH=CH-$$

s 8·32 *dd* 5·76 ($J = 5 + 1·6$) *dt* 3·59 ($J = 15 + 5$) *dm* 4·14 ($J = 1·5$) *d* 4·44 *dd* 3·33

$$CH=CH-(CH_2)_2-CO-CH_2CH_2CH_3 \qquad \textbf{180}$$

m 3·7–4·3 *m* 7·6 *t* 7·63 *tq* 8·41 *t* 9·1

$$\downarrow MnO_2$$

$$OCH-CH=CH-(C\equiv C)_2-CH=CH-CH=CH-(CH_2)_2-CO-C_3H_7 \qquad \textbf{182}$$

dd 0·45 ($J = 5 + 1·7$) *s* 3·44

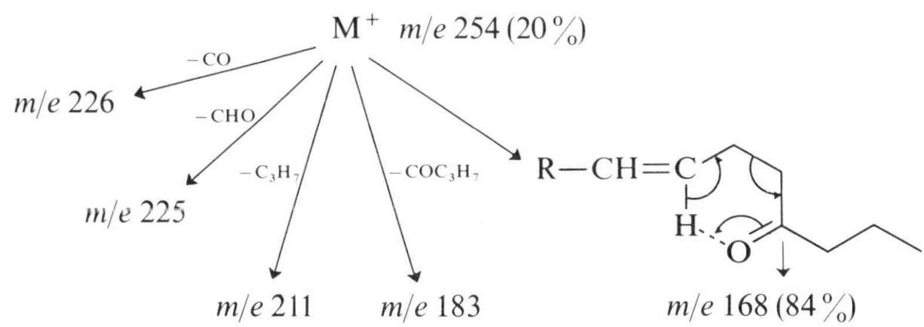

The strong tendency to form a McLafferty rearrangement with an olefinic proton is noteworthy, although there are enough saturated protons in **182**. A similar behaviour is observed by mass spectroscopic measurements of oenanthetone **147**. Like *Oenanthe crocata*, the water hemlock *Cicuta virosa*

has been known for many centuries to contain convulsant poisons and has been responsible for numerous fatalities amongst humans and animals. The root of *Cicuta virosa* is a bulbous structure containing many air cells. Its extraction, followed by chromatography, gave two pure main constituents. The toxic substance cicutoxin (**183**), isomeric with oenanthetoxin (**172**) and cicutol (**184**), isomeric with oenanthetol (**168**). Like oenanthetoxin, cicutoxin (**183**) is optically active too, but with opposite rotations ($[\alpha]_D^{36}$: $-14{\cdot}5$).

$$HO{-}CH_2CH_2CH_2{-}(C{\equiv}C)_2{-}(CH{=}CH)_3{-}\underset{\displaystyle OH}{\underset{|}{CH}}{-}C_3H_7 \quad \textbf{183}$$

cicutoxin

$$HO{-}CH_2CH_2CH_2{-}(C{\equiv}C)_2{-}(CH{=}CH)_3{-}CH_2{-}C_3H_7 \quad \textbf{184}$$

cicutol

172, **168**, **183**, and **184** form colourless low melting crystals which were transformed in air and light into brown insoluble resins. But they can be stored unchanged for some weeks below 0°C under oxygen-free nitrogen in the dark, and survive recrystallisation from warm solvents provided air and light were excluded.

In a manner analogous to **172**, cicutoxin was shown to contain the alternative triene-diyne chromophore **174**[219]. Oxidation of the maleic anhydride adduct of cicutoxin dimethyl ether (obtained by reaction of **183** with methyl iodide and silver oxide) gave γ-methoxy butyric acid, establishing the constitution of cicutoxin as all-*trans* heptadeca-8,10,12-triene-4,6-diyne-1,14-diol:

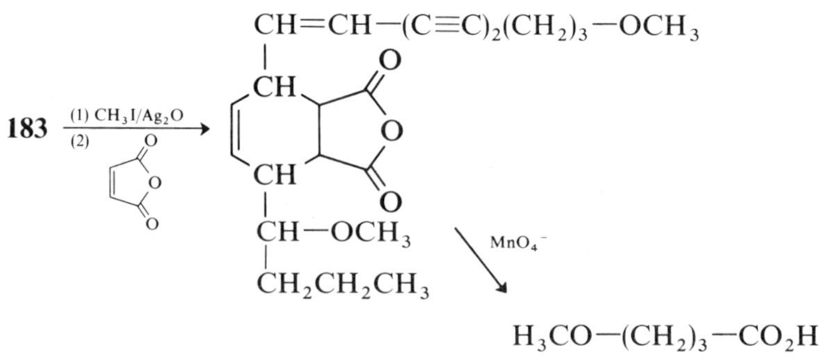

α-Methoxy-*n*-valeric acid, which would have been formed instead if cicutoxin had the structure **185**, was shown to be absent.

$$HO{-}(CH_2)_3{-}(CH{=}CH)_3{-}(C{\equiv}C)_2{-}\underset{\displaystyle OH}{\underset{|}{CH}}{-}C_3H_7 \quad \textbf{185}$$

In order to confirm the structure **183** was synthesised[222]. The method chosen is essentially the same as that described for the preparation of the model compound **178**. In the case of cicutoxin deca-2,4,6-trienal (**186**) was condensed with hepta-4,6-diyn-1-ol, the hydroxyl-group of the latter being protected as the tetrahydropyranyl ether **187**. This protecting group was removed under conditions which brought about the allylic rearrangement of the condensation product to give (±)-cicutoxin with light absorption almost identical with that of natural (−)-cicutoxin:

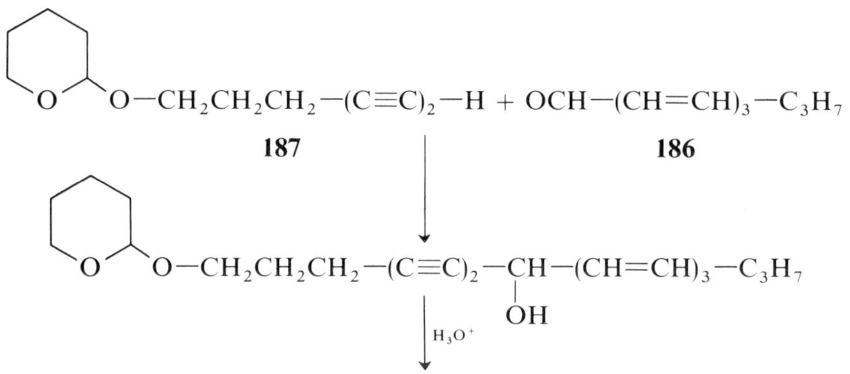

183 (anionotropic rearrangement)

For the preparation of deca-2,4,6-trienal (**186**), hex-2-en-1-al was chosen as the starting material. It was obtained by reducing hex-2-enoic acid with lithium aluminium hydride to hex-2-en-1-ol which was oxidised with manganese dioxide to hex-2-en-1-al. Reaction of the aldehyde with the lithium derivative of ethoxy-vinylacetylene[224] gave crude 1-ethoxy-deca-1,6-dien-3-yne-5-ol, which after partial hydrogenation with Lindlar's catalyst and subsequent treatment with acid, gave the trienal **186**.

Cicutol (**184**) was prepared by cross-coupling pent-4-yn-1-ol with dodeca-3,5,7-triene-1-yne (**188**). The latter compound was obtained by oxidation of pentenynol to the aldehyde **189** followed by condensation with the ylide **190** from hept-5-enyl-bromide[225]:

$$\underset{\mathbf{189}}{H-C\equiv C-CH=CH-CHO} + \underset{\mathbf{190}}{Ph_3P=CH-CH=CH-C_4H_9}$$

$$\downarrow$$

$$HO-(CH_2)_3-C\equiv C-H + H-C\equiv C-(CH=CH)_3-C_4H_9 \quad \mathbf{188}$$

$$\downarrow Cu_2Cl_2/O_2$$

cicutol **184**

Reinvestigation of roots from *Cicuta virosa* showed the presence of further polyacetylenes closely related to cicutoxin and cicutol[218]. Separation of the alcohol fraction gave two hydroxy ketones, which had to be conjugated ketones, as the alcohols obtained by sodium borohydride reduction yielded the natural hydroxy ketones, when oxidised with manganese dioxide.

On the basis of chemical and spectroscopic evidence, their structures were shown to be **191** and **192**:

$$HO-CH_2-CH_2-CH_2-(C\equiv C)_2-CH=CH-CH=CH-CH=CH-CO-CH_2-CH_2-CH_3 \quad \mathbf{191}$$

$$HO-CH_2CH_2CH_2-(C\equiv C)_2-(CH=CH)_2-CO-C_5H_{11} \quad \mathbf{192}$$

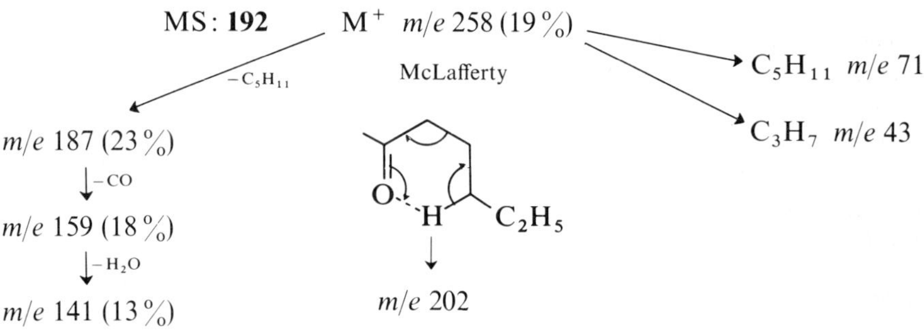

Sodium borohydride reduction of **191** yielded a diol, which was identical by spectral comparison with authentic material of cicutoxin (**183**). Otherwise a hydroxy ketone was obtained from cicutoxin by oxidation with manganese dioxide which proved to be identical with **191**.

The most polar fractions of *Cicuta virosa* contain besides cicutoxin a further diol, which could be separated after oxidation with manganese dioxide to a hydroxy ketone by TLC. The so obtained hydroxy ketone was identical with **192** in all spectroscopic properties, therefore, the structure of the diol has to be **193**:

$$HO-(CH_2)_3-(C\equiv C)_2-(CH=CH)_2-\underset{\substack{|\\ OH}}{CH}-C_5H_{11} \quad \mathbf{193} \xrightarrow{MnO_2} \mathbf{192}$$

2,3-Dihydro-oenanthetol (**194**) and the corresponding acetate **195** were obtained from the aerial parts of *Opopanax chironium* Kch.[215].

The acetate showed a carbonyl peak at 1750/cm in the i.r. spectrum and gave further bands consistent with the presence of triple bonds (2240, 2150/cm) and a *trans-trans*-diene-grouping (1640, 990/cm). The n.m.r.

spectrum confirmed the presence of the aceto-group and the conjugated *trans*-double bonds:

$$H_3CCO{-}O{-}CH_2CH_2CH_2{-}(C{\equiv}C)_2{-}$$

s 8·0 *t* 5·89 *m* 8·11 *t* 7·58

$$CH{=}CH{-}CH{=}CH{-}CH_2{-}(CH_2)_4{-}CH_3 \quad \mathbf{195}$$

d 4·58 *dd* 3·94 *dd* 3·41 *dt* 4·25 *m* 7·85 *m* 8,7 *t* 9·1

The n.m.r. data of the alcohol **194** only differs in the $HO{-}(CH_2)_3$ grouping. **195** was also converted to **194** by careful saponification. The n.m.r. of the alcohol so obtained was identical to that of the natural compound.

$$HO{-}CH_2{-}CH_2{-}CH_2{-}(C{\equiv}C)_2{-}(CH{=}CH)_2{-}C_6H_{13} \quad \mathbf{194}$$

s 8·7 *t* 6·31 *m* 7·95 *t* 7·55

The more polar fractions of *Oenanthe crocata* contain some polyacetylene compounds having obviously structural and functional relationships to oenanthetoxin (**172**) and cicutoxin (**183**), respectively[216]. The main difference between **172** and **183** is a reduced conjugation system caused by hydrogenation of one or two double bonds. Basing on n.m.r. and mass spectroscopic measurements the structures of the four compounds were assigned to be **196**, **197**, **198**, and **199**:

$$HOCH_2CH_2CH_2{-}(C{\equiv}C)_2{-}CH{=}CH{-}\underbrace{CH{=}CH}_{m\ 3\cdot7-4\cdot2}{-}$$

s 8·8 *t* 6·3 *tt* 8·24 *d* 4·55 *dd* 3·42

$$\underbrace{CH_2CH_2}_{m\ 7\cdot6}{-}CO{-}CH_2CH_2CH_3 \quad \mathbf{196}$$

t 7·7

$$HOCH_2CH_2CH_2{-}(C{\equiv}C)_2{-}CH{=}CH{-}CH{=}CH{-}$$

s 7·9

$$\underbrace{CH_2CH_2}_{m\ 8\cdot0}{-}\overset{m\ 6\cdot4}{CH}(OH){-}\underbrace{CH_2CH_2CH_3}_{m\ 8\cdot6} \quad \mathbf{197}$$

OH: *s* 7·9

$$AcO{-}(CH_2)_3{-}(C{\equiv}C)_2{-}(CH{=}CH)_2{-}CH_2CH_2{-}CH(OH){-}C_3H_7 \quad \mathbf{198} \xrightarrow{OH^-} \mathbf{197}$$

$$HO{-}CH_2CH_2CH_2{-}(C{\equiv}C)_2{-}CH{=}CH{-}CH_2{-}$$

d 4·55 *m* 3·8 *m* 7·6

$$(CH_2)_2{-}CH_2{-}CO{-}CH_2CH_2CH_3 \quad \mathbf{199}$$

m 8·6 *m* 7·6 *t* 7·7 *tq* 8·45 *t* 9·1

The occurrence and sometimes simultaneous occurrence in the plants of the polyacetylenic compounds described in Section 1.2.6 provides evidence

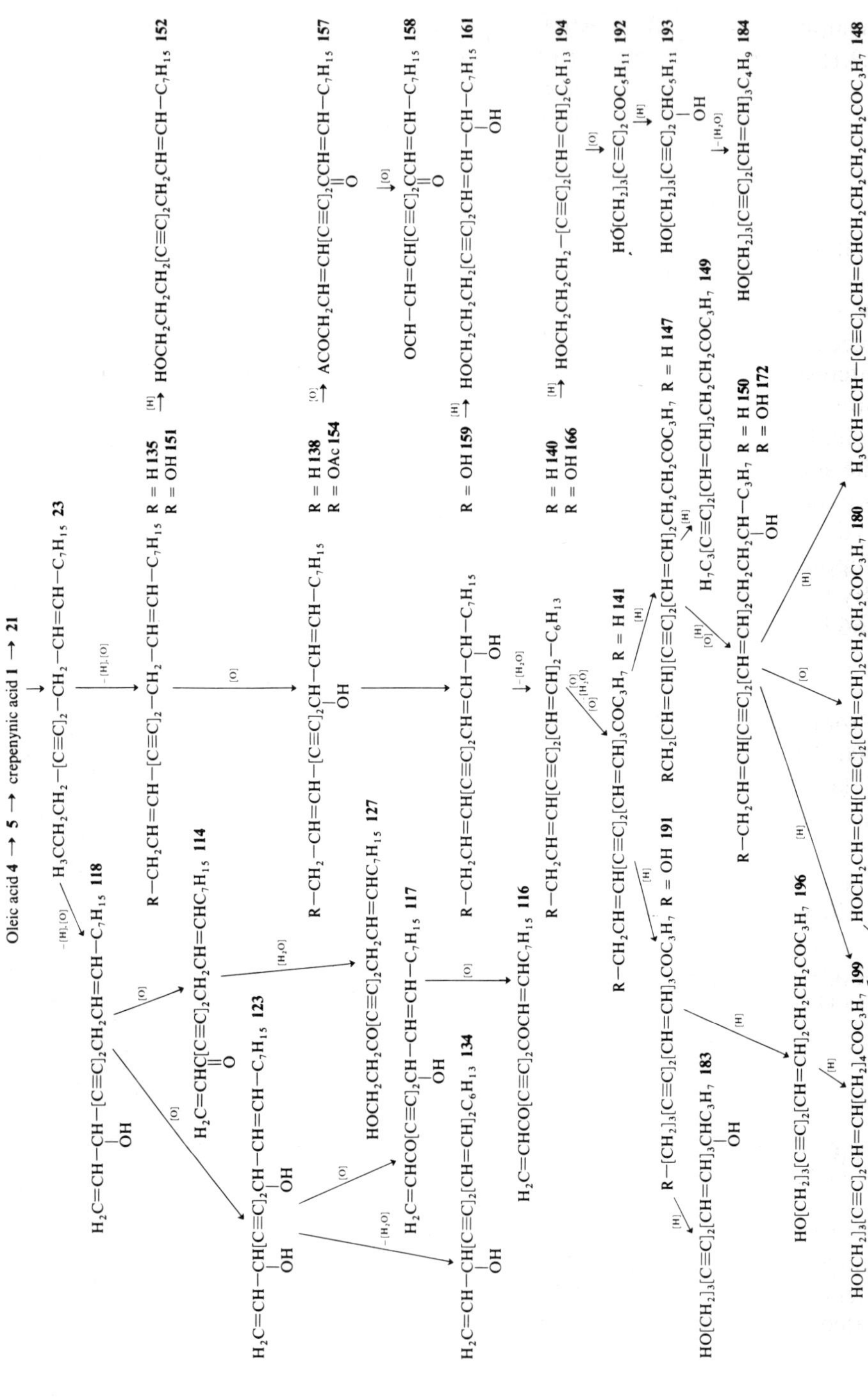

in support of the following biosynthetic reaction sequences (see also Section 1.2.1):

$$\text{oleic acid } \mathbf{4} \rightarrow \mathbf{5} \rightarrow \mathbf{1} \rightarrow \mathbf{21}$$
$$\downarrow$$
$$C_3H_7{-}(C{\equiv}C)_2{-}CH_2{-}CH{=}CH{-}R \quad \mathbf{23}$$

It is an open question, whether **23** consists of a C_{17}-chain ($R = -C_7H_{15}$) or possesses the intact C_{18}-chain length of the unsaturated acid ($R = -(CH_2)_7-CO_2R$). At which stage chain-shortening intervenes still needs to be established. Supposing that the former is the case, some very interesting reactions and rearrangements of fundamental nature must prevail in the plants (see the scheme on page 214). As yet there exists no definitive evidence for the exact mechanism, but it has been suggested that the simultaneous occurrence of **196** and **199**, or **147** and **148**, in the plants provides support for the existence of specific enzymes, which are able to hydrogenate double bonds to saturated groups in the umbelliferous plants. Support for this contention is found in the conversion of labelled **166** to **194**. (By tracer studies the authors showed[35] that **194** was wholly derived from **166** including hydrogenation of a double bond.)

1.2.7. C_{17}-compounds with oxygen-function at $C_{(1)}$

From the leaves of *Centaurea involucrata*[36], some homologous aldehydes **200**, **201**, and **202** were obtained and their structures elucidated by spectral data. The i.r. shows the presence of an aldehyde group (2730, 1740/cm) and the all-*trans* configuration of the double bonds (1640, 995, 960/cm). The mass spectrum exhibits peaks due to a homologous series of aldehydes with the chain length C_{15}, C_{16}, and C_{17}. Reduction of **201** with sodium borohydride yielded the C_{16}-alcohol **203**, already known from *Dahlia* species:

$$H_3C{-}CH{=}CH{-}(C{\equiv}C)_2{-}CH{=}CH{-}CH{=}CH{-}(CH_2)_n{-}CHO$$

dd 8·15 *dq* 3·82 *d* 4·52 *d* 4·52 *dd* 3·93 *dd* 3·43 *dt* 4·15 *m* 7·6–8·6 *t* 0·32

200: n = 5
201: n = 4
202: n = 3

$$H_3C{-}CH{=}CH{-}(C{\equiv}C)_2{-}(CH{=}CH)_2{-}CH_2{-}(CH_2)_3{-}CH_2OH \quad \mathbf{203}$$

dd 8·12 *dt* 7·83 *m* 8·55 *t* 6·41 *s* 9·07

Baccharis trinervis Pers. contains some C_{17}-esters, which perhaps are biogenetically important[226]. From the aerial parts of *Baccharis trinervis* Pers., three compounds with an ene-diyne-diene-chromophore were isolated. The least polar compound shows an ester grouping (1745/cm) and all-*trans* configurated double bonds (1638, 995, 958/cm) in the i.r. The mass spectrum exhibits the following fragmentation pattern:

M^+ *m/e* 270 $-OCH_3$ → *m/e* 239

→ $(H_3CCH{=}CH(C{\equiv}C)_2(CH{=}CH)_2CH_2^+)$

$-CH_2CO_2CH_3$ → *m/e* 197 *m/e* 155

$-CH_2CH_2CO_2CH_3$ → *m/e* 183

These data and, in particular, the n.m.r. data strongly indicate that the C_{17}-ester has structure **204**:

$H_3C{-}CH{=}CH{-}(C{\equiv}C)_2{-}CH{=}CH{-}CH{=}CH{-}$

dd 8·16 *dq* 3·8 *dm* 4·48 *d* 4·52 *dd* 3·42 *dd* 3·92 *dt* 4·25

$CH_2{-}(CH_2)_3{-}CH_2{-}CO_2CH_3$ **204**

dt 7·87 *m* 8·4 *t* 7·77 *s* 6·4

The most polar compound is a β-hydroxy ester (3580, 1760/cm). The mass spectrum shows a molecular ion peak due to $C_{18}H_{22}O_3$. In view of the n.m.r. data, it seemed probable that the unknown hydroxy ester is structurally related to **204**. Only the triplet at τ 7·77 is supplied by a multiplet at τ 7·65, which is typical for β-hydroxy esters. A multiplet at τ 6·12 could be ascribed to the proton on the same carbon atom as the hydroxyl group. The foregoing data lead to structure **205**:

$H_3C{-}CH{=}CH{-}(C{\equiv}C)_2{-}CH{=}CH{-}CH{=}CH{-}CH_2{-}$

dd 8·15 *d* 4·51 *dd* 3·33 *m* 3·8 *dt* 4·23 *dt* 7·85

m 6·12

$(CH_2)_2{-}CH(OH){-}CH_2{-}CO_2CH_3$ **205**

m 8·4 *m* 7·65 *s* 6·32

Finally, a compound was isolated in very small amounts, which had to be a methyl ester with an additional acetoxy group. In order to obtain it in a pure state, it was saponified and converted, using diazomethane, to the corresponding hydroxymethyl ester. The mass spectrum shows that the unknown compound is isomeric to **205**, probably the 17-hydroxy isomer of **205**. Only structure **206** could explain the cleavage of CH_2O in the mass spectrum of the alcohol:

$Ac{-}O{-}CH_2{-}CH{=}CH{-}(C{\equiv}C)_2{-}(CH{=}CH)_2{-}(CH_2)_5{-}CO_2CH_3$ **206**

A reasonable biosynthetic pathway for the C_{17}-aldehydes including α-oxidation is outlined in the scheme below:

$$-CH_2-CO_2H \xrightarrow{[O]} -\underset{\displaystyle OH}{\underset{|}{C}}H-CO_2H \rightarrow -CHO \xrightarrow{[H]} -CH_2OH$$

2. C_{16}- and C_{15}-acetylenes

2.1. C_{16}-acetylenes

Until recently C_{16}-acetylenic compounds have been very rare. A characteristic feature of the C_{16}-polyacetylenes is an aldehyde or alcohol function at $C_{(1)}$. As already mentioned, these compounds often occur with acetylenes of different chain lengths, which seems to be an indication that most of them are formed biogenetically by α-oxidation.

From the aerial parts of *Centaurea macrocephala* Puschk.[227], a mixture of two aldehydes was isolated, both containing an ene-diyne-chromophore. In order to obtain them in a pure state, they were converted using sodium borohydride to the corresponding alcohols and subsequently subjected to preparative TLC. The spectral data of the alcohols are only consistent with the structures **209** and **210**, and accordingly the original aldehydes have to be **207** and **208**:

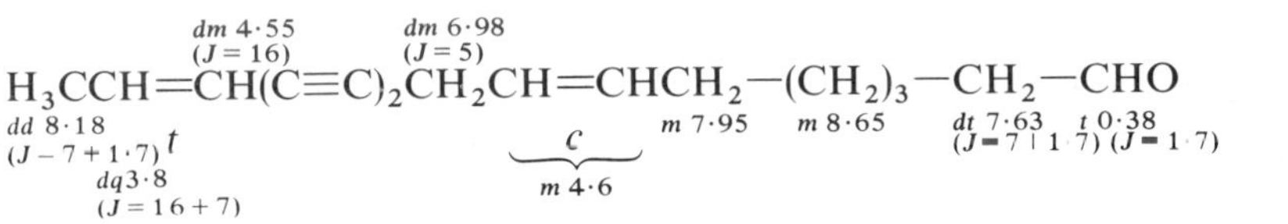

207

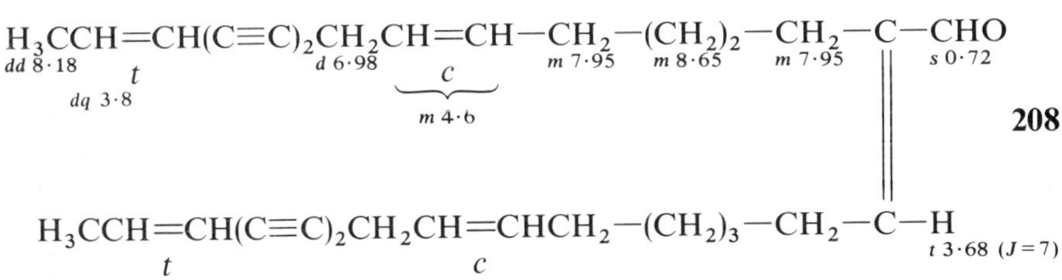

208

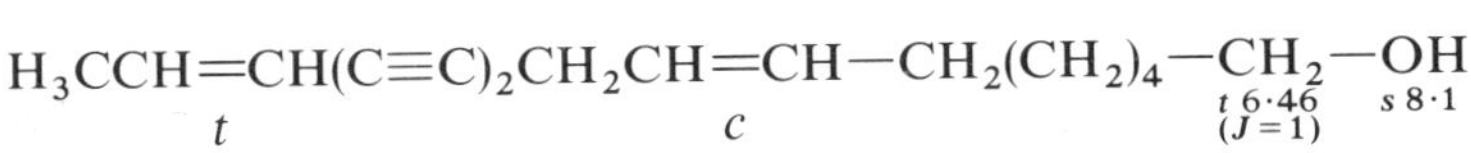

209

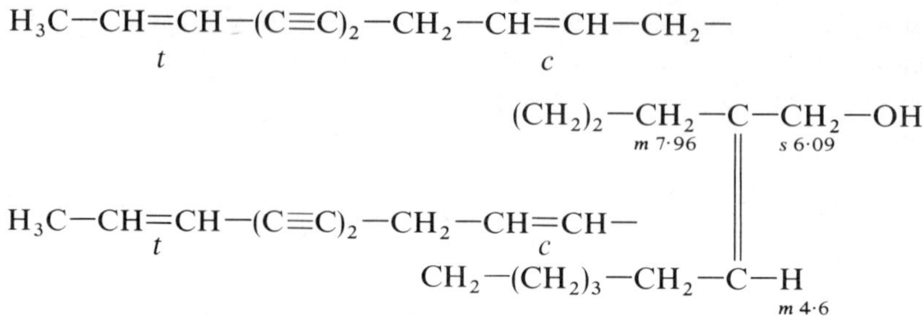

210

It is proposed that **208** is an artificial product arising from **207** by aldol condensation.

The more polar fractions of *Centaurea macrocephala* contain a compound with an ene-diyne-diene-chromophore (λ max: 336, 314, 295, 278, 265, 248 nm). Its i.r. spectrum displays bands for an acetate function (1750, 1250/cm). In the mass spectrum, a molecular ion peak at m/e 284 due to $C_{18}H_{20}O_3$ is observed. The structural assignments for the acetate **211** are mainly derived from its n.m.r. spectrum. The axial position of the *O*-acetyl-group is concluded from the line separation of proton (I):

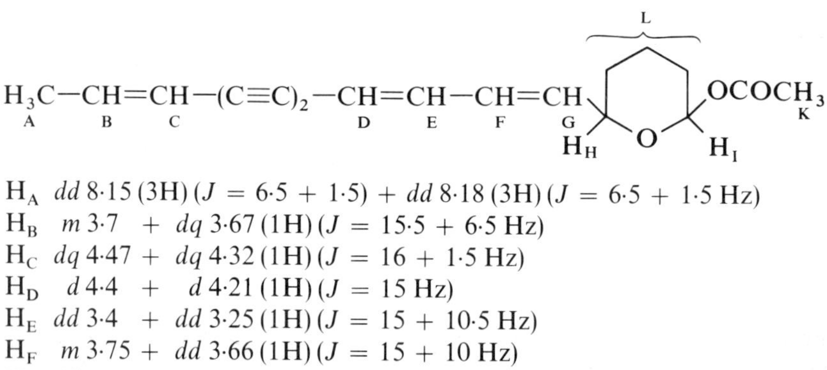

211

H_A *dd* 8·15 (3H) (J = 6·5 + 1·5) + *dd* 8·18 (3H) (J = 6·5 + 1·5 Hz)
H_B *m* 3·7 + *dq* 3·67 (1H) (J = 15·5 + 6·5 Hz)
H_C *dq* 4·47 + *dq* 4·32 (1H) (J = 16 + 1·5 Hz)
H_D *d* 4·4 + *d* 4·21 (1H) (J = 15 Hz)
H_E *dd* 3·4 + *dd* 3·25 (1H) (J = 15 + 10·5 Hz)
H_F *m* 3·75 + *dd* 3·66 (1H) (J = 15 + 10 Hz)
H_G *dd* 4·32 + *dd* 4·12 (1H) (J = 15 + 5 Hz)
H_H *dm* 5·7 + *dm* 5·58 (1H) (J = 11 Hz)
H_I *m* 3·9 + *m* 3·88 (1H)
H_K *s* 7·98 + *s* 7·97
H_L *m* 8·28 + *m* 8·30 (6H)

From the most polar fractions of *Centaurea macrocephala*, a mixture of two alcohols was obtained, which could not be separated. In view of the spectral results it seemed probable that the alcohols were structurally related to the former described acetate **211**. Obviously, it was the mixture of the two epimeric alcohols. Evidence for this was obtained by converting the alcohol mixture to the corresponding acetates. One of them proved to be identical with **211**. However, neither the alcohols nor the corresponding acetates could be separated. The structures must be **212** and **213**:

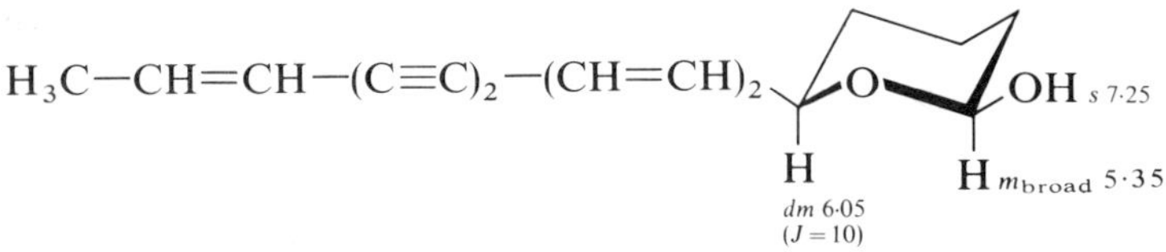

212

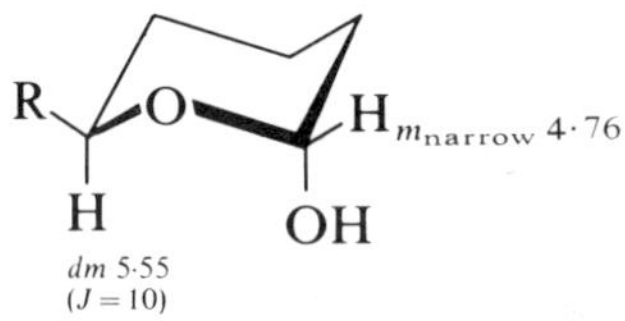

213

Obviously there exists a close relationship between **207** and the tetrahydropyran derivatives **212** and **213**. A reasonable biosynthetic route for these compounds is outlined in the scheme below:

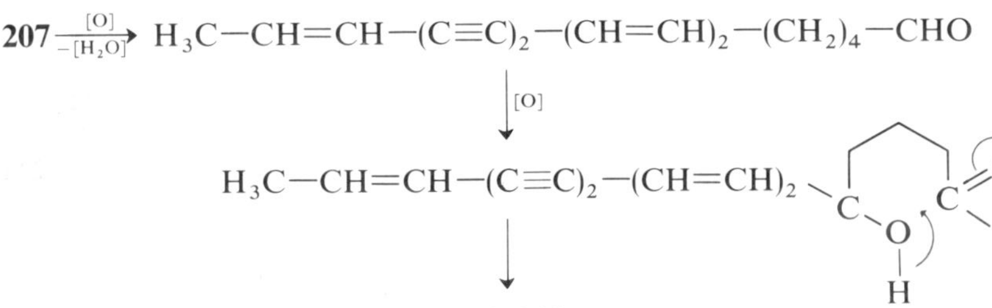

214

212, 213

The aerial parts of *Dahlia merckii* Lehm.[197] contain a complex mixture of acetates and some alcohols. Purification of the alcohol fraction by repeated column chromatography and TLC led to the isolation of a crystalline alcohol, which was identical with the saponification product of the also isolated acetate. The u.v. spectrum reveals an extended diene-yn-diene-chromophore (λ max: 337, 315 nm). The presence of a triple bond was shown by the i.r. only in high concentrations (2180, 2130/cm). In addition, the i.r. spectrum confirms the presence of a *trans-trans*-diene-grouping (1645, 988/cm). The n.m.r. spectrum consists of a complex multiplet in the range 3·15–4·5τ, due to eight olefinic protons. The methyl group adjacent to a diene grouping is indicated by a typical multiplet at τ 8·22. Furthermore, the spectrum includes a triplet at τ 6·35, attributable to a $-CH_2-O$-group, a multiplet for a methylene group adjacent to double bonds, and a multiplet at τ 8·55 for three methylene groups. Taking all these facts into consideration, only structure **216** and **217** can be suggested:

$$H_3C{-}\underset{t,t}{(CH{=}CH)_2}{-}C{\equiv}C{-}\underset{t,t}{(CH{=}CH)_2}{-}(CH_2)_4{-}CH_2{-}O{-}R$$

216: R = H
217: R = $COCH_3$

The structure was proved by hydrogenation of **216** to the corresponding saturated alcohol and formation of *n*-hexadecanol-(1)-acetate by acylation. The latter was identified and compared with authentical material by GLC.

The other two alcohols were not obtained in a pure state. Transformation into the corresponding azobenzoates led to crystalline products, which could be separated by repeated TLC. The less polar ester showed an u.v. maximum at 264 nm. Its i.r. spectrum exhibited bands due to a *trans-trans*-diene grouping and weak triple bond bands at 2250 and 2220/cm. The n.m.r. spectrum of the alcohol exhibited, *inter alia*, a three-hydrogen multiplet at τ 8·25, attributable to a methyl group adjacent to a *trans-trans*-diene grouping. A broadened doublet at τ 6·97 could be ascribed to the methylene group linking a triple bond with a further double bond. Furthermore, the n.m.r. showed a triplet at τ 5·7 indicating a $-CH_2-O$-group adjacent to a $-CH_2$-group, and a multiplet due to five methylene groups. The stereochemistry of the isolated double bond could be assigned on the basis of the i.r. spectrum, which lacked a corresponding *trans* band. The above evidences indicate that the alcohol has structure **218**:

$$H_3C{-}\underset{t,t}{(CH{=}CH)_2}{-}C{\equiv}C{-}CH_2{-}\underset{c}{CH{=}CH}{-}(CH_2)_5{-}CH_2{-}O{-}R$$

218: R = H
219: R = $COC_6H_4{-}N{=}N{-}C_6H_5$

Alkaline saponification of **219** yielded besides **218** the pentaene **220**, arisen by isomerisation.

$$\mathbf{219} \xrightarrow{OH^-} \mathbf{218} + H_3C{-}(CH{=}CH)_5{-}(CH_2)_4{-}CH_2{-}OH \quad \mathbf{220}$$

The more polar ester could be nearly obtained pure by repeated TLC. Besides the azobenzoate maximum the u.v. spectrum exhibits bands due to a diyne-ene-chromophore (λ max: 281, 265, 251 nm). The n.m.r. spectrum shows the presence of four olefinic protons as well as a signal attributable to an isolated methylene group. Furthermore, it confirms the presence of a methyl group adjacent to a double bond (*dd* 8·2τ (3H) (J = 7 + 1·5 Hz) and the $-CH_2O-$ grouping, and exhibits resonances due to five methylene groups in the range 7·8–8·7τ. Only one *trans*-disubstituted double bond is recognisable in the i.r. spectrum; therefore, the structure has to be **222**:

$$H_3C{-}\underset{t}{CH{=}CH}{-}(C{\equiv}C)_2{-}CH_2{-}\underset{c}{CH{=}CH}{-}(CH_2)_5{-}CH_2OR$$

222: R = H
221: R = $COC_6H_4{-}N{=}N{-}C_6H_5$

The alcohol **222** could not be obtained pure by saponification. The only product was the isomeric compound analogous to **220**.

Dahlia Merckii also produces the corresponding conjugated alcohol **223** in appreciable quantity[22]:

$$H_3C{-}CH{=}CH{-}(C{\equiv}C)_2{-}(CH{=}CH)_2{-}(CH_2)_4{-}CH_2OH \quad \mathbf{223}$$

From the aerial parts of *Serratula gmelini* Ldb. an aldehyde was isolated, which showed a dien-yne-diene chromophore in the u.v.[196]. The i.r. spectrum indicates the presence of an aldehyde group (2730, 1737/cm) and the existence of a *trans-trans*-diene-grouping (995/cm). These observations are consistent with structure **224**, which was proved by reduction of **224** to the known alcohol **216**:

$H_3C{-}CH{=}CH{-}CH{=}CH{-}C{\equiv}C{-}CH{=}CH{-}$

d 8·21 ($J = 6·5$) | *m* 3·7–4·4 | *dd* 3·57 ($J = 15+10$) *d* 4·49 ($J = 15$) | *d* 4·49 ($J = 15$) *dd* 3·57 ($J = 15+10$)

$CH{=}CH{-}CH_2{-}(CH_2)_2{-}CH_2{-}CHO$ **224**

m 3·9–4·4 | *m* 7·9 | *m* 8·7 | *m* 7·63 | *t* 0·31 ($J = 1·5$)

M$^+$ *m/e* 228 (30%) → $CH_3(CH{=}CH)_2C{\equiv}C(CH{=}CH)_2CH_2^+$ *m/e* 157 (100%)

M$^+$ *m/e* 228 (30%) → $CH_3(CH{=}CH)_2C{\equiv}C(CH{=}CH)_2CH_2CH_2^+$ *m/e* 171 (20%)

M$^+$ *m/e* 228 (30%) →

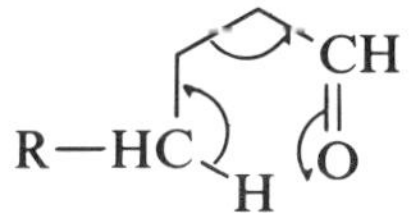

→ CH_3CHO^+ *m/e* 44 (100%)

The occurrence of dien-yne- and ene-diyne-compounds together is of biogenetical interest. As already discussed in Section 1.2.1, it seems to be probable that the acetylenes of this type are closely related. While in the case of the C_{17}-compounds the dehydrogenation of dien-ynes to ene-diynes has been confirmed, in the case of the C_{16}-acetylenes a similar pathway only can be supposed represented in the following scheme:

$$H_3C{-}(CH{=}CH)_2{-}C{\equiv}C{-}CH_2{-}\underset{cis}{CH{=}CH}{-}(CH_2)_6{-}OR \quad \mathbf{218}$$

$$\downarrow [O], -[H_2O]$$

$$H_3C{-}(CH{=}CH)_2{-}C{\equiv}C{-}(CH{=}CH)_2{-}(CH_2)_5{-}OR \quad \mathbf{216}$$

$$\swarrow -[H] \qquad\qquad\qquad \searrow [O]$$

$$\mathbf{223} \xrightarrow{[O]} \mathbf{214} \xrightarrow{[O]} \mathbf{212} \qquad\qquad \mathbf{224}$$

2.2. C_{15}-acetylenes

Structural analysis of the various members of the C_{15}-acetylenes reveals that these compounds may be related biosynthetically to the C_{17}-acetylenes described in Section 1.2. In many cases, structural features of special C_{17}-compounds are recognisable also in the C_{15}-series. Therefore, the biogenetic route of formation of the C_{15}-compounds seems to be relatively simple, including chain-shortening by β-oxidation in the "proximal" part of the C_{17}-molecules. For example, in the C_{15}-compound **225** most of the molecule of **127** is recognisable:

$$HOCH_2CH_2{-}CO{-}(C{\equiv}C)_2{-}CH_2{-}CH{=}CH{-}CH_2CH_2{-}CH{=}CH_2$$

225

225 was obtained from the aerial parts of *Cotula coronopifolia* L.[228]. It could not be obtained in a pure state and gave **226** on heating with acetanhydride.

$$\mathbf{225} \xrightarrow{Ac_2O} H_2C{=}CH{-}CO{-}(C{\equiv}C)_2{-}$$
$$CH_2{-}CH{=}CH{-}(CH_2)_2{-}CH{=}CH_2 \quad \mathbf{226}$$

The i.r. spectrum of **226** was nearly identical to that of dehydrofalcarinone (**52**). Besides **225**, a mixture of two alcohols was obtained from *Cotula coronopifolia* too. Both are showing a diyne-diene-chromophore. Hydrogenation and acylation of the mixture yielded *n*-pentadecanol-(3)-acetate. The two alcohols were assigned the structures **227** and **228**, largely on the basis of spectroscopic evidence:

$$\underset{m\,5{\cdot}02}{H_2C}{=}\underset{m\,4{\cdot}0}{CH}{-}\underset{OH}{CH}{-}(C{\equiv}C)_2{-}\underset{\substack{d\,4{\cdot}52\\(J=15)}}{CH}{=}\underset{\substack{dd\,3{\cdot}36\\(J=15+10)}}{CH}{-}\underbrace{CH{=}CH}_{m\,3{\cdot}7\text{—}4{\cdot}8}{-}\underset{m\,7{\cdot}82}{(CH_2)_2}{-}\underset{m\,4{\cdot}0}{CH}{=}\underset{m\,5{\cdot}02}{CH_2}$$

227

$$\underset{\substack{t\ 9{\cdot}0\\(J=7)}}{H_3C}-\underset{\substack{dq\ 8{\cdot}34\\(J=6{\cdot}8)}}{CH_2}-\overset{t\ 5{\cdot}67}{\underset{OH}{CH}}-(C{\equiv}C)_2-CH{=}CH-CH{=}CH-(CH_2)_2-CH{=}CH_2$$

228

The proposed structures were proved by synthesis[229]. Cadiot–Chodkiewicz coupling of **229** with **230** yielded the diyn-ene **231**. **231** was converted by manganese dioxide oxidation to the aldehyde **232**, which was condensed with the ylene **233** to the isomeric alcohols **234** and **235**. Main product was the *trans-cis* isomer **235**. The *trans-trans*-isomer could be enriched by u.v. isomerisation and thus be separated by repeated column chromatography and crystallisation. **234** was then oxidised to the highly unstable aldehyde **236**. The aldehyde **236** was treated with ethyl and vinyl magnesiumbromide to give the alcohols **228** and **227**:

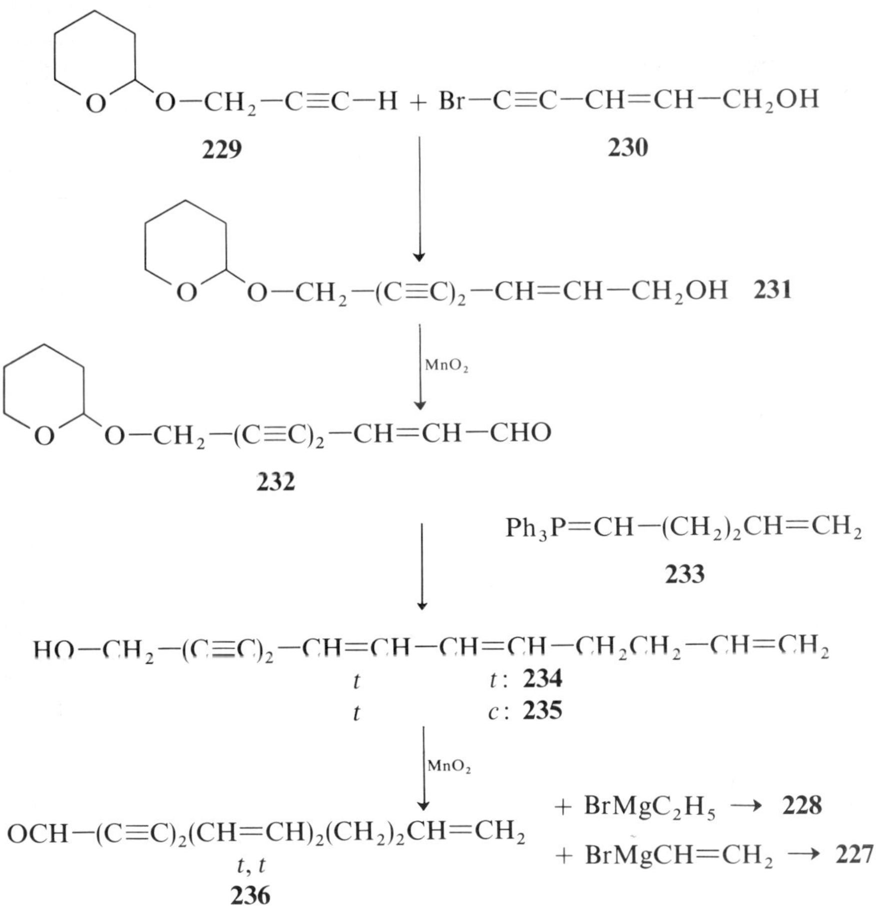

However, the two alcohols **228** and **227** were only obtained in small amounts. Their spectral properties were identical to those of the natural compounds. The main product of the Grignard reaction has been the carbinol **234** each time.

In addition, the carbinol **237** was obtained when **236** was treated with vinyl magnesiumbromide. A mechanistic route could be postulated, leading from **236** to **234** as a so-called Grignard reduction:

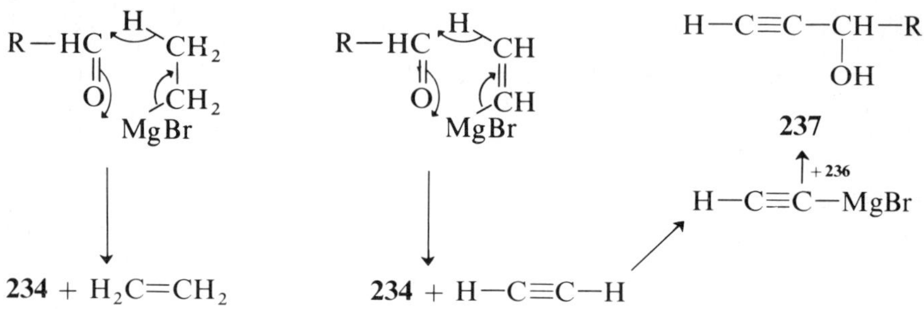

R = $-(C\equiv C)_2(CH=CH)_2(CH_2)_2CH=CH_2$

From the systematic point of view, the family Pittosporaceae is usually placed at the proximity of the Saxifragae. Special features, however, indicate the close relationship to the Araliaceae. Therefore, the roots of *Pittosporum buchanani* Hook. fil.[230] were investigated. (The aerial parts did not contain any polyacetylene.) The roots have been found to contain a very unstable conjugated polyyneketone shown to be **238** on the basis of spectroscopic evidence:

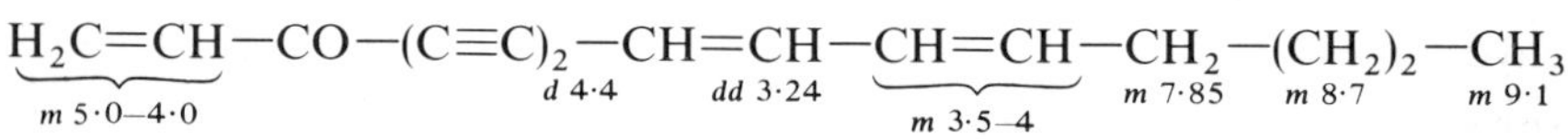

238

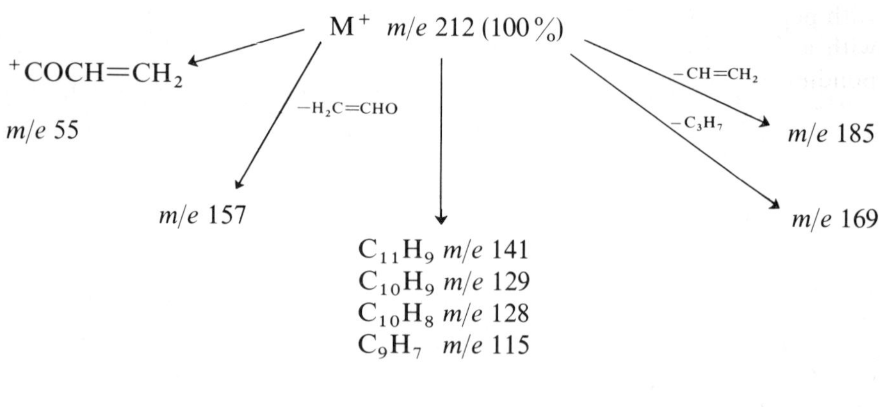

The more polar fractions of the plant extract gave a mixture of two alcohols, which could be separated only in parts. The i.r. characteristics indicate the following groupings: —OH, 3610/cm; —C≡C—, 2230/cm; *trans-trans*-diene, 990/cm; $-CH{=}CH_2$, 940/cm. In addition the u.v. spectrum revealed an extended diyne-diene-chromophore (λ max: 308, 294, 237·5, 228 nm). Together with the n.m.r. data, the compound was thus deduced to be the alcohol **239**:

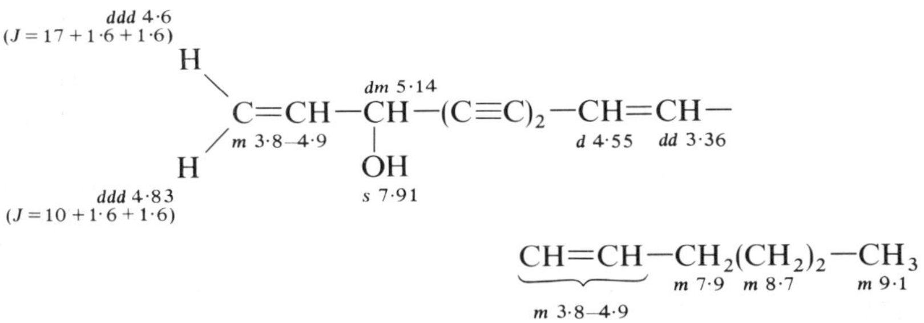

239

Oxidation with manganese dioxide yielded the ketone **238** described above, furthermore the ketone **114** (see Section 1.2.6). The accompanied alcohol, had to be the already mentioned C_{17}-alcohol falcarinol (**118**). The C_{17}-alcohol **134**, corresponding to **239**, has been isolated from *Oenanthe crocata* and *Opopanax chironium*.

As the investigation of the constituents of the plant gave only very small amounts of the described compounds, it seemed desirable to prove the structures of **239** and **238** by synthesis[231]. The route chosen involved preparation of the carbinol **240** by condensing pent-2-en-4-yn-1-al with pentylmagnesiumbromide to be followed by bromination with phosphortribromide. Dehydrohalogenation of **241** with 1,5-diaza-bicyclo-[5,4,0]-undecen-(5)[232] yielded the hydrocarbon **242**. Ten percent of the *cis*-isomer were separated by GLC. The alcohol **239** was finally obtained by Glaser coupling of **242** with pent-1-en-4-yn-3-ol. The alcohol obtained was identical in all respects with a natural sample. Manganese dioxide oxidation yielded the corresponding ketone **238**:

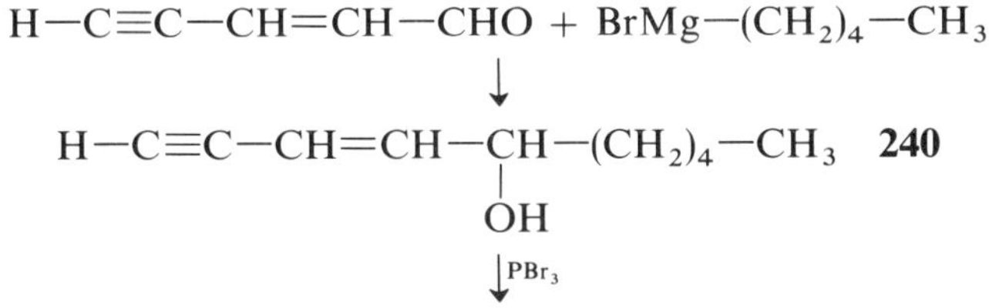

$$\downarrow PBr_3$$

$$H{-}C{\equiv}C{-}CH{=}CH{-}\underset{\displaystyle Br}{\underset{|}{C}H}{-}(CH_2)_4{-}CH_3 \quad \mathbf{241}$$

$$H_2C{=}CH{-}\underset{\displaystyle OH}{\underset{|}{C}H}{-}C{\equiv}C{-}H + \quad \downarrow$$

$$H{-}C{\equiv}C{-}CH{=}CH{-}CH{=}CH{-}(CH_2)_3{-}CH_3 \quad \mathbf{242}$$

$$\downarrow Cu_2Cl_2/O_2$$

$$\mathbf{239} \xrightarrow{MnO_2} \mathbf{238}$$

A hydrocarbon was obtained from the least polar fractions of *Pittosporum buchanani* Hook. fil.[230] in very small amounts. It has been shown to be **243** by spectroscopic measurements:

$$H_3C{-}CH{=}CH{-}(C{\equiv}C)_2{-}CH_2{-}CH{=}CH{-}C_5H_{11} \quad \mathbf{243}$$

The corresponding C_{17}-hydrocarbon with the same chromophore system has been isolated from *Oenanthe crocata*[216] (see Section 1.2.6, **135**).

From the roots of *Bupleurum ranunculoides* L., tribus Apioideae (Umbelliferae) an aldehyde was isolated, the structure of which was elucidated mainly on n.m.r. spectroscopic measurements[233]. The mass spectroscopic fragmentation pattern confirms the proposed structure **244**:

$$OCH{-}CH{=}CH{-}(C{\equiv}C)_2{-}CH_2{-}CH{=}CH{-}CH_2{-}(CH_2)_3{-}CH_3 \quad \mathbf{244}$$

OCH: *d* 0·08 (*J* = 7·5); CH=CH *cis*: *dd* 3·69 (*J* = 7·5 + 11), *dd* 3·43 (*J* = 11 + 0·8); CH_2: *dm* 6·91; CH=CH *cis*: *m* 4·6; CH_2: *m* 7·95; $(CH_2)_3$: *m* 8·7; CH_3: *t* 9·1 (*J* = 7)

$$M^+ \; m/e\ 214 \xrightarrow{-CHO} m/e\ 185$$

$$M^+ \; m/e\ 214 \xrightarrow{-C_4H_9} m/e\ 157 \xrightarrow{-CO} m/e\ 129$$

$$M^+ \; m/e\ 214 \xrightarrow{-C_5H_{11}} m/e\ 143 \xrightarrow{-C_2H_2} m/e\ 117$$

The corresponding alcohol **245**, together with the related C_{17}-alcohol **151**[215], was obtained from the more polar fractions. Its structure was proved by n.m.r. structural analysis as well as mass spectroscopic measurements.

$$HO{-}CH_2{-}CH{=}CH{-}(C{\equiv}C)_2{-}CH_2{-}CH{=}CH{-}CH_2{-}C_4H_9 \quad \mathbf{245}$$

$HOCH_2$: *dd* 5·68 (*J* = 6 + 1); CH=CH: *m* 3·9; CH_2: *d* 6·96 (*J* = 5); CH=CH: *m* 4·6; CH_2: *m* 7·97

$$M^+ \; m/e\ 216$$

$$\downarrow -C_4H_9$$

$$m/e\ 131 \xleftarrow{-CO} m/e\ 159 \xrightarrow{-H_2O} m/e\ 141$$

The last three compounds described above all possess an activated methylene group, which is known to be very oxidisable. Therefore, it is not surprising that there exists a compound with an oxygen function at the activated position. The previously mentioned compound **153**[217] (see Section 1.2.6) can easily be arranged into the probable biogenetic scheme.

$$AcO{-}CH_2{-}CH{=}CH{-}(C{\equiv}C)_2{-}\underset{\displaystyle OH}{\underset{|}{C}H}{-}CH{=}CH{-}C_5H_{11} \quad \textbf{153}$$

From the alcohol fraction of *Oenanthe crocata* L., a carbinol was separated which showed an ene-diyn-ene chromophore in the u.v. spectrum[216]. Oxidation with manganese dioxide gave a ketone, the structure of which was established by i.r., n.m.r. and mass spectroscopic measurements. These data indicate that the ketone is **247** and accordingly the original alcohol is **246**:

$$H_3C{-}CH{=}CH{-}(C{\equiv}C)_2{-}CH{=}CH{-}\underset{\displaystyle OH}{\underset{|}{C}H}{-}C_5H_{11} \quad \textbf{246}$$

$$\downarrow MnO_2$$

$$H_3C{-}CH{=}CH{-}(C{\equiv}C)_2{-}\underbrace{CH{=}CH}_{ABq\ 3{\cdot}42}{-}CO{-}CH_2{-}(CH_2)_3{-}CH_3 \quad \textbf{247}$$

dd 8·13 ($J = 6{\cdot}5 + 1{\cdot}7$); *dq* 3·68 ($J = 15 + 6{\cdot}5$); *dm* 4·45 ($J = 15$); ABq 3·42; *t* 7·56; *m* 8·7; *t* 9·1

From the aerial parts of *Cotula bipinnata* Thunb., a C_{15}-hydrocarbon was isolated, which showed a pure ene-diyne-diene chromophore in the u.v. spectrum. The compound absorbs in the i.r. region at 2200 and 2125/cm, the second band has about double the intensity of the first. The double bond regions give the following information: bands at 1845, 1640, and 912/cm indicate an isolated terminal ethylenic bond. A very strong band at 983/cm is assumed to contain also that of a conjugated diene grouping. Finally, a band at 948 cm of medium intensity exhibits a trans double bond adjacent to an acetylenic bond. *n*-Pentadecane was obtained by catalytic hydrogenation of the hydrocarbon. Ozonisation of the compound yielded a dominating dicarboxylic acid, which was shown to be succinic acid. These evidences are compatable with only two constitutional formula, **248** and **249**:

$$H_3C{-}CH{=}CH{-}(C{\equiv}C)_2{-}(CH{=}CH)_2{-}CH_2{-}CH_2{-}CH{=}CH_2 \quad \textbf{248}$$

$$H_3C{-}(CH{=}CH)_2{-}(C{\equiv}C)_2{-}CH{=}CH{-}CH_2{-}CH_2{-}CH{=}CH_2 \quad \textbf{249}$$

The orientation of this unsaturated system in the molecule was determined by ozonisation of the maleic anhydride adduct. The production of acetaldehyde (the presence of which could be demonstrated by formation of the

2,4-dinitrophenylhydrazone) shows the hydrocarbon to be **248**. To prove the structure, **248** was synthesised according to the following scheme:

$$H_2C{=}CH(CH_2)_3OH \qquad HOCH_2CH{=}CHC{\equiv}C{-}Br$$

$$+\ H{-}C{\equiv}CCH{=}CH{-}CH_3$$

$$\downarrow \qquad\qquad \downarrow$$

$$H_2C{=}CH(CH_2)_3Br \qquad HO{-}CH_2CH{=}CH{-}(C{\equiv}C)_2{-}CH{=}CH{-}CH_3$$

$$\downarrow \qquad\qquad \downarrow$$

$$H_2C{=}CH(CH_2)_2CH{=}PPh_3 + OCH{-}CH{=}CH{-}(C{\equiv}C)_2{-}CH{=}CH{-}CH_3$$

$$\downarrow$$

248

The i.r. spectrum of **248** so obtained showed that the reaction product was a mixture of the 10-cis- and 10-*trans*-isomers. Repeated and painstaking column chromatography led to pure *trans*-**248**. The spectral data were agreeing with those of the natural hydrocarbon[221].

The corresponding C_{17}-hydrocarbon, the so-called centaur X_4 (**101**), has been isolated from *Centaurea ruthenica*, as already mentioned in Section 1.2.5.

From the least polar fractions of *Oenanthe crocata* L.[216], the corresponding ene-diyne-diene with C_{15}-chain length was obtained besides **140** (see Section 1.2.6). The structure of the compound **250** was deduced by i.r., n.m.r. and mass spectroscopy. The all-*trans* configuration of the double bonds is shown by the i.r. and n.m.r. spectra:

$$H_3C{-}CH{=}CH{-}(C{\equiv}C)_2{-}CH{=}CH{-}CH{=}CH{-}CH_2{-}(CH_2)_2{-}CH_3 \quad \mathbf{250}$$

M^+ *m/e* 198

- $\xrightarrow{-C_3H_7}$ *m/e* 155
- $\xrightarrow{-C_4H_9}$ *m/e* 141 $\xrightarrow{-C_2H_2}$ *m/e* 115
- → *m/e* 129, *m/e* 128, *m/e* 127, *m/e* 91

From the roots of *Bupleurum ranunculoides* L.[233], three C_{15}-compounds were obtained all containing an ene-diyne-diene chromophore. They only differ in the nature of the oxygen function at $C_{(1)}$. Their structures were deduced by spectral properties to be **251**, **252** apd **253** (**253** has been obtained from the aerial parts). The C_{15}-alcohol **252** has been isolated too from *Oenanthe aquatica* (L.) Poit., but only in very small amounts[217].

$$OCH{-}CH{=}CH{-}(C{\equiv}C)_2{-}CH{=}CH{-}CH{=}CH{-}C_4H_9 \quad \mathbf{251}$$

d −0·08 (*J* = 7·5)

$$HO{-}CH_2{-}CH{=}CH{-}(C{\equiv}C)_2{-}CH{=}CH{-}\underbrace{CH{=}CH}{-}C_4H_9 \quad \mathbf{252}$$

s 8·4; *dd* 5·82 (*J* = 4·5 + 1·5); *dt* 3·68 (*J* = 16 + 4·5); *dt* 3·88 (*J* = 16 + 1·5); *d* 4·51 (*J* = 15·5); *dd* 3·39 (*J* = 15·5 + 9·5); *m* 4·0 − 4·35

$$AcO{-}CH_2{-}CH{=}CH{-}(C{\equiv}C)_2{-}CH{=}CH{-}CH{=}CH{-}C_4H_9 \quad \mathbf{253}$$

The allylic alcohol corresponding to **250** has been isolated from *Oenanthe crocata* L.[216]. The alcohol was oxidised with manganese dioxide to the corresponding ketone shown to be **255** on the basis of chemical and spectroscopic evidence. The alcohol was thus deduced to be **254**:

$$H_3C{-}CH{=}CH{-}(C{\equiv}C)_2{-}(CH{=}CH)_2{-}\underset{\displaystyle OH}{\underset{|}{CH}}{-}C_3H_7 \quad \mathbf{254}$$

$$\downarrow MnO_2$$

$$H_3C{-}CH{=}CH{-}(C{\equiv}C)_2{-}CH{=}CH{-}CH{=}CH{-}CO{-}CH_2{-}CH_2{-}CH_3 \quad \mathbf{255}$$

dd 8·16 (*J* = 6·7 + 1·6); *dq* 3·7 (*J* = 15 + 6·7); *dm* 4·44 (*J* = 15); *d* 3·99 (*J* = 15); *dd* 3·30 (*J* = 15 + 11); *dd* 2·93 (*J* = 15 + 11); *d* 3·85 (*J* = 15); *t* 7·56 (*J* = 7); *tq* 8·38 (*J* = 7 + 7); *t* 9·08 (*J* = 7)

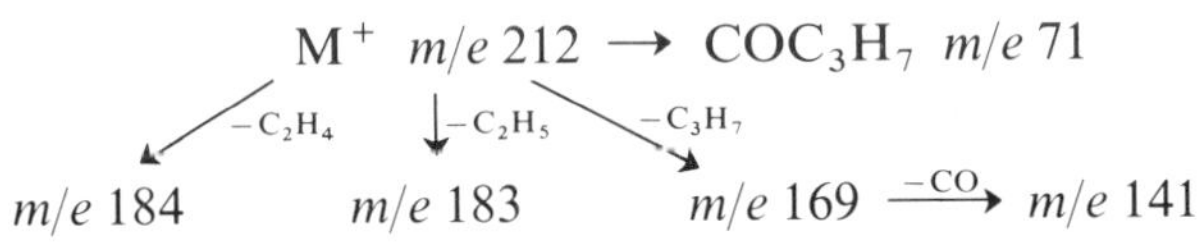

The proposed structure of **254** was confirmed by synthesis[231]. The route chosen involved Grignard reaction of dodeca-2,4,10-triene-6,8-diyn-1-al with propyl magnesium bromide:

$$H_3C{-}CH{=}CH{-}(C{\equiv}C)_2{-}(CH{=}CH)_2CHO + BrMgC_3H_7 \rightarrow \mathbf{254}$$

The spectral properties of the reaction product agreed with those of the natural alcohol.

The corresponding ketone with a shortened chromophore system (ene-diyn-ene: λ max: 311, 292, 275, 260 nm) was shown to occur together with **252** in *Oenanthe aquatica* (L.) Poir.[217]. In order to obtain it in a pure state it was converted, using sodium borohydride, to the corresponding alcohol and subsequently subjected to chromatography. Spectroscopic evidences indicate that the alcohol is **257** and accordingly the original ketone is **256**:

$$H_3C-CH=CH-(C\equiv C)_2-CH=CH-CH_2-CH_2-CO-C_3H_7 \quad \mathbf{256}$$

$$\downarrow NaBH_4$$

$$H_3C-CH=CH-(C\equiv C)_2-CH=CH-CH_2-CH_2-\underset{\displaystyle OH}{\underset{|}{CH}}-CH_2CH_2CH_3 \quad \mathbf{257}$$

Structural analysis of the various compounds described above reveals that these compounds may be related biosynthetically to the C_{17}-polyyne-family, occurring in *Oenanthe crocata* for instance (see Section 1.2.6). The postulated biogenetic route seems to be relatively simple starting with the hypothetical C_{15}-compound **258**, which could be obtained from the corresponding C_{17}-compound by β-oxidation:

$$(H_3C-CH=CH-(C\equiv C)_2-CH_2-CH=CH-C_5H_{11}) \quad \mathbf{258}$$

$$\downarrow [O]$$

$$\mathbf{246}$$

$$\downarrow -[H_2O]$$

$$\mathbf{250}$$

$$\mathbf{252} \xleftarrow{[O]} \mathbf{250} \xrightarrow{[O]} \mathbf{254} \xrightarrow{[O],\,[H]} \mathbf{256}$$

Homologous series of polyacetylenic aldehydes, alcohols and acetates were obtained from *Zoegea baldshuanica* C. Winkl.[192]. They all possess a dien-yne-diene chromophore. Separation of distinct compounds has been impossible. The chain length was determined by mass spectroscopic measurements. In part, the compounds were known from other plants and some of them are previously described in Section 1.1:

$$H_3C-(CH=CH)_2-C\equiv C-(CH=CH)_2-(CH_2)_n-CHO$$

41: n = 6
42: n = 5
43: n = 4
259: n = 3
260: n = 2

$$H_3C-(CH=CH)_2-C\equiv C-(CH=CH)_2-(CH_2)_n-CH_2OH$$

44: n = 6
45: n = 5
46: n = 4
261: n = 3
262: n = 2

$H_3C(CH{=}CH)_2{-}C{\equiv}C{-}(CH{=}CH)_2(CH_2)_n{-}CH_2OAc$ **263**: n = 3
264: n = 2
265: n = 1

The different chain-lengths of the aldehydes ($C_{18}{-}C_{14}$) and of the acetates ($C_{15}{-}C_{13}$) are remarkable. A reasonable biosynthetic pathway for these homologous compounds is outlined in Section 1.1.

3. Acetylenes related to the C_{14}-(12)-ene-(8,10)-diyne-(5)-*cis*-en-(1)-ol

3.1. Simple Acetylenes

As in the case of the methyl triyne compounds (see Chapter 2, Section 2), similar routes seem to operate in this series yielding the important precursors **A** and **B** with C_{14}-chain length in the following manner:

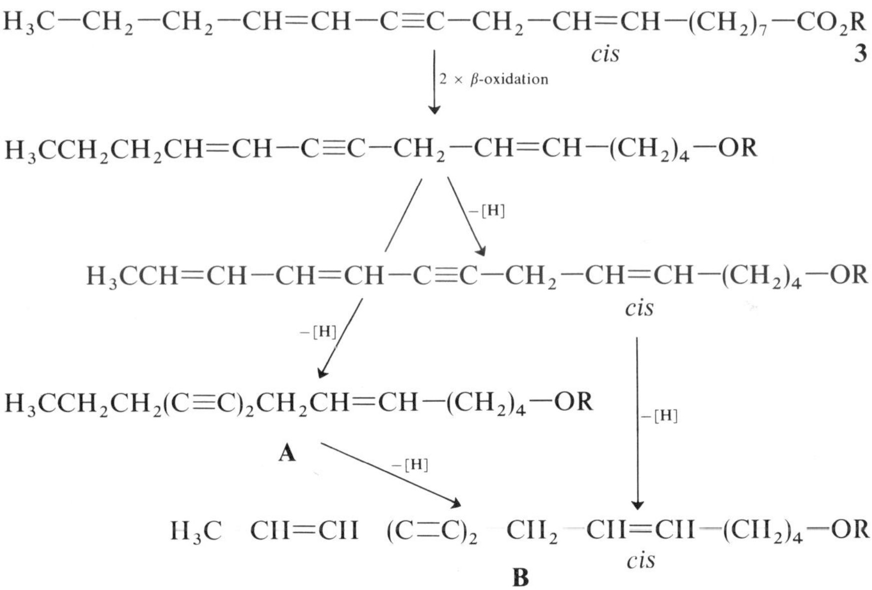

Both **A** and **B** are precursors of a considerable number of polyynes. In this section simple transformation products of **A** and **B** will be first discussed.

The Campanulaceae, generally agreed to be taxonomically close to the Compositae, have been investigated for polyacetylenes and most of them revealed the presence of polyacetylenic chromophores. The triols **266**, for example, **267** and **268** with a diyn-ene chromophore and an ene-diyn-ene chromophore, respectively[22]:

$$H_3C{-}CH_2{-}\underset{OH}{CH}{-}(C{\equiv}C)_2{-}CH{=}CH{-}\underset{OH}{CH}{-}(CH_2)_4{-}OH \qquad \mathbf{266}$$

$$HOCH_2{-}CH{=}CH{-}(C{\equiv}C)_2{-}CH{=}CH{-}\underset{OH}{CH}{-}(CH_2)_4{-}OH \qquad \mathbf{267}$$

$$H_2C{=}CH{-}\underset{OH}{CH}{-}(C{\equiv}C)_2{-}CH{=}CH{-}\underset{OH}{CH}{-}(CH_2)_4{-}OH \qquad \mathbf{268}$$

The rise of the different features of the propyl endgroup still needs to be established. Several mechanistic routes could be postulated which would lead from the saturated propyl endgroup eventually to the various unsaturated endgroups. In addition to the pathways already discussed some more possibilities have to be regarded.

The first of these, which we regard as more likely, would involve the attack of oxygen to form a "propargylic" alcohol. Dehydration would furnish the double bond conjugated to the triple bonds:

$$C_3H_7{-}C{\equiv}C{-} \xrightarrow{[O]} H_3C{-}CH_2{-}\underset{OH}{CH}{-}C{\equiv}C{-} \xrightarrow{-[H_2O]} H_3C{-}CH{=}CH{-}C{\equiv}C{-}$$

The second would involve β–γ-dehydrogenation, followed by oxidation of the allylic methylene group and, eventually, allylic rearrangement:

$$C_3H_7{-}C{\equiv}C{-} \xrightarrow{-[H]} H_2C{=}CH{-}CH_2{-}C{\equiv}C{-} \xrightarrow{[O]} H_2C{=}CH{-}\underset{OH}{CH}{-}C{\equiv}C{-} \xrightarrow{?} HOCH_2{-}CH{=}CH{-}C{\equiv}C{-}$$

The alternative route would involve ω-oxidation followed by dehydrogenation or dehydration:

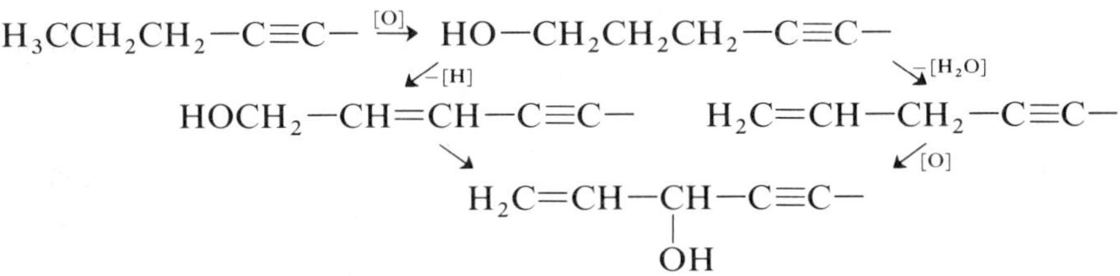

As previously remarked there are still open questions about the exact biosynthetic pathways. Possible routes are still being studied by tracer experiments and we hope to be able to give an answer quite soon.

It is proposed that the triols described above with hydroxyl functions at 1,5-position are precursors for the tetrahydropyranyl-polyacetylenic alcohols, occurring in *Campanula glomerata* L. (see Section 2.2). The majority of the C_{14}-compounds can be derived from a precursor of type **B** with a propenyl endgroup.

The roots of *Dahlia merckii* Lehm.[197] have been found to contain a C_{14}-alcohol shown to be **269** on the basis of chemical and spectroscopic evidence. The corresponding acetate **270** is shown to occur together with **269** in the same plant. The acetate **270** has been already isolated from the aerial parts of *Anthemis carpathica* Willd.[115]:

$$H_3C{-}CH{=}CH{-}(C{\equiv}C)_2{-}CH{=}CH{-}CH{=}CH{-}CH_2{-}CH_2CH_2{-}OR$$

dd 8·13
($J = 7 + 1·5$)

269: R = H
270: R = $COCH_3$

The position of the double bonds was established by the formation of a maleic anhydride adduct, which was still exhibiting a double doublet due to a methyl group adjacent to a double bond in the n.m.r. spectrum.

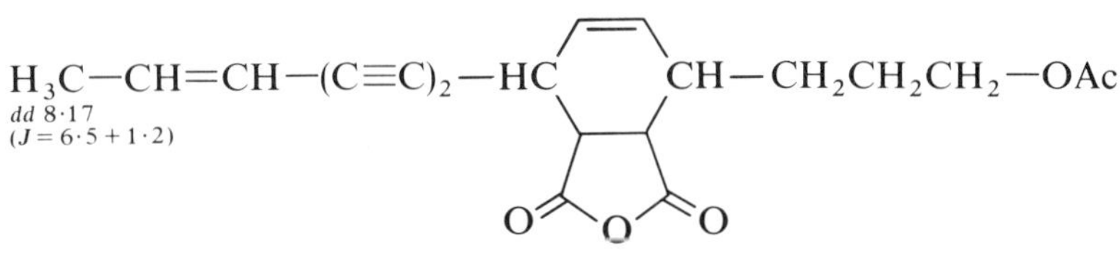

270 is also known from other plants like *Cotula coronopifolia*, *Coreopsis coronata* and *Coreopsis lanceolata*. In the latter, **270** is shown to occur with the corresponding C_{15}-acetate **271**:

$$H_3C{-}CH{=}CH{-}(C{\equiv}C)_2{-}(CH{=}CH)_2CH_2CH_2CH_2CH_2{-}OAc \quad \mathbf{271}$$

The most polar fractions of the aerial parts of *Coreopsis verticillata* L.[234] contain a compound with an ene-diyne-diene-chromophore and a molecular formula $C_{14}H_{16}O_2$ (M^+ *m/e* 216, 115). N.m.r. signals for two $-CH_2-O-$ groupings showed it to be a diol. In order to get a better defined n.m.r. spectrum, the diol was converted into the diacetate by acylation with acetanhydride. The n.m.r. of the diacetate was consistent with the proposed structure **273**. Accordingly, the original alcohol is **272**:

$$HO-CH_2-CH=CH-(C\equiv C)_2CH=CH-CH=CH-$$

s 7·18 *m* 5·82 *d* 4·11 (*J* = 15) *dm* 4·33 *dd* 3·76 (*J* = 15 + 10)
dt 3·54 (*J* = 15 + 5) *dd* 3·27 (*J* = 15 + 10) *dt* 4·04 (*J* = 15 + 5)

$$CH_2CH_2CH_2-OH \quad \textbf{272}$$

m 7·7 *m* 6·47 *s* 8·17
m 8·39

$$H_3C-CO-O-CH_2-CH=CH-(C\equiv C)_2-CH=CH-CH=CH-$$

s 7·97 *dd* 5·43 (*J* = 5·5 + 1·5) *dt* 3·76 *d* 4·21

$$CH_2CH_2CH_2-OCOCH_3 \quad \textbf{273}$$

t 6·0 *t* 8·02

Oxygen derivatives of **269** have been found in *Campanula*[22] and *Dahlia* species[235]. Their spectral properties support structure **274** and **275**:

$$H_3C-CH=CH-(C\equiv C)_2-\underset{\textstyle OH}{\underset{|}{CH}}-\underset{\textstyle OH}{\underset{|}{CH}}-CH=CH-CH_2CH_2CH_2-OH \quad \textbf{274}$$

$$H_3C-CH=CH-(C\equiv C)_2-CH=CH-\underbrace{CH-CH}_{O}-CH_2CH_2CH_2-OAc \quad \textbf{275}$$

Besides many C_{13}-polyacetylenes, some C_{14}-polyacetylenes were isolated from *Carthamus caeruleus* L., *C. lanatus* L. and *C. tinctorius* L.[236]. The diol **276** has been isolated from *Dahlia* species too[22]:

$$H_3C-CH=CH-(C\equiv C)_2-(CH=CH)_2-\underset{\textstyle OH}{\underset{|}{CH}}-CH_2-CH_2-OH \quad \textbf{276}$$

Purification and separation of the acetate fraction yielded two hydroxy acetates with the characteristic chromophore of an ene-diyne-diene system (λ max: 336, 314, 295, 278, 265, 248 nm). One of the two compounds is oxidisable with manganese dioxide. Careful saponification of both compounds gave the diol **276**. These experimental results suggest the interpretation that there exists the two isomer hydroxy acetates **277** and **278**. The n.m.r.

data agree upon the proposed structures. The double bonds are *trans* configurated as to be seen in the i.r.: 955, 990/cm:

$$H_3C{-}CH{=}CH{-}(C{\equiv}C)_2{-}(CH{=}CH)_2{-}\underset{\displaystyle OR}{\underset{|}{C}H}{-}CH_2{-}\underset{\displaystyle OR'}{\underset{|}{C}H_2}$$

277: R = Ac, R′ = H
278: R = H, R′ = Ac

The corresponding diacetate **279** has been also obtained from the same plants:

$$H_3C{-}CH{=}CH{-}(C{\equiv}C)_2{-}(CH{=}CH)_2{-}\underset{\displaystyle OAc}{\underset{|}{C}H}{-}CH_2{-}\underset{\displaystyle OAc}{\underset{|}{C}H_2} \quad \mathbf{279}$$

The n.m.r. data are identical for all the compounds. They only differ in the endgroup.

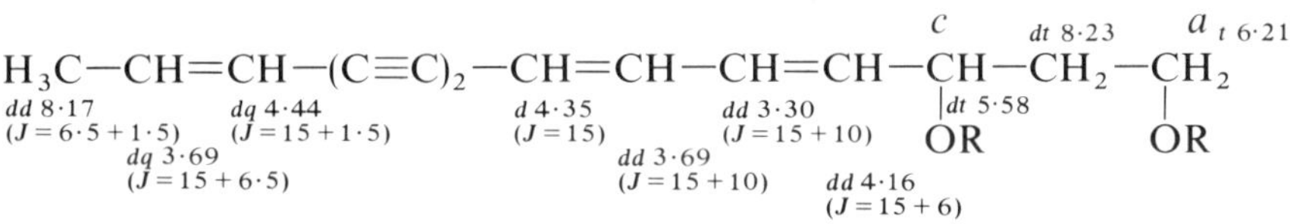

276

In **279**, the signals attributable to H_a and H_c are shifted to lower fields (*m* 5·98τ, *dt* 4·70τ). An analogous situation is given in **277** and **278**. Remarkable is the fact that the protons of the terminal $-CH_2OAc$ grouping in **279** and **278** are giving a multiplet, probably caused by the asymmetric C-atom (3).

From the aerial parts of *Anthemis carpathica* Willd.[115], very small amounts of a ketone were obtained, closely related to "artemisia ketone" [see Chapter 2, Section 2.2 (**44**)]. It could be separated from the latter by repeated TLC. The i.r. spectrum resembled those of artemisia ketone, but the u.v. showed an ene-diyn-ene chromophore. Hydrogenation gave tetradecanone-(3). Further structural assignments for the ketone were derived from its n.m.r. spectrum; the compound was thus deduced to be **280**:

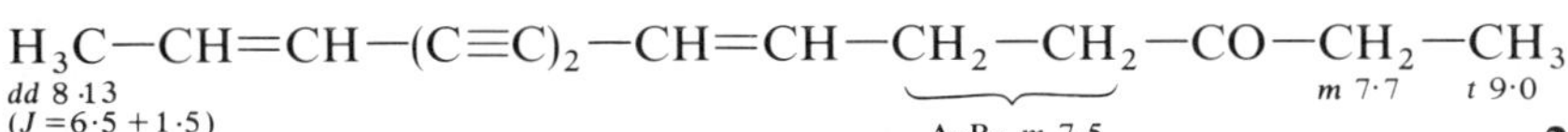

280

The proposed structure was confirmed by synthesis. The route chosen involved Cadiot–Chodkiewicz coupling of 1-bromo-pent-3-en-1-yne with the acetylenic ketone **281**. The obtained compound was identical with authentic material by spectral comparison:

$$CH_3{-}CH{=}CH{-}C{\equiv}C{-}Br$$

$$+\ H{-}C{\equiv}C{-}CH{=}CH{-}(CH_2)_2{-}CO{-}C_2H_5 \quad \textbf{281}$$

$$\downarrow$$

$$\textbf{280}$$

An alcohol closely related to **280** was obtained from the polar fractions of *Anthemis saguramica* Sosn.[57]. The structure was elucidated by spectral methods and seems to be **282**. This assignment has been somewhat difficult, as there also exists the artemisia alcohol **283** in great amounts in the same plant:

$$\underset{dd\ 8{\cdot}18\ (J=6{\cdot}5+1{\cdot}5)}{H_3C}{-}\underset{dq\ 3{\cdot}75\ (J=15+6{\cdot}5)}{CH}{=}\underset{d\ 4{\cdot}48\ (J=15)}{CH}{-}(C{\equiv}C)_2{-}\underset{dt\ 4{\cdot}35\ (J=15+1)}{CH}{=}\underset{dt\ 3{\cdot}80\ (J=15+6)}{CH}{-}\underset{m\ 7{\cdot}75}{CH_2}{-}\underset{m\ 8{\cdot}5}{CH_2}{-}\overset{m\ 6{\cdot}55}{\underset{\underset{s\ 8{\cdot}42}{OH}}{CH}}{-}\underset{m\ 8{\cdot}5}{CH_2}{-}\underset{t\ 9{\cdot}09}{CH_3} \quad \textbf{282}$$

$$H_3C{-}(C{\equiv}C)_3{-}CH{=}CH{-}CH_2{-}CH_2{-}\underset{OH}{CH}{-}C_2H_5 \quad \textbf{283}$$

From *Anthemis cupaniana* a keto ester was obtained, the structure of which could not be established[35]. The u.v. spectrum revealed an ene-diyn-ene chromophore. In order to obtain more information about the structure, it was carefully saponified. The saponification product showed an extended ene-diyn-en-one chromophore. In view of these results, it seemed probable that the unknown keto ester is structurally related to the compounds described above. Structure **284** could be suggested:

$$H_3C{-}CH{=}CH{-}(C{\equiv}C)_2{-}CH{=}CH{-}\underset{OR}{CH}{-}CH_2{-}CO{-}C_2H_5 \quad \textbf{284}$$

$$\downarrow OH^-$$

$$H_3C{-}CH{=}CH{-}(C{\equiv}C)_2{-}CH{=}CH{-}CH{=}CH{-}CO{-}C_2H_5$$

The investigation of the constituents of *Cotula coronopifolia* L.[228] gave, *inter alia*, a complex mixture of three polar compounds, which could be separated by repeated TLC. All of them contain a diyne-diene chromophore (λ max: 310, 294, 236, 226 nm). The most polar compound is a diol with a molecular ion peak at m/e 218. Its i.r. spectrum displays bands for a *trans-trans*-diene grouping (995/cm). The complex ill-defined n.m.r. spectrum could be analysed after partial spin-decoupling experiments. The two hydroxyl groups were determined to be of primary and secondary kind ($-CH_2-CH(OH)-C\equiv$: *t* 5·61 (1H) ($J = 6{\cdot}2$ Hz) and $-CH_2-CH_2-OH$: *t* 6·38τ (2H) ($J = 6{\cdot}5$). Spin decoupling of the protons adjacent to the $-CH_2OH$-group results in a triplet at τ 9·01 coupled to a doublet quartet at τ 8·27 for the $C_2H_5-CH(OH)-C\equiv$-grouping. Taking all these facts into consideration, structure **285** is strongly indicated:

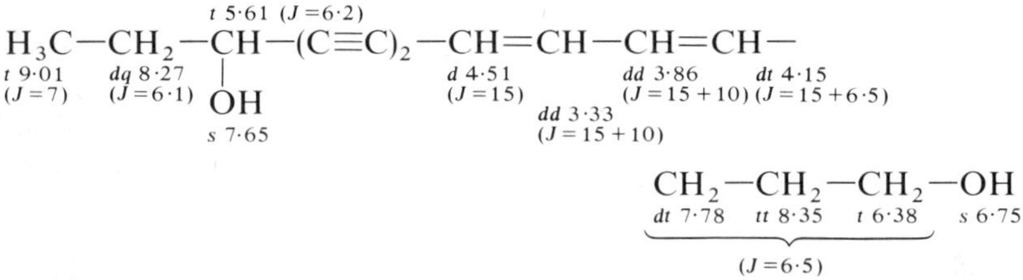

285

The i.r. spectrum of the second compound suggested the presence of a diacetate and the n.m.r. spectrum supported structure **286**:

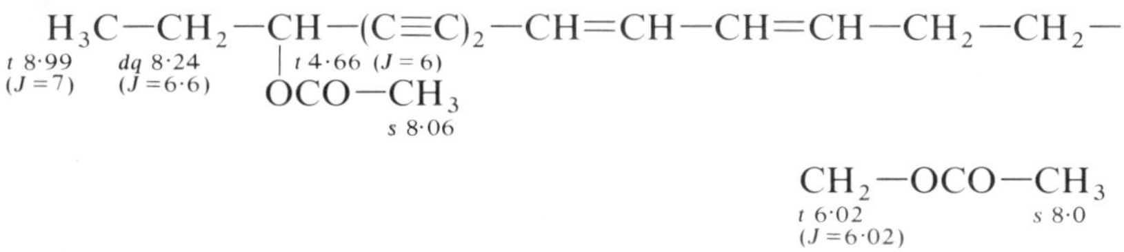

286

286 was converted on saponification into the diol **285**. The third diyne-diene was deduced to be the hydroxy acetate **287**. Acylation gave the diacetate **286**. Oxidation with manganese dioxide led to a diene-diyn-one chromophore:

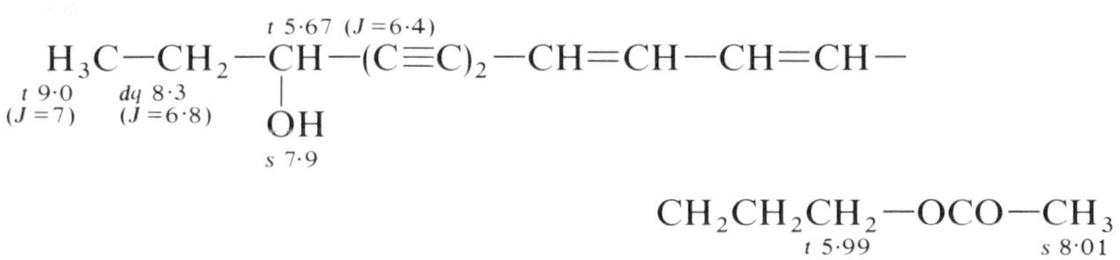

287

The structures of **285**–**287** have been further established by synthesis[239].

Structural analysis of the various members of the C_{14}-compounds reveals that these compounds may be related biosynthetically to the C_{16}-compounds. If this is the case, some rearrangements and degradation reactions must prevail. A possible pathway is the following:

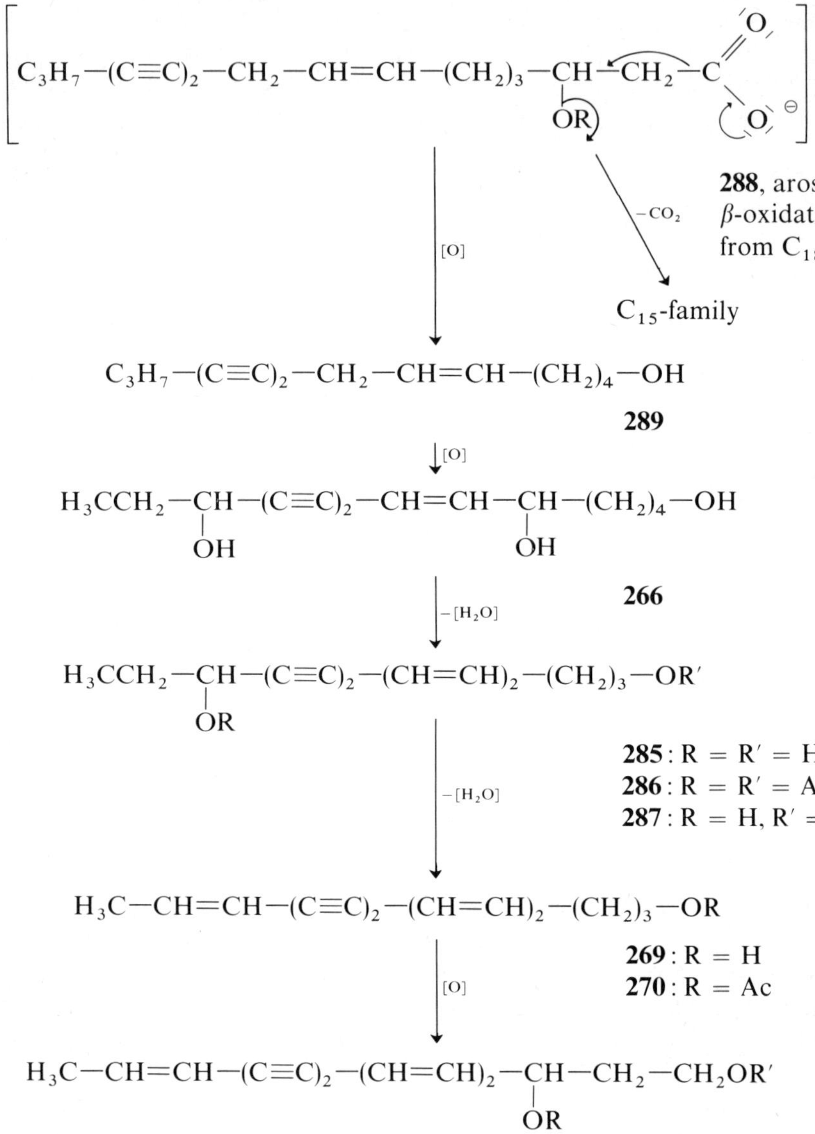

288, arose by β-oxidation from C_{18}

289

266

285: R = R′ = H
286: R = R′ = Ac
287: R = H, R′ = Ac

269: R = H
270: R = Ac

276: R = R′ = H
279: R = R′ = Ac
277: R = Ac, R′ = H
278: R = H, R′ = Ac

A further group of C_{14}-compounds obviously is related to **276**. In analogy to the triyne-ketone (**44**) (see Chapter 2, Section 2.2) it seems certain that **280** is formed in the following way:

$$\textbf{276} \xrightarrow{[O]} H_3C{-}CH{=}CH{-}(C{\equiv}C)_2{-}(CH{=}CH)_2{-}CO{-}(CH_2)_2OH \qquad \textbf{290}$$

$$\downarrow -[H_2O]$$

$$H_3C{-}CH{=}CH{-}(C{\equiv}C)_2{-}(CH{=}CH)_2{-}CO{-}CH{=}CH_2 \qquad \textbf{291}$$

$$\downarrow [H]$$

$$H_3C{-}CH{=}CH{-}(C{\equiv}C)_2{-}CH{=}CH{-}CH_2{-}CH_2{-}CO{-}CH_2{-}CH_3 \qquad \textbf{280}$$

$$\textbf{280} \xrightarrow{[H]} H_3CCH{=}CH(C{\equiv}C)_2CH{=}CH(CH_2)_2CH(OH)C_2H_5 \qquad \textbf{282}$$

$$\textbf{280} \xrightarrow{[O]} H_3CCH{=}CH(C{\equiv}C)_2CH{=}CH{-}CH(OR){-}CH_2{-}CO{-}C_2H_5 \qquad \textbf{284}$$

However, it should again be mentioned that all the biogenetical pathways discussed have perhaps to be revised in part, as the sequence of the proposed steps is not confirmed. Very probably, the formation of ene-diyne endgroups does not always follow the same scheme. Several feeding results with labelled precursors are strong indications in favour of this proposal.

The roots of *Sanvitalia procumbens* Lam.[237] have been found to contain a C_{14}-polyyne ester shown to be **292** on the basis of spectroscopic evidence. The u.v. spectrum reveals an extended ene-diyne-triene chromophore (λ max: 368, 348, 281, 275, 252 nm). The i.r. spectrum shows a carbonyl peak in the position characteristic of an unsaturated ester (1730/cm). The presence of triple bonds is shown by bands at 2240 and 2210/cm. In addition bands at 1627 and 1011/cm reveals an all-*trans*-triene grouping. The structural assignments for **292** are derived from its n.m.r. spectrum. A doublet at τ 8·06 coupled with 7 and 1·6 Hz to a doublet quartet at τ 4·46 ($J = 11 + 1{\cdot}6$ Hz) and a doublet quartet at τ 3·88 ($J = 11 + 7$ Hz) could be assigned to a *cis*-propylene grouping adjacent to a triple bond. A doublet doublet at τ 2·79 ($J = 15 + 10$ Hz) and a doublet at τ 4·16 ($J = 15$ Hz) could be ascribed to two further olefinic protons. The other four olefinic protons show resonances in a complex multiplet in the range of 3·1–3·7τ. Furthermore the n.m.r. shows a three-proton singlet at τ 6·32 indicating an ester methyl group. In view of these spectral results the proposed structure **292** seems to be very likely:

$$\underset{cis}{H_3C{-}CH{=}CH}{-}(C{\equiv}C)_2{-}\underset{t,t,t}{(CH{=}CH)_3}{-}CO_2CH_3 \quad \mathbf{292}$$

292 is the first C_{14}-polyacetylene ester to be isolated from a Compositae.

The biogenesis of **292** seems to be more or less established. A good incorporation into **292** has been observed[238] on feeding labelled **293**, presumably indicating the following pathway:

$$H_3C{-}CH{=}CH{-}(C{\equiv}C)_2{-}CH_2{-}\underset{cis}{CH{=}CH}{-}(CH_2)_4{-}OH \quad \mathbf{293}$$

$$\downarrow [O],\ -[H_2O]$$

$$H_3C{-}CH{=}CH{-}(C{\equiv}C)_2{-}(CH{=}CH)_2{-}CH_2CH_2{-}CO_2R$$

$$\downarrow [O]$$

$$H_3C{-}CH{=}CH{-}(C{\equiv}C)_2{-}(CH{=}CH)_2{-}\underset{OR}{CH}{-}CH_2{-}CO_2R$$

$$\downarrow -[H_2O]$$

$$H_3C{-}CH{=}CH{-}(C{\equiv}C)_2{-}(CH{=}CH)_3{-}CO_2CH_3 \quad \mathbf{292}$$

An acetylene ketone with a hitherto unknown feature has been isolated from the roots of *Centaurea pullata* L.[36]. Its i.r. spectrum indicates the presence of a terminal diyne-grouping (H—C≡C—C≡C— : 3318, 2235/cm) and of a carbonyl group (1728/cm). Reduction of the presumed ketone with sodium borohydride yielded *n*-tetradecanol-(2). Together with the n.m.r. data the ketone was deduced to be **294**. The mass spectrum confirms the proposed structure exhibiting ions due to α-cleavage of the ketone:

$$\underset{\substack{t\ 8\cdot22\\(J=1)}}{H}{-}C{\equiv}C{-}C{\equiv}C{-}\underset{\substack{d\ 7\cdot04\\(J=5)}}{CH_2}{-}\underset{\substack{cis\\m\ 4\cdot6}}{CH{=}CH}{-}\underset{m\ 8\cdot0}{CH_2}{-}\underset{m\ 8\cdot6}{(CH_2)_3}{-}\underset{t\ 7\cdot69}{CH_2}{-}CO{-}\underset{s\ 7\cdot99}{CH_3}$$

294

M^+ *m/e* 202 —($-CH_3COCH_3$)→ *m/e* 144

M^+ *m/e* 202 —($-CH_3$)→ *m/e* 187

M^+ *m/e* 202 → $H(C{\equiv}C)_2CH_2CH{=}CH(CH_2)_n^+$ + $COCH_3$

n = 1–5; $COCH_3$: *m/e* 43

m/e 103, 117, 131, 145, 159

Biogenetically **294** might be formed in the following way:

$$C_3H_7-(C\equiv C)_2-CH_2-CH=CH-(CH_2)_7-CO_2R \quad \mathbf{23}$$

$$\downarrow [O]$$

$$H_3C-CH_2-CO \wr (C\equiv C)_2-CH_2-\underset{cis}{CH=CH}-(CH_2)_5-CO-CH_2 \wr CO_2R$$

$$\downarrow [H_2O], -[CO_2]$$

294

A somewhat similar pathway has been confirmed in the case of the amides (see Section 6).

3.2. C_{14}-ring Compounds

As previously mentioned in Section 3.1, the triols **266**, **267** and **268** are very probably the precursors of the corresponding tetrahydropyranyl alcohols isolsted from Campanulaceae[22,240].

266 → 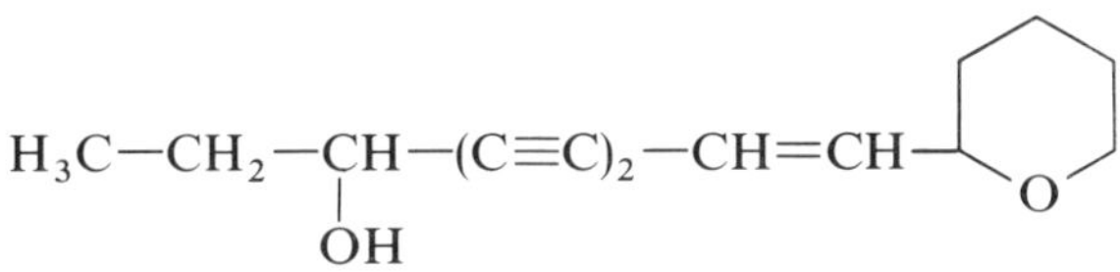 **295**

268 → 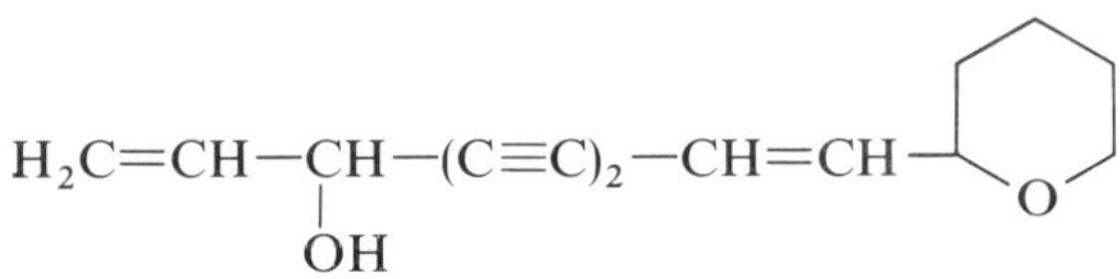 **296**

267 → 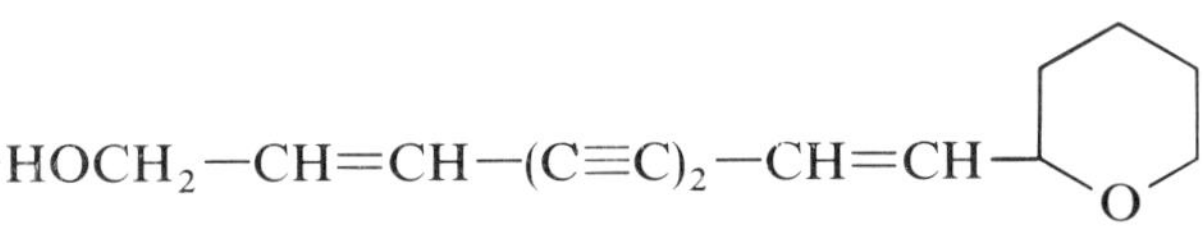 **297**

In particular, **297** has been isolated from the roots of *Campanula glomerata* L. (the clustered Bellflower). **297** gave a monoacetate and manganese dioxide converted it into the corresponding aldehyde showing matricarianal-like absorption:

$$H_3C-CH=CH-(C\equiv C)_2-CH=CH-CHO$$ matricarianal (see Section 7.1)

From the same plant, an ene-diyne alcohol inert to manganese dioxide oxidation was isolated. Spectral evidence was consistent with the corresponding dihydro-compound **298**:

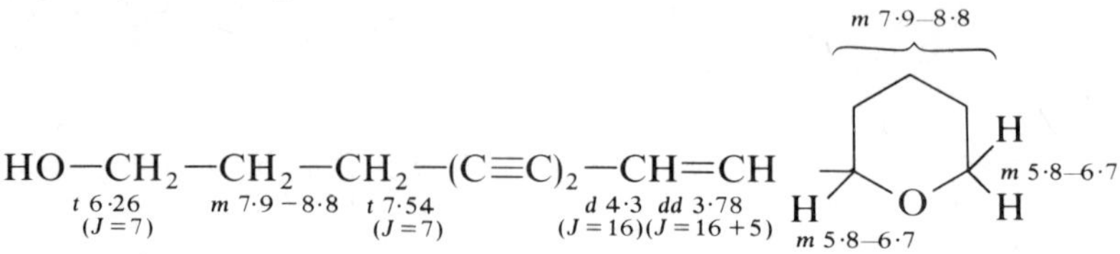

298

It seemed desirable to prove the structure of **297** and **298** by synthesis. The route chosen involved preparation of the C_9-diol **299**, obtained by a reaction of diacetylene with 2-hydroxy-tetrahydropyran and reduction of the diacetylene diol with lithium aluminium hydride:

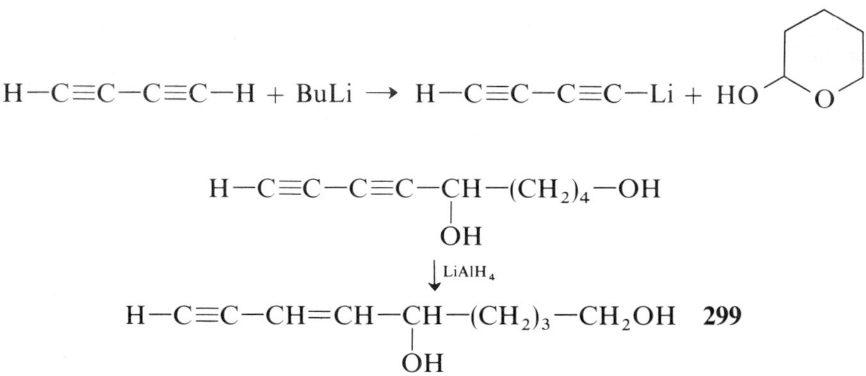

299 was cyclised in the presence of acid to the tetrahydropyran **300** and the latter was converted by Cadiot–Chodkiewicz coupling to the natural alcohols **297** and **298**:

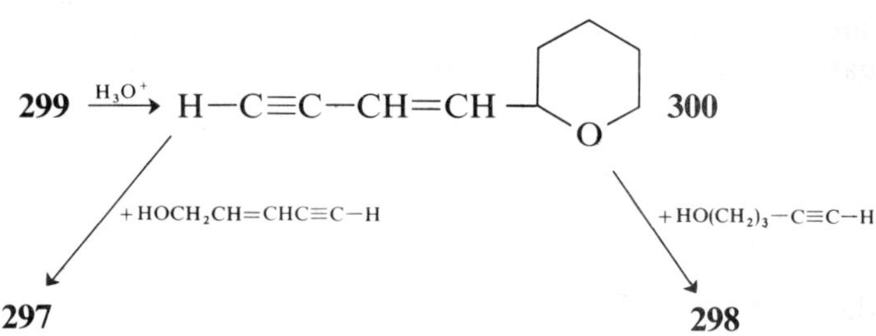

From the roots and aerial parts of *Trachelium caeruleum* L.[35], some very closely related compounds have been isolated besides **297**. As they occur only in very small amounts, it has proved very difficult to deduce the structures. Only the structure of **301** has been proved by n.m.r. structural analysis as well as by mass spectroscopic measurements:

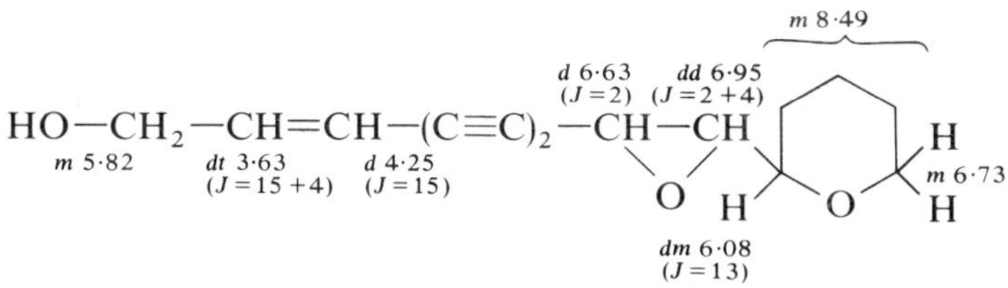

301

The coupling constant of about 2 Hz consists with the *trans*-configuration of the epoxide, while a *cis*-epoxide shows a coupling constant of about 4·2 Hz.

Furthermore, there has been isolated the corresponding epoxide aldehyde **302**, the alcohol **297** (described above) and the corresponding aldehyde of the latter **303**:

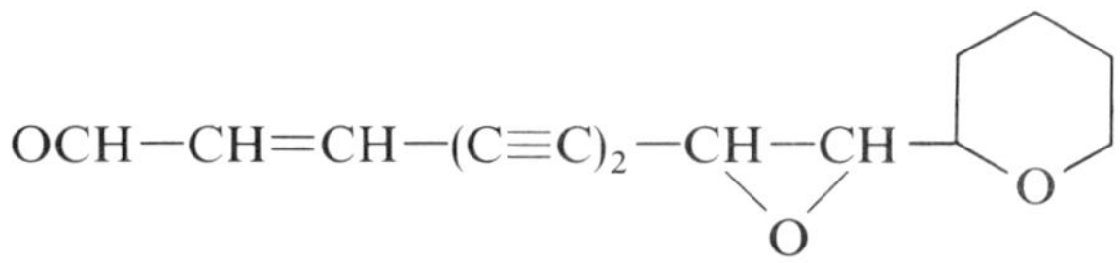

302

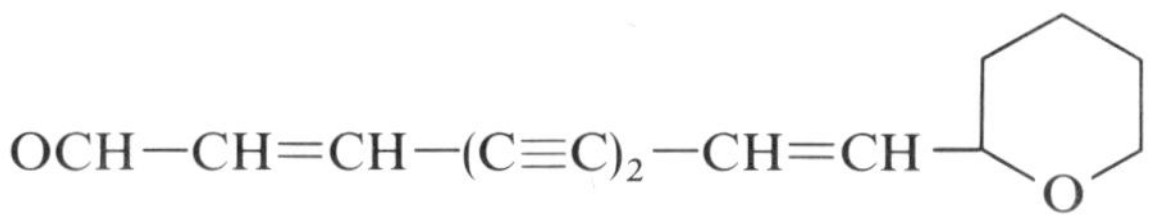

303

302 and **303** were converted, using sodium borohydride, to the corresponding alcohol derivatives. The spectral properties of the latter were shown to be identical with the known natural material.

From the roots of *Wahlenbergia gracilis* Schrad.[35] a tetrahydropyranyl alcohol with an ene-diyne chromophore (λ max: 284, 268, 254, 241 nm) was isolated. The mass spectrum with a molecular ion peak at *m/e* 218 furnished valuable confirmation of the proposed structure **304**:

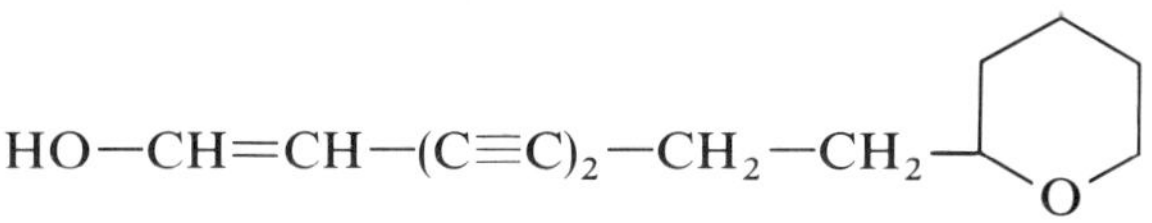

304

Manganese dioxide oxidation converted it into an aldehyde showing diyne-en-one-like absorption in the u.v. All the mentioned cyclic ethers are obviously closely related. The necessary biochemical transformations are already discussed.

As mentioned in Chapter 2, Section 2.3, two tetrahydropyran-derivatives with an ene-triyne chromophore (**59**, **60**) have been isolated from the leaves and flowers of *Dahlia coccinea* Cav. var. coccinea. However, the most striking structural feature of the compounds occurring in the roots of the same plant is the unversal presence of the terminal propenyl group in contrast to the corresponding propynyl group in the leaves and flowers. Two ene-diyn-ene-compounds isolated from the roots of *Dahlia coccinea* Cav. var. coccinea[60] were shown to be the alcohol **305** and its acetate **306**:

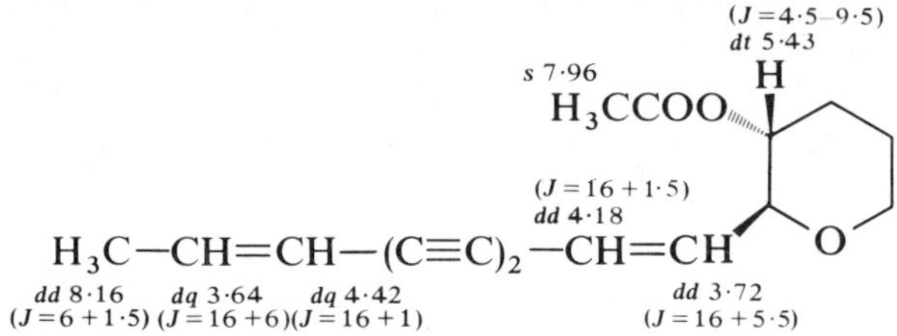

306

306 resembles the tetrahydropyranyl-acetate **59** of Chapter 2, Section 2.3 considerably in its i.r. spectrum and optical rotation; and the alcohol **305** shows corresponding similarities:

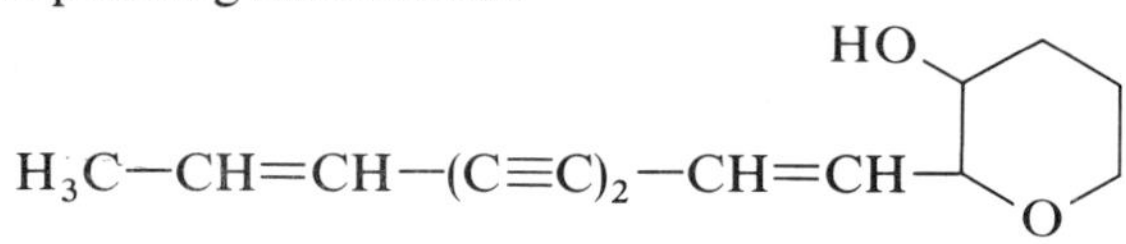

305

The mass spectrum of the alcohol shows a very interesting fragmentation pattern of the tetrahydropyranyl part of the ene-diyn-ene alcohol[59]:

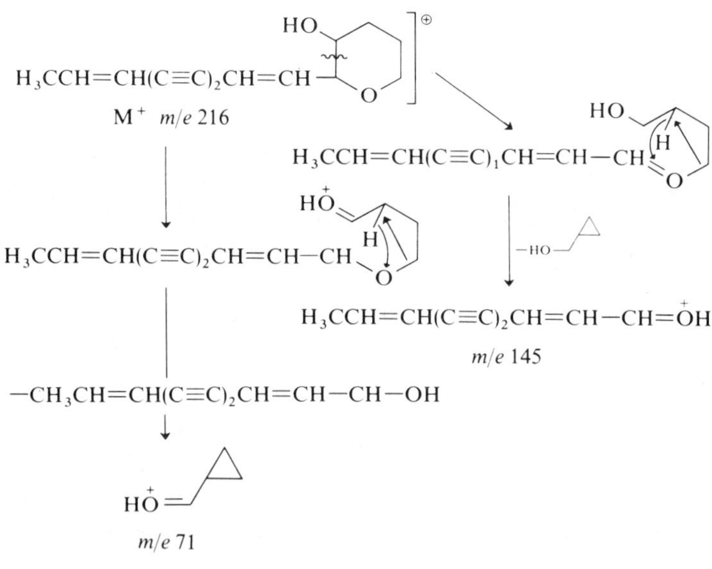

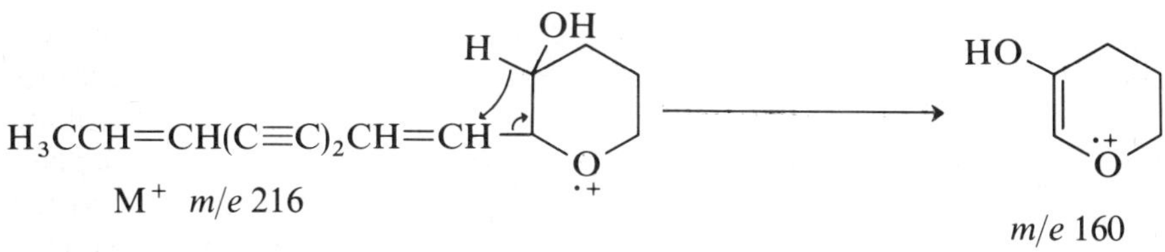

This fragmentation pattern, which is very similar to that of **60** (Chapter 2, Section 2.3), strongly confirms the relationship of the tetrahydropyranyl alcohols from *Dahlia* species. A tetrahydropyranyl derivative with a hitherto unknown feature was shown to occur in *Seseli hippomarathrum* Jacq.[241]. The acetylene ketone was assigned structure **307**, largely on the basis of spectroscopic evidence and some chemical transformations:

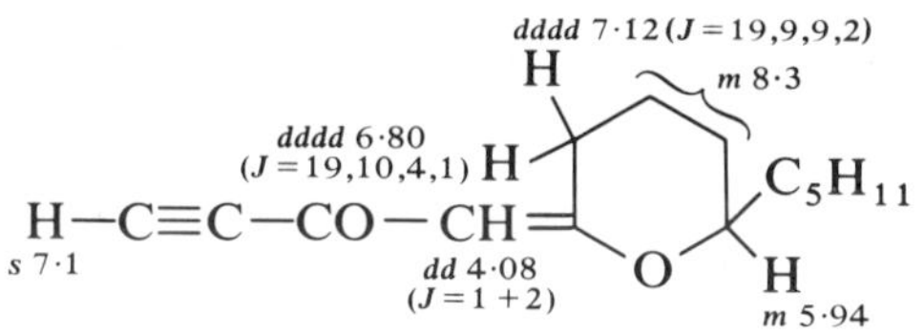

307

Partial hydrogenation of **307** yields the corresponding vinylogous ester, the spectral data of which confirms the proposed structure of the natural compound:

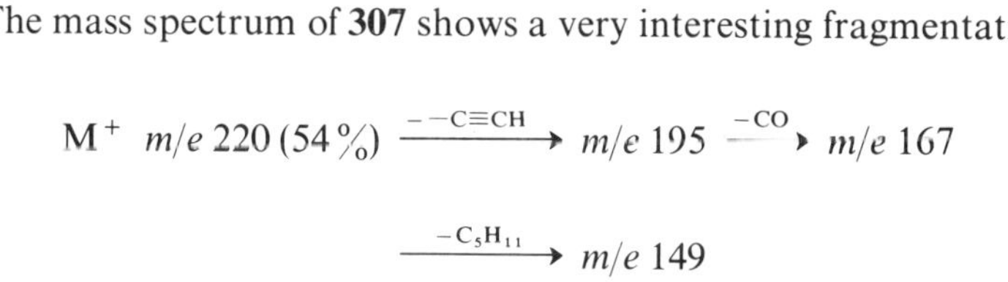

The mass spectrum of **307** shows a very interesting fragmentation pattern:

$$M^+ \; m/e\ 220\ (54\%) \xrightarrow{-C\equiv CH} m/e\ 195 \xrightarrow{-CO} m/e\ 167$$

$$\xrightarrow{-C_5H_{11}} m/e\ 149$$

$$\longrightarrow {}^{+}COC\equiv CH \; m/e\ 53$$

$$\longrightarrow m/e\ 95\ (100\%) \xrightarrow{-C_2H_2} m/e\ 69$$

The fragments 95 and 69 can be explained as following:

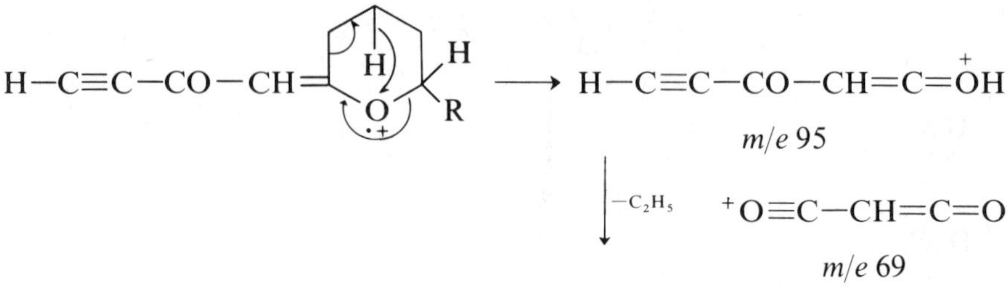

A reasonable biosynthetic pathway for this compound, starting with crepenynic acid (**4**) is outlined in the scheme below:

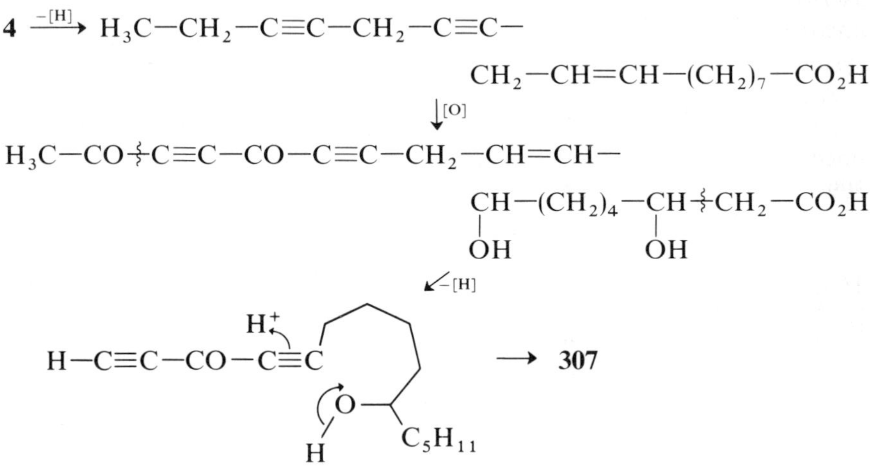

An antifungal acetylenic furanoid keto-ester, named wyerone, has been isolated from shoots of the broad bean *Vicia faba* L., family Papilionaceae[242]. Certain plants show resistance to fungal diseases and pieces of tissue or expressed sap from a large number of plants are toxic to fungi. Investigation of these fungitoxicity of the broad bean revealed the presence as fungitoxins, *inter alia*, of a lipophilic component of high activity. Its i.r. spectrum suggested the presence of an acetylenic keto-ester (2191, 1725, 1640/cm). The u.v. spectrum reveals an extended chromophore (λ max: 351, 291, 224 nm). The formation of a 2,4-dinitrophenylhydrazone and the presence of a strong i.r. band at 2192/cm, which disappeared after reduction with sodium borohydride. but was regenerated by manganese dioxide oxidation, points to a conjugated acetylenic carbonyl system. In addition the band at 1725/cm indicates the presence of a conjugated ester group. Oxidation with alkaline permanganate yields furan-2,5-dicarboxylic acid.

In view of these spectral and chemical results, it seemed probable that the unknown compound has structures of types **308** or **309**:

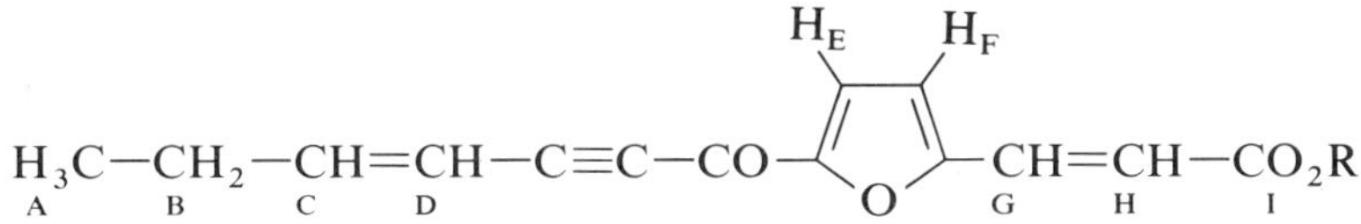

308

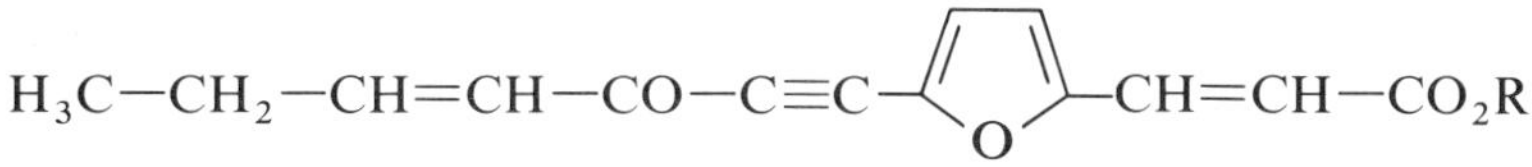

309

The n.m.r. spectrum of the natural product exhibits signals at τ 6·22 (3H) *s*, and τ 8·75 (3H) *t*, coupled to *dq* at τ 7·44 (2H) with 8 Hz, which could be ascribed to the protons of the methoxy and ethyl group. However, the resonances in the olefinic range of the 2,5-disubstituted furan and the olefinic systems were more difficult to interpret.

Assignments in this region were finally confirmed by spin–spin decoupling experiments. From these measurements the H_D-signal could be identified unequivocally, and its position at τ 4·29 supports structure **308** rather than **309**:

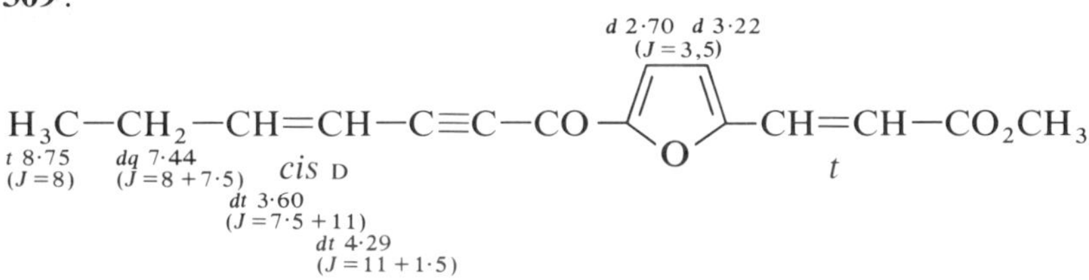

308

It seemed desirable to prove the proposed structure **308** by synthesis. The route chosen involved preparation of the hydroxy-ester **310** by the Wittig reaction with 5-hydroxymethylfurfurylaldehyde. **310** gave the formyl ester **311** when oxidised. *Cis*-hex-3-en-1-yne, prepared from propionaldehyde and prop-2-ynyl bromide, was coupled with the formyl acid **312**. Esterification gave the hydroxy ester **313**, the oxidation of which yielded the keto ester wyerone **308**, with spectra almost indistinguishable from those of the natural compound:

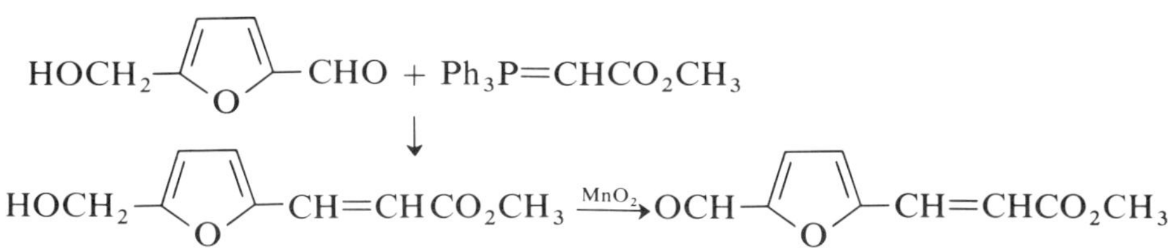

310 **311**

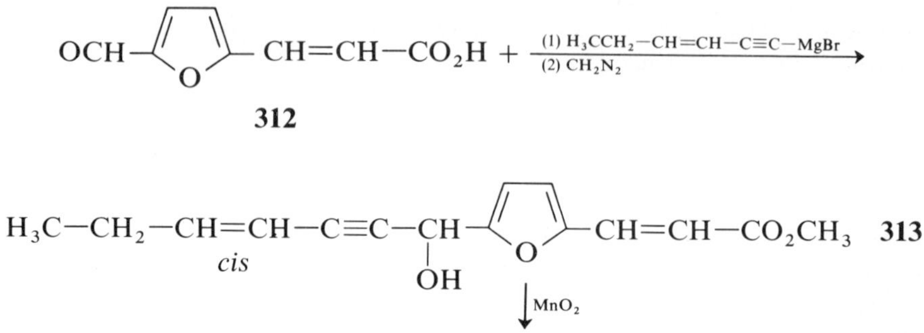

312

313

308

The mass spectrum of **308** shows a fragmentation pattern not unexpected for a compound of the proposed structure:

M^+ *m/e* 258 (85%)

– OCH_3 → *m/e* 227 (26%)

– CO_2CH_3 → *m/e* 199 (11%)

– $C_2H_5CH{=}CH{-}C{\equiv}C{-}$ → *m/e* 179 (21%) – CO → *m/e* 151 (100%)

A peak at *m/e* 260, much more intense in less pure samples, was attributed to the presence of a dihydro impurity **314**, with which the natural material is contaminated (about 2%):

314

Examinations of fractions containing more materials yielded very small amounts of the parent hydroxy ester **313**, identical with synthetic material:

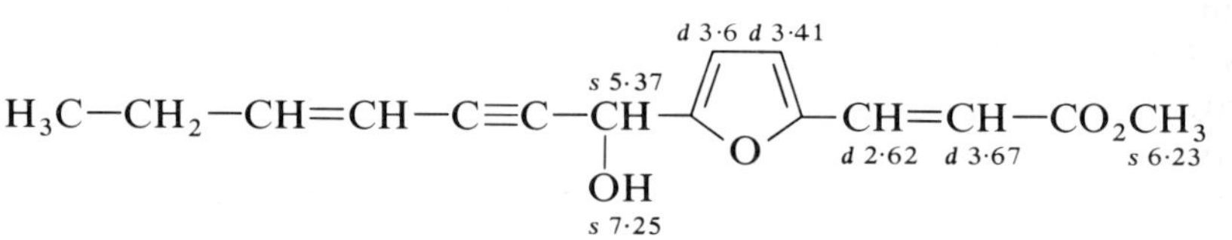

313

Investigation of the constituents of infected broad bean—incubated with *Botrytis cinerea*—gives very small amounts of an antifungal compound, which has been identified as wyerone acid **315**. The compound was extracted from infection droplets and diseased leaves, and was shown to form in response to infection of leaves by the virulent parasite *Botrytis fabae* as well as the avirulent *Botrytis cinerea*[243].

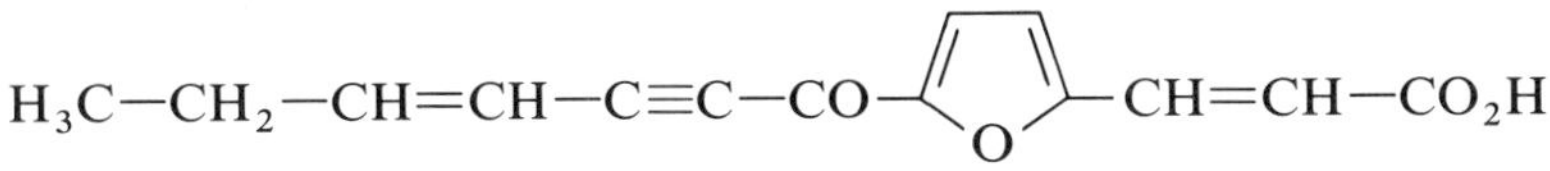 **315**

Wyerone acid is apparently absent in healthy tissue; possibly the acid is formed by hydrolysis of the methyl ester when cells are damaged by infection.

Wyerone itself is active against many fungi, but not against *Botrytis cinerea*. Conversion to the free acid **315** increases its activity against *Botrytis cinerea*. Even more active is the related acetylenic compound capillin (see Chapter 2, Section 3.3: **292**).

The biogenesis of these furan-acetylenes is still not established. Perhaps linoleic acid might be a precursor[242].

4. Aromatic C_{13}-acetylenes

From the flowering parts of *Coreopsis cardaminafolia*, an oily compound has been isolated, which was given the constitutional formula **316**[244]:

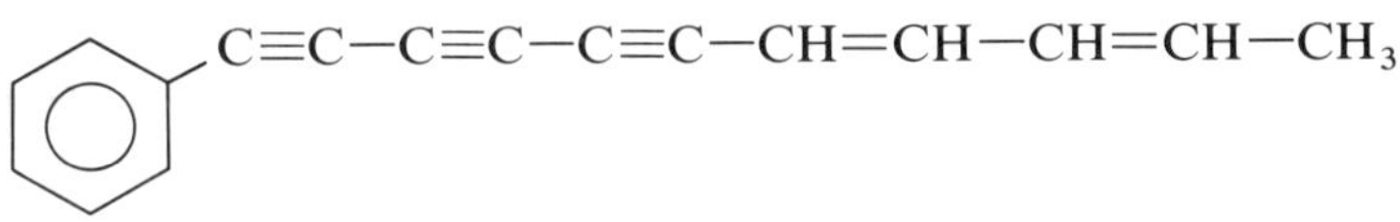 **316**

A synthesised *trans*-isomer of the so-called "3650-hydrocarbon," supposed to be **316**, was undoubtedly different in the u.v. spectra from the compound occurring in the investigated annual *Coreopsis* species[245]. The hypothesis that the naturally occurring compound might be one of the possible *cis*-isomers was, however, unlikely since the synthetic *trans*-isomer has the u.v. absorption maxima about 6·5 nm at longer wavelengths than the natural compound, whereas *cis*-polyen-ynic isomers mostly absorb at longer wavelengths than their corresponding *trans*-isomers.

As **316** was a liquid, the homogeneity was difficult to guarantee. The hydrogenation product, which should be 1-phenyl-*n*-undecane, did not give the correct melting point of the latter. Finally, the investigation of a melting point sample of the hydrogenation product in a vapour fractometer showed that the sample contained at least two products one of which might possibly be *n*-tridecane. Reinvestigation with narrower chromatographic cuts, yielded fractions which showed characteristic differences in their i.r. spectra. The characteristic phenyl peaks at 1494, 753 and 686/cm changed in intensity relative to the acetylenic band at 2200/cm and a characteristic peak at

775/cm. The suspicion hardened that **316** might be a mixture of two compounds, one of them containing no phenyl group. Repeated chromatography on deactivated alumina finally led to a separation of the mixture into two compounds, both of which could be crystallised. One component was deduced to be **317**:

$$H_3C{-}CH{=}CH{-}(C{\equiv}C)_3{-}CH{=}CH{-}CH{=}CH_2 \quad \mathbf{317}$$

The second, phenyl-containing compound was difficult to purify, because of its high solubility in all solvents. The purest crystallised fraction obtained had a melting point of 25–26°C. Its u.v. spectrum resembled the 1-phenyl-hepta-1,3-diyn-5-en-7-ol-acetate, isolated from all parts of *Coreopsis* species. For this reason the authors guessed on the structure 1-phenyl-hepta-1,3-diyn-5-ene **318**:

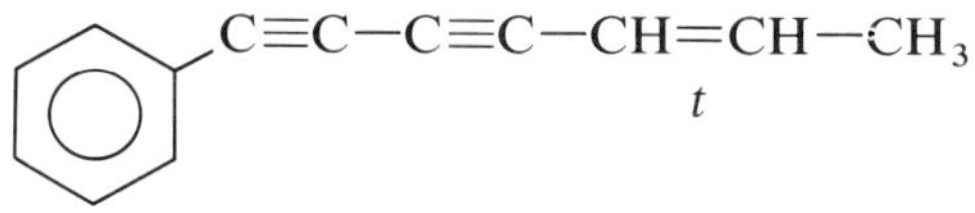

318

It seemed desirable to test this hypothesis by synthesis of the *cis*- and *trans*-isomers of **318**. The route chosen involved oxidative coupling of phenylacetylene with *cis*- and *trans*-pent-3-en-1-yne, respectively. The synthetic *trans*-isomer was identical by spectral comparison with authentic material from the plant.

The synthetic long known 7-phenyl-hept-2-ene-4,6-diyn-1-ol **319** has been isolated from *Coreopsis tintoria*[35]. In 1967 Jones *et al.*[246] showed that tritium labelled **318** was converted into the ω-alcohol **319** by *Dahlia* hybrids:

318 →

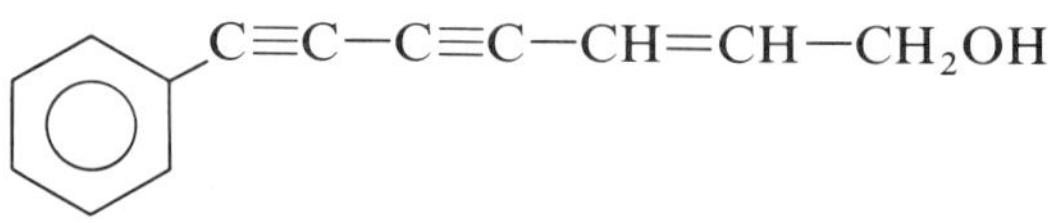

319

Such ω-oxidation seems to be a standard metabolic process for natural polyacetylenes. It has been established that aethusanol B (**320**) (see Section 5.3) is biosynthesised in *Aethusa cynapium* plants from the corresponding hydrocarbon:

$$HO{-}CH_2{-}CH{=}CH{-}(C{\equiv}C)_2{-}CH{=}CH{-}CH{=}CH{-}CH_2CH_3 \quad \mathbf{320}$$

and that the hydroxy acid **321** is produced from the parent methyl ester by fungal cultures of *Tricholoma paneolum* (see Chapter 2, Section 5.4):

$$HO{-}CH_2(C{\equiv}C)_3{-}CH{=}CH{-}CO_2H \quad \mathbf{321}$$

The acetate corresponding to **319**, 1-phenyl-*n*-hept-5-en-1,3-diyn-7-ylacetate **322** was shown to occur in all parts of *Coreopsis tinctoria* Nutt. in the garden varieties "*nigra speciosa*," "*nana radiata*" and *Coreopsis Drummondii* Torr and Gray, *Coreopsis cardaminefolia*, *C. atrosanguinea* and the perennial *Coreopsis verticillata* L.[247].

The compound shows strong absorption bands in the i.r. region at 1754 and 1225/cm, indicating the presence of an ester grouping. Furthermore, there occurs a sharp peak at 1490/cm mostly characteristic of phenyl residues. The u.v. spectrum reveals an extended chromophore: (λ max: 317·5, 298, 281, 263, 253, 245, 222 nm). Usually, in the u.v. spectra of polyenes, a phenyl group is more or less equivalent to $1\frac{1}{2}$ double bond, whilst in phenyl-polyynes it has been found equivalent to about one double bond. In view of these spectral results, the proposed structure **322** seems to be a reasonable one:

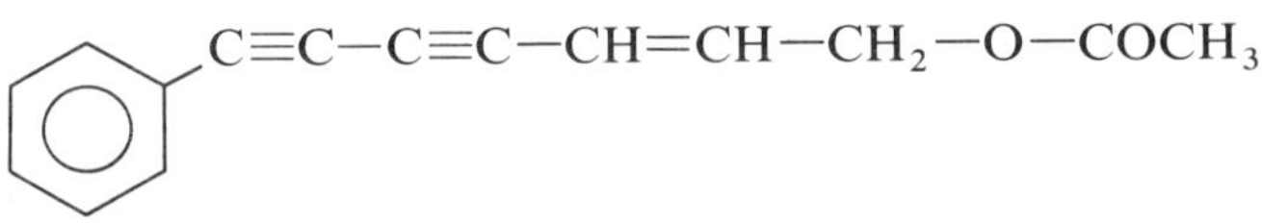

322

The structure of **322** was proved by synthesis[248]: **322** was readily obtained by a Glaser coupling of phenylacetylene with pent-2-en-4-yn-1-ol. The corresponding alcohol **319** gave a u.v. spectrum identical to that obtained from the alcohol from *Coreopsis* species. The alcohol was acetylated in pyridine with acetic anhydride, yielding **322**, which was identical by spectral comparison with authentic material.

From the roots of *Onopordon acanthium*[204], a compound was isolated in very small amounts, which could be crystallised after repeated TLC. Its u.v. spectrum exhibits bands at 346, 336, 322, 298, 265 and 253 nm. Its i.r. spectrum displays bands for an unsaturated aldehyde function (2740, 1700/cm), triple bonds (2220, 2145/cm) and a conjugated *trans* double bond (1608, 1586, 963/cm). A multiplet at τ 2·65 in the n.m.r. spectrum can be ascribed to the five protons of a monosubstituted phenyl ring. Furthermore, the n.m.r. shows an one proton double doublet at τ 0·47 (J = 5, 5 + 1, 5 Hz) indicating an aldehyde function adjacent to a double bond. In addition an AB quartet at τ 3·47 (2H) can be assigned to two olefinic protons. Taking all these facts into consideration, structure **323** must be suggested:

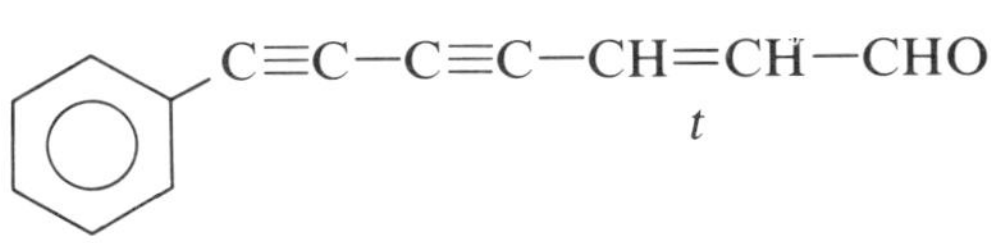

323

The occurrence of the phenyl-diyne **318** and the corresponding phenyl-triyne (see Chapter 2, Section 2.5, **86**) together in certain *Dahlia* hybrids seemed to offer an opportunity of learning something about the stage at which the formation of the terminal acetylenic bond takes place. The tritium-labelled hydrocarbon **318**, therefore, was inserted into the tuberous roots of two *Dahlia* hybrids. As previously mentioned, it found to be converted into **319**, but not into the phenyltriyne (**86**, Chapter 2, Section 2.5). **318** was synthesised from tritiated iodobenzene.

Investigation of the constituents of the *Dahlia* hybrid "Preference", yielded, *inter alia*, from the most polar fractions two active poly-alcohols, which seemed to be di- and trihydroxylated[246]. The two alcohols were separated by TLC and their structures elucidated by spectral methods. The spectral properties were consistent with structures **324** and **325**:

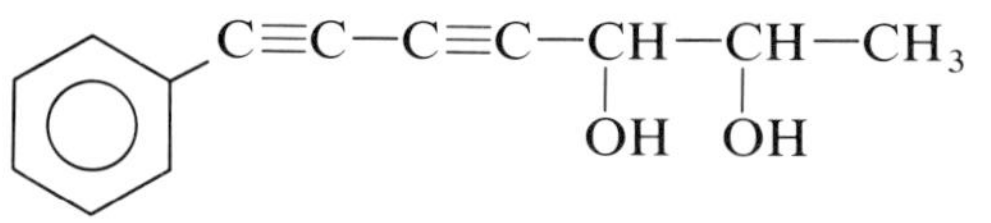

(*threo*) **324**

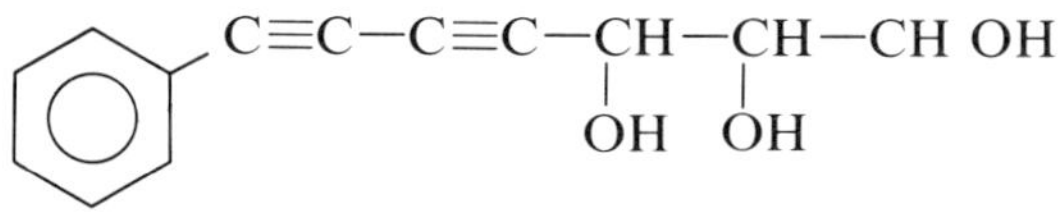

(*threo*) **325**

Periodate cleavage of the two alcohols followed by borohydride reduction revealed, that the C_{11}-alcohol **326** obtained contained most of the tritium activity:

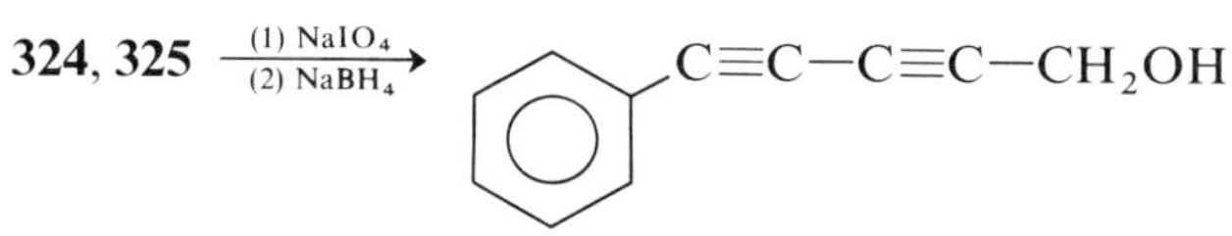

326

The hitherto not described threo-diol **324** was synthesised from the *trans* hydrocarbon **318** via osmium tetroxide hydroxylation[249]. Synthetic **324** was identical in all respects with a sample of natural diol:

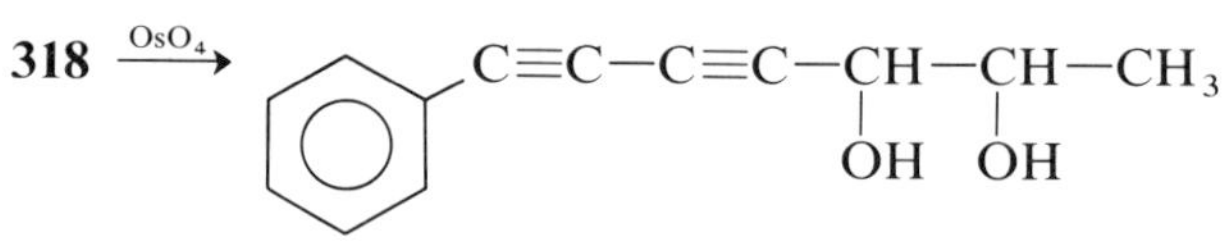

324

A reasonable biosynthetic pathway for these compounds is outlined in the scheme below following the one already established in the corresponding triynes (see Chapter 2, Section 2.5):

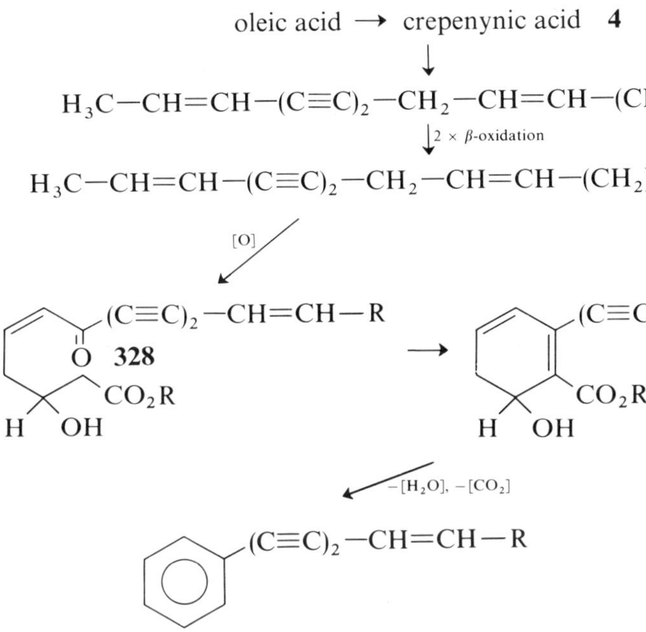

318: R = CH_3
323: R = CHO
319: R = CH_2OH
322: R = OAc

An alternative route only differs in the oxidation product of **327**. Instead of an hydroxy-ester (**328**) an oxo-ester **330** could arise, which would lead to 3-substituted phenylpolyyn-enes:

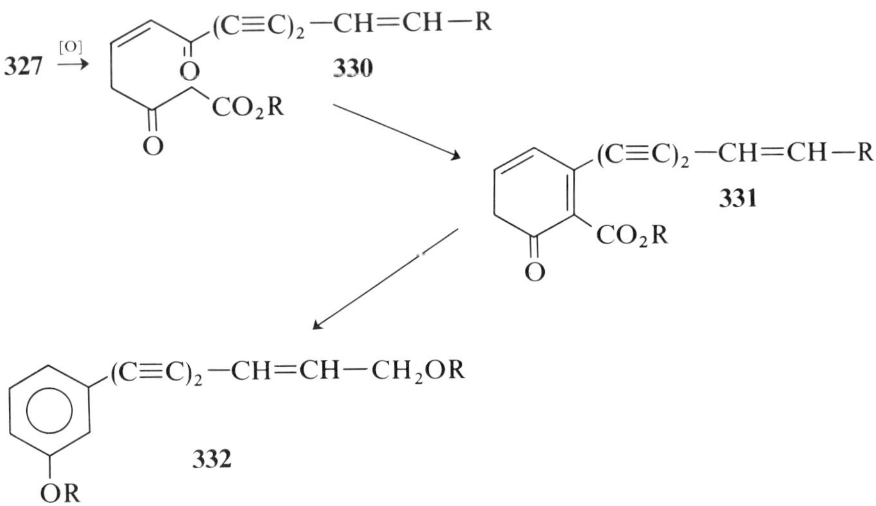

Representatives of **332** are the acetate **333** and diacetate **334**, respectively, isolated from *Coreopsis tinctoria* cv. tetra Goldteppich[250]:

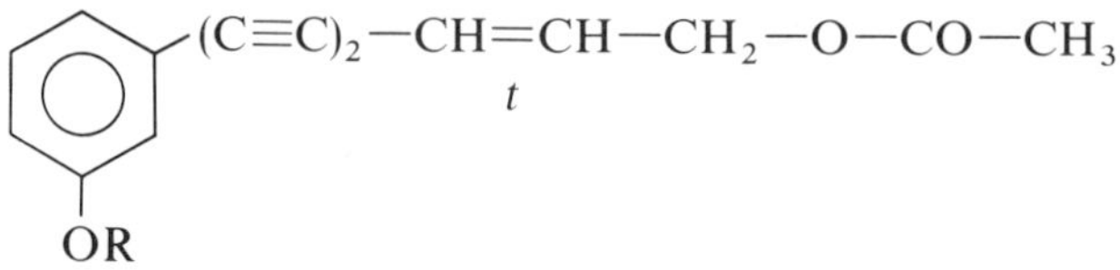

333: R = H
334: R = $COCH_3$

The n.m.r. spectrum of **334** shows two singlets at τ 7·78 and τ 7·98 arising from two acetoxy-groups. The existence of the $-C{\equiv}C-CH{=}CH-CH_2-OAc$-grouping is shown by the following signals: *dt* 4·22τ (J = 16 + 1·3); *dt* 3·71τ (J = 16 + 5·5) and *dd* 5·43τ (J = 5·5 + 1·3 Hz). The i.r. spectrum of **334** also indicates the presence of two *O*-acetyl groups (1783, 1760/cm). The structure of **334** was further proved by oxidative degradation. The degradation product was identified by spectral comparison with authentic material as *m*-methoxybenzoic acid.

The i.r. spectrum of **333** lacked the aromatic *O*-acetyl-absorption of **334**. The presence of a hydroxy group was shown by a band at 3620/cm in the i.r. spectrum. When treated with acetic anhydride, **333** was converted into **334**.

It seemed desirable to prove the structure of **334** by synthesis. The route chosen involved preparation of *m*-hydroxyphenylpropiolic acid **336**[117]. Therefore *m*-benzoyl-oxy-cinnamic acid **335** was brominated. The corresponding dibromide yielded after dehydrohalogenation and saponification **336**. Decarboxylation and acetylation of **336** gave **337**, the latter was coupled with 5-bromo-pent-4-yn-2-en-1-ol to **338**, which was acylated to **334**:

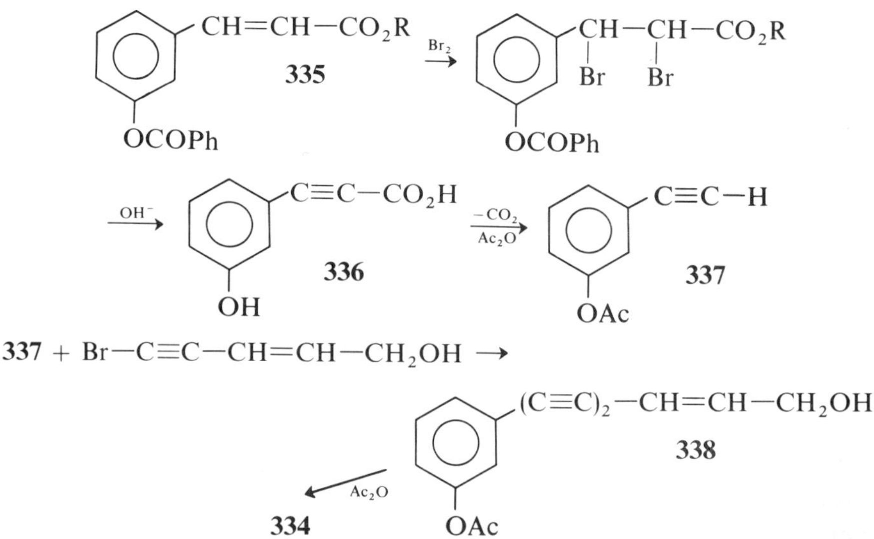

The *cis*-isomer of **333**, the acetate **339** is shown to occur in *Carlina vulgaris* L.[251]:

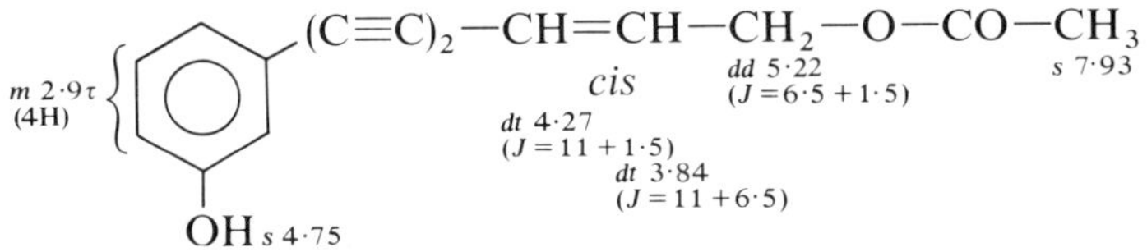

339

This structure is proved by n.m.r. structural analysis as well as mass spectroscopic measurements.

One of the longest-known, naturally occurring acetylenic compound is the so-called "carlina oxide" **340**, named by Semmler. It is the principal constituent of the essential oil obtained by steam distillation of the roots of the thistle *Carlina acaulis*. The oil had been investigated by Semmler in 1889[4] but it was not until 1906 that he reported the isolation of carlina oxide and assigned a structure to it[252]. Semmler's scepticism concerning the possible occurrence of acetylenic compounds in nature led him to prefer the allenic (**341**) rather than the acetylenic (**340**) constitution. In 1933 Gilman *et al.*[5] suggested the correct structure, benzyl-2-furylacetylene **340**, and two independent syntheses appeared soon after[5,253].

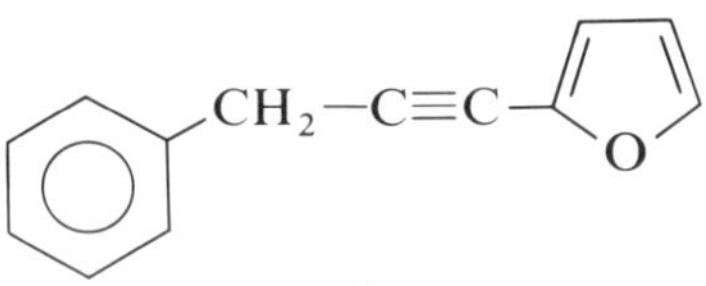

340

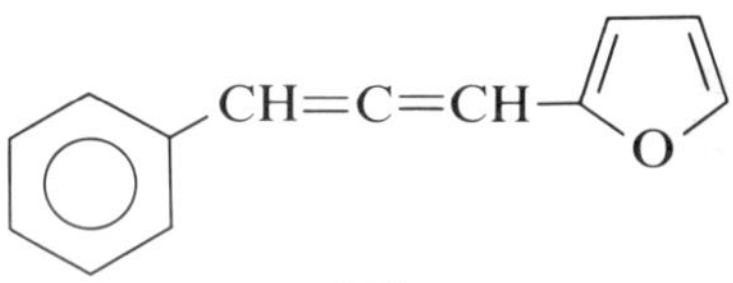

341

A synthesis scheme is outlined below:

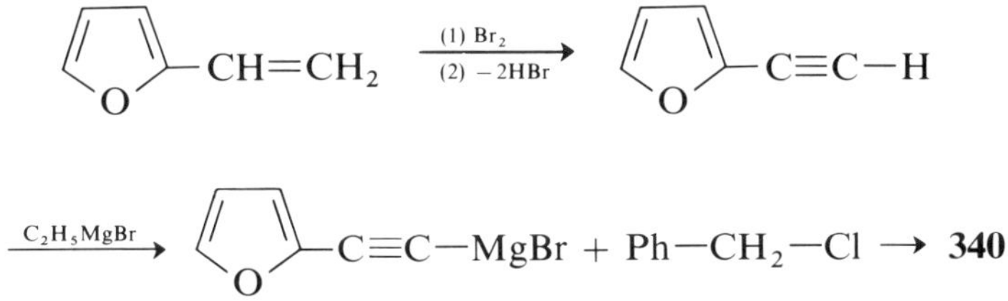

For a long time, no results have been available concerning the biosynthetic pathway of the formation of carlina oxide. A possible connection between **322** and carlina oxide has been pointed out by Jones and Whiting (see ref. 247) in the light of an observation made by Heilbron *et al.*[254] that the pentenynol **342** is readily converted in an acidic medium into α-methylfuran **343**:

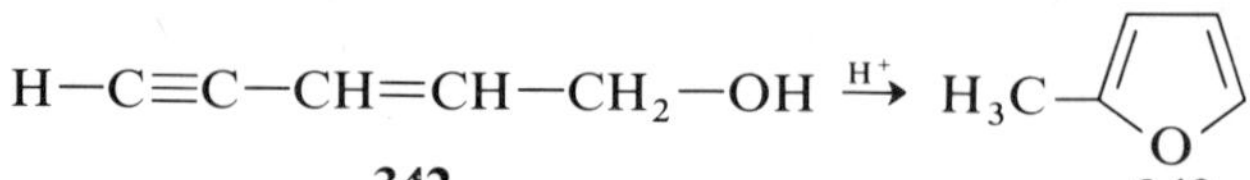

342 **343**

The conversion of **322** into **340** might have proceeded by an analogous mechanism:

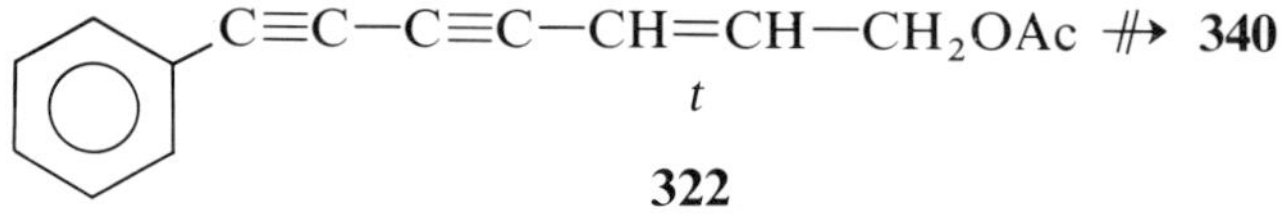

322

Several attempts by the Manchester school[254] to effect such a conversion have been without success. However, attempts with the corresponding *cis*-alcohol **344** have been successful. Tritium labelled **344** have been fed to *Carlina acaulis* L. and the radioactive carlina oxide isolated[255]:

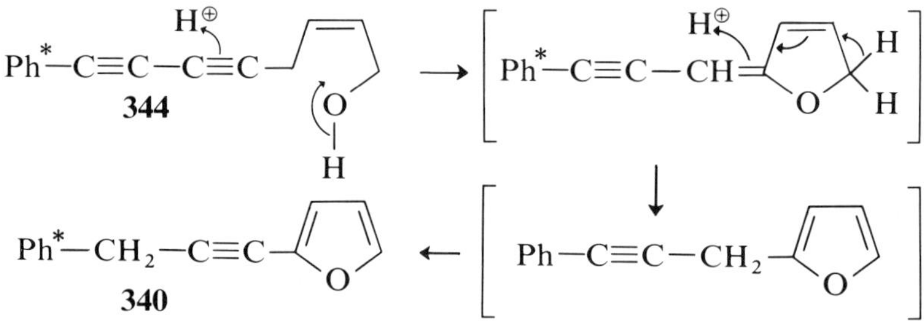

Oxidative degradation of the obtained labelled **340** revealed active benzoic acid. Furthermore, it could be shown that carlina oxide was wholly derived from labelled tetradec-5-ene-7,9-diyn-1-*O*-acetate **345** and tetradec-5,12-diene-7,9-diyn-1-ol **346**[35]:

$$C_3H_7-(C\equiv C)_2-CH_2-CH=CH-(CH_2)_4OAc \quad \mathbf{345} \rightarrow \mathbf{340}$$

$$CH_3-CH=CH-(C\equiv C)_2-CH_2-CH=CH-(CH_2)_4-OH \quad \mathbf{346} \rightarrow \mathbf{340}$$

As expected, **345** gave lower incorporation than **346**. These tracer studies supports our conception about the biogenesis of the C_{13}-phenyl-polyynes described above.

$$H_3C-CH=CH-(C\equiv C)_2-CH_2-CH=CH-(CH_2)_3-CO_2R \quad \mathbf{327}$$

327 →[H] **346**

327 → **328** → **329** → **318, 323, 319, 322**

5. Compounds related to the trideca-(2,9)-diene-(4,6)-diyne and the trideca-(2,4,8,10)-tetraen-(6)-yne

5.1. General Considerations to the Biogenesis of the C_{13}-polyacetylenes

C_{13}-polyacetylenic compounds are very widespread in the plant families of Compositae and Umbelliferae. The multiplicity of their features can be reduced to a simple biosynthetic scheme, which starts with a C_{14}-compound. The latter arise in the already mentioned way from oleic acid via its 12,13-dehydroderivative crepenynic acid. The necessary transformations involve further dehydrogenations and chain shortening (twice β-oxidation).

The C_{14}-alcohol **346** then will be converted by β-oxidation into the C_{13}-compound **347**, which is further transformed to the C_{13}-polyacetylenes isolated from *Aethusa cynapium* L. Similar pathways lead to the more unsaturated C_{13}-hydrocarbons as shown by feeding experiments using labelled **346**. The following scheme shows the main reactions:

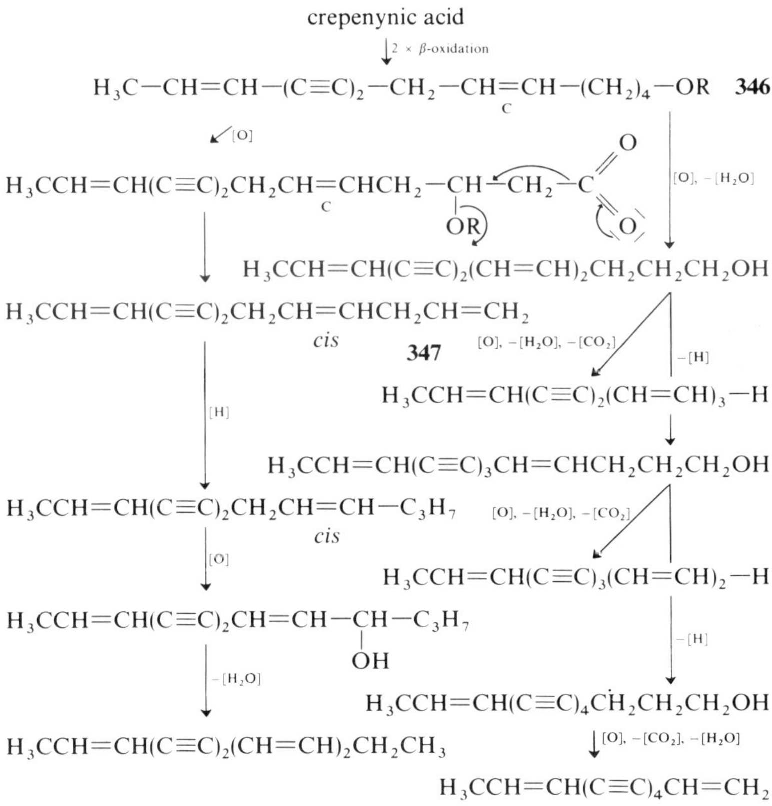

The resulting hydrocarbons themselves are again precursors for a considerable number of further natural acetylenes.

5.2. Compounds Containing an Ene-diyne-triene Chromophore

The essential oil obtained from the leaves and stems of some annual *Coreopsis* species contains, *inter alia*, a crystallising hydrocarbon (m.p.: 71°), which shows an ene-diyne-triene like absorption in the u.v. (λ max: 353·5, 330, 310, 274, 264 nm)[247]. The hydrocarbon is very unstable and is oxidised extremely rapidly. The presence of *trans* double bonds is indicated by the formation of a maleic anhydride adduct **349**, which exhibits the spectrum of a symmetrical diene-diyne. This evidence, coupled with the data from analysis and hydrogenation—the compound was hydrogenated to *n*-tridecane—points to an ene-diyne-triene, i.e., trideca-1,3,5,11-tetraene-7,9-diyne **348**:

$$H_3C{-}CH{=}CH{-}(C{\equiv}C)_2{-}CH{=}CH{-}CH{=}CH{-}CH{=}CH_2 \quad \textbf{348}$$

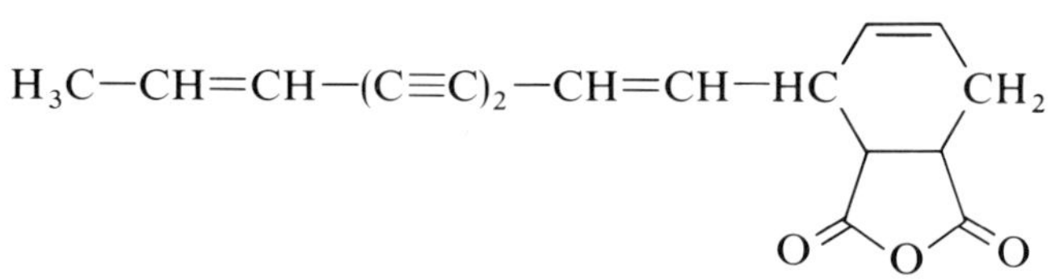

349

Some additional support of formula **348** can be obtained from the i.r. spectrum. The compound absorbs in the i.r. region at 905 and 1005/cm characteristic of conjugated and unconjugated *trans* double bonds, respectively. Since the solvent—CCl_4—covers most of the region where the *cis* double bond mostly absorb, a *cis* configuration is not excluded, but since the *trans* peaks are strong, most probably all double bonds are *trans*.

To confirm the structure, **348** has been synthesised, using the ylide compound **350** prepared by treating allylbromide with triphenylphosphine according to the method of Wittig[256]. This technique has been made good use of for the introduction of double bonds in the synthesis of polyenynes. In the present instance, the ylide **350** was condensed with matricarianal, obtained by oxidation of *trans–trans* matricarianol [see Section 7.1 (**514**)] with manganese dioxide:

$$H_3CCH{=}CH(C{\equiv}C)_2CH{=}CHCH_2OH \quad \textbf{514} \qquad Ph_3P + Br{-}CH_2CH{=}CH_2$$

$$\downarrow MnO_2 \qquad\qquad \downarrow$$

$$Ph_3P{-}CH_2{-}CH{=}CH_2]^+Br^-$$

$$\downarrow$$

$$\underset{t}{H_3C{-}CH{=}CH}(C{\equiv}C)_2{-}\underset{t}{CH{=}CH}{-}CHO \quad + Ph_3P{=}CH{-}CH{=}CH_2 \quad \textbf{350}$$

$$\downarrow$$

all-*trans* **348**

The identity of synthetic and natural material confirmed the view about the all-*trans* configuration of the latter. The previously discussed biogenesis of **348** has been confirmed by feeding labelled **346**[98].

The aerial parts of *Bidens ferulaefolia* Thuill. have been found to contain a triene-diyn-ene acetate, shown to be the allylic oxidation product **351** on the basis of chemical and spectroscopic evidence[47]:

$$H_3C{-}CO{-}O{-}CH_2{-}CH{=}CH{-}(C{\equiv}C)_2{-}(CH{=}CH)_3{-}H \quad \mathbf{351}$$

s 7·94   *d* 5·37 (*J* = 5·5)   *t*

Furthermore, the n.m.r. shows a poorly resolved multiplet in the olefinic range (*m* 2·6–4·9) for nine protons. The acetate **351** was isolated from *Carlina vulgaris* L. The n.m.r. spectrum of the crude acetate fraction, containing **351**, showed the presence of very small amounts of the corresponding Δ^2-*cis* isomer **352**[251]:

$$H_3C{-}CO{-}O{-}CH_2{-}CH{=}CH{-}(C{=}C)_2{-}(CH{=}CH)_3{-}H \quad \mathbf{352}$$

d 5·23 (*J* = 6 Hz)   *cis*

The corresponding alcohol **353** has been isolated from the roots of *Dahlia* species[22]:

$$HO{-}CH_2{-}CH{=}CH{-}(C{\equiv}C)_2{-}CH{=}CH{-}CH{=}CH{-}CH{=}CH_2 \quad \mathbf{353}$$

From the roots of *Centaurea pullata* a C_{13}-polyynene has been isolated, possessing a characteristic u.v. absorption at 379 and 359 nm[36]. Its i.r. spectrum indicates the presence of a triple bond (2188/cm), a vinyl group (914/cm), and a strong band at 1638/cm, which could be ascribed to an enol-ether. When treated with acid, the compound is readily converted into a compound with a trien-yne chromophore (λ max: 302, 289 nm) **355**. Taking all these facts into consideration structure **354** can be suggested:

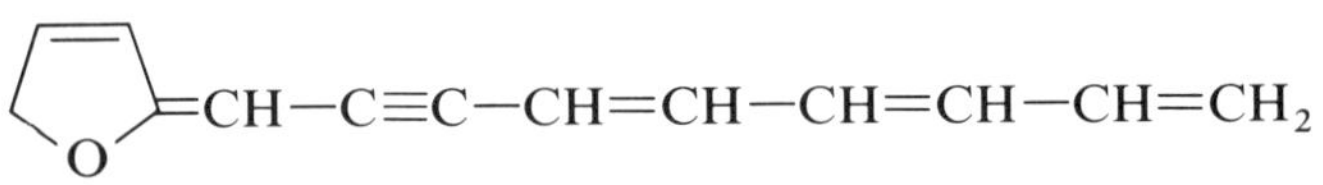

354

This agrees with a molecular ion at *m/e* 184 observed in the mass spectrum. The absence of protons resonating higher than τ 5 in the n.m.r. is consistent with the proposed structure:

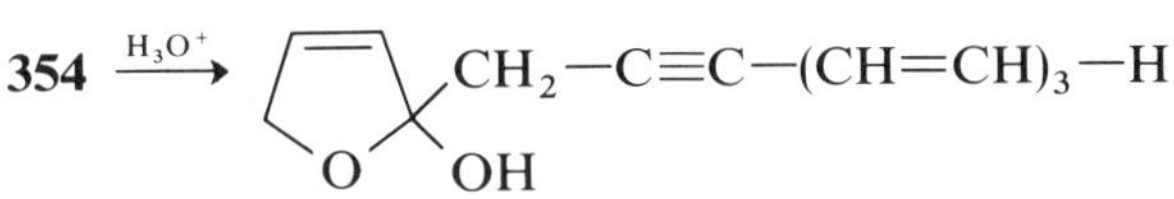

355

Structural analysis of the various constituents of *Centaurea pullata* reveals that **354** may be related biosynthetically to **348**, which is shown to occur with **354** in the same plant. A reasonable biosynthetic pathway would be allylic oxidation of **348** to the *cis*-alcohol **356** followed by ring closure:

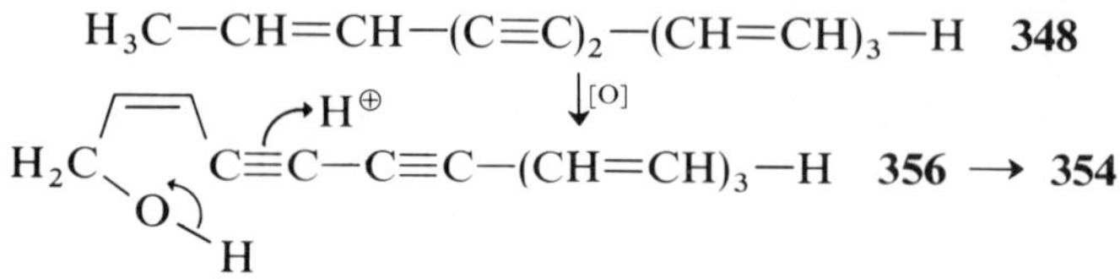

It seemed desirable to prove the structure of **354** by synthesis. The mono-tetrahydropyranylether **357** was converted by manganese dioxide oxidation and cleavage of the protecting group to the hydroxy aldehyde **358**. Treatment of the latter with potassium-tert.-butylate yielded, besides many resinification products, the aldehyde **359**, the structure of which was confirmed by n.m.r. spectroscopic measurements. Wittig reaction of **359** with an ylide compound, prepared by treating allyl bromide with triphenyl phosphine, gave an isomeric mixture of **354**:

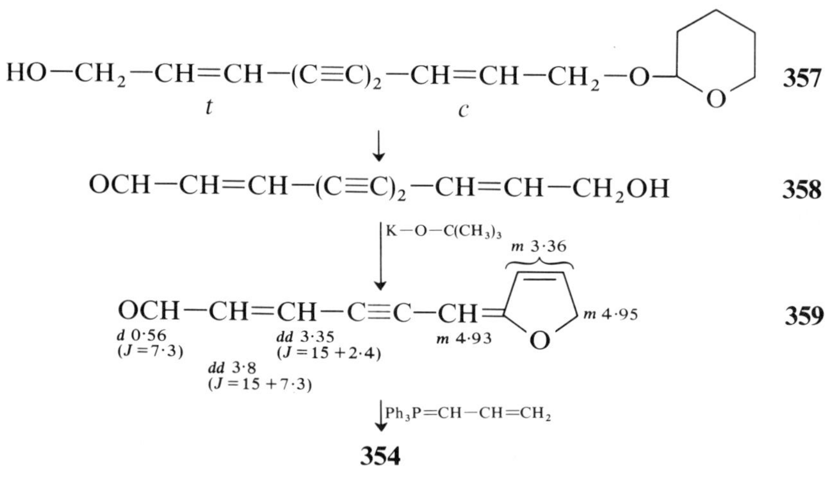

The u.v. spectrum of **359** shows an unusual longwave maximum at 365 nm. The synthesised isomeric **354** was nearly identical with the material from *Centaurea pullata*.

Two compounds, closely related to **348**, have been isolated from the roots of *Centaurea ruthenica* Lam.[100], showing an ene-diyne-diene chromophore in the u.v. The less polar compound **360** has the molecular constitution $C_{17}H_{18}O_4$ and the i.r. spectrum suggests the presence of an ester-carbonyl (1760/cm) as the only oxygen function. The i.r. spectrum of the saponification product exhibits bands attributable to hydroxy groups (3580, 3410/cm), but lacked the carbonyl absorption of the latter. Physical and spectral properties of the saponification product are identical with those derived from the more polar compound **361**, obtained from the same plant. **361** is cleaved with periodic acid to give a polyenyne aldehyde, the u.v. spectrum of which points to an ene-diyne-triene chromophore.

The next most likely form of **362** is unlikely, after the comparison with the known u.v. spectrum of a homologue of **362**[257]. Therefore, the aldehyde is deduced to be **363**. In view of these results, it seemed probable that the two unknown compounds have the proposed structures **360** and **361**:

$$H_3C-(CH=CH)_2-(C\equiv C)_2-CH=CH-CHO \qquad \mathbf{362}$$

$$H_3C-CH=CH-(C\equiv C)_2-(CH=CH)_2-\underset{\displaystyle OAc}{\underset{|}{CH}}-CH_2O-Ac \qquad \mathbf{360}$$

$$\downarrow OH^-$$

$$H_3C-CH=CH-(C\equiv C)_2-(CH=CH)_2-\underset{\displaystyle OH}{\underset{|}{CH}}-CH_2-OH \qquad \mathbf{361}$$

$$\downarrow HIO_4$$

$$H_3C-CH=CH-(C\equiv C)_2-(CH=CH)_2-CHO \qquad \mathbf{363}$$

The structure of **361** has been finally established by synthesis[101].

The aldehyde **365** was obtained by a vinyl–ether condensation with γ-acetoxy-crotonaldehyde-acetal (**364**). The aldehyde **365** was condensed with acetylene monomagnesiumbromide to give the diol **366**, which was coupled with 1-bromo-pent-3-en-1-yne to the diol **367**. Acid catalysed allylic rearrangement gave the methyl ether **368**, which was cleaved with acid to **361**:

$$AcO-CH_2-CH=CH-CH\begin{matrix}\diagup OR\\ \diagdown OR\end{matrix} \quad + H_2C=CH-OR/ZnCl_2$$

364

$$\downarrow$$

$$AcO-CH_2-CH=CH-CH=CH-CHO \qquad \mathbf{365}$$

$$\downarrow + BrMg-C\equiv C-H$$

$$HO-CH_2-CH=CH-CH=CH-\underset{\displaystyle OH}{\underset{|}{CH}}-C\equiv C-H \qquad \mathbf{366}$$

$$\downarrow + Br-C\equiv C-CH=CH-CH_3$$

$$\downarrow$$

$$H_3C{-}CH{=}CH{-}(C{\equiv}C)_2{-}CH(OH){-}CH{=}CH{-}CH{=}CH{-}CH_2{-}OH \quad \mathbf{367}$$

$$\downarrow \; CH_3OH/H_2SO_4$$

$$H_3C{-}CH{=}CH{-}(C{\equiv}C)_2{-}CH{=}CH{-}CH{=}CH{-}CH(OCH_3){-}CH_2OH \quad \mathbf{368}$$

$$\downarrow \; H_3O^+$$

361

The spectral properties of synthetic **361** were almost identical with that of the natural diol. A diacetate, isomer to **360** has been shown to occur in *Dahlia* species. Its structure was assigned to be **369** on the basis of i.r. and n.m.r. spectra and the elemental analysis[22]:

$$H_3C{-}CH{=}CH{-}(C{\equiv}C)_2{-}CH{=}CH{-}CH(OAc){-}CH(OAc){-}CH{=}CH_2 \quad \mathbf{369}$$

Biosynthetically, **369** is presumably derived from the hitherto unknown epoxide **370**:

$$\mathbf{348} \xrightarrow{[O]} [H_3C{-}CH{=}CH{-}(C{\equiv}C)_2{-}CH{=}CH{-}\underbrace{CH{-}CH}_{O}{-}CH{=}CH_2] \quad \mathbf{370}$$

$$\downarrow$$

369

An analogous mechanistic route could be postulated for **361** and **360**. The intermediate has to be the epoxide **371**:

$$\mathbf{348} \xrightarrow{[O]} H_3C{-}CH{=}CH{-}(C{\equiv}C)_2{-}(CH{=}CH)_2{-}\underbrace{CH{-}CH_2}_{O} \quad \mathbf{371}$$

$$\downarrow$$

361 → **360**

5.3. Compounds Containing an Ene-diyne-diene Chromophore

The C_{13}-hydrocarbon aethusin (**372**) has been isolated from *Aethusa cynapium* L.[257]:

$$H_3C{-}CH{=}CH{-}(C{\equiv}C)_2{-}CH{=}CH{-}CH{=}CH{-}CH_2{-}CH_3 \quad \mathbf{372}$$

aethusin

Its structure was elucidated by u.v. and i.r. spectroscopic measurements. The u.v. spectrum exhibits the presence of an ene-diyne-diene chromophore and the i.r. absorption bands suggest the presence of a *trans*-disubstituted double bond (950/cm), a *trans-trans* diene grouping (980/cm) and triple bonds (2200, 2120/cm). The presence of two conjugated *trans* double bonds was further indicated by the formation of a maleic anhydride adduct (**373**), which exhibits the u.v. spectrum of an ene-diyne. This evidence, coupled with the hydrogenation product—*n*-tridecane—points to the proposed structure **372**:

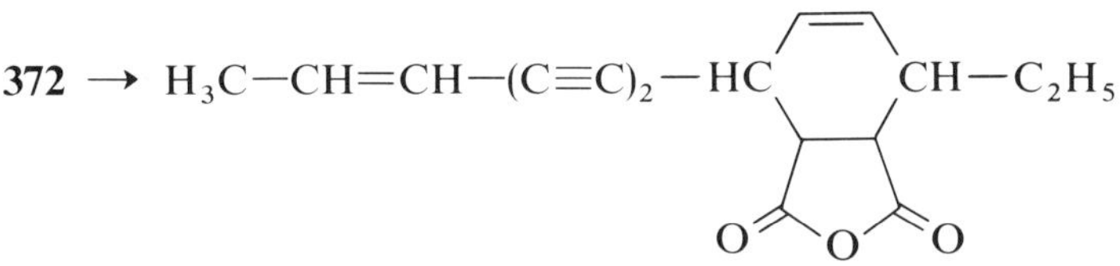

373

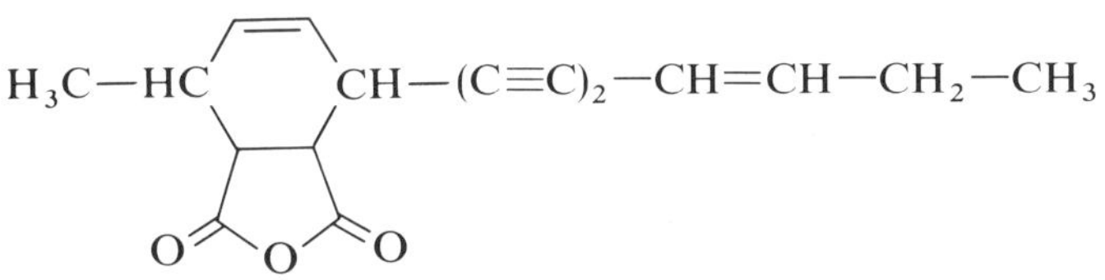

374

The orientation of the *trans* double bond system was determined by ozonolysis of the maleic anhydride adduct. The formation of acetaldehyde shows **372** to have the proposed structure, whilst **374** would give propionaldehyde. The structure of aethusin was finally confirmed by synthesis. The route chosen involved Grignard reaction of matricarianal (**513**) (see Section 7.1) with propyl magnesium bromide. The obtained racemic carbinol **383** (the so-called aethusanol A) is shown to occur with **372** in *Aethusa cynapium*; see below) was readily dehydrated to the hydrocarbon **372**:

$H_3C{-}CH{=}CH{-}(C{\equiv}C)_2{-}CH{=}CH{-}CHO$ **513**

$\downarrow + BrMg{-}CH_2{-}CH_2{-}CH_3$

$H_3C{-}CH{=}CH{-}(C{\equiv}C)_2{-}CH{=}CH{-}CH(OH){-}CH_2{-}CH_2{-}CH_3$ **383**

$\downarrow$ toluene-*p*-sulphonic acid

372

As already discussed in Section 5.1, aethusin biogenetically should be formed in the following way:

$$H_3C{-}CH{=}CH{-}(C{\equiv}C)_2{-}CH_2{-}CH{=}CH{-}(CH_2)_4{-}OH \quad \mathbf{346}$$

$$\downarrow [O],\ -[CO_2]$$

$$H_3C{-}CH{=}CH{-}(C{\equiv}C)_2{-}CH_2{-}CH{=}CH{-}CH_2{-}CH{=}CH_2 \quad \mathbf{347}$$

$$\downarrow [O],\ [H],\ -[H_2O]$$

372

By means of ^{3}[H]-tracer techniques, it has been shown that **372** is wholly derived from **346** and **347**, respectively[35]. These feeding experiments are a very good support for the proposed biosynthesis of the C_{13}-polyacetylenes, outlined in Section 5.1.

Aethusin epoxide **375** was shown to occur in very small amounts in *Aethusa cynapium* too[258]. Its u.v. spectrum reveals the presence of an ene-diyne-ene chromophore (λ max: 313, 293, 278 nm). **375** was treated with sulphuric acid to give the diol **376**, which has been isolated too from the more polar fractions of the plant. The glycol group of **376** was cleaved with periodic acid to give matricarianal (**513**) and propionaldehyde:

$$H_3C{-}CH{=}CH{-}(C{\equiv}C)_2{-}CH{=}CH{-}\underbrace{CH{-}CH}_{O}{-}CH_2{-}CH_3 \quad \mathbf{375}$$

$$\downarrow H_3O^+$$

$$H_3C{-}CH{=}CH{-}(C{\equiv}C)_2{-}CH{=}CH{-}\underset{OH}{\underset{|}{CH}}{-}\underset{OH}{\underset{|}{CH}}{-}CH_2{-}CH_3 \quad \mathbf{376}$$

$$\downarrow HIO_4$$

$$H_3C{-}CH{=}CH{-}(C{\equiv}C)_2{-}CH{=}CH{-}CHO \ + \ CH_3{-}CH_2{-}CHO$$

513

The corresponding 5,6-aethusin epoxide **377** also has been isolated from *Aethusa cynapium*[259]. In analogy to **375**, it has been transformed to the diol **378**, the periodate-cleavage of which gave *n*-octanol after hydrogenation and reduction of the reaction products:

$$H_3C{-}CH{=}CH{-}(C{\equiv}C)_2{-}\underbrace{CH{-}CH}_{O}{-}CH{=}CH{-}CH_2{-}CH_3 \quad \mathbf{377}$$

$$\downarrow$$

$$H_3C{-}CH{=}CH{-}(C{\equiv}C)_2{-}\underset{OH}{\underset{|}{CH}}{-}\underset{OH}{\underset{|}{CH}}{-}CH{=}CH{-}CH_2{-}CH_3 \quad \mathbf{378}$$

$$\downarrow HIO_4,\ H_2,\ NaBH_4$$

n-octanol

Aethusanol B (**320**), the corresponding acetate **379**, the aldehyde **380** and aethusanol B-epoxide **381** were shown to occur with the former described compounds in *Aethusa cynapium*[257,258]:

$$HO{-}CH_2{-}CH{=}CH{-}(C{\equiv}C)_2{-}CH{=}CH{-}CH{=}CH{-}C_2H_5 \quad \textbf{320}$$

aethusanol B

$$H_3C{-}CO{-}O{-}CH_2{-}CH{=}CH{-}(C{\equiv}C)_2{-}CH{=}CH{-}CH{=}CH{-}C_2H_5 \quad \textbf{379}$$

$$OCH{-}CH{=}CH{-}(C{\equiv}C)_2{-}CH{=}CH{-}CH{=}CH{-}C_2H_5 \quad \textbf{380}$$

$$HO{-}CH_2{-}CH{=}CH{-}(C{\equiv}C)_2{-}CH{=}CH{-}\underbrace{CH{-}CH}_{O}{-}C_2H_5 \quad \textbf{381}$$

320 reveals a u.v. spectrum nearly identical to those obtained from **372**. Furthermore, its i.r. spectrum suggests the presence of a hydroxy group. The allylic position of the latter was shown by oxidation with manganese dioxide to a diene-diyn-en-al **380**. Catalytic hydrogenation of **320** gave *n*-tridecanol. The position and orientation of the chromophore was deduced by using the maleic anhydride adduct method, mentioned earlier. When ozonolysed the adduct **382** did not yield propionaldehyde, thus confirming the proposed structure of **320**:

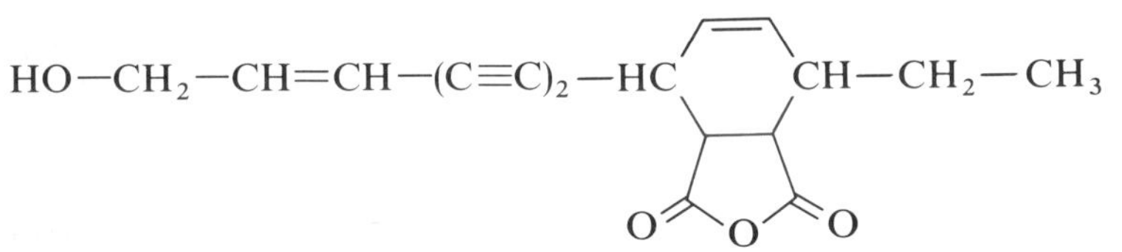
336

As already mentioned, aethusin (**372**) and aethusanol B (**320**) occur in very high concentrations in *Aethusa cynapium* L. A further main constituent of this plant is a secondary alcohol, called aethusanol A (**383**). It has been obtained in very high amounts from the polar fractions of *Aethusa cynapium* extracts[257]:

$$H_3C{-}CH{=}CH{-}(C{\equiv}C)_2{-}CH{=}CH{-}\underset{\displaystyle OH}{\underset{|}{C}H}{-}CH_2{-}CH_2{-}CH_3 \quad \textbf{383}$$

aethusanol A

The structure of **383** was elucidated by u.v. and i.r. spectroscopic measurements as well as some chemical transformations. The allylic position of the alcohol group was established by oxidation with manganese dioxide to a ketone. Hydrogenation gave a compound, which was identified to be sec.-tridecanol by GLC. Ozone cleavage of the corresponding azobenzoate yielded acetaldehyde. Its structure was finally established by synthesis. Matricarianal

(**513**) was condensed with *n*-propyl magnesium bromide. The obtained racemic carbinol was identical with natural aethusanol A. Together with the above described major constituents aethusin (**372**), aethusanol A (**383**) and aethusanol B (**320**), a lot of other C_{13}-compounds occur in *Aethusa cynapium*, biogenetically all closely related to the former. The compounds **384**–**388** occur in all parts of the plant though the proportions vary considerably:

$$H_3C-CH=CH-(C\equiv C)_2-CH_2-CH=CH-CH_2-CH_2-CH_3$$ **384**[259]

dd 8·21 (*J* = 7+2); *dq* 3·81 (*J* = 16+7); *dm* 4·6 (*J* = 16); *dm* 6·99 (*J* = 5); *cis*, *m* 4·6; *m* 7·9; *tq* 8·6 (*J* = 7+7); *t* 8·98 (*J* = 7)

$$H_3C-CH=CH-(C\equiv C)_2-CH_2-\underset{\diagdown O \diagup}{CH-CH}-CH_2-CH_2-CH_3$$ **385**[259]

$$H_3C-CH=CH-(C\equiv C)_2-CH=CH-\underset{OCOCH_3}{\underset{|}{CH}}-CH_2-CH_2-CH_3$$ **386**[258]

$$H_3C-CH=CH-(C\equiv C)_2-CH=CH-CO=CH_2-CH_2-CH_3$$ **387**[259]

$$H_3C-CH=CH-(C\equiv C)_2-CH_2-CH_2-CO-CH_2CH_2CH_3$$ **388**[259]

The occurrence of these compounds together in the same plant provides support for the proposed biogenetic scheme (see Section 5.1).

384 can well arise by hydrogenation of the terminal ethylenic bond of **347**:

$$H_3C-CH=CH-(C\equiv C)_2-CH_2-CH=CH-CH_2-CH=CH_2$$ **347**

$$\xrightarrow{[H]}$$ **384**

This step is usually observed at compounds from umbelliferous plants and has been confirmed by feeding experiments with labelled **347**[260]. Epoxidation of **384** gives **385**, whilst allylic oxidation followed by an allylic rearrangement yields aethusanol A (**383**). Dehydration of **383** yields aethusin (**372**), while oxidation of the allylic hydroxy group leads to the ketone **387**; **388** probably arises by hydrogenation of **387**. The structures of **384**–**388** were elucidated by spectral methods and confirmed by synthesis in parts. Besides aethusin (**372**) the corresponding dihydrocompound **389** was also obtained from *Aethusa cynapium*[259]:

$$H_3C-CH=CH-CH=CH-C\equiv C-CH=CH-CH=CH-CH_2-CH_3$$

d 8·23 (*J* = 6); *dq* 4·28 (*J* = 15+6); *dd* 3·96 (*J* = 15+10); *dd* 3·58 (*J* = 15+10); *d* 4·50 (*J* = 15); *d* 4·50; *dd* 3·58; *dd* 3·96; *dd* 4·28; *m* 7·9; *t* 8·98 (*J* = 7)

389

Its structure was established by spectral methods and finally confirmed by synthesis.

The reaction of the aldehyde **390** with propyl magnesium bromide leads to the carbinol **391**, the dehydration of which gave the all-*trans* **389**, with traces of Δ^{10}-*cis* **389**. The all-*trans* compound could be separated by chromatography and crystallisation.

$$\underset{\textbf{390}}{H_3C{-}\underset{t,t}{(CH{=}CH)_2}{-}C{\equiv}C{-}\underset{t}{CH{=}CH}{-}CHO} + BrMg{-}C_3H_7$$

$$\downarrow$$

$$H_3C{-}(CH{=}CH)_2{-}C{\equiv}C{-}CH{=}CH{-}\underset{OH}{\underset{|}{CH}}{-}CH_2CH_2CH_3 \quad \textbf{391}$$

$$\downarrow$$

$$\textbf{389}$$

The same dien-yne-diene chromophore was recognised by a compound isolated from *Zoegea baldshuanica* C. Winkl.[192]. The acetate **265** occurs together with the corresponding C_{14}- and C_{15}-compounds (**263**, **264**, see Section 2.2). The structures of **392**–**394** have been elucidated by spectral methods as well as some chemical reactions[259]:

$$H_3C{-}\underset{OH}{\underset{|}{CH}}{-}\underset{OH}{\underset{|}{CH}}{-}CH{=}CH{-}C{\equiv}C{-}CH{=}CH{-}CH{=}CH{-}C_2H_5 \quad \textbf{392}$$

$$OCH{-}CH{=}CH{-}C{\equiv}C{-}CH{=}CH{-}CH{=}CH{-}C_2H_5 \quad \textbf{393}$$

$$HO{-}CH_2{-}CH{=}CH{-}C{\equiv}C{-}CH{=}CH{-}CH{=}CH{-}C_2H_5 \quad \textbf{394}$$

The diol **392** is easily transformed by periodate cleavage to the corresponding aldehyde **393**.

The simultaneous occurrence of **392**–**394** in *Aethusa cynapium* provides circumstantial evidence in support of the following sequence of cellular reactions:

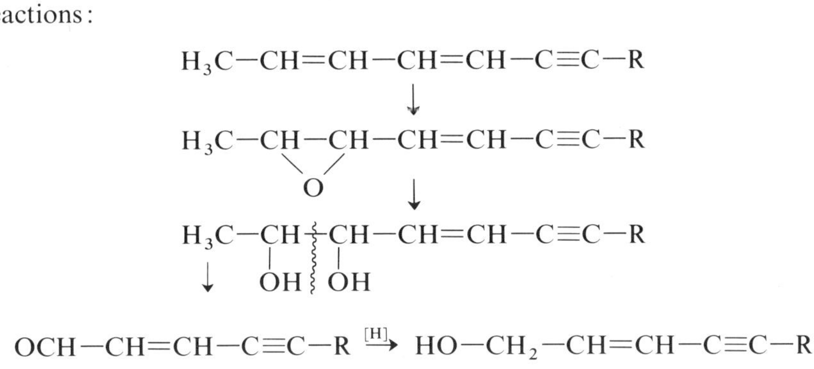

This elimination of two carbon atoms is one of the rare cases, while relatively often the elimination of one carbon atom can be observed (see Chapter 1, Section 4).

Three carbonyl compounds **395–397** were obtained in very small amounts from *Aethusa cynapium* too[259]. They could not be obtained in a pure state and their structure elucidation was rather difficult. The structures were deduced mainly by n.m.r. and mass spectroscopy:

$$H_3C{-}CH{=}CH{-}(C{\equiv}C)_2{-}CH{=}CH{-}CH_2{-}CO{-}CH_2{-}CH_3 \quad \mathbf{395}$$

$$H_3C{-}CH_2{-}(CH{=}CH)_2{-}C{\equiv}C{-}CH{=}CH{-}CO{-}CH_2{-}CH_3 \quad \mathbf{396}$$

$$H_3C{-}(CH{=}CH)_2{-}(C{\equiv}C)_2{-}CO{-}CH_2{-}CH_2{-}CH_3 \quad \mathbf{397}$$

The furane derivative **398** has been isolated from the roots of *Atractylis cancellata*[261].

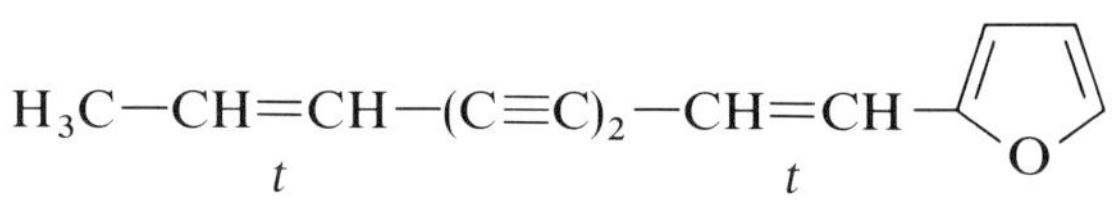

398

Biogenetically, **398** is presumably formed from a C_{13}-hydroxy-aldehyde in the following way:

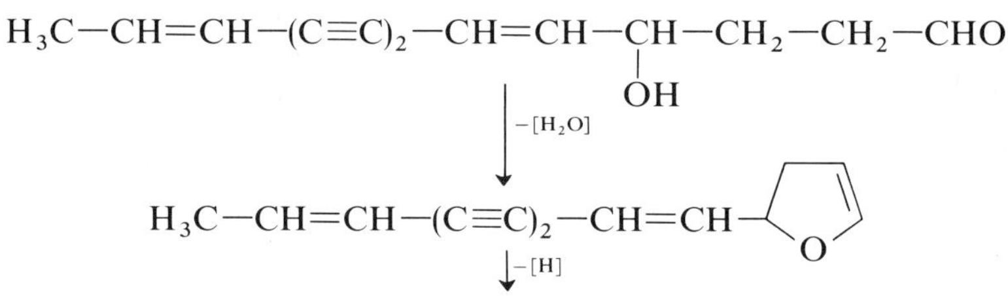

398

The corresponding triyne-derivative has been isolated from a *Chrysanthemum* species (see Chapter 2, Section 3.4).

5.4. Compounds Containing an Ene-triyne-diene Chromophore

As mentioned in Section 4, the so-called "3650"-hydrocarbon was recognised to be a mixture of 1-phenyl-hepta-1,3-diyn-5-ene (**318**) and the C_{13}-hydrocarbon **317**[244]:

$$H_3C{-}\underset{t}{CH{=}CH}{-}(C{\equiv}C)_3{-}\underset{c}{CH{=}CH}{-}CH{=}CH_2 \quad \mathbf{317}$$

317 could be crystallised from petroleum ether and showed a m.p. 51–56°C. The perhydrogenated compound crystallised readily in a refrigerator m.p.

$-7°C$. *n*-Tridecane has m.p. $-6{\cdot}5$ to $-6°C$. A 1:1 mixture showed no melting point depression. The fact which was overlooked by the authors was that the u.v. spectra of phenyl-triynes and en-triynes are very similar, especially in the longwave band group. The main difference is a small displacement of the first sharp maximum of the shortwave band group. The phenyltriynes have this maximum some 9 nm at longer wavelengths. As to the configuration of **317**, an accidental observation gave valuable clues. When both u.v. and i.r. spectra were taken on a same sample, surprisingly the subsequent u.v. spectrum demonstrated remarkable changes to earlier samples. The longwave band group had lost the characteristic double maxima; at the same time the main maxima had moved about 4 nm towards shorter wavelengths. This difference indicated that a *cis-trans*-isomerisation had taken place during the i.r. measurement.

Usually in polyenynes the *cis*-isomers absorb at somewhat longer wavelengths than the *trans*-isomers. However, the difference is normally very small. Therefore, the characteristic double maxima in natural **317** might be due to special constitutional details. The authors found it reasonable to suppose that the *cis-trans*-isomerism occurs at least at the C_{3-4} double bond. They assumed that a conjugated molecule tends to exist in a planar conformation. If at all possible, inspection of models revealed that **317** may exist in two planar conformations of rather different molecular shape. Model compounds should provide the answer as to whether the double maxima are due to the coexistence of two rotational isomers.

The all-*trans* isomer of **317** was obtained from the least polar fractions of *Centaurea macrocephala* Puschk.[206]:

$$\underset{}{H_3C}-\underset{t}{CH=CH}-(C\equiv C)_3-\underset{t}{CH=CH}-\underset{t}{CH=CH_2} \qquad \mathbf{399}$$

Its u.v. spectrum showed bands at 366·5, 352, 341, 329, 319, 299, 287, 272, 243·5, 234·5 and 224 nm. The u.v. and i.r. spectra were identical to those obtained from synthetic material. The route chosen for synthesis is outlined in the scheme below[262]:

$$H_3C-CH=CH-(C\equiv C)_3-CH=CH-CHO + IMg-CH_3$$

$$\downarrow$$

$$H_3C-CH=CH-(C\equiv C)_3-CH=CH-\underset{OH}{\underset{|}{CH}}-CH_3 \xrightarrow{-H_2O} \mathbf{399}$$

The all-*trans* alcohol **400** was obtained from the more polar fractions of *Cosmos sulphureus* Cav. and *Cosmos hybridus* Klondyke[263]. **400** shows the typical u.v. spectrum of an ene-triyne-diene chromophore (λ max: 367, 353, 342 nm) and is converted by hydrogenation to *n*-tridecanol.

$$\underset{}{HO{-}CH_2{-}}\underset{t}{CH{=}CH}{-}(C{\equiv}C)_3{-}\underset{t}{CH{=}CH}{-}\underset{t}{CH{=}CH_2} \quad \mathbf{400}$$

Acetylation of the alcohol gave an acetate, the spectral data of which were identical to those of synthetic material (see below, **408**).

The corresponding all-*cis* alcohol **401** has been isolated from the roots of *Carlina vulgaris* L.[251]:

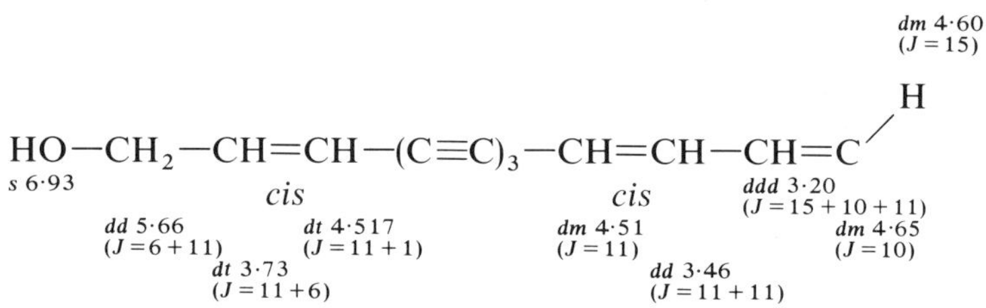

401

From the same plant, *Carlina vulgaris*, a polyyne has been isolated, which turned out to be closely related to **401**. The compound showed an u.v. maximum at 358 nm. Its i.r. spectrum showed bands due to acetylenic bonds (2200, 2130/cm) and a strong band at 1630/cm, which points to an enol-ether grouping. According to its elemental analysis and *m/e* value in the mass spectrum the compound has the composition $C_{13}H_{10}O$. Final information of the structure was obtained from the complex n.m.r., which exhibits signals in the range 3–5·2 and which is only in agreement with structure **402**:

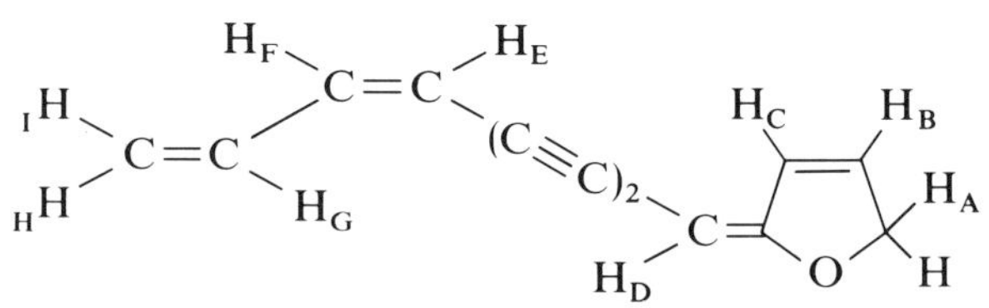

402

H_A AB*q* 4·97
H_B *ddt* 3·39 ($J_{AB} = 1{\cdot}8$, $J_{BC} = 5{\cdot}8$, $J_{BD} = 1{\cdot}5$)
H_C *dd* 3·33 ($J_{AC} = 1{\cdot}5$, $J_{BC} = 5{\cdot}8$)
H_D *m* 5·14
H_E *dm* 4·50 ($J_{EF} = 11$)
H_F *dd* 3·58 ($J_{EF} = 11$, $J_{FG} = 11$)
H_G *dddd* 3·16 ($J_{FG} = 11$, $J_{GH} = 10$, $J_{GI} = 17$, $J_{GE} = 0{\cdot}7$)
H_H *dm* 4·73 ($J_{GH} = 10$)
H_I *dm* 4·64 ($J_{GI} = 17$)

The mass spectrum of **402** shows predominant peaks at *m/e* 182, 181, 153 and 114:

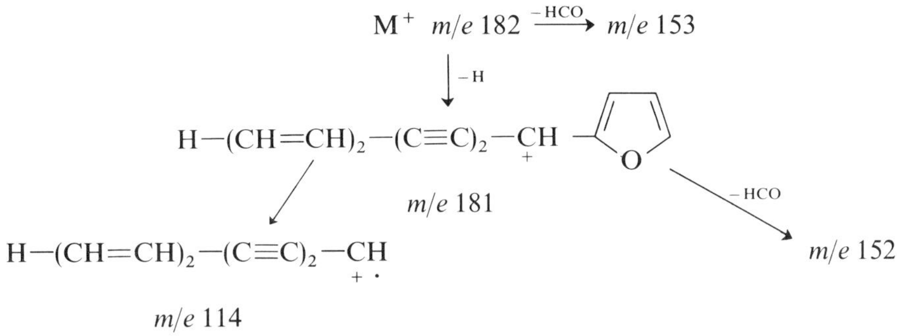

The stereochemistry of the enol-ether proton (H_D) could be assigned unambiguously on the basis of the observed position of 5·14τ, which is lower than the position of the corresponding *trans* compound (5·44τ). This assignment is in agreement with observations made by similar compounds[264]. The *cis* configuration of the disubstituted double bond could be assigned on the basis of the observed olefinic coupling constant (11 Hz) and causes the downfield shift of the signal of the vinylic proton H_G, which, in this special case, is influenced by the deshielding effect of the diyne grouping. The signals of the olefinic ring protons vary directly with the concentration. In high concentrations, a poorly resolved singlet occurs which splits in diluted solutions. Spin-decoupling of the protons H_A results in clearly interpretable signals for the two olefinic ring protons. The nature of the chromophore was confirmed by a rearrangement reaction, which could be observed when **402** was treated with acid. The rearrangement product exhibited an u.v. spectrum due to a diyne-diene system (λ max: 305, 288, 232·5, 222·5 nm). The structure of the produced compound was assigned on the basis of i.r. and n.m.r. spectra to be **403**:

403

H_A	*ddt* 2·75τ (J_{AB} = 1·8, J_{AC} = 0·8, J_{AD} = 0·4)
H_B	*dd* 3·76τ (J_{BC} = 3·2, J_{AB} = 1·8)
H_C	*ddt* 3·82τ (J_{BC} = 3·2, J_{AC} = 0·8, J_{CD} = 1·8)
H_D	*m* 6·30τ
$H_{E,F,H}$	*m* 3·5 – 4·8τ (4H)
H_G	*dddd* 3·13τ (J_{FG} = 11, J_{GH} = 16·5 + 10, J_{EG} = 0·7)

A reasonable pathway for the isomerisation is outlined in the scheme below:

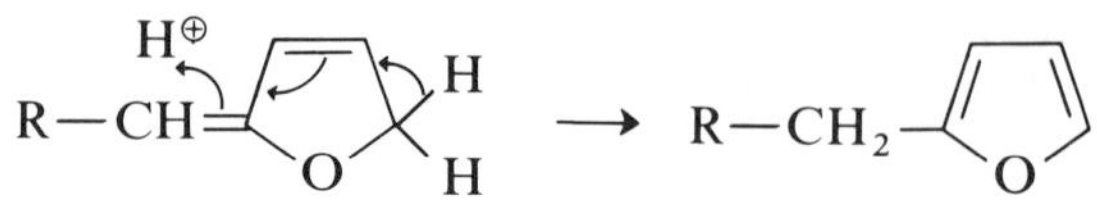

An analogous scheme has been supposed for the biosynthesis of carlina-oxide (**340**), (see Section 4), which occurs in *Carlina* species, too.

402 is formed from the cooccurring all-*cis* alcohol **401** in the plant, as shown by feeding labelled **401**[35]. The formation presumably proceeds through a cyclic intermediate involving intramolecular addition to the triple bond:

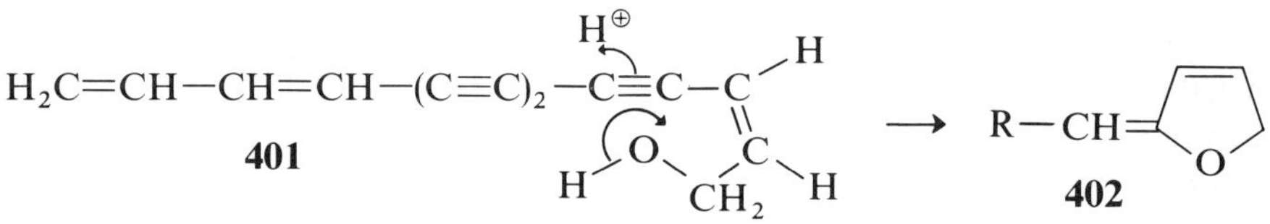

401 **402**

This assumption is supported by the isolation of **404** from the roots of *Carlina vulgaris* L.[251]:

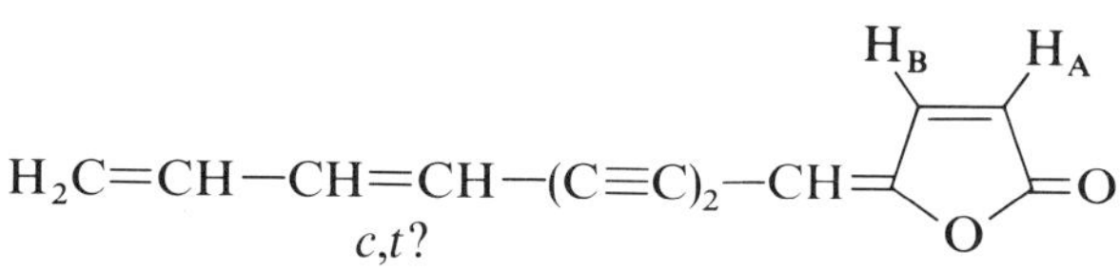

404

The butenolide is certainly formed by cyclisation of the acid **405**, which might be produced by oxidation of the alcohol **401**:

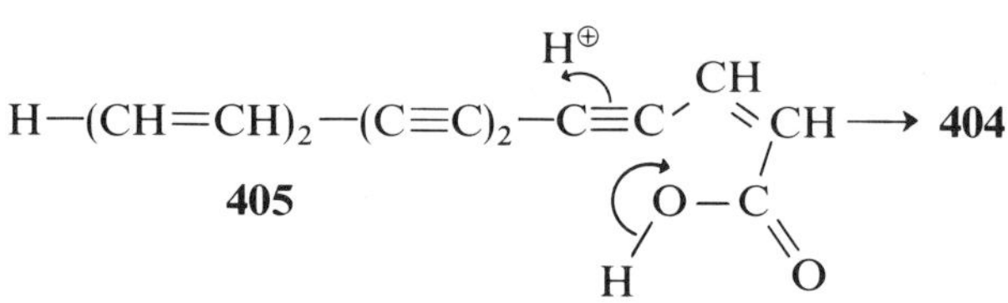

405

404 could not be obtained in a pure state. In spite of repeated TLC, there was still a mixture of isomers. Thus the stereochemistry of the double bonds is not yet clear. However, no doubts exist about the structure, which could be elucidated mainly on the basis of i.r. and mass spectroscopic measurements. Its i.r. spectrum displays bands for a butenolide derivative at 1870, 1800, 1110, 1075 and 885/cm; the mass spectrum shows prominent peaks at *m/e* 168, 114 and 82:

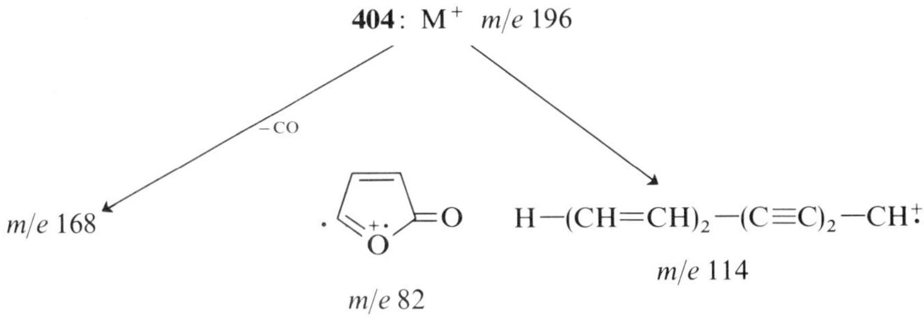

In the n.m.r. spectrum, two doublets [*d* 2·28τ (J = 5·5), *d* 2·64τ (J = 5·5)] attributed to the proton H_B establish the existence of a *cis* as well as a *trans* configurated enol-ether grouping. The stereochemistry of the second double bond could not be interpreted. The presence of a terminal ethylenic bond is significant (*m* 4·5–4·8τ). Synthetic **404** prepared from 2-*trans* **405** shows the same properties, especially the *cis-trans* isomerism on TLC separations[35].

The acetate corresponding to **401** has been isolated from common Carline thistle, *Carlina vulgaris* L. eu-vulgaris Holmboe[265]. Its i.r. spectrum, besides confirming the acetylenic nature of the compound, revealed that the substance was an acetate (sharp peak at 1750/cm, strong band at 1225/cm). The perhydrogenated acetate was saponified, the saponification product turned out to be *n*-tridecanol-(1), as shown by mixed melting point with authentic material. The compound reacted slowly with maleic anhydride. The u.v. maxima of the adduct agreed with those of a conjugated ene-triyne system of type R′—C≡C—C≡C—C≡C—CH=CH—R″.

Since the i.r. spectrum gives the information that the acetate ester is not a vinyl ester —R″ must be —CH_2OAc. Therefore the substance is given structure **406**:

$$H_3C{-}CO{-}O{-}CH_2{-}\underset{trans}{CH{=}CH}{-}(C{\equiv}C)_3{-}\underset{cis}{CH{=}CH}{-}CH{=}CH_2 \quad \mathbf{406}$$

The proposed structure has been finally established by synthesis[266]. The route chosen involved oxidative dimerisation of two acetylenic compounds with C_6- and C_7-chain length:

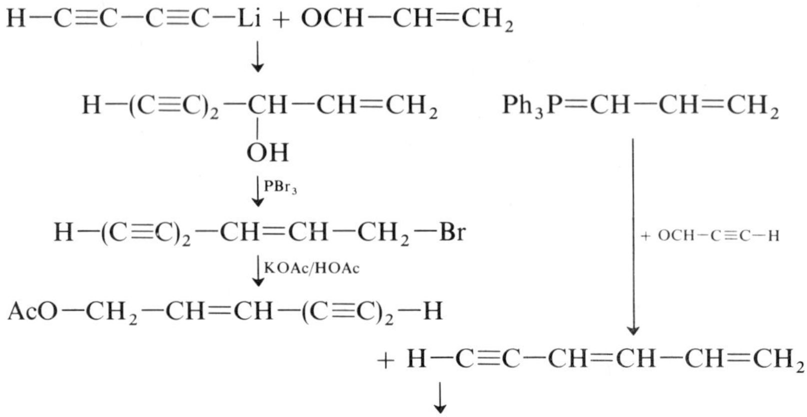

407

406

The u.v. spectrum of the synthesised acetate fully agreed with those obtained from natural material. The i.r. spectra of synthetic and natural material are also closely related but lacked the band of the latter at 750/cm. Further difference exists in the melting points. The natural acetate melts at −15°C, while the synthetic one melts at +31°C. In view of these observations, it seemed probable that there was a *cis-trans* diene grouping in the natural compound. Final confirmation about the exact stereochemistry of **406**, the so-called carlina acetate, was obtained by careful n.m.r. studies of a sample isolated from *Carlina vulgaris* L. in 1967[251]. These studies, and in particular spin-decoupling experiments, strongly indicate the *cis-cis* configuration of the carlina acetate **406**:

$$\underset{s\ 8{\cdot}03}{H_3C}-CO-O-\underset{A}{CH_2}-\underset{B}{CH}=\underset{cis}{}\underset{C}{CH}-(C\equiv C)_3-\underset{D}{CH}=\underset{cis}{}\underset{E}{CH}-\underset{F}{CH}=\underset{G,H}{CH_2} \quad \mathbf{406}$$

H_A	*dd* 5·24 (J_{AB} = 6·5, J_{AC} = 1·5)
H_B	*dt* 3·79 (J_{AB} = 6·5, J_{BC} = 11)
H_C	*dt* 4·51 (J_{BC} = 11, J_{AC} = 1·5)
H_D	*dm* 4·30 (J_{DE} = 11)
H_E	*dd* 3·42 (J_{DE} = 11, J_{EF} = 11)
H_F	*dddd* 3·15 (J_{FH} = 16·5, J_{FG} = 11, J_{EF} = 11, J_{DF} = 0·7)
H_G	*dm* 4·60 (J_{FG} = 11)
H_H	*dm* 4·56 (J_{FH} = 16·5)

The corresponding all-*trans* acetate **408** occurs in some *Bidens*, *Dahlia* and *Coreopsis* species[12]:

$$AcO-CH_2-\underset{t}{CH=CH}-(C\equiv C)_3-\underset{t}{CH=CH}-CH=CH_2 \quad \mathbf{408}$$

406 is accompanied in the plant by a mixture of two aldehydes both showing a diene-triyn-en-al chromophore in the u.v. (λ max : 387 nm). The structures were established by spectral methods and some chemical reactions. The two isomers **409** and **410** yielded, when reduced with sodium borohydride, the corresponding alcohols. The downfield shift for the aldehyde proton of the *cis-cis* isomer **409** must be due to the magnetic anisotropy effect of the triple bonds:

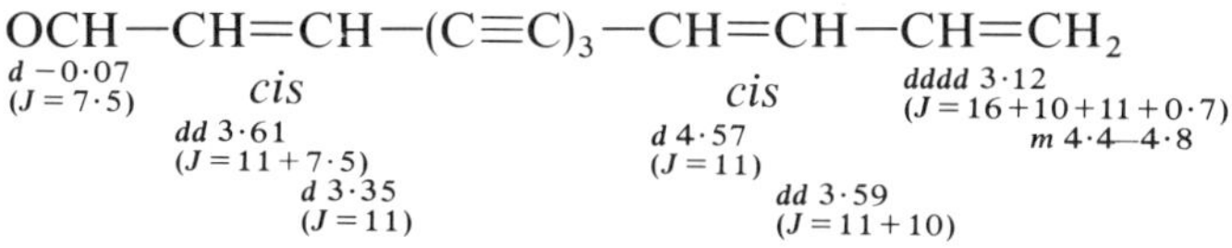

409

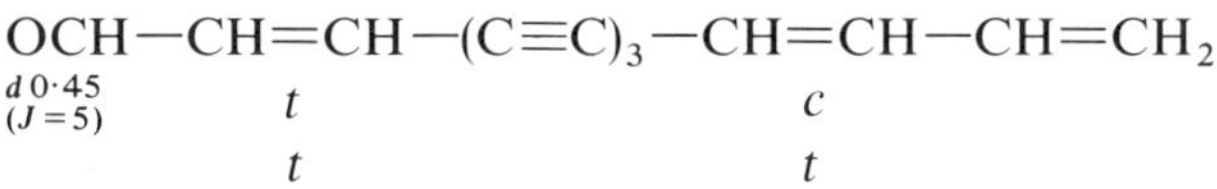

410

411

The corresponding all-*trans* aldehyde **411** has been isolated from *Cosmos sulphureus* Cav. and *Cosmos hybridus* Klondyke[263].

As mentioned in Chapter 1, Section 4, a usual oxidative reaction of polyynenes is the epoxidation which often is followed by secondary transformations.

The epoxide **412**, closely related to **399**, could be the hypothetical precursor for a number of bifunctional C_{13}-polyenynes, all possessing an ene-triyn-ene chromophore.

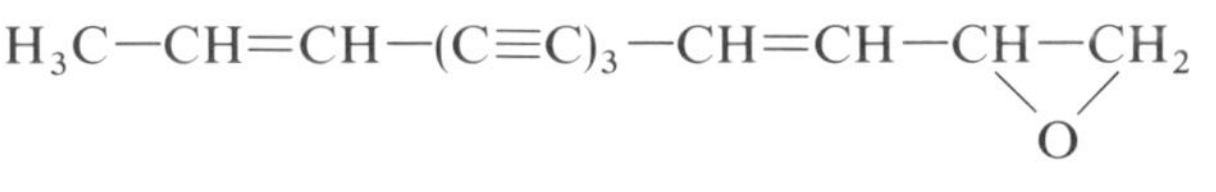

412

412 has been isolated in small amounts from the aerial parts of *Centaurea ruthenica* Lam.[100]. According to the scheme outlined in Chapter 1, Section 4, the biosynthesis of the following compounds, **413–418**, is rather simple to explain:

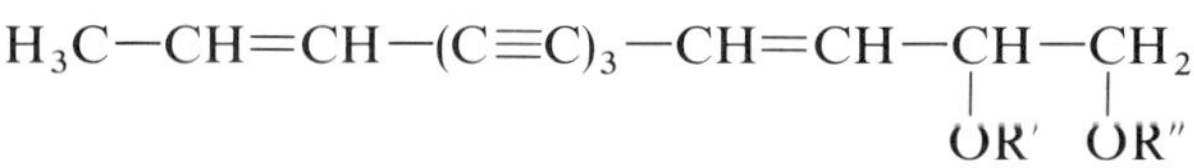

413: R′ = R″ = H
414: R′ = $COCH_3$, R″ = H

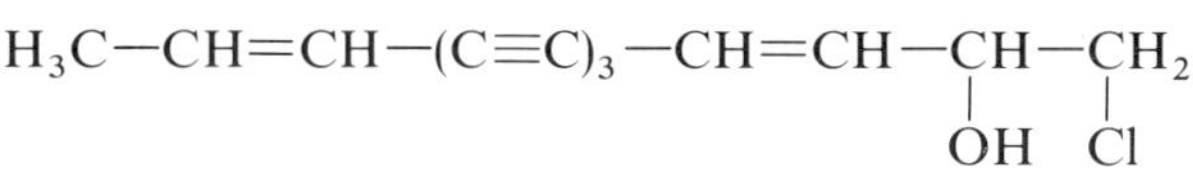

416 **415**: R′ = R″ = $COCH_3$

$$H_3C{-}CH{=}CH{-}(C{\equiv}C)_3{-}CH{=}CH{-}CH(Cl){-}CH_2OH$$

417

$$H_3C{-}CH{=}CH{-}(C{\equiv}C)_3{-}CH{=}CH{-}CH(Cl){-}CH_2OAc$$

418

All the compounds have been isolated from *Centaurea ruthenica* Lam.[100,206], except **416**, which has been isolated from *Carthamus caeruleus* L., *C. lanatus* L. and *C. tinctorius* L.[236]. The structures of the compounds have been elucidated by spectral methods as well as by chemical reactions and were confirmed by synthesis. For instance, the u.v. spectrum of **413** with λ max at 354 nm indicates that the substance is a polyacetylene with an ene-triyn-ene chromophore. The i.r. spectrum indicates the presence of aliphatic hydroxyl (3600, 3430, 1050/cm), *trans* ethylenic linkage (1620, 946/cm) and asymmetrically disubstituted acetylenic linkage (2190, 2170/cm). Absorption characteristic of *cis*-ethylenic linkage (700/cm) is not observed.

413 has been found to be the major antifungal compound from diseased *Carthamus tinctorius* L.[267]. A root and stem rot caused by the fungus *Phytophthora drechsleri* Tucker is one of the major diseases of *Carthamus tinctorius* L. The highest known level of resistance to *Ph. drechsleri* has been found in extracts of hypocotyls infected with *Ph. drechsleri*, showing highly fungi toxicity to the pathogen as compared with the antifungal activities of extracts from non-infected hypocotyls. The spectral data of the antifungal substance were identical with those of the already known **413**.

The optically active substance **414** ($[\alpha]_D^{21} = +39°$) melts at 86°C and shows an ene-triyn-ene like chromophore. The compound absorbs in the i.r. region at 3600 and 1750/cm, indicating the presence of a hydroxyl- and ester-carbonyl group. Saponification yielded **413**. As **414** could not be oxidised with manganese dioxide, the position of the hydroxyl group must be non-allylic:

$$\mathbf{414} \xrightarrow{H_3O^+} \mathbf{413}$$
$$\xrightarrow{MnO_2} \!\!\!\!\!\!/\!/$$

The route chosen for synthesis of some of the compounds started with the preparation of the acetylenic compounds **419** and **420**, respectively, by condensing γ-acetoxy-crotonaldehyde with lithium-acetylide or acetylene mono-magnesiumbromide. Oxidative coupling of **419** or **420** with hept-5-ene-1,3-diyne yielded the diol **421** and the hydroxy-acetate **422**. **421** is converted by diluted acid to **413**, including an allylic rearrangement. **415** is obtained by treatment of **413** with acetanhydride. Treatment of **422** with thionyl-chloride yielded the chloroacetate **418**:

$$\mathrm{AcO{-}CH_2{-}CH{=}CH{-}CHO + BrMg{-}C{\equiv}C{-}H}$$
$$\downarrow$$
$$\mathrm{RO{-}CH_2{-}CH{=}CH{-}\underset{\displaystyle OH}{\underset{|}{C}H}{-}C{\equiv}C{-}H}$$

419: R = H
420: R = $COCH_3$

419, **420** + $\mathrm{H{-}C{\equiv}C{-}C{\equiv}C{-}CH{=}CH{-}CH_3}$

$$\searrow$$
$$\mathrm{RO{-}CH_2{-}CH{=}CH{-}\underset{\displaystyle OH}{\underset{|}{C}H}{-}(C{\equiv}C)_3{-}CH{=}CH{-}CH_3}$$

421: R = H
422: R = $COCH_3$

$$\mathrm{HO{-}CH_2{-}CH{=}CH{-}\underset{\displaystyle OH}{\underset{|}{C}H}{-}(C{\equiv}C)_3{-}CH{=}CH{-}CH_3}\quad \mathbf{421} \xrightarrow{H_3O^+} \mathbf{413} \xrightarrow{Ac_2O} \mathbf{415}$$

$$\mathrm{CH_3COO{-}CH_2{-}CH{=}CH{-}\underset{\displaystyle OH}{\underset{|}{C}H}{-}(C{\equiv}C)_3{-}CH{=}CH{-}CH_3}\quad \mathbf{422}$$
$$\downarrow SOCl_2$$
$$\mathbf{418} \xrightarrow{H_3O^+} \mathbf{417}$$

The formation of the chloroacetate **418** presumably proceeds via

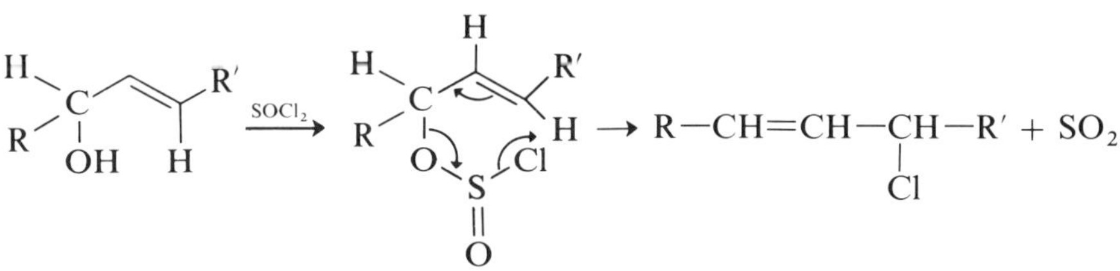

Finally, the chloroacetate **418** yielded when treated with acid the chlorohydrine **417**.

The proposed formation of **418** via **399** and **412** could be established by feeding labelled **399**[310].

In addition to the other compounds, **416** has been isolated from *Carthamus* species. Mass spectroscopic measurements showed that **416** had an empirical formula of $C_{13}H_{11}ClO$. When treated with alkali, **412** was obtained. The compound was finally identified with the aid of n.m.r. spectroscopy as the chlorohydrine **416**[236]:

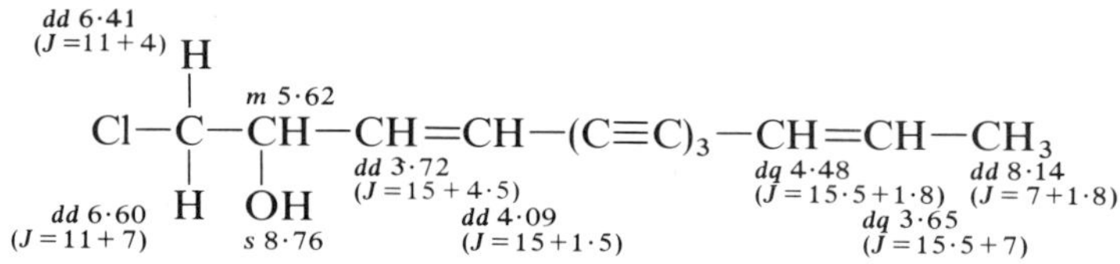

416

A cyclopropane derivative **423**, closely related to **418**, has also been isolated from the aerial parts of *Centaurea ruthenica*[268]. Its structure has been established with the aid of u.v., i.r., n.m.r. and mass spectroscopy. The structure has been finally confirmed by a synthesis via a photoreaction[269]. Irradiation of **418** with u.v. light yields in high amounts a *cis-trans* mixture of **423**, which is further converted with alkali to the enolethers **424** and **425**:

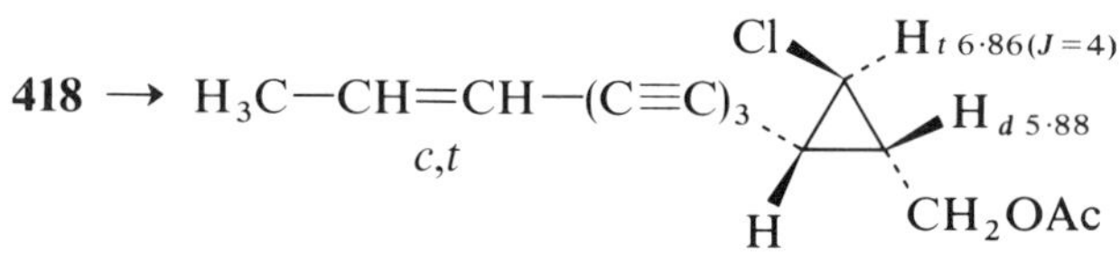

423

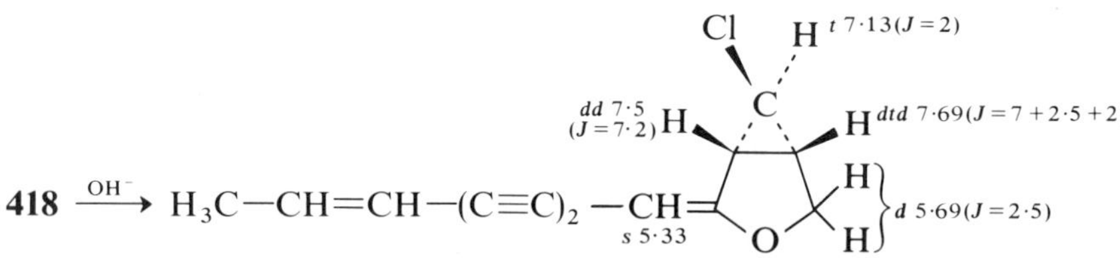

t **424**
c **425**

The chlorohydrine **416** might be the precursor of the enolether **426**, isolated from the roots of *Achyrocline satureioides*[35]:

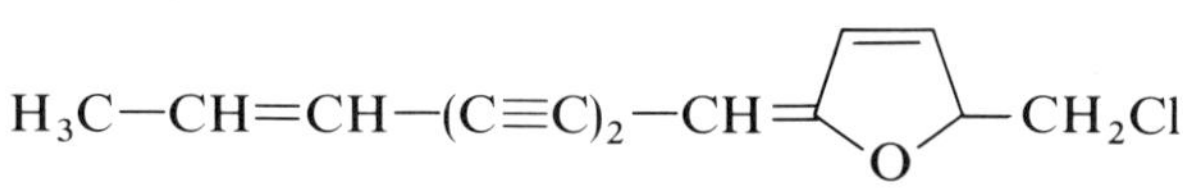

426

The aerial parts of *Coreopsis verticillata* L. contain besides Cosmen a non-polar compound with an u.v. maximum at 350 nm in very small amounts.

The same u.v. spectrum exhibits a more polar substance, which was obtained in larger amounts[234]. The i.r. spectrum of the latter displays bands of a hydroxy ester (3560, 1735/cm). Furthermore, the i.r. spectrum shows bands for a terminal ethylenic bond, a *trans* disubstituted double bond and a triple bond (915, 955, 2200/cm). According to its uncharacteristic u.v. spectrum, the substance might be a thiophene derivative. The n.m.r. spectrum of the compound exhibits, *inter alia*, a doublet at τ 2·99 (J = 4 Hz), attributable to a β-hydrogen of a 2,5-disubstituted thiophene derivative. High-resolution mass measurements show that the compound has an empirical formula of $C_{18}H_{20}O_3S$. A peak in the mass spectrum at m/e 199 ($C_{13}H_{11}S$) indicates the loss of $C_5H_9O_3$ to give an ion of high intensity. Therefore, the molecule presumably contains a C_5-hydroxy ester residue. The observation of an internal H-bonding of the OH-group in diluted CCl_4 solution points, together with the n.m.r. data, to a β-hydroxy isovalerate. Acetylation, which turns out rather difficult, produces an acetate, confirming the assumption. Saponification of the ester yields an alcohol, which was oxidised with manganese dioxide to the aldehyde **430**. In view of these spectral and chemical results, it seemed probable that the unknown ester has structure **427**; furthermore, the foregoing data lead to structure **428** for the acetate and **429** for the alcohol. The nonpolar compound with the same u.v. maximum might possess structure **431**. The proposed structure **427** has been finally established by synthesis[35]:

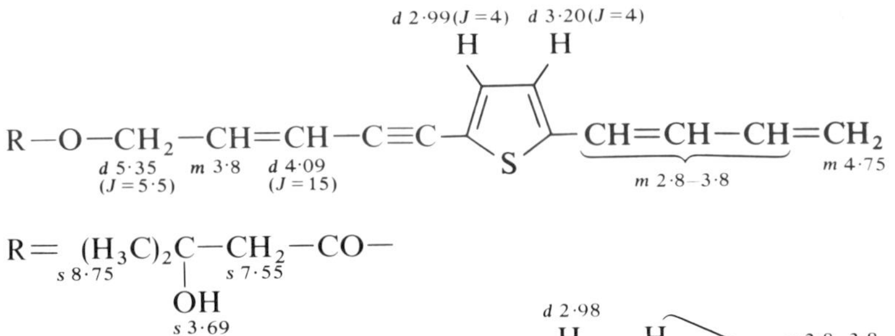

427

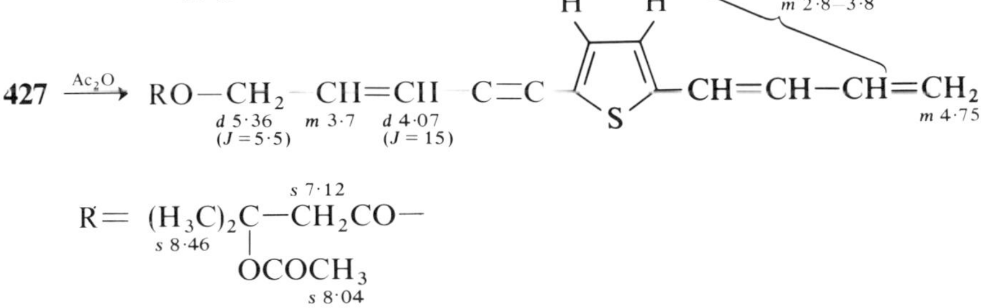

428

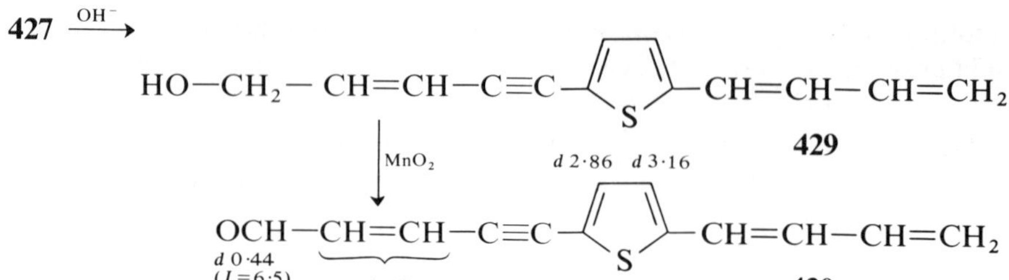

429

430

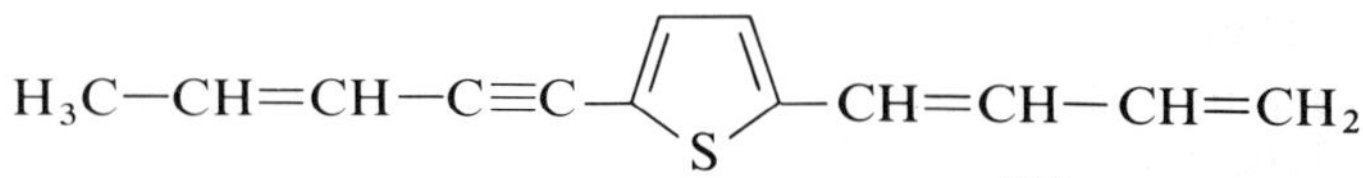

431

431 has also been isolated from *Centaurea uliginosa* Brot. and clearly shows the close relationship to the commonly occurring compound **317**. The biogenesis of such thiophenes is declared in Chapter 1, Section 4, including H_2S addition to the triyne system:

$$H_3C{-}CH{=}CH{-}C{\equiv}C{-}C{\equiv}C{-}C{\equiv}C{-}CH{=}CH{-}CH{=}CH_2 \;+\; H_2S$$

317 → **431**

The most polar fractions of extracts from the aerial parts of *Bidens bipinnatus* L. also contain a triol closely related to **413**. The triol was assigned structure **432**, largely on the basis of spectroscopic evidence[77]:

$$HO{-}CH_2{-}CH{=}CH{-}(C{\equiv}C)_3{-}CH{=}CH{-}\underset{OH}{CH}{-}\underset{OH}{CH_2} \quad \mathbf{432}$$

5.5. Compounds Containing an Ene-tetrayn-ene-Chromophore

The investigation of some common garden *Coreopsis* species—*Coreopsis cardaminefolia* var. atrosanguinea, *C. drummondii*, *C. tinctoria vars.* nana radiata and nigra speciosa—led to the discovery of some polyacetylenic compounds (see also Section 4). From the roots a hydrocarbon was isolated, first thought to be dodeca-1,11-diene-3,5,7,9-tetrayne **433**[247].

$$H_2C{=}CH{-}C{\equiv}C{-}C{\equiv}C{-}C{\equiv}C{-}C{\equiv}C{-}CH{=}CH_2 \quad \mathbf{433}$$

The evidence for this formula was mainly spectroscopic, supported by hydrogenation to "dodecane" and analysis. Jones, Thompson and Whiting[270] synthesised **433**, which is extremely unstable, by coupling vinyldiacetylene **434** prepared from the glycol **435**:

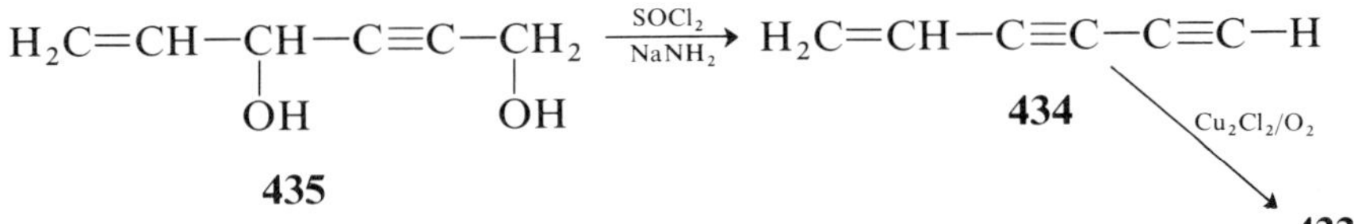

435 **434** **433**

Comparison of melting points and i.r. spectra indicated clearly the non-identity of the two compounds. Moreover, the maxima in the u.v. spectrum of the synthetic compound lie at wavelengths 1·6–2·8 nm shorter than those of the natural material. It was suggested from consideration of the spectrum that the compound is a monoalkyl-diene-tetrayne, probably trideca-1,11-diene-3,5,7,9-tetrayne **436**:

$$H_3C{-}\underset{c\ t}{CH{=}CH}{-}C{\equiv}C{-}C{\equiv}C{-}C{\equiv}C{-}C{\equiv}C{-}CH{=}CH_2 \quad \textbf{436}$$

This has been confirmed by synthesis of the *cis*- and *trans*-forms of **436**[76], both of which have now been shown to occur in *Coreopsis*. **436** has also been detected by means of its characteristic absorption spectrum, in the thistles *Carthamus lanatus* and *Silybium marianum*[247], all members of the tribe Cynareae as well as in many other Compositae.

The synthesis of symmetrically substituted acetylenic hydrocarbons via oxidative cross-coupling reaction is unsuitable because of the similarity of physical properties between the desired product and its symmetrical analogues. The removal of hydrogenchloride from terminal groupings in a polyacetylenic dichloride of type $R{-}CH(Cl){-}(C{\equiv}C)_n{-}CH(Cl){-}R'$ only proceeds in very low yields. The route chosen for synthesis of **436** involved therefore a modification of the latter, in which the two —C(Cl)H-groups were located in non-terminal positions. This required the preparation of an unsymmetrical glycol which could be prepared on a two-stage route from the two appropriate $\alpha\beta$-acetylenic aldehydes and acetylene:

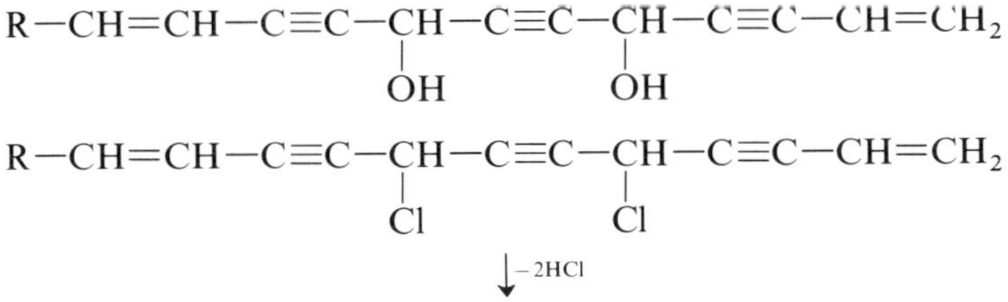

436

Some polyyn-epoxides, closely related to **436** have been isolated from the roots of *Centaurea deusta* Ten.[79]. Besides **436** three polyyn-epoxides have been obtained from the root extract, but they could not be separated. Therefore, the mixture was hydrolysed and the corresponding diol mixture separated by repeated TLC. The structural assignments for the diols were derived from their n.m.r. spectra, which supported the structures **440**, **441** and **442**. Therefore the structures of the natural epoxides have to be **437**, **438** and **439**:

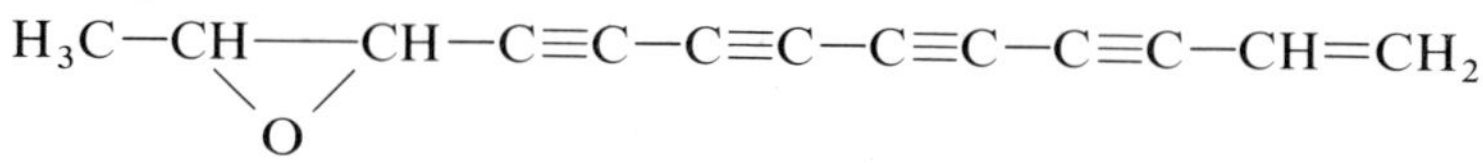

437

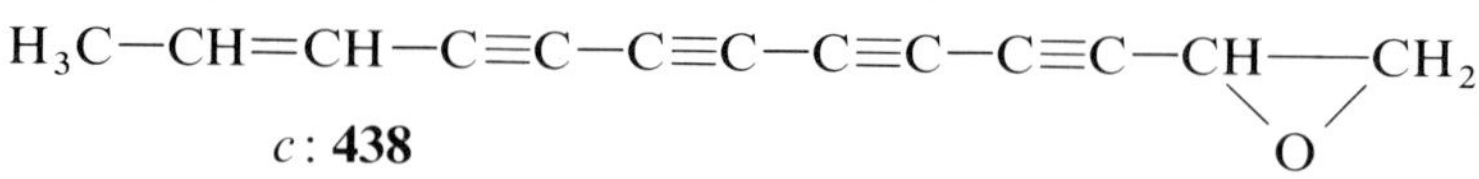

c: **438**
t: **439**

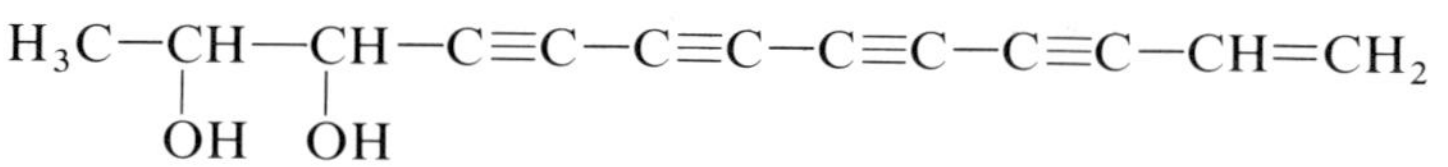

440

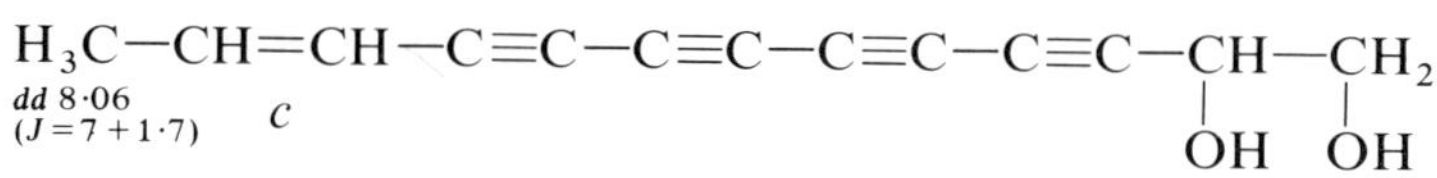

441

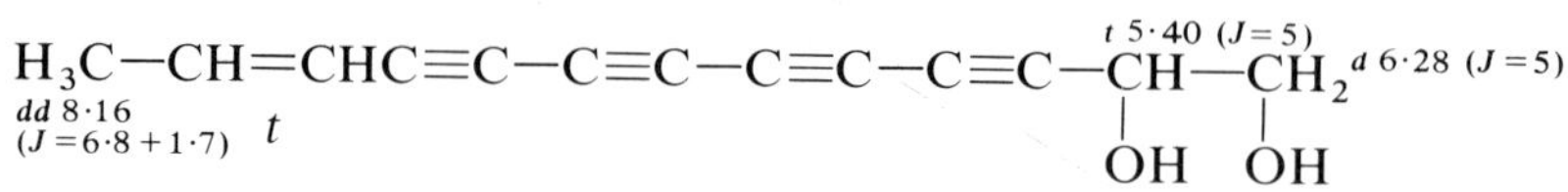

442

Subsequent reaction of the diols with periodate gave the highly unstable aldehydes **443**–**445**:

$$\mathbf{440} \rightarrow \mathrm{OCH{-}(C{\equiv}C)_4{-}CH{=}CH_2} \qquad \mathbf{443}$$

$$\mathbf{441}, \mathbf{442} \rightarrow \mathrm{H_3C{-}CH{=}CH{-}(C{\equiv}C)_4{-}CHO}$$

c: **444**
t: **445**

The u.v. spectra of **443** and **445** clearly indicates the non-identity of the two aldehydes. The maxima in the u.v. spectrum of **443** lie at wavelengths 5 nm shorter than those of **445**.

Some oxidation products of **436** have been isolated from the roots of *Bidens leucanthus*[271]. Besides **436** small amounts of the ene-tetrayn-ene-aldehyde **446**, the corresponding alcohol **447** and acetate **448** were obtained:

$$OCH{-}CH{=}CH{-}(C{\equiv}C)_4{-}CH{=}CH_2 \qquad \mathbf{446}$$

$$HOCH_2{-}CH{=}CH{-}(C{\equiv}C)_4{-}CH{=}CH_2 \qquad \mathbf{447}$$

$$H_3C{-}CO{-}OCH_2{-}CH{=}CH{-}(C{\equiv}C)_4{-}CH{=}CH_2 \qquad \mathbf{448}$$

The polyyn-aldehyde **446** shows a characteristic u.v. spectrum, indicating the presence of an ene-tetrayn-en-al-chromophore (λ max : 404, 373, 347 nm). The i.r. spectrum shows a carbonyl band in the position characteristic for an unsaturated carbonyl-compound (1700/cm). Furthermore the compound shows absorbance due to an aldehyde proton (2740/cm). Reduction with sodium borohydride yielded an alcohol with the typical u.v. spectrum of an ene-tetrayn-ene (λ max: 396, 366, 340, 318 nm). The obtained alcohol was shown to be identical with natural material obtained from the more polar fractions of the same plant. The acetate **448** gave an i.r. spectrum consistent with the presence of an OAc-grouping and triple bonds (1750, 2210/cm). Careful saponification of **448** yields an alcohol, identical with those obtained by reducing **446**.

The proposed structures have been finally established by synthesis. The route chosen involved cross-coupling of vinylacetylene with nona-1,3,5-triyn-7-en-9-ol. Purification by chromatography gave pure **447**, which had an absorption spectrum identical with that of the natural occurring alcohol.

The presence of the 2,3-dihydro-aldehyde **449** recently has detected in *Silybum marianum* Gaertn.[272]:

$$OCH{-}CH_2{-}CH_2{-}(C{\equiv}C)_4{-}CH{=}CH_2 \qquad \mathbf{449}$$

The aldehyde **449** was isolated from the root extract of the plant and identified with the aid of u.v., i.r., n.m.r. and mass spectroscopy. Its molecular formula of $C_{13}H_8O$ was confirmed by mass spectrometric studies; its n.m.r. spectrum contains characteristic signals of an aldehyde proton adjacent to a methylene group [t 0·17τ (J = 1·5 Hz); m 7·67–7·95τ (4H)]. Hydrogenation yielded *n*-tridecanal. The foregoing data lead to structure **449**.

Two polar compounds with an ene-tetrayne chromophore have been isolated from the aerial parts of *Bidens bipinnatus* L.[77]. The (presumed) polyalcohols were transformed by acetylation to a mixture of acetates and the pure acetates obtained by TLC. The n.m.r. spectrum of the less polar acetate shows a double doublet at 8·16τ (3H) (J = 7·3 + 1·3) arising from a methyl group adjacent to a double bond. The i.r. spectrum indicates the presence of —OAc groups (1760/cm) as well as of a disubstituted double bond. The foregoing data lead to structure **450**; therefore, the natural diol has structure **451**:

$$H_3C{-}\underset{t}{CH{=}CH}{-}(C{\equiv}C)_4{-}\underset{OR}{\underset{|}{CH}}{-}\underset{OR}{\underset{|}{CH_2}} \qquad \mathbf{451}: R = H \quad \mathbf{450}: R = COCH_3$$

The structure of the more polar acetate was assigned mainly on the basis of its n.m.r. spectrum:

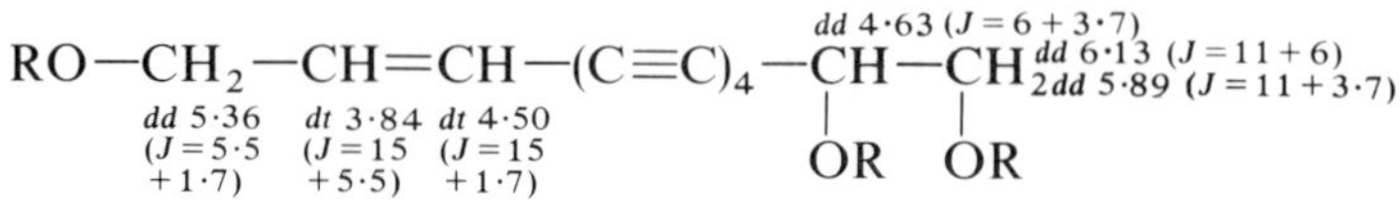

$R = COCH_3$: **452**
$R = H$: **453**

The natural triol must therefore be **453**.

The structures have been confirmed by syntheses of **453** and **452**[78]. Starting with diacetylene-monomagnesiumbromide the diacetate **454** was obtained by condensing the former with glycolaldehydeacetate **455** followed by acetylation. **454** was coupled with hept-1,3-diyn-5-ene and the acetate **456** to give the diacetate **450** and the triacetate **452**, respectively:

$$H{-}C{\equiv}C{-}C{\equiv}C{-}MgBr + OCH{-}CH_2{-}O{-}COCH_3 \quad \mathbf{455}$$

$$\downarrow Ac_2O$$

$$H{-}C{\equiv}C{-}C{\equiv}C{-}CH(OAc){-}CH_2OAc \quad \mathbf{454}$$

$$\mathbf{454} + H{-}C{\equiv}C{-}C{\equiv}C{-}CH{=}CH{-}CH_3 \rightarrow \mathbf{450}$$

$$\mathbf{454} + \underset{\mathbf{456}}{H{-}C{\equiv}C{-}C{\equiv}C{-}CH{=}CH{-}CH_2OAc} \rightarrow \mathbf{452}$$

The identity of synthetic and natural materials has been shown by comparison of the spectral properties. **453** has been detected as the aglycon of a glucoside obtained from *Bidens frondosa* flowers[273]. Its structure must be **457** since a tetra-ynenal-chromophore has been obtained on manganese dioxide oxidation and mass spectroscopy of the hydrogenated product gives the position of the glucoside[35].

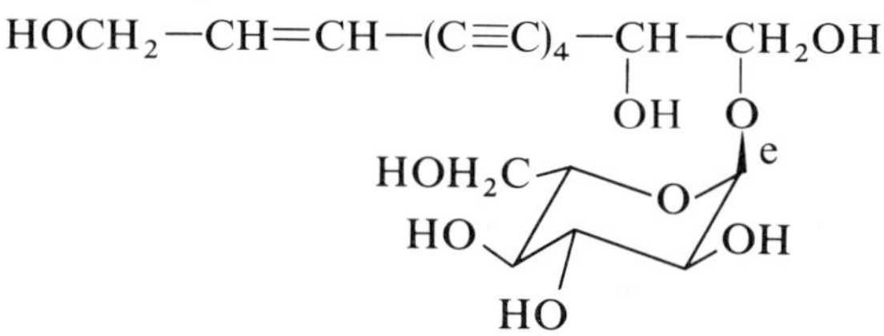

457

The oxygen derivatives described above all possess the skeleton of the ene-tetrayn-ene **436**, and it seems very probable that **436** is the biogenetic precursor for all of them. The simultaneous occurrence of **436** and the described oxygen derivatives provides evidence in support of this assumption. A biogenetic pathway, starting with crepenynic acid to **436**, is outlined in Section 5.1. The hydrocarbon **436** is also a precursor of a considerable number of thiophene derivatives, the formation of which has been described in Chapter 1, Section 4. Three possibilities of H_2S addition are:

(A) $H_3C-CH=CH-C\equiv C-C\equiv C-C\equiv C-C\equiv C-CH=CH_2$ **436**

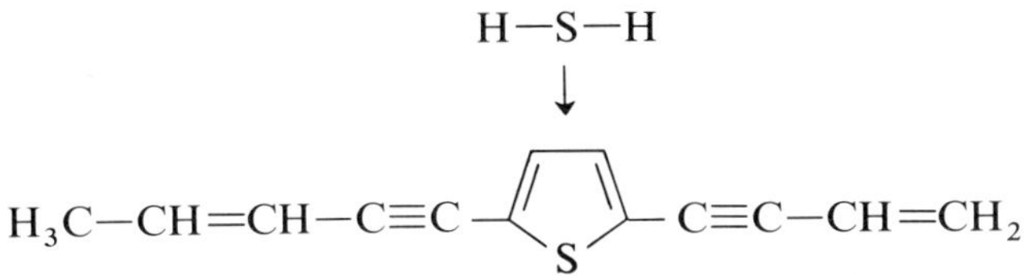

(B) $H_3C-CH=CH-C\equiv C-C\equiv C-C\equiv C-C\equiv C-CH=CH_2$

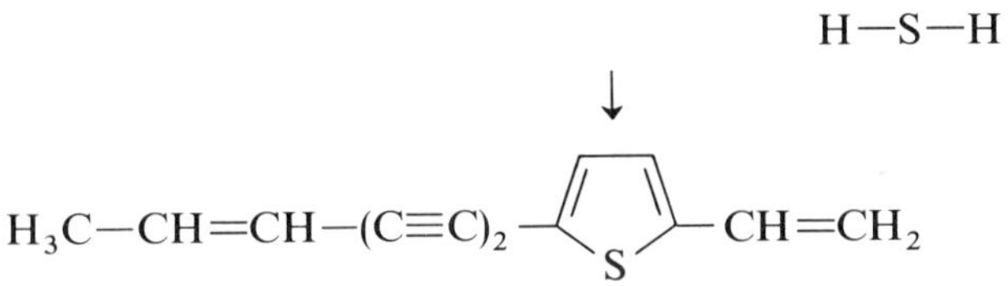

(C) $H_3C-CH=CH-C\equiv C-C\equiv C-C\equiv C-C\equiv C-CH=CH_2$

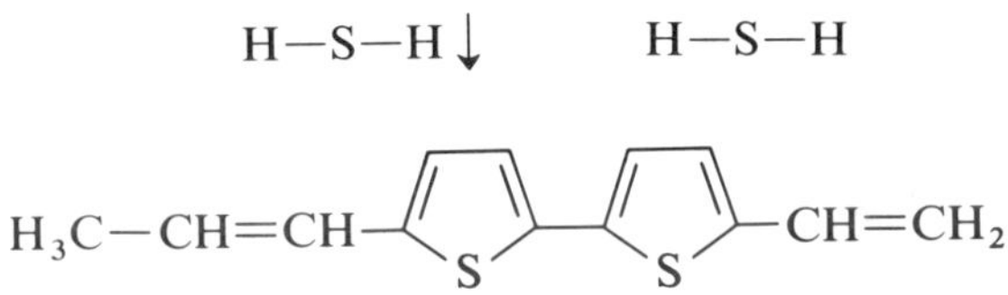

The most probable way of the thiophene formation is outlined in the scheme below:

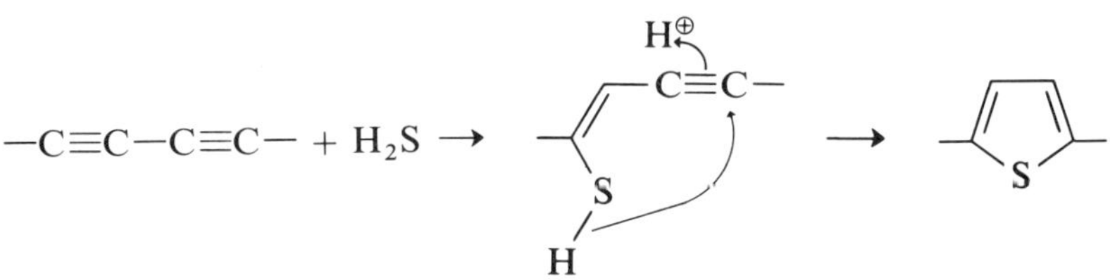

Members for all the three discussed possibilities are known. Extraction of the aerial parts of *Serratula radiata* M. Bieb. gave, *inter alia*, a yellow oil, which was obtained in the pure state by distillation (b.p.$_{0\cdot 001}$ = 80°C). The i.r. spectrum of the oil displays bands due to acetylenic linkage (2200/cm), a *trans* disubstituted double bond (1640, 954/cm), a terminal ethylenic bond

(970, 925/cm) and a 2,5-disubstituted thiophene residue (800/cm). The n.m.r. spectrum confirms the presence of a thiophene residue showing resonances at τ 3·04 (poorly resolved). The presence of a methyl group adjacent to a double bond is indicated by a double doublet at 8·11τ with coupling constants of 6 and 1·5 Hz. Hydrogenation of the compound gave 2-butyl-5-amyl-thiophene. Taking all these facts into consideration structure **458** must be suggested[50]:

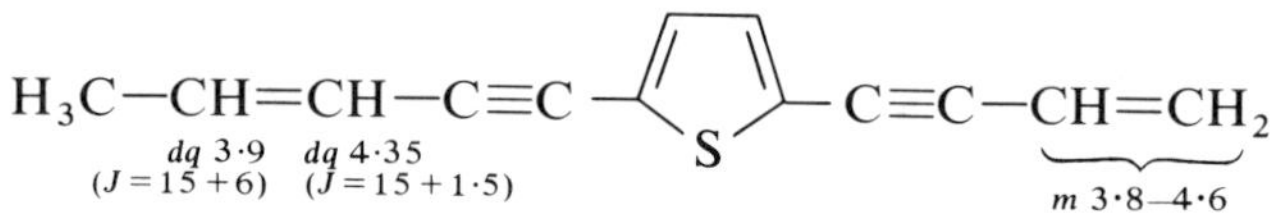
458

Confirmation of the proposed structure **457** was obtained by synthesis[274].

2,5-Diiodothiophene **459** was treated with one equivalent of the cuprous salt of propargyl aldehyde diethylacetal **460** to give the thiophene derivative **461**, which was condensed with a Wittig reagent from ethylbromide. The obtained thiophene derivative **464** was reacted with the cuprous salt of vinylacetylene to give **458**, identical with the natural material:

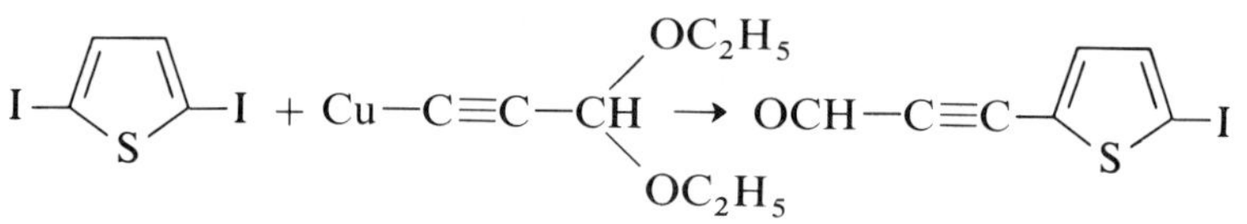
459 **460** **461**

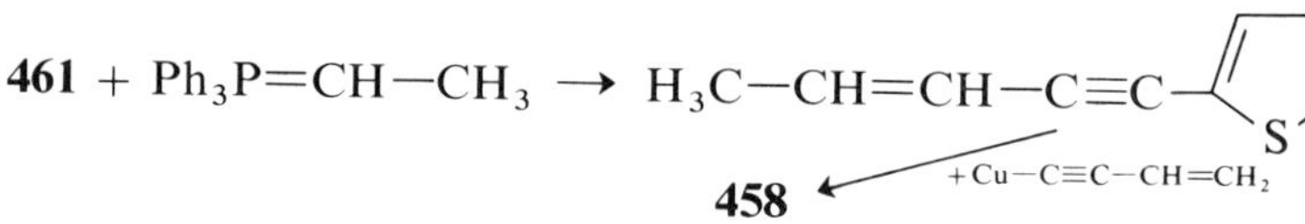
462

458 has been isolated from other sources too, e.g. *Centaurea nigra* L., *C. cineraria* L. and *C. pulcherrima* Wright; further on from *Bidens connatus* Mühlenbg. and *Lasthenia glaberimma* DC. Some oxygen derivatives, closely related to **458** have been isolated from the roots of *Baeria aristata* H. M. Hall[50]. The roots have been found to contain the alcohol **463**, the corresponding acetate **464** and the aldehyde **465**:

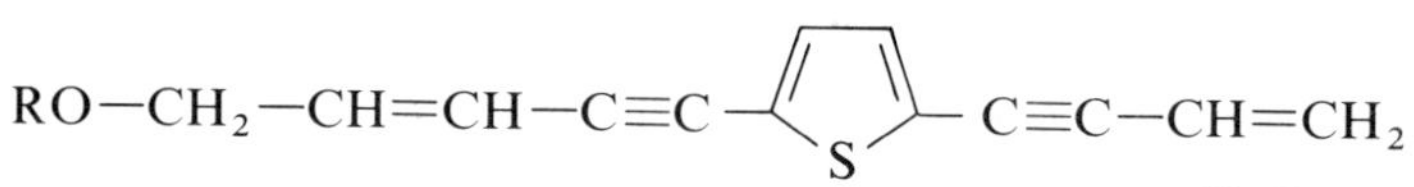
463: R = H
464: R = $COCH_3$

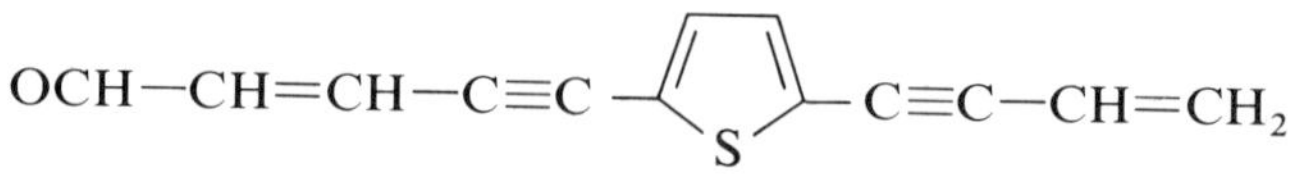
465

463 and **464** show nearly the same u.v. spectrum (λ max: 335 nm), while the spectrum of **465** lie at wavelengths 25 nm longer than those of the former.

The structures of the last three compounds have been finally established by synthesis[275]. The route chosen involved preparation of the iodothiophene derivative **466** by reaction of 2,5-diodothiophene with the cuprous salt of vinyl acetylene. Reaction of **466** with the cuprous salt of the pyranyl ether **467** gave after hydrolysation the alcohol **463**, which was identical in all respects with the natural compound. Acetylation of **463** gave the acetate **464**, while manganese dioxide oxidation yielded the aldehyde **465**:

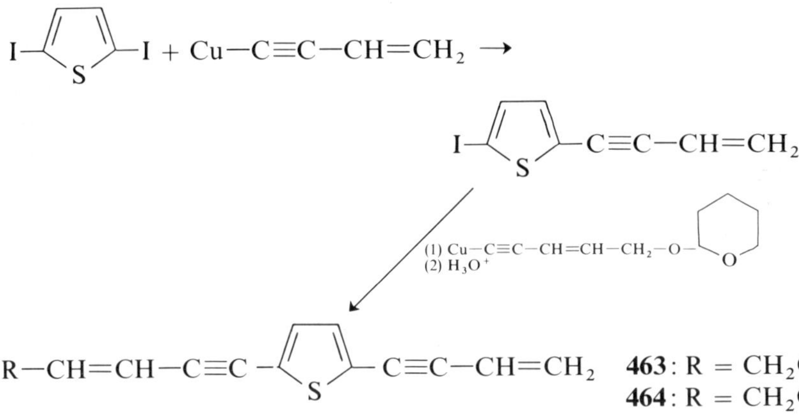

466

467

463: R = CH_2OH
464: R = CH_2OAc
465: R = CHO

The most polar portions of root extracts from *Saussurea pectinata* Bunge contain a mixture of diols, which could not be separated[193]. Periodate cleavage of the diols yielded the aldehydes **468** and **469**, the diols should be **470** and **471**:

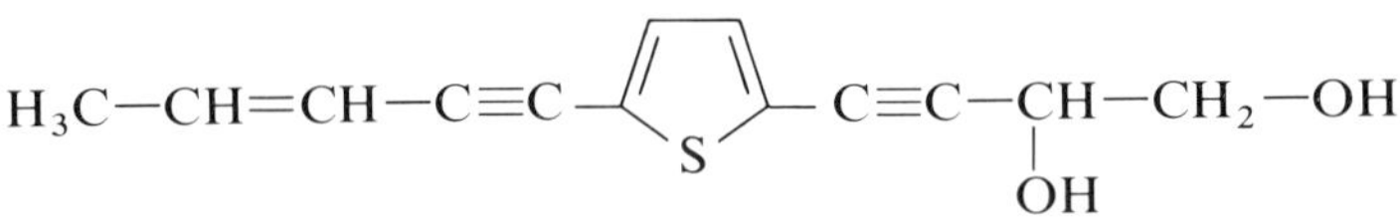

470

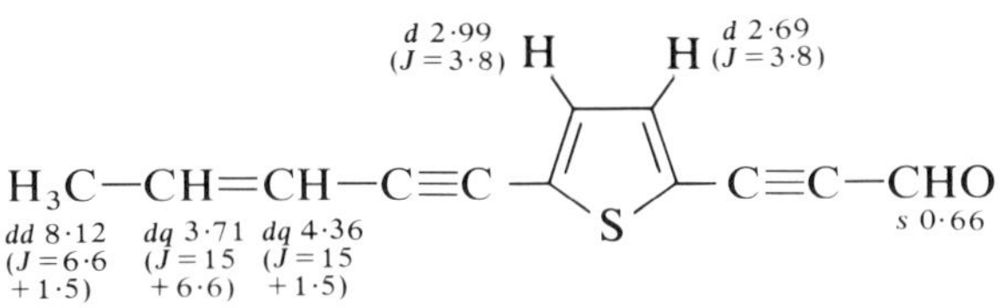

468

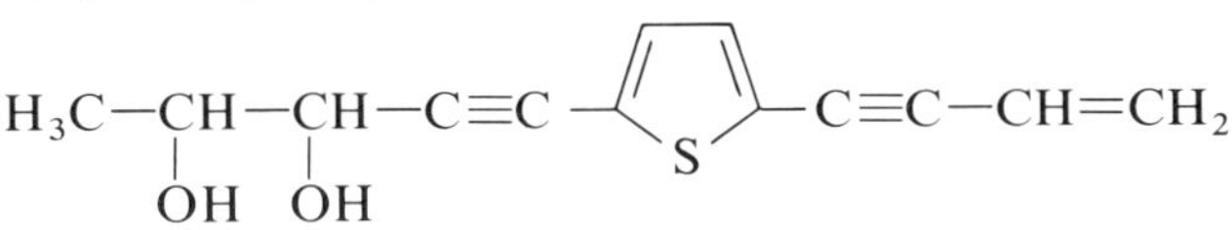

471

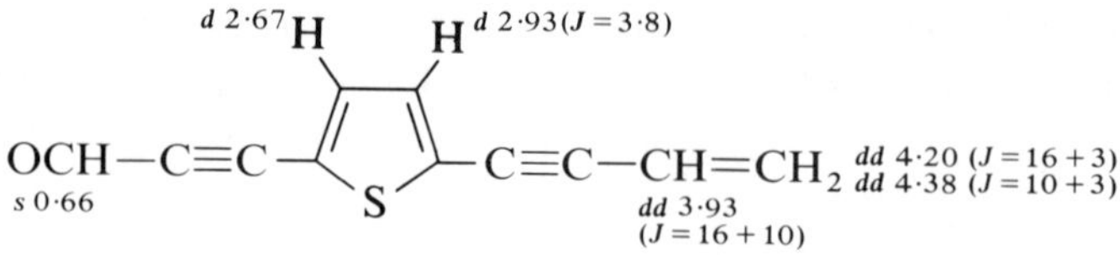 **469**

The aldehyde **469** has also been obtained from *Saussurea pectinata*. Sodium borohydride reduction of **469** gave the alcohol **472**, which has been obtained from extracts of the same plant too:

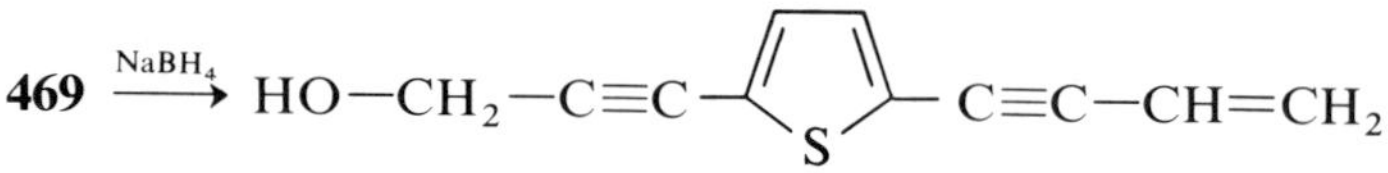 **472**

The maxima in the u.v. spectrum of the alcohol **472** (λ max: 316 nm) lie at wavelengths 23 nm shorter than those of the aldehyde.

It is proposed that **469** arises biogenetically from the diol **471** occurring in the same plant. This degradation is one of the rare cases of elimination of two carbon atoms in one step; a further sample has been already described in Section 5.3.

Three thiophene compounds represent the route "B" outlined in the scheme above.

The roots of *Serratula radiata* M. Bieb. have been found to contain a diol shown to be **473** on the basis of chemical and spectroscopic evidence[276]:

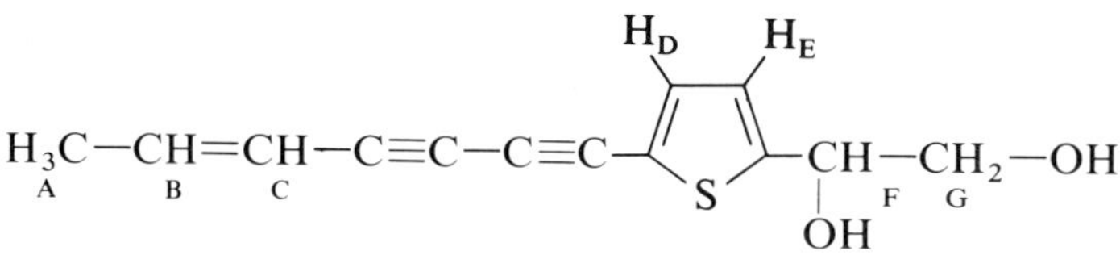 **473**

H_A	*dd* 8·19τ (J = 7 + 2 Hz)	H_E	*d* 2·86τ (J = 3·7 Hz)
H_B	*dq* 3·66τ (J = 16 + 7 Hz)	H_F	*dd* 5·02τ (J = 7 + 4 Hz)
H_C	*dq* 4·42τ (J = 16 + 2 Hz)	H_G	*m* 6·23τ
H_D	*d* 3·16τ (J = 3·7 Hz)		

The glycol group of **473** was cleaved with periodic acid to give the aldehyde **474**:

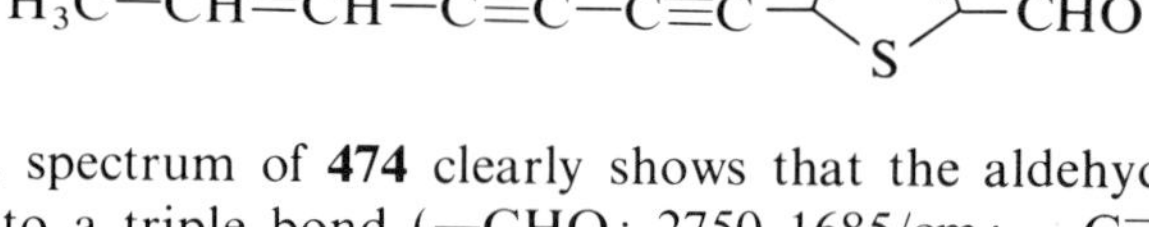 **474**

The i.r. spectrum of **474** clearly shows that the aldehyde group is not adjacent to a triple bond (—CHO: 2750 1685/cm; —C≡C—: 2210/cm, weaker intensity). The monoacetate **475** and the diacetate **476** are shown to occur with **474** in the same plant. Their structures have been established by spectral methods as well as some chemical reactions:

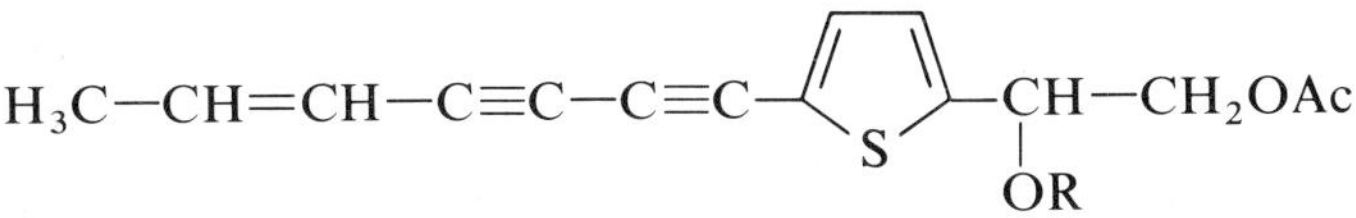

475: R = H
476: R = $COCH_3$

Like the previously described thiophene derivatives, **474–476** are also closely related to **436**. They differ, however, in the position of the thiophene ring formation. This is caused by a primary transformation of the terminal ethylenic bond of **436** to oxygen derivatives:

$$\mathbf{436} \xrightarrow{H_2S} \mathbf{458, 463, 464, 465}$$

$$\mathbf{436} \xrightarrow{[O]} H_3C{-}CH{=}CH{-}(C{\equiv}C)_4{-}CH(OR){-}CH_2OR \xrightarrow{H_2S} \mathbf{473\text{–}476}$$

The only dithienyl derivative representing route "C" is **477**, obtained from the roots of *Bidens connatus* Mühlenbg.[177]:

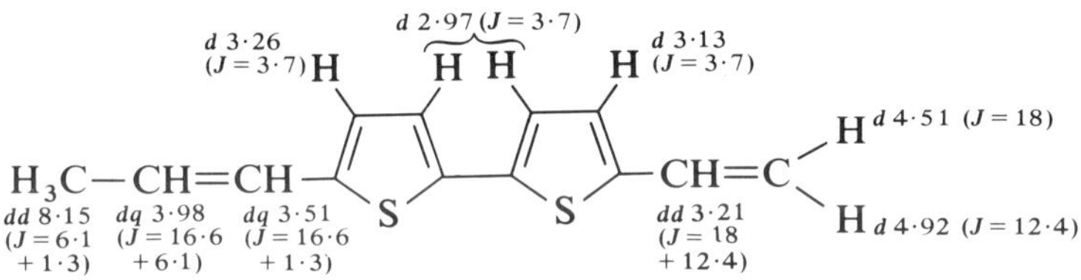

477

6. Amides

Lipid amides, especially isobutylamides, tyramides and phenylethylamides with characteristic olefinic and acetylenic unsaturation patterns, form a distinct group of natural products and have been isolated from members of the Compositae, Piperaceae and Rutaceae families. The polyene members especially frequently possess insecticide activity.

The biogenesis of these compounds is not yet clear. The amide group must certainly come from the corresponding peptide grouping, e.g. the isobutyl-group from a valine peptide:

$$R{-}CO{-}NH{-}\underset{\displaystyle CO_2H}{\underset{|}{CH}}{-}CH(CH_3)_2 \xrightarrow[-CO_2]{} R{-}CO{-}NH{-}CH_2{-}CH(CH_3)_2$$

During an investigation of the insecticidally active extracts of *Anacyclus pyrethrum* DC., Crombie[277] isolated the isobutylamide of a diacetylenic acid. The compound was identified as *N*-isobutyltetradeca-2,4-diene-8,10-diynamide(**478**), both double bonds being *trans*, since a maleic anhydride adduct is formed readily:

$$H_3C-CH_2-CH_2-C\equiv C-C\equiv C-(CH_2)_2-$$

$$\underset{t,t}{(CH=CH)_2}-\underset{\text{anacyclin}}{CONH-CH_2-CH(CH_3)_2} \quad \textbf{478}$$

The amide, named anacyclin, melts at 121°C and has weak insecticidal properties. Partial reduction of the triple bonds with the Lindlar catalyst gives the corresponding 2-*trans*-4-*trans*-8-*cis*-10-*cis*-tetraene, which proved to be a highly active insecticide.

The maleic anhydride adduct of anacyclin has been of great value in determining the nature and position of the remaining unsaturation. In the adduct, the 2,4-dienamide type of chromophore is eliminated and determination of the u.v. spectrum reveals the low intensity absorption characteristic of a conjugated diyne (λ max: 252, 237, 227 nm). The proposed structure of anacyclin has been confirmed by synthesis[278]. The route chosen is outlined in the scheme below. The Grignard reagent from hexa-1,5-diyne gave, after carboxylation, hepta-2,6-diynoic acid. The latter was esterified with diazomethane to its methyl ester **479**. The methyl ester was reduced with lithium aluminium hydride to hept-*trans*-2-en-6-yn-1-ol **480**. Oxidation of the alcohol with chromic acid yielded the aldehyde **481**, which was condensed with malonic acid to give nona-*trans*-2-*trans*-4-dien-8-ynoic acid **482**. The acid was converted, via the acid chloride, into *N*-isobutylnona-2,4-dien-8-ynamide **483**. Oxidative cross-coupling with pent-1-yne gave anacyclin, the spectroscopic and other data of which were consistent with natural material:

$$H-C\equiv C-(CH_2)_2-C\equiv C-H \xrightarrow[(2)\ CO_2]{(1)\ BrMgC_2H_5}$$

$$H-C\equiv C-(CH_2)_2-C\equiv C-CO_2H \xrightarrow{CH_2N_2}$$

$$H-C\equiv C-CH_2-CH_2-C\equiv C-CO_2CH_3 \quad \textbf{479}$$

$$\downarrow LiAlH_4$$

$$H-C\equiv C-(CH_2)_2-\overset{H}{\overset{|}{C}}=\underset{H}{\underset{|}{C}}-CH_2-OH \quad \textbf{480}$$

$$\downarrow CrO_3$$

$$\downarrow$$

$$\mathrm{H{-}C{\equiv}C{-}(CH_2)_2{-}\underset{}{\overset{H}{\overset{|}{C}}}{=}\underset{\underset{H}{|}}{C}{-}CHO} \qquad \mathbf{481}$$

$$\downarrow \mathrm{H_2C(COOH)_2}$$

$$\mathrm{H{-}C{\equiv}C{-}(CH_2)_2{-}CH{=}CH{-}CH{=}CH{-}CO_2H} \qquad \mathbf{482}$$

$$\downarrow$$

$$\mathrm{H{-}C{\equiv}C{-}(CH_2)_2{-}CH{=}CH{-}CH{=}CH{-}CONH{-}CH_2{-}CH(CH_3)_2} \qquad \mathbf{483}$$

$$\downarrow \mathrm{+\,H{-}C{=}C{-}C_3H_7/Cu_2Cl_2/O_2}$$

anacyclin **478**

A second synthesis, publicated simultaneously[279], involved cross-coupling of pent-1-yne with pent-4-yn-1-ol and conversion into the bromide **484**. The ylide **485** was prepared by treating the bromide with triphenyl phosphine. **485** was condensed with the aldehyde **486** obtained on oxidation of crotonic ester with selenium dioxide:

$$\mathrm{H_7C_3{-}C{\equiv}C{-}H + H{-}C{\equiv}C{-}(CH_2)_3{-}OH}$$

$$\downarrow$$

$$\mathrm{H_7C_3{-}(C{\equiv}C)_2{-}CH_2{-}CH_2{-}CH_2{-}OH}$$

$$\downarrow$$

$$\mathrm{H_7C_3{-}(C{\equiv}C)_2{-}(CH_2)_3{-}Br} \quad \mathbf{484} \qquad \mathrm{H_3C{-}CH{=}CH{-}CO_2CH_3}$$

$$\downarrow \mathrm{Ph_3P/LiBu} \qquad \swarrow \mathrm{SeO_2}$$

$$\mathrm{H_7C_3{-}(C{\equiv}C)_2{-}CH_2{-}CH_2{-}CH{=}PPh_3}$$

$$\mathrm{+\ OCH{-}CH{=}CH{-}CO_2CH_3} \quad \mathbf{486}$$

$$\mathbf{485} \qquad \swarrow$$

$$\mathrm{H_7C_3{-}(C{\equiv}C)_2{-}CH_2{-}CH_2{-}CH{=}CH{-}CH{=}CH{-}CO_2CH_3} \rightarrow \mathbf{478}$$

It is very probable that natural anacyclin is contaminated with small portions of **487** and **488**. As mentioned above, the 2,4-dienamide chromophore is eliminated in the maleic anhydride adduct; therefore, a low intensity absorption characteristic of a conjugated diyne is to be seen. But the spectrum of this grouping is not quite pure; some 4% of an ene-diyne is indicated (λ max: 281, 266, 252 nm)[277]:

$$\mathrm{H_3C{-}CH{=}CH{-}(C{\equiv}C)_2{-}CH_2CH_2{-}(CH{=}CH)_2{-}}$$

trans: **487**
cis: **488**

$$\mathrm{CONH{-}CH_2CH(CH_3)_2}$$

488 could be isolated as the only amide from the roots of *Anthemis fuscata* Brot.[121].

The structure has been confirmed by synthesis including cross-coupling of *N*-isobutylnona-2,4-dien-8-ynamide with *cis*- and *trans*-pent-2-en-4-yne. **487** and **488** were formed and may be related to the contaminant of anacyclin[278]. Until now it is not possible to say if either is the natural conjugated ene-diyne.

The structure of maleic anhydride adducts of *N*-isobutyl-trans-2-*trans*-4-dienamides is very interesting. The obvious structure **489** seems to be incorrect, for, in a series of such adducts no characteristic vibrations for NH, the amide carbonyl or the anhydride ring can be found. Observations of the i.r. spectrum near 2500/cm (broad band), however, suggests that a free carboxylic acid group is present in the abnormal adducts. **490** and **491** should be alternative formulations, the former being more consistent with the i.r. data:

$$\mathrm{R{-}CH{=}CH{-}CH{=}CH{-}CONH{-}CH_2CH(CH_3)_2} \rightarrow$$

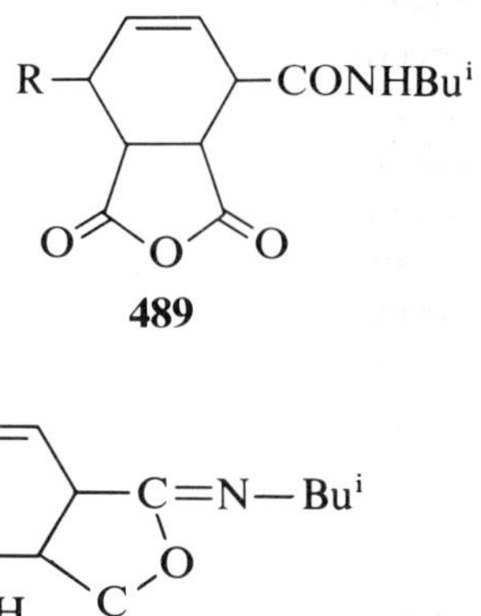

489

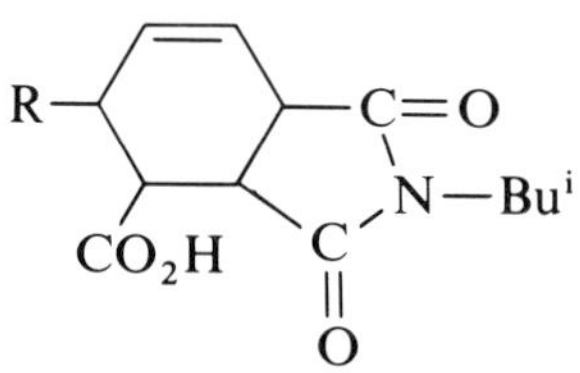

490

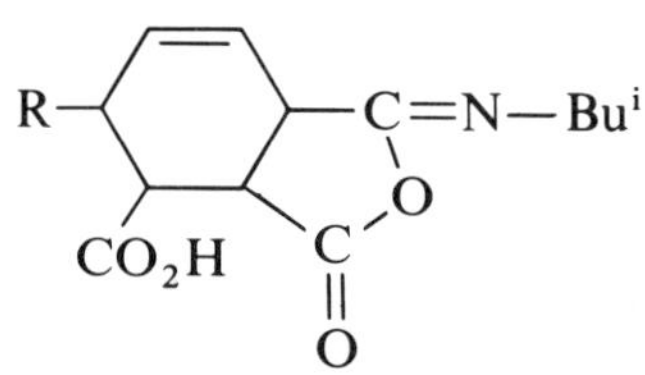

491

Formation of **490** is considered to occur through nucleophilic attack by the amide grouping at the carbon atom of the adjacent anhydride carbonyl grouping, either during the Diels-Alder reaction, or subsequently[278].

The roots of *Chrysanthemum frutescens* L. have been found to contain a C_{14}-isobutylamide shown to be **492** on the basis of chemical and spectroscopic evidence[280]. The u.v. spectrum reveals a dienamide like chromophore with a maximum at 251 nm and the i.r. spectrum confirms this assumption. The compound shows absorption at 3460, 1670 and 1560/cm due to a secondary amide. Bands at 1625 and 1005/cm are indicating a *trans-trans*-diene grouping. The foregoing data, together with the n.m.r. spectrum, lead to structure **492**:

$$H_3C-(CH_2)_7-CH_2-CH=CH-CH=CH-CO-NH-CH_2-CH(CH_3)_2 \quad \mathbf{492}$$

m 9·12; *m* 8·74; *m* 7·87; *m* 3·97; *dd* 2·83 (*J* = 15 + 10); *d* 4·13 (*J* = 15); *t* 3·52 (*J* = 6); *t* 6·87 (*J* = 6); *m* 8·2; *d* 9·08 (*J* = 6·5)

The structure of **492** has been finally established by synthesis[281].

A series of amides, with tyramine replacing isobutylamine, has been isolated from *Anacyclus pyrethrum* DC.[282]. From the root extract a crystalline fraction was obtained, at first thought to be homogeneous (TLC). The fraction shows an u.v. spectrum with a maximum at 259 nm. The i.r. spectrum shows it to be a monosubstituted amide (3300, 3070, 1635 and 1550/cm), which contains $\alpha\beta$-conjugated diene-unsaturation (1660, 1615/cm). The presence of a hydroxyl is indicated by a broad band at 3200–3400/cm. In addition, the u.v. spectrum of the maleic anhydride adduct shows a weak absorption maximum at 277 nm. The signal was identified and shifted to longer wavelengths on addition of alkali, indicating that the hydroxyl group was phenolic. This assumption was confirmed by acid hydrolysis of the amide which yielded tyramine hydrochloride. Analysis of the fraction gave no clear result and mass spectrometric measurements showed three parent prominent peaks at *m/e* 343, 315 and 287 in the ratio 4:5:1, indicating a mixture of homologues. As there was no success in separating the homologues **493**–**495**, the three structures were confirmed by synthesis:

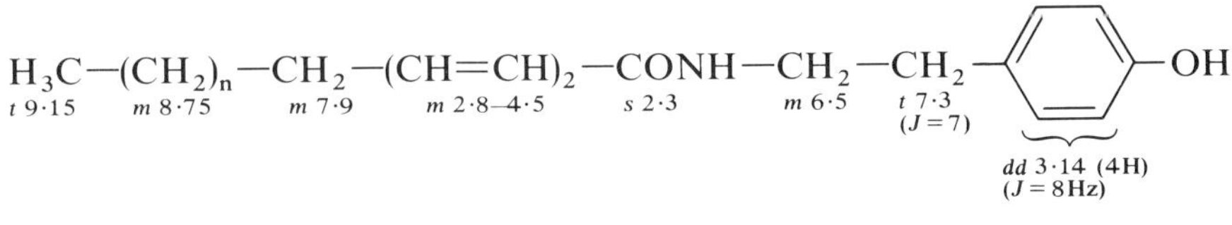

493: n = 7
494: n = 5
495: n = 3

The route chosen for synthesis involved the phosphonate method[283]. The phosphonate **496** was obtained by a Michaelis–Arbuzov reaction from methyl-4-bromo-crotonate and triethylphosphite. Base treatment of **496** gave the corresponding anion, which reacts readily with the aldehydes to give the methyl esters of different chain length. The esters were converted into the acid chlorides and then transformed on reaction with tyramine to the three tyramides:

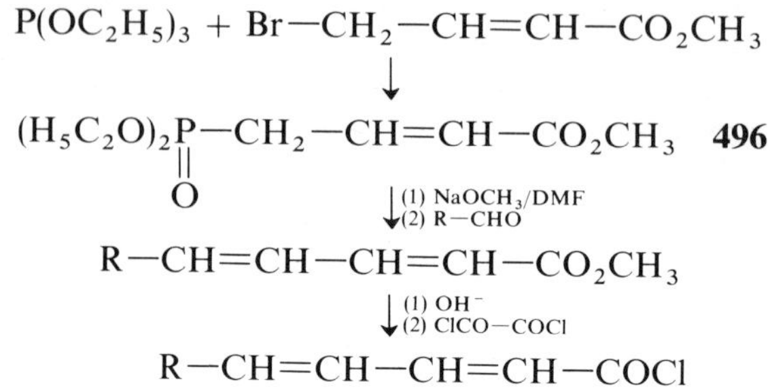

493: R = $CH_3(CH_2)_8$—
494: R = $CH_3(CH_2)_6$—
495: R = $CH_3(CH_2)_4$—

The investigation of the constituents of *Echinacea purpurea* Moench. and *E. angustifolia* DC. gave several compounds all possessing an isobutylamide grouping[284]. Two of them only differ in the chain length (C_{11} and C_{12}). One of the compounds, possessing a terminal acetylenic proton could be separated by transformation into the silver salt and was shown to be **497**. The second isobutylamide showed a methyl singulet in the n.m.r. (*s* 8·11τ) adjacent to triple bonds, instead of the signal of the acetylenic proton (broadened singlet 8·13τ) of the former. In view of the spectral results the structures have to be **497** and **498**. Of some interest is the *cis-trans*-configuration of the diene grouping:

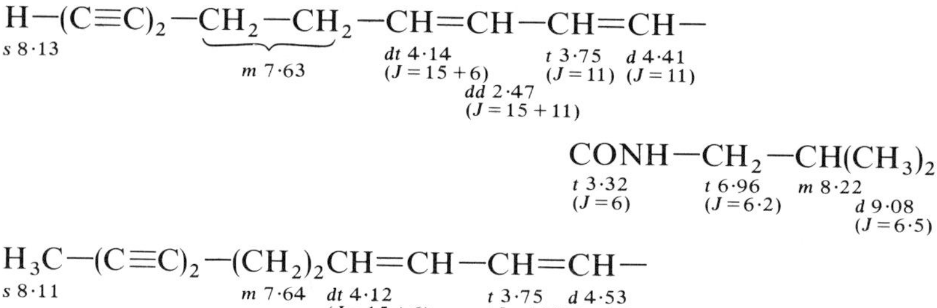

497

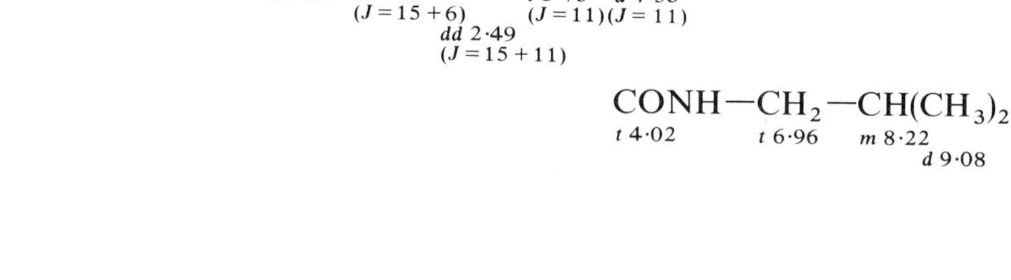

498

It seemed desirable to prove the structures of **497** and **498** by synthesis. The route chosen for **498** involved preparation of the aldehyde **499** by Cadiot–Chodkiewicz coupling of 1-bromo-prop-1-yne with pent-4-yn-1-ol, followed by oxidation with dimethyl sulphoxide to the corresponding aldehyde and chain lengthening by means of the ethoxy-acetylene method. The obtained $\alpha\beta$-unsaturated aldehyde **499** was treated with the ylide **500** to give a mixture of *cis-trans* isomeric **498**[281]:

$$H_3C{-}C{\equiv}C{-}Br + H{-}C{\equiv}C{-}(CH_2)_3{-}OH$$

$$\downarrow$$

$$H_3C{-}(C{\equiv}C)_2{-}(CH_2)_2{-}CH_2{-}OH$$

$$\downarrow \text{DMSO}$$

$$H_3C{-}(C{\equiv}C)_2{-}(CH_2)_2{-}CHO + BrMg{-}C{\equiv}C{-}OC_2H_5$$

$$\downarrow$$

$$H_3C{-}(C{\equiv}C)_2{-}(CH_2)_2{-}\underset{\displaystyle OH}{\underset{|}{C}H}{-}C{\equiv}C{-}OC_2H_5$$

$$\swarrow \text{LiAlH}_4$$

$$H_3C{-}(C{\equiv}C)_2{-}(CH_2)_2{-}CH\overset{}{\underset{t}{=}}CH{-}CHO$$

499

$$\downarrow \quad + Ph_3P{=}CH{-}CONH{-}CH_2{-}CH(CH_3)_2$$

500

$$H_3C{-}(C{\equiv}C)_2{-}(CH_2)_2{-}CH\underset{t\;/\;t}{=}CH{-}CH\underset{t\;/\;c}{=}CH{-}CONH{-}CH_2{-}CH(CH_3)_2$$

498

The isomeric amides could be separated by chromatography and **497** was obtained in a similar manner as outlined in the scheme below:

$$H{-}C{\equiv}C{-}(CH_2)_2{-}CHO \rightarrow H{-}C{\equiv}C{-}(CH_2)_2{-}\underset{\displaystyle OH}{\underset{|}{C}H}{-}C{\equiv}C{-}OC_2H_5$$

$$\swarrow \text{1) LiAlH}_4\ \text{2) H}_3\text{O}^+$$

$$H{-}C{\equiv}C{-}(CH_2)_2{-}CH{=}CH{-}CHO$$

$$\downarrow \quad + Ph_3P{=}CH{-}CONH{-}CH_2{-}CH(CH_3)_2 \quad \textbf{500}$$

$$H{-}C{\equiv}C{-}(CH_2)_2{-}CH{=}CH{-}CH{=}CH{-}CONH{-}CH_2{-}CH(CH_3)_2$$

501

The diyne grouping was introduced by coupling bromo-propiolic acid with **501** and decarboxylation of the coupling product:

$$\mathrm{HOOC{-}C{\equiv}C{-}Br} + \mathbf{501}$$

$$\downarrow$$

$$\mathrm{HOOC{-}C{\equiv}C{-}C{\equiv}C{-}(CH_2)_2{-}CH{=}CH{-}}$$
$$\mathrm{CH{=}CH{-}CONH{-}CH_2{-}CH(CH_3)_2} \quad \mathbf{502}$$

$$\downarrow -CO_2$$

$$\mathrm{H{-}C{\equiv}C{-}C{\equiv}C{-}(CH_2)_2{-}CH{=}CH{-}CH{=}CH{-}}$$

t *t* **503**
t *c* **497**

$$\mathrm{CONH{-}CH_2{-}CH(CH_3)_2}$$

The obtained *trans-trans*-isobutylamide **503** is known to occur in the roots of *Chrysanthemum frutescens* L.[280]. The spectral properties of natural and synthetic material are identical:

$$\mathrm{H{-}C{\equiv}C{-}C{\equiv}C{-}(CH_2)_2{-}CH{=}CH{-}CH{=}CH{-}}$$

s 8·12; *m* 7·66; *m* 3·9; *dd* 2·95 ($J = 10 + 15$); *d* 4·11 ($J = 15$)

$$\mathrm{CONH{-}CH_2{-}CH(CH_3)_2} \quad \mathbf{503}$$

t 3·22 ($J=6$); *t* 6·93 ($J=6$); *m* 8·21; *d* 9·09 ($J=6·5$)

An interesting feature of the structures of anacyclin **478** and the other described natural amides is the dimethylene between two unsaturated systems. This seems to be a characteristic feature of various insecticidal amides of plants.

Two further amides, a phenylethylamide corresponding to **503** and an isobutylamide related to dehydromatricariaester have also been found to occur in *Anacyclus pyrethrum* DC.[35] The structures **505** and **504** are proved by n.m.r. structural analysis as well as mass spectroscopic measurements. **505** has been already mentioned in Chapter 2, Section 5.1 (No: **385**).

$$\mathrm{H{-}(C{\equiv}C)_2{-}(CH_2)_2{-}(CH{=}CH)_2{-}CONH{-}CH_2{-}CH_2{-}}$$ 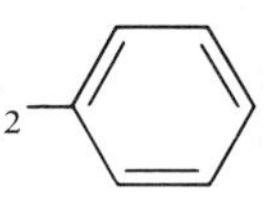 **504**

$$\mathrm{H_3C{-}(C{\equiv}C)_3{-}CH{=}CH{-}CONH{-}CH_2{-}CH(CH_3)_2} \quad \mathbf{505}$$

The roots of *Chrysanthemum frutescens* L. have been found to contain a thiophene compound shown to be **506** on the basis of chemical and spectroscopic evidence. The polar substance gave an i.r. spectrum consistent with the presence of a secondary amide grouping (3450, 1680, 1540/cm) and a conjugated diene grouping (1645, 1630/cm). The u.v. spectrum shows a maximum at 252 nm. According to its elemental analysis, the compound has the composition $C_{14}H_{19}NOS$. Its n.m.r. spectrum contains characteristic signals of an amide proton (τ 2·2) and an isopropyl grouping [*d* 9·1τ (6H), *m* 8·2τ (1H)]. Oxidation with permanganate gave thiophene-carboxylic acid-(2) and oxalic acid monoisobutylamide. Treatment with alkali yielded a compound with a thienyl-diene like chromophore (λ max: 306 nm). The latter gave a diene adduct with a u.v. spectrum due to a monosubstituted thiophene (λ max: 231 nm). In view of these results, the compound was assigned structure **506**:

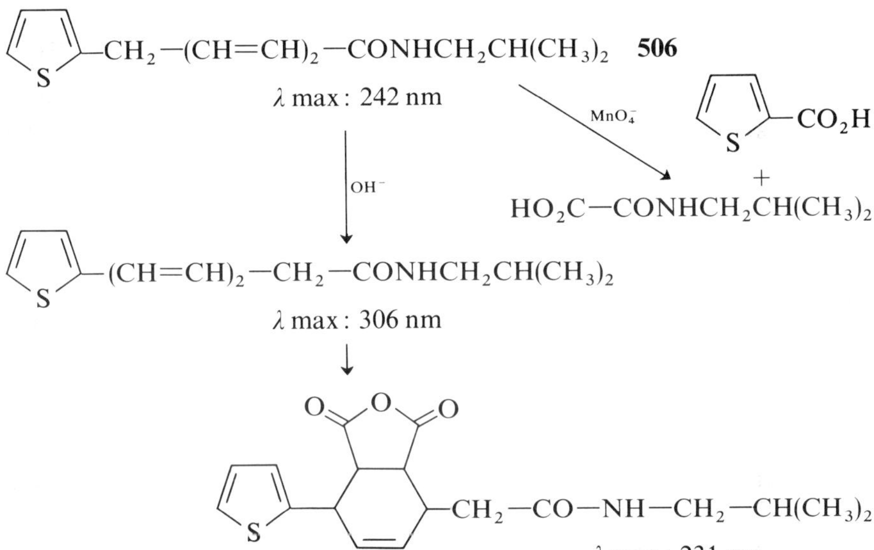

Two further highly insecticidal isobutylamides should be mentioned. Both, neoherculin (**507**) and affinin (**508**) are characterised by a dimethylene between two unsaturated systems and the lacking of triple bonds. The highly unstable neoherculin **507** has been found to be the insecticidal principle from *Zanthoxylum clavaherculis* L.[285] while the closely related affinin **508** has been isolated from *Heliopsis longipes* (A. Gray) Blake[286]:

$$H_3C—(CH{=}CH)_3—CH_2CH_2—CH{=}CH—CONH—CH_2—CH(CH_3)_2$$

neoherculin **507**

$$H_3C—(CH{=}CH)_2—CH_2CH_2—CH{=}CH—CONH—CH_2—CH(CH_3)_2$$

affinin **508**

The biogenesis of most of the amides is more or less clear as shown by tracer studies. It turns out that **498** is formed from the alcohol **289** (see Section 3.1) and from **478**[35]. Therefore, the following pathways seem to be possible:

$$H_3C{-}CH_2{-}CH_2{-}(C{\equiv}C)_2{-}CH_2{-}CH{=}CH{-}(CH_2)_4{-}OH \quad \textbf{289}$$

$\downarrow$ [O] *cis*

$$[H_3C{-}CH_2{-}CH_2{-}(C{=}C)_2{-}CH_2{-}CH{=}CH{-}CH_2{-}CH(OH){-}CH_2{-}CO_2R]$$

$\downarrow$

$$[H_3C{-}CH_2{-}CH_2{-}(C{\equiv}C)_2{-}CH_2{-}CH{=}CH{-}CH_2{-}CH{=}CH{-}CO_2R]$$

$\downarrow$

$$H_3C{-}CH_2{-}CH_2{-}(C{\equiv}C)_2{-}CH_2CH_2{-}(CH{=}CH)_2{-}CONHCH_2CH(CH_3)_2 \quad \textbf{478}$$

478 $\xrightarrow{[O]}$ $H_3C{-}CO \wr CH_2{-}(C{\equiv}C)_2{-}(CH_2)_2{-}(CH{=}CH)_2{-}CONHR$

478 $\xrightarrow{-[H]}$ **487** $\xrightarrow{-[H]}$ **507** $\xrightarrow{-[H]}$ **492**

$$H_3C{-}(C{\equiv}C)_2{-}(CH_2)_2{-}(CH{=}CH)_2{-}CONHR \quad \textbf{498}$$

498 $\xrightarrow{[O],\ -CO_2}$ **497**

498 $\xrightarrow{[H]}$ **508**

Perhaps the thiophene **506** is formed in the following way:

$$\textbf{289} \xrightarrow[-CO_2]{[O]} [H_7C_3{-}(C{\equiv}C)_2{-}CH_2{-}CH{=}CH{-}(CH_2)_3{-}OH]$$

$\downarrow$ [O]

$$[H_7C_3{-}(C{\equiv}C)_2{-}CH_2{-}CH{=}CH{-}CH(OH){-}CH_2{-}CO_2R]$$

$\downarrow$ $-[H_2O]$

$$H_7C_3{-}(C{\equiv}C)_2{-}CH_2{-}(CH{=}CH)_2{-}CONHR$$

$\downarrow$ [O]

$$H_3C{-}CO \wr CH_2{-}(C{\equiv}C)_2{-}CH_2{-}(CH{=}CH)_2{-}CONHR$$

$\downarrow$

$$H_3C{-}(C{\equiv}C)_2{-}CH_2{-}(CH{=}CH)_2{-}CONHR$$

$\downarrow$ (1) [O], $-CO_2$; (2) H_2S

506

7. Compounds Related to Matricariaester

7.1. Simple Derivatives

The Compositae is the largest and most widely dispersed family of flowering plants and is considered to mark the highest point in the evolution of the dicotyledons. It contains over 20,000 species divided into thirteen tribes. A preliminary investigation by Sörensen[287] of the essential oils of 94 members of the Compositae showed that 34 of these contained matricariaester (**509**). It is thus one of the most widely distributed acetylenic compound known; its structure has been established in 1941 by degradation and hydrogenation[7]. The action of alkali, for example, produces methyl vinyl ketone, while hydrogenation gives methyl decanoate with absorption of six moles of hydrogen:

$$H_3C{-}CH{=}CH{-}C{\equiv}C{-}C{\equiv}C{-}CH{=}CH{-}CO_2CH_3 \quad \text{matricariaester}$$

cis	*cis*	**509**
cis	*trans*	**509a**
trans	*trans*	**509b**

$$\mathbf{509} \xrightarrow{OH^-} H_3C{-}CH{=}CH{-}CH_2{-}CO\wr CH_2{-}CO{-}CH{=}CH\wr CO_2H$$

$$\downarrow$$

$$H_3C{-}CO{-}CH{=}CH_2$$

The name matricariaester is unfortunate since the compound occurs in part of the genus *Matricaria* which according to some botanists is atypical and should be separated as the genus *Tripleurospermum*[288]. Its presence does appear to be a characteristic of the genus *Erigeron* (tribe Astereae), the essential oil from some species of which is composed entirely of matricariaester. The occurrence of **509** in the thirteen tribes of the Compositae has been summarised earlier by Sörensen[289].

The roots of *Solidago virgaurea*, for instance, are one of the richest source known of matricariaester (0·8 %). *Matricaria inodora* from which **509** was first isolated in 1941 was found on reinvestigation to contain 2-*trans*-8-*cis*-matricariaester (**509a**). It was isolated in 1951 from the mother liquors after removal of **509** and also from another member of the Astereae, *Amellus strigosus*[287]. **509a** is difficult to crystallise and has a melting point of 2°C. It has been prepared by photoisomerisation of the all-*cis* isomer **509**, and was synthesised together with the all-*trans* isomer **509b** by a cross coupling of pent-2-en-4-yne with methyl *trans*-pent-2-en-4-ynoate[290]:

$$\underset{c,t}{H_3C{-}CH{=}CH{-}C{\equiv}C{-}H} + \underset{t}{H{-}C{\equiv}C{-}CH{=}CH{-}CO_2CH_3}$$

$$\downarrow Cu_2Cl_2/O_2$$

509a,b

In 1954, Sörensen obtained the first evidence for the existence of the *trans-trans*-isomer **509b** in nature, detecting it spectroscopically in *Bellis perennis*[291]. It was not in sufficient quantity for isolation and characterisation. It was finally obtained from the aerial parts of *Matricaria tenuifolia* Poir.[48].

The stereochemistry of the three isomers can be assigned unambiguously on the basis of the olefinic coupling constants of their n.m.r. spectra (see Table II below). The configuration of the double bonds can also be established by the

TABLE II
N.M.R. SIGNALS OF THE ISOMERIC MATRICARIAESTER

$$\underset{A}{H_3C}-\underset{B}{CH}=\underset{C}{CH}-(C\equiv C)_2-\underset{D}{CH}=\underset{E}{CH}-\underset{F}{CO_2CH_3}$$

	H_A	H_B	H_C	H_D	H_E	H_F
509:	*dd* 8·03 $J = 7 + 1{\cdot}7$	*dq* 3·81 $J = 11 + 7$	*dq* 4·41 $J = 11 + 1{\cdot}7$	*s* 3·85		*s* 6·28
509a:	*dd* 8·07 $J = 7 + 1{\cdot}8$	*dq* 3·81 $J = 11 + 7$	*ddq* 4·45 $J = 11, 1{\cdot}8 + 1$	*dd* 3·22 $J = 16 + 1$	*d* 3·75 $J = 16$	*s* 6·31
509b:	*dd* 8·14 $J = 7 + 1{\cdot}6$	*dq* 3·67 $J = 16 + 7$	*ddq* 4·44 $J = 16, 1 + 1{\cdot}6$	*dd* 3·25 $J = 16 + 1$	*d* 3·77 $J = 16$	*s* 6·31

different chemical shift, caused by the deshielding effects of the diyne or carbonyl grouping. The downfield shift for the proton H_E must be due to the diyne grouping in the Δ^2-*trans*-series; the shifting of H_D to lower fields is caused by the carbonyl group. In the *cis-cis*-isomer **509** the two groupings are of no effect. The electron withdrawal from the carbonyl- and the diyne-grouping is of the same strength, therefore H_D and H_E absorb together at the same τ-value. A similar situation is discussed for the Δ^8-double bond. When *trans*-configurated H_B is shifted down to lower fields, caused by the magnetic anisotropy of the triple bond, while when *cis*-configurated the methyl group is influenced by the deshielding sphere of the triple bond:

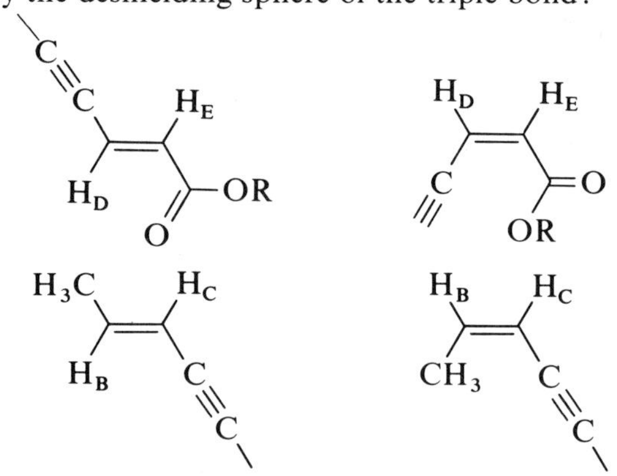

C_{10}-polyacetylenic compounds are formed—as mentioned in Chapter 1, Section 4—from C_{18}-acids by β-oxidation as outlined in the scheme below:

oleic acid **4** ⟶ crepenynic acid **1**

↓

$H_3CCH_2CH_2-(C{\equiv}C)_2-CH_2-CH{=}CH-(CH_2)_7-CO_2R$ **23**

↓ −[H], 4 × β-oxidation

509

Labelled **23** is incorporated very well in matricarianol (**514**). The terminal propenyl group is surely formed by dehydrogenation of the propyl group.

The free matricaria acid **510** has been found to occur in *Anthemis rudolfiana* hort.[85]:

$H_3C-CH{=}CH-(C{\equiv}C)_2-CH{=}CH-COOH$ **510**

cis *cis*

The aerial parts of *Grindelia humilis* Hook. contain, *inter alia*, three isomeric matricaria aldehydes **511**–**513**[292]

$H_3C-CH{=}CH-(C{\equiv}C)_2-CH{=}CH-CHO$

cis	*cis*	**511**
trans	*cis*	**512**
trans	*trans*	**513**

Trans-trans-matricarianal has also been isolated from *Aethusa cynapium* L.[258] and has been used for many syntheses, e.g. for the synthesis of **248**; see Section 2.2. The corresponding alcohols *cis-cis-* and *trans-cis*-matricarianol were obtained from *Diplopappus fruticolosus*[293] and some *Grindelia* species[292]. The *trans-trans*-alcohol has been found earlier to occur in *Grindelia arenicola*[289].

$H_3C-CH{=}CH-(C{\equiv}C)_2-CH{=}CH-CH_2OH$

trans	*trans*	**514**[289]
cis	*cis*	**515**[293]
cis	*trans*	**516**[292]
trans	*cis*	**517**[292]

The *trans-trans*-alcohol **514** is a solid substance (m.p.: 93°) while the others are oily compounds. The stereochemistry of the double bonds can be assigned on the basis of the observed olefinic coupling constants. The n.m.r. data of **515** are given below:

$H_3C{-}CH{=}CH{-}(C{\equiv}C)_2{-}CH{=}CH{-}CH_2{-}OH$ **515**

dd 8·06 (J=6·5+1·6); *dq* 3·88 (J=6·5+11); *dq* 4·46 (J=11+1·6); *dt* 4·40 (J=11+1·3); *dt* 3·81 (J=11+6); *dd* 5·63 (J=6+1·3); *s* 7·95

514 is formed in high yields by biochemical reduction of **509**[35].

10-hydroxy derivatives of matricarianal and matricarianol have been isolated from the aerial parts of *Diplopappus fruticolosus*[293]. The spectral properties of the two compounds were consistent with the proposed structures **518** and **519**:

$HO{-}CH_2{-}CH{=}CH{-}(C{\equiv}C)_2{-}CH{=}CH{-}CHO$ **518**

s 7·99; *dd* 5·55 (J=6·5+1·2); *dt* 3·65 (J=11+6·5); *ddt* 4·29 (J=11+1·2+1); *m* 3·38 (CH=CH); *dd* 0·41 (J=6+1·5)

$HO{-}CH_2{-}CH{=}CH{-}(C{\equiv}C)_2{-}CH{=}CH{-}CH_2OH$ **519**

s 5·90; *dt* 3·71 (J=11+6·5); *dt* 4·35 (J=11+1·2); *dd* 5·67 (J=6·5+1·2)

M^+ *m/e* 162

$\xrightarrow{-H_2O}$ *m/e* 144

$\xrightarrow{-CHO}$ *m/e* 133 $\xrightarrow{-CO}$ *m/e* 115

The same plant, *Diplopappus fruticolosus*, contains the *cis-cis*-matricarianol-acetate **520**. The corresponding *trans-trans*-acetate **521** has been isolated first from the leaves and roots of *Grindelia arenicola* by Holme *et al.*[289], while the *cis-trans*-isomer has been found to occur in an *Erigeron* species[9,55]:

$H_3C{-}CH{=}CH{-}(C{\equiv}C)_2{-}CH{=}CH{-}CH_2{-}OCOCH_3$

cis	*cis*	**520**
trans	*trans*	**521**
cis	*trans*	**522**

Some special esters of matricarianol have been isolated from *Carthamus tinctorius* L.[294] (**523**), *Grindelia humilis* Hook.[292] (**524**), *Grindelia camporum* Greene[293] (**525**) and *Bellium bellidioides* Desf.[35] (**526**). The structures were assigned on the basis of i.r. and n.m.r. spectra as well as mass spectroscopic measurements:

$H_3C{-}CH{=}CH{-}(C{\equiv}C)_2{-}CH{=}CH{-}CH_2{-}OCO{-}CH_2{-}CH(CH_3)_2$ **523**

dd 8·07; *cis*: *dq* 3·85, *dm* 4·50; *t*: *dm* 4·21, *m* 3·8; *dd* 5·42; *m* 7·89; *m* 8·45; *d* 9·04

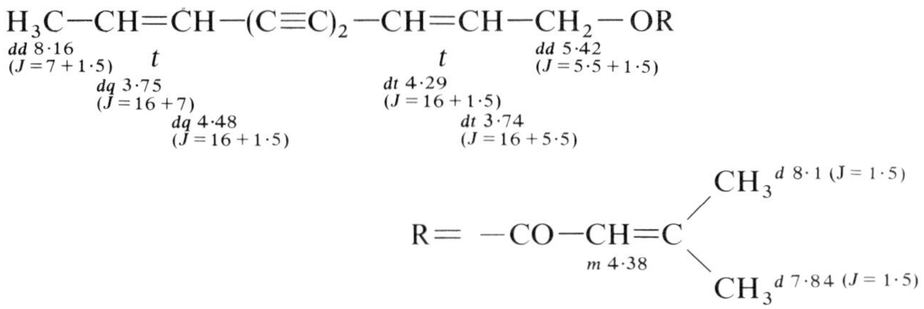

524

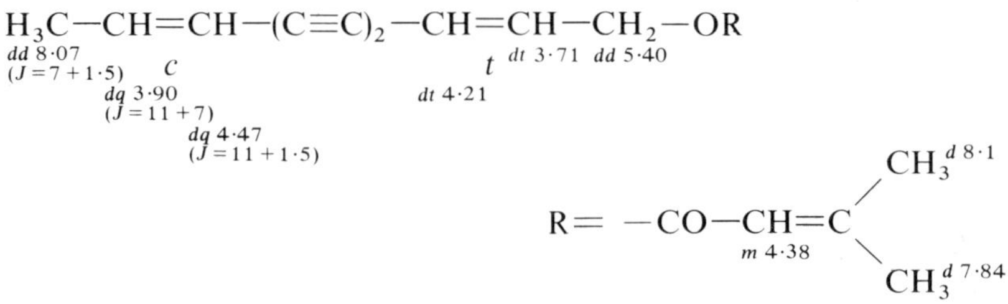

525

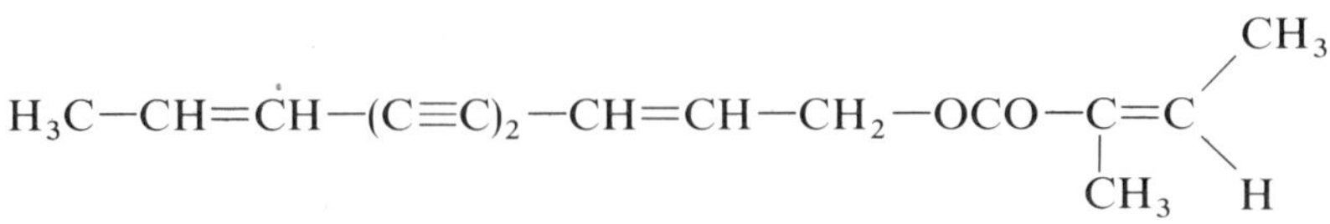

526

7.2. Addition Products and Ring Compounds

In 1950 Sörensen and Stavholt described a compound isolated from the leaves of *Matricaria inodora* L.[295] for which they proposed the formula **527**, the so-called "Composit-Cumulene I":

$$H_3C-(CH_2)_4-CH=C=C=CH-CO_2CH_3 \quad \mathbf{527}$$

527 was isolated by chromatography of the mother liquors of the essential oil after removal of the main component matricaria ester. Five years later beside matricaria ester small amounts of a crystalline compound with one broad absorption band at the same wavelength as **527** was isolated from the roots of *Chrysothamnus viscidiflorus* (Hooker) Nuttal. Its i.r. spectrum revealed the interesting fact that the new compound contained both a disubstituted acetylenic bond and an unsaturated γ-lactone grouping. These findings indicated that **527** might be a mixture of compounds with the lactone from *Chrysothamnus* as the main constituent. During the work on the synthesis of matricaria ester, it was found that when oxidising the alcohol **528** with chromic acid, a lactone was formed during the isolation of the product, when the acidic and neutral fractions were separated via sodium bicarbonate solution:

$$R{-}C{\equiv}C{-}C{\equiv}C{-}CH{=}CH{-}CH_2OH \xrightarrow{CrO_3}$$

528

The spectral properties of the lactone showed striking resemblances to those of the Composit-Cumulene I and to the *Chrysothamnus* lactone. When *cis-cis*-matricaria acid was treated with sodium bicarbonate, a lactone was formed, identical in all respects to those obtained from *Chrysothamnus*. The crystals melted at 37·5°C. Under special conditions, the authors[296] finally had success in crystallising the so-called Composit-Cumulene I. This material also gave the same crystalline lactone identical with the compound obtained by lactonisation of the *cis-cis*-matricaria acid. Therefore, the formula first proposed (**527**) is completely wrong; the correct formula is **529**:

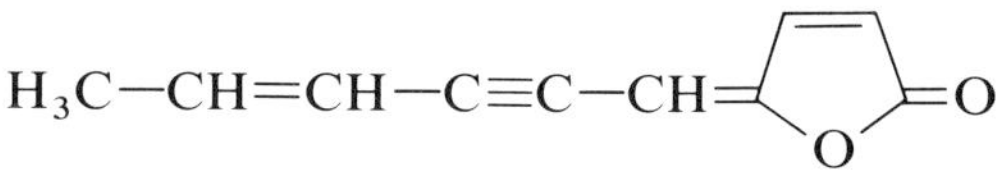

cis *trans* **529**
cis *cis* **530**

The isomer **530** is shown to occur beside the *cis-trans* matricaria lactone **529** in the aerial parts of *Matricaria caucasica* Poir.[48].

The investigation of the constituents of some *Conyza*, *Boltonia* and *Erigeron* species gave small amounts of a compound, closely related to **529**. It is the first natural cumulene to be isolated from a plant and was identified as **531**[297]:

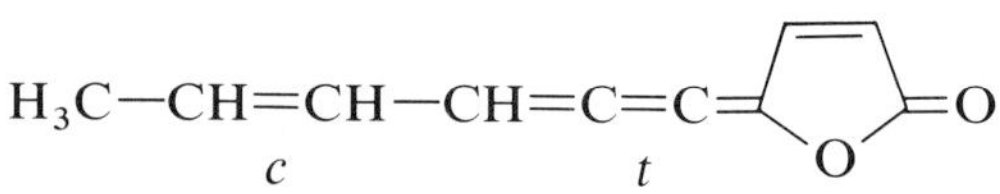

531

c *t*
c *c*

The highly unstable cumulene shows an u.v. spectrum with λ max: 380 nm. Its i.r. spectrum displays bands for an unsaturated five-ring lactone (1795, 1760, 1630, 1100, 1065, 880/cm) and weak bands due to the cumulene grouping at 2092, 2045, and 1960/cm.

Formula **532** could be excluded since the i.r. lacked bands typical for deformation vibrations of the proton located at the exocyclic double bond (940 or 720/cm):

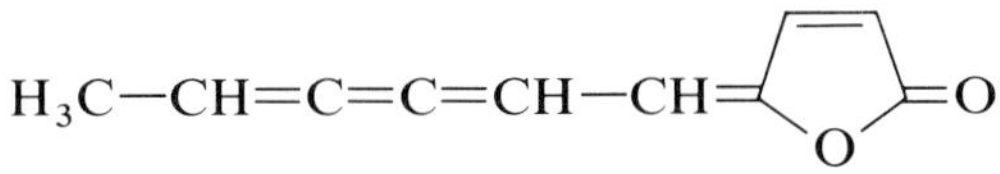 **532**

A further confirmation for the proposed structure was obtained by the n.m.r. spectrum. It shows a three proton double doublet at τ 8·14 indicating a methyl group adjacent to a double bond (J = 6·4 + 1·1 Hz). Methyl adjacent to an allene grouping should resonate at higher fields. In order to clarify this question **531** was partial hydrogenated with deuterium and subsequently irradiated to give the all-*trans* compound **533**:

531 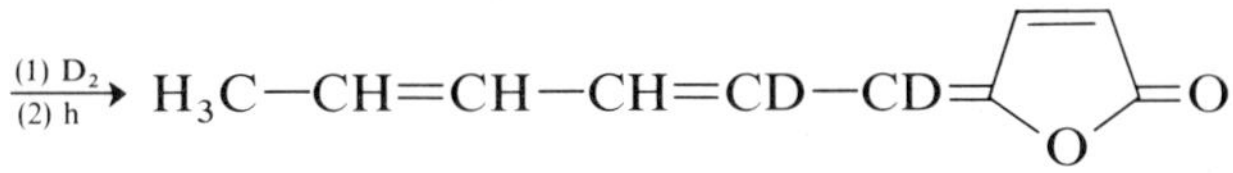 **533**

The presence of a CH_3—CH=CH-grouping was again indicated by a doublet doublet at τ 8·1 (J = 6·8 + 1·2 Hz), while a CH_3—CH=CD—CD=CH-grouping should give another splitting. Biogenetically **531** derives from matricaria acid and corresponds to the matricaria lactone **529**:

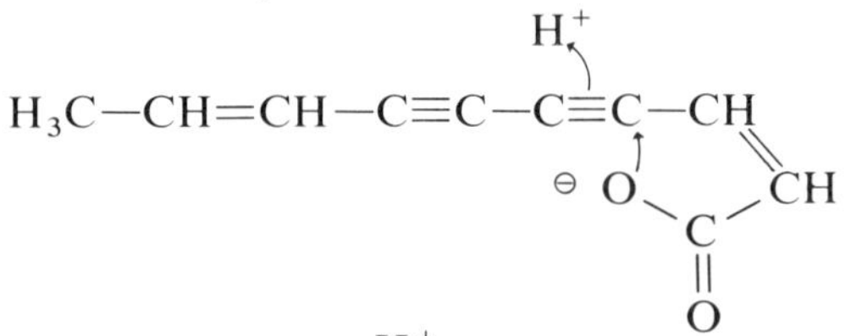 → **529**

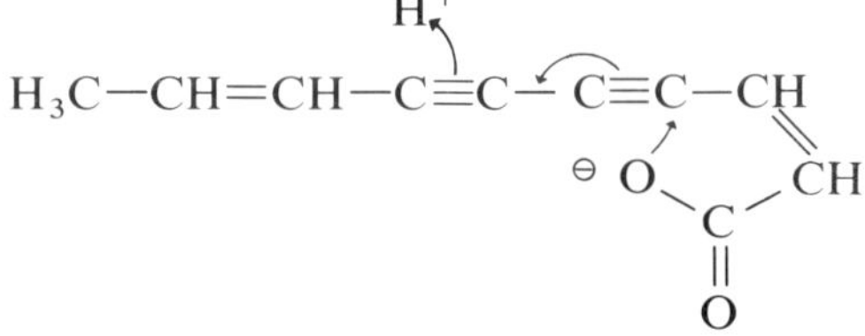 → **531**

The aerial parts of *Diplopappus fruticolosus* have been found to contain an aldehyde, shown to be **534** on the basis of chemical and spectroscopic evidence[293]:

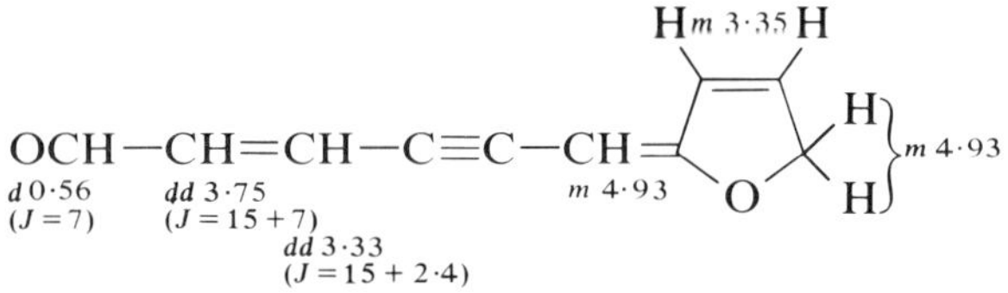 **534**

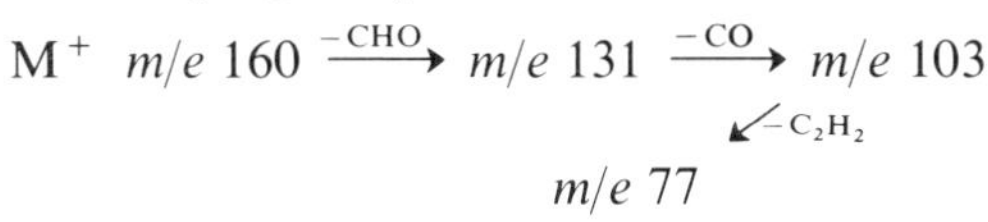

Characteristic constituents of members of the *Anthemis* L. species are dehydromatricaria ester and the related sulphur compounds (see Chapter 2, Sections 5.1 and 5.2). However, the investigation of the constituents of *Anthemis austriaca* Jacq. shows that dehydromatricaria ester is absent and that the main constituent is matricariaester **509** and some closely related sulphur compounds[165]. From the root extract, a mixture of sulphur containing compounds was obtained, which could not be separated. Using monoperphthalic acid, two sulphones were obtained together with a third compound which had not reacted. The latter showed a broad u.v. maximum at 345 nm. The i.r. spectrum shows some bands in the position characteristic of matricaria ester, but lacked the acetylene absorption of the latter. The structure **535** is proved by n.m.r. structural analysis as well as mass spectroscopic measurements:

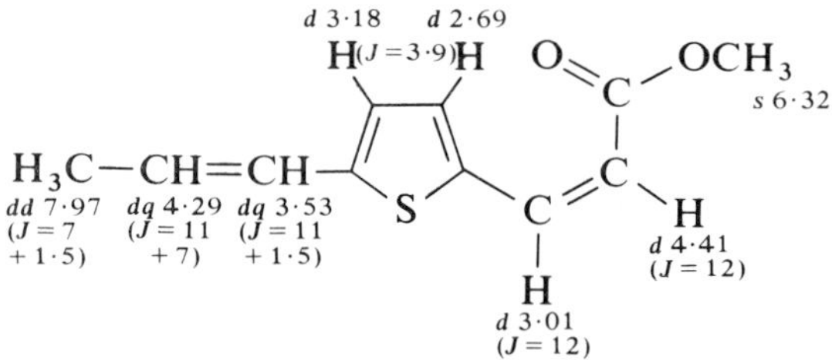

535

The downfield shift in **535** for one of the thiophene protons must be due to the carbonyl group of the ester grouping. The obtained sulfones could be separated by chromatography. Their structures were elucidated by n.m.r. spectroscopy. In view of these data the original thioethers have the structure **536** and **537**:

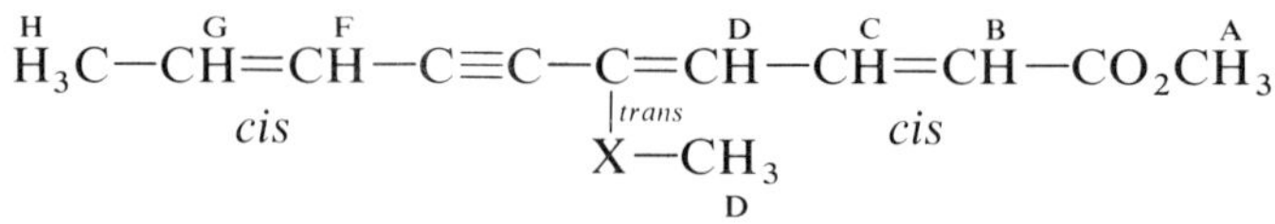

536 X = S

n.m.r. signals:

	536: X = S		**538**: X = SO_2
H_A	*s* 6·32	H_A	*s* 6·29
H_B	*d* 4·20 (J_{BC} = 11)	H_B	*d* 4·07 (J_{BC} = 11)
H_C	*dd* 2·91 (J_{BC} = 11, J_{CD} = 12)	H_C	*dd* 2·13 (J_{BC} = 11, J_{CD} = 12·5)
H_D	*d* 2·23 (J_{CD} = 12)	H_D	*d* 1·80 (J_{CD} = 12·5)
H_E	*s* 7·60	H_E	*s* 6·99
H_F	*dq* 4·3 (J_{FG} = 11, J_{FH} = 1·5)	H_F	*dq* 4·32 (J_{FG} = 11, J_{FH} = 1·5)
H_G	*dq* 3·9 (J_{FG} = 11, J_{GH} = 6·5)	H_G	*dq* 3·83 (J_{FG} = 11, J_{GH} = 6·5)
H_H	*dd* 8·11 (J_{GH} = 6·5, J_{FH} = 1·5)	H_H	*dd* 8·01 (J_{GH} = 6·5, J_{FH} = 1·5)

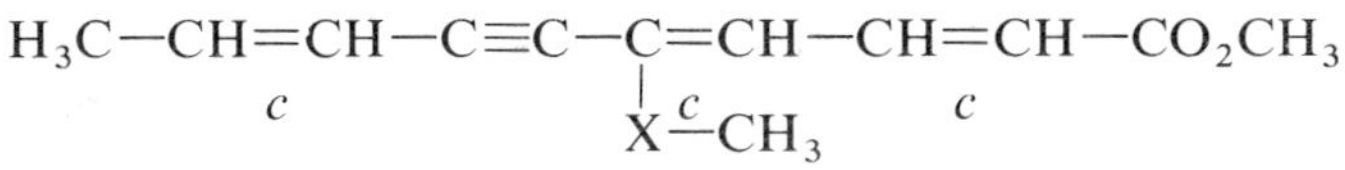

537: X = S

539: X = SO_2

536 and **537** obviously are methyl mercaptan addition products of **509**. The proposed structures and configurations have been confirmed by synthesis, using the route outlined in the scheme below[275]:

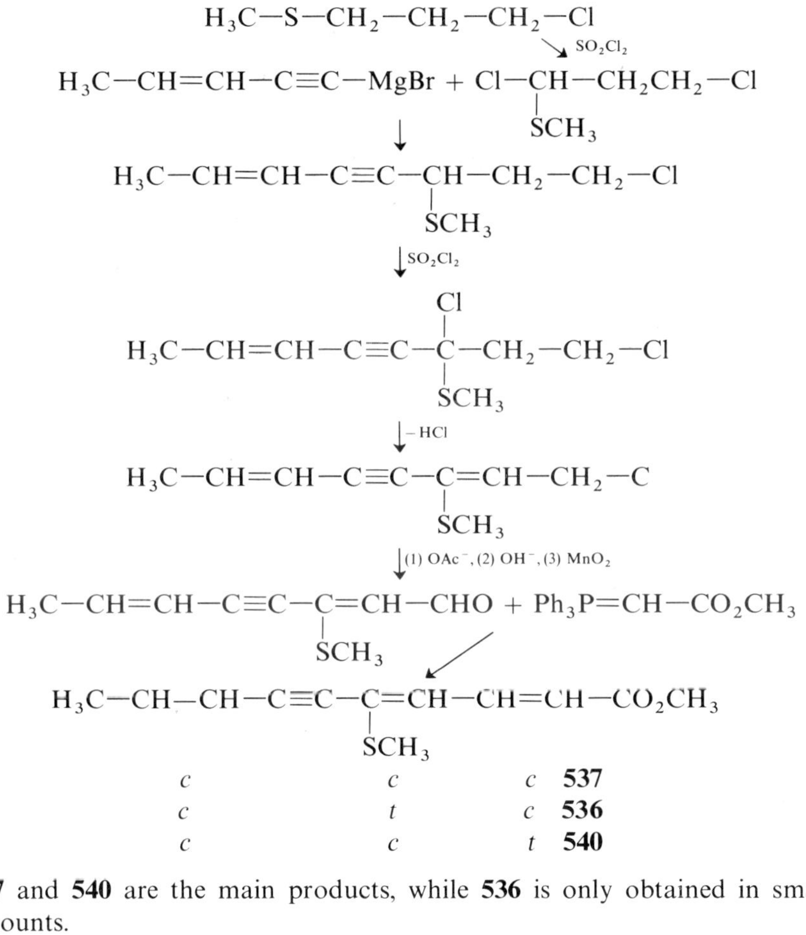

c	*c*	*c*	**537**
c	*t*	*c*	**536**
c	*c*	*t*	**540**

537 and **540** are the main products, while **536** is only obtained in small amounts.

7.3. Compounds Related to Lachnophyllum Ester

As mentioned in the introduction lachnophyllum ester is the longest known natural diacetylenic compound. It was isolated in 1935 by Wiljams, Smirnow and Goljmow[6] from *Lachnophyllum gossipinum*, a member of the

Compositae. The ester can be regarded as a dihydro-matricaria ester, its one double bond having the *cis* configuration. Its structure was determined from analytical and degradative evidence and since the action of alkali produces maleic acid it is apparent that the double bond has the *cis*-configuration:

$$H_3C{-}CH_2{-}CH_2{-}(C{\equiv}C)_2{-}\underset{cis}{CH{=}CH}{-}CO_2CH_3 \quad \mathbf{541}$$

lachnophyllum ester

In 1954, Holme and Sörensen[291] reported the presence of *trans*-lachnophyllum ester (**542**) in the roots of *Bellis perennis* and identified it with a synthetic sample, prepared from pent-1-yne and *trans*-pent-2-en-4-yn-1-ol by a mixed coupling reaction[298]:

$$H_7C_3{-}C{\equiv}C{-}H + H{-}C{\equiv}C{-}\underset{t}{CH{=}CH}{-}CH_2OH$$

$$\downarrow Cu_2Cl_2/O_2$$

$$H_7C_3{-}C{\equiv}C{-}C{\equiv}C{-}\underset{t}{CH{=}CH}{-}CH_2{-}OH$$

$$\downarrow CrO_3,\ CH_2N_2$$

$$\mathbf{542}$$

The 8-hydroxy-lachnophyllum ester (**543**) is shown to occur in *Aster radula*[299]. The angelica ester of **543** is shown to occur with **541** in *Aster Novi Belgii* L.[300]:

$$H_3C{-}CH_3{-}\underset{OH}{\underset{|}{CH}}{-}C{\equiv}C{-}C{\equiv}C{-}CH{=}CH{-}CO_2CH_3 \quad \mathbf{543}$$

The structure of the angelica ester **544** is proved by n.m.r. structural analysis as well as mass spectroscopic measurements and finally has been confirmed by synthesis[298,301]:

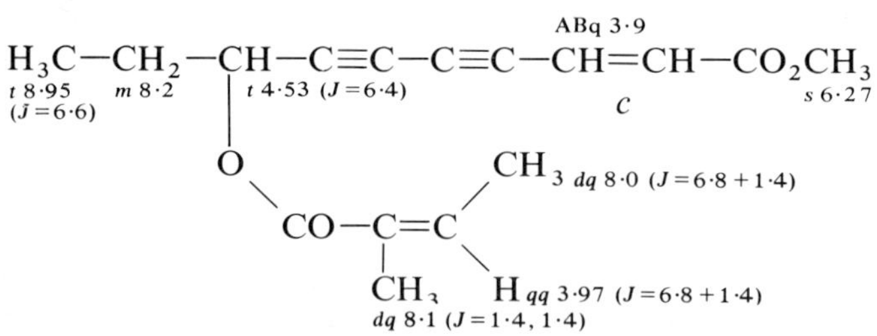

544

MS:

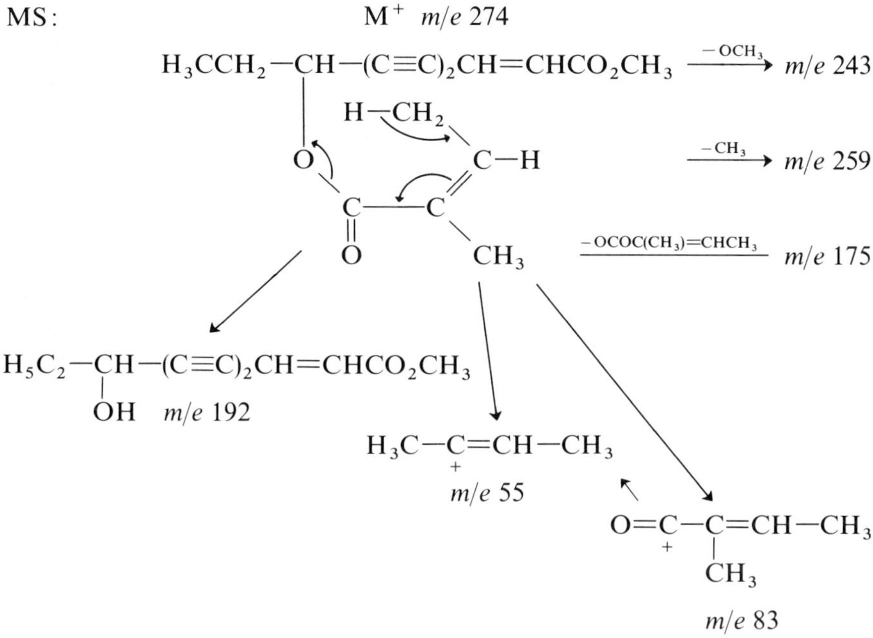

The constituents of *Aster Novi Belgii* vary seasonally. In Spring **544** has been mainly found, while in Autumn matricaria ester instead of **544** has been isolated. Therefore it has been proposed that there exists a direct biosynthetic route from Lachnophyllum ester to matricaria ester as outlined in the scheme below:

$$H_3C{-}CH_2{-}CH_2(C{\equiv}C)_2{-}CH{=}CH{-}CO_2CH_3$$

$$\downarrow [O]$$

$$H_3C{-}CH_2{-}\underset{\displaystyle OR}{\underset{|}{CH}}{-}(C{\equiv}C)_2{-}CH{=}CH{-}CO_2CH_3$$

$$\downarrow -ROH$$

$$H_3C{-}CH{=}CH{-}(C{\equiv}C)_2{-}CH{=}CH{-}CO_2CH_3$$

However, feeding experiments with labelled lachnophyllum ester and 8-hydroxy-lachnophyllum ester gave negative results. Further tracer studies have shown that the propenyl-grouping is formed directly by a dehydrogenation step[302]:

$$H_3C{-}CH_2{-}CH_2{-}C{\equiv}C{-} \xrightarrow{-H} H_3C{-}CH{=}CH{-}C{\equiv}C{-}$$

An aldehyde with a conjugated *trans* double bond, closely related to lachnophyllum ester, has been isolated from the aerial parts of *Grindelia humilis*

Hook.[292]. The structure of the aldehyde was elucidated by n.m.r. structural analysis. These data lead to structure **545**:

$$\underset{F}{H_3C}-\underset{E}{CH_2}-\underset{D}{CH_2}-(C{\equiv}C)_2-\underset{C}{CH}{=}\underset{B}{CH}-\underset{A}{CHO} \quad \mathbf{545}$$

The n.m.r. data as measured are:

	A	B	C	D	E	F
CCl_4	*dd* 0·46 J = 5 + 2·5	ABq 3·51		*t* 7·63 J = 6·7	*tq* 8·37 J = 7 + 6·7	*t* 8·96 J = 7
$(D_3C)_2CO$	*d* 0·36 J = 7·5	*dd* 3·5 J = 15·5 + 7·5	*dt* 3·09 J = 15·5 + 0·7	*t* 7·57 J = 6·7	*tq* 8·41 J = 7 + 6·7	*t* 9·01 J = 7

Trans-lachnophyllol (**546**) is shown to occur in small amounts in the roots of *Callistephus chinensis*[299]. **546** is cooccurring with the acetate **547** and the angelica ester **548** in the same plant. The senecio ester **549** has been isolated from the roots of *Grindelia camporum* Greene[292]:

$$H_3C-CH_2-CH_2-(C{\equiv}C)_2-CH\underset{t}{=}CH-CH_2OH \quad \mathbf{546}$$

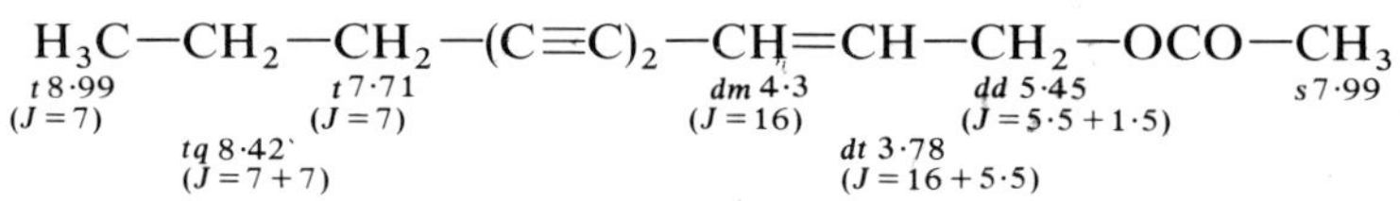

547

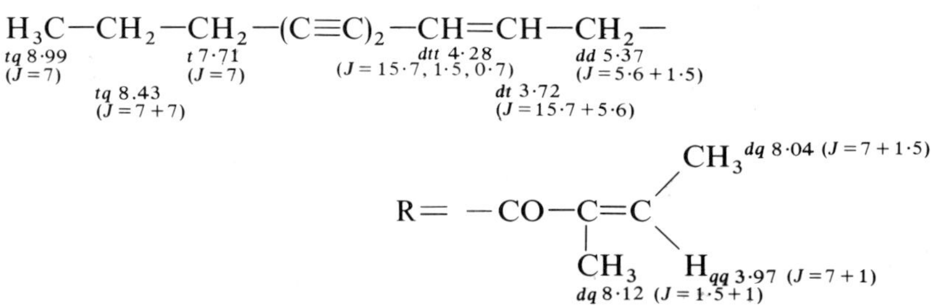

548

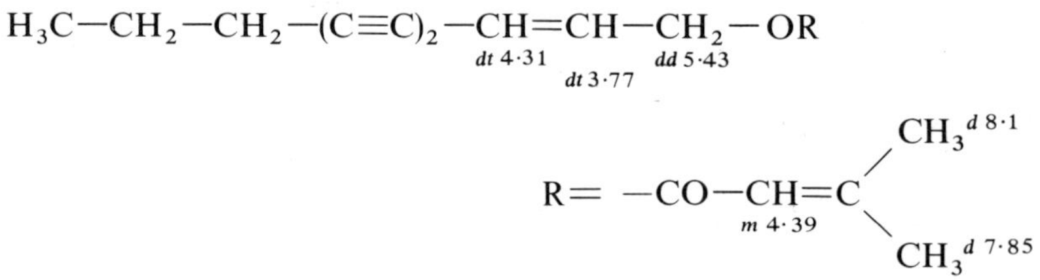

549

The lactone **550**, occurring in many *Aster*, *Boltonia* and *Conyza* species[112] may well arise from lachnophyllum ester:

541 → 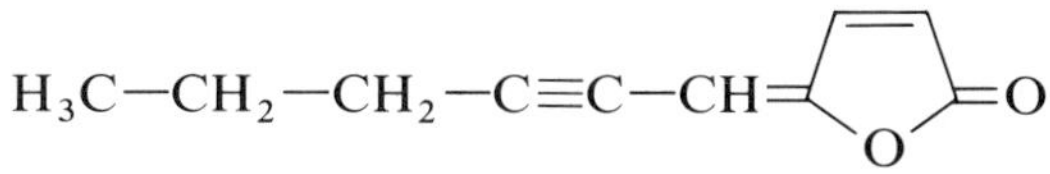 **550**

A cumulene, closely related to **531**, recently has been isolated from the aerial parts of *Erigeron canadensis* L.[303]. Its physical and spectroscopic properties resemble **531**. Its u.v. spectrum shows a maximum at 337 nm. The n.m.r. data, and in particular the mass spectrometric fragmentation pattern, which shows significant differences to **550**, strongly indicates that the new cumulene is a derivative of lachnophyllum ester and has structure **551**:

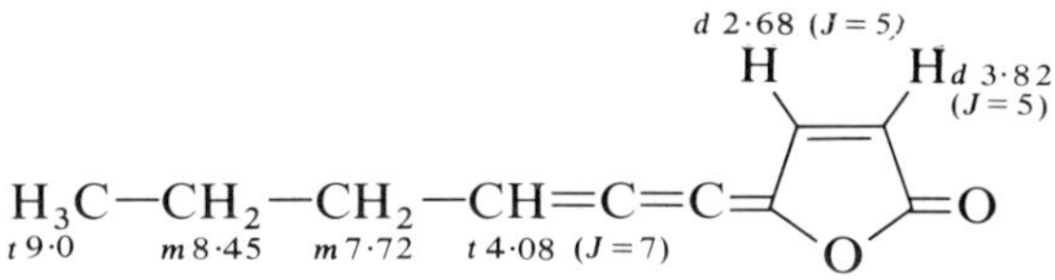

551

The mass spectrum of **551** shows a prominent peak at *m/e* 120 obviously formed by a cyclic fragmentation, while **550** is showing a base peak at *m/e* 82:

551 M^+ *m/e* 162 (70%) $\xrightarrow{-C_2H_5}$ *m/e* 133· $\xrightarrow{-CO}$ *m/e* 105

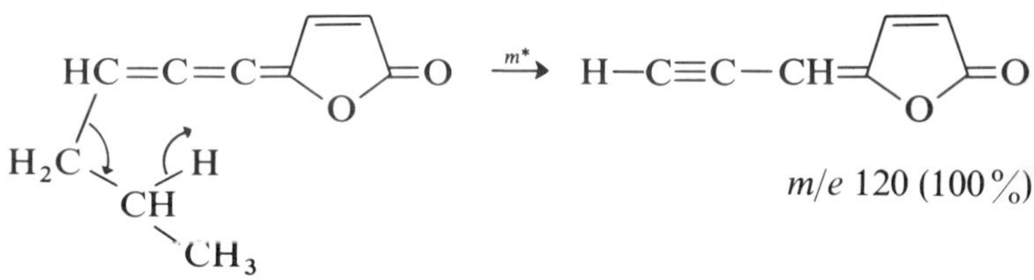

550 M^+ *m/e* 162 (80%)

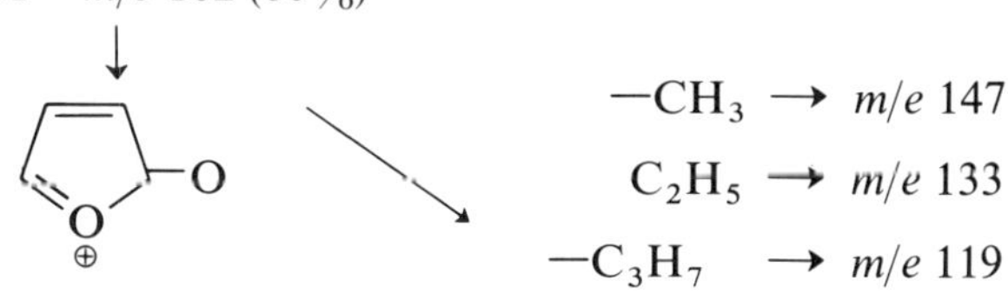

m/e 82 (100%)

7.4. Compounds Related to C_{10}-2,6,8-trien-4-yne

Matricaria ester **509**, lachnophyllum ester **541** and matricarianol acetate **520** are very widespread in plants belonging to the tribus Astereae. In some

Grindelia species, however, they only occur in small amounts. From *Grindelia robusta* Nutt.[304], a new acetate has been isolated which shows a u.v. spectrum at 307 nm. An alcohol with the same u.v. spectrum was also obtained from the same plant. Finally an aldehyde, too, with λ max at 342 nm. Saponification of the acetate yields an alcohol, identical in all respects with the natural material. Manganese dioxide oxidation of the latter furnishes an aldehyde, identical with the natural one.

According to its elemental analysis, the alcohol has the composition $C_{10}H_{12}O$. I.r. and u.v. spectrum indicate an en-yn-diene chromophore. Such a compound with a *cis*-double bond (**552**) has been already described by Sörensen[9]:

$$\underset{trans}{H_3C{-}CH{=}CH}{-}C{\equiv}C{-}\underset{cis}{CH{=}CH}{-}\underset{trans}{CH{=}CH}{-}CH_2OR \quad \mathbf{552}$$

The assumption, that the isolated compound would be the all-*trans* isomer of **552** was refuted, as synthesised all-*trans* **552** did not show any resemblance to the natural material.

The natural alcohol was finally assigned structure **553**, largely on the basis of n.m.r. spectroscopic evidence. The n.m.r. spectrum of **552** shows a double doublet at τ 8·2 arising from the $CH_3{-}CH{=}CH{-}C{\equiv}$-grouping, while the methyl group of the natural alcohol gives rise to a complex multiplet caused by the coupling with the diene grouping. Taking these facts into consideration, structure **553** is most likely:

$$H_3C{-}\underset{t,t}{(CH{=}CH)_2}{-}C{\equiv}C{-}CH{=}CH{-}R \quad \begin{array}{l}\mathbf{553}: R = CH_2OH\\ \mathbf{554}: R = CHO\\ \mathbf{555}: R = CH_2OAc\end{array}$$

The corresponding angelica ester has been isolated from the roots of *Gutierrezia gymnospermoides*[299]. It also occurs in the plant with **553** and **555** and its spectral data points to structure **556**:

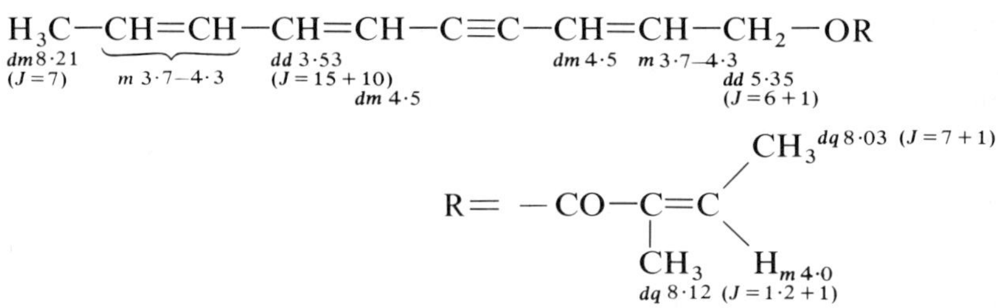

556

$$M^+ \; m/e\ 230 \xrightarrow{-CH_3} m/e\ 215$$

$$\xrightarrow{-COR} m/e\ 147 \xrightarrow{-CO} m/e\ 119$$

$$\longrightarrow m/e\ RCO^+ \; m/e\ 83\ (100\%)$$

553 is the alcohol component of a senecio ester, which is shown to occur together with **524** in the aerial parts of *Grindelia humilis* Hook.[292]. Its spectral properties are consistent with structure **557**:

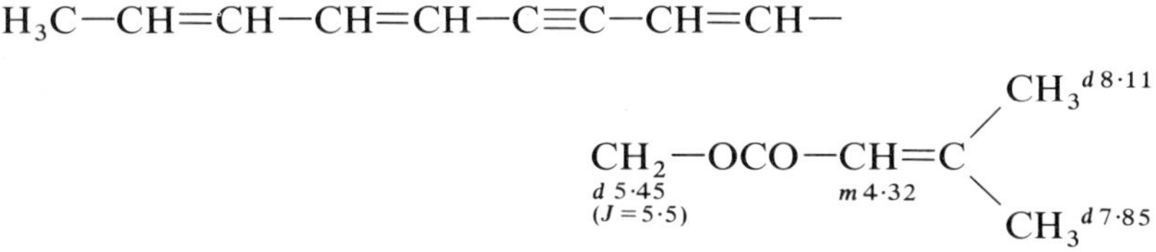

557

The aerial parts of *Matricaria caucasica* Poir. have been found to contain a lactone shown to be **558** on the basis of chemical and spectroscopic evidence[48]:

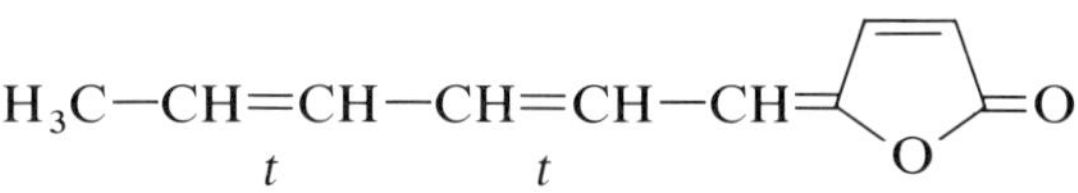

558

558 already has been obtained by partial hydrogenation and u.v. isomerisation of the cumulene **531** (see Section 7.2).

7.5. More or Less Hydrogenated C_{10}-acetylenic Compounds

After removal of 2-*trans*-8-*cis* matricaria ester from *Matricaria inodora* and *Amellus strigosus* extracts extensive chromatography of the mother liquors gave a mixture of acetylenic esters, the absorption spectrum of which indicated the presence of an en-diyne chromophore[287]. One compound was finally separated as an oil from which a pure acid (m.p. 64·5°C) was obtained after saponification under nitrogen at 0°C. It was shown to be a dihydro-matricaria ester (**559**) by comparison with the synthetic *cis*- and *trans*-forms. Pent-2-en-4-yne was coupled with methyl pent-4-ynoate and the asymmetrical product saponified[305]. The *trans* acid melted at 144°C but the *cis*-isomer melted at 60·5°C and was identical with that from the natural product which must therefore be 8-*cis*-2,3-dihydro matricaria ester **559**, biogenetically formed by hydrogenation of **509** as shown by feeding experiments:

$$H_3C{-}CH{=}CH{-}C{\equiv}C{-}H + H{-}C{\equiv}C{-}CH_2{-}CH_2{-}CO_2CH_3$$

$$\downarrow$$

$$\underset{cis}{H_3C{-}CH{=}CH}{-}C{\equiv}C{-}C{\equiv}C{-}CH_2{-}CH_2{-}CO_2CH_3 \quad \mathbf{559}$$

Furthermore the presence of **559** has been detected in *Cephalophora aromatica* and *Felicia tenella*.

A mixture of two esters was shown to occur with matricaria and lachnophyllum ester in the roots of *Bellis perennis* L.[299]. The structures of the esters **560** and **561** were assigned on the basis of n.m.r. and mass spectra:
and natural ester:

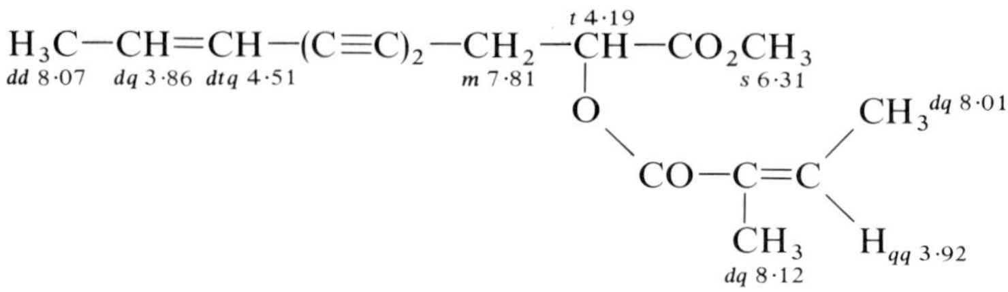

560

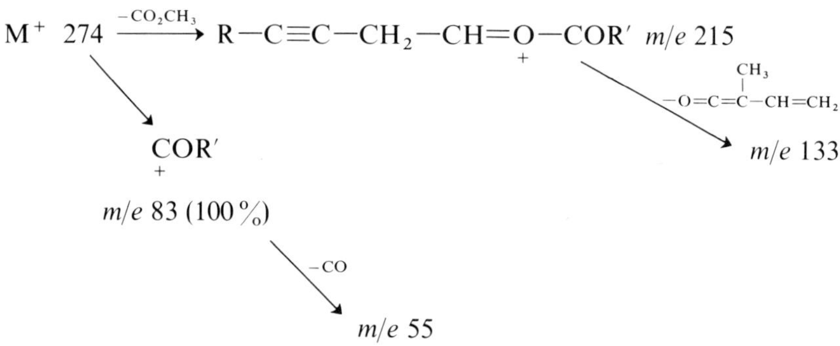

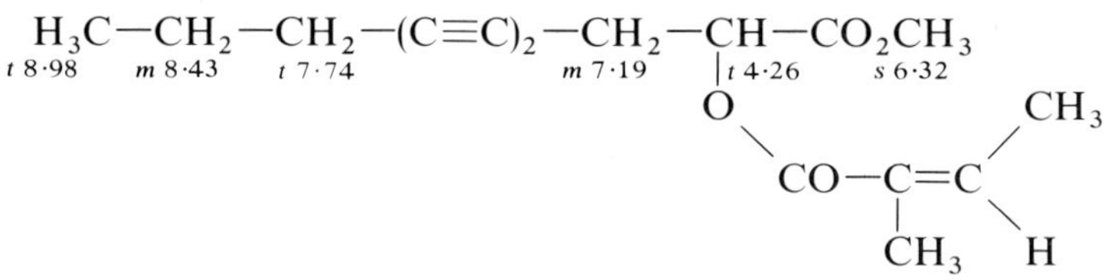

561

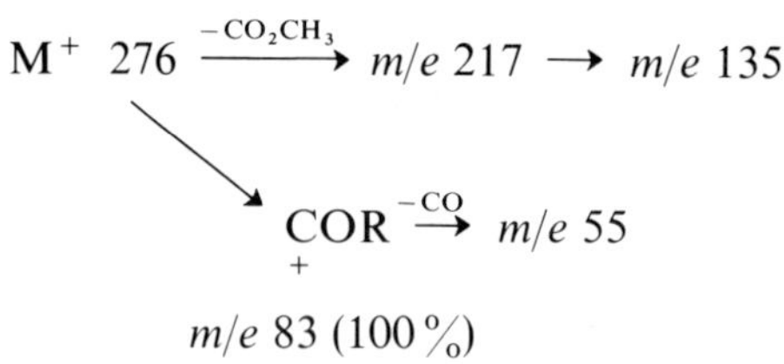

The position of the angeloyl residue was not established; therefore, it seemed desirable to prove the structures by synthesis[306]. An alternative structure would be **562** and **563**, respectively. The route chosen for synthesis of **561** involved preparation of the diester **565**, which was obtained by treatment of 2-hydroxy-pent-4-ynoate **564** with angeloyl-chloride. Coupling of the diester with pent-1-yne yielded the presumed ester **561**. The n.m.r. spectrum of the latter clearly showed the difference between the synthesised and natural ester:

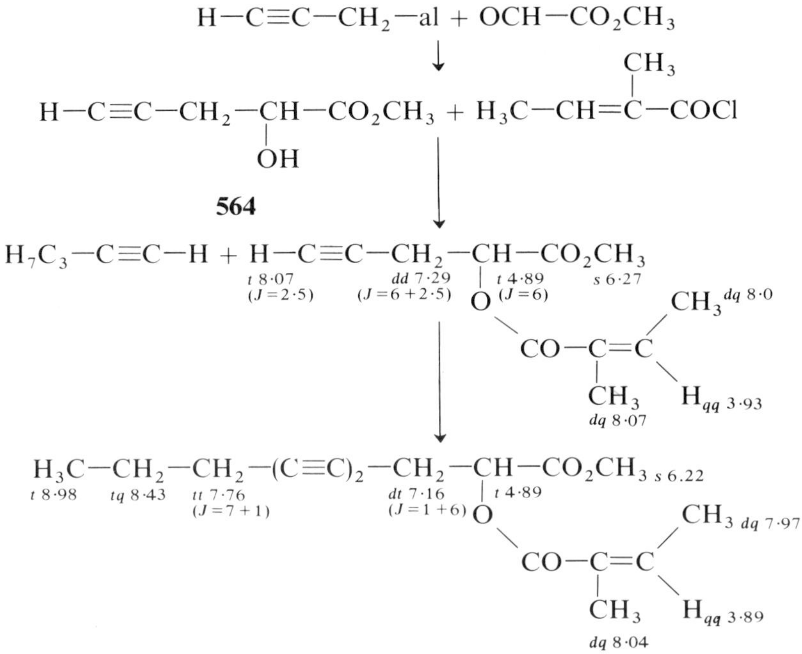

564

565

561

The alternative structure **563** was proved by synthesis in the following way:

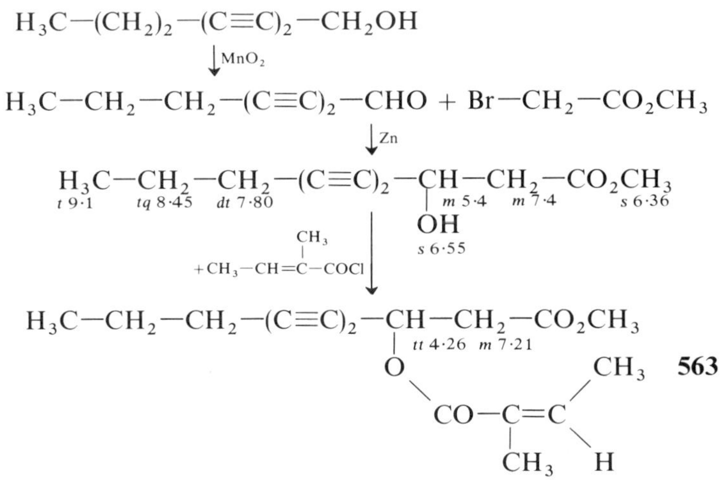

563

The ester **563** obtained was identical in all spectral and chemical properties with the natural compound. Therefore, the first proposed structure of the natural compound **561** is wrong, and so presumably is the structure **560**, which has to be replaced by **562** as the right one with the angeloyl residue at C_3:

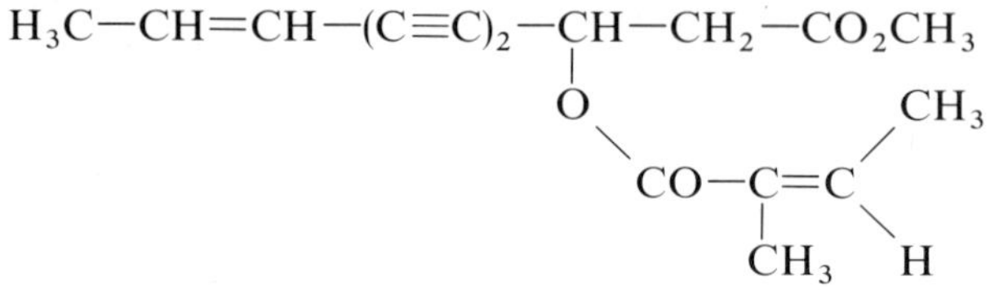

562

The alcohol **566** and the corresponding acetate **567**, both closely related to **559**, have been isolated from the aerial parts of *Chrysanthemum maximum* Ramond[202]. I.r. and u.v. spectra indicate the presence of a *cis*-ene-diyne-grouping and the n.m.r. spectrum confirms the proposed structures:

$H_3C{-}CH{=}CH{-}(C{\equiv}C)_2{-}CH_2{-}CH_2{-}CH_2{-}OR$ **566**: R = H; **567**: R = $COCH_3$

H_3C: *dd* 8·09 (J=6·8+1·2); CH: *dq* 3·96 (J=11+6.8); CH: *d* 4·55 (J=11); CH_2: *t* 7·56; CH_2: quintet 8·26; CH_2: *t* 6·32

The aerial parts of *Carthamus tinctorius* L.[294] have been found to contain a further derivative of **566**, shown to be **568** on the basis of chemical and spectroscopic evidence:

$H_3C{-}CH{=}CH{-}(C{\equiv}C)_2{-}CH_2{-}CH_2{-}CH_2{-}OCO{-}CH_2{-}CH(CH_3)_2$

dd 8·08, *dq* 3·95, *dm* 4·55, *t* 7·57, *m* 8·12, *t* 5·87, *m* 7·89, *m* 8·45, *d* 9·04

568

The finding of **568** in *C. tinctorius* L. is very interesting, as plants of the tribus Cynarea are usually lacking C_{10}-compounds.

An ester containing an en-yn-ene chromophore has been isolated from *Aster schreberi* Ch. G. Nees.[194]. The structure **569** is proved by n.m.r. structural analysis as well as mass spectroscopic measurements:

$H_3C{-}CH_2{-}CH_2{-}CH{=}CH{-}C{\equiv}C{-}CH{=}CH{-}CO_2CH_3$ **569**

H_3C: *t* 9·07; CH_2: *tq* 8·47; CH_2: *ddt* 7·88 (J=7,7+1·2); CH: *dt* 3·80 (J=15·5+7); CH: *ddt* 4·43 (J=15·5,1·2,2·2); CH: *dd* 3·32 (J=15·5+2·2); CH: *d* 3·92 (J=15·5); CH_3: *s* 6·31

$$M^+\ 178\ m/e \xrightarrow{-C_2H_5} m/e\ 149$$
$$\xrightarrow{-OCH_3} m/e\ 147$$
$$\xrightarrow{-CO_2CH_3} m/e\ 119$$

The same chromophoric system (en-yn-ene) has been found to occur in three compounds, which has been isolated from the aerial parts of *Grindelia humilis* Hook.[292]. The three closely related compounds were shown to be the aldehyde **570**, the alcohol **571** and the corresponding acetate **572**:

$$H_3C{-}CH_2{-}CH_2{-}CH{=}CH{-}C{\equiv}C{-}CH{=}CH{-}CHO \quad \mathbf{570}$$

t 9·05 *tq* 8·51 *dt* 7·82 *dt* 3·76 *dm* 4·36 *dd* 3·38 *d* 3·60 *d* 0·48

$$H_3C{-}CH_2{-}CH_2{-}CH{=}CH{-}C{\equiv}C{-}CH{=}CH{-}CH_2OH \quad \mathbf{571}$$

dm 4·20 *m* 3·80 *d* 5·85 *s* 6·80

$$H_3C{-}CH_2{-}CH_2{-}CH{=}CH{-}C{\equiv}C{-}CH{=}CH{-}CH_2OCOCH_3 \quad \mathbf{572}$$

dm 4·26 *dt* 3·95 *d* 5·46 *s* 7·99

The 9,10-dihydro derivative of **568** has been isolated from *Carthamus tinctorius* L. too[292]. Its spectroscopic properties are consistent with the proposed structure **573**:

$$H_3C{-}CH_2{-}CH_2(C{\equiv}C)_2{-}CH_2CH_2CH_2OCO{-}CH_2{-}CH(CH_3)_2 \quad \mathbf{573}$$

Careful saponification of **573** yielded the alcohol **574**, the structure of which was established by spectral methods:

$$H_3C{-}CH_2{-}CH_2{-}(C{\equiv}C)_2{-}CH_2{-}CH_2{-}CH_2{-}OH \quad \mathbf{574}$$

t 9·0 *tq* 8·45 *t* 7·79 *t* 7·66 *tt* 8·29 *t* 6·36 *s* 6·73

A conjugated acetylene ketone, closely related to **574**, has been isolated from *Erodiophyllum elderi* F. Müller[200]. Its u.v. spectrum shows maxima at 282, 267 and 253 nm. The i.r. spectrum suggests the presence of an acetylene ketone (2245, 1680/cm) and, in addition, a hydroxyl group (3645/cm). The n.m.r. spectrum of the compound is consistent with the proposed structure **575**:

$$H_3C{-}CH_2{-}CO{-}(C{\equiv}C)_2{-}CH_2{-}CH_2{-}CH_2{-}OH \quad \mathbf{575}$$

t 8·87 *q* 7·45 *t* 7·50 *tt* 8·22 *t* 6·31 *s* 8·3

The structure of **575** has been finally established by synthesis. The route chosen involved Cadiot–Chodkiewicz coupling of 1-bromo-pent-1-yn-3-ol with the tetrahydro-pyranyl ether **576**. Cleavage of the pyranyl ether and manganese dioxide oxidation of the propargylic alcohol yielded the ketone, the spectroscopic properties of which were identical with the natural product:

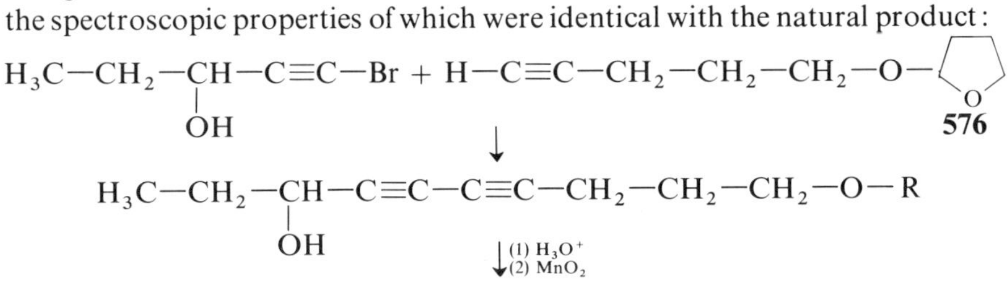

576

575

Until now, it is an open question whether the partial hydrogenated C_{10}-compounds arise from distinct C_{18}-acids with the corresponding chromophore, or from matricaria and lachnophyllum ester, which must be hydrogenated by specific enzymes.

8. Further Acetylenes from Micro-organisms

As mentioned in Chapter 2, Section 5.4, polyacetylenic compounds derived from micro-organisms differ in a special way from those obtained from higher plants. Those occurring in plants are relatively nonpolar (usually hydrocarbons, esters and acetates), while the micro-organism polyacetylenes are generally more polar (monohydric and polyhydric alcohols and acids). It is a remarkable feature of polyacetylenes obtained from fungi that the presence of any oxygen function next to a triple bond is not precluded, as it obviously is in the case in plants.

In this chapter, the compounds described are mainly closely related to matricaria ester and lachnophyllum ester, respectively, both of which are known as important and widespread constituents of higher plants (see Section 3.7). The most important biosynthetic relationships have already been described in Chapter 2, Section 5.4; they are more or less valid for the following compounds too.

The fungus *Polyporus anthracophilus* produces a complex and variable mixture of several acetylenic compounds, all of which are closely related[307]. The most important acetylene derivative produced is matricaria ester **577**—already described in Section 7.1, **509**. Against that the corresponding free acid **578** only occurs in comparatively small amounts:

$$H_3C{-}CH{=}CH{-}(C{\equiv}C)_2{-}CH{=}CH{-}CO_2R$$

577: R = CH_3 (**509**: 3.7.1)
578: R = H (**510**: 3.7.1)

In view of the spectroscopic evidence, they were identified as *trans-trans* matricaria ester and free matricaria acid. The principal constituent of rather old cultures (150 days) has been the corresponding *trans-trans* matricarianol **579**, the structure of which was confirmed by direct comparison with a synthetic sample and with material obtained by reduction of **577** with lithium aluminium hydride:

$$\underset{\qquad\qquad\quad t}{H_3C{-}CH{=}CH}{-}(C{\equiv}C)_2{-}\underset{t\qquad\qquad}{CH{=}CH}{-}CH_2OH$$ **579** (**514**: 3.7.1)

The hydroxy ester **580**, together with the corresponding acid **581**, has also been obtained from the culture. Both show up in the u.v. spectrum of a diyn-ene-carbonyl-hydroxy-chromophore. The i.r. spectrum of the former resembles those of *trans*-lachnophyllum ester, but, in addition, the compound shows absorbance in the i.r. due to a hydroxy group. Hydrogenation of the acid followed by chromic acid oxidation, gave sebacic acid, which proved the structure **580** for the natural ester. This was therefore synthesised by oxidative coupling of pent-4-yn-1-ol with methyl pent-*trans*-2-en-4-ynoate. The obtained product was identical with the natural ester and on hydrolysis gave the acid **581**:

$$HO{-}CH_2{-}CH_2{-}CH_2{-}C{\equiv}C{-}H + H{-}C{\equiv}C{-}CH{=}CH{-}CO_2CH_3$$

$$\downarrow$$

$$HOCH_2{-}CH_2{-}CH_2{-}(C{\equiv}C)_2{-}CH{=}CH{-}CO_2CH_3 \quad \mathbf{580}$$

$$\downarrow OH^-$$

$$HOCH_2{-}CH_2{-}CH_2{-}(C{\equiv}C)_2{-}CH{=}CH{-}COOH \quad \mathbf{581}$$

$$\downarrow (1)\ H_2;\ (2)\ CrO_3$$

$$HOOC{-}(CH_2)_8{-}COOH$$

The mixture of acids produced by *P. anthracophilus* have been found to contain the diacid **582**. Both the free acid and its dimethyl ester **583** occur in the cultures:

$$RO_2C{-}CH{=}CH{-}(C{\equiv}C)_2{-}CH{=}CH{-}CO_2R \quad \mathbf{582}: R = H; \quad \mathbf{583}: R = CH_3$$

The diester **583** also has been isolated as a metabolite of the Basidiomycete *Drosophila* (*Psathyrella*) *subatrata*[147].

The separation of another dimethyl ester from the above diester **583** was only achieved with some difficulty. Its i.r. spectrum shows only an $\alpha\beta$-unsaturated ester peak, but lacked an acetylenic band. Its u.v. spectrum resembles that of a polyene ester, but the intensity is much lower than would be expected. Such a lowering of the extinction coefficient is a characteristic effect of replacing a double by a triple bond. The proposed structure **584** has been finally confirmed by synthesis:

$$H_3CO_2C{-}\underset{t}{CH{=}CH}{-}C{\equiv}C{-}\underset{t}{CH{=}CH}{-}CO_2CH_3 \quad \mathbf{584}$$

The route chosen for synthesis is outlined below:

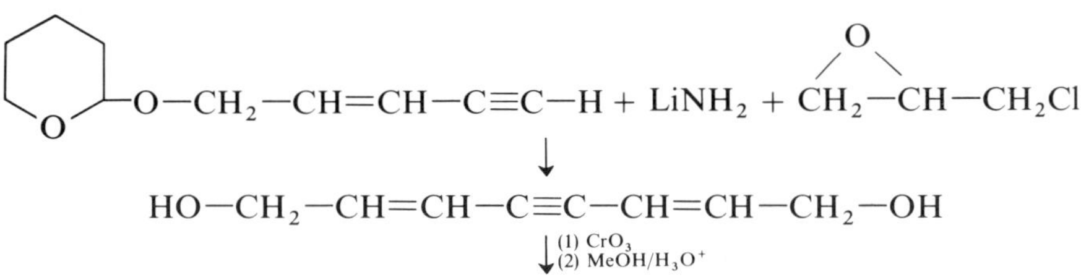

584

In **584** the triple bond is symmetrically placed so that its stretching vibration is i.r. inactive.

A further diacid shows an u.v. spectrum closely related to **581**. Comparison of the i.r. spectra, however, showed that the corresponding ester contains no hydroxyl group. On the other hand it contains two different ester groups (1740, 1720/cm). In view of these spectral results, it seems probable that the diacid has structure **585**, the corresponding diester **586**, respectively:

$$RO_2C{-}CH_2{-}CH_2{-}(C{\equiv}C)_2{-}CH{=}CH{-}CO_2R \quad \textbf{585}: R = H \quad \textbf{586}: R = CH_3$$

The nature of the mixed esters **583** and **588** was first deduced from their absorption spectra. The ester **587** gave an u.v. spectrum almost identical with that obtained by summation of the spectral curves for the methyl ester **577** and the alcohol **579**, whilst in its i.r. spectrum the intensity of the carbonyl band was nearly one-half of that in the spectrum of **577**. Similarly, the u.v. spectrum of the diester **588** could be duplicated by summation of the curves for the ester **577** and the hydroxy ester **580**:

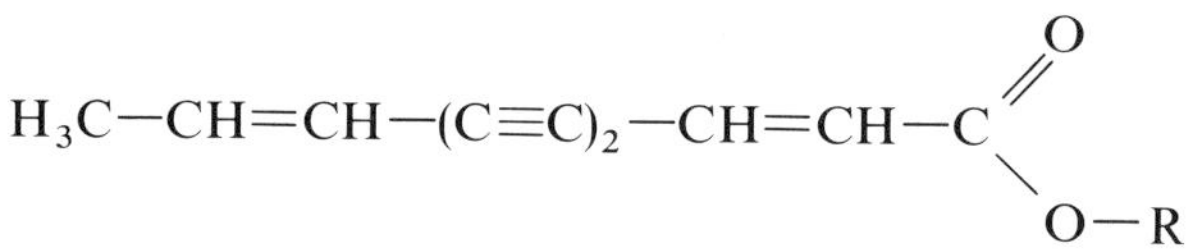

$$R = {-}CH_2{-}CH{=}CH{-}(C{\equiv}C)_2{-}CH{=}CH{-}CH_3 \quad \textbf{587}$$

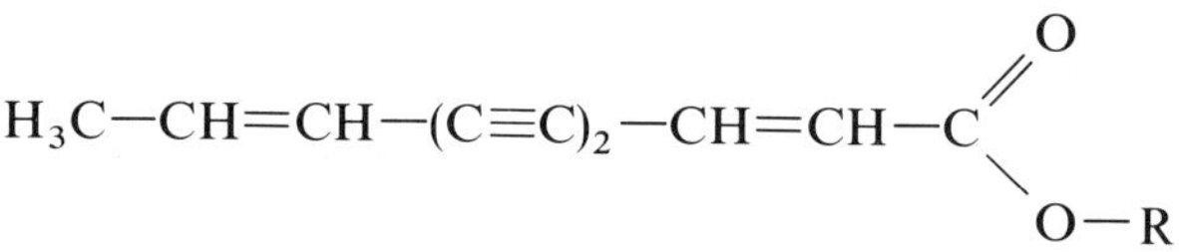

$$R = {-}(CH_2)_3{-}(C{\equiv}C)_2{-}CH{=}CH{-}CO_2CH_3 \quad \textbf{588}$$

Both structures were fully confirmed by synthesis of **587** and **588** from the chloride of the acid **578** and the hydroxy-compounds **579** and **580**, respectively.

There are indications for further compounds occurring in the fungus. A compound which behaves as an alcohol and showing an ene-diyne chromophore might well be **589**. Similarly a compound with an u.v. spectrum resembling those of **577** is probably the hydroxy ester **590**. These two compounds only occur in very small amounts, and hitherto it was impossible to confirm their structures:

$$HO{-}CH_2{-}CH_2{-}CH_2{-}(C{\equiv}C)_2{-}CH{=}CH{-}CH_2OH \quad \textbf{589}$$

$$HOCH_2{-}CH{=}CH{-}(C{\equiv}C)_2{-}CH{=}CH{-}CO_2CH_3 \quad \textbf{590}$$

A striking feature of the compounds described above is the *trans* or *trans-trans*-configuration of the double bonds, this contrasts sharply with the Compositae series where *cis* configuration is more common (see Section 7).

The Basidiomycete fungus *Fistulina hepatica* (Huds.) Fr. produces, *inter alia*, an alcohol, which gave *n*-tridecanol on catalytic hydrogenation[138]. Oxidation with manganese dioxide yielded a triyn-ene aldehyde, which had an i.r. band at 1115 cm typical of the —C≡C—CH=CH—CHO grouping. Oxidation with permanganate/periodate gave *n*-valeric acid. Thus, the compound must be **591** and this was confirmed by synthesis:

$$HO{-}CH_2{-}CH{=}CH{-}(C{\equiv}C)_3{-}CH_2{-}CH_2{-}CH_2{-}CH_3 \quad \mathbf{591}$$

The most polar fractions of extracts from the culture medium of *Drosophila* (*Psathyrella*) *subatrata*[147] have been found to contain the C_{11}-diacid **592**:

$$HO_2C{-}CH_2{-}CH_2{-}CH_2{-}(C{\equiv}C)_2{-}CH{=}CH{-}CO_2H \quad \mathbf{592}$$

To prove the structure, **592** was synthesised from hexynoic acid and *trans*-bromo-pentenynoic acid.

2-*cis*-8-*trans*-matricarianol **593**—also known from *Grindelia* species, see Section 7.1, **517**—has been found to occur in cultures of *Polyporus guttulatus*[153].

$$\underset{trans}{H_3C{-}CH{=}CH}{-}(C{\equiv}C)_2{-}\underset{cis}{CH{=}CH}{-}CH_2{-}OH \quad \mathbf{593} \quad (\mathbf{517}: 3.7.1)$$

An exception to the generalisation that polyacetylenic fungal metabolites containing an odd number of carbon atoms also have an ethynyl grouping (see Chapter 2, Section 5.4) is represented by the diol **594**, isolated from the Basidiomycete fungus *Poria sinuosa* Fr.[134]. Its u.v. spectrum shows an ene-diyne chromophore and hydrogenation followed by GLC showed it to be a C_9-diol. The ready conversion into a hydroxy aldehyde on oxidation with activated manganese dioxide showed that one alcoholic group was allylic. Then the diol could only be **594** and it was synthesised from but-3-yn-1-ol and *trans*-5-bromo-pent-2-en-4-yn-1-ol:

$$HO{-}CH_2{-}CH_2{-}C{\equiv}C{-}C{\equiv}C{-}CH{=}CH{-}CH_2{-}OH \quad \mathbf{594}$$

$$HO{-}(CH_2)_2{-}C{\equiv}C{-}H + Br{-}C{\equiv}C{-}CH{=}CH{-}CH_2OH$$

$$\downarrow$$

$$\mathbf{594} \xrightarrow{H_2} HO{-}(CH_2)_9{-}OH$$

$$\searrow MnO_2$$

$$HO(CH_2)_2{-}(C{\equiv}C)_2{-}CH{=}CH{-}CHO$$

The principal polyacetylenic metabolite of *Polyporus selecta* Karst. ex Rom. was the corresponding acid **595**[308]:

$$HO_2C{-}CH_2{-}(C{\equiv}C)_2{-}CH{=}CH{-}CH_2OH \quad \mathbf{595}$$

595 is a further exception to the generalisation that fungal polyacetylenes containing an odd number of carbon atoms possess a terminal ethynyl group.

Two further C_9-polyacetylenic metabolites have been found in cultures from *Clitocybe rhizophora* Velen[309]. The major constituent is a very stable acetylenic triol, shown to be a disubstituted polyhydroxy-diacetylene. Periodate cleavage yielded formaldehyde and a compound have a u.v. spectrum characteristic of a diyne-aldehyde; thus establishing the presence of a terminal 1,2-glycol grouping adjacent to a diacetylene grouping. The hydrogenated product, when cleaved with periodate followed by chromic acid oxidation, gave 6-oxo-octanoic acid, thereby confirming the structure **596**:

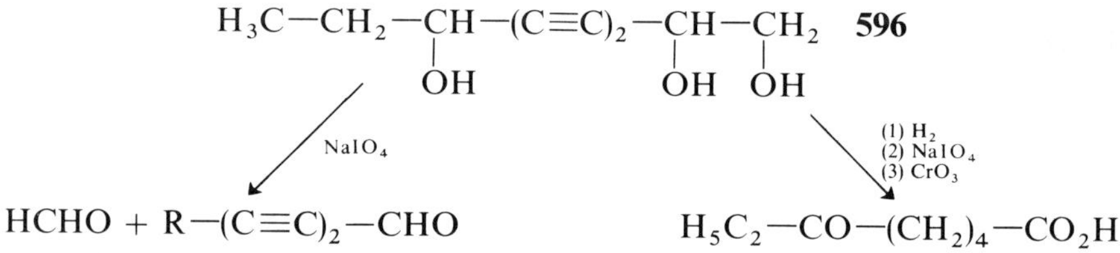

596 is insoluble in CCl_4, but when converted to the isopropylidene derivative **597**, it became soluble and permitted n.m.r. examination in CCl_4. The n.m.r. spectrum exhibits the presence of an ethylcarbinol group, confirmed by the data of pent-1-yn-3-ol:

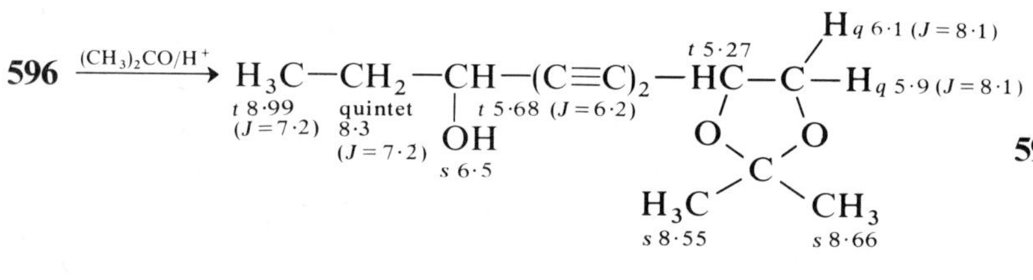

597

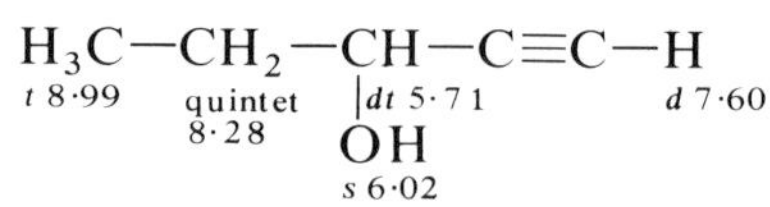

The resonance signals from the ketal ring clearly show the non-equivalence of the *gem*-dimethyl group and ketal ring protons. The *gem*-dimethyl protons give two lines and the ring protons form a 10-line ABX-system.

Closely related to **596** is the second compound obtained from *C. rhizophora.* It is the first polyacetylenic ketone to be isolated from a micro-organism. The ketone **598**, which has an intense u.v. spectrum (λ max: 282·5, 267, 253·5, 241 and 227·5 nm) was isolated and identified by way of its isopropylidene derivative **599**:

$$H_5C_2-CO-(C{\equiv}C)_2-\underset{OH}{\underset{|}{CH}}-\underset{OH}{\underset{|}{CH_2}} \xrightarrow{(CH_3)_2CO/H^+} \mathbf{599}$$

598

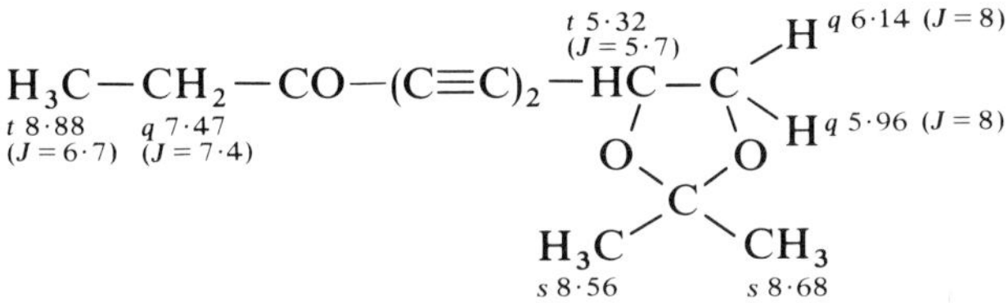

599

These data were confirmed through examinations of the spectrum of 1,2-*O*-isopropylidene-but-3-yn-1,2-diol **600**:

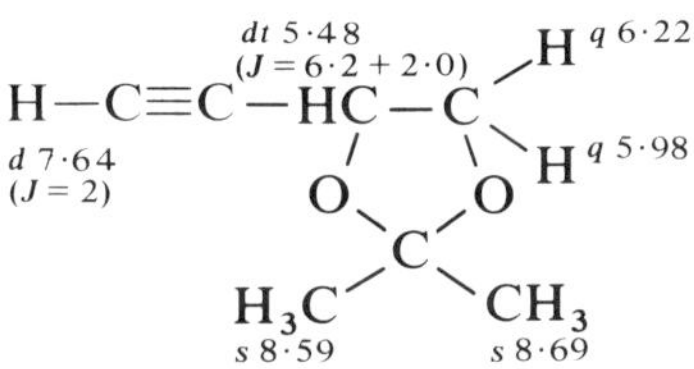

600

The two compounds are seen to be closely related:

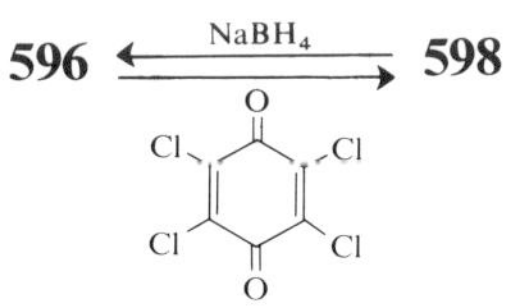

An epoxy-mono-ol (**601**) also has been detected in cultures from *Clitocybe rhizophora* Velen[311], its structure being elucidated by spectral methods and finally confirmed by synthesis:

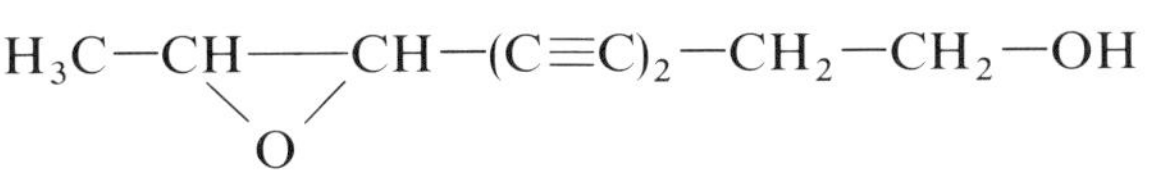

601

A hitherto unknown enediynol, possibly the keto-diyne-mono-ol (**602**) (u.v. spectrum) also has been found in very small amounts, but its structure was not established.

$$H_3C-CH_2-CO-(C{\equiv}C)_2-CH_2-CH_2-OH \quad \mathbf{602}$$

By feeding experiments with labelled diethyl malonate, methyl oleate, methyl linoleate, and methyl crepenynate, it could be shown[311] that all of them were readily incorporated into the triol **596** by cultures of *C. rhizophora*. For instance, the fungus incorporated [2-^{14}C]-malonate into the triol shown by the labelling pattern in **596a**. Almost equal activities at carbon atoms 1, 3, 5, and 7 suggest that the triol is formed from a C_{10}-precursor derived from an acetate primer and four malonate units followed by the loss of a carboxy-carbon atom:

$$H_3C{-}CH_2{-}\underset{\displaystyle OH}{\overset{*}{C}H}{-}C{\equiv}\overset{*}{C}{-}C{\equiv}\overset{*}{C}{-}\underset{\displaystyle OH}{CH}{-}\overset{*}{C}H_2{-}OH \quad \textbf{596a}$$

Furthermore, it could be shown that labelled lachnophyllum ester (**541**) (see Section 7.3) is a precursor of the triol **596**

$$\underset{\textbf{541}}{H_3C{-}CH_2{-}CH_2{-}(C{\equiv}C)_2{-}CH{=}CH{-}CO_2CH_3} \rightarrow \textbf{596}$$

These feeding experiments can be summarised in the following scheme; the last stages, however, of the scheme still need to be established:

$$H_3C{-}(CH_2)_7{-}CH{=}CH{-}(CH_2)_7{-}CO_2R$$

$$\downarrow$$

$$H_3C{-}(CH_2)_4{-}CH{=}CH{-}CH_2{-}CH{=}CH{-}(CH_2)_7{-}CO_2CH_3$$

$$\downarrow$$

$$H_3C{-}(CH_2)_4{-}C{\equiv}C{-}CH_2{-}CH{=}CH{-}(CH_2)_7{-}CO_2R$$

$$\downarrow$$

$$[H_3C{-}(CH_2)_2{-}CH{=}CH{-}C{\equiv}C{-}CH_2{-}CH{=}CH{-}(CH_2)_7{-}CO_2R]$$

$$\downarrow$$

$$[H_3C{-}(CH_2)_2{-}C{\equiv}C{-}C{\equiv}C{-}CH_2{-}CH{=}CH{-}(CH_2)_7{-}CO_2R]$$

$$\downarrow$$

$$[H_3C{-}CH_2{-}CO{-}C{\equiv}C{-}C{\equiv}C{-}CH_2{-}CH{=}CH{-}(CH_2)_7{-}CO_2R]$$

$$\downarrow$$

$$[H_3C{-}CH_2{-}CO{-}(C{\equiv}C)_2{-}CH{=}CH{-}CO_2R]$$

$$\downarrow$$

$$[H_3C{-}CH_2{-}CO{-}(C{\equiv}C)_2{-}CH(OH){-}CH_2{-}OH] \quad \textbf{598}$$

$$\downarrow$$

$$H_3C{-}CH_2{-}CH(OH){-}(C{\equiv}C)_2{-}CH(OH){-}CH_2OH \quad \textbf{596}$$

Finally, a few polyacetylenic compounds should be mentioned, all of which are obtained from fungi[22]. We are giving only the constitutional formula, since further details are lacking:

		Obtained from:
$H_7C_3{-}(C{\equiv}C)_3{-}CH{=}CH{-}\underset{OH}{\underset{\mid}{C}H}{-}\underset{OH}{\underset{\mid}{C}H_2}$	**603**	*Fistulina pallida* (Berk.a.Rav.)
$H_7C_3{-}(C{\equiv}C)_2{-}CH{=}CH{-}CO_2CH_3$	**604**	
$H_7C_3{-}(C{\equiv}C)_2{-}CH{=}CH{-}CH_2OH$	**605**	*Polyporus anthracophilus*
$H_3CCH{=}CH(C{\equiv}C)_2(CH_2)_2CO_2CH_3$	**606**	
$HOCH_2CH{=}CH(C{\equiv}C)_2CH{=}CHCH_2OH$	**607**	*P. anthracophilus*
$H_5C_2{-}\underset{OH}{\underset{\mid}{C}H}{-}(C{\equiv}C)_2{-}CH{=}CH{-}CH_2OH$	**608**	*Tricholoma grammo-podium*

Nearly all the C_{10}-acetylenes from fungi described above are more or less closely related to matricaria ester (**577**) and lachnophyllum ester (**604**), which also are very important precursors in the higher plants (see Sections 7.1 and 7.3).

The last two compounds can be derived from oleic acid (via linoleic acid) and crepenynic acid by further dehydrogenation and chain shortening.

A remarkable feature of the C_{10}-polyacetylenes from fungi is the predominant oxygen function at C_{10}. The relationships are cleared in the following scheme:

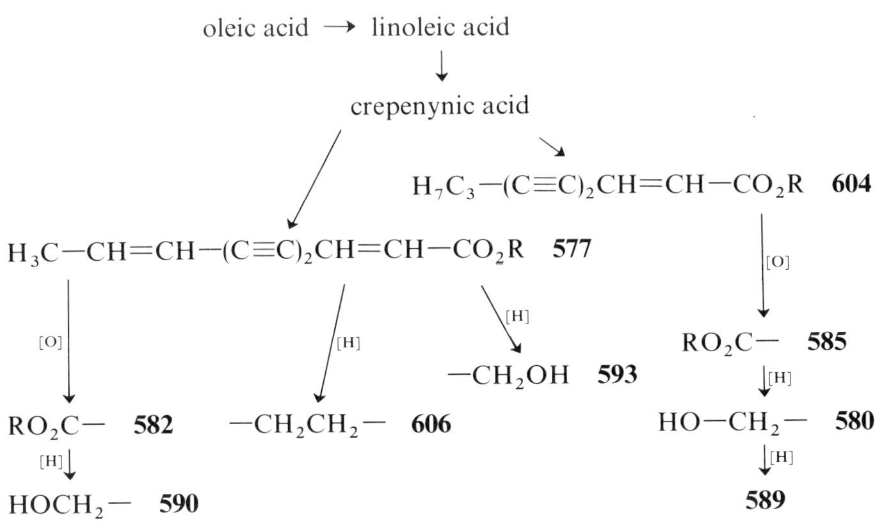

4

Acetylenes from Triglycerides

The occurrence of acetylenic fatty acids in seed oils of different origin had been already recognised in the last century. The first compound of this type was isolated in 1892 from the seed fats of the Guatemalan tree *Picramnia tariri* DC.—family Simarubaceae. Ten years later, the structure **1** for the so-called tariric acid could then be assigned on the oxidation to lauric and adipic acid[312]. The structure has been further established by synthesis[313].

$$H_3C{-}(CH_2)_{10}{-}C{\equiv}C{-}(CH_2)_4{-}CO_2H \quad \mathbf{1}$$

$$\downarrow [O]$$

tariric acid

$$H_3C{-}(CH_2)_{10}{-}CO_2H + HO_2C{-}(CH_2)_4{-}CO_2H$$

The work on tariric acid was contemporary with the first investigations of the Isano acids[314], its structures only being elucidated many years later (**11**–**18**). In the last twenty years, several further acetylenic fatty acids have since been isolated from the seed oils and from the lipids of the whole plants, most of them closely related to stearolic acid (**2**), which itself also has been isolated from the seed fats of *Pyrularia pubera* Michx.—family Santalaceae[315]:

$$H_3C{-}(CH_2)_7{-}C{\equiv}C{-}(CH_2)_7{-}CO_2H \quad \mathbf{2}$$

stearolic acid

A dehydro derivative of **2** is present in several members of the family Olacaceae. By means of oxidative degradation, the structure has been elucidated as **3**[316]:

$$H_3C{-}(CH_2)_5{-}CH{=}CH{-}C{\equiv}C{-}(CH_2)_7{-}CO_2H \quad \mathbf{3}$$

ximenynic acid

The corresponding hydroxy acid **4** is present together with **3** in the seeds of a *Ximenia* species[317], while in other parts of this plant and in related species, small amounts of fat are present which contain more highly unsaturated acids[318]. Especially rich sources of such acids are the seedlings of *Santalum accuminatum*, where the C_{18}-acids **5–9** have been identified[319]:

$$H_3C{-}(CH_2)_5{-}CH{=}CH{-}C{\equiv}C{-}\underset{\displaystyle OH}{\underset{|}{C}H}{-}(CH_2)_6{-}CO_2H \quad \mathbf{4}$$

$$H_3C{-}(CH_2)_3{-}(CH{=}CH)_2{-}C{\equiv}C{-}(CH_2)_7{-}CO_2H \quad \mathbf{5}$$

$$H_3C{-}(CH_2)_3{-}CH{=}CH{-}(C{\equiv}C)_2{-}(CH_2)_7{-}CO_2H \quad \mathbf{6}$$

$$H_5C_2{-}(CH{=}CH)_2{-}(C{\equiv}C)_2{-}(CH_2)_7{-}CO_2H \quad \mathbf{7}$$

$$H_5C_2{-}CH{=}CH{-}(C{\equiv}C)_3{-}(CH_2)_7{-}CO_2H \quad \mathbf{8}$$

$$H_2C{=}CH{-}CH{=}CH{-}(C{\equiv}C)_3{-}(CH_2)_7{-}CO_2H \quad \mathbf{9}$$

Structures **7**–**9** are not yet fully established since only the chromophoric systems are definite, the position of the unsaturated bonds being placed only by analogy. Although the structure of **3** could be established by synthesis, the structure of **7** has to be reinvestigated, as synthetic **7** was not identical with the natural dien-yn acid[320]. Primarily results indicate that the positions of the unsaturated bonds have to be shifted most probably to **10**, as on oxidation of the maleic anhydride adduct mainly suberic acid has been obtained:

$$H_3C{-}(CH_2)_2{-}\underset{t,t}{(CH{=}CH)_2}{-}(C{\equiv}C)_2{-}(CH_2)_6{-}CO_2H \quad \mathbf{10}$$

The fatty acids of the Isano oil previously mentioned have been reinvestigated by modern methods. It turns out that the fatty acid mixture consists of nine acetylenic acids (75 % total acids), 20 % "normal" acids and small amounts of *cis*-9,10-eopxy-stearic acid and 9,10-dihydroxy-stearic acid[321]. By chromatography of the corresponding methyl esters all compounds could be separated. Oxidative degradation led to the following structures:

$$H_3C{-}(CH_2)_3{-}\underset{cis}{CH{=}CH}{-}(C{\equiv}C)_2{-}(CH_2)_7{-}CO_2R \quad \mathbf{11}$$

$$H_2C{=}CH{-}(CH_2)_2{-}\underset{cis}{CH{=}CH}{-}(C{\equiv}C)_2{-}(CH_2)_7{-}CO_2R \quad \mathbf{12}$$

$$H_3C{-}(CH_2)_5{-}(C{\equiv}C)_2{-}(CH_2)_7{-}CO_2R \quad \mathbf{13}$$

$$H_2C{=}CH{-}(CH_2)_4{-}(C{\equiv}C)_2{-}(CH_2)_7{-}CO_2R \quad \mathbf{14}$$

isanic acid

$$H_3C{-}(CH_2)_5{-}(C{\equiv}C)_2{-}\underset{\displaystyle OH}{\underset{|}{C}H}{-}(CH_2)_6{-}CO_2R \quad \mathbf{15}$$

$$H_3C{-}(CH_2)_3{-}CH{=}CH{-}(C{\equiv}C)_2{-}\underset{OH}{\underset{|}{CH}}{-}(CH_2)_6{-}CO_2R \qquad \mathbf{16}$$

$$H_2C{=}CH{-}(CH_2)_4{-}(C{\equiv}C)_2{-}\underset{OH}{\underset{|}{CH}}{-}(CH_2)_6{-}CO_2R \qquad \mathbf{17}$$

$$H_2C{=}CH{-}(CH_2)_2{-}CH{=}CH{-}(C{\equiv}C)_2{-}\underset{OH}{\underset{|}{CH}}{-}(CH_2)_6{-}CO_2R \qquad \mathbf{18}$$

Only the structure of one of the isolated acids, **12**, is not elucidated. Perhaps **12** is identical with **6**, where the double configuration is not established. The old structure of the so-called isanolic acid (**19**)[322] must therefore be eliminated:

$$H_3C{-}(CH_2)_2{-}CH{=}CH{-}(C{\equiv}C)_2{-}CH_2{-}\underset{OH}{\underset{|}{CH}}{-}(CH_2)_6{-}CO_2H \qquad \mathbf{19}$$

Several further acetylenic fatty acids have been isolated from the seed fats from a member of the family Santalaceae-*Acanthosyris spinescens* Griseb. Using counter-current distribution, the corresponding methyl esters could be separated and the structures have been elucidated by spectral as well as by chemical methods. Together with the acids **3** and **4** already known, the following new ones could be identified[323]:

C_{18}:

$$H_2C{=}CH{-}(CH_2)_6{-}C{\equiv}C{-}(CH_2)_7{-}CO_2R \qquad \mathbf{20}$$

$$H_2C{=}CH{-}(CH_2)_4{-}CH{=}CH{-}C{\equiv}C{-}(CH_2)_7{-}CO_2R \qquad \mathbf{21}$$

$$H_2C{=}CH{-}(CH_2)_4{-}CH{=}CH{-}C{\equiv}C{-}\underset{OH}{\underset{|}{CH}}{-}(CH_2)_6{-}CO_2R \qquad \mathbf{22}$$

C_{17}:

$$H_3C{-}(CH_2)_5{-}CH{=}CH{-}C{\equiv}C{-}(CH_2)_6{-}CO_2R \qquad \mathbf{23}$$

$$H_3C{-}(CH_2)_5{-}CH{=}CH{-}C{\equiv}C{-}\underset{OH}{\underset{|}{CH}}{-}(CH_2)_5{-}CO_2R \qquad \mathbf{24}$$

$$H_2C{=}CH{-}(CH_2)_4{-}CH{=}CH{-}C{\equiv}C{-}(CH_2)_6{-}CO_2R, \qquad \mathbf{25}$$

$$H_2C{=}CH{-}(CH_2)_4{-}CH{=}CH{-}C{\equiv}C{-}\underset{OH}{\underset{|}{CH}}{-}(CH_2)_5{-}CO_2R \qquad \mathbf{26}$$

Interesting is the cooccurrence of C_{18}- and C_{17}-acids. The position of the unsaturated bonds clearly indicates that the C_{17}-acids are formed by α-oxidation of the C_{18}-acids.

Perhaps the most important acetylenic fatty acid is the so-called crepenynic acid (mentioned above) isolated first from the seed oil of *Crepis foetida* L.[185] (see Chapter 3, Section 1.1). By routine analysis of the fatty acid methyl esters, the main component, which shows the consumption of three moles of hydrogen, was not identical with linolenic acid methyl ester. Structure determinations of the relatively unstable compounds show the presence of an en-yn ester having structure **27**. The corresponding dehydro compound **28** has been isolated from micro-organism[324]. Further on the interesting hydroxy-acid **29** from the triglycerides of *Helichrysum brateatum* seeds is to mention[187] (see Chapter 3, Section 1.1):

$$\mathrm{H_3C{-}(CH_2)_4{-}C{\equiv}C{-}CH_2{-}\underset{\textit{cis}}{CH{=}CH}{-}(CH_2)_7{-}CO_2R} \qquad \mathbf{27}$$

$$\mathrm{H_3C{-}(CH_2)_2{-}\underset{\textit{cis}}{CH{=}CH}{-}C{\equiv}C{-}CH_2{-}\underset{\textit{cis}}{CH{=}CH}{-}(CH_2)_7{-}CO_2R} \qquad \mathbf{28}$$

$$\mathrm{H_3C{-}(CH_2)_4{-}C{\equiv}C{-}\underset{\textit{trans}}{CH{=}CH}{-}\underset{\displaystyle |\atop \displaystyle OH}{CH}{-}(CH_2)_7{-}CO_2R} \qquad \mathbf{29}$$

Sterculia species normally have the cyclopropene acid **30** as the main constituent of their fatty acids. *Sterculia alata* also contains an acetylenic derivative, its structure being **31**[325]:

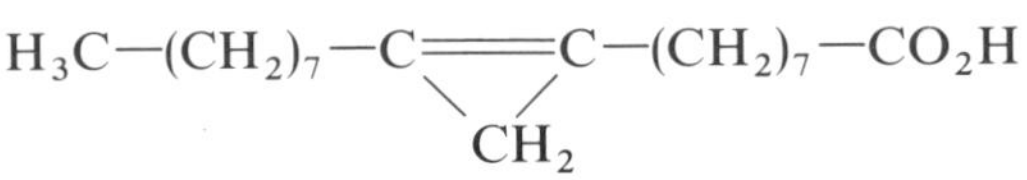 **30**

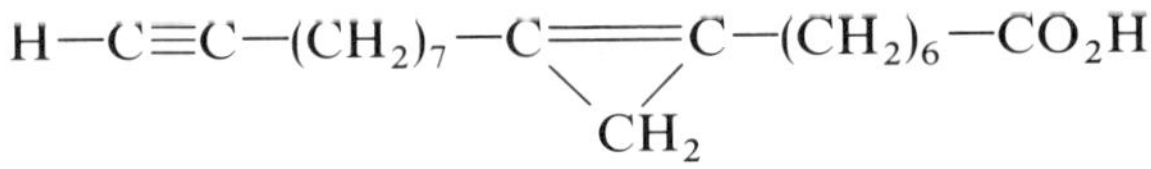 **31**

31 is the only acid of this type, but one further acid with an ethynyl endgroup is known in the C_{11}-compound **32**[326], probably formed by dehydrogenation of the cooccurring vinyl acid:

$$\mathrm{H_2C{=}CH{-}(CH_2)_8{-}CO_2H \xrightarrow{-[H]} H{-}C{\equiv}C{-}(CH_2)_8{-}CO_2H} \qquad \mathbf{32}$$

The biogenesis of the acetylenic fatty acids already known is not established. However, some common lines seem to be very probable if we consider the acids in separate groups. The most important scheme is probably the dehydrogenation pathway starting with oleic acid. Two different routes are possible. Firstly, the one already discussed, where oleic acid is transformed to crepenynic acid via linolic acid:

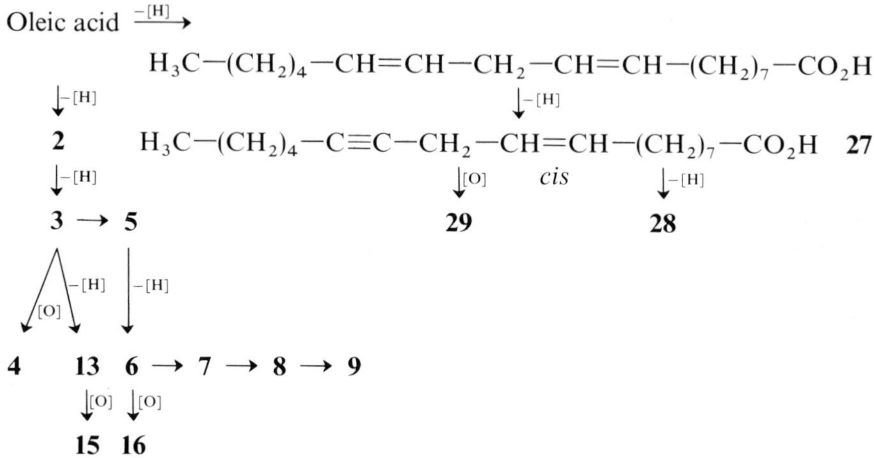

The second pathway is the direct dehydrogenation of oleic to stearolic acid (**2**) followed by further dehydrogenation to **3**, **5**, **6**, **7**, **8**, and **9**. Several of these acids are further transformed by oxidation in the propargylic position yielding the hydroxy-acids **4**, **15**, and **16**. Another modification of these pathways is the introduction of a vinyl endgroup, probably preceding by ω-oxidation and elimination of water:

$$H_3C{-}CH_2{-}R \xrightarrow{[O]} HO{-}CH_2{-}CH_2{-}R \xrightarrow{-H_2O} H_2C{=}CH{-}R$$

An exception to this common pathway is the formation of tariric acid (**1**), which is probably formed by dehydrogenation of petroselinic acid. As already mentioned, the C_{17}-acids **23**–**26** are surely formed by α-oxidation of the corresponding C_{18}-acids.

5

Further Acetylenes

In addition to the various types of naturally occurring acetylenes already discussed, a considerable number of compounds with acetylenic bonds have been isolated in the last ten years from very different sources and, in part, clearly not derived from fatty acids, e.g. the compounds with branched carbon chains related to terpenoic precursors.

1. Compounds Derived from Terpenoid Precursors

From a fungus, *Helminthosporium siccans* g 207, the hydroquinone derivative **1** has been isolated[327]. This compound most probably is formed by dehydrogenation of the corresponding diene:

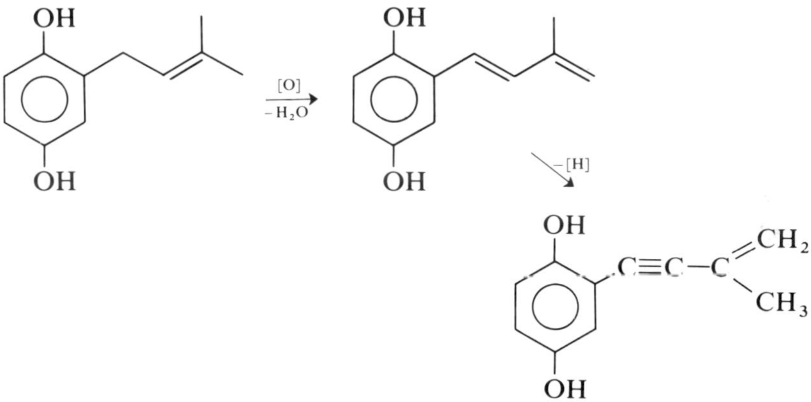

1

A sesquiterpene is clearly the precursor of freelingine (**2**) isolated from the wood oil of the angiosperm *Eremophilia freelingini* (Family Myoparaceae[328]):

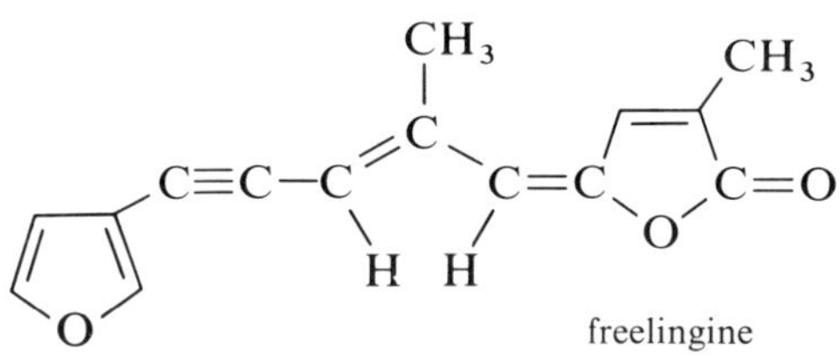

freelingine

A very similar compound (**3**) has been isolated from *Ursina* species (family Compositae): perhaps **2** is formed by oxidation cyclisation and dehydrogenation of **3**:

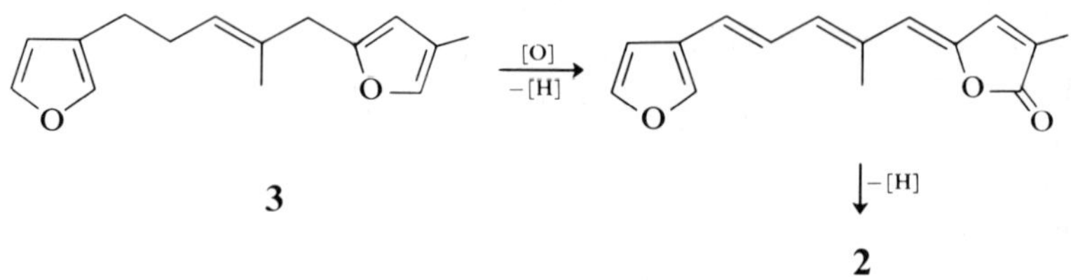

The compounds **5**–**11**[329] with a free ethynyl group isolated from Benihi tree, *Chamaecyparis formosensis* Matsum (family Cupressaceae), are certainly formed from the sesquiterpenoid compound **4** also present:

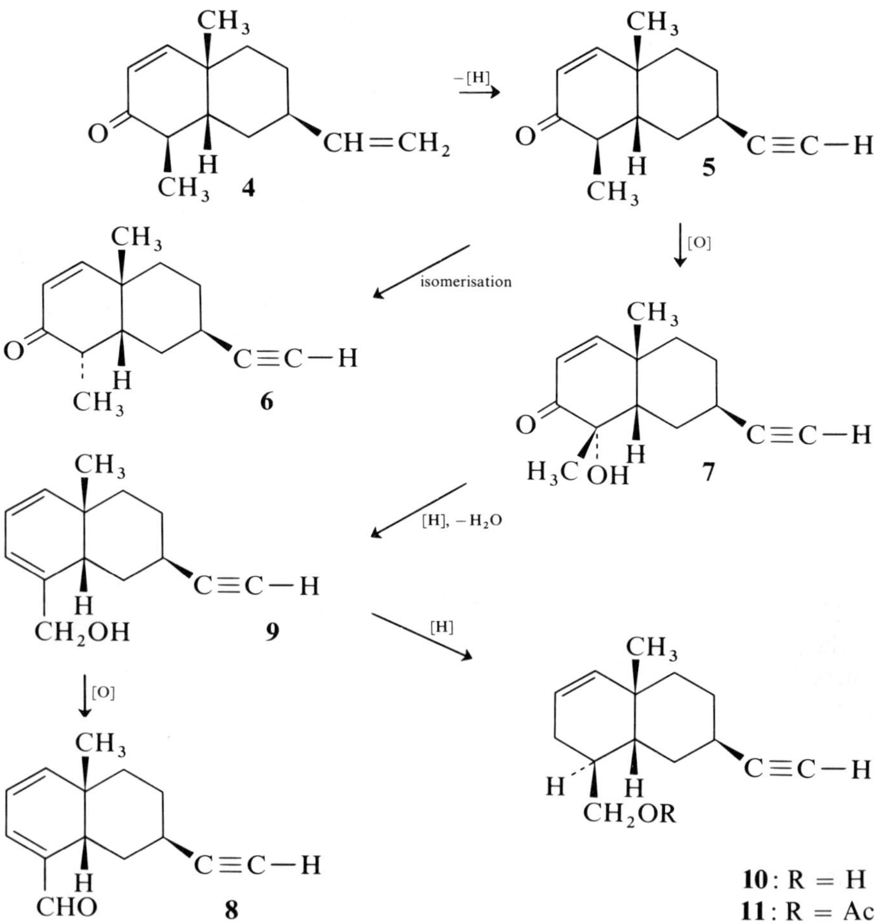

The sequence of the proposed biochemical transformations is not established. The formation of **4** itself being an aposesquiterpene has not yet been investigated.

A number of carotenoids isolated from algae and marine organisms also have an acetylenic bond always in the position 7,8. The first one fully characterised was Alloxanthin (**16**). In all cases the corresponding normal carotenoid also was found. However also allenic carotenoids as **24**[330] are present, which aroused interest in possible biochemical interconversion of the corresponding carotenoids in the following way:

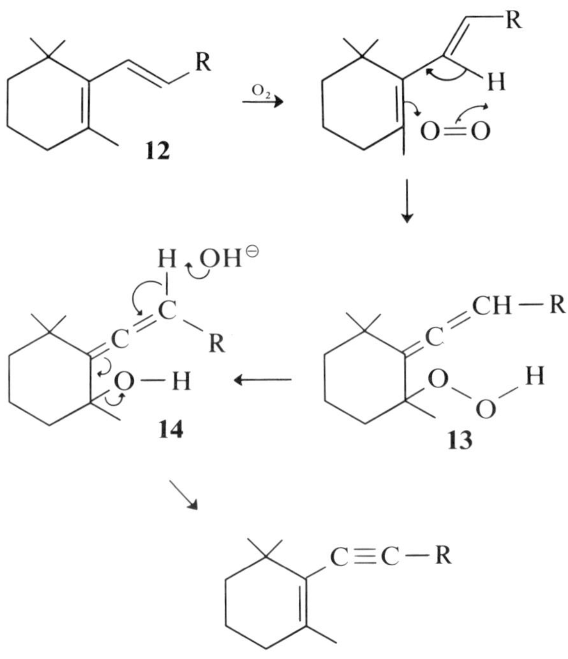

The transformation of **12** to **13** has been shown by irradiation of *β*-jonole in the presence of oxygen and a sensitiser[331]. This *in vitro* experiment and the unestablished pathway **14** ⟶ **15** have to be confirmed by further investigations. The following compounds are already known:

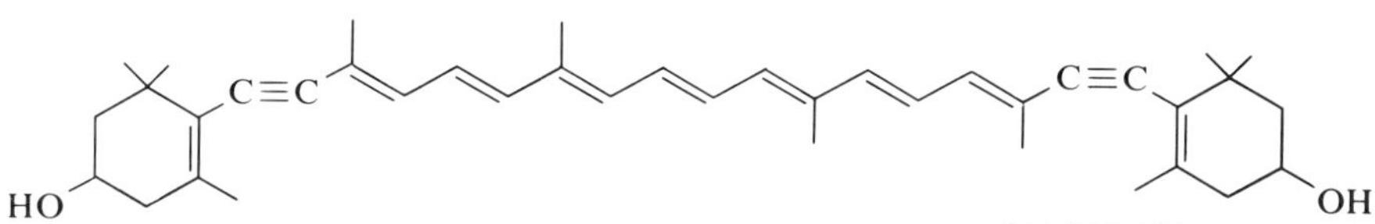

16[332,333,334]

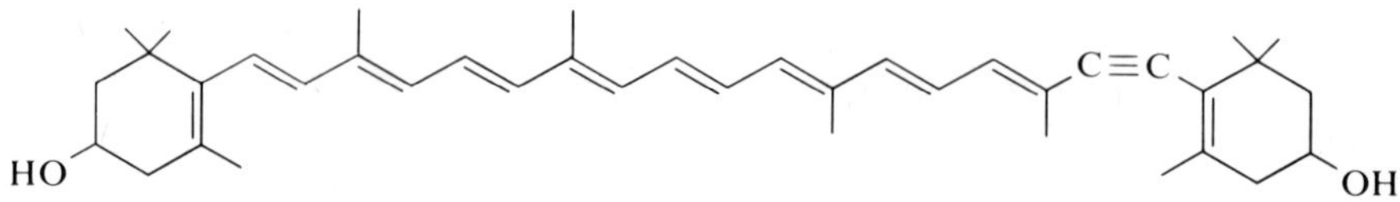

17[332,335]

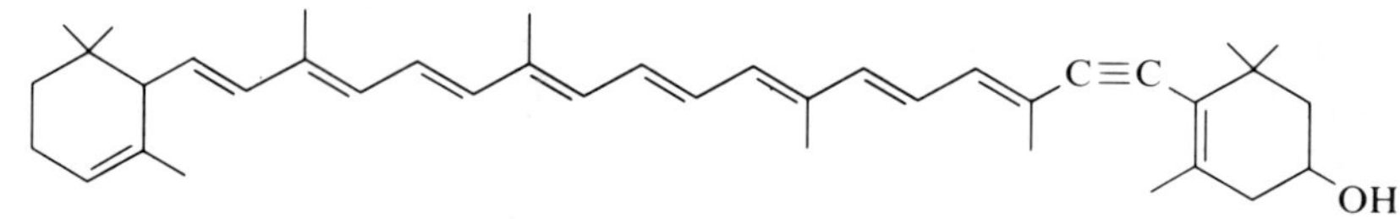

18[332]

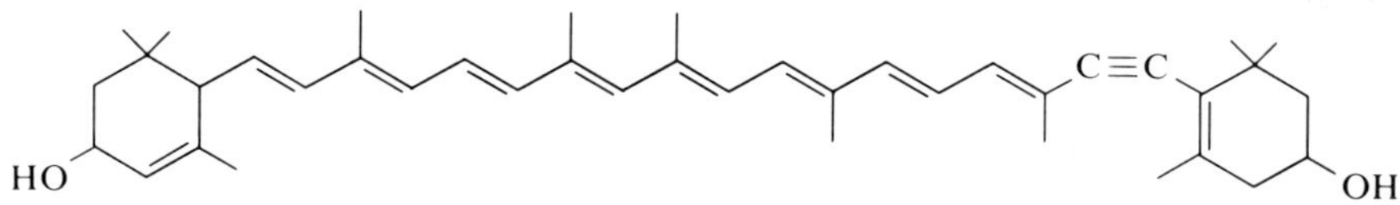

19[332]

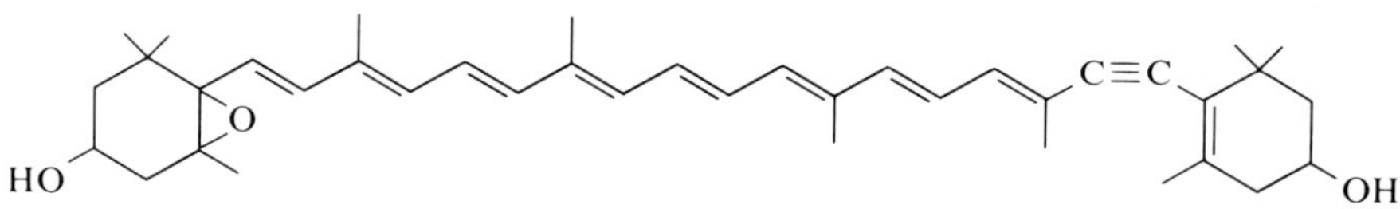

20[333]

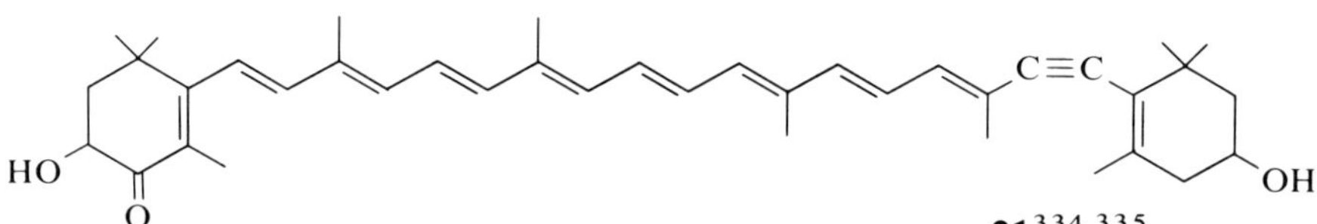

21[334,335]

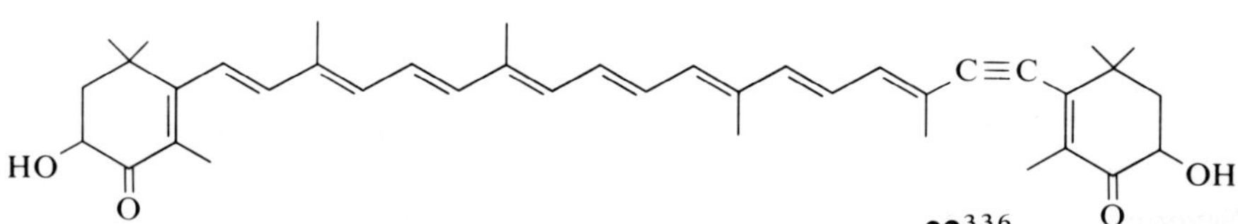

22[336]

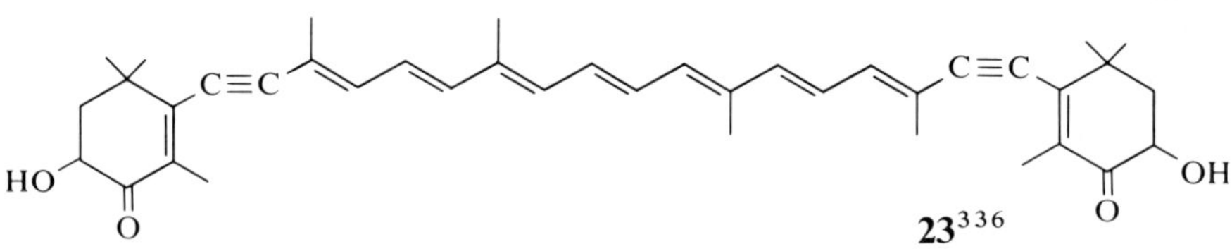

23[336]

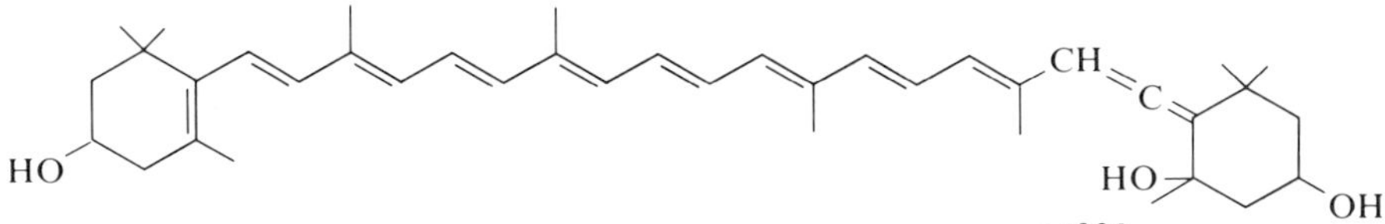

24[330]

All these compounds occur together with the corresponding normal carotenoids and also with the previously mentioned allenic compounds, except the diallenes corresponding to **16** and **23**.

The absolute configuration of most of these compounds have also been elucidated, mainly by chemical transformations and by synthesis of the allenic ketone isolated from grasshoppers in the following way:

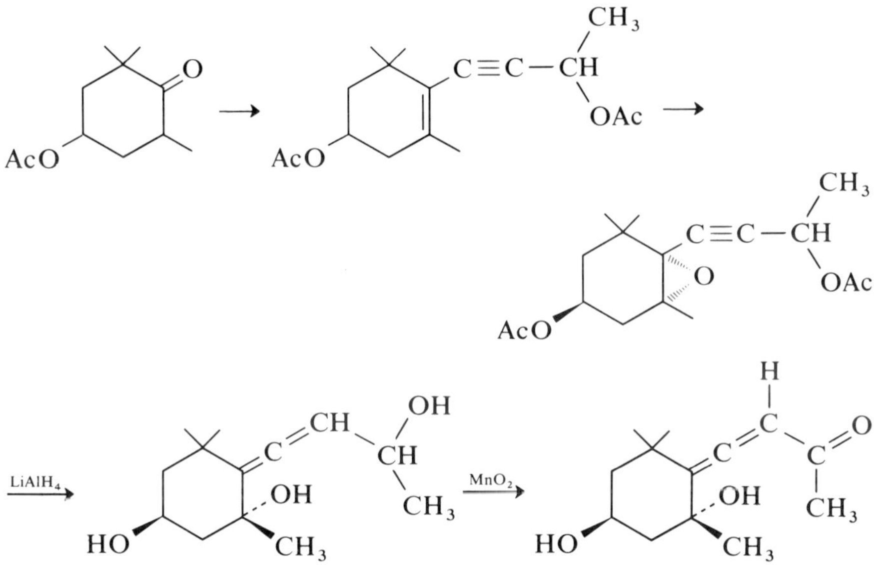

2. Compounds Derived from Unbranched Chains

Several more unique acetylenes, all having only one triple bond, have been found in very different organisms.

The essential oil of the tree *Litsea oderifera* Val. (Family Lauraceae) contains the methoxy compounds **25** and **26**[337]:

$$\underset{\displaystyle \underset{OCH_3}{|}}{H_3C-CH}-(CH_2)_n-C{\equiv}C-H \qquad \mathbf{25}: n = 7 \quad \mathbf{26}: n = 9$$

No biogenetic considerations are yet possible and the same is true for acetylenes isolated from the seeds of the avocado pear, *Persea gratissima* Gaertn. (Family Lauraceae)[338]. However, in this case, all the corresponding vinyl compounds are also present and therefore the acetylenes are most likely formed by dehydrogenation of the vinyl compounds. All constituents have a C_{17}-carbon chain which may indicate a possible relationship to C_{18}-fatty acids. However, the substitution pattern is difficult to correlate with known compounds:

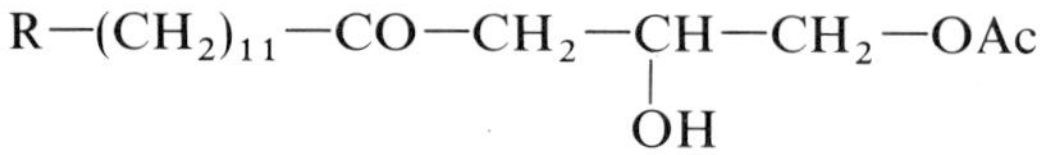

27: R = H—C≡C—
28: R = H_2C=CH—

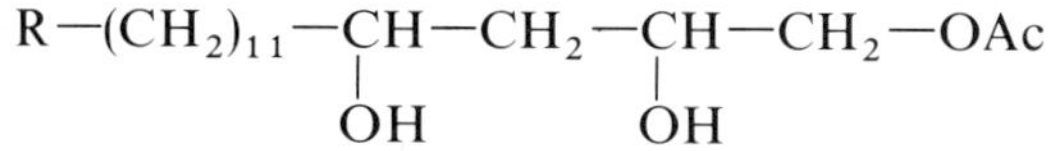

29: R = H—C≡C—
30: R = H_2C=CH—

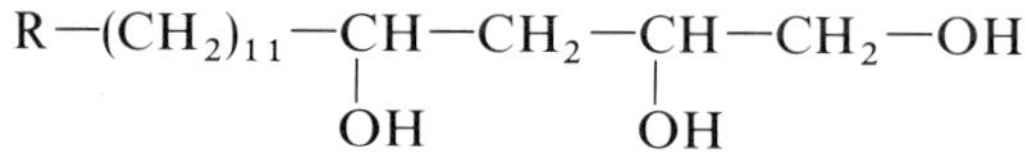

31: R = H—C≡C—
32: R = H_2C=CH—

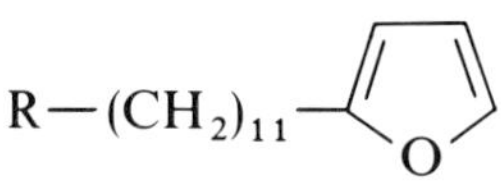

33: R = H—C≡C—
34: R = H_2C=CH—

33 and **34**, respectively, are clearly formed by cyclisation of **27** and **28** with subsequent elimination of water.

A further compound already mentioned, ethynyl, together with the corresponding vinyl compound, has been isolated from the yeast *Rhodotorula glutinus* var. lusitanica[326], while simple propiolic acid has been found in *Bacterium coli*[339] and the diamide in *Streptomyces griseus*[340]:

$H{-}C{\equiv}C{-}(CH_2)_8{-}CO_2H$	**35**	$H_2C{=}CH{-}(CH_2)_8{-}CO_2H$	**36**
$H{-}C{\equiv}C{-}CO_2H$	**37**	$H_2N{-}CO{-}C{\equiv}C{-}CO{-}NH_2$	**38**

An interesting group of acetylenes has been isolated from the red algae *Laurentia nipponica* Yamada and *L. glandulifera* Kutting[341]. The structures and the configuration of these eight-membered oxygen ring compounds are fully established, mainly by spectroscopic methods. Biogenetically these compounds probably are formed by the following sequence starting with the hexadeca-4,7,10,13-tetraenoate (**39**) isolated from Chlorophyta[342]:

$$H_3C(CH_2{-}CH\underset{c}{=}CH)_4{-}(CH_2)_2{-}CO_2R \quad \mathbf{39}$$

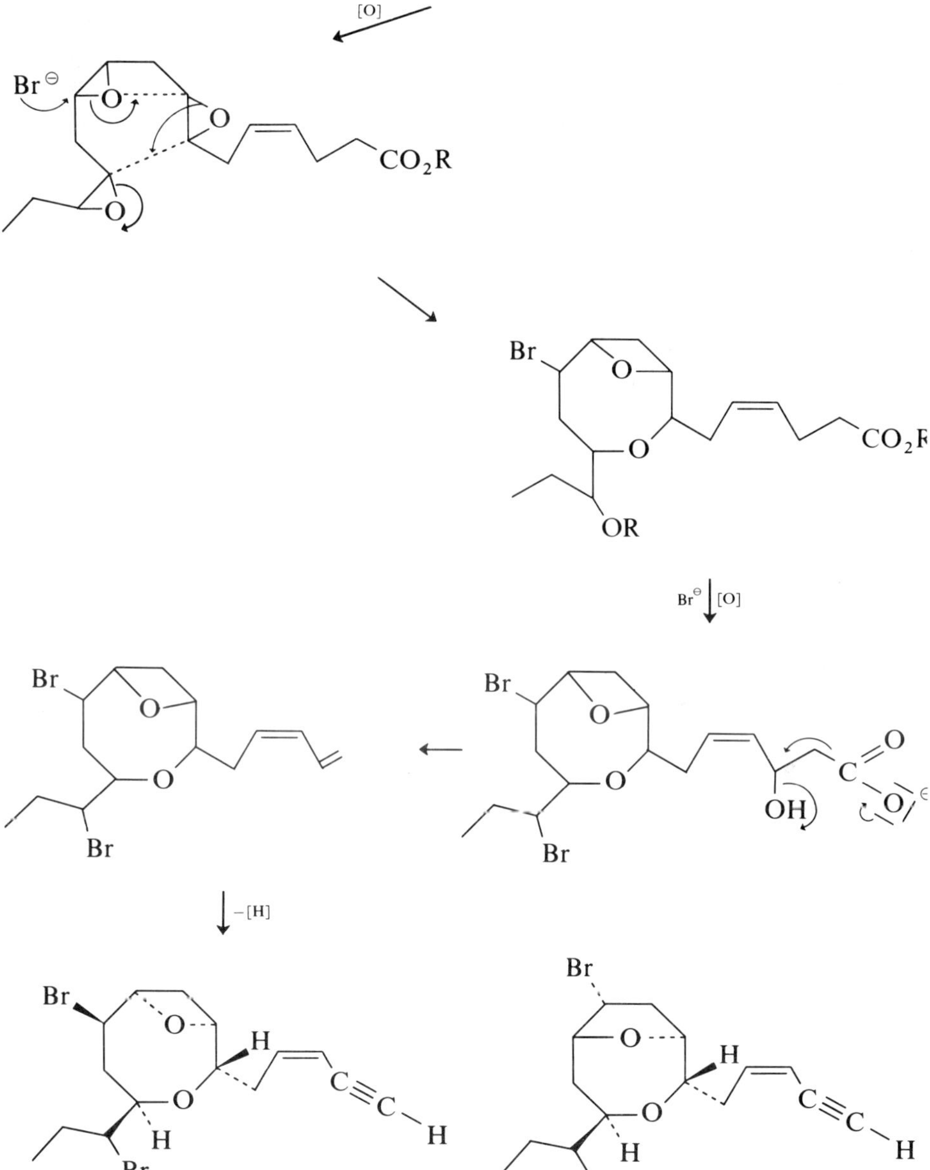

40 **41**

The isomeric compound **41** can clearly be formed in an analogous manner, while **42** must be formed via the diepoxide followed by the same pathway:

39 →
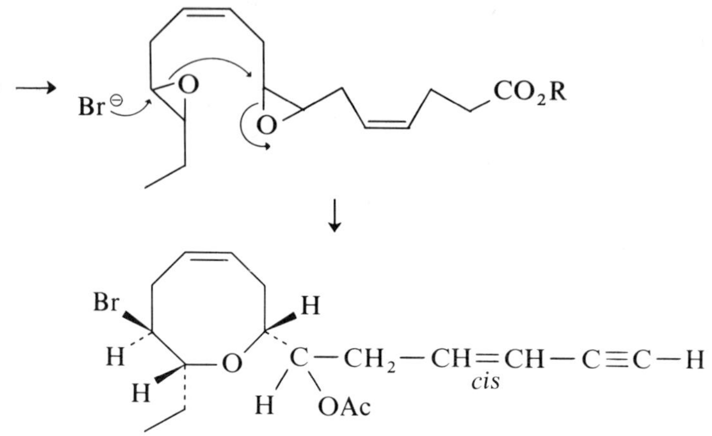
42

From an *Anona* species—*Alphonsea ventricosa*—(Family Anonaceae), three acetylenic triols have been isolated, their structures not being fully established up to now. However, the primary results indicate a possible structural relationship to the acetylenic fatty acids of the Santalales group. Although the position of the unsaturation is not clear, all results agree best with the following structures[35]:

$$H_2C{=}CH(CH_2)_2CH{=}CH{-}(C{\equiv}C)_2(CH_2)_7\underset{OH}{CH}CH_2\underset{OH}{CH}{-}\underset{OH}{CH_2} \qquad \mathbf{43}$$

$$n\text{-}C_4H_9{-}CH{=}CH{-}(C{\equiv}C)_2{-}(CH_2)_7{-}\underset{OH}{CH}{-}CH_2{-}\underset{OH}{CH}{-}\underset{OH}{CH_2} \qquad \mathbf{44}$$

$$n\text{-}C_6H_{13}{-}(C{\equiv}C)_2{-}(CH_2)_7{-}\underset{OH}{CH}{-}CH_2{-}\underset{OH}{CH}{-}\underset{OH}{CH_2} \qquad \mathbf{45}$$

A possible pathway for the biogenesis of the triols (e.g. **43**) would be the following one:

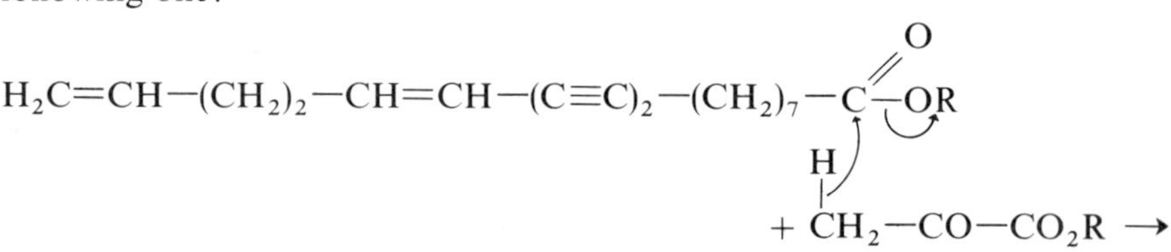

$$H_2C{=}CH(CH_2)_2CH{=}CH{-}(C{\equiv}C)_2{-}(CH_2)_7COCH_2COCO_2R \xrightarrow{[H]} \mathbf{43}$$

This would indicate really a close relationship to the Santalales compounds.

3. Special Structures

Besides the natural acetylenes already discussed, only a few further compounds need to be mentioned. Three compounds isolated from the seed oil of *Euphoria longan* (Family Sapindaceae) are amino-acids, its structures being elucidated as **46–48**[343]:

$$H{-}C{\equiv}C{-}\underset{\displaystyle CH_2{-}R}{\underset{|}{CH}}{-}CH_2{-}\underset{\displaystyle NH_2}{\underset{|}{CH}}{-}CO_2H \qquad \mathbf{46}: R = H \quad \mathbf{47}: R = OH$$

$$H{-}C{\equiv}C{-}\underset{\displaystyle OH}{\underset{|}{CH}}{-}CH_2{-}\underset{\displaystyle NH_2}{\underset{|}{CH}}{-}CO_2H \qquad \mathbf{48}$$

These acetylenic acids are isolated together with homoisoleucine proposed as a possible precursor.

The number of known monoacetylenes has increased very rapidly in the last decade. Perhaps many more compounds of this type may be present in plants and micro-organisms as so far all these constituents with acetylenic bonds have been discovered more or less by chance.

The first acetylenes isolated from animals were those from the Colombian frog, *Dendrobates histrionicus*[344]. Chiefly from X-ray analysis, these toxic compounds have structures **49** and **50**; therefore, these acetylenes are also the first acetylenic alkaloides:

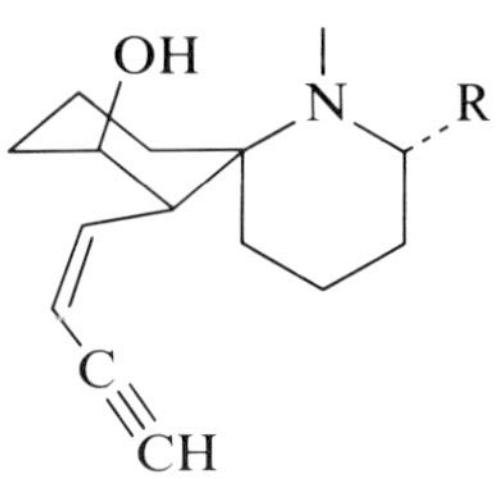

49: $R = {-}CH_2{-}CH{=}CH{-}C{\equiv}C{-}H$
cis
histrionicotoxin

50: $R = {-}CH_2{-}CH_2{-}CH{=}C{=}CH_2$

6

Distribution of Acetylenes

Acetylenes have been found in 19 families of the higher plants with one exception belonging to the Dicotyledons. However, only the families Compositae and Umbelliferae have yet been fully investigated. Up to now, more than 1100 species out of 267 genera have been studied from the Compositae and 195 species out of 89 genera from the Umbelliferae.

From other families, only a few species have been investigated. More detailed studies have already been achieved in the field of micro-organisms, especially with those belonging to the Basidiomycetes. Acetylenes have been found in 54 species out of 22 genera. Only a few further micro-organisms have also been found to produce acetylenes.

1. Compositae

As already mentioned, the Compositae are the best studied family and most of the known natural acetylenes have been isolated there. They are present in both subfamilies and in all 13 tribus. However, the amount of structure variation is extremely wide. Especially rich in acetylenes are the tribus Heliantheae, Anthemideae and Cynareae.

1.1. Tribus Vernonieae and Eupatorieae

More than 50 species of these two, possibly oldest, tribus of the family Compositae have been investigated up to now. In general the content of acetylenes is very minute. Most widespread is the pentaynene (**94**) (see Chapter 2, Section 2.6) isolated from members of 15 genera out of 19 already examined. The amount of **94** is very small (in the range 10^{-3}–10^{-5}% of the fresh root weights). In several cases, where only small amounts of material were available, nothing could be detected. Therefore, it is not possible to say whether **94** is present in all species or not. Besides **94**, only a few other acetylenes have been detected in plants belonging to these tribus: the tetrayne (**436**) (see Chapter 3, Section 5.5), two thiophene acetylenes (**112**) (see Chapter 2, Section 6.3) and (**460**) (see Chapter 2, Section 5.3), and only in one case

TABLE III
SUMMARY OF THE SPECIES INVESTIGATED

Family		Genus	Species
Annonaceae		1	1
Araliaceae		4	7
Campanulaceae		9	32
Compositae*	I/II	19	52
	III	36	167
	IV	36	108
	V	57	193
	VI	13	41
	VII	32	273
	VIII	12	43
	IX	4	19
	X	11	41
	XI	22	194
	XII	9	12
	XIII	18	29
Cupressaceae		1	1
Euphorbiaceae		1	1
Lauraceae		2	2
Loranthaceae		2	2
Myoporaceae		1	1
Olacaceae		2	5
Opiliaceae		1	1
Papilionaceae		1	1
Pittosporaceae		1	1
Santalaceae		8	19
Sapindaceae		1	1
Simarubaceae		1	4
Sterculiaceae		1	2
Umbelliferae		89	195
Valerianaceae		1	1
Micro-organisms		33	64

* Tribus following Engler–Prantl.

the dithio-compound (**128**) (Chapter 2, Section 6.3). The investigated species are summarised in Tables IV and V.

TABLE IV
TRIBUS VERNONIEAE[35]

1. Vernonieae-Vernoninae	Constituents (roots)
Sparganophorus sparganophora	$H_3C\equiv_5=$
Ethulia conyzoides L.	$H_3C\equiv_5=$, $H_3C{=}\equiv_4=$, $H_3C\equiv_2$ 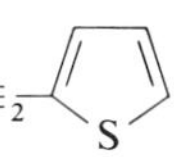 $\equiv=$
Centhratherum camporum Malme.	$H_3C\equiv_5=$
C. punctatum Cass.	$H_3C\equiv_5=$

TABLE IV—Continued

1. Vernonieae-Vernoninae	Constituents (roots)
Vanillosmopsis discolor Benth. *et* Hook.	$H_3C{\equiv}_5{=}$
Vernonia anthelmintica Willd.	$H_3C{\equiv}_5{=}$
V. brevifolia Less.	$H_3C{\equiv}_5{=}$
V. cinerea Less.	$H_3C{\equiv}_5{=}$
V. cistifolia O. Hoffm.	—
V. cognata Less.	—
V. echioides Less.	$H_3C{\equiv}_5{=}$
V. megapotamica Spreng.	$H_3C{\equiv}_5{=}$
V. novobaracensis Willd.	$H_3C{\equiv}_5{=}$
V. senegalensis Less.	$H_3C{\equiv}_5{=}$
Corymbium villosum	$H_3C{\equiv}_5{=}$, $H_3C{=}{\equiv}_4{=}$
Hoplophyllum spinosum DC.	$H_3C{=}{\equiv}_4{=}$

2. Vernonieae-Lychnophorinae[35]	Constituents (roots)
Elephantopus corilianus Willd.	$H_3C{\equiv}_5{=}$, $H_3C{=}{\equiv}_4{=}$
E. mollis H.B. *et* K.	$H_3C{=}{\equiv}_4{=}$

TABLE V
TRIBUS EUPATORIEAE

1. Eupatorieae-Piquerinae	Constituents (roots)
Piqueria trinervia Cav.[35]	$H_3C{\equiv}_5{=}$
Adenostemma lavenia[35]	$H_3C{\equiv}_5{=}$

2. Eupatorieae-Ageratinae[35]	Constituents (roots)
Carelia cistifolia Less.	$H_3C{\equiv}_5{=}$
Ageratum conyzoides L.	—
A. houstonianum Mill.	—
A. mexicanum Sims.	—
Stevia lanceolata Lag.	—
S. ovata Lag.	$H_3C{\equiv}_5{=}$
S. paniculata Lag.	—
S. purpurea Pers.	—
Symphyopappus halschbachii	—
Eupatorium altissimum L.	$H_3C{\equiv}_5{=}$
E. altrinubis	—
E. aromaticum L.	—
E. asclepiadeum DC.	—
E. betonicaeforme Baker	—
E. cannabnum C.	$H_3C{\equiv}_5{=}$
E. confertum Gardn.	—
E. glandulosum H.B. *et* K.	—
E. hecatanthum Baker	—
E. macrocephalum Less.	—
E. maracayense	—
E. numeralis	—
E. rugosum H.B. *et* K.	—

TABLE V—Continued

2. Eupatorieae-Ageratinae[35]	Constituents (roots)
E. subcapitalum	—
E. tanacetifolium Gill.	—
E. urticaefolium L.	—
Mikania officinalis Mart.	$H_3C{\equiv}_5{=}$
M. scandens (L.) Willd.	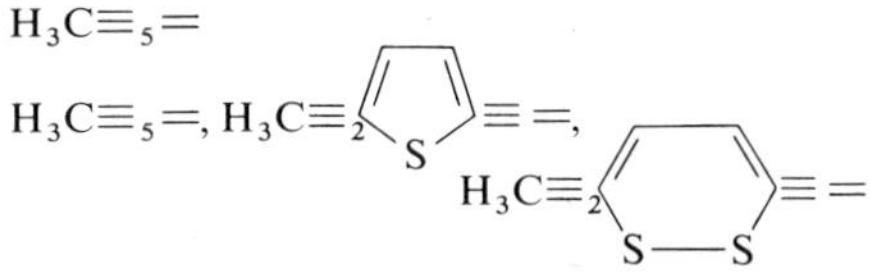

3. Eupatorieae-Adenostylinae[35]	Constituents (roots)	
Adenostyles alliariae Kern.	$H_3C{\equiv}_5{=}$	
Brickellia guatemaliensis	—	
Liatris scariosa (L.) Willd.	—	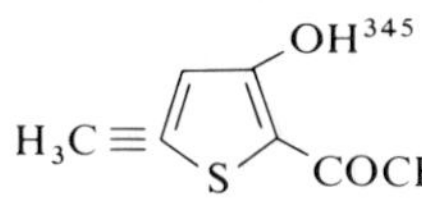
L pycnostachya Michx.	$H_3C{\equiv}_5{=}$,	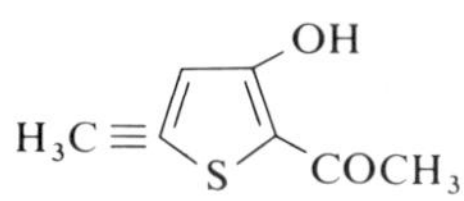
L spicata (L.) Willd.	$H_3C{\equiv}_5{=}$,	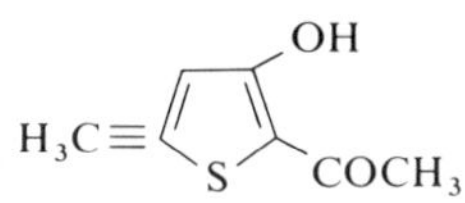

Besides the sequiterpene lactones common in these two tribus, some other compounds have been isolated.

From *Piqueria trinervia* (−)-Santalal, from the *Carelia* as well as from the *Brickellia* species derivatives of dehydro-nerolidol and again from *Carelia*, a benzofurane derivative closely related to those isolated from *Ageratum* and *Eupatorium* species, but also from *Encelia* and *Helianthella* species belonging to the tribus Heliantheae.

1.2. Tribus Astereae

This tribus has been investigated more extensively as the previous ones. Up to now more than 150 species have been examined. The variations in acetylenic structures are clearly larger, but most common are closely related C_{10}-compounds. In some genera or sections C_{17}-acetylenes have been found probably formed from precursors of the C_{10}-compounds in an early biogenetical stage. Very interesting is the fact that compounds related to all the possible unsaturated C_{18}-acids discussed in Chapter 1, Section 4, are present in this tribus as shown in the series of isolated C_{10}-acetylenes:

$$H_3C-CH_2-CH_2-CH=CH-C\equiv C-CH=CH-CO_2CH_3$$

$$H_3C-CH_2-CH_2-C\equiv C-C\equiv C-CH=CH-CO_2CH_3$$

$$H_3C-CH=CH-C\equiv C-C\equiv C-CH=CH-CO_2CH_3$$

$$H_3C-C\equiv C-C\equiv C-C\equiv C-CH=CH-CO_2CH_3$$

$$H_3C-CH=CH-CH=CH-C\equiv C-CH=CH-CO_2CH_3$$

In Table VI the investigated species are listed showing the main acetylenic constituents:

TABLE VI
TRIBUS ASTEREAE

1. Astereae-Solidagininae	Constituents
Xanthocephalum gymnospermoides Benth. *et* Hook.[35]	$C_3H_7\equiv_2=CO_2CH_3$ (*LE*)
Grindelia arenolica L.[35,289]	$H_3C=\equiv_2=CO_2CH_3$ (*ME*) $H_3C=\equiv_2=CH_2OAc$
G. camporum Greene[292]	*ME*, $H_3C=\equiv_2=R$ R = CH_2OH, CHO, CH_2OAc, $CH_2OCOCH=C(CH_3)_2$ $H_3C-CH_2CH_2\equiv_2=CH_2OR$ R = H, Ac, $COCH=C(CH_3)_2$
G. humilis Hook et Arn.[292]	*ME*, *LE*, $H_3C=\equiv_2=R$ R = CH_2OH, CHO, CH_2OAc, $CH_2OCOCH=C(CH_3)_2$ $H_3CCH_2CH_2\equiv_2=CH_2OR$ R = H, Ac, $COCH=C(CH_3)_2$ $H_3C=_2\equiv=R$ R = CH_2OH, CHO, CH_2OAc $H_3CCH_2CH_2=\equiv=R$ R = CH_2OH, CHO, CH_2OAc $H_3CCH_2CH_2\equiv_2=CHO$
G. integrifolia DC.[35]	*ME*, $H_3C=\equiv_2=R$ $H_3C=_2\equiv=R$ R = CH_2OH, CH_2OAc, CHO, $CH_2OCOCH=C(CH_3)_2$
G. oregana A. Gray[35]	*ME*, $H_3C=\equiv_2=R$ R = CH_2OH, CH_2OAc, CHO, $CH_2OCOCH=C(CH_3)_2$
G. robusta Nutt.[304,292] *G. squarrosa* Dum.[304] *G. stricta* DC.[35,289]	*ME*, $H_3C=\equiv_2=R$ R = CH_2OH, CHO, CH_2OAc $H_3C=_2\equiv=R$ R = CH_2OH, CHO, CH_2OAc
Guttierrezia gymnospermoides A. Gray[299]	$H_3C=_2\equiv=R$ R = CH_2OH, CH_2OAc, $CH_2OCOCH=C(CH_3)_2$

TABLE VI—Continued

1. Astereae-Solidagininae	Constituents
Xanthisma texanum DC.[35]	*LE*
Chrysopsis villosa Nutt.[35]	$R-=\equiv_2=_2(CH_2)_4=$ $R = CH_3, CH_2OH, CH_2OAc$
Hysterionica pinifolia Beth et Hook.[35]	*LE*, 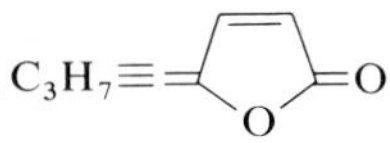 (*LL*), 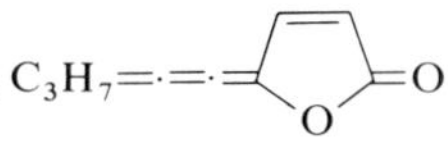
Solidago arguta Ait.[35]	—
S. chinensis Spr.	$H_3C\equiv_3=CO_2CH_3$ (*DME*)
S. glabra Desf.[35]	—
S. virgaurea L.[203]	*ME*, *LL*, 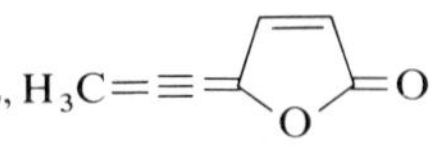 (*ML*) $H_3C=\equiv_2CH(R)=(CH_2)_5=$ (R = H, OH) 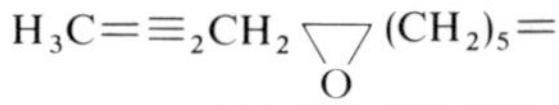
S. sempervirens L.[9]	$H_3C=\equiv_2CH_2CH_2CO_2CH_3$
Happlopappus brandegei A. Gray[35]	*ME*, *LE*, *LL*, 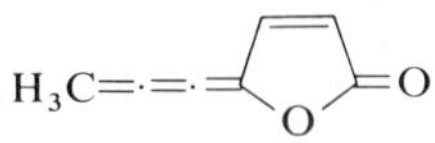
H. clementis Clarke[35]	*ME*, 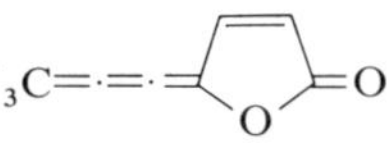
H. croceus A. Gray[35]	*ME*, $H_3C=\equiv_2=CH_2OAc$
H. fructicosus[293]	*ME*, *ML*, $H_3C=\equiv_2=R$ $R = CH_2OH, CH_2OAc$
H. integrifolius Porter[35]	*ME*, *LE*
Lessingia hololeuca Greene[35]	trace $-C\equiv C-$
Pteronia pallens L.[35]	—
P. paniculata Thunb.[35]	—

2. Astereae-Granginae	Constituents
Grangea maderaspatana Poir.[35]	—
Dichrocephala latifolia DC.[35]	$-C\equiv C-$

3. Astereae-Bellidinae	Constituents
Erodiophyllum elderi F. v. Müll.[200]	$C_3H_7\equiv_2=CH(OH)(CH_2)_5=$, $C_3H_7\equiv_2CO=(CH_2)_5=$ $C_3H_7\equiv_2CH(OH)=(CH_2)_5=$, $C_3H_7\equiv_2CH(OH)(CH_2)_7=$ $H_3CCH_2CO\equiv_2CH_2CH_2CH_2OH$

TABLE VI—Continued

3. Astereae-Bellidinae	Constituents
Myriactis nepalensis Less.[35]	$=\underset{\displaystyle OH}{\underset{\vert}{C}}H\equiv_2CH_2=(CH_2)_5=$
Brachycome iberidifolia Benth.[289]	*ME*
Bellis bernardi Boiss.[35]	*ME*, $H_3C=\equiv_2=CH_2O-CO\underset{\displaystyle CH_3}{\underset{\vert}{C}}=CH-CH_3$
B. perennis L.[306]	*ME, LE*
B. rotundifolia Boiss. *et* Reut.[35]	$H_3C=\equiv_2\underset{\displaystyle O-\text{Angeloyl}}{\underset{\vert}{C}}HCH_2CO_2CH_3$
B. caerulescens Cass.[35]	$C_3H_7\equiv_2\underset{\displaystyle O-\text{Angeloyl}}{\underset{\vert}{C}}HCH_2CO_2CH_3$

4. Astereae-Asterinae	Constituents
Amellus annuus Willd.[35]	*ME*, *ML*, $H_3C=\equiv_2CH_2CH_2CO_2CH_3$
A. strigosus Less.[35,287,289]	*ME*, $H_3C=\equiv_2CH_2CH_2CO_2CH_3$, Cosmen
Townsendia formosa Greene[35]	*ME*, *LE*, $H_3C\equiv\equiv\equiv$ 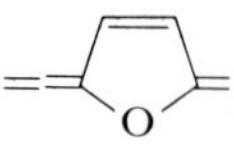 $=O$
T. parryi Eaton[289]	*ME*
Bellium bellidioides Desf.[35]	$H_3C=\equiv_2=R$ (R = CH_2OH, CH_2OAc, CH_2O-Angeloyl)
B. minutum L.[35]	*ME*, $H_3C=\equiv_2=CH_2O$Angeloyl
Boltonia asteroides L'Hérit[35]	*LE, LL*
B. diffusa Ell.[35]	*LL*
B. latisquama A. Gray[35]	*LL*, *ME*, $H_3C\equiv\equiv\equiv$ 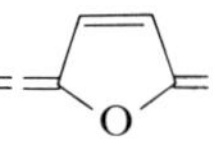 $=O$
Asteromea lautureana Hand-Mazz.[35]	*ME*
A. mongolica Kitamura[35]	*ME, LE*
Callistephus chinensis (L.) Nees.[299]	$H_3C=_2\equiv=CH_2O$Angeloyl $C_3H_7\equiv_2=R$ (R = CH_2OH, CH_2OAc, CH_2O-Angeloyl)
Charieis heterophylla Cass.[35]	polyenes
Mairia coriacea Bolus[35]	*ME*
Vittadinia triloba hort.[35]	—
Podocoma hieracifolia Cass.[35]	$H_3C=\equiv_2=CH_2OAc$
Olearia haastii Hook.[35]	—
Felicia aethiopica (Burm.) Bolus et Wolley Dod[35]	—
F. amelloides (L.) Voss.[289]	—
F. bergeriana (Sprengl.) O. Hoffm.[35]	—
F. filifolia Burtt. Davy[35]	*LE*
F. fragilis Cass[35]	*ME*
F. rotundifolia C. G. Taylor[35]	*ME*
F. tenella DC.[35]	*ME*, $H_3C=\equiv_2CH_2CH_2CO_2CH_3$
F. echinata Nees.[35]	—
F. ovata (Thb.) Compton[35]	—
Lachnophyllum gossypinum Bge.[6]	*LE*

TABLE VI—Continued

4. Astereae-Asterinae	Constituents					
	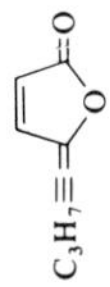	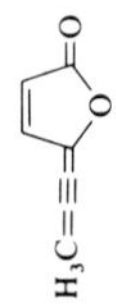	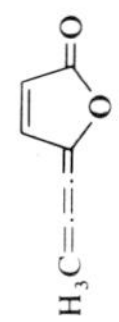	$H_3C{=}{\equiv}_2{=}CO_2CH_3$	$H_3CCH_2CH_2{\equiv}_2{=}CO_2CH_3$	Cosmen
Erigeron acer L.[35]	+	+	++	+		+
E. acris L.[43]				+	+	
E. annuus Pers.[35,346]	++	++	+	+	+	
E. arcticus Rouy[35]	+	+	++	++	+	++
E. armerifolius Turcz.[35]	+		++		+	++
E. atticus Vill.[35,346]			++	+	+	++
E. aurantiacus Rgl.[35,346]	+		++	+	+	++
E. bellidifolius Muhl.[35]	+		++		+	++
E. borealis Vierh. *et* Simm.[346]				+	+	
E. canadensis L.[303,346]	+		+	+	+	
E. candidus Widder[346]				+	+	
E. compositus Pursh.[35,346]	++	+	++	++	+	++
E. conyzoides F. Muell.[33]	+	+		+	+	
E. coulterii Porter *et* Coult.[35]	?	?	?	+	+	?
E. crispus Pour.[35]	+		++	+		++
E. droebachiensis O. F. Muell.[35]	+	+	++	+	+	++
E. eliator Greene[346]				+	+	
E. eriocephalus J. Vahl.[35,346]	?	?	?	+	+	?
E. eucephaloides Greene[35]	+		++		+	++
E. flagellaris A. Gr.[35,346]		+	++	+	+	++
E. gauetiana[35]	+		++		+	++
E. glabellus Nutt.[35,346]	?	?	?	+	+	?
E. glaucus Ker.[35,346]	+	+	++	+	+	++
E. glandiflorus Hook.[35]	+		++		+	++
E. karwinskianus DC.[35]			++	+		++
E. linifolius Willd.[35,346]	+	+	++	+	+	++
E. maximus Otto.[35]		+				
E. mairei Braun-Blanq.[346]				+	+	
E. montevidensis Bak.[35,346]	?	?	?	+	+	?
E. nematophyllus Rydb.[346]				+	+	
E. philadelphicus L.[35,346]	?	?	?	+	+	?
E. polymorphus Scop.[35,346]	+	+		+	+	
E. politus Hieron.[346]				+	+	
E. speciosus DC.[346]				+	+	
E. superbus Greene[346]				+	+	
E. tweediei Hook. *et* Arn.[35]		+		+		
E. uniflorus L.[35,346]	?	?	?	+	+	?
E. khorassanicus Bouss.[346]	$H_3C{=}{\equiv}_2{=}CH_2OAc$					
E. pappochroma Labill.[346]	$H_3C{=}{\equiv}_2{=}CH_2OR$ (R = H, Ac), $C_3H_7{\equiv}_2{=}CH_2OH$					

TABLE VI—Continued

4. Astereae-Asterinae	Constituents
Aster linosyris (L.)[35] Bernh.	$H_3C=\equiv_2=CH_2OAc$
A. purdomii Hutch.[35]	$H_3C=\equiv_2=CH_2OAc$
A. sibiricus L.[35]	$H_3C=\equiv_2=CH_2OAc$, $C_3H_7\equiv_2=CH_2OAc$, LE
A. tanacetifolius H.B. *et* K.[35]	$H_3C=\equiv_2=CH_2OAc$, $H_3C=\equiv_2=CHO$
A. tripolium L.[35]	$H_3C=\equiv_2=CH_2OAc$, ME
Biotia corymbosa DC.[35] *B. latifolia* DC.[194] *B. macrophylla* DC.[194] *Aster schreberi* Nees.[194]	$C_3H_7=\equiv=_2(CH_2)_4=$ $C_3H_7=\equiv=CO_2CH_3$ $C_3H_7=\equiv=_2(CH_2)_3=CH_2OH$ $C_3H_7=\equiv=_2(CH_2)_3\underset{\mid \atop OH}{CH}=$
Aster caucasicus Willd.[35] *A. macrophyllus* L.[35] *A. bellidiastrum*[35] *A. brachytrichus* Franch.[35] *A. lateriflorus* Britt.[35]	$H_3C=\equiv_2=_2(CH_2)_4=$ $H_3C=\equiv_2CH_2=(CH_2)_5=$ $H_3C=\equiv_2=\underset{\mid \atop OH}{C}H(CH_2)_5=$ *ME*, ML, $H_3C=\!=\cdot\!=\cdot\!=$ 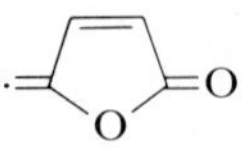
A. thomsonii C. B. Clarke[35]	*LE*
A. amethystinus Nutt.[35]	*LE*
A. alpinus L.[35]	*LE*
A. farreri W. W. Smith[35]	*LE*
A. modestus Lindl.[35]	*LE*
A. moranensis H.B.[35]	*LE*
A. conspicus Lindl.[35]	*LE*, $H_3C=\equiv_2\underset{\mid \atop O\text{-Angeloyl}}{C}H-CH_2CO_2CH_3$
A. bellidiastrum Scop.[208]	*LE*, *LL*, $H_3C=\equiv_2CH_2CH_2CO_2CH_3$, $H_3C=\equiv_2\underset{\mid \atop O\text{-Angeloyl}}{C}HCH_2CO_2CH_3$
A. salicifolius Scholler[35]	*LE*, $H_3CCH_2\underset{\mid \atop O\text{Angeloyl}}{C}H\equiv_2=CO_2CH_3$
A. novi-belgii L.[300]	*LE*, $H_3CCH_2\underset{\mid \atop O\text{Angeloyl}}{C}H\equiv_2=CO_2CH_3$
A. yunnanensis Franch.[35]	*LE*, $H_3CCH_2\underset{\mid \atop O\text{Angeloyl}}{C}H\equiv_2=CO_2CH_3$, coumarins
A. grandiflorus L.[35]	*LE*, $H_3CCH_2\underset{\mid \atop O\text{Angeloyl}}{C}H\equiv_2=CO_2CH_3$, coumarins
A. incisus Fisch.[35]	*LE*, $H_3CCH_2\underset{\mid \atop O\text{Angeloyl}}{C}H\equiv_2=CO_2CH_3$, coumarins
A. mackii Regel[35]	*LE*, $H_3CCH_2\underset{\mid \atop O\text{Angeloyl}}{C}H\equiv_2=CO_2CH_3$, coumarins
A. sericeus Vent.[35]	*LE*, $H_3CCH_2\underset{\mid \atop O\text{Angeloyl}}{C}H\equiv_2=CO_2CH_3$, coumarins

TABLE VI—Continued

4. Astereae-Asterinae	Constituents
A. depauperatus Fernald.[35]	*LE*, $H_3CCH_2\underset{\mid \atop \text{OAngeloyl}}{C}H{\equiv}_2{=}CO_2CH_3$, coumarins
A. radula Ait.[35]	*LE*, $H_3CCH_2\underset{\mid \atop \text{OAngeloyl}}{C}H{\equiv}_2{=}CO_2CH_3$, $H_5C_2\underset{\mid \atop \text{OH}}{C}H{\equiv}_2{=}CO_2CH_3$
A. cordifolius L.[35]	*LE*, coumarins
A. floribundus Willd.[35]	*LE*, coumarins
A. foliaceus Lindl.[35]	*LE*, coumarins
A. gracilipis Alex.[35]	*LE*, coumarins
A. herveyi A. Gray[35]	*LE*, coumarins
A. johannensis Fernald.[35]	*LE*, coumarins
A. lanceolatus Willd.[35]	*LE*, coumarins
A. lautureanus Franch.[35]	*LE*, coumarins
A. likiangensis Franch.[35]	*LE*, coumarins
A. pattersonii A. Gray[35]	*LE*, coumarins
A. salignus Willd.[35]	*LE*, coumarins
A. tradescanti L.[35]	*LE*, coumarins
A. umbellatus Mill.[35]	*LL*
A. canus W. *et* K.[35]	*LL*
A. novi-angliae L.[35]	*ME*, *ML*
A. ptarmicoides Torr. *et* Gray[299,289,35]	*ME*, *DME*
A. amellūs L.[35]	ML, 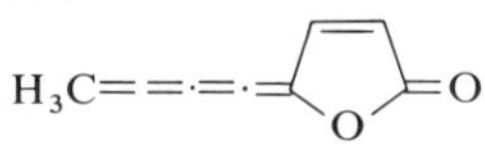
A. rotundifolius Thunb.[35]	polyenes
A. bigelowii A. Gray[35]	—
A. amelloides DC.[35]	—
A. tongolensis Franch.[35]	—
A. laevis Vahl.[35]	—
Diplopappus fructicolosus Less.[293]	ME, ML, $H_3C{=}{\equiv}_2{=}CH_2OR$ (R = H, Ac) $HOCH_2{=}{\equiv}_2{=}CH_2OH$, $OCH{=}{\equiv}_2{=}CH_2OH$, 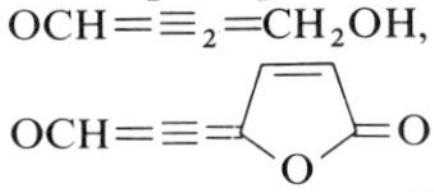
D. filifolius DC.[35]	ME, $ROCH_2{=}{\equiv}_2{=}CH_2OR'$ (R=R′=H, R=R′=Ac, R=H, R′=Ac)

5. Astereae-Conyzinae	Constituents
Microglossa angolensis Oliver *et* Hiern.[35] *Nidorella auriculata* DC.[35]	$-(C{\equiv}C)_2-CH{=}CH-$ $={CO}{\equiv}_2CH_2{=}(CH_2)_5{=}$, $=\underset{\mid \atop \text{OH}}{C}H{\equiv}_2CH_2{=}(CH_2)_5{=}$
Conyza bonariensis Cronq.[297]	ML, LL, 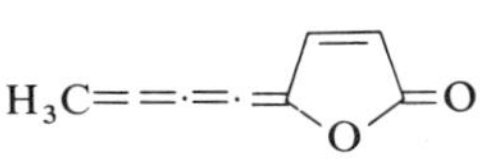

TABLE VI—Continued

5. Astereae-Conyzinae	Constituents
C. crispa Rupr.[35]	*LE*, *LL*, $H_3C{\equiv}{\equiv}{\equiv}$ 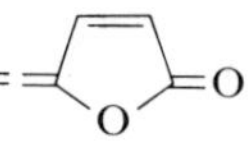
C. setosa Malme[35]	*ME*
Chrysocoma coma-aurea L.[35]	*LE*, *LL*
C. tenuifolia Jacq.[35]	—
C. virginiana Hort. ex DC.[35]	—

6. Astereae-Baccharidinae	Constituents
Baccharis pilularis DC.[35]	*ME*, *LE*, $C_2H_5\underset{|}{C}H{\equiv}_2{=}CO_2CH_3$ OAngeloyl
B. trinervis Pers.[226]	*ME*, *ML*, $H_3C{=}{\equiv}_2{=}_2(CH_2)_5CO_2CH_3$ $H_3C{=}{\equiv}_2{=}_2(CH_3)_3{-}\underset{|}{C}H{-}CH_2CO_2CH_3$ OH $AcOCH_2{=}{\equiv}_2{=}_2(CH_2)_5CO_2CH_3$

Although the investigations are still more or less of a preliminary nature, some groups are already visible; this may be an indication to botanists in difficult systematic situations. The variations, e.g. in the genus Aster are partly in agreement with the botanical systematics. The C_{17}-acetylenes are only typical for the sections *Biotia* and *Alpigena*, while for the section *Aster*, the occurrence of lachnophyllum ester (**541**) (see Chapter 3, Section 7.3) and the corresponding angelica-ester (**544**) (see Chapter 3, Section 7.3), mostly together with coumarins seems to be characteristic. Again, in the first two sections (*Lynosyris* and *Macheranthera*), simple matricarianol derivatives seems to be typical as again in the genus *Grindelia*. Further relationships can be found looking more detailed at the tables, some of them perhaps are of interest in connection with phylogenetic view point too.

Besides coumarins, no other constituents seem to be typical for this tribus. Sesquiterpene lactones, very widespread in the family Compositae, are absent in this tribus.

1.3. Tribus Inuleae

This large tribus with species distributed all over the world is botanically related to the tribus Astereae as well as to Heliantheae. The last relationship is much more pronounced, if we look at the acetylenic compounds isolated from members of this tribus. The same is true for the sesquiterpene-lactones. The variation in structure of the isolated acetylenes in the tribus Inuleae are not very large. The main compound is again the pentayn-ne (**94**) (see Chapter 2, Section 2.6) as shown in Table VII.

TABLE VII
TRIBUS INULEAE

1. Inuleae-Tarchonanthinae	Constituents
Tarchonanthus camphoratus L.[50]	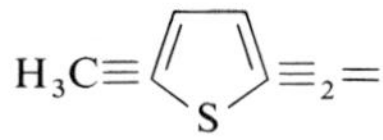

2. Inuleae-Plucheinae	Constituents
Stenachaenium macrocephalum Benth.[35]	$H_3C{\equiv}_5{=}$
Blumea lacera DC.[35]	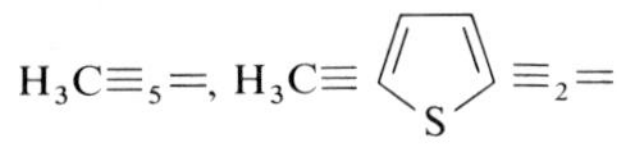
B. laciniata DC.[35,75]	$H_3C{\equiv}_5{=}$
Pluchea dioscorides DC.[35]	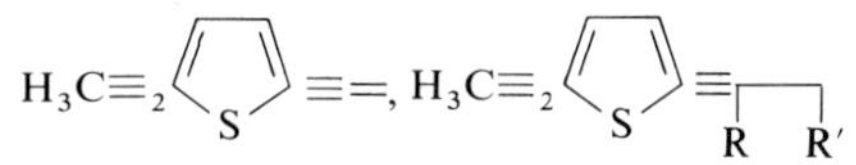 (R = R′ = OH; R = Cl, R′ = OH; R = Cl, R′ = OAc)
P. indica (L.) Less.[35]	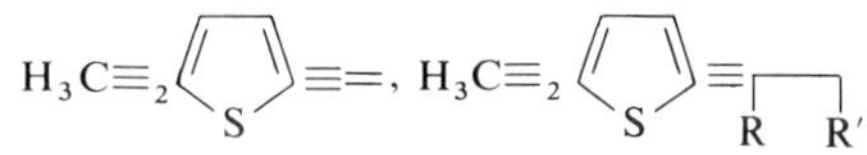 (R = R′ = OH; R = Cl, R′ = OH; R = Cl, R′ = OAc)
Sphaeranthus indicus L.[35]	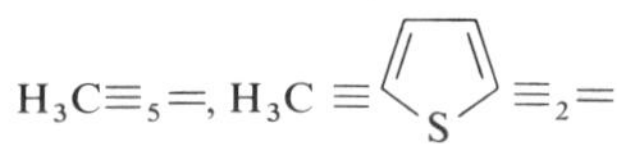

3. Inuleae-Filagininae	Constituents
Micropus supinus L.[35]	$H_3C{\equiv}_5{=}$, $H_3C{=}{\equiv}_4{=}$
Evax pygmea Brot.[35]	$H_3C{\equiv}_5{=}$
Filago arvensis L.[35]	$H_3C{\equiv}_5{=}$, $H_3C{=}{\equiv}_4{=}$, $H_3C{\equiv}_3{=}_2(CH_2)_3OH$
F. germanica L.[35]	$H_3C{\equiv}_5{=}$, $H_3C{=}{\equiv}_4{=}$, $H_3C{\equiv}_3{=}_2(CH_2)_3OH$
F. spathula Presl.[35]	$H_3C{\equiv}_5{=}$, $H_3C{=}{\equiv}_4{=}$, $H_3C{\equiv}_3{=}_2(CH_2)_3OH$

4. Inuleae-Gnaphalinae	Constituents
Antennaria alpina Gaertn.[35]	$H_3C{\equiv}_5{=}$, $H_3C{\equiv}_3{=}_2(CH_2)_3OH$
A. aprica Greene[35]	$H_3C{\equiv}_5{=}$, $H_3C{\equiv}_3{=}_2(CH_2)_3OH$
A. decipiens Greene[35]	$H_3C{\equiv}_5{=}$, $H_3C{\equiv}_3{=}_2(CH_2)_3OH$
A. dioeca Gaertn.[35]	$H_3C{\equiv}_5{=}$, $H_3C{\equiv}_3{=}_2(CH_2)_3OH$
A. gaspensis[35]	$H_3C{\equiv}_5{=}$, $H_3C{\equiv}_3{=}_2(CH_2)_3OH$
A. hansii Kern.[35]	$H_3C{\equiv}_5{=}$, $H_3C{\equiv}_3{=}_2(CH_2)_3OH$
A. howellii Greene[35]	$H_3C{\equiv}_5{=}$, $H_3C{\equiv}_3{=}_2(CH_2)_3OH$
A. imbricata E. Nelson[35]	$H_3C{\equiv}_5{=}$, $H_3C{\equiv}_3{=}_2(CH_2)_3OH$
A. intermedia Porslid.[35]	$H_3C{\equiv}_5{=}$

TABLE VII—Continued

4. Inuleae-Gnaphalinae	Constituents
A. microphylla Rydb.[35]	$H_3C\equiv_5=$
A. nagellanica[35]	$H_3C\equiv_5=$
A. neglecta Greene[35]	$H_3C\equiv_5=$, $H_3C\equiv_3=_2(CH_2)_3OH$
A. parvifolia Nutt.[35]	$H_3C\equiv_5=$
A. plataginea R. Br.[35]	$H_3C\equiv_5=$
A. rosea Greene[35]	$H_3C\equiv_5=$, $H_3C\equiv_3=_2(CH_2)_3OH$
Leontopodium alpinum Cass.[35]	$H_3C\equiv_5=$
L. calocephalum Beauverd.[35]	$H_3C\equiv_5=$
L. jacotianum Beauverd.[35]	$H_3C\equiv_5=$
L. kamschaticum Komarov.[35]	$H_3C\equiv_5=$
Anaphalis cinnamomea C. B. Clarke[35]	$H_3C\equiv_5=$, 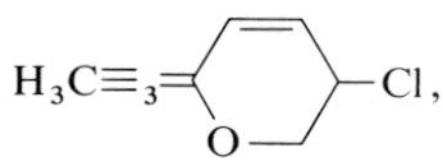 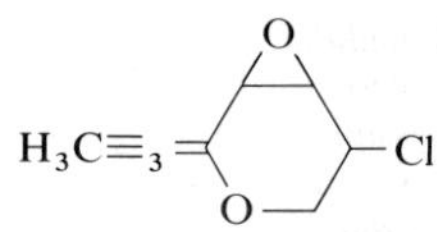
A. margaritacea B. *et* H.[102]	$H_3C\equiv_5=$, 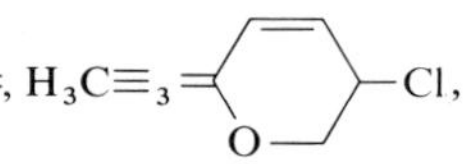 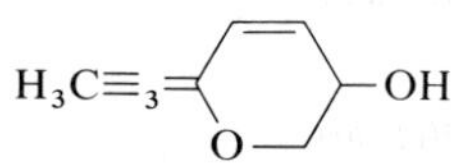
A. triplinervis Benth.[103]	$H_3C\equiv_5=$, 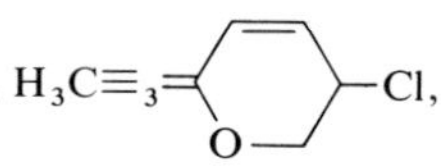 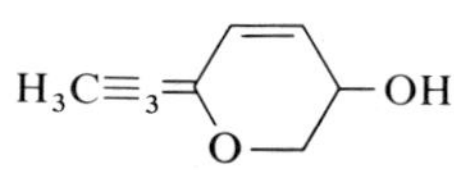
A. yeodensis Maxim.[35]	$H_3C\equiv_5=$, 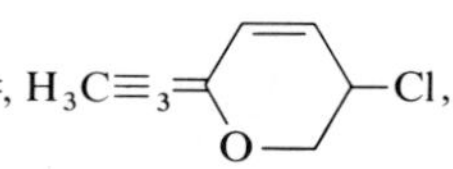 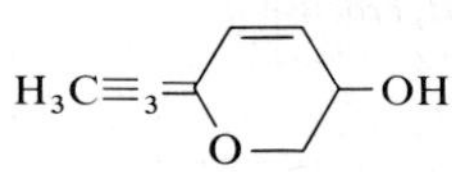
Phagnalon saxatile Cass.[35]	—
P. sordidum Reichb.[35]	—
Achyrocline satureioides DC.[35]	$H_3C\equiv_5=$, 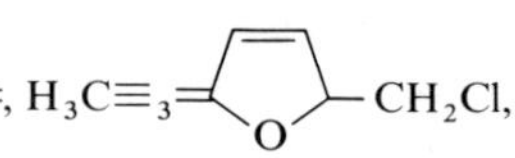 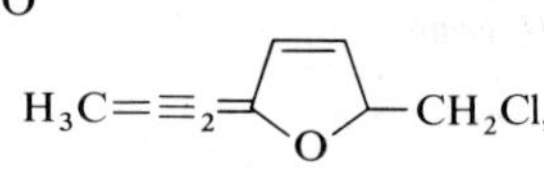
Gnaphalium falcatum Lam.[35]	—
G. gaūdichaūdianum DC.[35]	—

TABLE VII—Continued

4. Inuleae-Gnaphalinae	Constituents
G. luteo-album L.[75]	$H_3C{\equiv}_5{=}$
G. purpureum L.[35]	$H_3C{\equiv}_5{=}$
G. uliginosum L.[35]	$H_3C{\equiv}_5{=}$, $H_3C{\equiv}_3{=}_2(CH_2)_3OH$
G. obtusifolium L.[103]	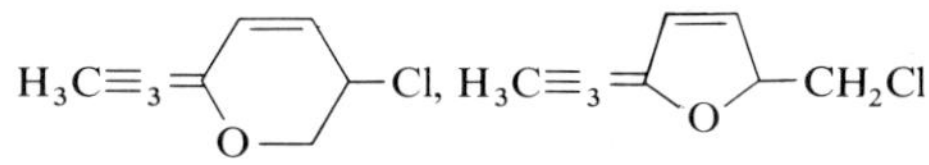
G. silvaticum L.[35]	—
G. sprengelii Hook. *et* Arn.[35]	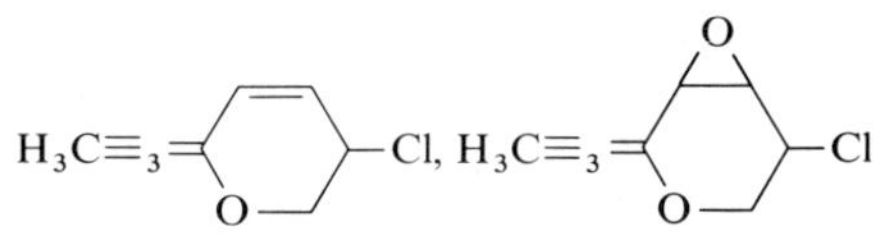
G. undulatum L.[35]	$H_3C{\equiv}_5{=}$, $H_3C{\equiv}_3{=}_2(CH_2)_3OAc$
G. vaticum[35]	$H_3C{\equiv}_5{=}$
Cassinia longifolia R. Br.[35]	$H_3C{\equiv}_5{=}$, $H_3C{\equiv}_3{=}_2(CH_2)_3OAc$
C. spectabilis R. Br.[35]	$H_3C{\equiv}_5{=}$, $H_3C{=}{\equiv}_4{=}$
C. vauvilliersii Hook.[35]	$H_3C{\equiv}_5{=}$, $H_3C{\equiv}_3{=}_2(CH_2)_3OAc$
Helipterum manglesii F. v. Müller[75]	$H_3C{\equiv}_5{=}$
H. roseum Benth.[35,75]	$H_3C{\equiv}_5{=}$, $H_3C{\equiv}_3{=}_2(CH_2)_3OH$
H. sandfordii Hook.[75]	$H_3C{\equiv}_5{=}$
Helichrysum allioides Less.[35]	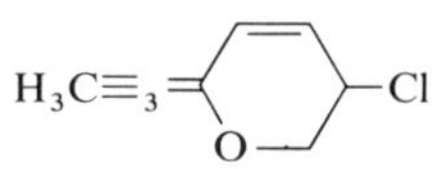
H. arenarium DC.[35]	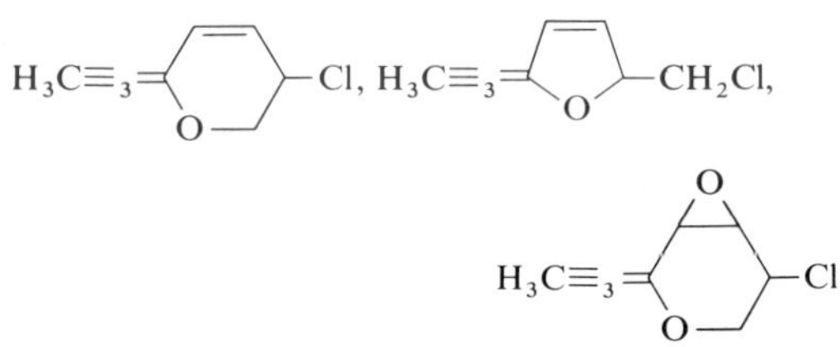
H. bracteatum Willd.[35]	$H_3C{\equiv}_5{=}$, $H_3C(CH_2)_4C{\equiv}CCH{=}CHCH(OH)(CH_2)_7CO_2R$
H. diosmaefoliūm Sweet.[35]	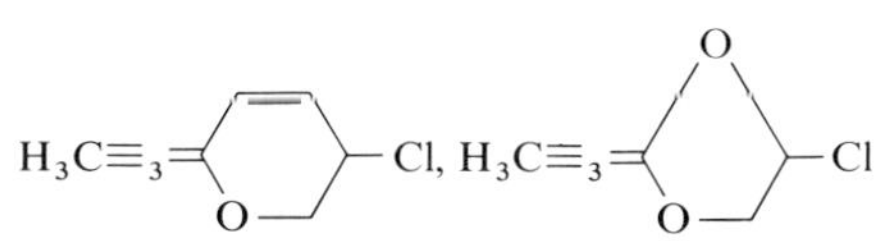
H. foetidum Moench.[35]	$H_3C{\equiv}_5{=}$
H. grandiflorum D. Don.[35]	$H_3C{\equiv}_5{=}$
H. lanatum DC.[35]	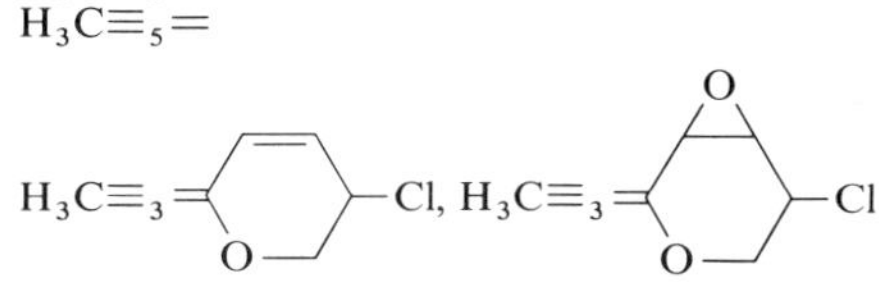

TABLE VII—Continued

4. Inuleae-Gnaphalinae	Constituents
H. latifolium Less.[35]	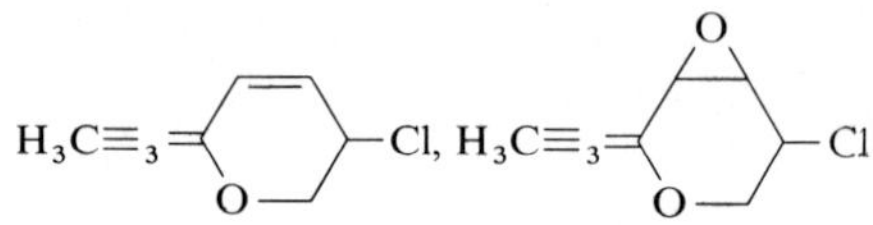
H. serotinum Boiss.[35]	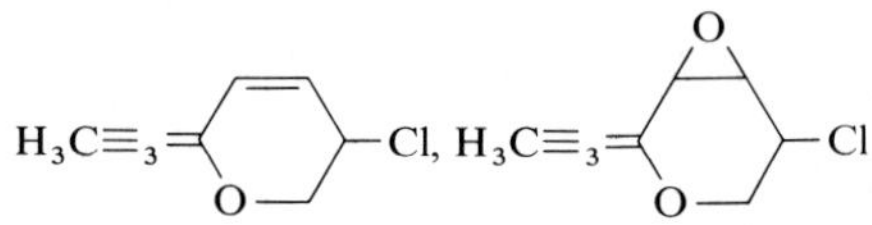
H. stoechas DC.[35]	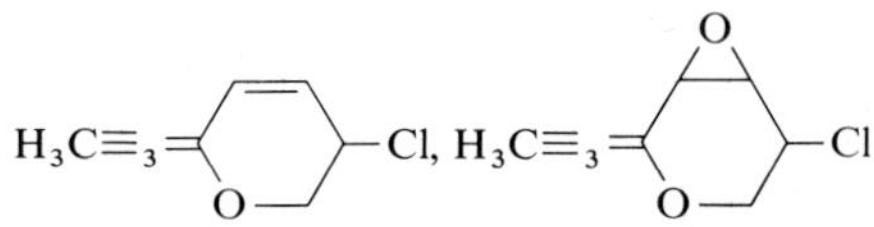
H. tianschanicum Regel.[35]	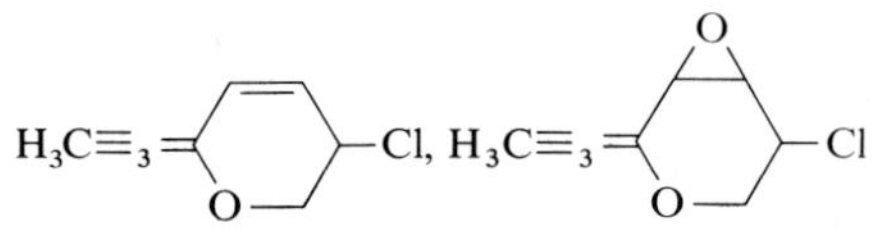
H. argenteum Thunb.[35]	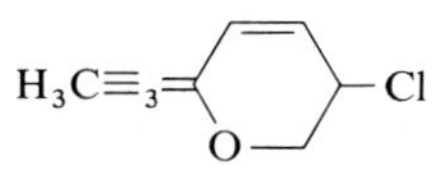
H. argyrophyllum DC.[35]	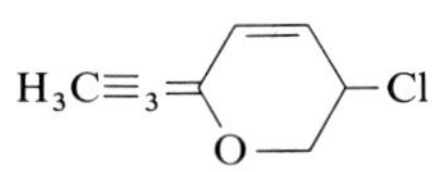
H. nudiflolium (L.) Less.[35]	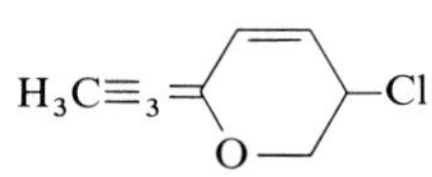
H. odoratissimum Sweet.[35]	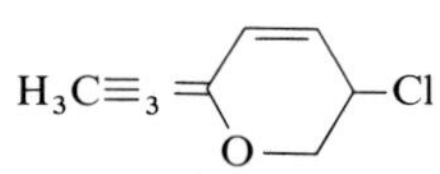
H. paniculatum Willd.[35]	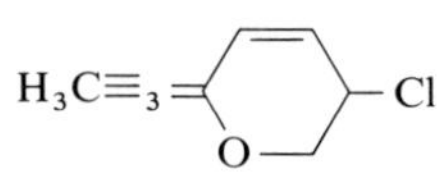
H. petiolatum DC.[35]	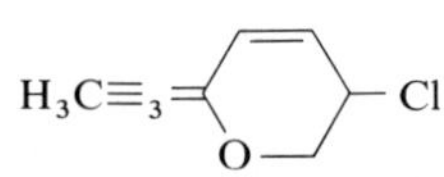
Ammobium alatum R. Br.[35]	$H_3C\equiv_5=$, $H_3C\equiv_3=_2(CH_2)_3OR$ (R = H, Ac)

TABLE VII—Continued

5. Inuleae-Agianthinae	Constituents
Caesulia axillaris Roxb.[35]	$H_3C\equiv_5=$
Myriocephalus guerinae F. v. Müller[35]	$H_3C\equiv_3=_2(CH_2)_3OH$
M. stuartii Benth.[35]	$H_3C\equiv_5=$
Calocephalus citreus Less.[354]	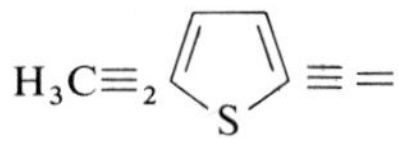

6. Inuleae-Relhaninae	Constituents
Metalasia muricata R. Br.[35]	$H_3C\equiv_5=$
Relhania quinquenervis Thunb.[35]	$H_3C\equiv_5=$
R. sessiliflora Thunb.[35]	$H_3C\equiv_5=$

7. Inuleae-Athrixinae	Constituents
Heterolepis aliena (L.f.) Druce[35]	$H_3C\equiv_5=$
Podolepis canescens Grah.[35]	—

8. Inuleae-Inulinae	Constituents
Inula ensifolia L.[35]	$H_3C\equiv_5=$
I. helenium L.[35]	$H_3C\equiv_5=$
I. hirta L.[35]	$H_3C\equiv_5=$
I. latifolia DC.[35]	$H_3C\equiv_5=$
I. magnifica[35]	$H_3C\equiv_5=$
I. royleana DC.[35]	$H_3C\equiv_5=$
I. spiraeifolia L.[35]	$H_3C\equiv_5=$
I. salicina L.[35]	$H_3C\equiv_5=$
I. vulgaris Trev.[35]	$H_3C\equiv_5=$
Humea elegans Sm.[35]	$H_3C\equiv_5=$
Vieraea laevigata Webb.[35]	$H_3C\equiv_5=$, $H_3C\equiv_3=_2(CH_2)_3OH$
Jasonia perennis[35]	$H_3C\equiv_5=$
J. tuberosa DC.[35]	$H_3C\equiv_5=$, $H_3C\equiv_3=_2(CH_2)_3OH$
Allagopappus dichotomus Cass.[35]	$H_3C\equiv_5=$, $H_3C\equiv_2\overset{}{=}\equiv_2=$ (with $SOCH_3$ substituent)
Pulicaria dysenterica Bernh.[35,347]	$H_3C\equiv_5=$, $H_3C=\equiv_4=$
P. odora Reichb.[35,347]	$H_3C\equiv_5=$, $H_3C=\equiv_4=$
P. sicula Moris.[35,347]	$H_3C\equiv_5=$, $H_3C=\equiv_4=$
P. uliginosa Stev.[35,347]	$H_3C\equiv_5=$, $H_3C=\equiv_4=$
P. vulgaris Gaertn.[35,75,347]	$H_3C\equiv_5=$, $H_3C=\equiv_4=$
Carpesium abrotanoides L.[35]	$H_3C\equiv_3=$
C. cernuum L.[35]	$H_3C\equiv_5=$, $H_3C\equiv_3=_2(CH_2)_3OAc$

9. Inuleae-Buphthalminae	Constituents
Pallenis spinosa Cass.[35]	$H_3C\equiv_5=$, $H_3C=\equiv_4=$, $H_3C=\equiv_3=_2$, $H_3C=\equiv_2=_3$
Buphthalmum grandiflorum L.[91]	$H_3C\equiv_5=$, 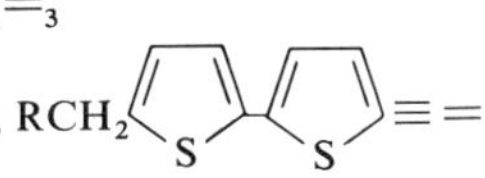 (R = H, OH, AcO)

TABLE VII—Continued

9. Inuleae-Buphthalminae	Constituents
B. salicifolium L.[91,108]	$H_3C\equiv_5=$, RCH_2 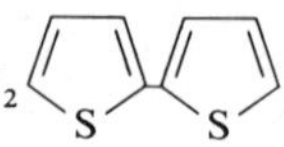 $\equiv=$ (R = H, OH, AcO) $H_3C\equiv_3=_3$, $H_3C\equiv_2\urcorner=_3$ (with SCH_3)
Telekia speciosa Baumg.[91]	$H_3C\equiv_5=$
T. cordifolia DC.[35]	$H_3C\equiv_5=$
Odontospermum maritimum Sch. Bip.[35]	$H_3C\equiv_5=$

As already mentioned, most species belonging to the tribus Inuleae contain the pentayne (**94**) (see Chapter 2, Section 2.6). Only in the subtribus Inuleae-Gnaphalinae are the chloro-enolethers (**183**–**188**) more common (see Chapter 2, Section 7.1). However, even here are several genera with the pentaynene (**94**) as the main compound very often occurring together with the precursor (**36**) (see Chapter 2, Section 2.1).

Only in a few cases have thiophene acetylenes also been isolated. These compounds are, however, typical for the genus *Buphthalmum* except for those species now combined in *Telekia*, where once again, only the pentaynene is present. The constituents of *Buphthalmum* species are almost identical with those of *Wedelia forsteriana* belonging to the tribus Heliantheae and showing the close relationship which is also suggested by botanical reasons.

1.4. Tribus Heliantheae

This large tribus is divided, from the botanical viewpoint, into ten subtribus. Again most of them have species with pentaynene (**94**) (see Chapter 2, Section 2.6) as a typical constituent. However, in some subtribus, other compounds are characteristic.

In the subtribus Heliantheae-Coreopsidinae, the enetetraynene (**436**) (see Chapter 3, Section 3.5) and its derivatives, together with several other acetylenes, are very common and with only a few exceptions. In the subtribus H.-Galinsoginae, the dehydrofalcarinone (**52**) (see Chapter 3, Section 1.2.5) and other C_{17}-acetylenes seem to be characteristic. But some C_{17}-compounds have been isolated in parts of other tribus. Even on the level of the genera, clear differentiations can be seen in several cases. The importance of these observations however has to be further established by further chemical and botanical work.

The first subtribus has only one genus showing the presence of dehydrofalcarinone which is otherwise only found in the genera *Jaegeria*, *Helianthus* and *Tithonia* belonging to subtribus H.-Verbessinae, in the subtribus H.-Galinsoginae and in an Iva species.

In the next four subtribus, the polyynes are not very characteristic. The pentaynene (**94**) is very common, sometimes occurring together with sulphur derivatives. Only in the genus Silphium could no acetylenes be detected.

The subtribus H.-Zinninae seems not to be homogenous. The genus *Heliopsis* has constituents more common in the subtribus H.-Coreopsidinae, while *Sanvitalia* has a very specific acetylene and the genus *Zinnia* only very small amounts of the pentayne (**94**).

The botanically complicated subtribus H.-Verbesinae is also not very uniform. Again, several genera contain the pentayne or simple sulphur derivatives, while others are more complicated, e.g. Eclipta, where a very large amount of structure variation can be seen. In addition, the genus Rudbeckia, already divided into sections, contains very different acetylenes. Most common are thiophenes derived from the pentayne and also from the tetraynediene. However, the section *Echinacea* has amides as typical constituents; these are also typical for *Spilanthes*. *Wedelia*, as already mentioned, has the same constituents as *Buphthalmum* belonging to the tribus Inuleae.

The most complex subtribus is H.-Coreopsidinae. The first genera shows still some relationship to the preceding one (*Guizotia* and *Synedrella*) as the pentayne (**94**) and its derivatives are still dominant. However, the occurrence of less unsaturated compounds typical for the following genera can also be observed. The large genus *Coreopsis* contains many different acetylenes. Typical of these are the ene-tetrayn-ene and its derivatives together with phenylacetylenes and less unsaturated compounds and there is a relatively high concentration of acetylenes in the leaves. This is also true for the genera *Dahlia*, *Bidens* and *Heterospermum*, which seem to be very closely related to *Coreopsis*. The occurrence of acetylenes having longer carbon chains is also typical for dahlias. This is also characteristic for the related genera *Glossocardia* and *Isostigma*, which have only C_{17}-acetylenes. In the genus *Cosmos* two groups can be differentiated, one related to *Bidens* and a second one having no acetylenes.

The occurrence of dehydrofalcarinone in the subtribus H.-Galinsoginae has been mentioned before. However, in the genus *Tridax* we have many other C_{17}-acetylenes, several of them being unique. The genus *Calea* however, does not have constituents in agreement with the other genera already investigated.

The last subtribus H.-Madinae again does not contain very typical acetylenes but the pentayne and its sulphur derivatives are common.

TABLE VIII
TRIBUS HELIANTHEAE

1. Heliantheae-Lagascinae	Constituents (roots)
Lagascea mollis Cav.[105]	$H_3C{\equiv}_5{=}$, ${=}CO{\equiv}_2CH_2{=}(CH_2)_5{=}(C_{17})$

2. Heliantheae-Millerinae	Constituents (roots)
Elvira biflora DC.[351]	$H_3C{\equiv}_5{=}$
Elvira martyni Cass.[35]	$H_3C{\equiv}_5{=}$
Milleria quinqueflora L.[35]	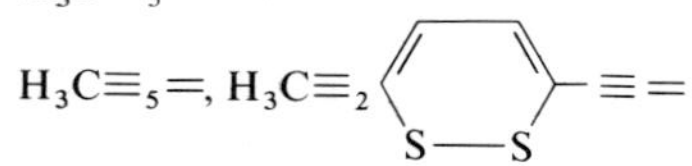

3. Heliantheae Melampodinae	Constituents (roots)
Ichthyothere terminalis Spreng.[59]	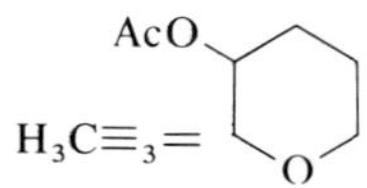
Polymnia connata Blake[35]	$H_3C{\equiv}_5{=}$
P. uvedalia L.[35]	$H_3C{\equiv}_5{=}$
Melampodium divaricatum DC.[50,89]	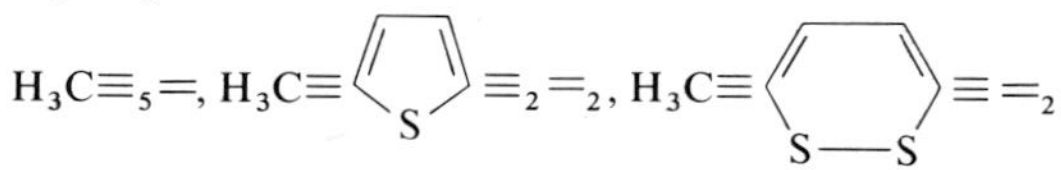
M. longifolium Brouss. *et* Willd.[50,89]	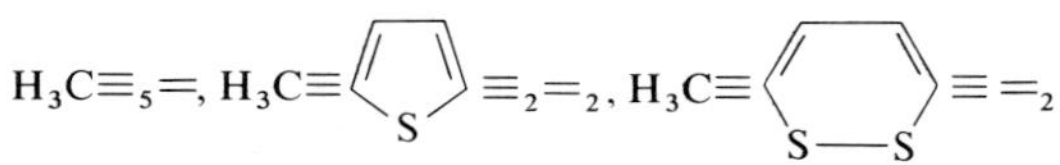
M. rhomboideum DC.[35]	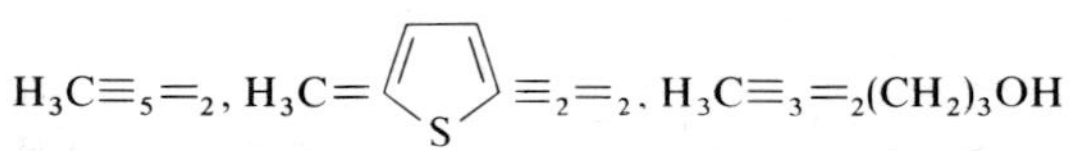
M. perfoliatum H.B. *et* K.[35]	$H_3C{\equiv}_5{=}$, — $H_3C{\equiv}_3{=}_2(CH_2)_3OH$

M. paludosum H.B. *et* K.[35]	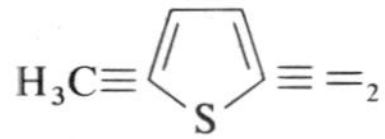
Acanthospermum hispidum DC.[35]	$H_3C{\equiv}_5{=}$
A. humile DC.[35]	$H_3C{-}_5{=}$, $H_3C{\equiv}_2\underset{SOCH_3}{|}{=}{\equiv}_2{=}$
Silphium asteriscus L.[35]	—
S. scaberrimum Ell.[35]	—
S. integrifolium Michx.[35]	—
S. hornemannii Schrad. ex DC.[35]	—
S. perfoliatum L.[35]	—
Berlandiera texana DC.[35]	—
Lindheimera texana A. Gray.[35]	$H_3C{\equiv}_5{=}$
Parthenium integrifolium L.[35]	$H_3C{\equiv}_5{=}$

4. Heliantheae-Ambrosinae	Constituents (roots)
Iva xanthiifolia Nutt.[50,89,105]	$H_3C{\equiv}_5{=}$, $H_3C{=}{\equiv}_4{=}$, ${=}CO{\equiv}_2CH_2{=}(CH_2)_5{=}$, 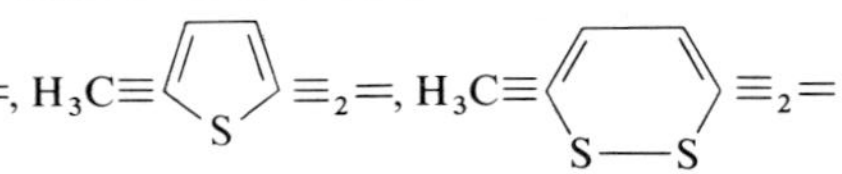
Ambrosia eliator L.[89]	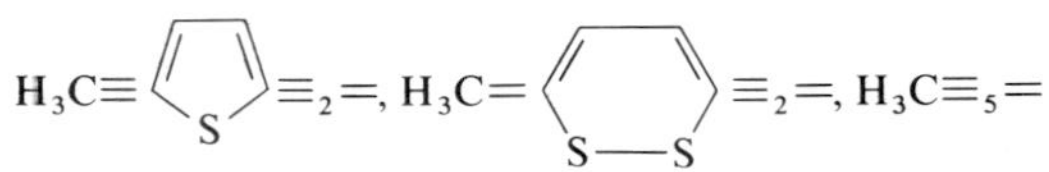
A. artemisiifolia L.[50,89]	$H_3C{\equiv}_5{=}$, 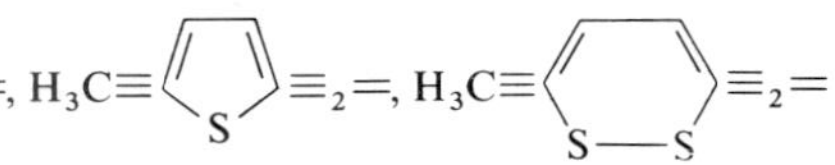
A. trifoliata[50,89]	$H_3C{\equiv}_5{=}$, 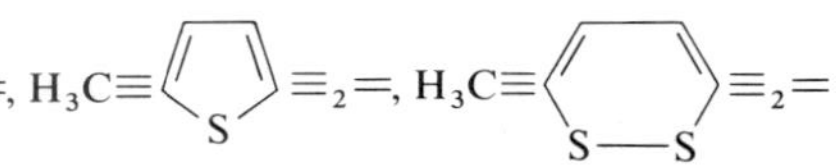
A. trifida L.[50,89]	$H_3C{\equiv}_5{=}$, 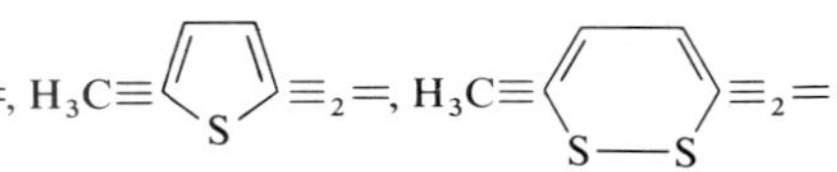

TABLE VIII—Continued

4. Heliantheae-Ambrosinae	Constituents (roots)
Xanthium canadense A. Gray[35]	$H_3C{\equiv}_5{=}$
X. orientale Best. var. italicum[35]	$H_3C{\equiv}_5{=}$, $H_3C{\equiv}_3{=}_3$
X. strumarium L.[35,75]	$H_3C{\equiv}_5{=}$
5. Heliantheae-Petrobinae	**Constituents (roots)**
6. Heliantheae-Zinninae	
Zinnia angustifolia H.B. *et* K.[35]	$H_3C{\equiv}_5{=}$
Z. elegans Jacq.[35]	$H_3C{\equiv}_5{=}$
Z. haageana Regel[35]	$H_3C{\equiv}_5{=}$
Z. linearis Benth.[35]	$H_3C{\equiv}_5{=}$
Z. peruviana L.[35]	$H_3C{\equiv}_5{=}$
Z. pumila A. Gray[35]	$H_3C{\equiv}_5{=}$
Z. tenuifora Jacq.[35]	—
Z. verticillata And.[35]	—
Sanvitalia procumbens Lam.[237,75]	$H_3C{\equiv}_5{=}$, $H_3C{=}{\equiv}_2{=}_3CO_2CH_3$, $H_3C{\equiv}_3{=}_2(CH_2)_3OH$
Heliopsis grandiflora[35] *H. helianthoides* Sweet.[35] *H. scabra Torr. et* Gray.[35]	$H_3C{\equiv}_5{=}$, $H_3C{=}{\equiv}_4{=}$, $OCH{=}{\equiv}_4{=}$ Leaves: $Ph{\equiv}_3CH_3$ $Ph{\equiv}_3CH_2R$ (R = H, OH)
7. Heliantheae-Verbesininae	**Constituents (roots)**
Siegesbeckia flosculosa L'Hérit[35]	$H_3C{\equiv}_5{=}$
S. orientalis L.[35]	$H_3C{\equiv}_5{=}$
Jaegeria hirta Less.[35]	Leaves: ${=}CO{\equiv}_2CH_2{=}(CH_2)_5{=}$
Eclipta alba (L.) Hassk.[50,86]	$H_3C{\equiv}_5{=}$, $H_3C{\equiv}_2$ 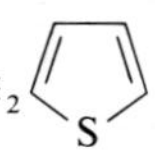 ${\equiv}{=}$, and others
E. martynii Cass.[35]	$H_3C{\equiv}_5{=}$

E. erecta L.[50,86]	Roots: $H_3C{\equiv}_5{=}$, $H_3C{\equiv}_2\underset{SCH_3}{C}{=}{\equiv}_2{=}$, $H_3C{\equiv}_3{=}_2(CH_2)_3OH$, $H_3C{\equiv}_3{=}_2\underset{OAc}{C}HCH_2CH_2OAc$,
E. prostata L.[86]	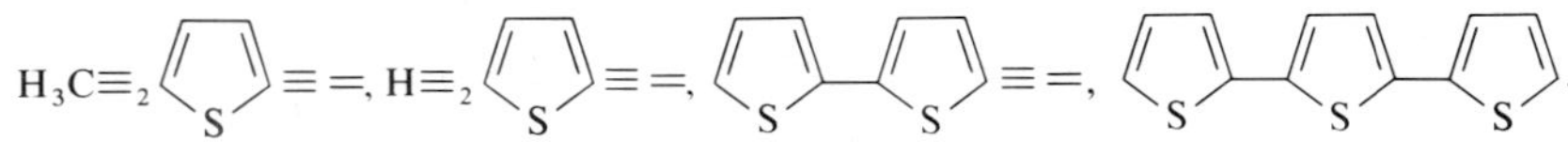 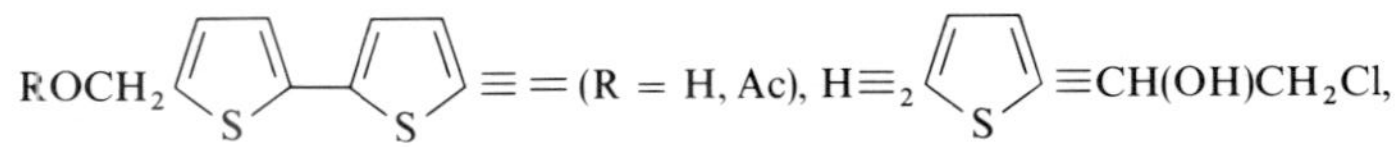 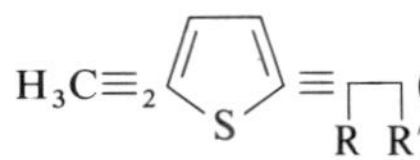 (R = OH, R′ = OAc; R = OH, R′ = Cl; R = OAc, R′ = Cl) 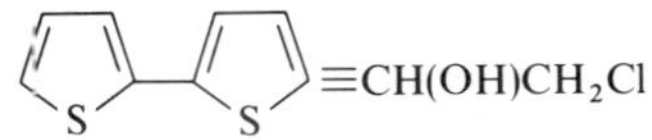 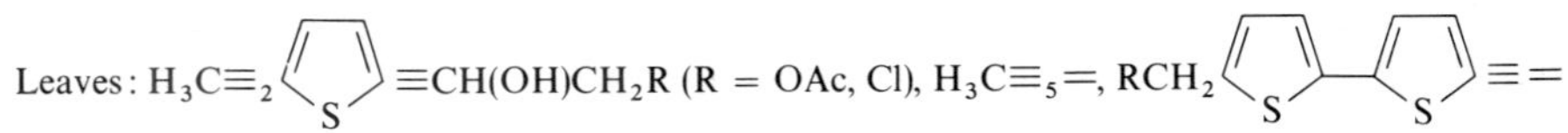 (R = H, OH, OAc, O-Ang., O-Tigl., $OCOCH(CH_3)_2$, 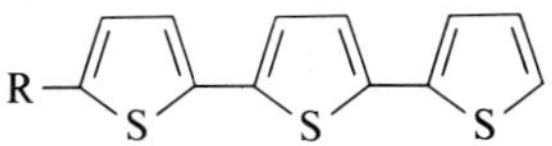 (R = H, CH_2OH, CHO, CH_3, CH_2OAc, CH_2O-Ang., CH_2O-Tigl., $CH_2OCOCH(CH_3)_2$)
Sclerocarpus africanus Jacq.[35]	—
S. unisersialis Benth. *et* Hook.[35]	—
Rudbeckia	
Section *Eurudbeckia*	
Rudbeckia bicolor Nutt.[50,89]	$H_3C{\equiv}_4{=}_2$, 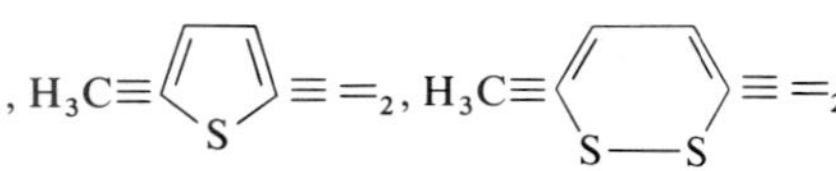
R. digitata Mill.[35]	$H_3C{\equiv}_5{=}$
R. flava Greene[35]	—

TABLE VIII—Continued

Section *Eurudbeckia*	
R. fulgida Ait.[35]	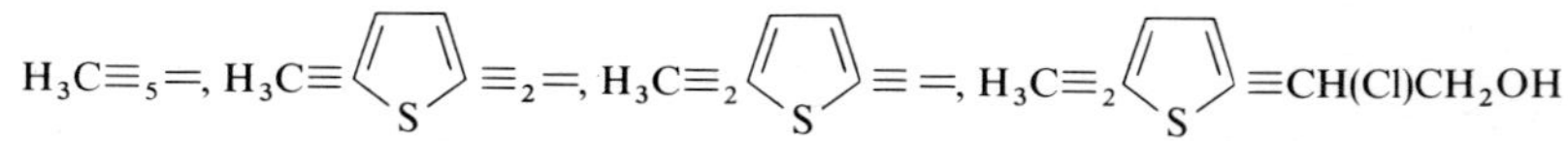
R. hirta L.[89]	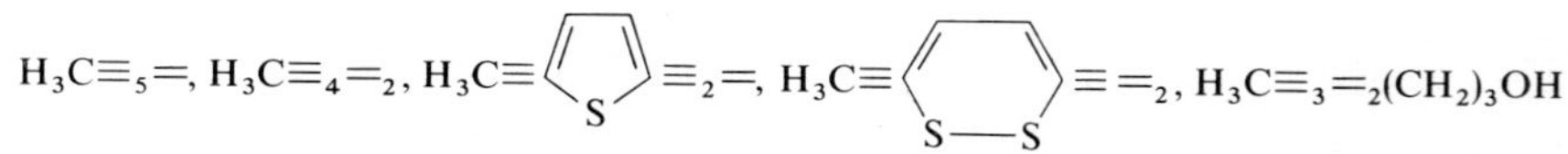
R. laciniata L.[35]	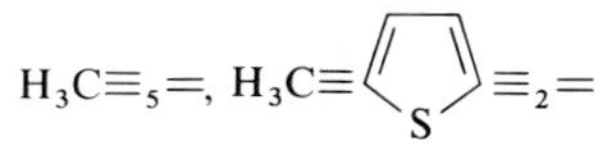
R. nitida Nutt.[35]	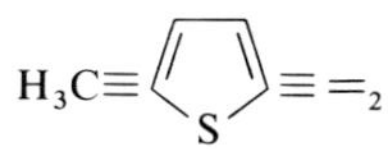
R. neumanni Steud.[35]	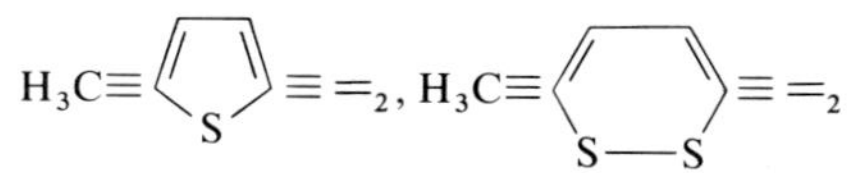
R. speciosa Lk.[89]	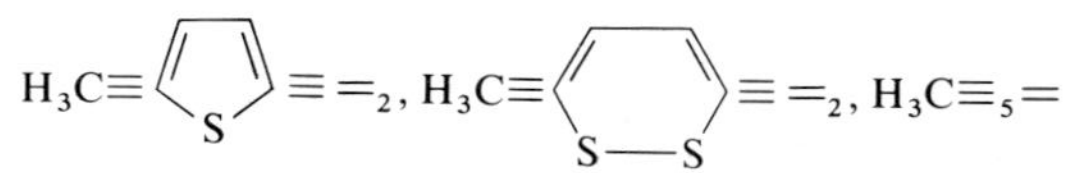
R. sullivantii Boyton[35]	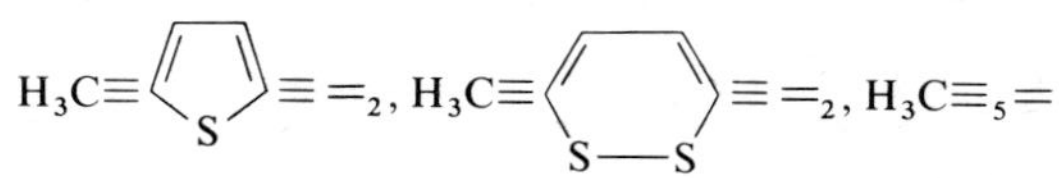
R. triloba L.[35]	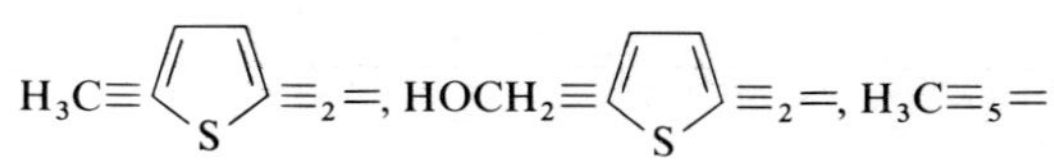

Section *Dracopxis*	
R. amplexicaulis Vahl.[50,74]	Roots: $H_3C{\equiv}_5{=}$, $H_3C{\equiv}_3{=}_2(CH_2)_3OH$
	Leaves: R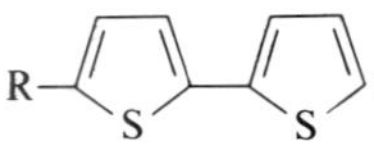$\equiv\,=$, (R = CH_3, CH_2OH, CH_2OAc)
	H_3C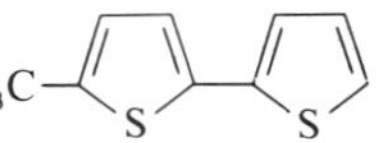$=_2$, $H_3C{\equiv}_3{=}_2CH(OAc)CH_2CH_2OAc$
Section *Echinacea*	
Echinacea angustifolia DC.[284,348] *E. purpurea* Moench.[284,348]	$H_3C{\equiv}_5{=}$, $H_3C{\equiv}_2CH_2CH_2{=}_2CONHCH_2CH(CH_3)_2$, $H_3C{=}{\equiv}_4{=}$, $H_3C{\equiv}_4{=}_2$, $H_3C{\equiv}_3{=}_3$, $H_3C{\equiv}_3{=}$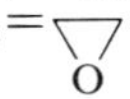$=$
Section *Lepachys Raf.*	
Ratibida pinnata Barnh.[35]	$H_3C{\equiv}_5{=}$
Blainvillea rhomboidea Cass.[35]	$H_3C{\equiv}_5{=}$, $H_3C{\equiv}_3{=}_2(CH_2)_3OH$
Wedelia forsteriana Endl.[108]	Roots: $H_3C{\equiv}_5{=}$
	Leaves: $H_3C{\equiv}_5{=}$, $H_3C{=}{\equiv}_4{=}$, $H_3C{\equiv}_2$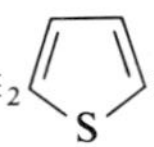$\equiv\,=$,
	R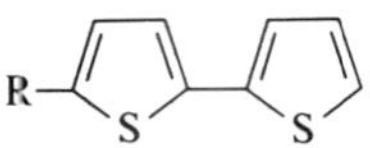$\equiv\,=$ (R = CH_3, CH_2OAc, CH_2OH), $H_3C{\equiv}_2{=}(SCH_3){=}_3$
W. paludosa DC.[35]	$H_3C{\equiv}_5{=}$, $H_3C{\equiv}$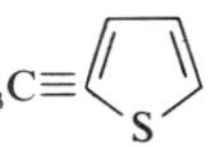$\equiv_2{=}$
Wyethia angustifolia Nutt.[35]	$H_3C{\equiv}_5{=}$

TABLE VIII—Continued

Section *Lepachys Raf.*	Constituents (roots)
Aspilia montevidensis Kuntze[35]	$H_3C\equiv$ 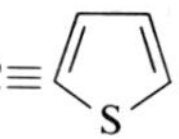 $\equiv_2=$
Tithonia rotundifolia Blake[35]	$H_3C\equiv_5=$, $=CO\equiv_2CH_2=(CH_2)_5=$
T. speciosa Hook.[105]	$H_3C\equiv_5=$, $H_3C=\equiv_4=$
T. tagetifloria Desf.[35]	$H_3C\equiv_5=$, $H_3C=\equiv_4=$
Viguiera grandiflora Gardn.[35]	—
Helianthus annuus L.[105]	$=CO\equiv_2CH_2=(CH_2)_5=$
H. californicus DC.[105]	$=CO\equiv_2CH_2=(CH_2)_5=$
H. decapetalus L.[35]	$=CO\equiv_2CH_2=(CH_2)_5=$, $H_3C\equiv_5=$
H. giganteus L.[35]	$=CO\equiv_2CH_2=(CH_2)_5=$, $H_3C\equiv_5=$
H. helianthoides[35]	$=CO\equiv_2CH_2=(CH_2)_5=$
H. lactiflorus Pers.[35]	$=CO\equiv_2CH_2=(CH_2)_5=$, $H_3C\equiv_5=$
H. macrophyllus Willd.[105]	$=CO\equiv_2CH_2=(CH_2)_5=$
H. maximiliani Schrad.[35]	$=CO\equiv_2CH_2=(CH_2)_5=$
H. mollis Lam.[35]	$H_3C\equiv_5=$
H. nobilis[35]	$\equiv_2=$
H. occidentalis Riddell.[35]	$H_3C\equiv_5=$
H. orgialis DC.[105]	$-(C\equiv C)_2-CH=CH-$
H. scaberrimus Ell.[35]	$=CO\equiv_2CH_2=(CH_2)_5=$, $H_3C\equiv_5=$
H. strumosus L.[35]	$=CO\equiv_2CH_2=(CH_2)_5=$
H. tuberosus L.[105]	$=CO\equiv_2CH_2=(CH_2)_5=$
Flourensia cernua DC.[35]	—
Melanthera deltoidea Michx.[35]	$H_3C\equiv_5=$
Spilanthes acmella Murr.[35]	—
S. americana Hieron.[35]	$R\equiv_2CH_2CH_2=_2CONHCH_2CH(CH_3)_2$ ($R = CH_3, H$)
S. mauritiana DC.[35]	$R\equiv_2CH_2CH_2=_2CONHCH_2CH(CH_3)_2$ ($R = CH_3, H$)
S. oleracea Jacq.[35]	$R\equiv_2CH_2CH_2=_2CONHCH_2CH(CH_3)_2$ ($R = CH_3, H$)
S. stolonifera DC.[35]	$R\equiv_2CH_2CH_2=_2CONHCH_2CH(CH_3)_2$ ($R = CH_3, H$)
Encelia farinosa A. Gray[35]	$H_3C\equiv_5=$
Helianthella quinquenervis A. Gray[35]	$H_3C\equiv_5=$

H. uniflora Torr. *et* A. Gray[350]	$H_3C{\equiv}_5{=}$ + chromenes and benzofuranes	
Podachaenium eminens Benth. *et* Oerst.[35]	$H_3C{\equiv}_5{=}$	
Actinomeris helianthoides Nutt.[35]	$H_3C{\equiv}_5{=}$, $H_3C{\equiv}_3{=}_3$	
A. nemorosa[35]	$H_3C{\equiv}_5{=}$, $H_3C{=}{\equiv}_4{=}$	
A. tetraptera DC.[35]	$H_3C{\equiv}_5{=}$	
Verbesina alata L.[35]	$H_3C{\equiv}_5{=}$, $H_3C{=}{\equiv}_4{=}$, $H_3C{\equiv}_2$ 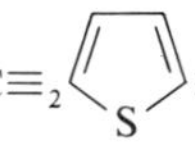 $\equiv=$, $H_3C{\equiv}_2$ 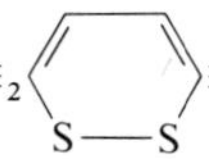 $\equiv=$	
V. alternifolia Brith.[35]	$H_3C{\equiv}_5{=}$	
V. enceloides Benth. *et* Hock.[35]	$H_3C{\equiv}_5{=}$, $H_3C{=}{\equiv}_4{=}$	
V. helianthoides H.B. *et* K.[35]	$H_3C{\equiv}_5{=}$	
8. Heliantheae-Coreopsidinae	Constituents (roots)	
Guizotia abyssinica Cass.[35]	Roots: $H_3C{\equiv}_5{=}$, $H_3C{\equiv}_2{=}{\equiv}_2{=}$ (with SCH_3 substituent), $H_3C{\equiv}_3{=}_2(CH_2)_3OH$	
	Leaves: $H_3C{=}{\equiv}_3{=}_2$, $H_3C{=}{\equiv}_2{=}_3$, $H_3C{=}{\equiv}_4{=}$, $H_3C{=}{\equiv}$ 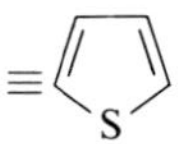 $\equiv=$	
G. oleifera DC.[74]	Roots: $H_3C{\equiv}_5{=}$, $H_3C{=}{\equiv}_4{=}$, $H_3C{\equiv}_3{=}_2(CH_2)_3OH$	
	Leaves: $H_3C{=}{\equiv}_4{=}$, $H_3C{=}{\equiv}$ 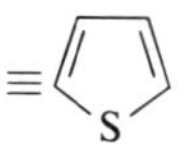 $\equiv=$	
Synedrella nudiflora Gaertn.[75]	$H_3C{\equiv}_5{=}$	
S. uniflora[35]	$H_3C{\equiv}_5{=}$, $H_3C{\equiv}_3{=}_3$, $H_3C{\equiv}_3{=}_2(CH_2)_3OH$	
	Constituents	
	Roots	Leaves
Coreopsis auriculata L.[35]	$R{=}{\equiv}_4{=}(R = CH_3, CHO)$ $H_3C{\equiv}_3{=}_2(CH_2)_3OH$	$Ph{\equiv}_3CH_3$ $H_3C{\equiv}_3{=}_2(CH_2)_3OR$ (R = H, Ac)

TABLE VIII—Continued

8. Helliantheae-Coreopsidinae	Constituents: Roots	Constituents: Leaves
C. atrosanguinea[35]	$R=\equiv_4=(R = CH_3, CH_2OH)$ 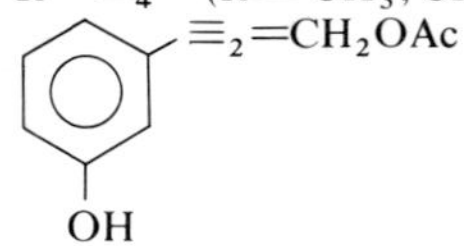	$H_3C=\equiv_2=_3, H_3C=\equiv_3=_2, H_3C=\equiv_4=$
C. bicolor Bosse[35]	$H_3C=\equiv_4=$	$H_3C=\equiv_4=, H_3C=\equiv_3=_2, H_3C=\equiv_2=_3$
C. basalis[35]	$R=\equiv_4=(R = CH_3, CHO, CH_2OH, CH_2OAc)$ $Ph\equiv_3CH_3$	$Ph\equiv_3CH_2R(R = H, OH)$ $H_3C\equiv_5=, H_3C=\equiv_4=$
C. bigelowii H. M. Hall.[35]	$R=\equiv_4=$ $(R = CH_3, CHO, CH_2OH, CH_2OAc)$, $H_3C=\equiv_2=_3$, 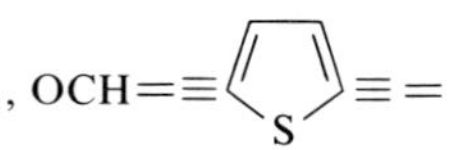	as roots
C. calliopsidea A. Gray[35]	$R=\equiv_4=$ $(R = CH_3, CH_2OH, CH_2OAc, CHO)$	as roots
C. coronata Hook.[35]	$R=\equiv_4=$ $(R = CH_3, CH_2OH, CH_2OAc, CHO)$	as roots
C. cardaminefolia Torr. *et* Gray[244]	$H_3C=\equiv_4=, H_3C=\equiv_3=_2, H_3C=\equiv_2=_3$ $Ph\equiv_3CH_3, Ph\equiv_2=CH_2R (R = H, OAc)$ 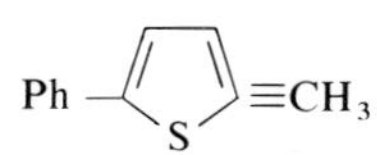	
C. delphinifolia Lam.[35]	$H_3C\equiv_5=, R=\equiv_4=$ $(R = CH_3, CHO, CH_2OAc)$ $H_3C\equiv_3=_2(CH_2)_3OR (R = H, Ac)$	$H_3C\equiv_3=_2(CH_2)_3OR (R = H, Ac)$
C. douglasii H. M. Hall.[35]	$H_3C\equiv_5=, H_3C=\equiv_4=, Ph\equiv_3CH_2R$ $(R = H, OAc), H_3C\equiv_3=_2(CH_2)_3OH$	$Ph\equiv_3CH_2R (R = H, OH)$ $H_3C\equiv_3=_2(CH_2)_3OH$

C. drummondii Torr. *et* Gray[247,69]	as *C. cardaminefolia*	
C. elegans hort.[35]	$H_3C{=}{\equiv}_2{=}_3$, $R{=}{\equiv}_4{=}$ (R = CH_3, CHO, CH_2OH, CH_2OAc)	$H_3C{=}{\equiv}_2{=}_3$, $H_3C{=}{\equiv}_3{=}_2$
C. gigantea H. M. Hall.[199]	$RCH_2{=}{\equiv}_2{=}_2(CH_2)_4{=}$(R = H, OH) $H_3C{=}{\equiv}_2{=}_2(CH_2)_3OH$	—
C. grandiflora Hogg.[69]		$Ph{\equiv}_3CH_2R$ (R = H, OH, OAc)
	$H_3C{=}{\equiv}_4{=}$, $H_3C{=}{\equiv}_3{=}_2$, $Ph{\equiv}_3CH_3$	$H_3C{\equiv}_3{=}_2(CH_2)_3OR$ (R = H, Ac)
C. lanceolata L.[69]	$H_3C{\equiv}_5{=}$, $RCH_2{=}{\equiv}_4{=}$ (R = H, OH, OAc) $H_3C{=}{\equiv}_3{=}_2$, $H_3C{\equiv}_3{=}_3$	$Ph{\equiv}_3CH_3$, $H_3C{\equiv}_3{=}_2(CH_2)_3OAc$
C. maritima Hook.[35]	$R{=}{\equiv}_4{=}$(R = CH_3, CHO, CH_2OH)	$RCH_2{=}{\equiv}_4{=}$(R = H, OH)
C. nuecensis A. Heller[73]	$H_3C{\equiv}_5{=}$, $RCH_2{=}{\equiv}_4{=}$ (R = H, OH, OAc) $Ph{\equiv}_3CH_3$, $Ph{\equiv}_2{=}CH_2OAc$	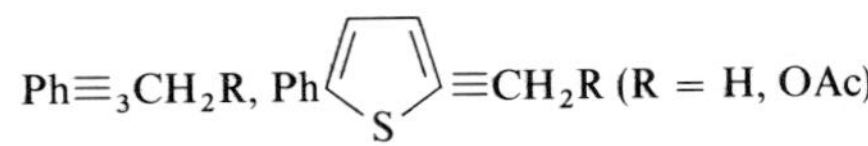 $Ph{\equiv}_2{=}CH_2OH$, $H_3C{\equiv}_3{=}_2(CH_2)_3OR$ (R = H, Ac) $H_3C{=}{\equiv}_4{=}$
C. palmata Nutt.[35]	$RCH_2{=}{\equiv}_4{=}$(R = H, OH, OAc)	$Ph{\equiv}_3CH_2R$ (R = H, OAc)
C. pubescens Ell.[35]	$H_3C{\equiv}_5{=}$, $RCH_2{=}{\equiv}_4{=}$(R = H, OH, OAc)	$H_3C{\equiv}_3{=}_2(CH_2)_3OR$ (R = H, Ac).
C. radiata Mill.[35]	$RCH_2{=}{\equiv}_4{=}$(R = H, OH, OAc) $Ph{\equiv}_2{=}CH_2OAc$	$H_3C{=}{\equiv}_4{=}$, $H_3C{\equiv}_3{=}_2(CH_2)_3OR$ (R = H, Ac) $H_3C{=}{\equiv}_2{=}_2(CH_2)_3OR$ (R = H, Ac) $H_3C{\equiv}_3{=}_2CH(OR)(CH_2)_2OR$ (R = H, Ac)
C. saxicola Alex[69]	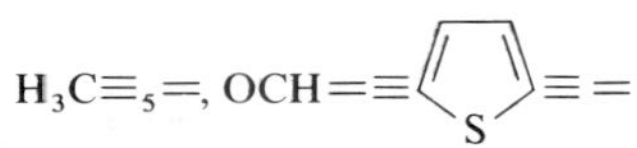	$Ph{\equiv}_3CH_2R$ (R = H, OH, Ac)
	$R{=}{\equiv}_4{=}$ (R = CH_3, CHO, CH_2OH, CH_2OAc)	$H_3C{\equiv}_3{=}_2(CH_2)_3OR$ (R = H, Ac)
C. stuhlmannii O. Hoffm.[35]	$R{=}{\equiv}_4{=}$ (R = CH_3, CHO, CH_2OH, CH_2OAc)	as roots

TABLE VIII—Continued

8. Heliantheae-Coreopsidinae	Constituents: Roots	Constituents: Leaves
C. tinctoria var. purpurea[250]	$R{=}{\equiv}_4{=}$ (R = CH_3, CHO, CH_2OH, CH_2OAc) $H_3C{=}{\equiv}_2{=}_3$, $Ph{\equiv}_2{=}CH_2OR$ (R = H, Ac) 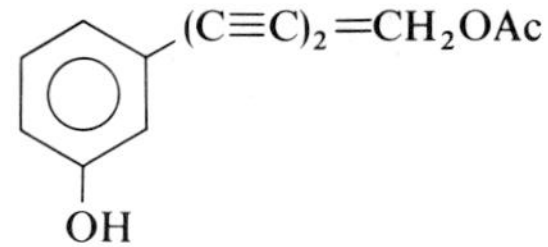	$H_3C{=}{\equiv}_2{=}_3$, $Ph{\equiv}_2{=}CH_2OAc$ $HOCH_2{=}{\equiv}_4{=}$
C. tinctoria Nutt.[244,247]	$R{=}{\equiv}_4{=}$ (R = CH_3, CHO, CH_2OH, CH_2OAc) $H_3C{=}{\equiv}_2{=}_3$	$H_3C{=}{\equiv}_2{=}_3$, $HOCH_2{=}{\equiv}_4{=}$
C. tinctoria cv. Tetra Goldteppich[250]	$H_3C{=}{\equiv}_4{=}$, $Ph{\equiv}_2{=}CH_2OH$ 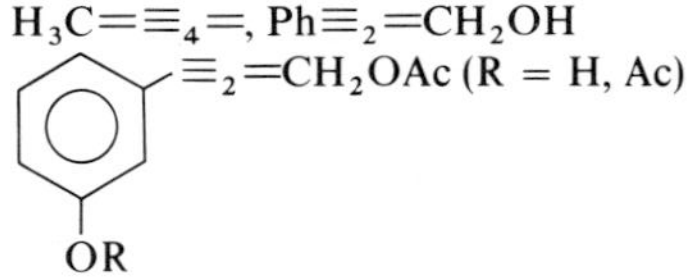	$RCH_2{=}{\equiv}_2{=}_3$ (R = H, OAc) $ROCH_2{=}{\equiv}_3{=}_2$ (R = H, Ac) $HOCH_2{=}{\equiv}_4{=}$
C. tinctoria cv. Radiata Nana[250]	$RCH_2{=}{\equiv}_4{=}$ (R = H, OH, OAc) $H_3C{=}{\equiv}_2{=}_3$, $Ph{\equiv}_2{=}CH_2OR$ (R = H, Ac)	$H_3C{=}{\equiv}_4{=}$, $H_3C{=}{\equiv}_3{=}_2$, $H_3C{=}{\equiv}_2{=}_3$, $HOCH_2{=}{\equiv}_3{=}_2$, $Ph{\equiv}_2{=}CH_2OAc$
C. verticillata L.[234,69,247]	(1)[247,69] $H_3C{=}{\equiv}_4{=}$, $H_3C{=}{\equiv}_3{=}_2$, $H_3C{=}{\equiv}_2{=}_3$, $Ph{\equiv}_3CH_3$ $Ph{\equiv}_2{=}CH_2R$ (R = H, OAc)	
	(2)[234] $H_3C{=}{\equiv}_4{=}$ (R = CH_3, CHO, CH_2OH, CH_2OAc)	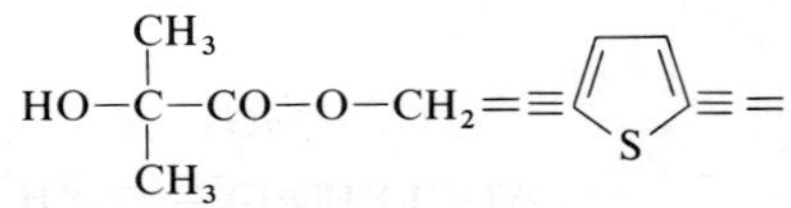 $HOCH_2{=}{\equiv}_2{=}_2(CH_2)_3OH$

Dahlia coccinea Cav. var. coccinea[60]	$RCH_2=\equiv_4=$ (R = H, OH, OAc), $RCH_2=\equiv_3=_2$ (R = H, OH, OAc) $H_3C=\equiv_2=_2\underset{\mid}{C}HCH_2CH_2OAc$, (OAc on CH) $H_3C=\equiv_2=_2(CH_2)_4=$ $H_3C=\equiv_2=_2CH(CH_2)_3=$, (OAc on CH) 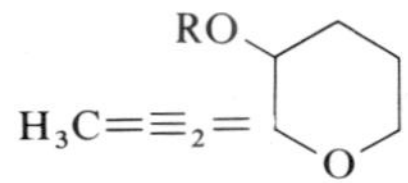 (R = H, Ac)	$Ph\equiv_3CH_3$, $H_3C\equiv_3=_2(CH_2)_3OR$ (R = H, Ac) $H_3C\equiv_3=_2CHCH_2CH_2OAc$ (OAc on CH) 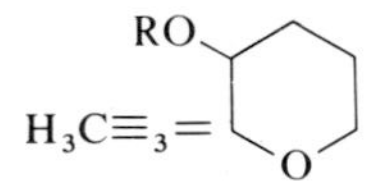 (R = H, Ac)
D. merckii Lehm.[197]	$H_3C=\equiv_4=$, $H_3C=\equiv_2=_2(CH_2)_3OR$ (R = H, Ac)	$H_3C=_2\equiv=_2(CH_2)_5OR$ (R = H, Ac) $H_3C=_2\equiv CH_2=(CH_2)_6OH$, $H_3C=\equiv_2CH_2=(CH_2)_6OH$ $H_3C=\equiv_2=_2(CH_2)_3OR$ (R = H, Ac)
D. rosea Cav.[35]	—	—
D. pinnata Cav.[22]	$H_3C=\equiv_4=$, $Ph\equiv_2=CH_2$ $H_3C\equiv_5=$, $H_3C=\equiv_4=$, $H_3C=\equiv_2=_3$, $H_3C=\equiv_3=_2$, $H_3C\equiv_4=_2$, $Ph\equiv_3CH_2R$ (R = H, OAc) $H_3C\equiv_3=_2CHCH_2CH_2OR$ (R = H, Ac) (OR on CH) $H_3C=\equiv_2=_2CHCH_2CH_2OR$ (R = H, Ac) (OR on CH) $H_3C=\equiv_2=CH(OAc)CH(OAc)=$, $Ph\equiv_2=R$ (R = CH_3, CHO, CH_2OH, CH_2OAc) $H_3C\equiv_3=_2(CH_2)_4=$	as roots
D. variabilis Desf.[35]		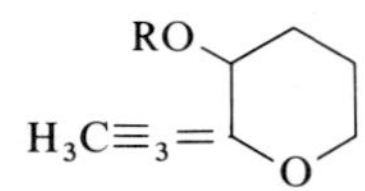 (R = H, Ac)
Glossocardia linearifolia Cass.[35]	$HOCH_2=\equiv_2=_2(CH_2)_4=$	$H_3C\equiv_3=_2(CH_2)_3OH$ $RCH_2=\equiv_4=$ (R = H, OAc)

TABLE VIII—Continued

8. Heliantheae-Coreopsidinae	Constituents Roots	 Leaves
Isostigma peucedanifolium Less.[71]	$RCH_2{=}{\equiv}_2{=}_2(CH_2)_4{=}$(R = H, OH)	
Heterospermum xanti A. Gray[35]	$RCH_2{=}{\equiv}_4{=}$(R = H, OH, OAc)	$H_3C{=}{\equiv}_4{=}$, $AcOCH_2{=}{\equiv}_3{=}_2$ $RCH_2{=}{\equiv}_2{=}_3$ (R = H, OH, OAc)
Thelesperma burridgeanum Blake[35]	$H_3C{=}{\equiv}_4{=}$ $HCO{=}{\equiv}_4{=}$ $HOCH_2{=}{\equiv}_4{=}$ $AcOCH_2{=}{\equiv}_4{=}$	as roots
T. ficifolium Gray[35]		as roots
T. trifidum Voss.[35]		as roots
Bidens bipinnatus L.[77]	$R{=}{\equiv}_4{=}$ (R=CH_3, CHO, CH_2OH, CH_2OAc)	$H_3C{\equiv}_5{=}$, $RCH_2{=}{\equiv}_4$–CH(OH)–CH(OH)– (R = H, OH)
B. biternata Merill. *et* Sherff.[35]	$R{=}{\equiv}_4{=}$ (R=CH_3, CHO, CH_2OH, CH_2OAc)	$Ph{\equiv}_3CH_3$, $AcOCH_2{=}{\equiv}_3{=}_2$
B. chrysanthemoides Michx.[35]	$H_3C{=}{\equiv}_4{=}$	$Ph{\equiv}_3CH_2R$ (R = H, OH)
B. connatus Muehl.[50,77]	$RCH_2{=}{\equiv}_4{=}$(R = H, OH, OAc)	$H_3C{=}{\equiv}_4{=}$, 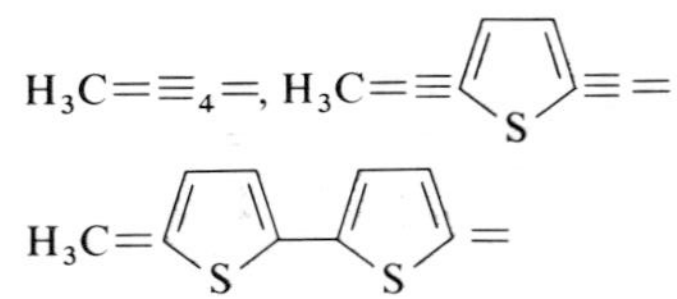
B. cernuus L.[35]	$Ph{\equiv}_3CH_3$	
B. dahlioides S. Wats.[77]	$RCH_2{=}{\equiv}_4{=}$(R = H, OH, OAc) $Ph{\equiv}_2{=}R$ (R = CHO, CH_2OH, CH_2OAc) 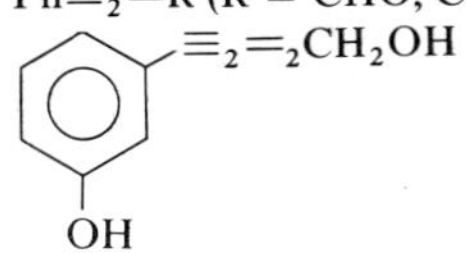	$Ph{\equiv}_3CH_2R$ (R = H, OH) $Ph{\equiv}_2{=}CH_2OH$, $H_3C{\equiv}_5$–CH(OH)–CH(OH)– $H_3C{\equiv}_4{=}$–CH(OH)–CH(OH)–
B. ferulaefolia Thuill.[349,47]	$RCH_2{=}{\equiv}_4{=}$ (R = H, OH, OAc)	$H_3C{=}{\equiv}_4{=}$, $RCH_2{=}{\equiv}_2{=}_3$ (R = H, OH, OAc) $Ph{\equiv}_2{=}CH_2R$ (R = H, OH, OAc) $H_3C{=}{\equiv}_2{=}_2CH(OAc)CH_2CH_2{-}OAc$

B. frondosa L.[35]	$R{=}{\equiv}_4{=}$ (R = CH_3, CHO, CH_2OH, CH_2OAc)	$H_3C{=}{\equiv}_4{=}$, $H_3C{=}$ 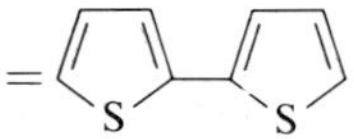 $=$
B. frutelifolius (DC.) Jacq.[35]	$RCH_2{=}{\equiv}_4{=}$ (R = H, OH, OAc)	$RCH_2{=}{\equiv}_2{=}_3$ (R = H, OH, OAc) $H_3C{=}{\equiv}_3{=}_2$, $H_3C{=}{\equiv}_2{=}_2$ (OAc OAc)
B. grandiflorus Balb.[35]	$R{=}{\equiv}_4{=}$ (R = CH_3, CHO, CH_2OH) $H_3C{=}{\equiv}_2{=}_3$	$RCH_2{=}{\equiv}_4{=}$ (R = H, OH)
B. humilis H.B.[35]	$R{=}{\equiv}_4{=}$ (R = CH_3, CHO, CH_2OH, CH_2OAc)	$RCH_2{=}{\equiv}_4{=}$ (R = H, OH), $H_3C{=}{\equiv}_3{=}_2$ $RCH_2{=}{\equiv}_2{=}_3$ (R = H, OAc)
B. leucanthus Poepp.[271,77]	$R{=}{\equiv}_4{=}$ (R = CH_3, CHO, CH_2OH, CH_2OAc) $Ph{\equiv}_2{=}CH_2OAc$	$RCH_2{=}{\equiv}_4{=}$ (R = H, OH)
B. melanocarpus K. M. Wieg.[35]	$R{=}{\equiv}_4{=}$ (R = CH_3, CHO, CH_2OH)	$RCH_2{=}{\equiv}_4{=}$ (R = H, OH)
B. pilosus L.[71]	$RCH_2{=}{\equiv}_4{=}$ (R = H, OH, OAc) $Ph{\equiv}_2{=}CH_2OR$ (R = H, Ac)	$H_3C{\equiv}_5{=}$, $Ph{\equiv}_3CH_2R$ (R = H, OH, OAc)
B. radiatus Thuill.[35,349]	$RCH_2{=}{\equiv}_4{=}$ (R = H, OH, OAc)	$H_3C{=}{\equiv}_4{=}$, $H_3C{=}$ 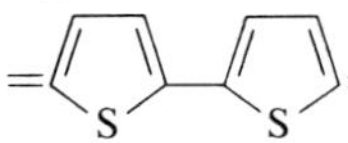 $=$
B. tripartitus L.[271]	$R{=}{\equiv}_4{=}$(R = CH_3, CHO, CH_2OH, CH_2OAc) $Ph{\equiv}_2{=}CH_2OAc$	
B. vulgatus Greene[35]	$H_3C{=}{\equiv}_4{=}$	$H_3C{=}{\equiv}_4{=}$
B. warszewicziana Regel[35]	$H_3C{\equiv}_5{=}$, $RCH_2{=}{\equiv}_4{=}$(R = H, OH, OAc)	$RCH_2{=}{\equiv}_2{=}_3$(R = H, OH, OAc)
Cosmos diversifolius Otto[263]	$R{=}{\equiv}_4{=}$(R = CH_3, CHO, CH_2OH, CH_2OAc)	
C. hybridus[263]	$R{=}{\equiv}_3{=}_2$(R = CHO, CH_2OH) $H_3C{=}{\equiv}_2{=}_3$	$H_3C{=}_2{\equiv}{=}_2(CH_2)_4COCH_2CH_2OAc$
C. sulphureus Cav.[263,189]	$R{=}{\equiv}_3{=}_2$(R = CH_3, CHO, CH_2OH) $H_3C{=}{\equiv}_2{=}_3$	$H_3C{=}_2{\equiv}{=}_2(CH_2)_4COCH_2CH_2OAc$
C. bipinnatus Cav.[263]	—	Cosmen
C. caudatus H.B. *et* K.[263]	—	Cosmen

	Constituents	
9. Heliantheae-Galinsoginae	Roots	Leaves
Calea reticulata Gardn.[35]	$H_3C{\equiv}_5{=}$, $H_3C{=}{\equiv}_4{=}$	—
C. phyllolepis Baker[35]	$H_3C{\equiv}_5{=}$	—

TABLE VIII—Continued

9. Heliantheae-Galinsoginae	Constituents: Roots	Leaves
C. pilosa Baker[35]	$H_3C\equiv_5=$	—
Tridax balbisoides A. Gray[35]	$=CO\equiv_2CH_2=(CH_2)_5=$	$R=\equiv_2=(CH_2)_4=$ (R = CHO, CH_2OH)
T. trilobata Hemsl.[105,209,154]	$=CO\equiv_2CH_2=(CH_2)_5=$ $=CO\equiv=_2(CH_2)_6=$ $=CO=\equiv=_2(CH_2)_4=$	$R=_2\equiv_2(CH_2)_4=$ (R = CHO, CH_2OH) $=CO\equiv_2CH_2=(CH_2)_5=$, $R=_2\equiv=_2(CH_2)_4=$ (R = CHO, CH_2OH),
T. procumbens L.		$R=\equiv_2=_2(CH_2)_4=$ (R = CHO, CH_2OH) $H_3C\equiv_3(CH_2)_3OH$ $H_3C\equiv_3=_2(CH_2)_3OH$
Galinsoga aristulata E. P. Pickn.[35]	$=CO\equiv_2CH_2=(CH_2)_5=$	as roots
G. ciliata Blake[35]	$=CO\equiv_2CH_2=(CH_2)_5=$	as roots
G. parviflora Cav.[105]	$=CO\equiv_2CH_2=(CH_2)_5=$	as roots
G. quadriratiata Ruiz. *et* Pav.[35]	$=CO\equiv_2CH_2=(CH_2)_5=$	as roots

10. Heliantheae-Madinae	Constituents: Roots	Leaves
Madia dissitiflora Torr. *et* Gray[35]	$H_3C\equiv_5=$	—
M. elegans D. Don[35]	$H_3C\equiv_5=$	—
M. sativa Mol.[35]	$H_3C\equiv_5=$	—
Hemizonia corymbosa Torr. *et* Gray[50]	$H_3C\equiv_5=$, $H_3C\equiv$ 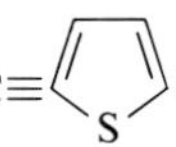 $\equiv_2=$	—
H. luzulaefolia DC.[35]	$H_3C\equiv_5=$	—
Layia chrysanthemoides A. Gray[35]	$H_3C\equiv_5=$	—
L. elegans Torr. *et* Gray[50]	$H_3C\equiv_5=$, $H_3C\equiv$ 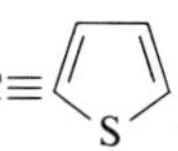 $\equiv_2=$	—
L. platyglossa A. Gray[50]	$H_3C\equiv_5=$, $H_3C\equiv$ 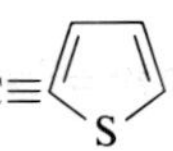 $\equiv_2=$	—

Achyrachaena mollis Schauer[50]	$H_3C\equiv_5=, H_3C\equiv_3=_2(CH_2)_3OH$	$H_3C\equiv_5=, H_3C\equiv_2$ 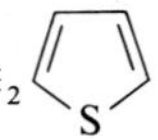 $\equiv=$
Calycadenia ciliosa (= Hemizonia)[35]	$H_3C\equiv_5=, H_3C\equiv_3=_2(CH_2)_3OH$	—

1.5. Tribus Helenieae

TABLE IX

HELENICAE

1. Helenieae-Jauminae	Constituents (roots)
Jaumea congensis O. Hoffm. ex Th.[35]	$H_3C\equiv_5=$
2. Helenieae-Heleninae	
Baeria aristata H. M. Hall.[50]	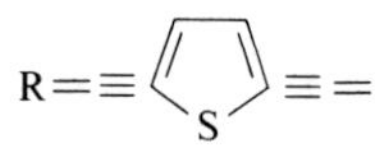 (R = CH_3, CH_2OH, CHO, CH_2OAc)
B. chrysostoma Fisch. *et* Mey.[50]	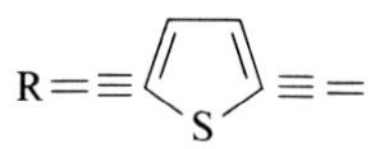 (R = CH_3, CH_2OH, CHO, CH_2OAc)
B. coronaria A. Gray[50]	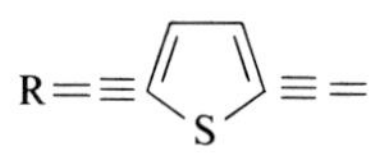 (R = CH_3, CH_2OH, CHO, CH_2OAc)
B. maritima A. Gray.[35]	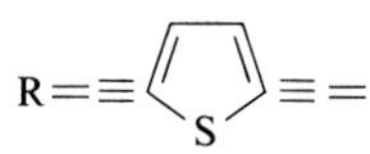 (R = CH_3, CH_2OH, CHO, CH_2OAc) $H_3C=\equiv_4=$
Lasthenia glaberrima DC.[50]	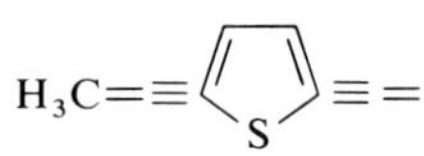
Monolopia lanceolata Nutt.[35]	$H_3C\equiv_5=$
Flaveria australasica Hook.[35]	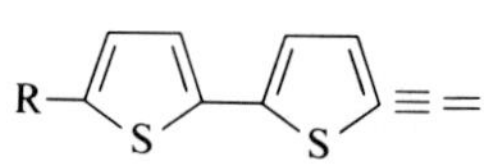 (R = H, CH_2OH, CH_2OAc)
F. repanda Lag.[81,83] *F. trinervata* Baill.[35]	$H_3C\equiv_5=$, $H_3C\equiv_3=_3$, $H_3C\equiv_2\ulcorner=\equiv_2=$ (SCH_3) $H_3C\equiv_2\ulcorner=\equiv_2=$ ($SOCH_3$), AcOCH$_2$ 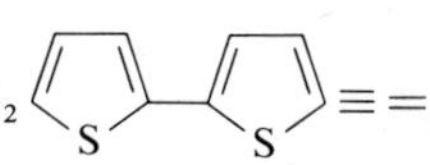 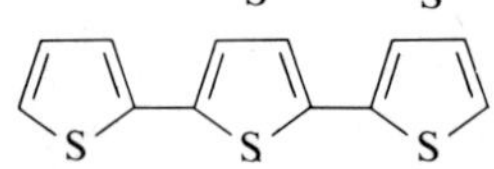

TABLE IX—Continued

2. Helenieae-Heleninae	Constituents (roots)
Schkuhria abrotanoides Roth.[35]	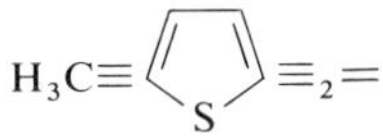
S. advena Thell.[50,89]	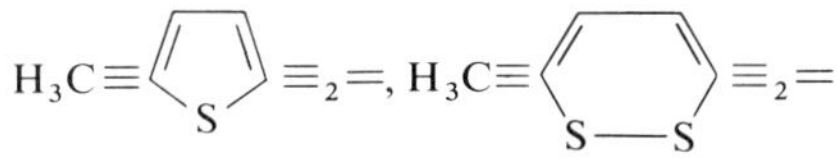
S. pinnata O. Ktze.[50,89]	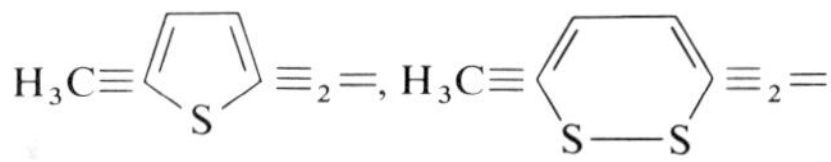
S. senecioides Nees.[50,89]	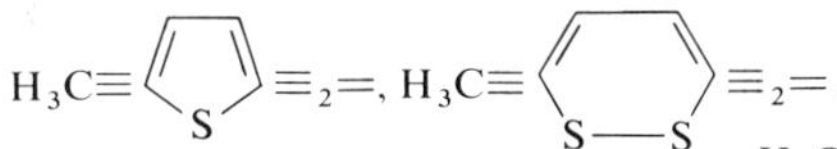 $H_3C\equiv_5=$ $H_3C=\equiv_4=$
Palafoxia texana DC.[35]	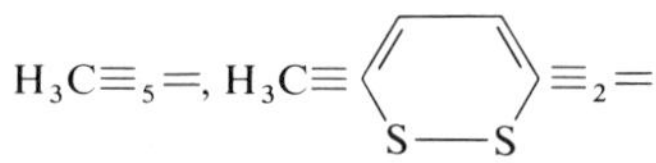
P. hookeriana Torr. *et* Gray[35]	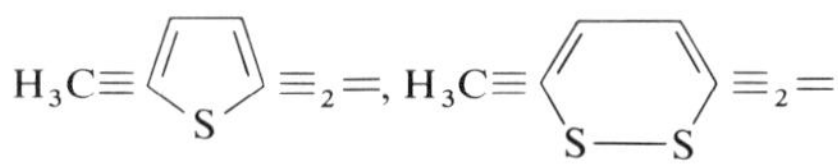
Eriophyllum caespitosum Dougl.[89,352]	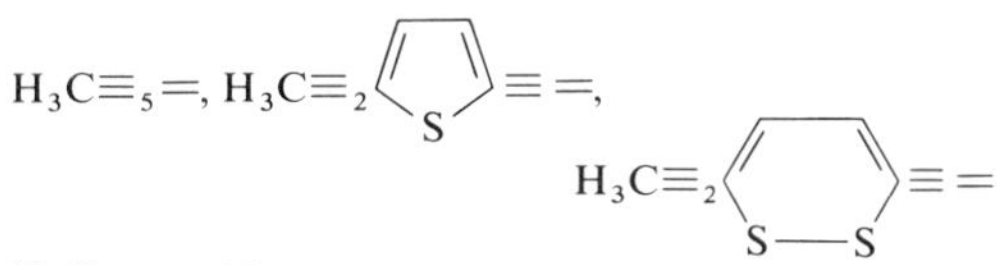 $H_3C\equiv_3=_2(CH_2)_3OH$
E. lanatum Forb.[35]	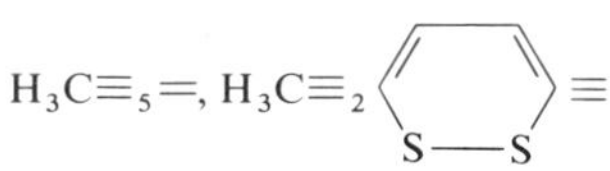
Chaenactis glabriuscula DC.[35]	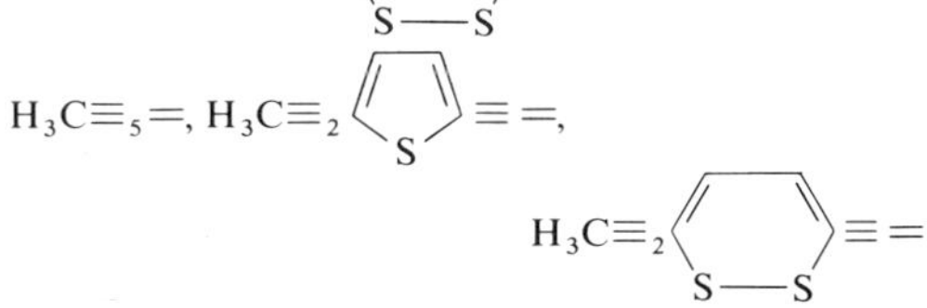 $H_3C\equiv_3=_2(CH_2)_3OH$
Actinella richardsonii Nutt.[35]	$H_3C\equiv_5=$
Helenium autumnale L.[35]	$H_3C\equiv_5=$
H. aromaticum Bailey[83] =*Cephalophora aromatica*	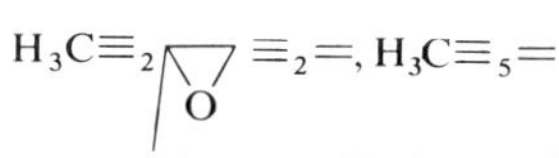 SO_2CH_3, ($H_3C=\equiv_2CH_2CH_2CO_2CH_3$(?)[287]
H. californicum Link.[83]	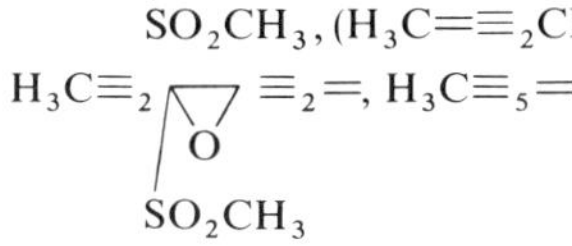

TABLE IX—Continued

2. Helenieae-Heleninae	Constituents (roots)
H. hoopesii A. Gray[83]	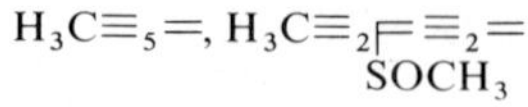
H. nudiflorum Nutt.[35]	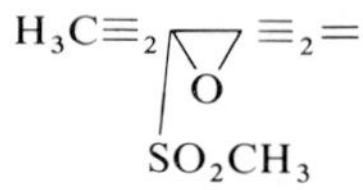
H. tenuifolium Nutt.[83]	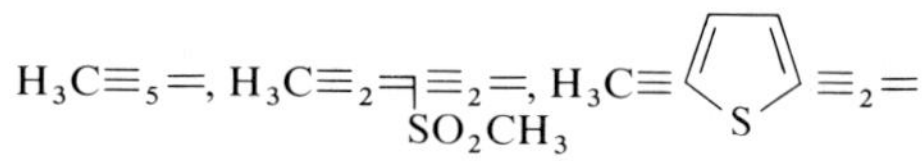
Gaillardia amblyodon J. Gay.[83]	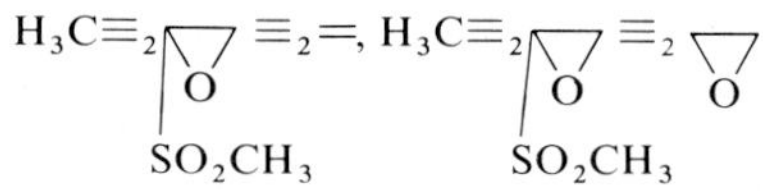
G. aristata Pursh.[83]	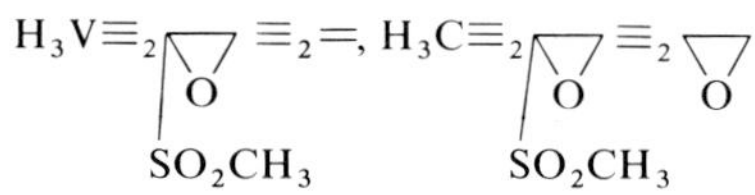
G. drummondii DC.[35]	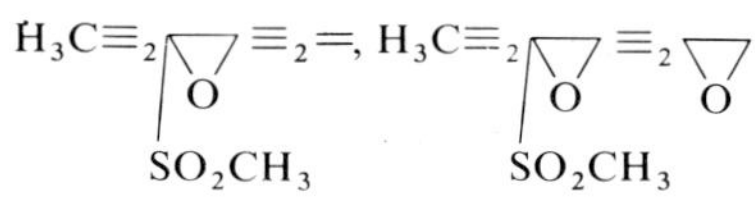
G. lanceolata Michx.[35]	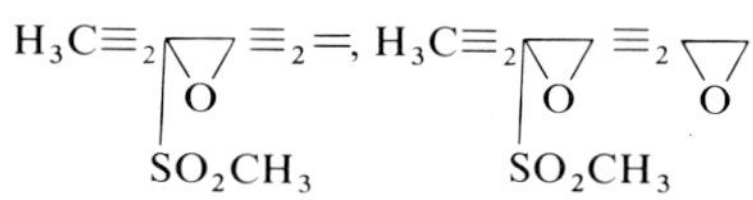
G. lutea Greene[35]	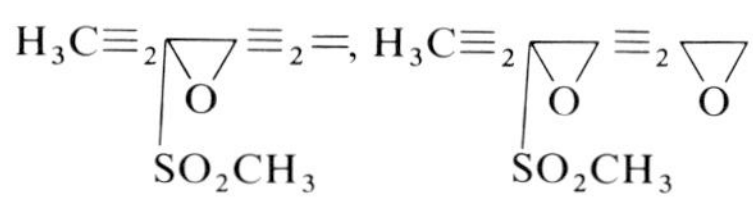
G. pulchella Fouq.[35]	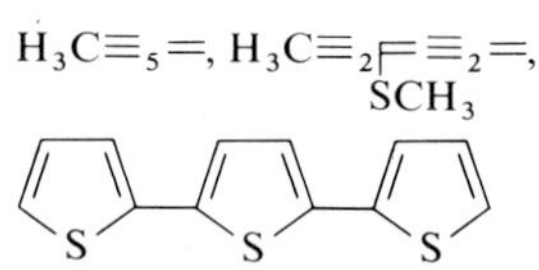

4. Helenieae-Tagetininae	Constituents (roots)
Tagetes erectes L.[74,93]	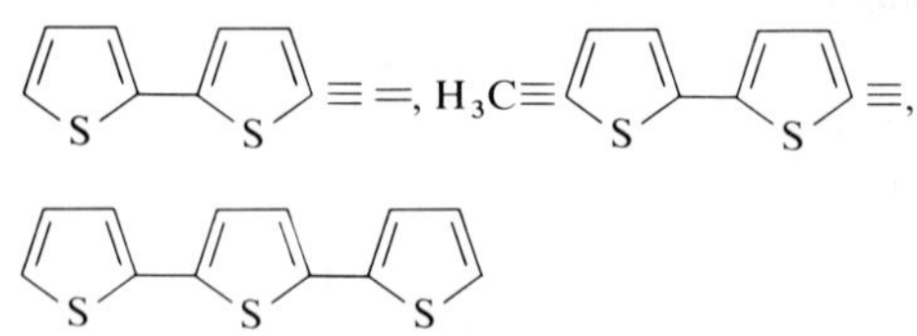

TABLE IX—Continued

4. Helenieae-Tagetininae	Constituents (roots)
T. glandulifera Schrank.[35]	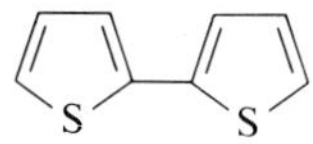 $\equiv =$,
	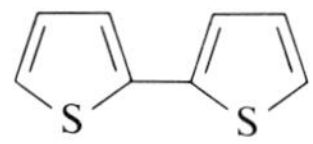 $\equiv CH_2CH_2OR$ (R = H, Ac)
T. lucidus Cav.[35]	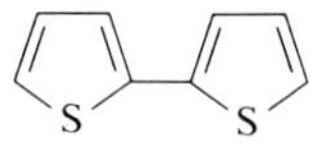 $\equiv =$,
	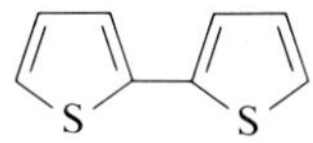 $\equiv CH_2CH_2OR$ (R = H, Ac)
	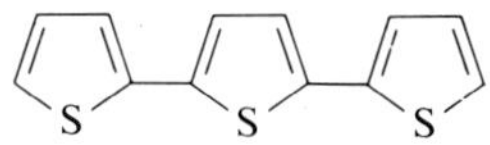
T. incida[35]	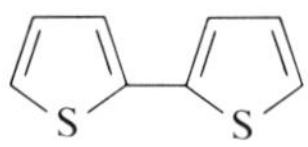 $\equiv =$,
	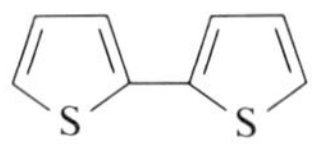 $\equiv CH_2CH_2OR$ (R = H, Ac)
	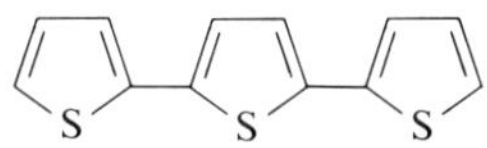
T. minutus L.[35]	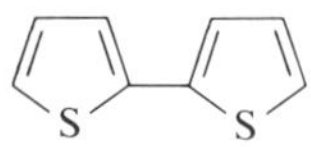 $\equiv =$,
	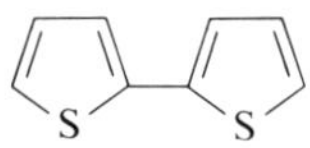 $\equiv CH_2CH_2OAc$
T. tenuifolius H.B. *et* K.[35]	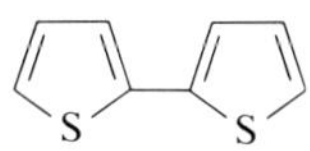 $\equiv =$,
	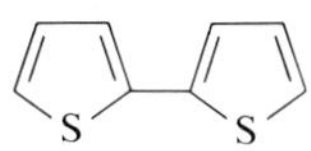 $\equiv CH_2CH_2OAc$
T. pauciloba DC.[35]	$H_3C\equiv$ 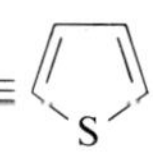 $\equiv_2 =$
T. patulus L.[93]	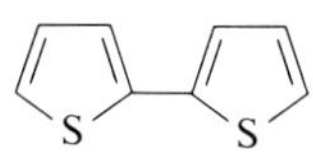 $\equiv =$,
	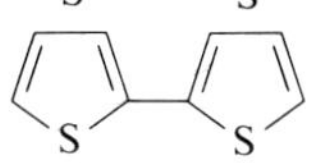 $\equiv CH_2CH_2OAc$,
	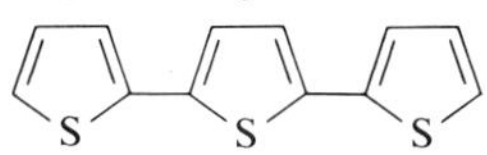

TABLE IX—Continued

4. Helenieae-Tagetininae	Constituents (roots)
T. signatūs Bartl.[35]	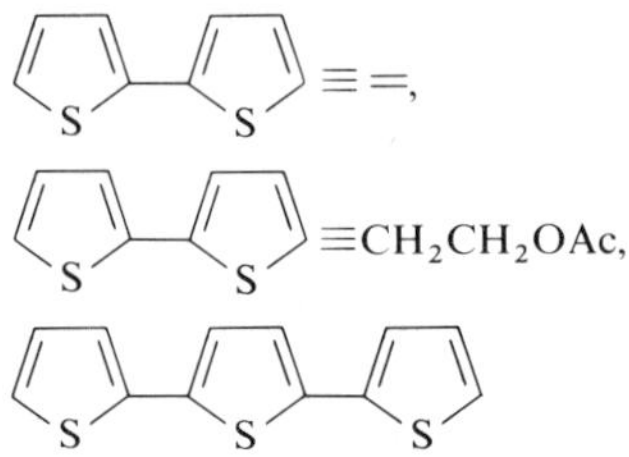

Surveying the constituents of tribus Helenieae, it appears once again that the pentaynene (**94**) (see Chapter 2, Section 2.6) is present. However, in most cases sulphur derivatives of **94** are more common. Only in the genera *Baeria* and *Lasthenia* is ene-tetrayn-ene the precursor of the observed acetylenes. The similarity of botanically related genera is very pronounced, e.g. in the genus *Helenium* and *Gaillardia*: only the annual *G. pulchella* does not contain the typical highly oxygenated polyynes.

The relationship between *Helenium* and *Gaillardia* is also clear if we look at the other constituents. The sesquiterpene-lactones and also thymol derivatives of very similar constitution are typical. The genera *Schkuhria*, *Palafoxia* and *Erodiophyllum* are also very closely related, while the genus *Flaveria* seems to be a transition of subtribus H.-Heleninae to H.-Tagetininae. No sign of the relationship to the tribus Anthemideae suggested by the botanical viewpoint can be observed. However, close relationships to the tribus Heliantheae are obvious, but the differences are relatively clear especially for the better investigated genera.

1.6. Tribus Anthemideae

This tribus is the most complicated both chemically and botanically. Several large genera do not appear very uniform and further subdivision seems to be necessary. The differences in the constituents in the two subtribus are not very clear. Many relationships between genera of both groups can be observed. *Eriocephalus* dehydrofalcarinone seems to be typical of the genus, but the same acetylene is also present in one group of *Artemisia* species and in the genus Cotula, as well as in some genera of the tribus Heliantheae. The genera *Lasiospermum*, *Athanasia*, and *Eumorphia* do not contain acetylenes at all: furane sesquiterpenes are as characteristic here as in *Asaemia* and *Phymaspermum*—belonging to the second subtribus—and also Ursinia, which now is attributed to this tribus. *Gonospermum* and *Santolina* show some relationship to special groups of *Chrysanthemum* and the same is true of *Lonas*.

TABLE X
SUBTRIBUS ANTHEMIDEAE-ANTHEMIDINAE

	Constituents	
	Roots	Leaves
Eriocephalus africanus L.[35]	—	$=CO\equiv_2CH_2=(CH_2)_5=$
E. racemosus L.[105]	$=CO\equiv_2CH_2=(CH_2)_5=, H_3C\equiv_3=_3$	—
Lasiospermum bipinnatum Druce[35]	—	furan-sesquiterpenes
L. radiatum Trev.[35]	—	furan-sesquiterpenes
L. brachyglossum DC.[35]	—	furan-sesquiterpenes
Athanasia parviflora Thunb.[35]	furan-sesquiterpenes	furan-sesquiterpenes
A. crithmifolia L.[35]	—	furan-sesquiterpenes
A. incisa Harv.[35]	—	furan-sesquiterpenes
A. trifurcata L.[35]	—	furan-sesquiterpenes
Eumorphia dregeana DC.[35]	—	furan-sesquiterpenes
Gonospermum canariense Less.[35]	Caryophyllene, *DME*, $H_3C\equiv_3=_2(CH_2)_4=$ 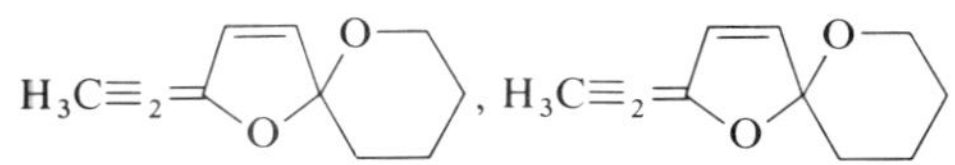	—
G. revolutum Sch. Bip.[35]	$H_3C\equiv_3\equiv_2(CH_2)_4=, H_3C\equiv_2=$ 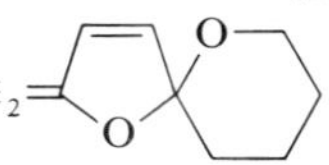	—
Santolina africana Jord. *et* Fourc.[35]	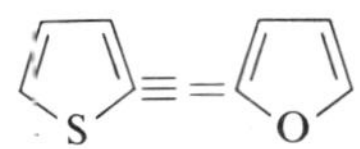	—
S. chamaecyparissus L.[132,133]	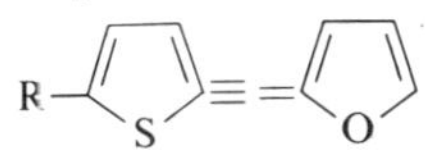 (R = H, CH_2OH, CH_2OAc, CHO) 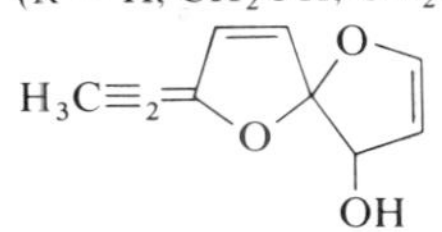	$H_3C\equiv_3=_3, H_3C\equiv_2=$ 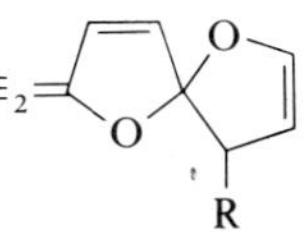 (R = H, OH, OAc)

TABLE X—Continued

	Constituents: Roots	Constituents: Leaves
S. neapolitana Jord.[35]	$H_3C\equiv_3$, $=_2(CH_2)_4=$, 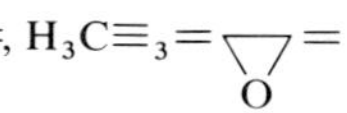 $H_3C\equiv_3=CH_2CH_2COC_2H_5$ 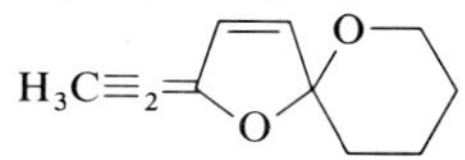	$H_3C\equiv_3=_3$, 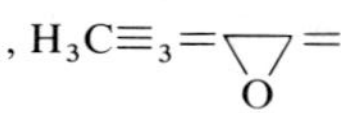 $H_3C\equiv_3=CH_2CH_2COC_2H_5$ $H_3C\equiv_3=CH_2CH_2CH(OH)C_2H_5$ 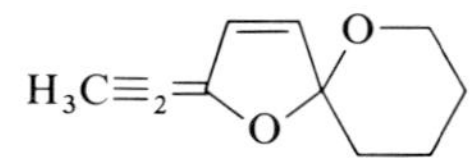
S. pinnata Viv.[35]	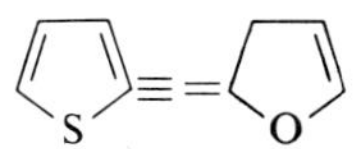	—
S. rosmarinifolia L.[35]	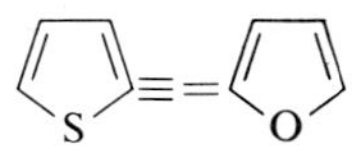	—
S. scariosa Ball.[35]	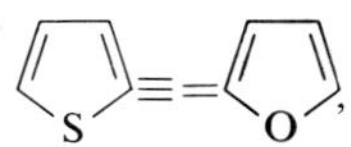, $H_3C\equiv_3=_2(CH_2)_4=$	—
Lonas inodora (L.) Gaertn.[129] (= L. annua)	$Ph-CO-\equiv_2CH_3$ $Ph-CH(OAc)\equiv_2CH_3$	$PhCO\equiv_2CH_2R$ (R = H, OH, O-Senecioyl)

TABLE XI
ANTHEMIS

Subgenus *Anthemis*	Constituents: Roots	Constituents: Leaves
Anthemis carpathica Willd.[115,155,156]	$H_3C{\equiv}\underset{SCH_3}{\ulcorner\!=}{\equiv}{=}CO_2CH_3$, *DME* $H_3C{\equiv}_2\underset{SCH_3}{\ulcorner\!=}{=}CO_2CH_3H_3H{\equiv}_3{=}_3$	$H_3C{\equiv}_3{=}\ CH_2CH_2COC_2H_5(AK)$ $H_3C{=}{\equiv}_2{=}_2CH_2CH_2COC_2H_5$
A. montana L.[155,156]	$H_3C{\equiv}\underset{SCH_3}{\ulcorner\!=}{\equiv}{=}CO_2CH_3$, *DME* 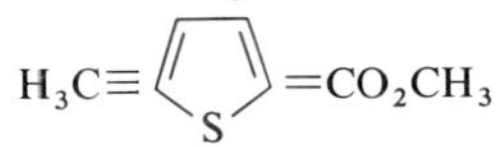	$H_3C{=}{\equiv}_2{=}_2CH_2CH_2CH_2OAc$ $H_3C{=}{\equiv}_2{=}\ CH_2CH_2COC_2H_5$
A. mucronulata Reichb.[35]	$H_3C{\equiv}\underset{SCH_3}{\ulcorner\!=}{\equiv}{=}CO_2CH_3$, *AK* $H_3C{\equiv}_2\underset{SCH_3}{\ulcorner\!=}{=}CO_2CH_3H_3C{\equiv}_3{=}CH_2CH_2\underset{OH}{C}HC_2H_5$ $H_3C{=}{\equiv}_2{=}CH_2CH_2COC_2H_5$	*AK*, $H_3C{=}{\equiv}_2{=}_2CH_2CH_2COC_2H_5$
A. petraea Ten.[35]	$H_3C{\equiv}\underset{SCH_3}{\ulcorner\!=}{\equiv}{=}CO_2CH_3$, *DME* $H_3C{=}_2\underset{SCH_3}{\ulcorner\!=}{=}CO_2CH_3$, *AK*	*AK*, $H_3C{=}{\equiv}_2{=}_2CH_2CH_2COC_2H_5$ $H_3C{=}{\equiv}_2{=}_2CH_2CH_2\underset{OH}{C}HC_2H_5$
A. styriaca Vest.[35]	$H_3C{\equiv}\underset{SCH_3}{\ulcorner\!=}{\equiv}{=}CO_2CH_3$, *DME* 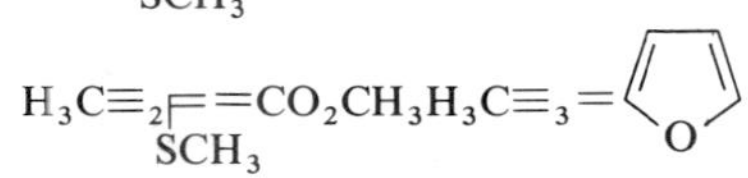 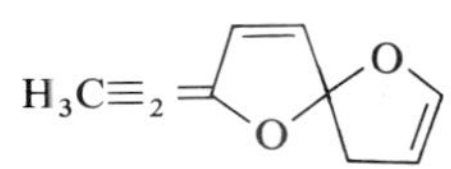 $H_3C{=}{\equiv}_2{=}_2CH_2CH_2\underset{OH}{C}HC_2H_5$	$H_3C{=}{\equiv}_2{=}_2CH_2CH_2COC_2H_5$ $H_3C{=}{\equiv}_2{=}_2CH_2CH_2CH_2OH$

TABLE XI—Continued

Subgenus *Anthemis*	Constituents: Roots	Constituents: Leaves
A. cupaniana Tod.[155]	$H_3C{\equiv}\underset{SCH_3}{\ulcorner}{\equiv}{=}CO_2CH_3$, *DME*	$H_3C{=}{\equiv}_2{=}_2CH_2CH_2COC_2H_5$
	$H_3C{\equiv}_2\underset{SCH_3}{\ulcorner}{=}CO_2CH_3$ 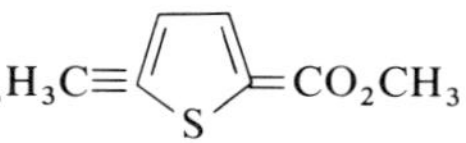	$H_3C{=}{\equiv}_2{=}_2CH_2CH_2CH_2OH$
	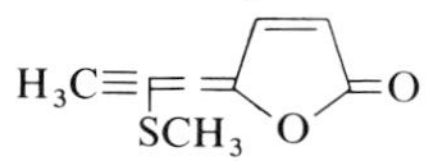 $H_3C{=}{\equiv}_2{=}\underset{OH}{C}H(CH_2)_5{=}$	
A. punctata Vahl.[156]	$H_3C{\equiv}\underset{SCH_3}{\ulcorner}{\equiv}{=}CO_2CH_3$, *DME*	$H_3C{=}{\equiv}_2{=}_2CH_2CH_2COC_2H_5$
	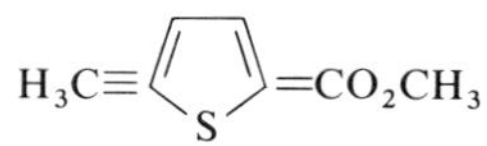	
	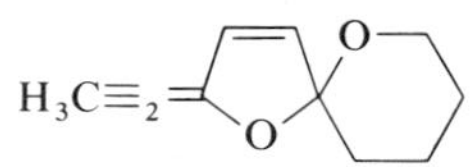 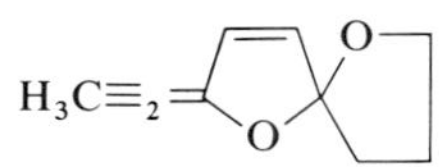	
A. arvensis L.[155,156]	$H_3C{\equiv}_2\underset{SCH_3}{\ulcorner}{=}CO_2CH_3$, *DME*	*AK*, *ME*, *ML*
	$H_3C\underset{SCH_3}{\ulcorner}{\equiv}_2{=}CO_2CH_3$ 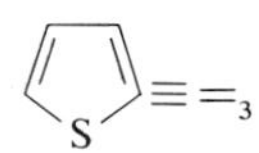	$H_3C{=}_2{\equiv}{=}_2(CH_2)_3OAc$
	$H_3C{\equiv}{=}\underset{SCH_3}{\ulcorner}{=}CO_2CH_3$	$H_3C{=}{\equiv}_2{=}_2(CH_2)_3OR$ (R = H, Ac)
A. punctata var. sicula[35]	$H_3C{\equiv}\underset{SCH_3}{\ulcorner}{\equiv}{=}CO_2CH_3$, *DME*	

	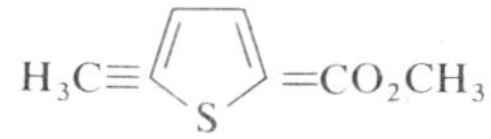	
	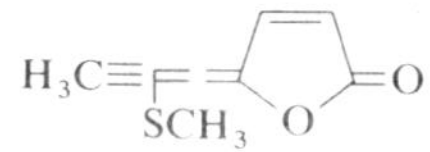	
A. ruthenica Bieb.[155,156]	$H_3C{\equiv}{=}{\equiv}{=}CO_2CH_3$, *DME*, *AK* SCH_3	$H_3C{\equiv}_3{=}_3$
	$H_3C{\equiv}_2{=}{=}CO_2CH_3$ $H_3C{\equiv}_3{=}CH_2CH_2CHC_2H_5$ SCH_3 OH	
	$H_3C{\equiv}{=}{=}{=}CO_2CH_3$ $H_3C{\equiv}_3{=}_2(CH_2)_3OAc$ SCH_3	
A. maritima L.[155,156]	$H_3C{\equiv}{=}{\equiv}{=}CO_2CH_3$, *DME* SCH_3	—
	$H_3C{\equiv}_2{=}{=}CO_2CH_3$ SCH_3	
	$H_3C{=}{\equiv}_2{=}CO_2CH_3$ SCH_3	
	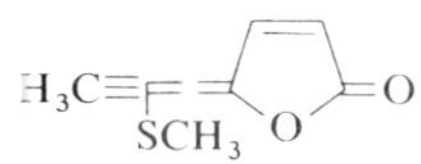	
A. cotula L.[155,156]	$H_3C{\equiv}{=}{\equiv}{=}CO_2CH_3$, *DME* SCH_3	$H_3C{=}{\equiv}_2{=}_2(CH_2)_3OAc$
	$H_3C{\equiv}_2{=}{=}CO_2CH_3$ SCH_3	
	$H_3C{=}{\equiv}{=}{=}CO_2CH_3$ SCH_3	

TABLE XI—Continued

Subgenus *Cota*	Constituents: Roots	Constituents: Leaves
A. altissima L.[105,155,156] (= A. cota L.)	$H_3C{\equiv}\underset{SCH_3}{=}{\equiv}{=}CO_2CH_3$ *DME, AK* $H_3C{\equiv}_3{=}_3$, 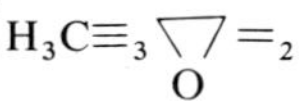	—
A. austriaca Jacq.[155,156,160,165]	$H_3C{\equiv}\underset{SCH_3}{=}{\equiv}{=}CO_2CH_3$, *ME, ML* $H_3\underset{SCH_3}{C}{=}{\equiv}_2{=}CO_2CH_3$ 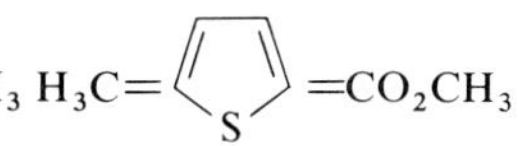 $H_3C{=}{\equiv}\underset{SCH_3}{=}{=}CO_2CH_3$, 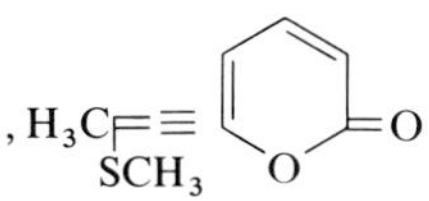 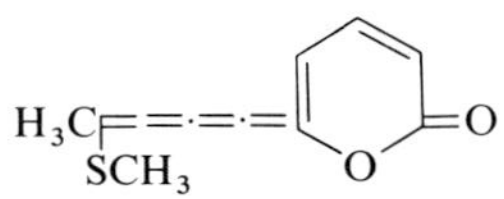 $H_3\underset{SCH_3}{C}{=}{=}{\cdot}{=}{\equiv}{=}CO_2CH_3$	—
A. brachycentros F. Gray[156]	$H_3C{\equiv}\underset{SCH_3}{=}{\equiv}{=}CO_2CH_3$, *DME, AK* 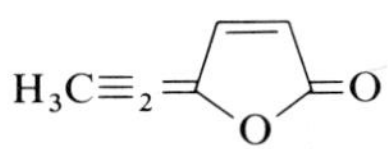 $H_3C{\equiv}_2\underset{SCH_3}{=}{=}CO_2CH_3$ 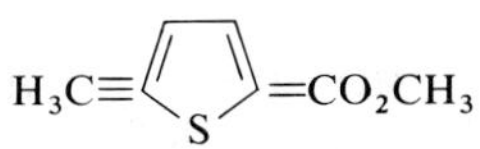 $H_3C{=}{\equiv}_2{=}_2CH_2CH_2COC_2H_5$	

A. cairica Vis.[155,156]	$H_3C{\equiv}\underset{SCH_3}{\ulcorner\!{=}}{\equiv}{=}CO_2CH_3$ $H_3C{\equiv}_3{=}_3$, *AK* $H_3C{\equiv}_3{=}_2(CH_2)_3OH$	$H_3C{\equiv}_3{=}_2(CH_2)_3OAc$
A. rigescens Willd.[156]	$H_3C{\equiv}\underset{SCH_3}{\ulcorner\!{=}}{\equiv}{=}CO_2CH_3$ *DME* 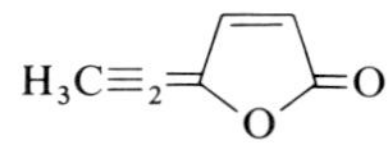 $H_3C{\equiv}_2\underset{SCH_3}{\ulcorner\!{=}}{=}CO_2CH_3$ $H_3C{\equiv}{=}\underset{SCH_3}{\ulcorner\!{=}}{=}CO_2CH_3$ $H_3C\underset{SCH_3}{\ulcorner\!{=}}{\equiv}_2{=}CO_2CH_3$ 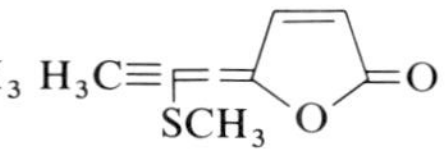 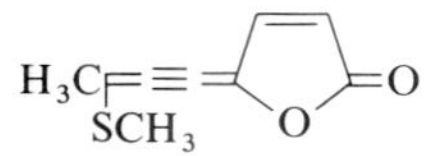	$H_3C{\equiv}_3{=}_2(CH_2)_3OAc$
A. melanoloma Trautv.[35]	$H_3C{\equiv}{=}\underset{SCH_3}{\ulcorner\!{=}}{=}{=}CO_2CH_3$, *DME* $H_3C{\equiv}_2\underset{SCH_3}{\ulcorner\!{=}}{=}CO_2CH_3$ $H_3C\underset{SCH_3}{\ulcorner\!{=}}{\equiv}_2{=}CO_2CH_3$ 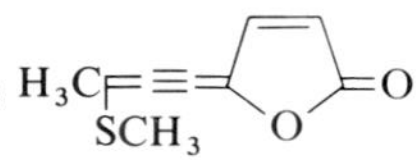	
A. tinctoria L.[155,156,162]	$H_3C\underset{SCH_3}{\ulcorner\!{=}}{\equiv}_2{=}CO_2CH_3$, *DME* $H_3C{\equiv}{=}\underset{SCH_3}{\ulcorner\!{=}}{=}CO_2CH_3$ 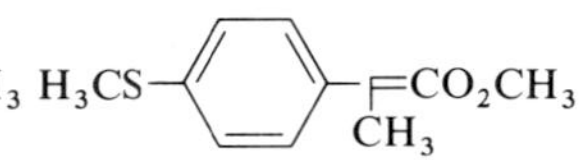	$H_3C{\equiv}_3{=}_2(CH_2)_3OAc$
A. tinctoria (Bot. Garden Bucharest)[85]	$H_3C{\equiv}\underset{SCH_3}{\ulcorner\!{=}}{\equiv}{=}CO_2CH_3$ $HOCH_2{=}{\equiv}_2{=}_2(CH_2)_4{=}$ $H_3C\underset{SCH_3}{\ulcorner\!{=}}{\equiv}_2{=}CO_2CH_3$ $HOCH_2{=}{\equiv}_2CH_2{=}(CH_2)_5{=}$ $H_3C{\equiv}{=}\underset{SCH_3}{\ulcorner\!{=}}{=}CO_2CH_3$ $H_3C{\equiv}_2\underset{SCH_3}{\urcorner}{=}CO_2CH_3$	$H_3C{\equiv}_3{=}_2(CH_2)_3OAc$ $HOCH_2{=}{\equiv}_2{=}_2(CH_2)_4{=}$

TABLE XI—Continued

Subgenus *Anthemis*	Constituents: Roots	Constituents: Leaves
A. dumetorum Sosn.[35]	$H_3C\!=\!\equiv_2\!=\!CO_2CH_3$, *DME* SCH_3 $H_3C\equiv=\,=\,=CO_2CH_3$ SCH_3	—
A. woronowii Sosn.[156]	$H_3C\equiv=\equiv=CO_2CH_3$, *DME* SCH_3 $H_3C\!=\!\equiv_2\!=\!CO_2CH_3$ $H_3C\equiv_3=_2(CH_2)_3OAc$ SCH_3 $H_3C\equiv=\,=\,=CO_2CH_3$ SCH_3	$H_3C\equiv_3=_2(CH_2)_3OAc$
A. zyghia Woron.[156]	$H_3C\!=\!\equiv_2\!=\!CO_2CH_3$, *DME* SCH_3 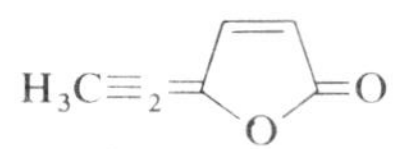	—
A. saguramica Sosn.[57,155,156]	 $\equiv=R$ (R = $CH_2CH_2COC_2H_5$) $CH_2CH(OCOR)COC_2H_5$, $-CHO$, $CH(OCOR)CH_2COC_2H_5$, $=COC_2H_5$ $CH_2CH_2CH(OH)C_2H_5$, $=CH(OH)C_2H_5$ *DME*, $H_3C\equiv_3=_3$, *AK*	*ME*, *ML*, *AK* $H_3C\equiv_3=_3$ $H_3C\equiv_3=CH_2CH_2CHC_2H_5$ OH
A. Kelevays[156]	*DME*, $H_3C\equiv=\,=\,=CO_2CH_3$, $H_3C\equiv=\equiv=CO_2CH_3$ SCH_3 SCH_3 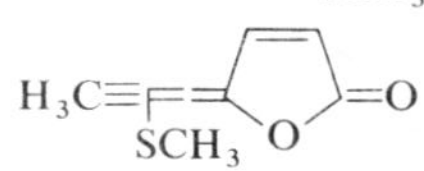	

Chamaemelum	Constituents: Roots	Constituents: Leaves
A. nobilis L.[68,105,156,167]	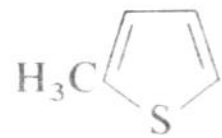 $\equiv=CO_2CH_3$ *DME*	—
	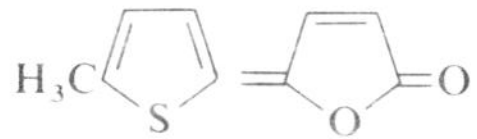 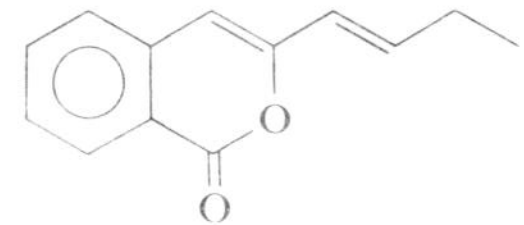	
	$H_3C\equiv_3=_3$	
A. fuscata Brot.[45,121,130,156]	 $\equiv=CO_2CH_3$, *DME*	 $\equiv=CO_2CH_3$
	 $\equiv=CH_2CH_2COC_2H_5$	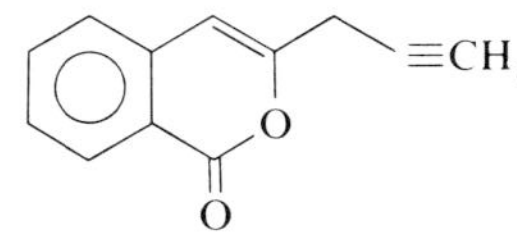
	$H_3C=\equiv_2CH_2CH_2=_2CONHR$	
	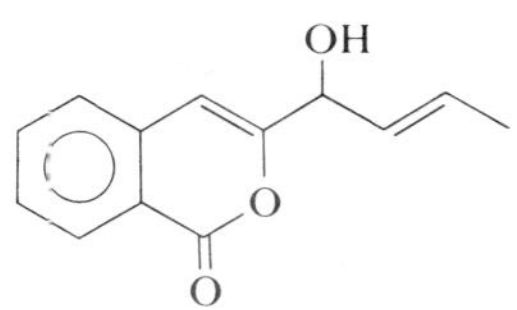 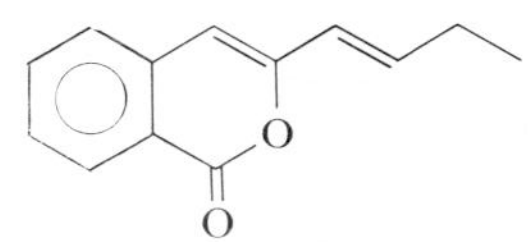	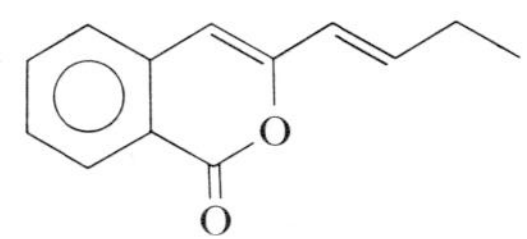

TABLE XI—Continued

Chamaemelum	Constituents: Roots	Constituents: Leaves
A. nobilis var. discoides (Boiss.) P. Silve *et* G. Sobr.[35]	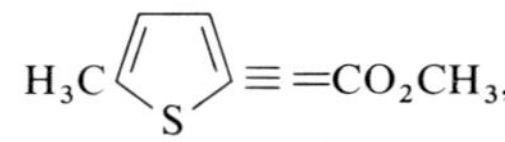, *DME*	—
	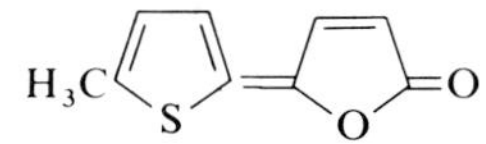 $H_3C\equiv_3=_3$	
	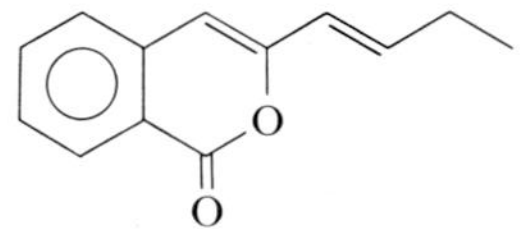	
A. scariosa DC.[156]	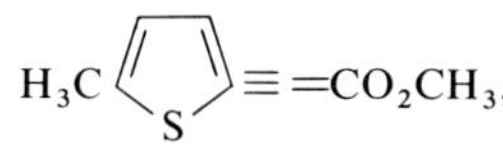, *DME*	—
	$H_3C\equiv_3=_2(CH_2)_4=$	
A. santolinoides Munby[35]	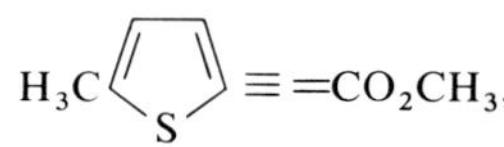, *DME*	—
	$H_3C\equiv_3=_3$	
A. mixta L.[155,156] (= Ormenis mixta)	$H_3C\equiv_3=CONHCH_2CH(CH_3)_2$, *DME*	as roots
	$H_3C\equiv_3=_3$, 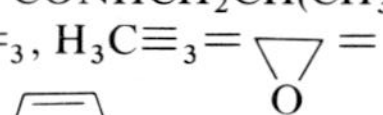	
	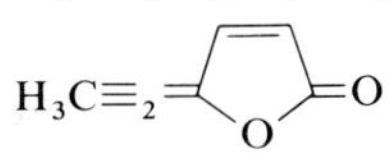	
A. ridolfiana Hort.	*ME*, *ML*, $H_3C=\equiv_2CH(OH)$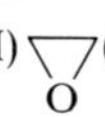$(CH_2)_5=$	

Lepidophorum	Constituents: Roots	Constituents: Leaves
L. grisleyi Samp.[35]	$PhCO{\equiv}_3CH_3$, $PhCH(OAc){\equiv}_2CH_3$	$PhCO{\equiv}_2CH_3$ 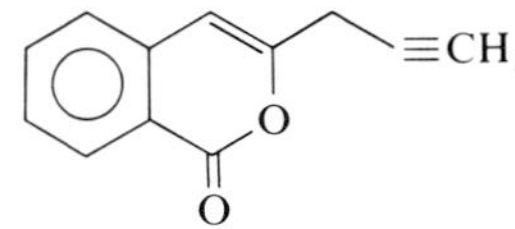 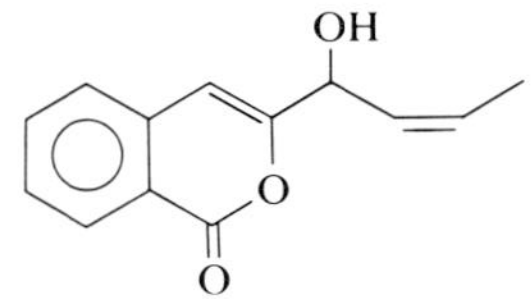

Anacyclus	Roots	Leaves
I. Anacyclus pyrethrum DC.[35,45]	$H_3C{\equiv}_3{=}_3$, *DME*, *AK* $H_3C{\equiv}_3{=}CONHR$, $C_3H_7{\equiv}_2CH_2CH_2{=}_2CONHR$ $H_3C{=}{\equiv}_2CH_2CH_2{=}_2CONHR$ $H{\equiv}_2CH_2CH_2{=}_2CONHCH_2CH_2Ph$	—
A. officinarum Hayne[35]	*DME*, *AK*	—
A. marrocanus Ball.[35]	$H_3C{\equiv}_3{=}_3$, *DME*, *AK* $H_3C{\equiv}_3{=}CONHR$, $C_3H_7{\equiv}_2CH_2CH_2{=}_2CONHR$	—
II. A. clavatus Pers.[35]	$H_3C{\equiv}_3{=}_3$, *DME*, *AK* 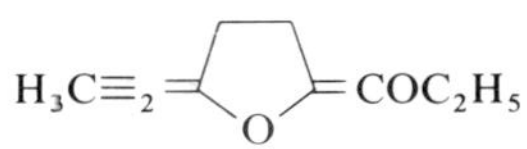	$H_3C{\equiv}_3{=}_3$, *DME*, *AK* 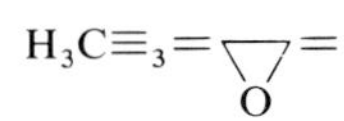 $H_3C{\equiv}_3{=}CH_2CH_2CH(OH)C_2H_5$

TABLE XI—Continued

Anacyclus	Constituents Roots	 Leaves
A. depressus Ball.[35]	$H_3C\equiv_3=_3$, *DME, AK*	$H_3C\equiv_3=_3$, *DME*
	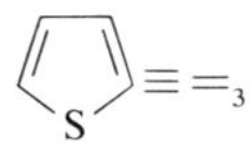	$H_3C=\equiv_2=CH_2CH_2COC_2H_5$
A. valentinus L.[35]	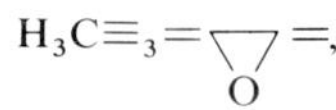, *DME, AK*	$H_3C\equiv_3=_3$
A. radiatus Loisel[58]	$H_3C\equiv_3=_3$, DME, AK	$H_3C\equiv_3=_3$, DME, AK
	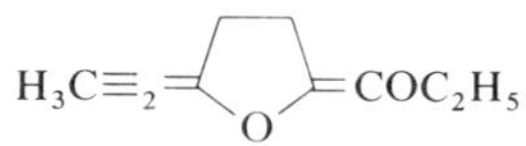, 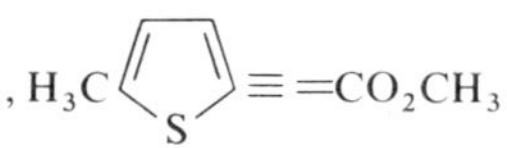	$H_3C\equiv_3=_2(CH_2)_3OAc$ $H_3C\equiv_3=CH_2CH_2CH(OH)C_2H_5$

The section *Chamaemelum* now is separated as a genus from the large genus *Anthemis*; the same is true for Lepidophorum. The remaining species, divided into four sections, are relatively uniform as regards the main acetylenic compounds. The methyl thioenolethers derived from dehydro-matricaria ester are clearly typical, but the variations in the different sections are not very distinct. However, the differentiation with *Chamaemelum* is very apparent. Here (the thiophenes are typical), the occurrence with capillarine and related compounds might be a hint of a relationship with *Chrysanthemum* and also with *Tripleurospermum. Lepidophorum*, in which aromatic acetylenes are found, is clearly separated and seems to be a link with *Agyranthemum.*

The last two genera of the first subtribus to be investigated—*Diotis* and *Cladanthus*—have only the epoxide typical for the genus *Achillea*; but this is also present in other genera.

The genus Achillea is also divided into several sections and is relatively uniform when considering the triyne-triene and the corresponding epoxide, but the occurrence of amides in some species is important. Amides are also typical of several species of the genus *Anacyclus*, but do not occur in the last section; this is clearly different from the other sections.

TABLE XII
ACHILLEA

1. Section *Ptarmica*	Constituents (roots)
Achillea ageratifolia Sibeth. *et* Sm.[105,107]	$H_3C\equiv_3=_3$, $H_3C\equiv_3=\underset{O}{\bigtriangledown}=$
A. atrata L.[105,107]	$H_3C\equiv_3=_3$, $H_3C\equiv_3=\underset{O}{\bigtriangledown}=$
A. clavenae L.[105,107]	$H_3C\equiv_3=_3$, $H_3C\equiv_3=\underset{O}{\bigtriangledown}=$, *DME*, $H_3C\equiv_3=_2(CH_2)_3OH$
A. clusiana Tausch.[105,107]	$H_3C\equiv_3=_3$, $H_3C\equiv_3=\underset{O}{\bigtriangledown}=$
A. impatiens L.[105,107]	$H_3C\equiv_3=_3$, $H_3C\equiv_3=\underset{O}{\bigtriangledown}=$, *DME*, $H_3C\equiv_3=CONHCH_2CH(CH_3)_2$
	$H_3C\equiv_3=_2(CH_2)_3OR$ (R = H, Ac)
A. ptarmica L.[105,107]	$H_3C\equiv_3=_3$, $H_3C\equiv_3=\underset{O}{\bigtriangledown}=$, *DME*
A. schurii Sch. Bip. = *Anthemis tenuifolius* Schr.[105]	$H_3C\equiv_3=_3$, $H_3C\equiv_3=\underset{O}{\bigtriangledown}=$
A. serbica Nym.[105,107]	$H_3C\equiv_3=_3$, $H_3C\equiv_3=\underset{O}{\bigtriangledown}=$
A. sibirica Ledeb.[105,107]	$H_3C\equiv_3=_3$, $H_3C\equiv_3=\underset{O}{\bigtriangledown}=$, *DME*, $H_3C\equiv_3=CONHCH_2CH(CH_3)_2$
	$H_3C\equiv_3=_2(CH_2)_3OR$ (R = H, Ac)
A. umbellata Sibeth.[105,107]	$H_3C\equiv_3=_3$, $H_3C\equiv_3=\underset{O}{\bigtriangledown}=$

2. Section *Millefolium*	Constituents (roots)
Achillea asplenifolia Vent.[105,107]	$H_3C\equiv_3=_3$, *DME*
A. chrysocoma Friv.[107]	$H_3C\equiv_3=_3$, *DME*
A. collina J. Beck.[35]	$H_3C\equiv_3=_3$, *DME*
A. distans W. *et* K.[105,107]	$H_3C\equiv_3=_3$, *DME*
A. ligustica All.[105,107]	$H_3C\equiv_3=_3$, *DME*
A. millefolium Kelw.[107]	$H_3C\equiv_3=_3$, *DME*
A. millefolium L.[105,107]	$H_3C\equiv_3=_3$, *DME*, $H_3C\equiv_3=\underset{O}{\triangledown}=$
A. setacea Waldst. *et* Kit.[105]	*DME*
A. sudetica Boiss.[105,107]	$H_3C\equiv_3=_3$, *DME*, $H_3C\equiv_3=\underset{O}{\triangledown}=$
A. tanacetifolia All.[105,107]	$H_3C\equiv_3=_3$, *DME*, $H_3C\equiv_3=\underset{O}{\triangledown}=$

3. Section *Filipendulanae*	Constituents (roots)
Achillea ageratum L.[107]	*DME*
A. clypeolata S. Sm.[105,107]	$H_3C\equiv_3=_3$, $H_3C\equiv_3=\underset{O}{\triangledown}=$
A. compacta Lam.[105,107]	$H_3C\equiv_3=_3$, $H_3C\equiv_3=\underset{O}{\triangledown}=$, *DME*, $H_3C\equiv_3=_2(CH_2)_3OH$
A. filipendulina var. coronarium Lam.[107]	$H_3C\equiv_3=_3$, *DME*, $H_3C\equiv_3=_2(CH_2)_3OH$
A. pectinata Willd.[105,107]	$H_3C\equiv_3=_3$, $H_3C\equiv_3=\underset{O}{\triangledown}=$
A. taygetea Boiss.[105,107]	$H_3C\equiv_3=_3$, $H_3C\equiv_3=\underset{O}{\triangledown}=$, *DME*
A. vandassii Vel.[105,107]	$H_3C\equiv_3=_3$, $H_3C\equiv_3=\underset{O}{\triangledown}=$, *DME*, $H_3C\equiv_3=_2(CH_2)_3OH$

TABLE XII—Continued

4. Section *Santolinaideae*	Constituents (roots)
Achillea santolina L.[105,107]	$H_3C{\equiv}_3{=}_3$, $H_3C{\equiv}_3{=}\underset{O}{\bigtriangledown}{=}$, $H_3C{\equiv}_3{=}_2(CH_2)_3OR$ (R = H, Ac)
A. santolinoides Lag.[35]	$H_3C{\equiv}_3{=}_3$, *DME*
A. sulphurea Boiss.[105,107]	$H_3C{\equiv}_3{=}_3$, *DME*
A. tomentosa L.[107]	$H_3C{\equiv}_3{=}\cdots OH$

Further species	Constituents (roots)
Achillea conescens L.[105]	$H_3C{\equiv}_3{=}_3$, $H_3C{\equiv}_3{=}\underset{O}{\bigtriangledown}{=}$, *DME*, $H_3C{\equiv}_3{=}_2(CH_2)_3OH$
A. conjugata[35]	$H_3C{\equiv}_3{=}_3$, $H_3C{\equiv}_3{=}\underset{O}{\bigtriangledown}{=}$, *DME*
A. grandifolia Friv.[35]	*DME*
A. kitaibeliana[35]	*DME*
A. nabelikii Heimerl. ex Nabelek[35]	$H_3C{\equiv}_3{=}\underset{O}{\bigtriangledown}{=}$, DME
A. odorata L.[25]	$H_3C{\equiv}_3{=}_3$, $H_3C{\equiv}_3{=}\underset{O}{\bigtriangledown}{=}$, DME
Diotis candidissima Koch[35]	$H_3C{\equiv}_3{=}\underset{O}{\bigtriangledown}{=}$
D. maritima (L.) Cass.[35]	$H_3C{\equiv}_3{=}_3$, $H_3C{\equiv}_3{=}\underset{O}{\bigtriangledown}{=}$, $H_3C{\equiv}_3{=}_2(CH_2)_3OAc$
Cladanthus arabicus Cass.[35]	$H_3C{\equiv}_3{=}\underset{O}{\bigtriangledown}{=}$, *DME*, *AK*

Considering the subtribus Anthemidea-Chrysantheminae, the genus *Matricaria* is also non uniform in the connection with the chemical constituents. But also botanically subdivisions into the genera *Matricaria*, *Tripleurospermum* and *Pentzia* seem to be excepted being in agreement with the chemotaxonomical findings. The South African genera *Lidbeckia* and *Thaminophyllum* botanically placed near *Matricaria* however contain C_{17}-acetylenes not found in *Matricaria* species.

TABLE XIII
MATRICARIA

	Constituents	
	Roots	Leaves
Matricaria recutita L.[35]	*DME*, *AK*, 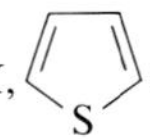 $\equiv=_3$ $H_3C\equiv_2=$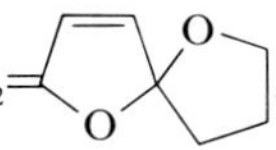, $H_3C\equiv_2=_3(CH_2)_2OAc$	*ME*, *ML*, $H_3C\equiv_3=_3$ $H_3C\equiv_2=$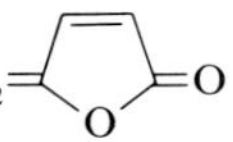
M. matricarioides Porter[62]	$H_3C\equiv_2=_3CH_2CH_2OAc$	$H_3C\equiv_2=$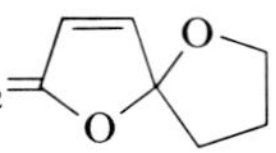
M. capensis L.[48]	$H_3C\equiv_2=$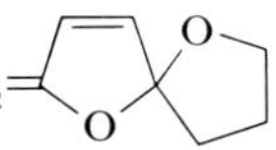	$H_3C\equiv_2=$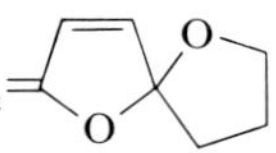
M. africana Berg.[48]	$H_3C\equiv_2=$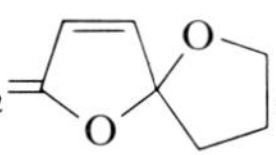, $H_3C\equiv_3=_3$	$H_3C\equiv_2=$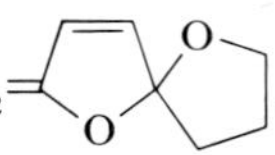, $H_3C\equiv_2=$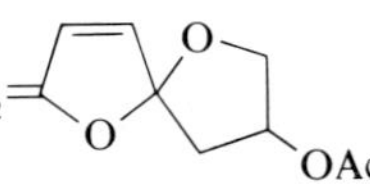
M. nigellaefolia DC.[48]	*DME*, phenacetine $H_3C\equiv_3=_2(CH_2)_3OAc$ $H_3C=\equiv_2=_2(CH_2)_3OR$ (R = H, Ac)	*DME*, Custonolide $H_3C\equiv_3=_3$, $H_3C=\equiv_2=_2(CH_2)_3OAc$ $H_3C=\equiv_2=$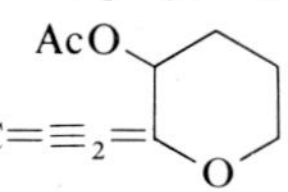

Lidbeckia pectinata Berg.[35]	*DME*, $-{=}{\equiv}_2-$
	$=\underset{\mid \atop OH}{C}H{\equiv}_2CH_2{=}(CH_2)_5{=}$
Thaminophyllum multiflorum Harv.[35]	$=\underset{\mid \atop OH}{C}H{\equiv}_2CH_2{=}(CH_2)_5{=}$
Phymaspermum parvifolium Benth. *et* Hook.[35]	furan sesquiterpenes
P. schroeteri Compton[35]	furan sesquiterpenes

TABLE XIV
PENTZIA

	Constituents	
	Roots	Leaves
Pentzia pilulifera Fourc.[35,106]	$H_3C\equiv_3=_3$, $H_3C\equiv_3=$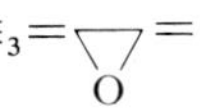$=$	$H_3C\equiv_2=$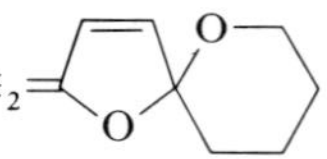
	$H_3C\equiv_3=$$=$, $H_3C\equiv_2=$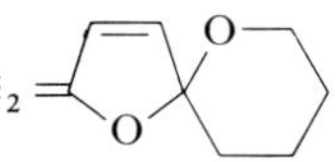	$H_3C\equiv_2=$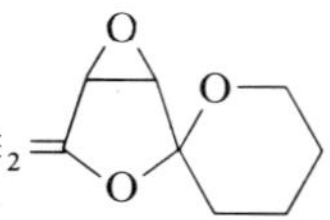
	$H_3C\equiv_3=_2(CH_2)_4=$	$H_3C\equiv_3=$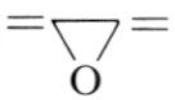$=$
P. globifera (= *M. globifera* Fenzl.)[48]	$H_3C\equiv_2=$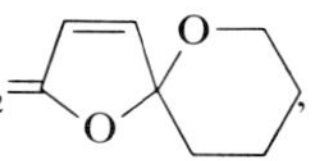, $H_3C\equiv_3=_2(CH_2)_4=$	$H_3C\equiv_3=_2(CH_2)_4=$
	$H_3C\equiv_3=$$=$, $H_3C\equiv_3=$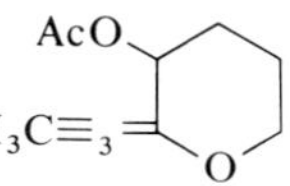	$H_3C\equiv_2=$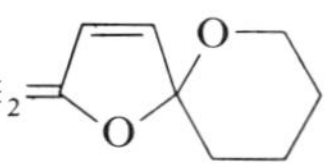
		$H_3C\equiv_2=$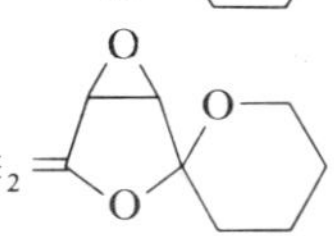
P. elegans DC. (= *M. multiflora*)[35]	$H_3C\equiv_3=$$=$, *AK*, $H_3C\equiv_3=_2(CH_2)_3OR$	$H_3C\equiv_3=_2(CH_2)_4=$
	$H_3C\equiv_3=_2(CH_2)_4=$	$H_3C\equiv_3=$$=$

P. grandiflora Hutch.[106]

$H_3C{\equiv}_3{=}_3$, 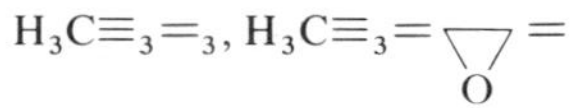

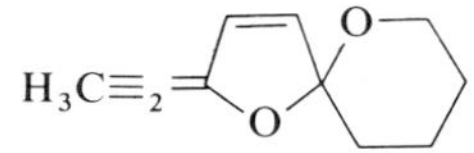

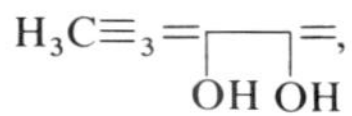, 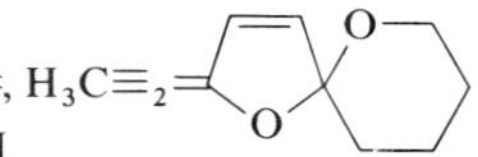

$H_3C{\equiv}_3{=}CH_2CH_2CH(OH)C_2H_5$

$H_3C{\equiv}_3{=}_2(CH_2)_3OR$ (R = H, Ac)

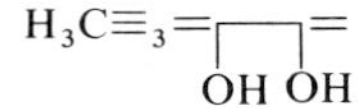

$H_3C{\equiv}_3{=}CH_2CH_2CH(OH)C_2H_5$

TABLE XV
TRIPLEUROSPERMUM

	Constituents: Roots	Constituents: Leaves
M. inodora L.[7,48,149]	*ME*, *ML*, *DME*, *AK*	$H_3C{\equiv}_3{=}_3$
	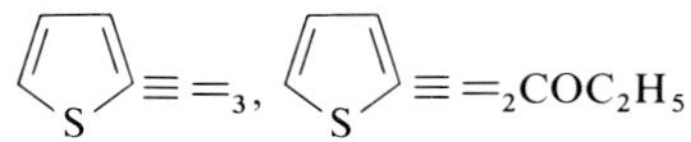	
	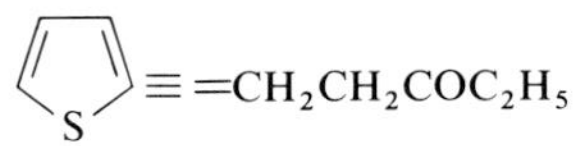	
M. tennusfolia Poir.[48]	*ME*, *ML*, *DME*, *AK*	
	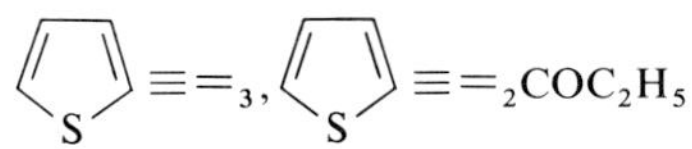	t,t-*ME*
	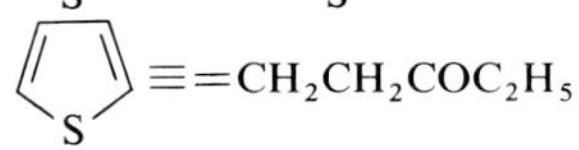	
M. oreades Boiss.[47,48,149]	*ML*, *DME*, *AK*	$H_3C{\equiv}_3{=}_3$, *ME*, *ML*
	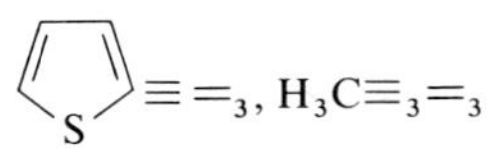	$H_3C{\equiv}_3{=}_2CH(OAc)CH_2CH_2OAc$
	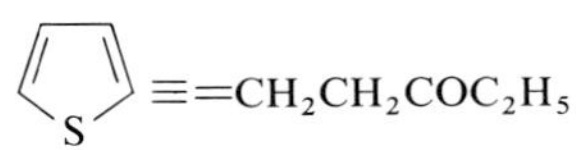	
M. grandiflora Fenzl.[48]	*DME*, *AK*, *ML*	$H_3C{\equiv}_3{=}_3$, *ME*, *ML*, *AK*
	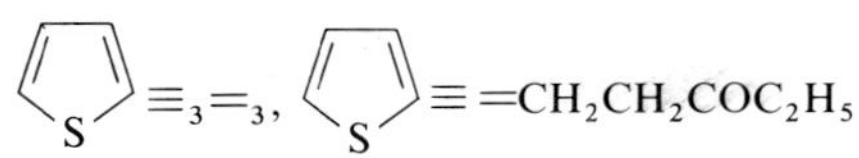	$H_3C{\equiv}_3{=}_2(CH_2)_3OH$

M. caucasica Poir.[48]		*DME*, *ME*, *ML*, *AK*, phenacetine
	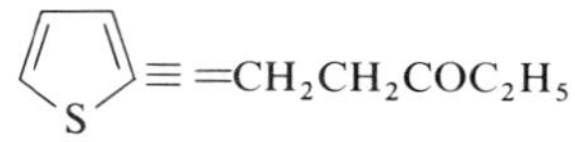	$H_3C{\equiv}_3{=}_3$
	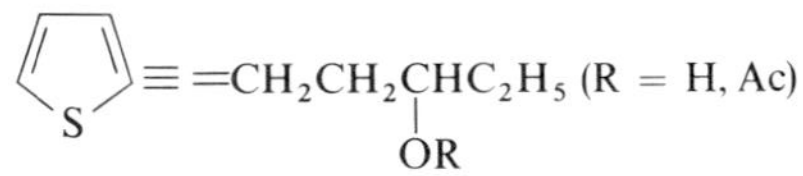	$H_3C{=}{\equiv}_2{=}_2(CH_2)_3OR$ (R = H, Ac)
	$H_3C{\equiv}_3{=}CH_2CH_2CH(OH)C_2H_5$	$H_3C{\equiv}_3{=}_2(CH_2)_3OR$ (R = H, Ac)
		$H_3C{=}{\equiv}_2{=}_2CH(OR)CH_2CH_2OR$
		$H_3C{\equiv}_3{=}_2CH(OR)CH_2CH_2OR$
		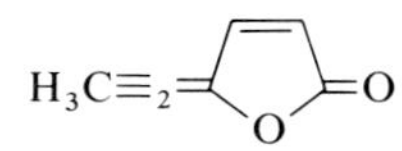
		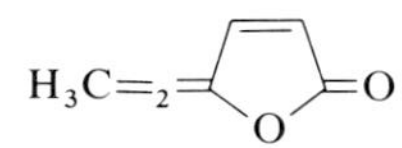
M. maritima L.[48]	*DME*	*ME*, $H_3C{\equiv}_3{=}_3$
M. tschihatchewii Boiss.[105]	*DME*, $H_3C{\equiv}_3{=}_3$	—

The genus Chrysanthemum is very complicated from the botanical point of view. Several partially very different subdivisions have been proposed. It is therefore impossible to use one botanically accepted system. In the following tables (XVI–XXIV) the investigated species are listed following the isolated acetylenes and partly in botanically proposed groupings. The subdivision of this large genus into the genera *Agyranthemum, Chrysanthemum, Pinardia, Leucanthemum, Kremeria, Dendranthema, Pyrethrum and Tanacetum* seems also to be logical from chemotaxonomical considerations.

TABLE XVI
AGYRANTHEMUM

	Roots	Leaves
Chrysanthemum frutescens L.[16,36,65,170,198]	$PhCO\equiv_2CH_3$, $PhCH(OAc)\equiv_2CH_3$, 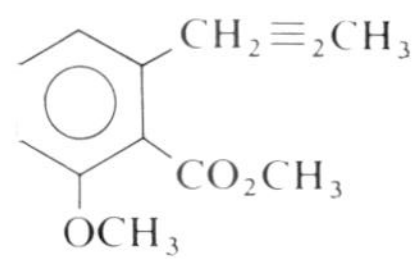 (and **270**, **272**, **274**, **276**) 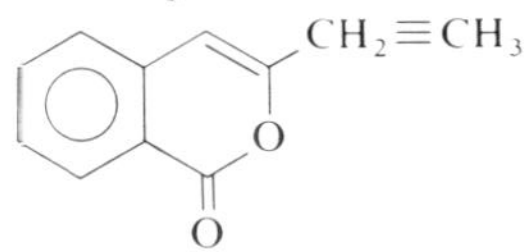 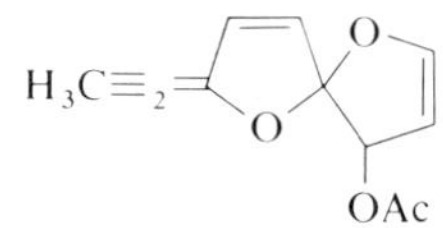 $R\equiv_2CH_2CH_2=_2CONHR'$ (R = H, CH_3) 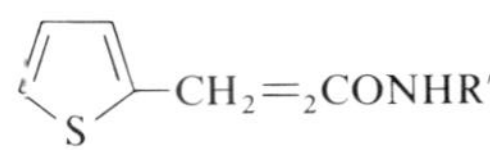	—
Chr. foeniculaceum DC.[65]	$PhCO\equiv_2CH_3$, $PhCH(OAc)\equiv_2CH_3$, 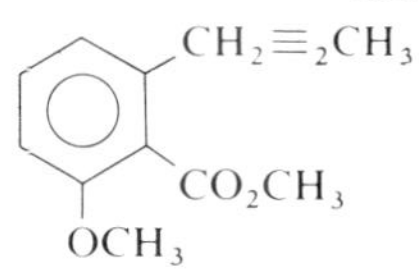 (and **270**, **272**, **274**, **276**) 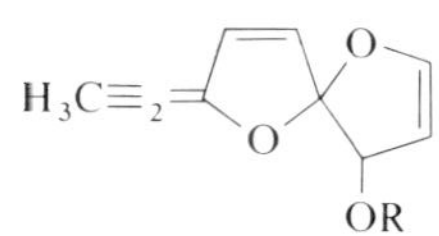 (R = H, Ac)	—

TABLE XVI—Continued

	Roots	Leaves
Chr. gracile Masf.[35]	$PhCO\equiv_2CH_3$, $H_3C\equiv_2$= 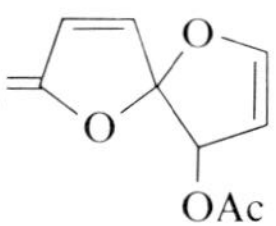	—
	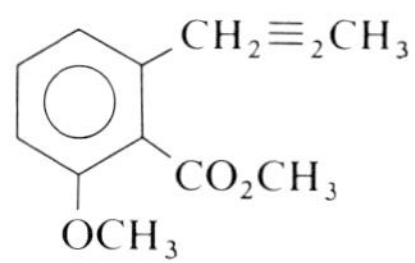 (R = CH_3, H)	
	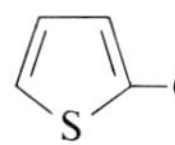 $-CH_2=_2CONHR'$	
	$H\equiv_2CH_2CH_2=_2CONHR'$	
Chr. fasciculatum[35]	$H_3C\equiv_2$= 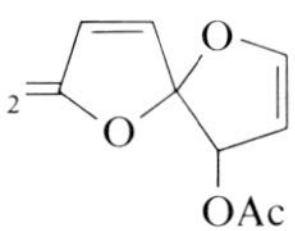, $H_3C\equiv_2$= 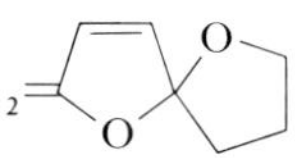	—
	$H_3C\equiv_2CH_2$ 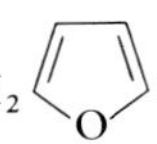 =CHO	
Chr. dugourii Christ.[35]	$H_3C\equiv_2$= 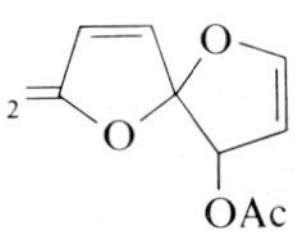	—

Chr. callichrysum Svent.[35]	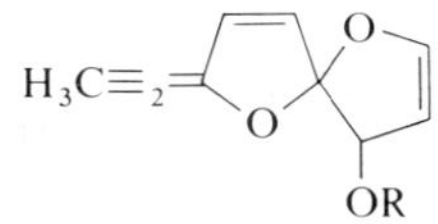 (R = H, Ac)	—
Chr. anethifolium Brouss.[35]	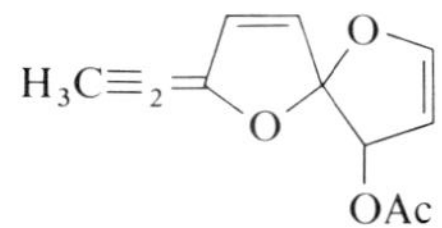	—
Chr. broussinetii Balb.[35]	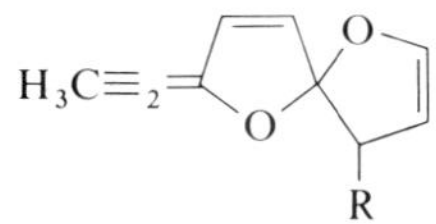 (R = H, OAc)	—
Chr. filifolium Christ.[35]	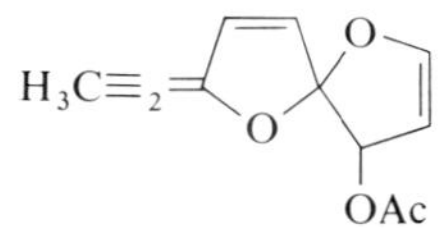	—

TABLE XVII
CHRYSANTHEMUM

	Roots	Leaves
Chr. carinatum Schousb.[65,67]	$PhCO{\equiv}_2R$ (R = H, CH_3) 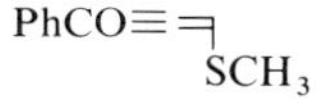 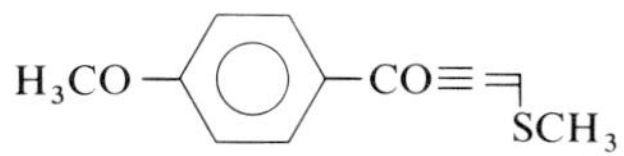 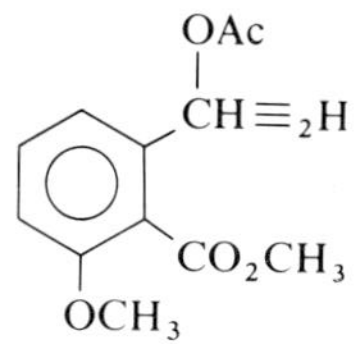 , $H_3C{\equiv}_3{=}_3$, 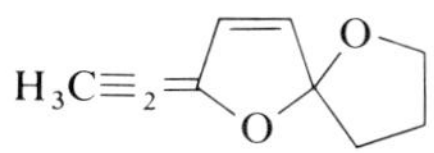 , 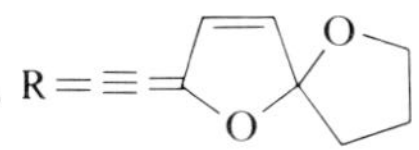 (R = SCH_3, H_3CSO) 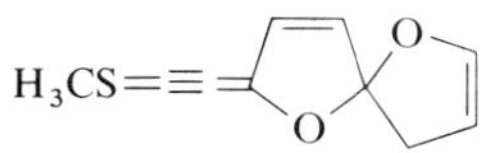	$PhCO{\equiv}_2R$ (R = H, CH_3) 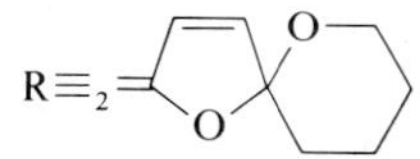 (R = H, CH_3) 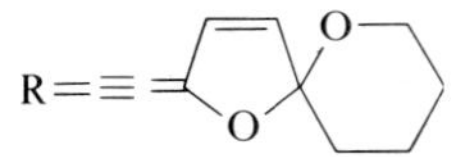 (R = H_3CS, H_3CSO, H_3CSO_2), 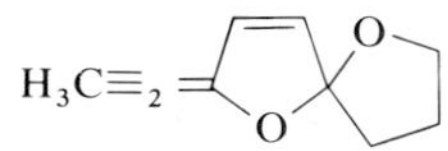 , 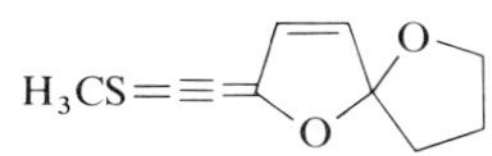 , 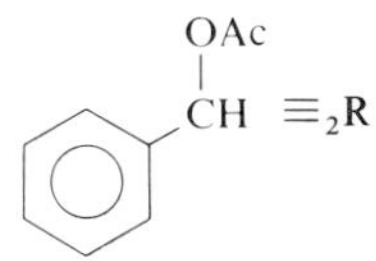 (R = H, CH_3)
Chr. coronarium L.[65]	$PhCO{\equiv}_2R$ (R = H, CH_3), $PhCO{\equiv}{=}SCH_3$, $PhCH_2{\equiv}{=}SCH_3$, $H_3C{\equiv}_2$ 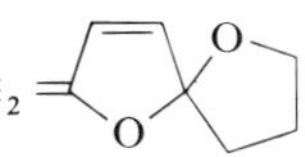 ,	$H_3C{\equiv}_3{=}_3$, $H_3C{\equiv}_2$ 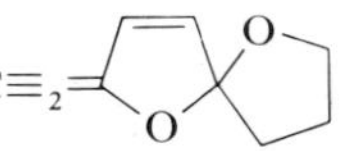 , 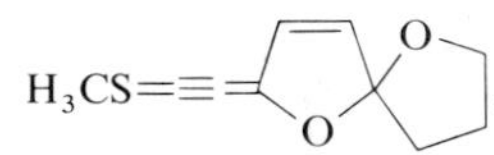 ,

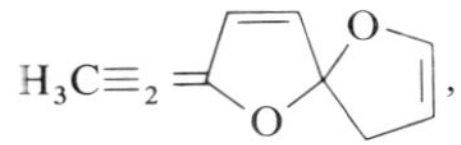

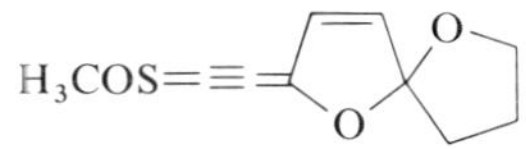

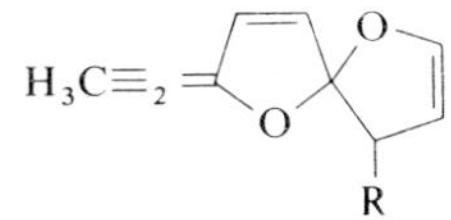 (R = H, OAc)

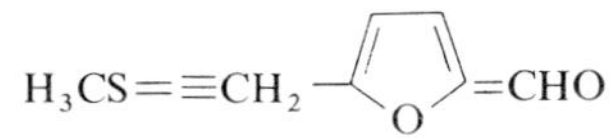

Chr. segetum L.[63,65]

PhCO≡$_2$R (R = H, CH_3),
PhCH$_2$≡$_2$H, PhCH$_2$≡ =SCH_3,
PhCO≡ =SCH_3,

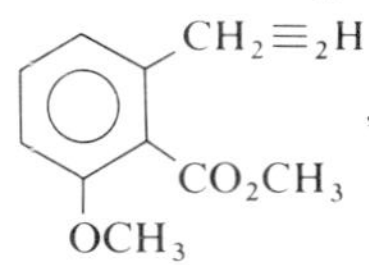 , 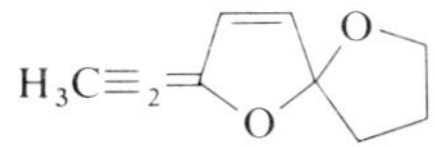 ,

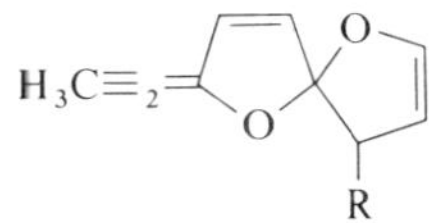 (R = H, OAc)

PhCH$_2$≡$_2$H

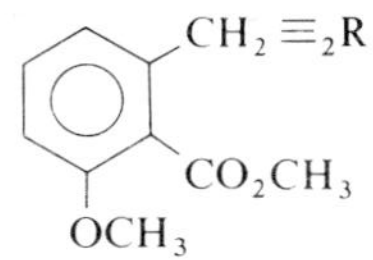 (R = H, CH_3)

Chr. spatiosum[67]

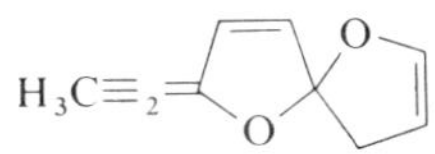 , 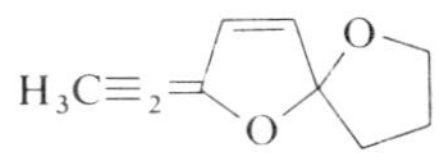

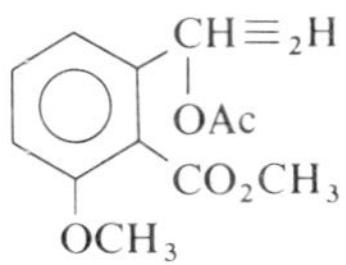

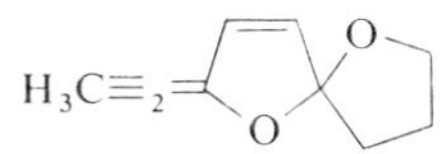 , 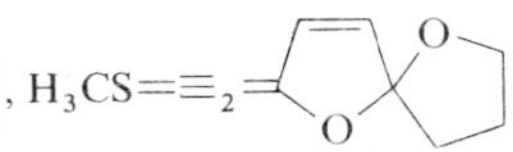

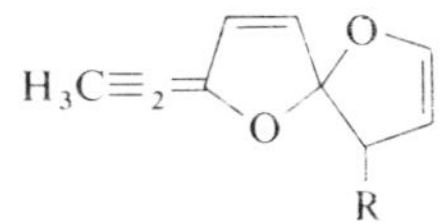 (R = H, OAc)

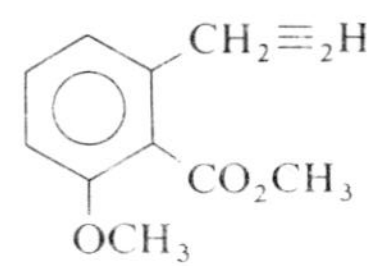

TABLE XVII—Continued

	Roots	Leaves
Chr. multicolor[35]	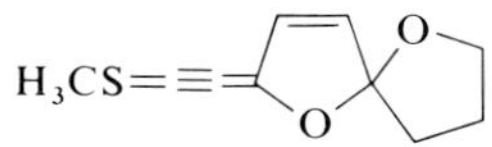	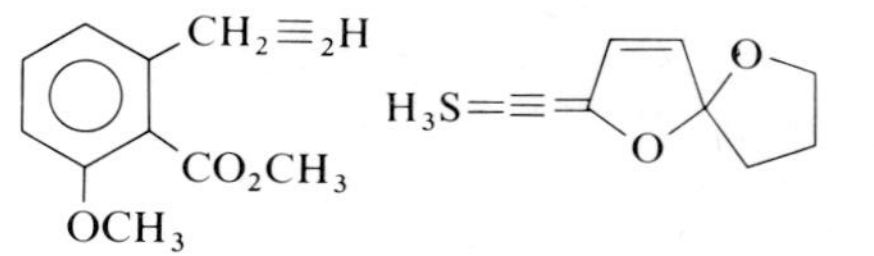
Chr. spectabilis[35]	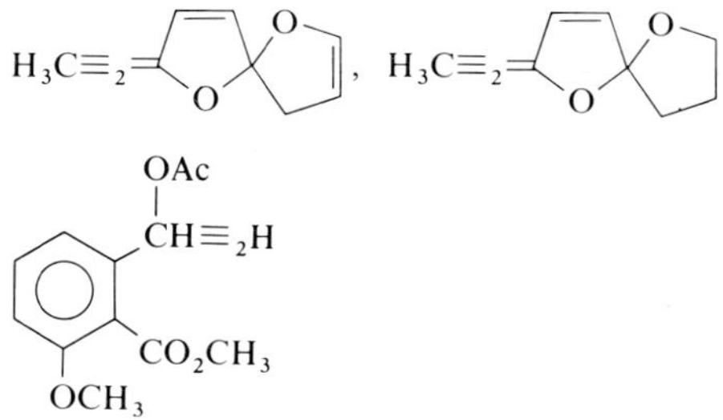	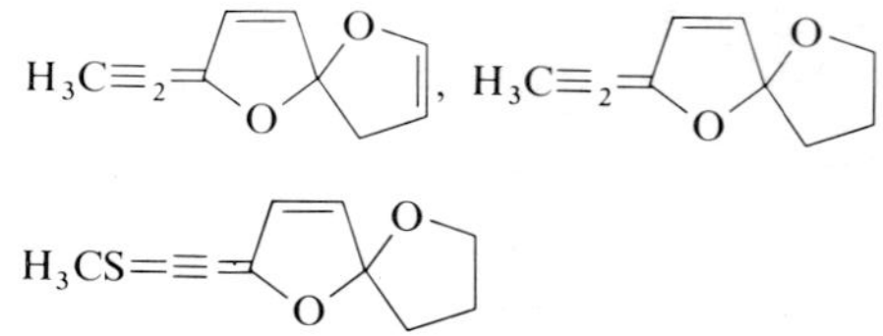

TABLE XVIII
PINARDIA

	Roots	Leaves
Chr. viscidi-hirtum Thell.[65]	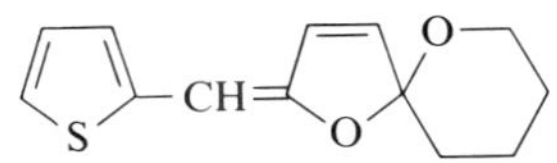	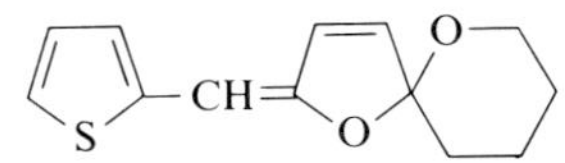
Chr. corynaeflorum[65]	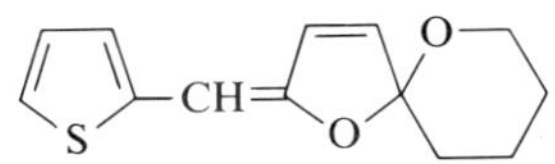	—

TABLE XIX
LEUCANTHEMUM

	Roots	Leaves
Chr. leucanthemum L.[45,62,65,115]	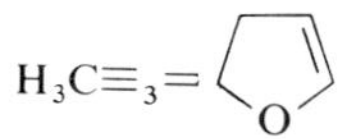	$H_3C\equiv_3=_2(CH_2)_2OAc$
	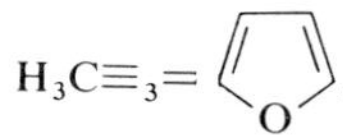, $H_3C\equiv_3=_2(CH_2)_2OAc$	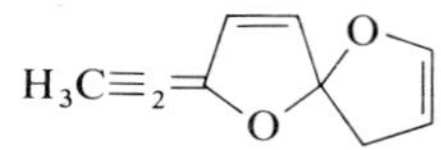
	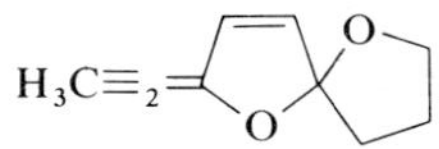 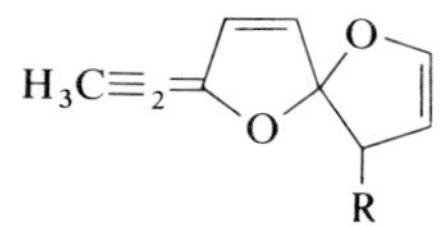	
	(R = H, OAc), 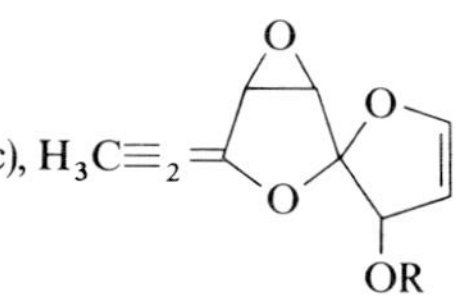 (R = COCH$_3$, Ang.)	
Chr. ircutianum Turcz.[37,65]	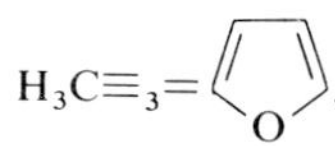, 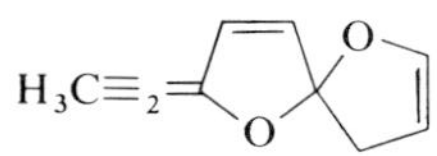 (R = H, Ac)	$H_3C\equiv_3=_2(CH_2)_3=$, DME $H_3C\equiv_3=_2CH_2CH_2OR$ (R = H, Ac) 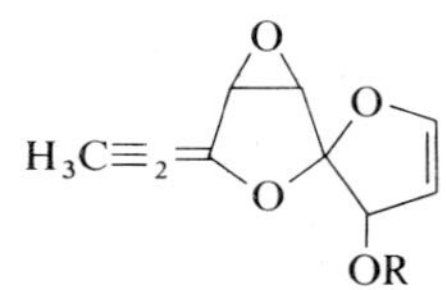 (R = H, Ang.)
Chr. maximum Ramond.[202]	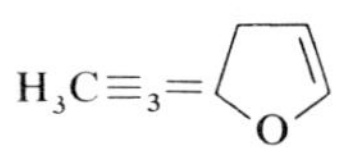	$H_3C=\equiv_2CH_2=(CH_2)_5=$, $H_3C\equiv_3=_2(CH_2)_3=$ $H_3\equiv_3=_2CH_2CH_2OR$, $H_3C=\equiv_2CH_2CH_2CH_2OR$

Chr. lacustre Brot.[65]	$H_3C\equiv_2=$ 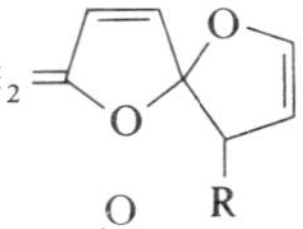 (R = H, OAc) $H_3C\equiv_2=$ 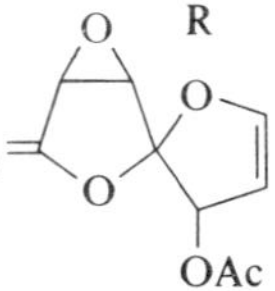	$H_3C\equiv_3=_2(CH_2)_3=$ $H_3C\equiv_3=_2CH_2CH_2OAc$ $H_3C\equiv_2=$ 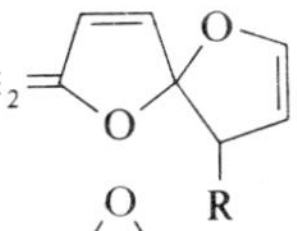 (R = H, OAc) $H_3C\equiv_2=$ 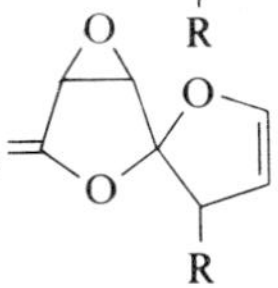 (R = H, Ac, Ang.)
Chr. discoideum All.[35]	$H_3C\equiv_2=$ 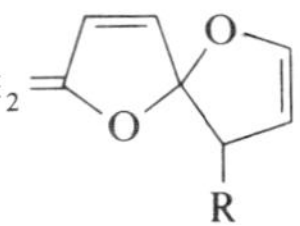 (R = H, OAc)	$H_3C\equiv_3=_2(CH_2)_3=$ $H_3C\equiv_3=_2CH_2CH_2OR$ (R = H, Ac)
Chr. atratum Jacq.[65]	$H_3C\equiv_3=$ 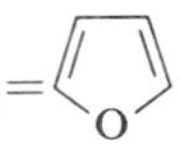, $H_3C\equiv_2=$ 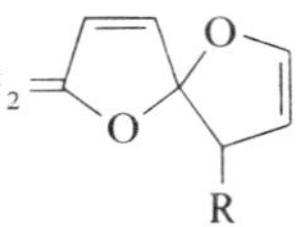 (R = H, OH, OAc) $H_3C\equiv_2=$ 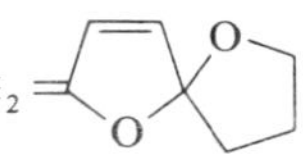, $H_3C\equiv_3=_2(CH_2)_2OR$ (R = H, Ac)	$H_3C\equiv_3=_2(CH_2)_3=$, $H_3C\equiv_2=$ 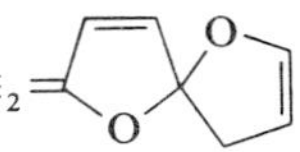 $H_3C\equiv_2=$ 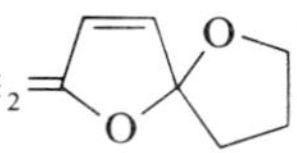, $H_3C\equiv_3=_2(CH_2)_2OR$ (R = H, Ac)
Chr. croaticum Horwatic[35]	$H_3C\equiv_3=_3$, $H_3C\equiv_2=$ 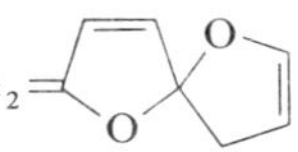	$H_3C\equiv_3=_2(CH_2)_3=$, $H_3C\equiv_2=$ 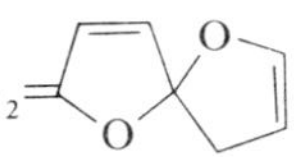 $H_3C\equiv_3=_2(CH_2)_2OR$ (R = H, Ac) $H_3C\equiv_3CH_2=(CH_2)_3OAc$

TABLE XIX—Continued

	Roots	Leaves
Chr. sylvaticum Hoffmg. *et* LK.[35]	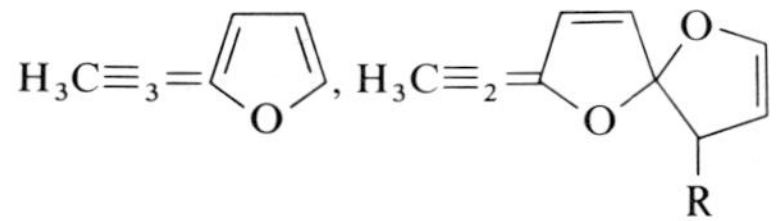 (R = H, OAc)	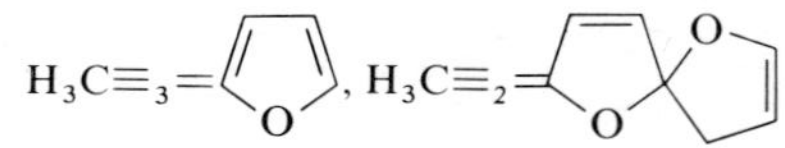 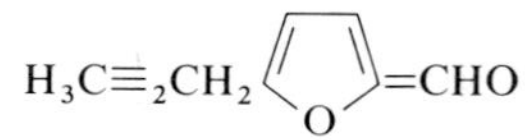
	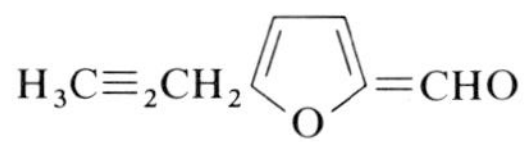	
Chr. montanum All.[35]	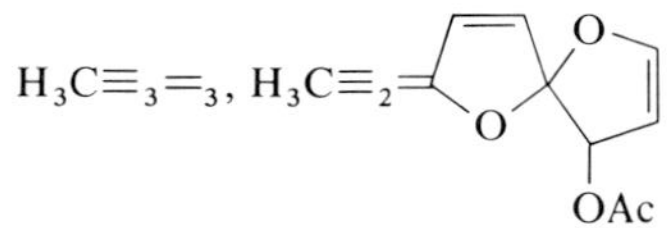	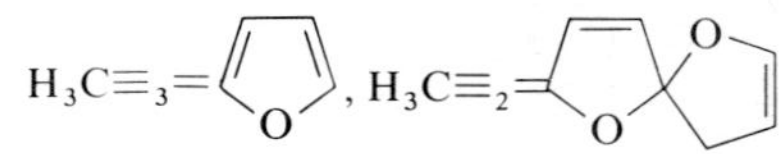
Chr. coronopifolium Vill.[35]	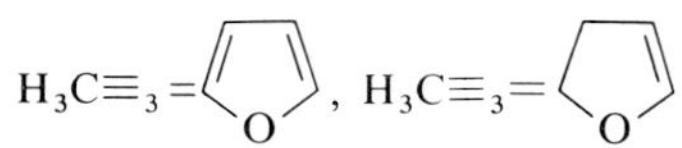	$H_3C\equiv_3=_2(CH_2)_n=$ (n = 3, 4)
	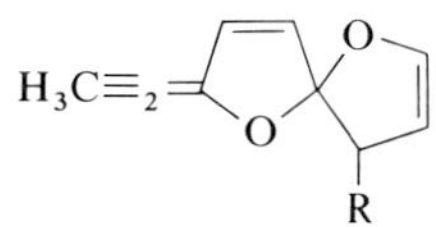 (R = H, OAc)	
	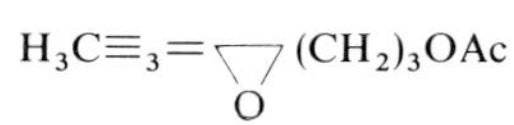	
	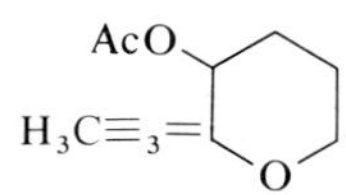	

Chr. rotundifolium Wald. *et* Kit.[35]

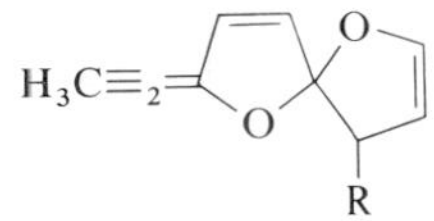 (R = H, OAc)

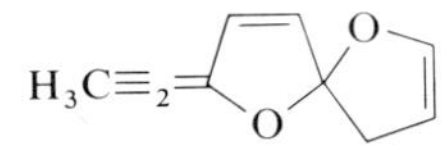

Chr. nivellii Braun-Blanquet[35,65]

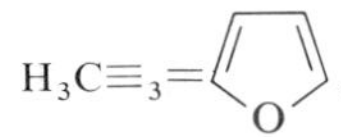,

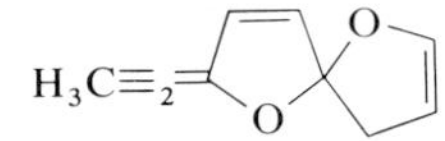

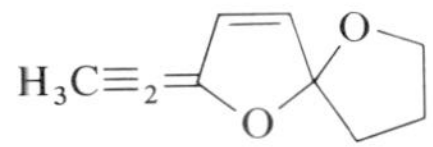

$H_3C\equiv_3=_2(CH_2)_3OR$ (R = H, Ac)

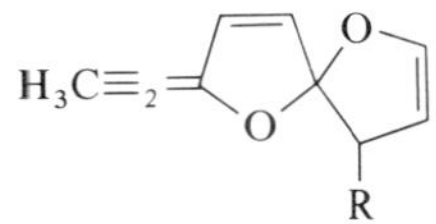 (R = H, OAc)

Chr. monspeliensis L.[106]

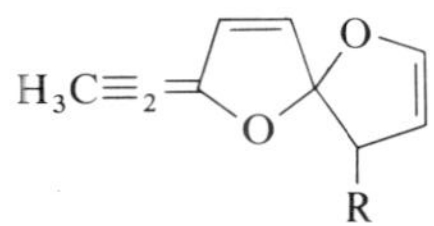

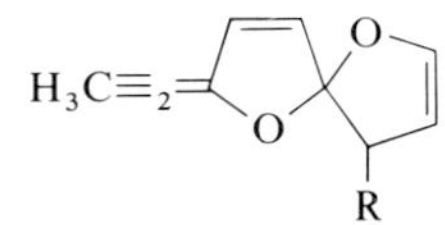

(R = H, OH, OAc, O-i-Val.)

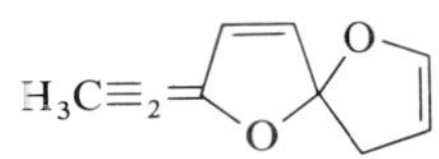, 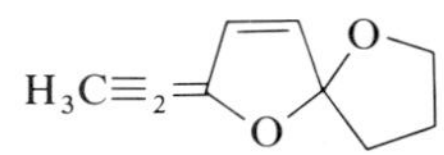

(R = H, OH, OAc, O-Sen., O-Ang., O-i-Val.)

Chr. glabrum DC.[35]

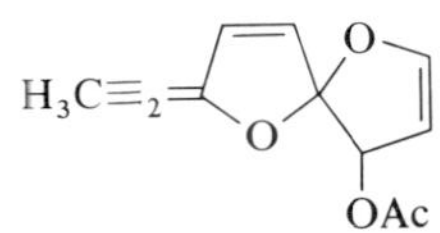

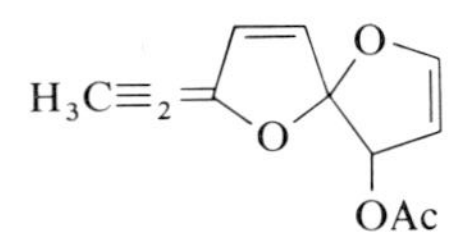

TABLE XIX—Continued

	Roots	Leaves
Chr. graminifolium L.[35]	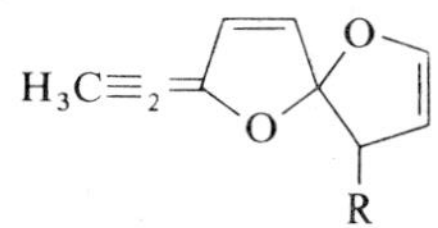 (R = H, OAc)	—
Chr. goyanum Ball.[35]	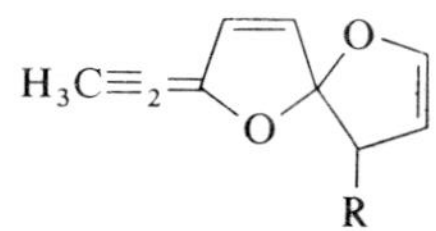 (R = H, OAc)	—

TABLE XX
KREMERIA

	Roots	Leaves
Chr. myconis L.[65,105]	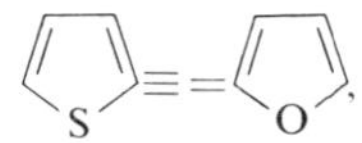, 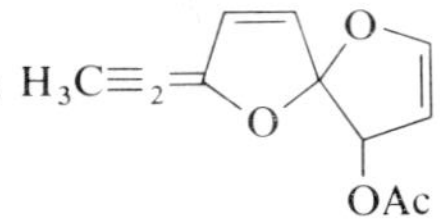	$H_3C\equiv_3=_2(CH_2)_3=$, 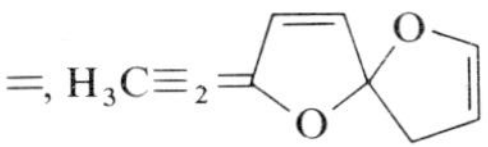
Chr. multicaule Desf.[65]	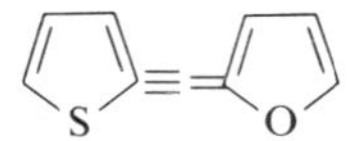	$H_3C\equiv_3=_2(CH_2)_3=$, *DME* $H_3C\equiv_3=_2CH_2CH_2OR$ (R = H, Ac)
Chr. hybridum Guss.[35]	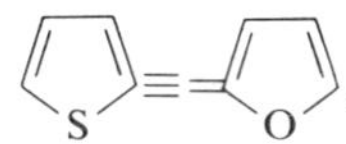, 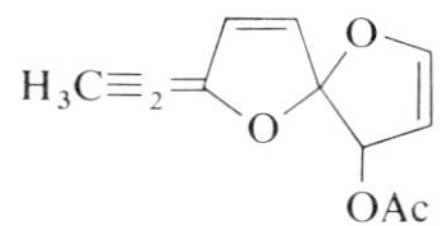	$H_3C\equiv_3=_2(CH_2)_3=$, 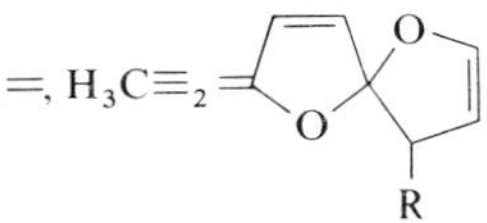 (R = H, OAc), 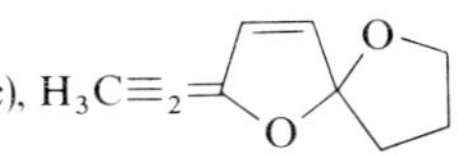 $H_3C\equiv_3=_2CH_2CH_2OR$ (R = H, Ac) *AK*, *DME*
Chr. grande Hook.[35]	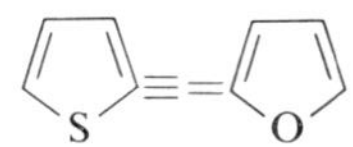	—
Chr. flosculosum L.[65]	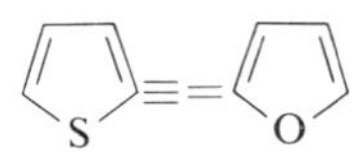 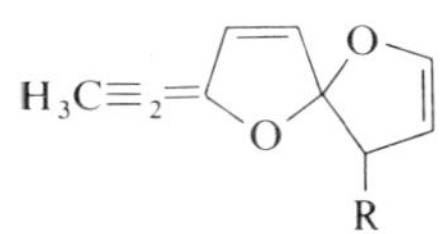 (R = H, OAc) 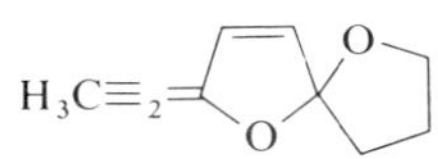	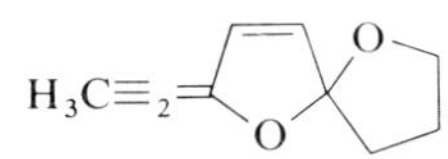, 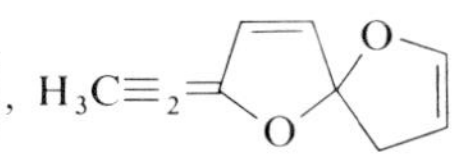 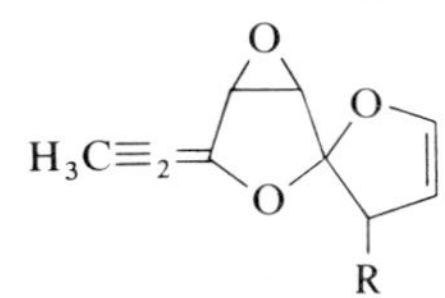 (R = H, OAc) $H_3C\equiv_3=_2CH_2CH_2OH$

TABLE XXI
DENDRANTHEMA

	Roots	Leaves
Chr. arcticum L.[63,65]	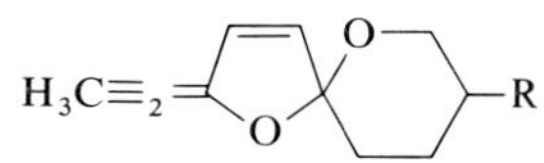 (R = H, OAc)	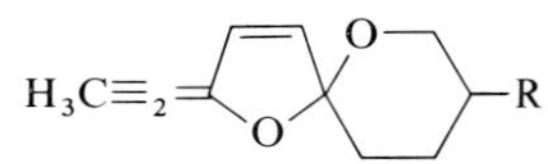 (R = H, OAc)
Chr. coreanum Nakai[65]	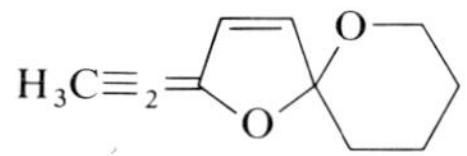 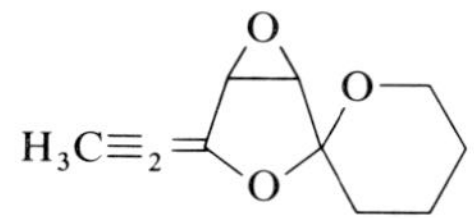	—
Chr. indicum L.[65]	$H_3C\equiv_3=_2(CH_2)_4=$ 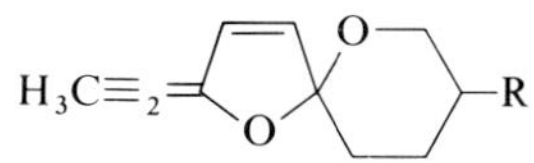 (R = H, OAc)	—
Chr. yezoense Maekawa[65]	$H_3C\equiv_3=_2(CH_2)_4=$ 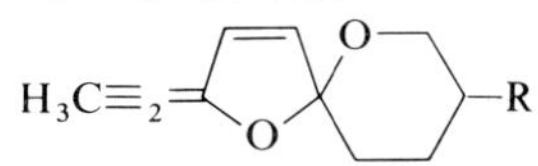 (R = H, OAc)	$H_3C\equiv_3=_2(CH_2)_4=$ 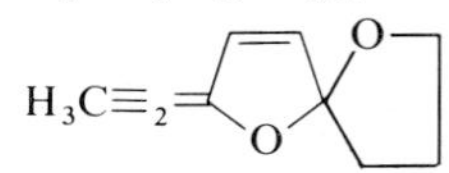
Chr. zawadskii Herbich[35]	$H_3C\equiv_3=_2(CH_2)_4=$ 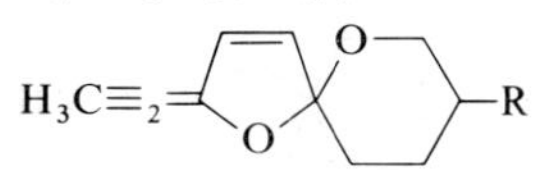 (R = H, OAc)	as roots

Chr. weyrichii Miyabe *et* Miyake[65]	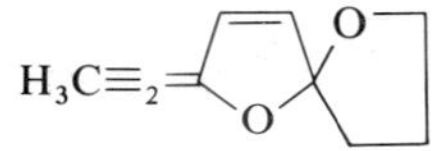	—
Chr. koraiense Nakai[35]	trace 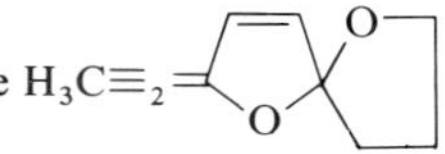	—
Chr. nactongense[35]	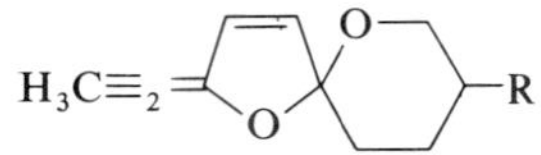 (R = H, OH, OAc) 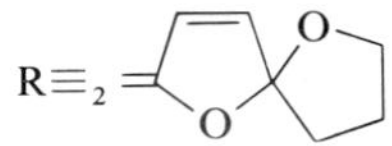 (R = H, CH_3)	—

TABLE XXII
PYRETHRUM

	Roots	Leaves
I. *Chr. macrophyllum* W. *et* K.[65]	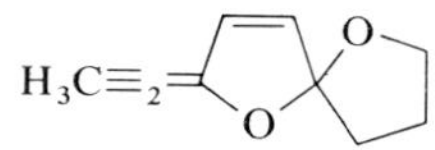	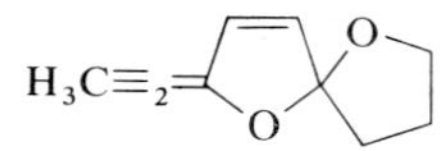
Chr. parthenicum Bernh.[45,65]	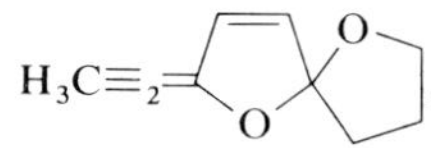, $H_3C{\equiv}_3{=}_3$, *DME* 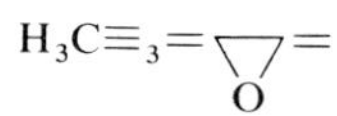	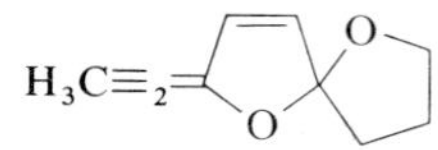
Chr. mawii Hook.[65]	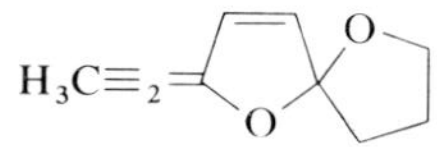	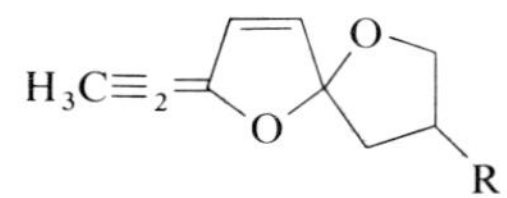 (R = H, OAc)
Chr. corymbosum L.[65]	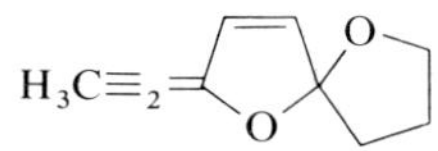	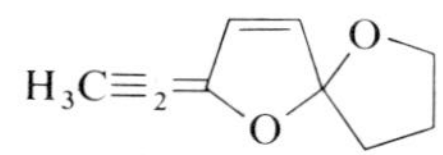
Chr. partheniifolium Pers.[37,65]	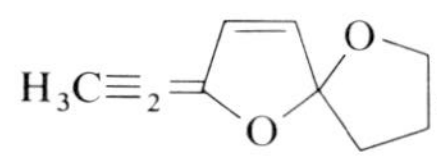, 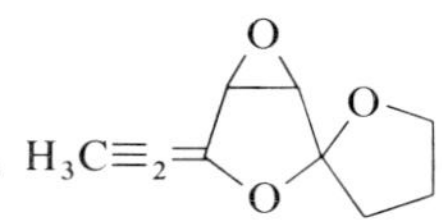	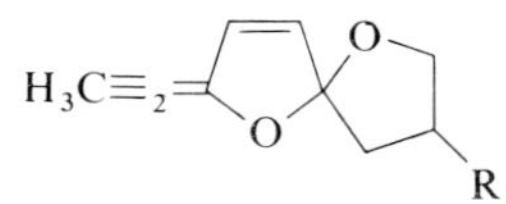 (R = H, OAc)
Chr. palustre Pers.[35]	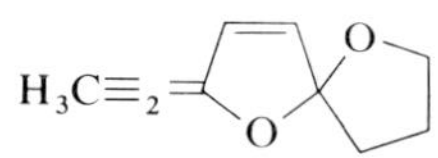	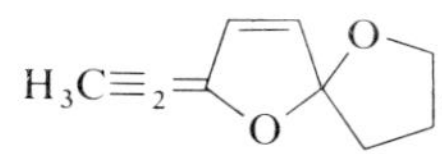, 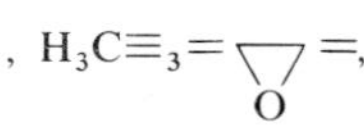, *DME*
Chr. marschallii Aschers.[65]	—	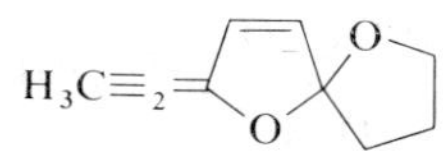

Chr. niveum Lag.[65]	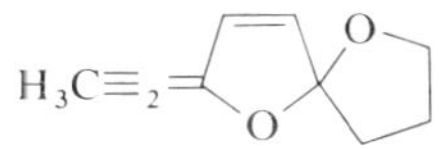 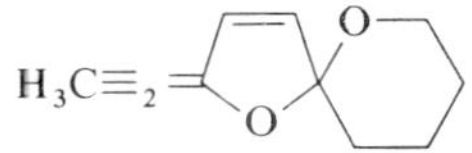	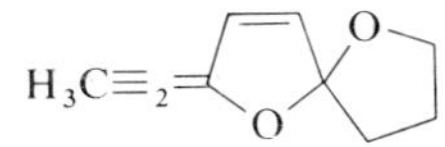
Chr. prealtum Vent.[35]	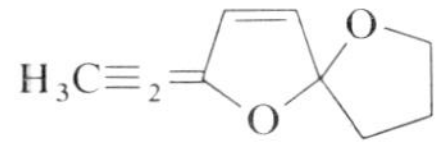 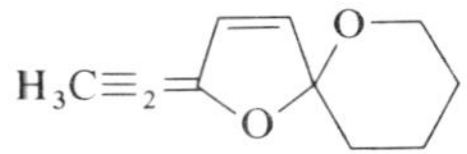	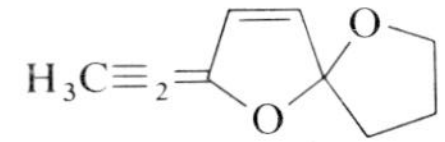
Chr. leptophyllum DC.[106]	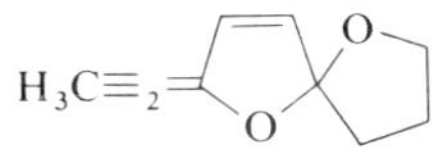 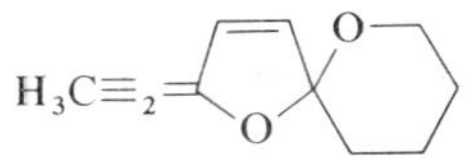 $H_3C\equiv_2=_3CH_2CH_2OH$, $H\equiv_2=_3CH_2CH_2OH$	—
II. *Chr. rubellum* Sealy[65]	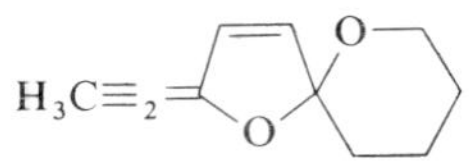	—
Chr. densum-rubellum Geyig.[35]	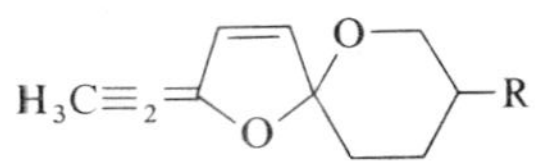 (R = H, OAc) 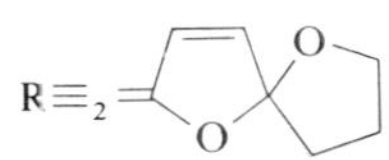 (R = H, CH_3)	$H_3C\equiv_3=_2(CH_2)_4=$, *AK*, 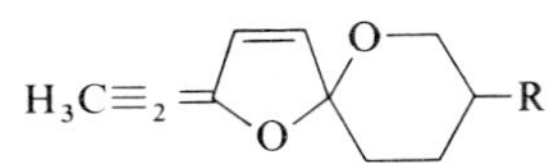 (R = H, OAc)

TABLE XXII—Continued

	Roots	Leaves
Chr. clusii Kreutz[65]	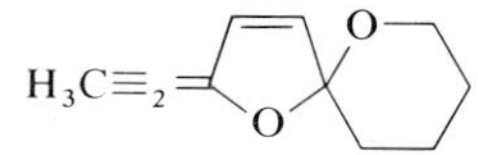	—
Chr. serotinum L.[40,65,105]	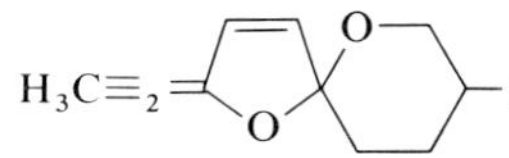 R (R = H, OAc)	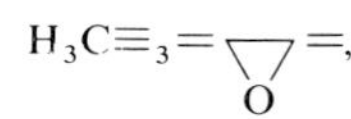, $H_3C\equiv_3=_2(CH_2)_3OAc$
	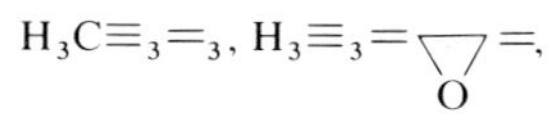, *DME*	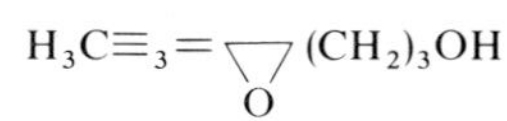
	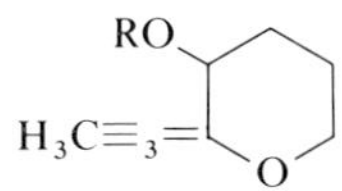 (R = H, Ac)	$H_3C\equiv_3CH_2=(CH_2)_4OAc$
		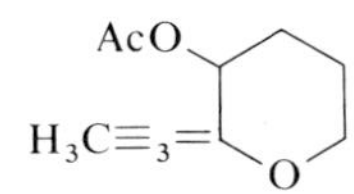
Chr. ochroleucum Masf.[35]	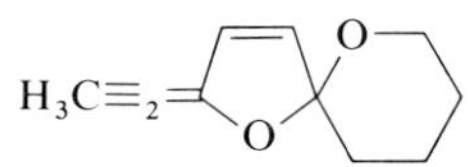	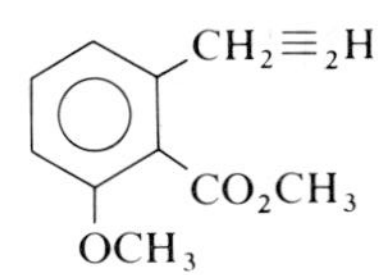
Chr. nipponicum Masf.[65]	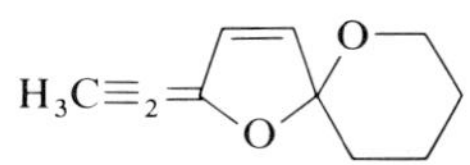	$H_3C\equiv_3=_2(CH_2)_4=$, 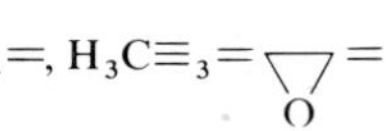
	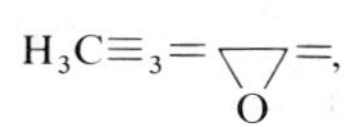, *AK*	$H_3C\equiv_3=_2(CH_2)_3OH$

Chr. douglasii Hulten[35]	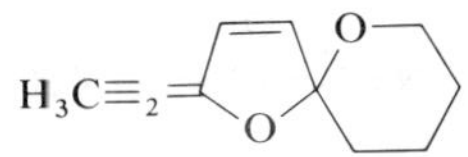	—
	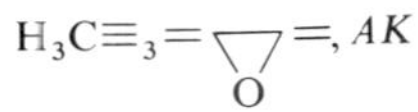	
	$H_3C\equiv_3=CH_2CH_2CHC_2H_5$, *DME* (CH with OH)	
Chr. coccineum Willd.[65]	—	ev. 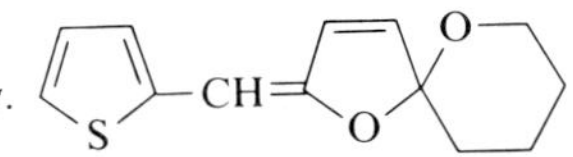
Chr. balsamita L.[65]	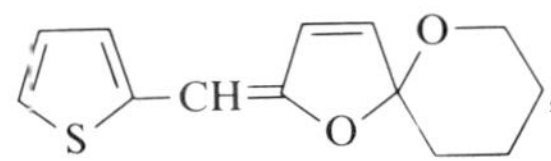, *AK*, *DME*	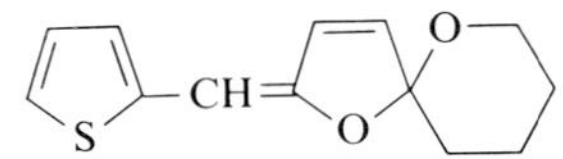
	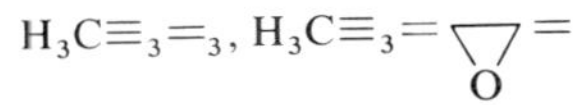	
Chr. roseum Adam.[35]	trace $-C\equiv C-$	as roots
Chr. alpinum L.[35]	$H_3C=\equiv_2=_2(CH_2)_3OAc$	as roots

TABLE XXIII
TANACETUM

	Roots	Leaves
I. *T. vulgare* L.[35,43,62,355]	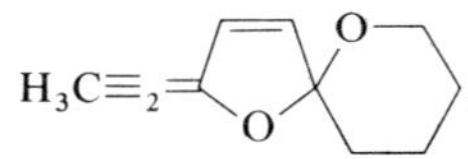, $H_3C{\equiv}_3{=}_3$	*DME*
	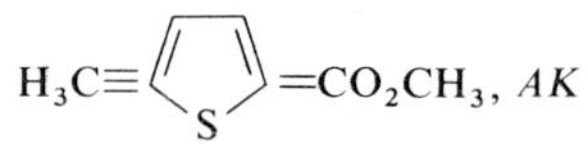, *AK*	
	$H_3C{\equiv}_3{=}_2(CH_2)_3OAc$	
T. annuum L.[35]	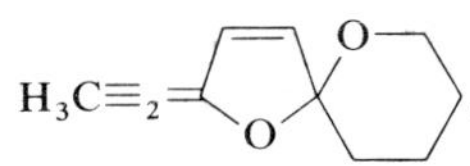, *AK*	—
T. pseudoachillea C. Winkl.[35]	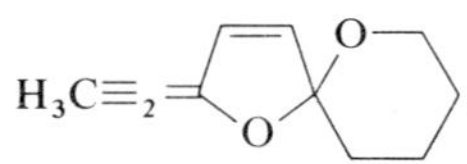, $H_3C{\equiv}_3{=}_3$	$H_3C{\equiv}_3{=}_2(CH_2)_4{=}$, *DME*
	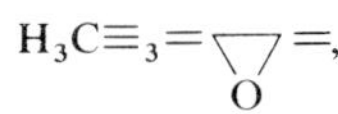, *DME, AK*	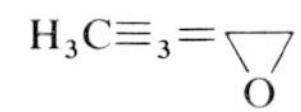 $(CH_2)_3OAc$
	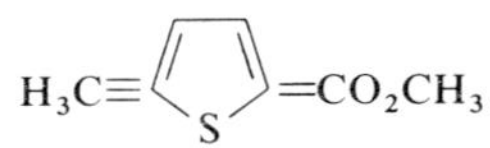, $H_3C{\equiv}_3{=}_2(CH_2)_3OAc$	
	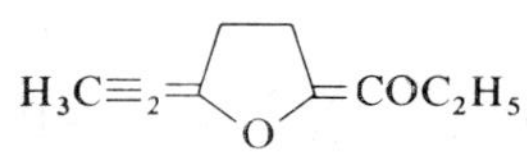	
	$H_3C{\equiv}_3{=}CH_2CH_2CH(OH)C_2H_5$	

T. huronense Nutt. (= *T. camphoratum* Less.)[35]	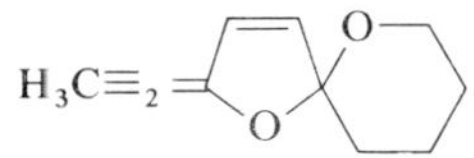, $H_3C{\equiv}_3{=}_3$, *AK*	$H_3C{\equiv}_3{=}_2(CH_2)_4{=}$
	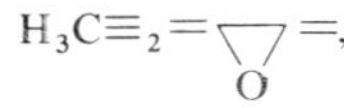, $H_3C{\equiv}_3{=}_2(CH_2)_3$, OR (R = H, Ac)	
T. boreale Fisch.[65]	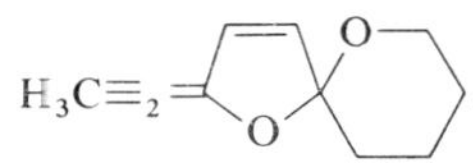, $H_3C{\equiv}_3{=}_3$	$H_3C{\equiv}_3{=}_2(CH_2)_3OAc$
	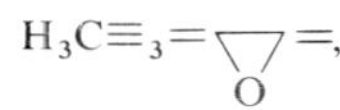, *AK*, *DME*	
	$H_3C{\equiv}_3{=}_2(CH_2)_3OAc$	
	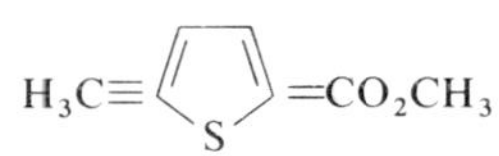	
II. *T. bipinnatum* Sch. Bip.[35]	$H_3C{\equiv}_3{=}_2(CH_2)_4{=}$, *DME*, *AK*	$H_3C{\equiv}_3{=}_2(CH_2)_4{=}$, *DME*
T. tomentosum Loisel[35]	$H_3C{\equiv}_3{=}_2(CH_2)_4{=}$, *DME*, *AK*	$H_3C{\equiv}_3{=}_2(CH_2)_4{=}$, *DME*
T. millefoliatum Fisch. *et* Mey.[35]	$H_3C{\equiv}_3{=}_2(CH_2)_4{=}$, *DME*, *AK*	$H_3C{\equiv}_3{=}_2(CH_2)_4{=}$, *DME*
	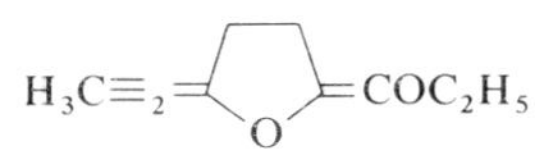,	*AK*, $H_3C{\equiv}_3{=}_2(CH_2)_3OAc$
	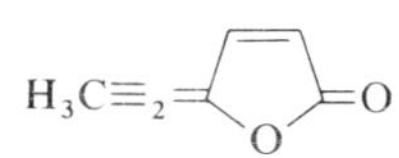	

The *Tanacetum* group, as shown by Table XXII, is very homogenous. Nevertheless, the whole *Chrysanthemum* group is more or less well characterised by the very common occurrence of spiroketalenolethers. This clearly shows the close relationship to *Matricaria* and to some *Artemisia* species.

Many other constituents have been isolated from *Chrysanthemum* species besides acetylenes, but up to now the number of these fully investigated is too small to be of chemotaxonomical value. Worth a mention is the coumarin Herniarine isolated from several *Chrysanthemum* as well as from *Matricaria* and *Artemisia* species, while sesquiterpene-lactones are rare, although they are very common in the genus *Artemisia*.

TABLE XXIV
COTULA

	Roots	Leaves
I. *C. anthemoides* L.[35]	$C_{17}{=}CO{\equiv}_2CH_2{=}(CH_2)_5{=}$	as roots
C. australis Hook.[35]	$C_{17}{=}CO{\equiv}_2CH_2{=}(CH_2)_5{=}$	as roots
C. barbata DC.[35]	$C_{17}{=}CO{\equiv}_2CH_2{=}(CH_2)_5{=}$	as roots
C. bipinnata Thunb.[35,221]	—	$C_{15}H_3C{=}{\equiv}_2{=}_2CH_2CH_2{=}$ $C_{14}H_3C{=}{\equiv}_2{=}_2(CH_2)_3OAc$
C. dioica Hook.[105]	$C_{17}{=}CO{\equiv}_2CH_2{=}(CH_2)_5{=}$	—
C. filicula Benth.[221]	$C_{17}{=}CO{\equiv}_2CH_2{=}(CH_2)_5{=}$	—
C. microglossa[35]	$C_{17}{=}CO{\equiv}_2CH_2{=}(CH_2)_5{=}$ $H_3C{\equiv}_3{=}_2(CH_2)_4{=}$, $HOCH_2{=}{\equiv}_2C_3H_7$	—
C. plumosa Hook.[221]	$C_{17}HOCH_2{=}{\equiv}_2{=}_2C_6H_{13}$	—
C. squalida Hook.[105]	$C_{17}{=}CO{\equiv}_2CH_2{=}(CH_2)_5{=}$	—
C. turbinata L.[105]	$C_{17}{=}CO{\equiv}_2CH_2{=}(CH_2)_5{=}$	—
C. reptans Benth.[221]	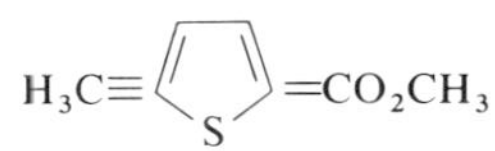	—
C. alpina Hook.[221]	*DME*	—
II. *C. coronopifolia* L.[228]	$C_{15}{=}CH(OH){\equiv}_2{=}_2CH_2CH_2{=}$ $C_2H_5CH(OH){\equiv}_2{=}_2CH_2CH_2{=}$ $HOCH_2CH_2CO{\equiv}_2CH_2{=}(CH_2)_3{=}$ $C_{14}H_3C{=}{\equiv}_2{=}_2(CH_2)_3OAc$, *AK*, *DME*	$C_{14}H_3C{=}{\equiv}_2{=}_2(CH_2)_3OR$ (R = H, Ac) $C_2H_5CH(OR){\equiv}_2{=}_2(CH_2)_3OR'$ (R = R′ = H, R = H, R′ = Ac, R = R′ = Ac) C_{15} as roots, *DME*, Estafiatn

The genus *Cotula* is characterised by the occurrence of C_{17}-acetylenes, especially dehydrofalcarinone; *C. coronopifolia* L., belonging to the section *Eucotula*, is an exception. Only C_{15}-acetylenes with very similar structure occur together with the main compounds having C_{14}-chains. The same situation is true for *C. bipinnata*, while *C. alpina* and *C. reptans* have only dehydromatricariaester and the corresponding thiophene.

The next three genera seem to have some relations to the large genus *Artemisia*.

TABLE XXV
SOLVIA, PHALOCROCARPUM AND HIPPIA

	Constituents	
	Roots	Leaves
Soliva anthemifolia R. Br.[35]	*DME*	*DME*
Phalacrocarpum oppositifolium (Brot.) Willk.[35]	*AK*, $H_3C\equiv_3=_3$	—
Hippia frutescens L.[35]	$H_3C\equiv_3=_3$, 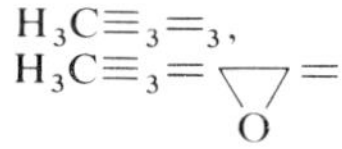 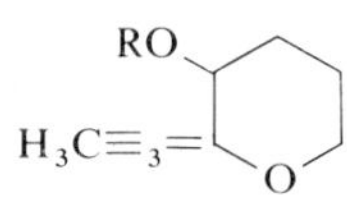	$H_3C=\equiv_2=_2(CH_2)_3$-OAc
H. pilosia Hutch.[35]	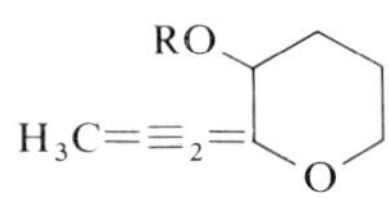	

In the genus *Artemisia*, the botanical situation is one again very complicated and no common expected scheme is available for a subdivision of this large genus.

In Table XXVI the species are placed according to their acetylenic constituents.

TABLE XXVI
ARTEMISIA

	Constituents: Roots	Constituents: Leaves
(1) *Artemisia atrata* Lam.[105,201]	$=CO\equiv_2CH_2=(CH_2)_5=$ $=CH(OH)\equiv_2CH_2=(CH_2)_5=$	—
A. armeniaca Lam.[35]	$=CO\equiv_2CH_2=(CH_2)_5=$	—
A. campestris L.[105,201]	$=CO\equiv_2CH_2=(CH_2)_5=$	$=CO\equiv_2CH_2=(CH_2)_5=$ $=CO\equiv_2CO=(CH_2)_5=$
A. capillaris Thunb.[35] *	$=CO\equiv_2CH_2=(CH_2)_5=$	—
A. crithmifolia L.[201]	—	$=CO\equiv_2CH(OH)=(CH_2)_5=$
A. laciniata Willd.[35]	} $=CO\equiv_2CH_2=(CH_2)_5=$ } $=CH(OH)\equiv_2CH(R)=(CH_2)_5=$ (R = H, OH)	—
A. macilenta Krasch.[35]	(as above)	$=CO\equiv_2CH_2=(CH_2)_5=$
A. scoparia W. *et* K.[35]	$=CO\equiv_2CH_2=(CH_2)_5=$ $HOCH_2CH_2CO\equiv_2CH_2=(CH_2)_5=$ $PhCH_2\equiv_2CH_3$, $PhCH_2\equiv_2H$	—
(2) *Artemisia abrotanum* L.[35]	$H_3C\equiv_3=_2(CH_2)_4=$	$H_3C\equiv_3=_2$, $H_3C\equiv_3=$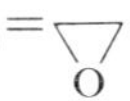$=$
A. alba Turra.[35]	$H_3C\equiv_3=_2(CH_2)_4=$	—
A. austriaca Jacq.[35]	*DME*, *AK*, $H_3C\equiv_2=$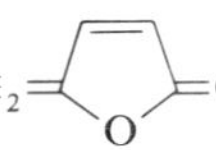$=O$ $H_3C\equiv_3=_2(CH_2)_4=$, H_3C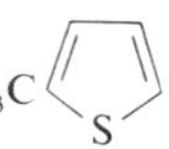$\equiv=CO_2CH_3$ $H_3C\equiv_3=CH_2CH_2CH(OH)C_2H_5$ $H_3C\equiv_3=_2(CH_2)OR$ (R = H, Ac)	$H_3C\equiv_3=_2(CH_2)_4=$

* Japanese workers have isolated aromatic acetylenes $Ph-CO-(C\equiv C)_2-CH_3$ and others[125].

TABLE XXVI—Continued

	Constituents: Roots	Constituents: Leaves
A. camphorata Vill.[35]	$H_3C\equiv_3=_2(CH_2)_4=$	—
A. cana Pursh.[35]	*AK*, $H_3C\equiv_3=_2(CH_2)_4=$ $H_3C\equiv_3=$ 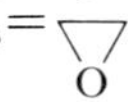 $=$	—
A. commutata Bess.[35]	*DME*, *AK*, $H_3C\equiv_3=_2(CH_2)_4=$ $H_3C\equiv_3=_2(CH_2)_3OR$ (R = H, Ac) $H_3C\equiv_3=CH_2CH_2CH(OH)CH_2CH_3$	—
A. persica Boiss.[35]	*DME*, *AK*, $H_3C\equiv_3=_2(CH_2)_4=$ $H_3C\equiv_3=_2(CH_2)_3-(OR)$(R = H, Ac) $H_3C\equiv_3=CH_2CH_2CH(OH)C_2H_5$, $H_3C\equiv_2=$ 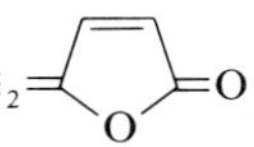	—
A. selengensis auct. (Turcz.)[32]	*DME*, *AK* $H_3C\equiv_3=_2(CH_2)_4=$, $H_3C\equiv_3=_2CH(OH)(CH_2)_3=$ $H_3C\equiv_3=CH(OH)(CH_2)_5=$, $H_3C\equiv_3=_3$, $H_3C\equiv_2=$ 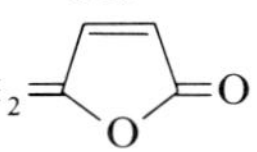 $H_3C\equiv_3=_2(CH_2)_3-OR$ (R = H, Ac) $H_3C\equiv_3=CH_2CH_2CH(OH)C_2H_5$	trace $H_3C\equiv_3=_2(CH_2)_4=$

A. superba Pampan.[35]	*DME*, *AK*, $H_3C{\equiv}_3{=}_2(CH_2)_4{=}$ $H_3C{\equiv}_3{=}_2CO(CH_2)_3{=}$ $H_3C{\equiv}_3{=}CH_2CH_2CH(OH)C_2H_5$ 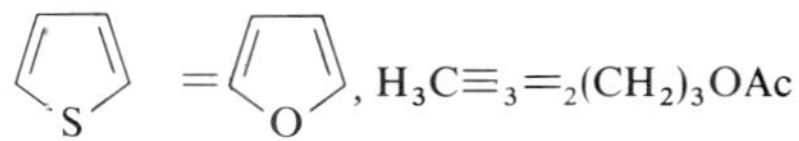	trace $H_3C{\equiv}_3{=}_2(CH_2)_4{=}$
A. verlotorum Lam.[35]	*DME*, *AK*, $H_3C{\equiv}_3{=}_2(CH_2)_4{=}$ 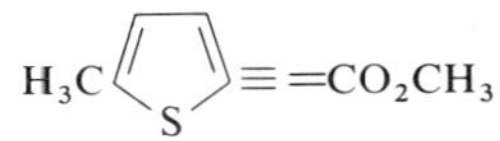 $H_3C{\equiv}_3{=}CH_2CH_2CH(OH)C_2H_5$ $H_3C{\equiv}_3{=}_2(CH)_3OR$ (R = H, Ac) 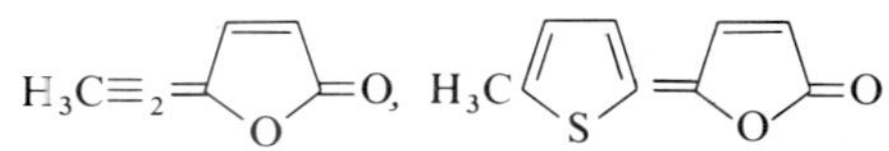	—
A. vulgaris L.[9,34,35,68,28]	*DME*, *AK*, $H_3C{\equiv}_3{=}_2(CH_2)_4{=}$ 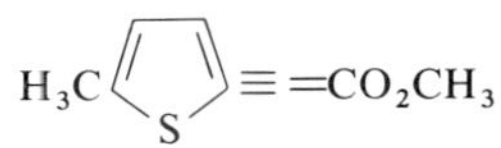 $H_3C{\equiv}_3{=}_2(CH_2)_3OR$ (R = H, Ac)	—
(3) *Artemisia annua* L.[35]	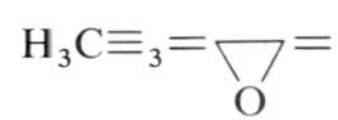	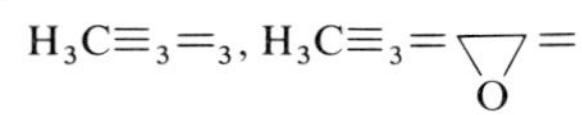
A. mandschurica Korr.[105]	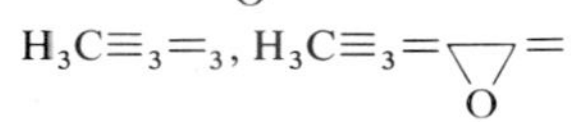	—
A. pontica L.[43,35,105]	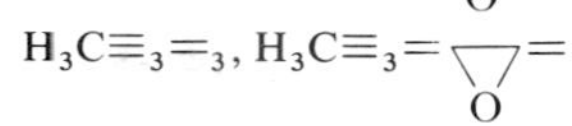	$H_3C{\equiv}_3{=}_2(CH_2)_3OAc$
A. sacrorum Ledeb. var. viridis[35]	*DME*, *AK*, $H_3C{\equiv}_3{=}_3$ 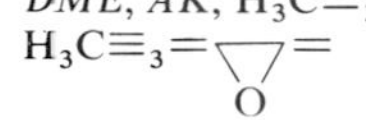	$H_3C{\equiv}_3{=}_2(CH_2)_3OR$ (R = H, Ac)

TABLE XXVI—Continued

	Constituents: Roots	Constituents: Leaves
A. senjavinensis Bess.[105]	*DME*, *AK*, $H_3C{\equiv}_3{=}_3$	—
A. taurica Willd.[35]	*DME*, $H_3C{\equiv}_3{=}_3$	*DME*, $H_3C{\equiv}_3{=}_3$
A. afra Jacq.[35]	$H_3C{\equiv}_3{=}_3$, $H_3C{\equiv}_3{=}$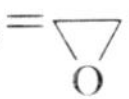$=$, $H_3C{\equiv}_3{=}_2(CH_2)_3OAc$, $H_3C{\equiv}_3{=}_2CH(OH)CH_2CH_2OAc$ 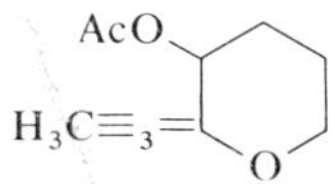	—
(4) *Artemisia integrifolia* Falk.[35]	*AK*, $H_3C{\equiv}_3{=}_3$, 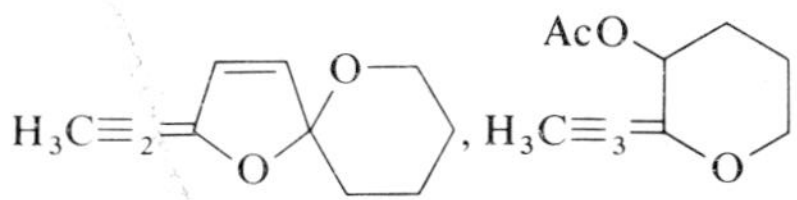 $H_3C{\equiv}_3{=}_2(CH_2)_3OH$, $H_3C{\equiv}_3{=}CH_2CH_2CH(OH)C_2H_5$ $H_3C{=}{\equiv}_2{=}_2CH(OAc)CH_2CH_2OAc$	trace *DME*, *AK*
A. laxa Fritsch.[35]	*DME*, *AK*, $H_3C{\equiv}_3{=}_3$, $H_3C{\equiv}_3{=}$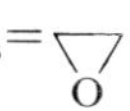$=$, $H_3C{\equiv}_3{=}_2(CH_2)_3OH$ 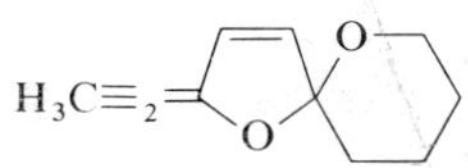	*DME*, 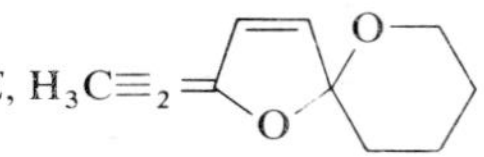

A. lobelii All.[35]	$H_3C\equiv_3=_3$, $H_3C\equiv_2=$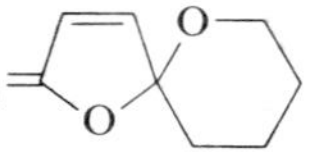	—
A. ludoviciana Nutt.[68]	*AK*, *DME*, $H_3C\equiv_3=_3$ 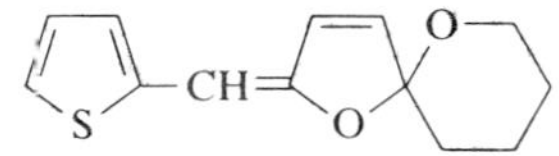	—
A. mutellina Vill.[35]	*DME*, $H_3C\equiv_3=_3$, $H_3C\equiv_3=$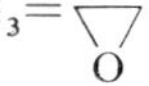$=$, $H_3C\equiv_3=_2(CH_2)_3OAc$ $H_3C\equiv_2=$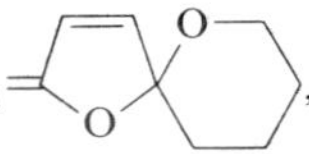, $H_3C\equiv_2=$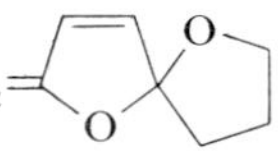	*DME*, $H_3C\equiv_3=_3$, $H_3C\equiv_2=$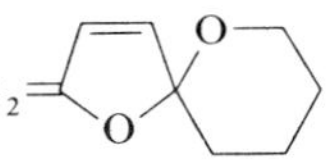
A. princeps Pampan.[35]	*AK*, $H_3C\equiv_3=_3$, $H_3C\equiv_3=$$=$, $H_3C\equiv_2=$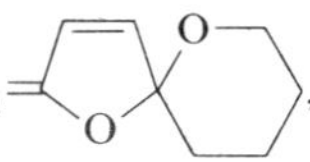, $H_3C\equiv_3=_2(CH_2)_3OH$, $\equiv=CH_2CH_2CH(OH)C_2H_5$	—
A. suksdorfii Piper[35]	*DME*, *AK* $H_3C\equiv_3=_3$, $H_3C\equiv_3=$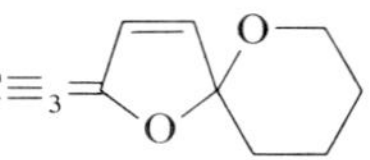	—
(5a) *Artemisia alpina* DC.[35]	$H_3C\equiv_2=$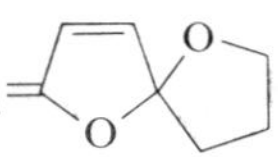	—
A. japonica Thunb.[35]	$H_3C\equiv_2=$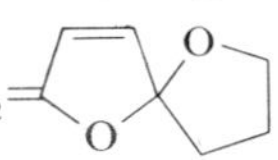	—

TABLE XXVI—Continued

	Constituents: Roots	Leaves
A. sericea Weber[35]	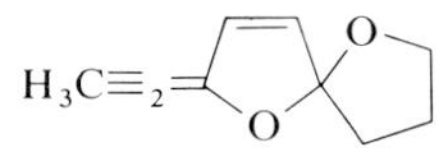	—
A. tournefortiana Reichb.[35]	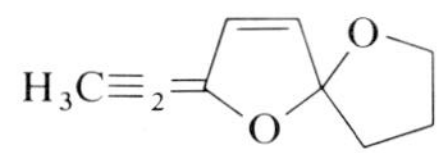	—
(5b) *A. argentea* Birds.[35]	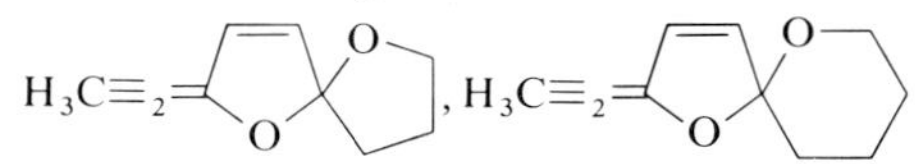	—
A. biennis Willd.[35]	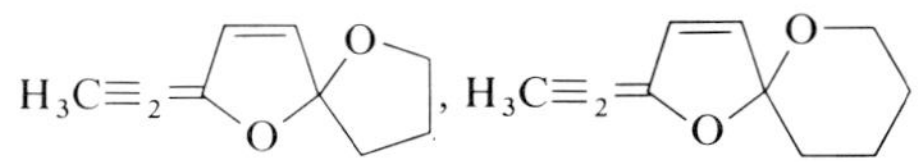	as roots
A. caucasica Willd.[35]	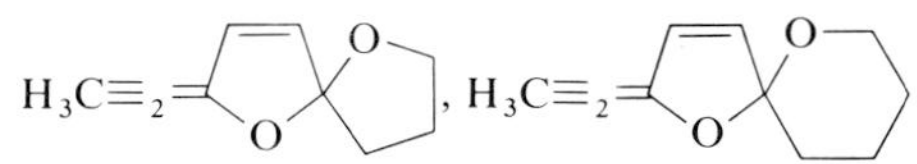 *DME*	—
A. freyniana Krasch.[35]	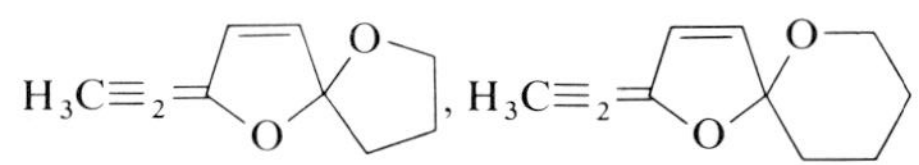 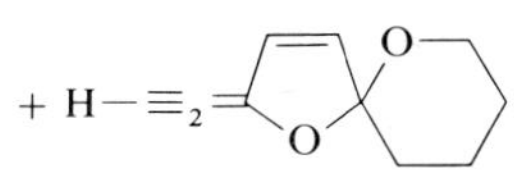	—
Artemisia maritima L.[35]	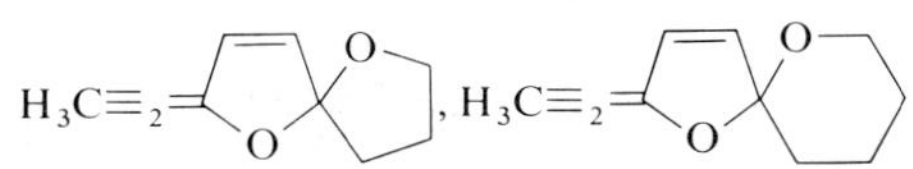	—

A. pedemontana Balb.[66]

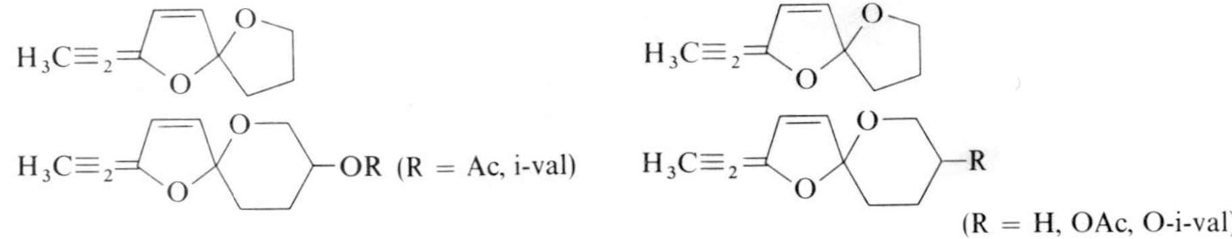

A. petrosa Jan.[35]

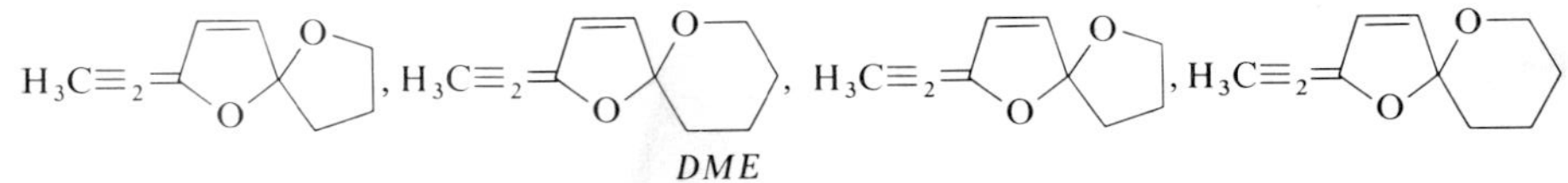

A. splendens Willd.[35]

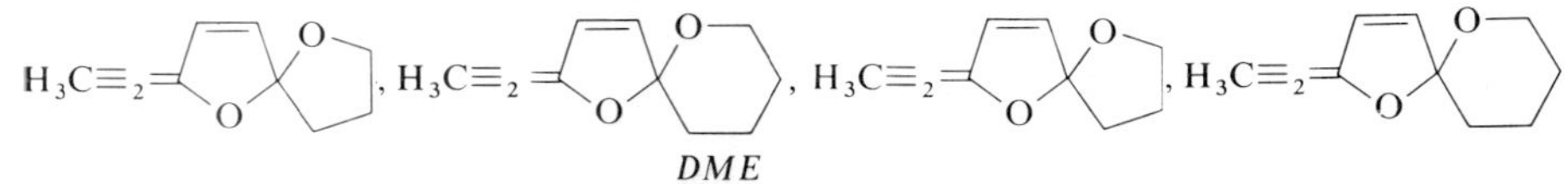

(5c) *A. douglasiana* Bess.[35]

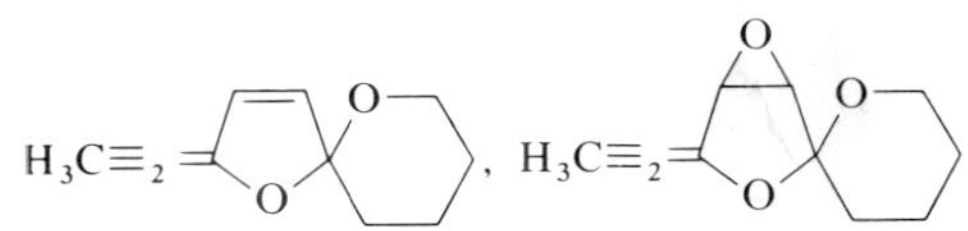

—

A. glacialis L.[35]

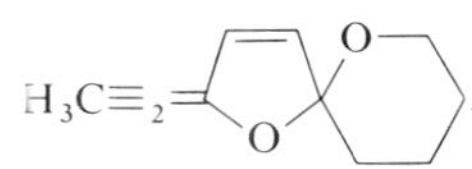

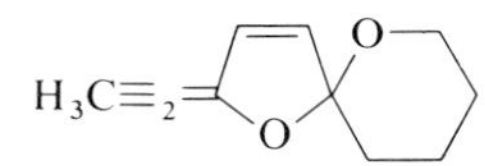

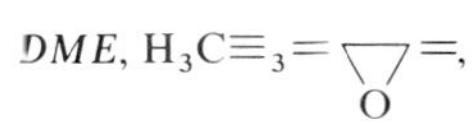

$H_3C{\equiv}_3{=}_2(CH_2)_3OAc$

$H_3C{\equiv}_3{=}_2(CH_2)_3OAc$

A. latifolia Ledeb.[35]

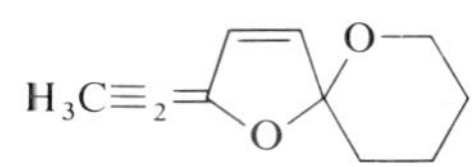

Sesamin derivatives

TABLE XXVI—Continued

	Constituents: Roots	Constituents: Leaves
A. sieversiana Ehrh.[35]	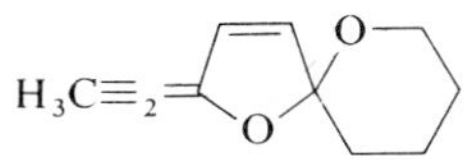	—
A. vallesiaca All.[35]	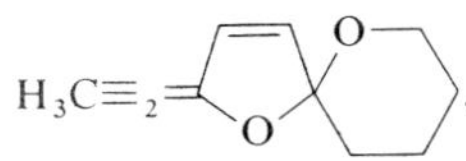, DME	—
A. spicata Wulf.[35]	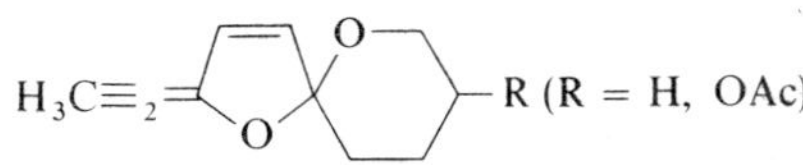	as roots
(5d) *A. orientalis* Willd.[35]	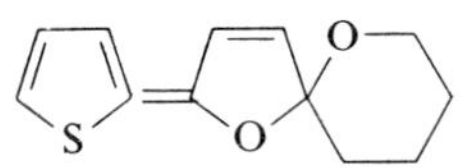	—
A. purshiana Bess.[35]	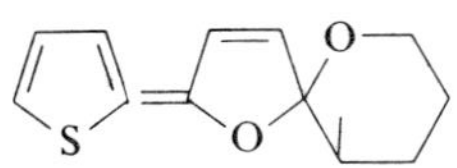	—
A. stelleriana Bess.[35]	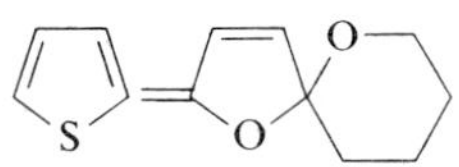	—
(6) *A. dracunculus* L.[120]	$PhCH_2\equiv_2R$ (R = CH_3, H) $PhCO\equiv_2CH_3$, 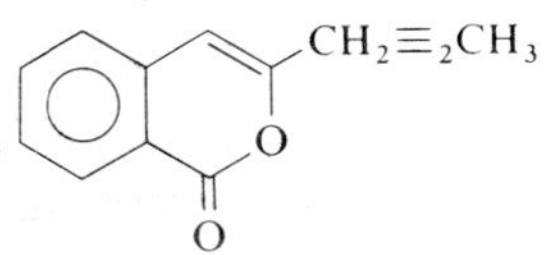	—

(7) *A. frigida* Willd.[35]	*AK*	—
A. genipi Stechm.[35]	*AK*, $H_3C\equiv_3{=}CH_2CH_2CH(OH)C_2H_5$	*AK*
A. lactiflora Wall. (?)[35]	*ME*, *LE*	—
A. tridentata Nutt.[35]	*AK*	—
A. turanica[35]	*DME*	—
A. variabilis Ten.[35]	*DME*	—
A. arborescens L.[170]	*DME*, $H_3C\equiv$ 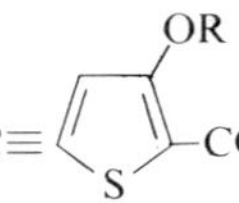 $-COCH_3$ ($R = H, CH_3$)	—
(8) *A. absinthium* L.[35]	—	—
A. canariensis Less.[35]	—	—
Asaemia axillaris[35]	—	furan sesquiterpenes

The compounds obtained from *Artemisia* species are, in part, very different from the chemical viewpoint. The first group seems to have a special position as only dehydrofalcarinone and related acetylenes are present. Only *A. scoparia* has also aromatic acetylenes as published for *A. capillaris* Thunb. however, this could not be established with a species grown in Europe.

The next group also has C_{17}-acetylenes, but only Centaur X_3 and related compounds occur together with other acetylenes very common in the tribus Anthemideae.

The third group is characterised by the occurrence of the triyne-triene, while the next also has spiroketalenolethers. These compounds are predominant in the next group, divided according to the occurrence of the 5-membered, 5- and 6-, 6-membered and finally thiophene spiroketalenolethers. In contrast to *Matricaria* and some groups of *Chrysanthemum*, the spiroketalenolethers are normally only present in the roots and only traces of them occur in the aerial parts of the plants.

A. dracunculus is the only species with aromatic acetylenes besides the already mentioned sample in the first group. The seventh group does not contain typical acetylenes; only *A. arborescens* contains the thiophenes also present in *Liatris* species. The last two species do not contain acetylenes at all, although they have been investigated several times.

The genus *Artemisia* also contain many other compounds, but up to now, investigations of other types of constituents have not been very extensive. Only the sesquiterpene lactones have been studied more precisely, but unfortunately the number of species investigated, both for sesquiterpenes and acetylenes, is still too small to establish any similarities. Only the special situation of *A. dracunculus* agrees very well with the results in both fields. In the whole tribus Anthemideae there are partial signs of homogeneity if we consider the structures of the isolated acetylenes. Spiroketalenolethers are only present in this tribus, but dehydromatricariaester and the "Artemisia-ketone" are also typical; several others show the transitions to other tribus.

1.7. Senecioneae, Calendulaceae and Arctotideae

These three tribus are not very interesting as regards structure variations in the occurring acetylenes. In the tribus Senecioneae, only the genera *Arnica*, *Doronicum* and *Gamolepis* contain acetylenes, while in the first two, the pentayn-ene (**94**) (see Chapter 2, Section 2.6) seems to be typical and two unique acetylenes are present in *Psilothonna tagetes* (belonging to Gamolepis). The typical genera of this tribus do not contain acetylenes. The difference between *Arnica* and *Doronicum* and the rest of the tribus is also found in the sesquiterpenoide constituents. The occurrence of alkaloids is also typical of many genera. Therefore, it seems likely that the main part of this tribus holds a unique position in the family.

In the tribus Calendulaceae, only small concentrations of the pentayn-ene (**94**) have been observed, sometimes together with the ene-tetrayn-ene.

The tribus Arctotideae does not appear to be very homogeneous. The species of the first subtribus and of some genera of the second one (Gorteria, *Gazania*, *Hirpicium*), only contain the pentayne (**94**) in small concentrations as before, while the genera *Berkheya*, *Didelta*, *Cullumia*, and *Cuspidia* are characterised by thiophene acetylenes, and in part by very unique ones. However, several are identical with those typical for the genus *Echinops* showing a possible relationship of this group to the tribus Cynareae which is also indicated from the botanical point of view.

A special position seems to be held by the South African genus *Ursinia*. All species do not contain acetylenes but furane sesquiterpenes are present. The genus is now considered to be in the tribus Anthemideae, containing the genera *Lasiospermum* and *Athanasia* also containing furane sesquiterpenes but no acetylenes. Probably, these genera are closely related and more or less different from the other members of the tribus Anthemideae. We can possibly assume that these genera might be a link to Senecioneae, where furane sesquiterpenes are also common.

TABLE XXVII
SENECIONEAE

(1) Senecioneae-Liabinae	Constituents
(2) Senecioneae-Senecioninae	
Tussilago farfara L.[35]	—
Petasites officinalis Mnch.[35]	—
Homogyne alpina (L.) Cass.[35]	—
Erechthites arguta DC.[35]	—
E. hieracifolius Raf.[35]	—
Gynura aurantica DC.[35]	$H_3C\equiv_5=?$
G. crepidioides Benth.[35]	—
G. rubens (Juss. *et* Jacq.) S. Moore[35]	—
Cineraria maritima L.[35]	Trace $\equiv_3=_2$
Cacalia hastata L.[35]	—
C. suaveolens L.[35]	—
Senecio aquaticus Ten.[35]	—
S. brasiliensis Less.[35]	—
S. burchellii DC.[35]	—
S. coccineus Klatt.[35]	—
S. elegans L.[35]	—
S. flammeus DC.[35]	—
S. fuchsii Gmel.[35]	—
S. jacobaea L.[35]	—
S. leptoschizus Bong.[35]	—

TABLE XXVII—Continued

(2) Senecioneae-Senecioninae	Constituents
S. niveus Willd.[35]	—
S. pseudo-arnica Less.[35]	—
S. rupester Waldst. *et* Kit.[35]	—
S. sagittatus O. Hoffm.[35]	—
S. sonchifolius Moench.[35]	—
S. squalidus L.[35]	—
S. umbellatus L.[35]	—
S. viscosus L.[35]	—
S. westermannii Dusen.[35]	—
Ligularia divorum[35]	—

Arnica and *Doronicum*	Constituents
Arnica alpina Olin.[35]	$H_3C\equiv_5=$, $H_3C\equiv_3=_2(CH_2)_3OH$
A. chamissonis Less.[35]	$H_3C\equiv_5=$, $H_3C\equiv_3=_2(CH_2)_3OH$
A. foliosa Nutt.[35]	$H_3C\equiv_5=$, $H_3C\equiv_3=_2(CH_2)_3OH$
A. longifolia Eaton[35]	$H_3C\equiv_5=$
A. montana L.[35,80]	$H_3C\equiv_5=$, $H_3C=\equiv_4=$
A. sachalinensis A. Gray.[35]	$H_3C\equiv_5=$, $H_3C\equiv_3=_2(CH_2)_3OH$, 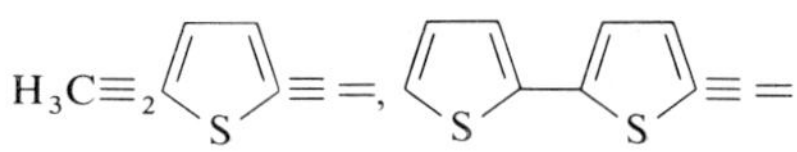
A. whitneyi Fernald.[35]	$H_3C\equiv_5=$, $H_3C\equiv_3=_2(CH_2)_3OH$
Doronicum austriacum Jacq.[35]	$H_3C\equiv_5=$
D. corsicum Poir.[35]	$H_3C\equiv_5=$
D. grandiflorum Lam.[35]	$H_3C\equiv_5=$
D. pardalianches L.[35]	$H_3C\equiv_5=$

(3) Senecioneae-Othonninae	Constituents
Gamolepis chrysanthemoides DC.[35]	$H_3C\equiv_5=$
G. tagetes (L.) DC.[109] (= genus *Psilothonna* E. Mey.)	$H_3C\equiv_3=_2$, $H_3C\equiv_3CH_2CH_2=$

Calenduleae[35]	Constituents
Calendula aegyptica Desf.	$H_3C\equiv_5=$
C. algeriensis Boiss.	$H_3C\equiv_5=$
C. balansae Boiss.	$H_3C\equiv_5=$
C. hortensis	$H_3C\equiv_5=$
C. lusitanica Boiss.	$H_3C\equiv_5=$
C. maritima Guss.	$H_3C\equiv_5=$
C. microphylla Lange	$H_3C\equiv_5=$
C. officinalis L.	$H_3C\equiv_5=$
Dimorphotheca pluvialis Moench.	$H_3C\equiv_5=$
D. sinuata DC.	$H_3C\equiv_5=$
Osteospermum amplectens Norlindh.	$H_3C\equiv_5=$, $H_3C=\equiv_4=$
O. clandestinum Norlindh.	$H_3C\equiv_5=$, $H_3C=\equiv_4=$

TABLE XXVII—Continued

Calenduleae	Constituents
O. ecklonis Norlindh.	$H_3C\equiv_5=$
O. hyseroides Norlindh.	$H_3C\equiv_5=$, $H_3C=\equiv_4=$
O. muricatum E. Mey.	$H_3C\equiv_5=$
O. pinnatifidum L'Hérit	$H_3C\equiv_5=$
O. vaillantii Norlindh.	$H_3C\equiv_5=$
Tripteris cheiranthifolia Sch. Bip. ex Hochst.	$H_3C\equiv_5=$, $H_3C=\equiv_4=$
T. glandulosa Turcz.	$H_3C\equiv_5=$

TABLE XXVIII
ARCTOTIDEAE

1. Arctotodeae-Arctotidinae	Constituents
Ursinia anthemoides Gaertn.[35]	
U. calenduliflora Benth. *et* Hook.[35]	
U. montana DC.[35]	furan sesquiterpenes
U. speciosa DC.[35]	
U. versicolor[35]	
Arctotis acaulis L.[35]	$H_3C\equiv_5=$, $H_3C\equiv_3=_2(CH_2)_3OH$
A. angustifolia L.[35]	$H_3C\equiv_5=$, $H_3C=\equiv_4=$
A. calendula (L.) Levyns[35]	$H_3C\equiv_5=$, $H_3C\equiv_3=_2(CH_2)_3OH$
A. decurrens Jacq.[35]	$H_3C\equiv_5=$, $H_3C\equiv_3=_2(CH_2)_3OH$
A. fastuosa Jacq.[35]	$H_3C\equiv_5=$, $H_3C=\equiv_4=$
A. grandis Thunb.[35]	$H_3C\equiv_5=$, $H_3C=\equiv_4=$
A. laevis Thunb.[35]	$H_3C\equiv_5=$, $H_3C=\equiv_4=$
A. populifolia (Berg.) T. Norl.[35]	$H_3C\equiv_5=$
A. stoechedifolia Berg.[35]	$H_3C\equiv_5=$, $H_3C=\equiv_4=$
A. venusta Norl.[35]	$H_3C\equiv_5=$, $H_3C=\equiv_4=$

2. Arctotideae-Gorterinae	Constituents
Gorteria diffusa Thb. var. calendulacea (DC.) Roessler[35]	$H_3C\equiv_5=$, $H_3C\equiv_3=_2(CH_2)_3OH$
Gazania hybridus[35]	$H_3C=\equiv_4=$, $H_3C\equiv_3=_2(CH_2)_3OH$
G. krebsiana Less.[35]	—
G. lichtensteinii Less.[35]	—
G. linearis (Thb.) Druce[35]	—
G. rigens R. Br.[35]	$H_3C\equiv_5=$
G. splendens Hort.[35]	$H_3C\equiv_5=$
Dymondia margaretae Compton[35]	$H_3C=\equiv_4=$
Hirpicium alienatum (Thb.) Druce[35]	—
H. integrifolium Less.[35]	—
Berkheyopsis gossweileri S. Moore[35]	—
Cullumia bisulca (Thb.) Less.[35]	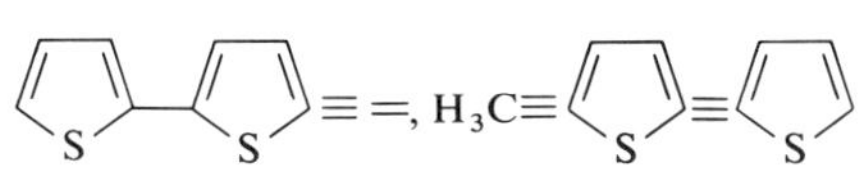

TABLE XXVIII—Continued

2. Arctotideae-Gorterinae	Constituents
C. setosa (L.) R. Br.[35]	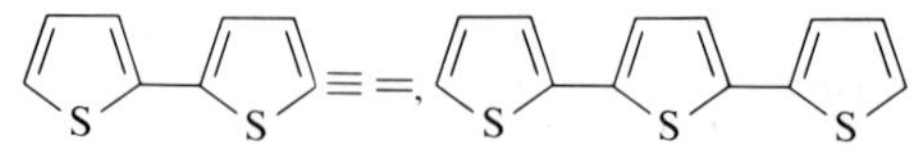 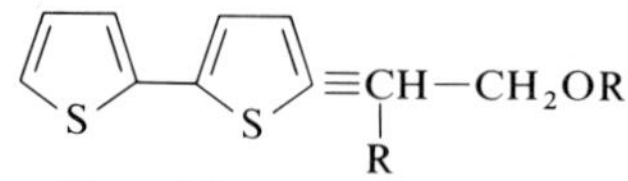 (R = OAc, R′ = Ac; R = H, R′ = Ac; R = H, R′ = H)
Berkheya macrocephala J. M. Wood.[35]	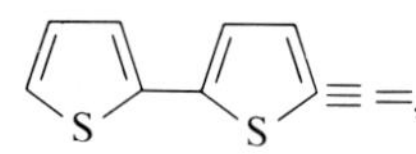 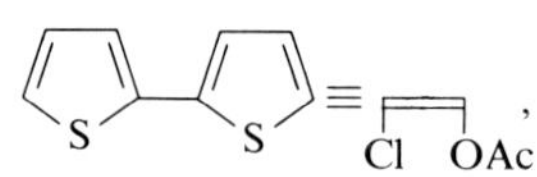 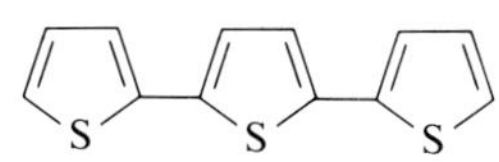
B. onopordifolia Burtt. Davy[35]	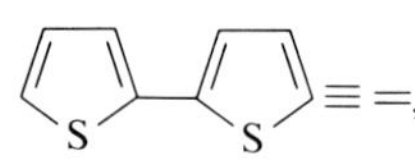 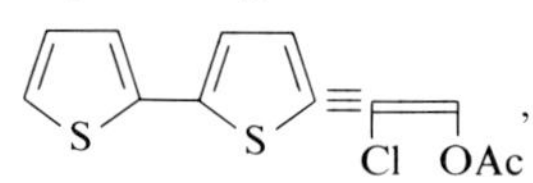 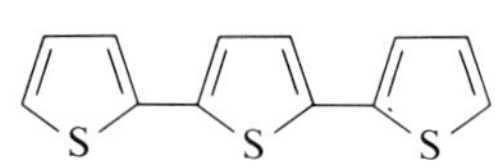
B. radula (Harv.) de Willd.[95] = *B. adlami* Hook.[95]	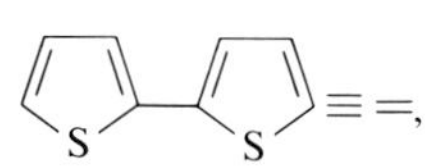 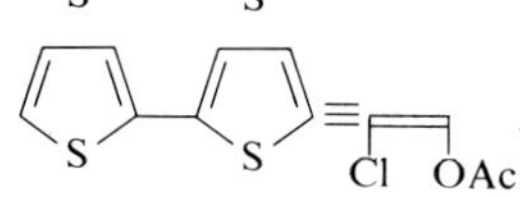 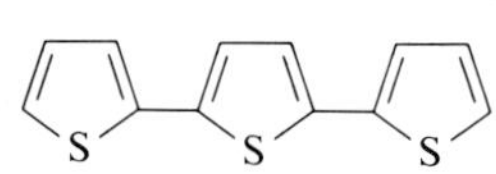
B. fruticosa (L.) Ehrb.[35]	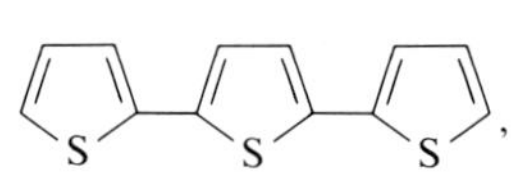 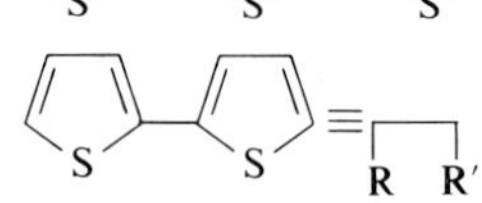 (R = R′ = OAc; R = H, R′ = OAc)

TABLE XXVIII—Continued

2. Arctotideae-Gorterinae	Constituents
B. barbata (L.f.) Hutch.[35]	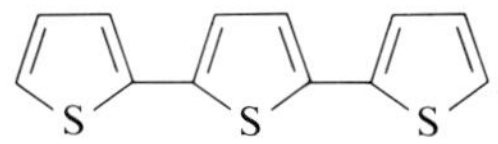, 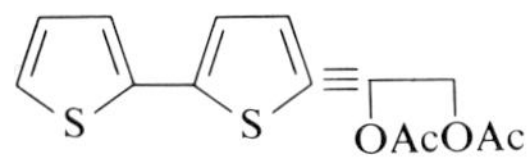 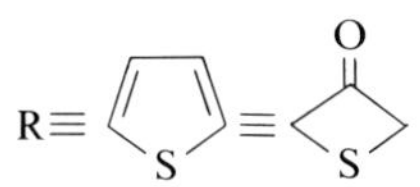 (R = CH_3, H) 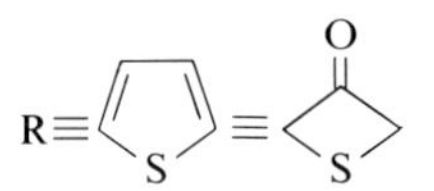 (R = CH_3, H)
B. heterophylla (Thb.) O. Hoffm.[35]	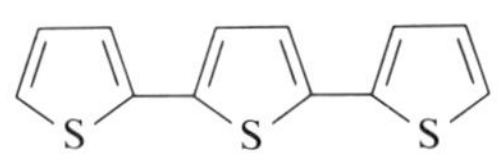, 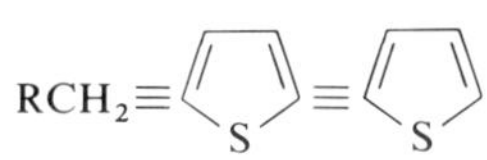 (R = H, OH) 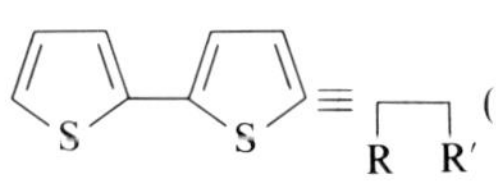 (R = R′ = OAc; R = H, R′ = OAc; R = H, R′ = OH)
B. armata (Vahl.) Druce[35] *B. decurrens* (Thb.) Willd.[35] *B. herbacea* (L.f.) Druce[35] *B. purpurea* (DC.) Masters[35]	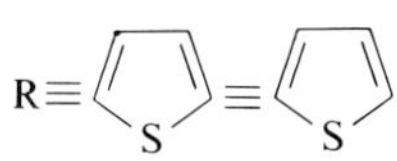 (R = H_3C, $AcOCH_2$, $HOCH_2$, OCH) 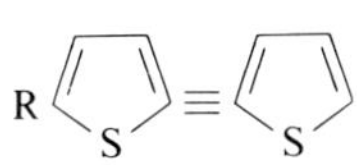 (R = $AcOCH_2$, $HOCH_2$, OCH, H_3CCO, $H_3C\underset{\mid}{C}H$ OH) $H_3C\equiv_3=_2CHCH_2CH_2OH$ (OH on CH)

TABLE XXVII—Continued

2. Arctotideae-Gorterinae	Constituents
Cuspidia cernua Burtt.[35]	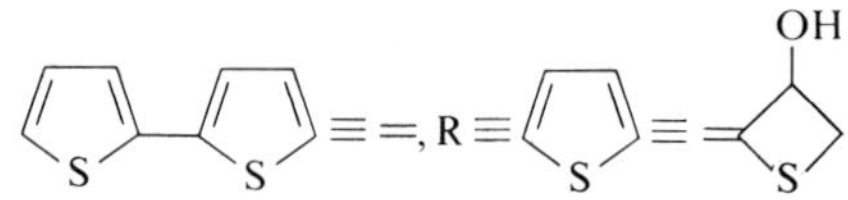 R = H, CH_3 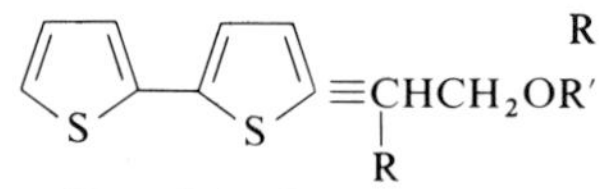 (R = OAc, R′ = Ac; R = H, R′ = Ac; R = H, R′ = H)
Didelta carnosa Ait.[35]	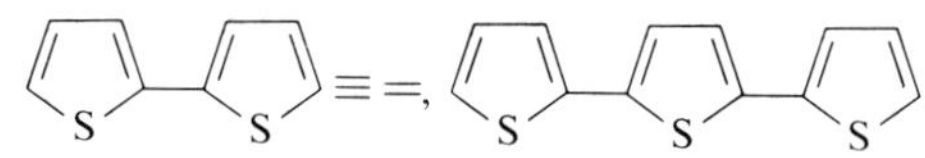 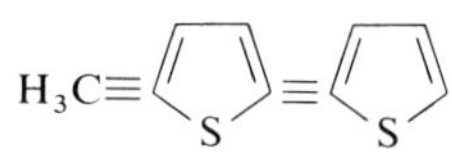 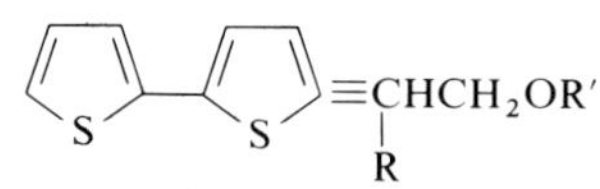 (R = OAc, R′ = Ac; R = H, R′ = Ac; R = H, R′ = H)
D. spinosa Ait.[35]	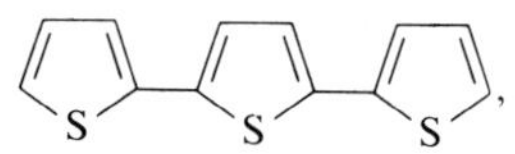 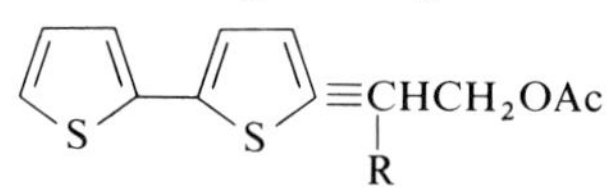 (R = H, R = OAc)

1.8. Cynareae

This large tribus is divided into four subtribus and is much more differentiated. In the first subtribus with two genera, only *Echinops* species have been investigated. The thiophenes are typical and occur in considerable variation. Two groups can be separated since the monothiophenes are missing in one part of the species.

The second subtribus has very different constituents. While special thiophenes are characteristic of the genus *Xeranthemum* which are also found in the tribus Anthemideae, *Carlina* and *Atractylis* contain furan derivatives as typical constituents as well as a number of C_{13}-acetylenes with a different degree of unsaturation and different functional groups.

TABLE XXIX
TRIBUS CYNAREAE

1. Cynareae-Echinopsidinae	Constituents (roots)
Echinops bannaticus Roch.[46] *E. cornigerus* DC.[46] *E. niveus* Wall.[46]	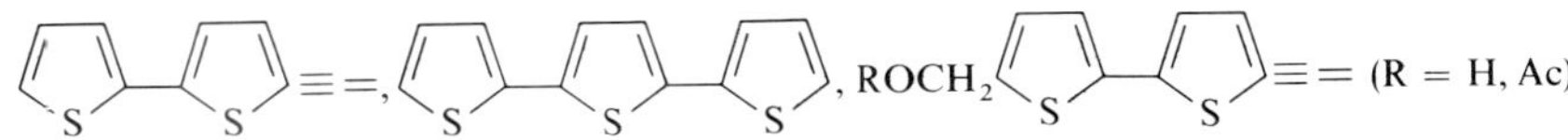 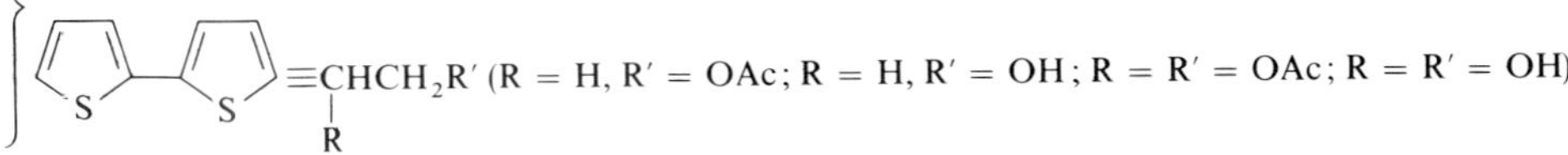
E. champtavicus[35] *E. commutatus* Jar.[35] *E. dahuricus* Fisch.[46] *E. exaltus* Rchb. *et* auct.[35] *E. horridus* Desf.[46] *E. humilis* Bieb.[35] *E. persicus* Fisch.[46] *E. ritro* L.[46] *E. sphaerocephalus* L.[46] *E. strigosus* L.[46] *E. viscosus* L.[35]	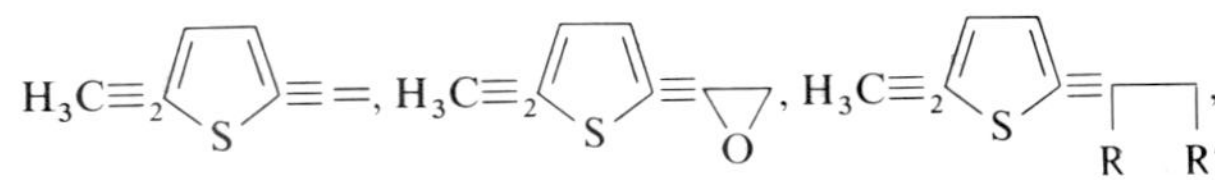 (R = H, R′ = OH; R′ = OAc, R = H; R = Cl, R′ = OH; R = Cl, R′ = OAc) 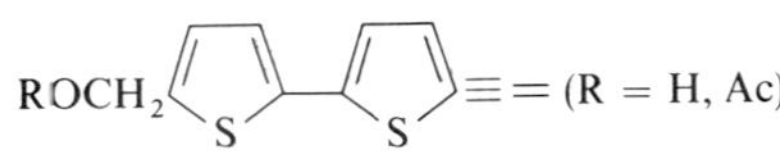 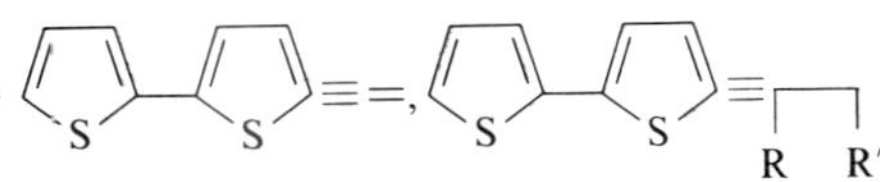 (R = R′ = OH; R = R′ = OAc; R = OAc, R′ = OH; R = OH, R′ = OAc, R = H, R′ = OAc; R = H, R′ = OH) 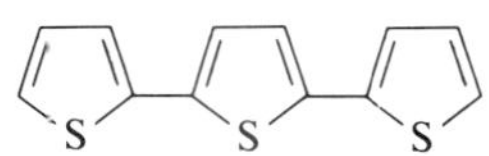 $H_3C{\equiv}_3{=}_2CH{-}CH(OH)CH_2CH_2OH$

TABLE XXIX—Continued

2. Cynareae-Carlininae	Constituents (roots)
Xeranthemum annuum L.[50] *X. cylindraceum* Sibth.[35] *X. foetidum* Moench.[35] *X. inapertum* (L.) Willd.[35]	$H_3C\equiv_5=$  $\equiv=_3$  $\equiv=_2CH_2CH_2OR$ (R = H, Ac)
X. longepapposum Fisch. *et* Mey.[35] *X. orientale* Mill.[35] *X. squarrosum* Boiss.[35]	$H_3C\equiv_5=$
Chardinia xeranthemoides Desf.[35]	$H_3C\equiv_5=$
Carlina acaulis L.[251] *C. salicifolia* Cav.[251]	$PhCH_2\equiv$ 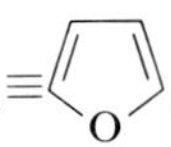, $=_2\equiv_2=$ 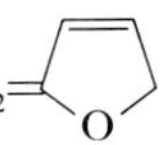 $R=\equiv_3=_2$ (R = CH_2OH, CH_2OAc, CHO) $AcOCH_2=\equiv_2=_3$
C. acanthifolia Haussk.[251]	$=_2\equiv_2=$ 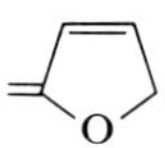, $HOCH_2=\equiv_3=_2$
C. racemosa L.[251]	$=_2\equiv_2=$ 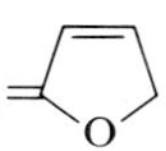, $R=\equiv_3=_2$ (R = CHO, CH_2OAc), $AcOCH_2=\equiv_2=_3$

C. vulgaris L.[251] *C. lanata* L.[35] *C. caulescens* Lam.[35]	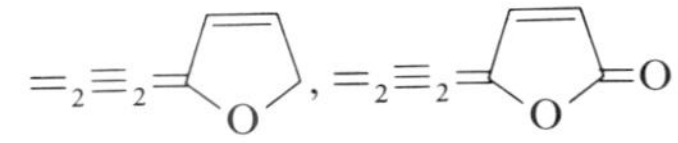 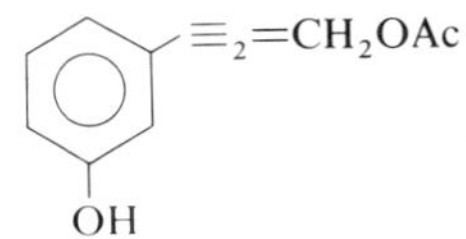 , $R{=}{\equiv}_2{=}_3$ (R = CH_3, CH_2OAc)
Atractylis cancellata L.[35] *A. gummifera* L.[35]	$R{=}{\equiv}_3{=}_2$ (R = CH_2OH, CH_2OAc, CHO) $R{=}{\equiv}_3{=}_2$ (R = CH_2OH, CH_2OAc, CHO) $AcOCH_2{=}{\equiv}_3{=}_2$
Atractylodes[261]	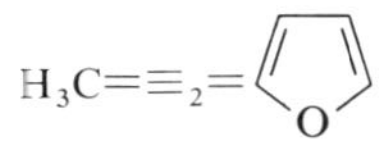

3. Cynareae-Carduinae	Constituents (roots)
Arctium lappa L.[204] *A. major* Schrank.[204] *A. minus* Schrank.[204] *A. nemorosum* Lej.[204] *A. tomentosum* Mill.[204]	$H_3C{=}{\equiv}_4{=}$ $H_3C{\equiv}_5{=}$ 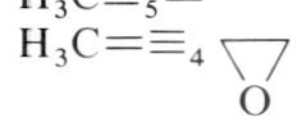 (traces)
Cousinia ambigens Juzupczuk.[35]	$H_3C{\equiv}_5{=}$, $H_3C{=}{\equiv}_4{=}$
C. coronata Franch.[35]	$H_3C{\equiv}_5{=}$, $H_3C{=}{\equiv}_4{=}$
C. hystrix C. A. May[204,205]	$H_3C{\equiv}_5{=}$, $H_3C{=}{\equiv}_4{=}$, $H_3C{=}{\equiv}_2{=}CH(OH)(CH_2)_5{=}$
C. integrifolia Franch.[35]	$H_3C{\equiv}_5{=}$, $H_3C{=}{\equiv}_4{=}$
C. microcarpa Boiss.[35]	$H_3C{\equiv}_5{=}$, $H_3C{=}{\equiv}_4{=}$
C. pseudoarcticum Bornm.[35]	$H_3C{\equiv}_5{=}$, $H_3C{=}{\equiv}_4{=}$
C. radians Bunge[204]	$H_3C{\equiv}_5{=}$, $H_3C{=}{\equiv}_4{=}$, $Ph{\equiv}_2{=}CH_2OAc$ (leaves), $H_3C{\equiv}_3{=}_2(CH_2)_3OH$ (leaves)
C. sewerzrowi Regel[35]	$H_3C{\equiv}_5{=}$, $H_3C{=}{\equiv}_4{=}$
Saussurea albescens Hook. f. *et* Thom.[193,204]	$H_3C{\equiv}_5{=}$, $H_3C{=}{\equiv}_4{=}$
S. alpina DC.[193,204]	$H_3C{\equiv}_5{=}$, $H_3C{=}{\equiv}_4{=}$, $H_3C{=}_2{\equiv}_2{=}_2(CH_2)_3OR$ (R = H, Ac) (leaves)
S. candicans C. B. Clarke[193,204]	$H_3C{\equiv}_5{=}$, $H_3C{=}{\equiv}_4{=}$
S. discolor DC.[193,204]	$H_3C{\equiv}_5{=}$, $H_3C{=}{\equiv}_4{=}$

TABLE XXIX—Continued

3. Cynareae-Carduinae	Constituents (roots)
S. japonica DC.[193,204]	$H_3C{\equiv}_5{=}$, $H_3C{=}{\equiv}_4{=}$
S. pectinata Bunge.[193]	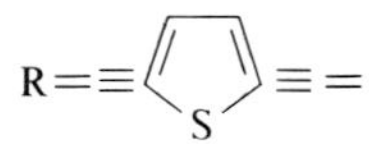 (R = CH_3, CH_2OH, CH_2OAc, CHO)
	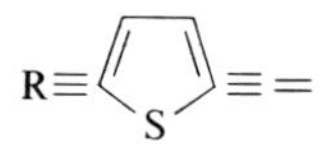 (R = CH_2OH, CHO)
	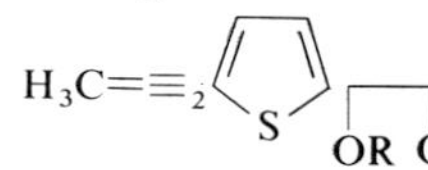 (R = H, Ac)
	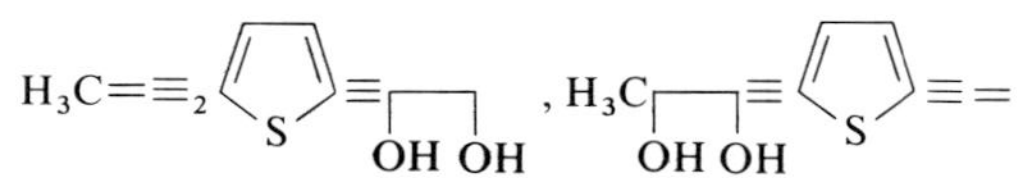
	Leaves: 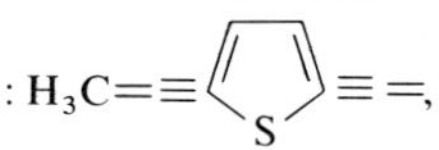, $H_3CCH_2{=}_4(CH_2)_nCOCH_2CH_2OR$ (n = 3, 4, 5; R = H, Ac)
Jurinea alata Cass.[196,204]	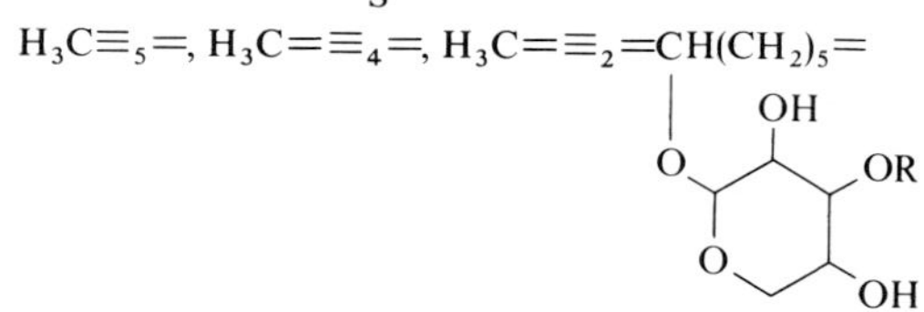 (R = H, Ac)
J. arachnoidea Bunge[35]	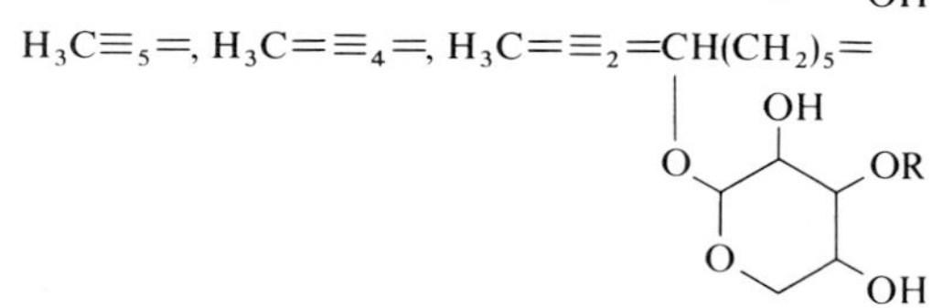 (R = H, Ac)

Species	Compounds
J. transsilvanica Schur.[35]	$H_3C\equiv_5=$, $H_3C=\equiv_4=$, $H_3C=\equiv_2=CH(CH_2)_5=$ 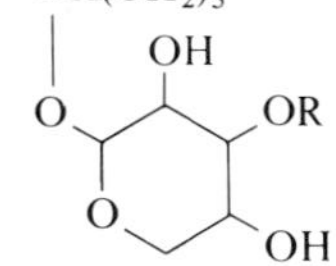 (R = H, Ac)
J. cyanoides Rchb.[196] *J. mollis* (L.) Rchb.[196]	$H_3C\equiv_5=$, $H_3C=\equiv_4=$, $H_3C=\equiv_2=CH(CH_2)_5=$ 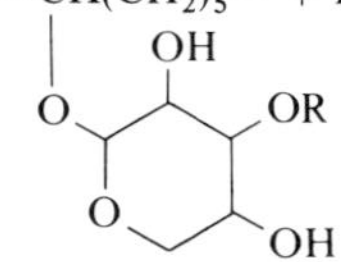 (R = H, Ac) + $H_3C=\equiv_2=CH(OR)(CH_2)_5=$ (R = H, Ac)
Carduus argyroa Bir.[204]	$H_3C\equiv_5=$, $OCH=\equiv=_2(CH_2)_4=$
C. collinus W. *et* K.[204]	$H_3C\equiv_5=$, $OCH=\equiv=_2(CH_2)_4=$
C. glaucus Baumg.[204]	$H_3C\equiv_5=$
C. leiophyllus Petrov.[204]	$H_3C\equiv_5=$
C. leucophyllus Turcz.[204]	$H_3C\equiv_5=$, $H_3C\equiv_3=_2(CH_2)_3OH$
C. nigrescens Vill.[204]	$H_3C\equiv_5=$, $H_3C=\equiv_4=$
C. nutans L.[204]	$H_3C\equiv_5=$
C. personata Jacq.[204]	$H_3C\equiv_5=$, $H_3C=\equiv_4=$
C. pycnocephalus L.[204]	$H_3C\equiv_5=$
Galactites tomentosa Moench.[204]	$H_3C\equiv_5=$, $H_3C=\equiv_4=$
Cirsium acaule (L.) Web.[204]	$H_3C\equiv_5=$, $H_3C=\equiv_4=$
C. arvene Scop.[204]	$H_3C\equiv_5=$
C. candolleanum Naeg.[204]	$H_3C\equiv_5=$, $H_3C=\equiv_4=$
C. canum (L.) All.[204]	$H_3C\equiv_5=$, $H_3C=\equiv_4=$
C. carnolicum Scop.[204]	$H_3C\equiv_5=$
C. decussatum Janka[204]	$H_3C\equiv_5=$, $H_3C=\equiv_4=$
C. eriophorum (L.) Scop.[204]	$H_3C\equiv_5=$
C. erisithales Scop.[204]	$H_3C\equiv_5=$, $H_3C=\equiv_4=$
C. ferox DC.[204]	$H_3C\equiv_5=$
C. heterophyllum All.[204]	$H_3C\equiv_5=$, $H_3C\equiv_3=_2(CH_2)_3OH$
C. lanceolatum Mill.[204]	$H_3C\equiv_5=$, $H_3C=\equiv_4=$, $H_3C\equiv_3=_2(CH_2)_3OH$

TABLE XXIX—Continued

3. Cynareae-Carduinae	Constituents (roots)
C. montanum Spr.[204]	$H_3C{\equiv}_5{=}$, $H_3C{=}{\equiv}_4{=}$, $H_3C{\equiv}_3{=}_2(CH_2)_3OH$
C. occidentalis Jepson[204]	$H_3C{\equiv}_5{=}$, $H_3C{=}{\equiv}_4{=}$
C. olevaceum Scop.[204]	$H_3C{\equiv}_5{=}$, $H_3C{=}{\equiv}_4{=}$, $H_3C{\equiv}_3{=}_2(CH_2)_3OH$
C. palustre Scop.[204]	$H_3C{\equiv}_5{=}$, $H_3C{=}{\equiv}_4{=}$, $H_3C{\equiv}_3{=}_2(CH_2)_3OH$
C. pannonicum Gaud.[204]	$H_3C{\equiv}_5{=}$, $H_3C{=}{\equiv}_4{=}$
C. spinosissimum Scop.[204]	$H_3C{\equiv}_5{=}$, $H_3C{=}{\equiv}_4{=}$
C. syriacum Grtn.[204]	$H_3C{\equiv}_5{=}$, $H_3C{=}{\equiv}_4{=}$
C. tuberosum All.[204]	$H_3C{\equiv}_5{=}$, $H_3C{=}{\equiv}_4{=}$, $H_3C{\equiv}_3{=}_2(CH_2)_3OH$
Cynara cardunculus L.[204] *C. scolymus* L.[204]	$H_3C{\equiv}_5{=}$, $H_3C{=}{\equiv}_4{=}$, $R{=}{\equiv}_2{=}_2(CH_2)_4{=}$ (R = CH_3, CHO, CH_2OH)
Silybum eburneum Coss *et* Dur.[204]	$H_3C{\equiv}_5{=}$, $H_3C{=}{\equiv}_4{=}$, $H_3C{=}{\equiv}_2\underset{\displaystyle OH}{\underset{\vert}{CH}}{=}(CH_2)_5{=}$
S. marianum Gaertn.[204]	$H_3C{\equiv}_5{=}$, $H_3C{=}{\equiv}_4{=}$, $H_3C{=}{\equiv}_2\underset{\displaystyle OH}{\underset{\vert}{CH}}{=}(CH_2)_5{=}$, $H_3C{\equiv}_4{=}_2$, $H_3C{=}{\equiv}_3{=}_2$, $H_3C{=}{\equiv}_2CH_2{=}(CH_2)_5{=}$, $H_3C{\equiv}_3{=}(CH_2)_3OAc$, $OCHCH_2CH_2{\equiv}_4{=}$
Tyrimnus leucographus (L.) Cass.[35]	$H_3C{\equiv}_5{=}$, $H_3C{\equiv}_5\underset{O}{\bigtriangledown}$, $H_3C{\equiv}_3{=}_2\underset{\displaystyle OAc}{\underset{\vert}{C}}HCH_2CH_2OAc$
Onopordon acaule Willd.[204]	$H_3C{=}{\equiv}_4{=}$, $R{=}{\equiv}_2{=}_2(CH_2)_4{=}$ (R = CH_3, CHO, CH_2OH)
O. algeriense Pormel.[204]	$H_3C{=}{\equiv}_4{=}$, $R{=}{\equiv}_2{=}_2(CH_2)_4{=}$ (R = CH_3, CHO, CH_2OH)
O. acanthium L.[204]	$H_3C{=}{\equiv}_4{=}$, $R{=}{\equiv}_2{=}_2(CH_2)_4{=}$ (R = CH_3, CHO, CH_2OH) + $Ph{\equiv}_2{=}CH_2OH$
O. arabicum L.[204]	$H_3C{=}{\equiv}_4{=}$, $R{=}{\equiv}_2{=}_2(CH_2)_4{=}$ (R = CH_3, CHO, CH_2OH)
O. bracteatum Boiss.[204]	$H_3C{=}{\equiv}_4{=}$, $R{=}{\equiv}_2{=}_2(CH_2)_4{=}$ (R = CH_3, CHO, CH_2OH) (leaves: ditto)
O. illyricum L.[204]	$H_3C{=}{\equiv}_4{=}$, $R{=}{\equiv}_2{=}_2(CH_2)_4{=}$ (R = CH_3, CHO, CH_2OH)
O. tauricum Willd.[204]	$H_3C{=}{\equiv}_4{=}$, $R{=}{\equiv}_2{=}_2(CH_2)_4{=}$ (R = CH_3, CHO, CH_2OH) (leaves: ditto)

4. Cynareae-Centaureinae	Constituents Roots	 Leaves
Zoegea baldshuanica C. Winkl.[192]	$H_3C{\equiv}_5{=}$, $H_3C{=}{\equiv}_4{=}$, $Ph{\equiv}_3CH_3$, $Ph{\equiv}_2{=}CH_2R$ (R = H, OH)	$H_3C{\equiv}_5{=}$, $H_3C{=}{\equiv}_4{=}$, $Ph{\equiv}_3CH_3$ $Ph{\equiv}_2{=}CH_3$, $C_2H_5{=}_4(CH_2)_nCHO$ (n = 5 – 7), $H_3C{=}{\equiv}_2{=}_2(CH_2)_3OAc$ $H_3C{=}_2{\equiv}{=}_2(CH_2)_nOR$ (n = 2–5) (R = H, Ac) $H_3C{=}_2{\equiv}{=}_2(CH_2)_nCHO$ (n = 2–5) $H_3C{=}{\equiv}_2{=}_2(CH_2)_3CHO$
Crupina crupinastrum Vis.[35]	$H_3C{=}{\equiv}_4{=}$	$H_3C{=}{\equiv}_4{=}$, $C_2H_5{=}_4(CH_2)_nR$ $H_3C{=}{\equiv}_2{=}(CH_2)_nR$, $H_3C{\equiv}_3{=}_2(CH_2)_nR$ (n = 2 – 5) (R = CHO, CH_2OH)
C. vulgaris Cass.[36]	$H_3C{=}{\equiv}_4{=}$, $H_3C{=}{\equiv}_4$ $H_3C{=}{\equiv}_2{=}CH(OH)CH_2CH_2OAc$	as *C. crupinastrum* + $H_3C{\equiv}_3{=}_2(CH_2)_4{=}$ $H_3C{=}{\equiv}_2{=}_2CH(OH)CH_2CH_2OAc$
Serratula coronata L.[35] *S. heterophylla* Desf.[196]	} $H_3C{=}{\equiv}_4{=}$	$H_3C{=}{\equiv}_4{=}$, $C_2H_5{=}_4(CH_2)_nCHO$ $H_3C{=}_2{\equiv}{=}_2(CH_2)_nCHO$ $H_3C{=}{\equiv}_2{=}_2(CH_2)_nCHO$ (n = 3–5)
S. gmelini Ldb.[196]	$H_3C{=}{\equiv}_4{=}$, $H_3C{\equiv}_5{=}$, $H_3C{=}{\equiv}_2{=}CH(OH)(CH_2)_5{=}$, $H_3C{=}{\equiv}_2{=}CH(CH_2)_5{=}$, 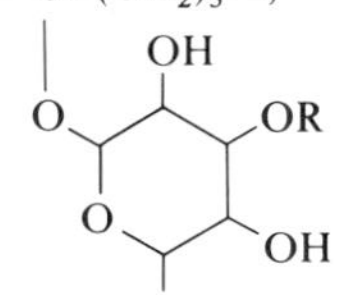	$C_2H_5{=}_4(CH_2)_nCHO$, $H_3C{=}_2{\equiv}{=}_2(CH_2)_nCHO$ $H_3C{=}{\equiv}_2{=}_2(CH_2)_nCHO$ $H_3C{\equiv}_3{=}_2(CH_2)_nCHO$ $H_3C{=}{\equiv}_2{=}_2(CH_2)_nCH_2OAc$ $H_3C{=}_2{\equiv}{=}_2(CH_2)_nCH_2OH$ (n = 3–6)

TABLE XXIX—Continued

4. Cynareae-Centaureinae	Constituents: Roots	Constituents: Leaves
	$H_3CCH_2CH_2{\equiv}_2{=}CH(CH_2)_5{=}$ 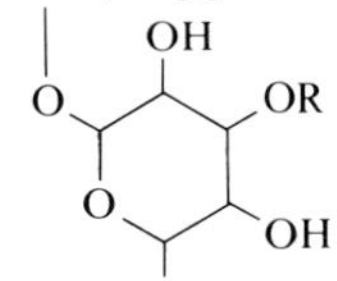 (R = H, Ac)	
	$C_3H_7{\equiv}_2{=}CH(OH)(CH_2)_5{=}$	
S. pinnatifida Poir.[196]	$H_3C{=}{\equiv}_4{=}$	—
S. quinquefolia Bieb.[35]	$H_3C{\equiv}_5{=}$, $H_3C{=}{\equiv}_4{=}$	—
S. radiata Bieb.[276,50]	$H_3C{=}{\equiv}_4{=}$, $R{=}{\equiv}$ 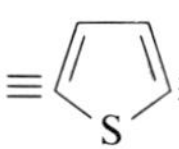 ${\equiv}{=}$ (R = CH_3, CH_2OAc, CH_2OH, CHO)	—
	$H_3C{=}{\equiv}_2$ 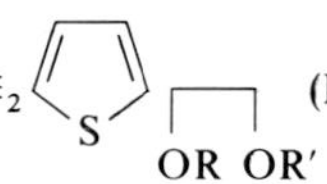 (R = R′ = Ac; R = R′ = H; R = H, R′ = Ac)	
S. tinctoria L.[35]	$H_3C{=}{\equiv}_4{=}$	—

The third subtribus is characterised by the occurrence of the ene-tetrayn-ene normally only together with traces of the pentayn-ene. However, in several genera less common acetylenes have also been observed. For most of the investigated genera, the absence of acetylenes in the aerial parts is typical. For the genus *Jurinea*, glycosides are characteristic, while in the genera *Sylibum* and *Onopordon* C_{17}-acetylenes are common. A transition to the next subtribus seems to be the genus *Zoegea* as much more structure differentiation can already be observed and the occurrence of acetylenes in the aerial parts typical for *Centaurea* species has been recognized.

The next genus, *Centaurea*, is a very large one. The botanical limits are not well established. Following the systematic of Engler-Prantl 41 sections can be drawn up. If we look at the isolated acetylenic constituents only a few show clear separations. Perhaps several of the listed species may be hybrids, however the more characteristic ones are well estimated. Only sections with characteristic acetylenes will be listed separately. Really a special group is the section *Centaurium* showing as main constituents the chlorohydrines and -acetates. These are not found in any other section. However we find them again in the genera *Carthamus* and *Carduncellus*, while the genus *Cnicus* is more related to the other *Centaurea* sections. Typical for these sections is the occurrence of a number of different unsaturated aldehydes with chain lengths from C_{14}–C_{17} in the aerial parts, while in the roots always the ene-tetrayn-ene is present together with several other acetylenes. Characteristic are further the bifunctional ene-triyn-enes and ene-diyne-dienes. The section *Melanoloma* also differs from the other sections as some unique acetylenes are present in the roots. In contrast to the other subtribus of Cynareae the content of acetylenes in the aerial parts is typical.

TABLE XXX
CENTAUREA

1. Section Amberboa DC.	Constituents: Roots	Leaves
C. moschata L.[38]	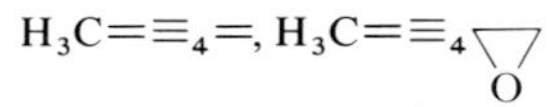 $H_3C{=}{\equiv}_2{=}_2(CH_2)_3OAc$	$H_3C{\equiv}_3{=}_2(CH_2)_4{=}$, $C_2H_5{=}_4(CH_2)_nCHO$ $H_3C{=}{\equiv}_2{=}_2(CH_2)_nCHO$ (n = 3–5) $H_3C{\equiv}_3CH_2{=}(CH_2)_4OH$, $H_3C{\equiv}_3{=}_2(CH_2)_3OH$ 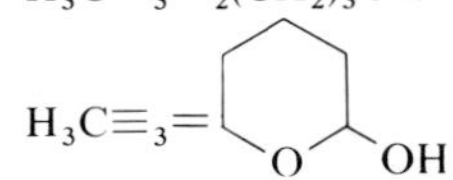
C. lippii L.[36]	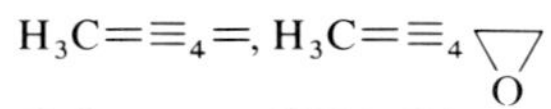 $H_3C{=}{\equiv}_2{=}_2(CH_2)_3OAc$	$H_3CCH_2{=}_4(CH_2)_nR$ (R = CH_2OH, CHO)

2. Section Centaurium Cass.	Roots	Leaves
C. alpina L.[35]	$H_3C{=}{\equiv}_3{=}$[Cl OR] (R = H, Ac)	—
C. ruthenica Lam.[100,206,269]	$H_3C{=}{\equiv}_2{=}_2(CH_2)_4{=}$ $H_3C{=}{\equiv}_3{=}$[Cl OR] (R = H, Ac) $H_3C{=}{\equiv}_3{=}$[OR OR′] (R = R′ = H; R = R′ = Ac) (R = Ac, R′ = H) $H_3C{=}{\equiv}_2{=}_2$[OR OR′] (R = R′ = H; R = R′ = Ac) 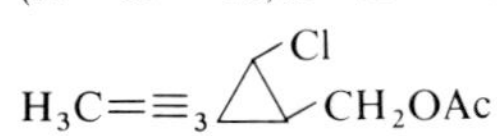	$H_3C{=}{\equiv}_3{=}$[Cl OR] (R = H, Ac) $H_3C{\equiv}_4{=}$[Cl OR] (R = H, Ac) 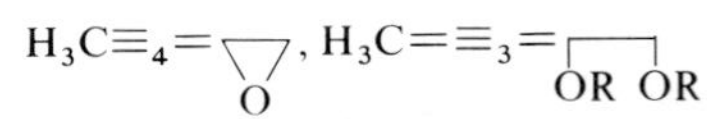 (R = H, Ac)
C. tagana Brot.[35]	$H_3C{=}{\equiv}_3{=}$[Cl OR] (R = H, Ac)	

3. Section Cynastrum Cass.	Roots	Leaves
C. muricata L.[41]	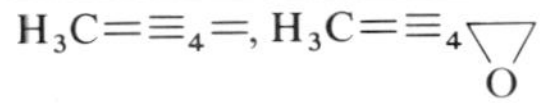	$C_2H_5=_4(CH_2)_nCHO$, $H_3C=\equiv_2=_2(CH_2)_nCHO$ $H_3C=_2\equiv=_2(CH_2)_nCHO$ $H_3C\equiv_3=_2(CH_2)_nCHO$ $H_3C\equiv_3=CH(OH)(CH_2)_4OR$ (R = H, Ac) 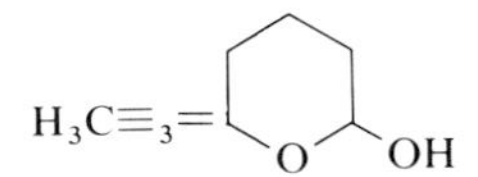 $H_3C\equiv_3=_2CH(OAc)(CH_2)_2OH$

4. Section Melanoloma Cass.	Roots	Leaves
C. pullata L.[36]	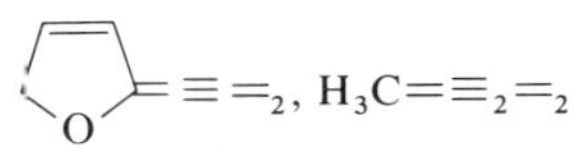 $H_3C=\equiv_4=$, $H_3C=\equiv_2=_2(CH_2)_3OAc$ $H_3C=\equiv_2=_2CH(OAc)CH_2CH_2OH$ $H\equiv_2CH_2=(CH_2)_5COCH_3$	$C_2H_5=_4(CH_2)_4CHO$ $H_3C=\equiv_2=_2(CH_2)_nCHO$ $H_3C\equiv_3=_2(CH_2)_nCHO$ (n = 3–5) $H_3C=\equiv_2=_2(CH_2)_4=$

5. Further species	Roots	Leaves
C. amara L.[35]	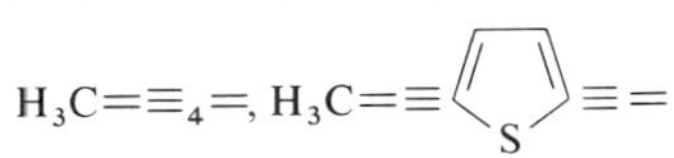 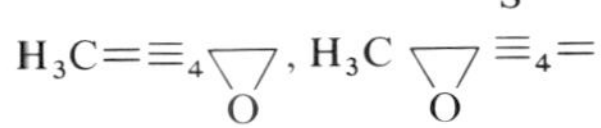	$H_3C\equiv_3=_2(CH_2)_4=$, $H_3C=\equiv_2=_2(CH_2)_4=$ $C_2H_5=_4(CH_2)_nCHO$, $H_3C=\equiv_2=_2(CH_2)_nCHO$ $H_3C\equiv_3=_2(CH_2)_nCHO$ (n = 3–5) and the alcohols

TABLE XXX—Continued

5. Further species	Roots	Leaves
C. angustifolia Miller.[36]	$H_3C{=}{\equiv}_4{=}$, 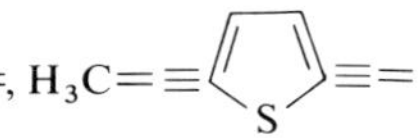	$H_3C{=}{\equiv}_2{=}_2(CH_2)_nCHO$
C. aspera L.[35]	$H_3C{=}{\equiv}_4{=}$	$H_3C{\equiv}_3{=}_2(CH_2)_4{=}$, $C_2H_5{=}_4(CH_2)_nCHO$ $H_3C{\equiv}_3{=}_2(CH_2)_nCHO$
C. atropurpurea Waldst. *et* Kit.[35]	$H_3C{=}{\equiv}_4{=}$	$C_2H_5{=}_4(CH_2)_nCHO$ $H_3C{=}{\equiv}_2{=}_2(CH_2)_nCHO$ $H_3C{\equiv}_3{=}_2(CH_2)_nCHO (n = 3–5)$ $H_3C{=}{\equiv}_4{=}$
C. austriaca Willd.[35]	$H_3C{=}{\equiv}_4{=}$	—
C. babylonica L.[35]	$H_3C{=}{\equiv}_4{=}$	Aldehydes
C. bella Trautv.[35]	$H_3C{=}{\equiv}_4{=}$, 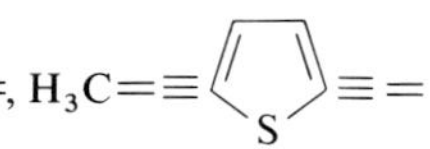	—
C. calcitrapa L.[35]	$H_3C{=}{\equiv}_4{=}$	$H_3C{=}{\equiv}_4{=}$, 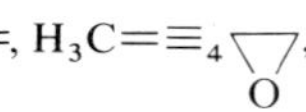, $C_2H_5{=}_4(CH_2)_4CHO$
C. canariensis Willd.[35]	$H_3C{=}{\equiv}_4{=}$, 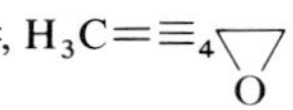	$H_3C{=}{\equiv}_2{=}_2(CH_2)_4{=}$ $H_3C{=}{\equiv}_2{=}_2(CH_2)_nR$ (R = CHO, CH_2OH)
C. cineraria L.[35]	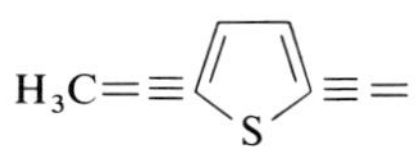	$C_2H_5{=}_4(CH_2)_nCHO$
C. chrithmifolia Vis.[35]	$H_3C{=}{\equiv}_4{=}$, $H_3C{\equiv}_5{=}$ 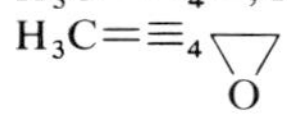	$H_3C{\equiv}_3{=}_2(CH_2)_4{=}$, $H_3C{=}{\equiv}_2{=}_2(CH_2)_4{=}$ Aldehydes
C. collina L.[35]	$H_3C{\equiv}_5{=}$, $H_3C{=}{\equiv}_4{=}$, 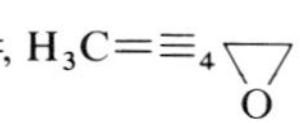	Aldehydes

C. cristata Bartl.[36]	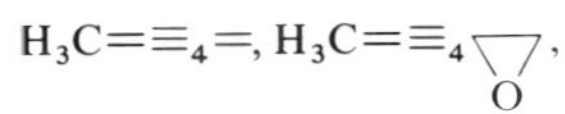	Aldehydes
	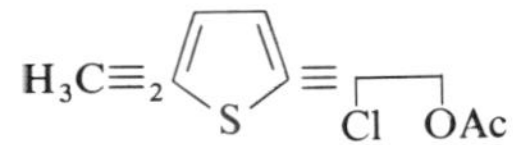	
C. crocodylium L.[35]	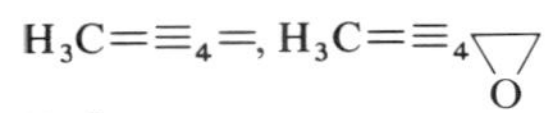	$H_3C{\equiv}_3{=}_2(CH_2)_4{=}$, Aldehydes
C. corymbosa Pour.[35]	$H_3C{=}{\equiv}_4{=}$	$C_2H_5{=}_4(CH_2)_nCHO$
C. csatoi Borb.[35]	$H_3C{=}{\equiv}_4{=}$	$C_2H_5{=}_4(CH_2)_nCHO$
C. cassia Boiss.[35]	$H_3C{=}{\equiv}_4{=}$	Aldehydes
C. cynaroides Lk.[35]	$H_3C{=}{\equiv}_4{=}$	Aldehydes
C. cyanus L.[34,35,206]	$H_3C{=}{\equiv}_4{=}$, $H_3C{=}{\equiv}_3{=}_2$	$H_3C{\equiv}_3{=}_2(CH_2)_4{=}$
	$H_3C{\equiv}_3{=}_2CH(OAc)CH_2CH_2OAc$	$H_3C{=}{\equiv}_2{=}_2(CH_2)_4{=}$
	$H_3C{=}{\equiv}_2{=}_2CH(OAc)CH_2CH_2OAc$	
C. dalmantica L.[35]	$H_3C{=}{\equiv}_4{=}$	—
C. dealbata Willd.[35]	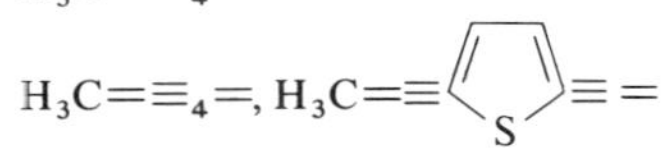	$H_3C{=}{\equiv}_2{=}_2(CH_2)_4{=}$, Aldehydes
C. depressa Bieb.[35]	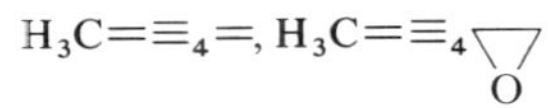	$H_3C{\equiv}_3{=}_2(CH_2)_4{=}$, $H_3C{=}{\equiv}_2{=}_2(CH_2)_4{=}$ Aldehydes
C. deusta Ten.[79]	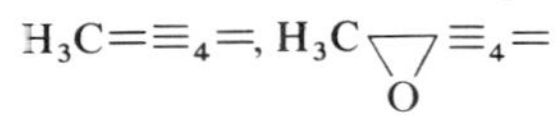	Aldehydes
	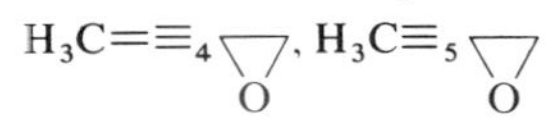	
	$H_3C{=}{\equiv}_2{=}_2CH(OAc)CH_2CH_2OAc$	
	$H_3C{\equiv}_3{=}_2CH(OAc)CH_2CH_2OAc$	
	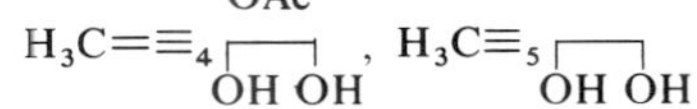	

TABLE XXX—Continued

5. Further species	Roots	Leaves
C. dichroantha Kern.[35]	$H_3C{=}{\equiv}_4{=}$, $H_3C{=}{\equiv}_4$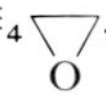, $H_3C{\equiv}_5{=}$	Aldehydes
C. diffusa Lam.[36]	$H_3C{=}{\equiv}_4{=}$, $H_3C{=}{\equiv}_4$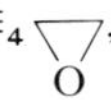, $H_3C{\equiv}_5{=}$	$C_2H_5{=}_4(CH_2)_nCHO$ $H_3C{\equiv}_3{=}_2(CH_2)_4{=}$
C. dilutata Ait.[36]	$H_3C{=}{\equiv}_4{=}$, $H_3C{=}{\equiv}_2{=}_2\underset{\mid \atop R}{C}HCH_2CH_2OR'$ (R = H, R′ = Ac; R = OAc, R′ = Ac; R = OAc, R′ = H)	$H_3C{\equiv}_3{=}_2(CH_2)_4{=}$ Aldehydes
C. dubia subsp. eudubia subvar. candolei Koch[35]	$H_3C{=}{\equiv}_4{=}$, $H_3C{=}{\equiv}_4$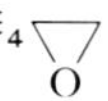 $H_3C{=}{\equiv}$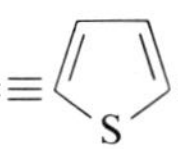${\equiv}{=}$	$H_3C{\equiv}_3{=}_2(CH_2)_4{=}$, $H_3C{=}{\equiv}_2{=}_2(CH_2)_4{=}$ Aldehydes and corresponding alcohols
C. divaricata Wall. ex Bornm.[35]	$H_3C{=}{\equiv}_4{=}$, $H_3C{\equiv}_5{=}$	—
C. eriophora L.[36]	$H_3C{=}{\equiv}_4{=}$	Aldehydes
C. ferox Desf.[36]	$H_3C{=}{\equiv}_4{=}$, $H_3C{\equiv}_5{=}$	Aldehydes
C. friderici Vis.[35]	$H_3C{=}{\equiv}_4{=}$, $H_3C{\equiv}_5{=}$	$H_3C{\equiv}_2{=}_2(CH_2)_4{=}$, $H_3C{=}{\equiv}_2{=}_2(CH_2)_4{=}$ Aldehydes
C. glastifolia L.[35]	$H_3C{=}{\equiv}_4{=}$, $H_3C{\equiv}_5{=}$, $H_3C{=}{\equiv}$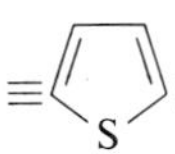${\equiv}{=}$	$H_3C{=}{\equiv}_4{=}$, Aldehydes $H_3C{\equiv}_3{=}_2(CH_2)_4{=}$
C. gymnocarpa Moris.[35]	$H_3C{=}{\equiv}_4{=}$	—
C. iberica Trev.[35]	$H_3C{=}{\equiv}_4{=}$, $H_3C{=}{\equiv}_2{=}_2(CH_2)_3OAc$	$H_3{=}{\equiv}_4{=}$, Aldehydes $H_3C{=}{\equiv}_2{=}_2(CH_2)_4{=}$
C. indurata Janka[35]	$H_3C{=}{\equiv}_4{=}$, $H_3C{=}{\equiv}$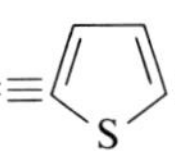${\equiv}{=}$	$H_3C{\equiv}_3{=}_2(CH_2)_4{=}$, Aldehydes and corresponding alcohols
C. involucrata Desgl.[36]	$H_3C{=}{\equiv}_4{=}$, $H_3C{=}{\equiv}_4$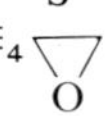	$H_3C{=}{\equiv}_4{=}$, Aldehydes

C. jacea L. subsp. jacea[36] *C. jacea* subsp. macroptilon[35]	$H_3C{=}{\equiv}_4{=}$ 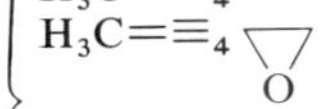	$H_3C{\equiv}_3{=}_2(CH_2)_4{=}$, $H_3C{=}{\equiv}_2{=}_2(CH_2)_4{=}$ Aldehydes and corresponding alcohols
C. jacea subsp. jacea var. pectinata[35]	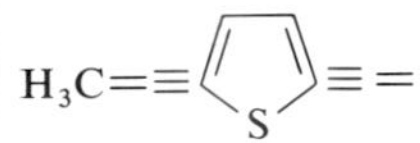	
C. kosanini Hayek[35]	$H_3C{=}{\equiv}_4{=}$	—
C. kotschyana Koch.[35]	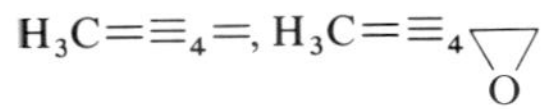	Aldehydes
C. leucophylla Bieb.[35]	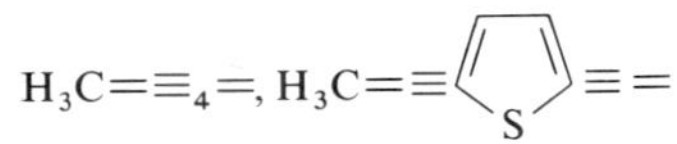	$H_3C{=}{\equiv}_2{=}_2(CH_2)_4{=}$, Aldehydes
C. lusitanica Boiss.[35]	$H_3C{=}{\equiv}_4{=}$, $H_3C{=}{\equiv}_3{=}_2$ 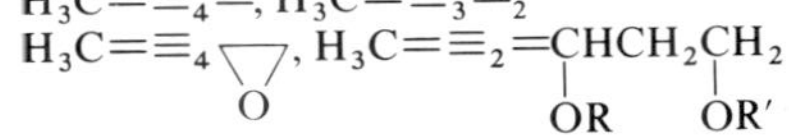 (R = R′ = Ac; R = R′ = H; R = Ac, R′ = H)	Aldehydes, $H_3C{=}{\equiv}_4{=}$
C. maculosa Lam.[36]	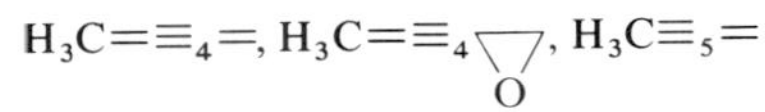 $H_3C{=}{\equiv}_2{=}_2CH(OAc)CH_2CH_2OAc$, $H_3C{\equiv}_3{=}_2CH(OAc)CH_2CH_2OAc$	Aldehydes
subsp. rhenana[35]	$H_3C{=}{\equiv}_4{=}$, $H_3C{\equiv}_5{=}$	Aldehydes
C. macrocephala Puschk.[205,227]	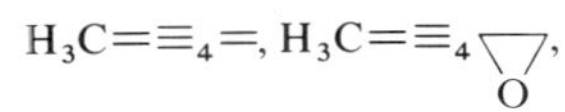 $H_3C{=}{\equiv}_3{=}_2$ $H_3C{=}{\equiv}_2{=}_2CH(OAc)CH_2CH_2OAc$	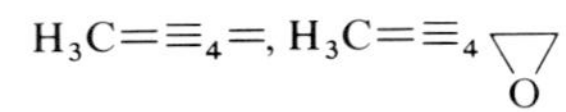 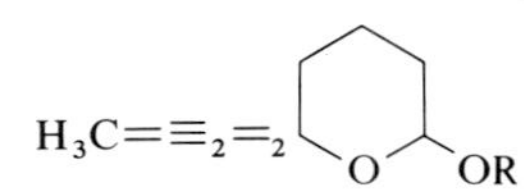 (R = H, Ac) $H_3C{=}{\equiv}_2CH_2{=}(CH_2)_5CHO$

TABLE XXX—Continued

5. Further species	Roots	Leaves
C. macroptilon Hay[35]	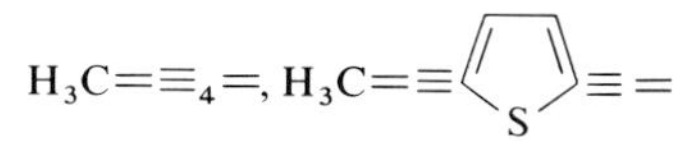	Aldehydes and corresponding alcohols
	$H_3C{\equiv}_3{=}_2CH(OAc)CH_2CH_2OAc$	$H_3C{=}{\equiv}_4{=}$
	$H_3C{=}{\equiv}_2{=}_2CH(OAc)CH_2CH_2OAc$	
C. melitensis L.[36]	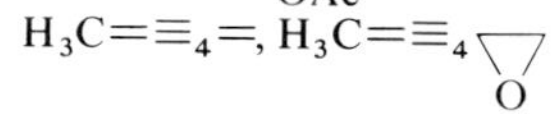	$H_3CCH_2{=}_4(CH_2)_nCHO$
C. montana L.[35]	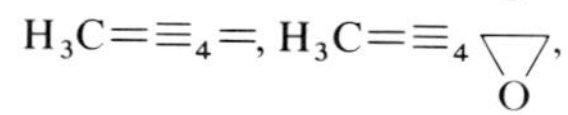	Aldehydes and alcohols
	$H_3C{=}{\equiv}_2{=}_2CH(R)CH_2CH_2OR'$ (R = H, R′ = Ac; R = OAc, R′ = Ac; R = OH, R′ = H) (R = H, R′ = H; R = OAc, R′ = H)	$H_3C{\equiv}_3{=}_2(CH_2)_4{=}$, $H_3C{=}{\equiv}_4{=}$
C. napifolia L.[36]	$H_3C{=}{\equiv}_4{=}$	$H_3C{\equiv}_3{=}_2(CH_2)_4{=}$, $C_2H_5{=}_4(CH_2)_nCHO$
C. nana Desfr.[35]	$H_3C{=}{\equiv}_4{=}$	$C_2H_5{=}_4(CH_2)_nCHO$
C. nigrescens Willd.[35]	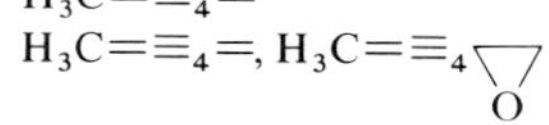	Aldehydes
	$H_3C{=}{\equiv}_2{=}_2CH(OR)CH_2CH_2OR$	
C. nicaeensis All.[35]	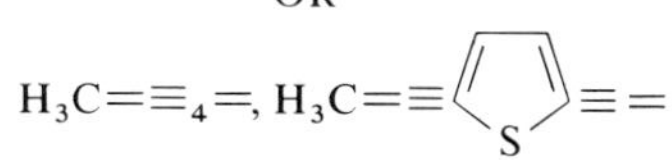	Aldehydes
C. orientalis L.[35]	$H_3C{=}{\equiv}_4{=}$	Aldehydes
C. ornata Willd.[35]	$H_3C{=}{\equiv}_4{=}$	—
C. ossica C. Koch[35]	$H_3C{=}{\equiv}_4{=}$	—

C. oxylepis Wim. *et* Grab.[35]	$H_3C=\equiv_4=$	—
C. phrygia L. subsp.	$H_3C=\equiv_4=$, $H_3C=\equiv$ 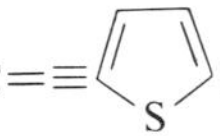 $\equiv=$	$H_3C=\equiv_4=$
pseudophrygia Gugl.[35]	$H_3C\equiv_5=$, $H_3C=\equiv_4$ 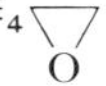	$H_3C\equiv_3=_2(CH_2)_4=$
subsp. stenolepis[35]	$H_3C=\equiv_4=$, $H_3C=\equiv_4$ 	Aldehydes
C. salmantica L.[35]	$H_3C=\equiv_4=$	$C_2H_5=_4(CH_2)_nCHO$ and Alcohols
C. scabiosa L. subsp. scabiosa[35]	$H_3C=\equiv_4=$, $H_3C=\equiv_4$ 	Aldehydes
C. sempervirens L.[36]	$H_3C=\equiv_4=$	—
C. solstitalis L.[36]	$H_3C=\equiv_4=$, $H_3C=\equiv_4$ 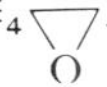, $H_3C\equiv_5=$	$C_2H_5=_4(CH_2)_nCHO$
C. sphaerocephala L.[35]	$H_3C=\equiv_4=$	Aldehydes
C. sulphurea Willd.[36]	$H_3C=\equiv_4=$	Aldehydes
C. trinervia Stev. ex Willd.[35]	$H_3C=\equiv_4=$	Aldehydes
C. triumfettii All.[35]	$H_3C=\equiv_4=$	Aldehydes
C. transalpina Schleich.[35]	$H_3C=\equiv_4=$, $H_3C\equiv_5=$, $H_3C=\equiv$ 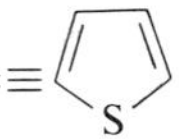 $\equiv=$	Aldehydes, $H_3C=\equiv_4=$
C. uliginosa Brot.[35]	$H_3C=\equiv_4=$, $H_3C=\equiv$ 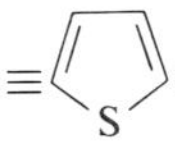 $\equiv=$, $H_3C=\equiv_3=_2$ $H_3C=\equiv_2=_2(CH_2)_3OR$ (R = H, Ac)	Aldehydes
C. uniflora L.[35]	$H_3C=\equiv_4=$	—
Carthamus caeruleus L.[236]	$H_3C=\equiv_2=_3$, traces $H_3C=\equiv_4=$, $H_3C\equiv_5=$ $H_3C=\equiv_3=$  (R = H, Ac) $H_3C=\equiv_2=_2CH(OR)CH_2CH_2OR'$ (R = R′ = H; R′ = H, R = Ac; R = H, R′ = Ac; R = R′ = Ac)	$H_3C=\equiv_3=_2$ $H_3C=\equiv_2=_3$ $H_3C=\equiv_3=$ 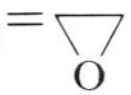 $H_3C=\equiv_2=CH(OAc)CH_2CH_2OH$

TABLE XXX—Continued

5. Further species	Roots	Leaves
C. glaucus Bieb.[35]	$H_3C{=}{\equiv}_4{=}$	$H_3C{=}{\equiv}_4{=}$, $H_3C{=}{\equiv}_3{=}_2$ $H_3C{\equiv}_4{=}_2$, $H_3C{=}{\equiv}_3{=}\underset{O}{\triangledown}$ $H_3C{=}{\equiv}_3{=}\underset{Cl\quad OH}{\sqcap}$, $H_3C{=}{\equiv}_3{=}\underset{OH\;OH}{\sqcap}$
C. lanatus L.[236]	$H_3C{=}{\equiv}_4{=}$ $H_3C{=}{\equiv}_3{=}\underset{OR\;OR}{\sqcap}$	$H_3C{=}{\equiv}_3{=}_2$, $H_3C{=}{\equiv}_4{=}$ $H_3C{=}{\equiv}_3{=}\underset{O}{\triangledown}$ $H_3C{=}{\equiv}_3{=}\underset{R\quad R'}{\sqcap}$ (R = R′ = OH; R = R′ = OAc; R = Cl, R′ = OH; R = OH, R′ = Cl; R = OAc, R′ = OH)
C. tinctorius L.[236,294]	$H_3C{=}{\equiv}_4{=}$ $H_3C{=}{\equiv}_4\underset{O}{\triangledown}$ $H_3C{=}{\equiv}_3{=}\underset{O}{\triangledown}$ $H_3C{=}{\equiv}_3{=}\underset{OR\;OR}{\sqcap}$	$H_3C{=}{\equiv}_3{=}_2$, $H_3C{\equiv}_3{=}_3$, $H_3C{=}{\equiv}_4\underset{O}{\triangledown}$ $H_3C{=}{\equiv}_3{=}\underset{O}{\triangledown}$, 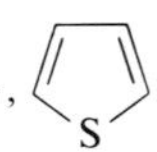$\equiv{=}_3$, $H_3C{\equiv}_5\underset{O}{\triangledown}$ $H_3C{=}{\equiv}_3{=}\underset{R\quad R'}{\sqcap}$ (R = R′ = OH; R = Cl, R′ = OH; R = OH, R′ = Cl; R = R′ = OAc, R = Cl, R′ = OAc) $H_3C{=}{\equiv}_2CH_2CH_2CH_2OCOCH_2CH(CH_3)_2$ $C_3H_7{\equiv}_2CH_2CH_2CH_2OCOCH_2CH(CH_3)_2$
Carduncellus coeruleus Less.[35]	$H_3C{=}{\equiv}_4{=}$ $H_3C{=}{\equiv}_4\underset{O}{\triangledown}$	$H_3C{=}{\equiv}_4{=}$, $H_3C{\equiv}_4{=}_2$, $H_3C{=}{\equiv}_3{=}_2$ $H_3C{=}{\equiv}_2{=}_3$, $H_3C{=}{\equiv}_3{=}\underset{O}{\triangledown}$

	$OCH{=}{\equiv}_2{=}_2(CH_2)_4{=}$ $H_3C{=}{\equiv}_2{=}_2CH(OAc)CH_2CH_2OAc$	$H_3C{\equiv}_4{=}$(epoxide, O), $H_3C{=}{\equiv}_3{=}CH(R)CH_2R'$ (R = R′ = OAc; R = Cl, R′ = OH; R = OH, R′ = Cl) $H_3C{\equiv}_4{=}CH(Cl)CH_2OH$ $H_3C{=}{\equiv}_2{=}_2CH(OAc)CH_2CH_2OAc$
Cnicus benedictus L.[35]	$H_3C{=}{\equiv}_4{=}$, $H_3C{=}{\equiv}_3{=}_2$ $H_3C{\equiv}_5{=}$ $H_3C{=}{\equiv}_2{=}_2CH(OAc)CH_2CH_2OAc$	$H_3C{=}{\equiv}_4{=}$, $H_3C{\equiv}_5{=}$ $C_2H_5{=}_4(CH_2)_nR$ (R = CHO, CH_2OH)
C. diacantha Desf.[35]	$H_3C{=}{\equiv}_4{=}$, $H_3C{\equiv}_5{=}$	—
C. syriacus Roth.[35]	$H_3C{=}{\equiv}_4{=}$, $H_3C{\equiv}_5{=}$	—

1.9. Mutisieae and Cichorieae

In species of the tribus Mutisieae already investigated, only small amounts of acetylenes have been isolated; the pentayn-ene, sometimes together with the ene-tetrayn-ene is typical. Only the genus *Dicoma* does not contain these hydrocarbons, while from the genera *Perezia* and Mutisia, only trace amounts of unidentified acetylenes have been isolated.

In the tribus Cichorieae subtribus some acetylenes have been found only in species belonging to the last subtribus. However, only in the genus *Lactuca* could several compounds be identified.

The seeds of *Crepis foetida* contain the crepenynic acid, a common precursor of natural acetylenes. This may be important as no acetylenes have been detected in the whole plant, showing that perhaps even in cases where no acetylenes can be isolated some of them might still be present as precursors with a very rapid interconversion.

TABLE XXXI
TRIBUS MUTISIEAE

1. Mutisieae-Gochnatinae	Constituents (roots)
Dicoma zeyheri Sond.[356]	$H_3C{=}{\equiv}_3{=}\underset{\displaystyle Cl}{\underset{\vert}{C}}HCH_2OR$ (R = H, Ac) $H_3C(CH_2)_{10}CH{=}C{=}CH(CH_2)_3CO_2R$ (R = H, CH_3)

2. Mutisieae-Mutisinae	Constituents (roots)
Mutisia clematis L.[35]	$H_3C{\equiv}_5{-}R$?
Trichocline incana Less.[35]	$H_3C{\equiv}_5{=}$
Gerbera anandria L.[35]	$H_3C{\equiv}_5{=}$, $H_3C{=}{\equiv}_4{=}$
G. kunzeana A. Br. *et* Aschers[35]	$H_3C{\equiv}_5{=}$, $H_3C{=}{\equiv}_4{=}$
G. nivea Sch. Bip.[35]	$H_3C{\equiv}_5{=}$, $H_3C{=}{\equiv}_4{=}$, $H_3C{\equiv}_3{=}_2(CH_2)_3OH$
G. natalensis Sch. Bip.[35]	$H_3C{=}{\equiv}_4{=}$
Chapitalia arediavaletae Areth.	$H_3C{\equiv}_5{=}$, $H_3C{=}{\equiv}_4{=}$

3. Mutisieae-Nassauvinae	Constituents (roots)
Perezia multiflora Less.[35]	${\equiv}_3{=}$?
Trixis eryngioides Cabrera[35]	$H_3C{\equiv}_5{=}$
Moscharia pinnatifida Ruiz. *et* Pav.[35]	$H_3C{\equiv}_5{=}$
Schlechtendalia luzulaefolia Less.[35]	$H_3C{\equiv}_5{=}$

TABLE XXXII
TRIBUS CICHORIEAE

1. Cichorieae-Scolyminae	Constituents (roots)
Scolymus hispanicus L.[35]	—

2. Cichorieae-Dendroseridinae 3. Cichorieae-Cichorinae	Constituents (roots)
Tolpis barbata Gaertn.[35]	—
Hypochoeris maculata L.[35]	—
Leontodon autumnalis L.[35]	—
L. hispidus L.[35]	—
Picris hieracioides Sibth. *et* Sm.[35]	—
P. echioides L.[35]	—
Tragopogon heterospermum Schweigg.[35]	—
Scorzonera laciniata L.[35]	—

5. Cichorieae-Crepidinae	Constituents (roots)
Andryala integrifolia L.[35]	—
Taraxacum officinale Web.[35]	—
Launaea procumbens Amin.[35]	Traces $\equiv_3=_2$, $\equiv_2=$
Sonchus arvensis L.[35]	—
Lactuca macrocephala[35]	$H_3C\equiv_3\underset{\displaystyle R}{CH}CH_2CH_2OH$ (R = H, OH); leaves also
L. plumieri Gren. *et* Godr.[152]	$H_3C\equiv_3CO=$, $H_3C\equiv_3CO$ , $H_3C\equiv_3=CHO$ $H_3C\equiv_3=CH_2OH$, $H_3C\equiv_3COCH_2CH_2OH$ $H_3C\equiv_3CH_2CH_2CH_2OH$, $H_3C\equiv_2=$ 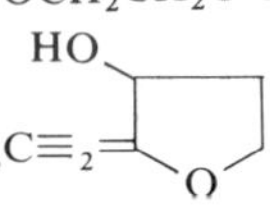 $H_3C\equiv_3\underset{\displaystyle OH}{CH}CH_2CH_2OH$; leaves;
L. sativa L.[152]	— Leaves: $H_3C\equiv_3=_2(CH_2)_4=$
L. alpina L.[152]	$\equiv_3=$
L. seriola L.[35]	—
L. tatarica C. A. Mey.[35]	—
L. tenerrima Pourr.[152]	$\equiv_3=_2$
Reichardia tigitana L. Pan.[35]	Traces $\equiv_3=$ (leaves)
R. ligulatum var. iliciola[35]	—
Troximon pterocarpus[35]	—
Crepis biennis L.[35]	— Leaves: traces $=\equiv_2=$
C. foetida L.[35,185]	— Seeds: $H_3C(CH_2)_4\equiv CH_2=(CH_2)_7CO_2R$
C. sibirica L.[35]	—
Hieracium umbellatum L.[35]	—
Urospermum dalechampii F. W. Schmidt[152]	$\equiv_3=_2$
Catananche caerulea L.[152]	$\equiv_3=_2$

2. Umbelliferae

Besides Compositae, the Umbelliferae is the family which has been the most investigated for acetylenes among the higher plants. Up to now about 200 species out of 90 genera have been investigated and more than 90% found to contain acetylenes. Other very common constituents are coumarins, terpenes and several aromatic compounds such as myristicine and eugenole-like ethers.

Unfortunately, the number of species investigated for all these compounds is limited and therefore, their importance as regards botanical classification is still not clear. The most common acetylenic compounds in this family are falcarinone and its closely related compounds. These constituents are found in all three subfamilies and in nearly all tribus except Coriandreae and the uninvestigated Echinophorae (see Tables XXX–XXXII).

TABLE XXXIII
I. SUBFAMILY HYDROCOTYLOIDEAE

1. Tribus Hydrocotyleae	Constituents (roots)	
Hydrocotyle ramiflora Maxim.[35]	$=CO\equiv_2CH_2=C_7H_{15}$	
H. vulgaris L.[36]	$=CO\equiv_2CH_2=C_7H_{15}$	
Didiscus coeruleus DC.[35]	$=CO\equiv_2CH_2=C_7H_{15}$	
Didiscus pilosus Domin.[35]	$=CO\equiv_2CH_2=C_7H_{15}$	
	Leaves: also + $=\underset{OH}{CH}\equiv_2CH_2=C_7H_{15}$	
Trachymene australis Benth.[35]	$=CO\equiv_2CH_2=C_7H_{15}$	
T. glauca[35]	$=CO\equiv_2CH_2=C_7H_{15}$	Leaves: as roots
T. pilosa Sm.[35]	$=CO\equiv_2CH_2=C_7H_{15}$	
2. Tribus Mulinae	**Constituents (roots)**	
Bowlesia inacana Ruiz. *et* Pav.[35]	$=CO\equiv_2CH_2=C_7H_{15}$	
B. tenella Meyen.[35]	$=CO\equiv_2CH_2=C_7H_{15}$	
B. tenera Spr.[35]	$=CO\equiv_2CH_2=C_7H_{15}$	
Azorella trifurcata (Gaertn.) Pers.[217]	$HOCH_2=\equiv_2\underset{OH}{CH}=C_7H_{15}$,	
	$AcOCH_2=\equiv_2\underset{OH}{CH}=R$ ($R = C_7H_{15}, C_5H_{11}$)	

TABLE XXXIV
II. SUBFAMILY SANICULIDEAE

3. Tribus Saniculeae	Constituents (roots)
Sanicula europeae L.[35]	Trace $-C\equiv C-$
Hacquetia epipactis DC.[35]	Trace $-C\equiv C-$

TABLE XXXIV—Continued

3. Tribus Saniculeae	Constituents (roots)
Astrantia carniolica Wulff.[35]	—
A. trifida Hoffm.[35]	— 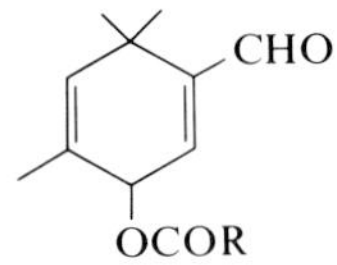
A.1. Eryngium alpinum L.[357]	$=CO\equiv_2\underset{\mid\atop R}{CH}=C_7H_{15}$ (R = H, OH), $=\underset{\mid\atop OH}{CH}\equiv_2\underset{\mid\atop R}{CH}=C_7H_{15}$ (R = H, OH)
	$AcOCH_2=\equiv_2\underset{\mid\atop OH}{CH}=C_7H_{15}$ 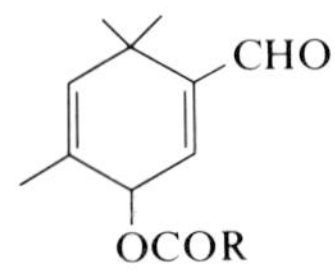
E. bromeliifolium de La Roche[357]	$=CO\equiv_2\underset{\mid\atop R}{CH}=C_7H_{15}$ (R = H, OH) 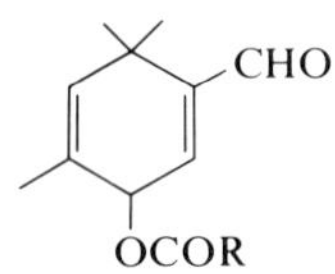
E. campestre L.[357]	$=CO\equiv_2\underset{\mid\atop R}{CH}=C_7H_{15}$ (R = H, OH), $AcOCH_2=\equiv_2\underset{\mid\atop OH}{CH}=C_7H_{15}$
E. caucasicum Fisch.[357]	As *E. alpinum*
E. caeruleum Lk.[357]	As *E. alpinum*
E. giganteum M.B.[357]	As *E. alpinum*
E. planum L.[337,338]	As *E. alpinum*
A.2. E. amethystinum L.[357]	— 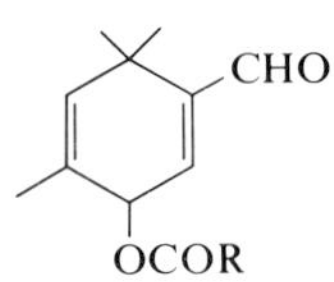 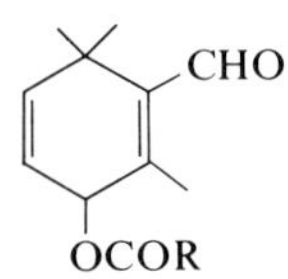
E. serbicum Panc.[357]	As *E. amethystinum*
E. Eryngium agavifolium Griseb.[357]	$=CO\equiv_2CH_2=C_7H_{15}$, $=CO\equiv_2\underset{\mid\atop OH}{CH}=C_7H_{15}$

4. Tribus Lagoecieae	Constituents (roots)
Lagoecia cuminoides L.[35]	$=CO\equiv_2CH_2=C_7H_{15}$
Arctopus echinata L.[35]	$H_3C=\equiv_2=_2C_2H_5$

TABLE XXXV
III. SUBFAMILY APIOIDEAE

	Constituents: Roots	Constituents: Leaves
5. Tribus Echinophoreae		
6. Tribus Scandiceae		
Physocaulis nodosus (L.) Tausch.[35]	$=CO\equiv_2CH_2=C_7H_{15}$	$=CO\equiv_2\underset{\mid \atop OH}{CH}=C_7H_{15}$
Chaetosciadium trichospermum Boiss[358]	$-C\equiv C-$	—
Chaerophyllum aromaticum L.[35]	Coumarins	Trace $-C\equiv C-$
C. temulum L.[35]	$=CO\equiv_2CH_2=C_7H_{15}$	—
Anthriscus cerefolium Hoffm.[35]	—	Trace $-C\equiv C-$
Scandix balansae Reut.[35]	—	$=CO\equiv_2CH_2=C_7H_{15}$
S. brachycarpa Guss.[35]	Trace $-C\equiv C-$	—
S. pecten-veneris L.[35]	Trace $-C\equiv C-$	—
Osmorhiza aristata Makino[358]	$-C\equiv C-$	—
Myrrhis odorata Scop.[35]	$\equiv_2=$	—
Molopospermum cicutarium DC.[35]	$=\underset{\mid \atop OH}{CH}\equiv_2\underset{\mid \atop OH}{CH}=C_7H_{15}$	—
Torilis arvensis Link.[35,358]	$=CO\equiv_2CH_2=C_7H_{15}$	—
T. helvetica C. Gmel.[35]	$=CO\equiv_2CH_2=C_7H_{15}$, $=\underset{\mid \atop OH}{CH}\equiv_2CH_2=C_7H_{15}$	—
T. infesta Clairv.[35]	$=CO\equiv_2CH_2=C_7H_{15}$, $=\underset{\mid \atop OH}{CH}\equiv_2CH_2=C_7H_{15}$	—
T. japonica DC.[35,358]	$=CO\equiv_2CH_2=C_7H_{15}$, $=\underset{\mid \atop OH}{CH}\equiv_2CH_2=C_7H_{15}$	—
T. nodosa Gaertn.[35]	$-C\equiv C-$	—
T. rubella Moench.[35]	$=CO\equiv_2\underset{\mid \atop R}{CH_2}=C_7H_{15}$ (R = H, OH) $=\underset{\mid \atop OH}{CH}\equiv_2CH_2=C_7H_{15}$	—

Astrodaucus orientalis Drude.[358]	$-C\equiv C-$	—
Turgenia latifolia Hoffm.[35,358]	$=CO\equiv_2\underset{OH}{CH}=C_7H_{15}$	—
Caucalis anthriscus Huds.[35]	$=CO\equiv_2\underset{R}{CH}=C_7H_{15}$ (R = H, OH)	$=CO\equiv_2CH_2=C_7H_{15}$
C. daucoides L.[35]	$H_5C_2CO\equiv_2CH_2=C_7H_{15}$	—
C. lappula Grande[35]	Trace $-C\equiv C-$	—
C. platycarpos L.[35]	Trace $-C\equiv C-$	—
Orlaya daucorlaya Murb.[358]	$-C\equiv C-$	—
O. grandiflora Hoff.[35]	$=CO\equiv_2\underset{R}{CH}=C_7H_{15}$ (R = H, OH)	—
7. Tribus Coriandreae	Roots	Leaves
Coriandrum sativum L.[35]	—	—
Bifora radians M. Bieb.[35]	Trace $-C\equiv C-$	—
8. Tribus Smyrnieae	Roots	Leaves
Conium maculatum L.[217]	$=CO\equiv_2\underset{R}{CH}=C_7H_{15}$ (R = H, OH) $H_5C_2CO\equiv_2CH_2=C_7H_{15}$, Coumarins	$=CO\equiv_2CH_2=C_7H_{15}$
Arracacia purpusii Rose.[35]	Trace $-C\equiv C-$	—
Hladnikia athamanticum[35]	Trace $-C\equiv C-$, 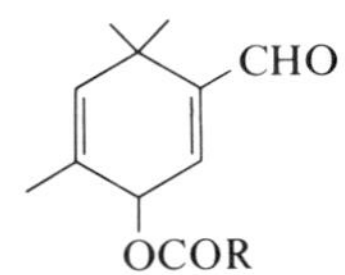	—
Pleurospermum austriacum Hoffm.[35]	$=CO\equiv_2\underset{R}{CH}=C_7H_{15}$ (R = H, OH)	

TABLE XXXV—Continued

9a. Ammineae-Carinae	Constituents: Roots	Constituents: Leaves
Heteromorpha trifoliata Eckl. *et* Zeyh.[35]	$=CO\equiv_2\underset{R}{CH}=C_7H_{15}$ (R = H, OH) $=\underset{OH}{CH}\equiv_2\underset{R}{CH}=C_7H_{15}$ (R = H, OH)	—
Bupleurum gerardii All.[233]	$HOCH_2=\equiv_2=_2C_5H_{11}$, $OCH=\equiv_2CH_2=C_5H_{11}$	—
B. longiradiatum Turcz.[35]	$-C\equiv C-$	—
B. multinerve DC.[233]	$HOCH_2=\equiv_2=_2C_4H_9$ $HOCH_3=\equiv_2CH_2=C_5H_{11}$ $OCH=\equiv_2CH_2=C_5H_{11}$ $OCH=\equiv_2=_2C_4H_9$	—
B. baldense Turra.[35]	$OCH=\equiv_2CH_2=C_5H_{11}$ $HOCH_2=\equiv_2CH_2=C_5H_{11}$	—
B. ranunculoides L.[233]	As *B. multinerve* DC.	—
B. rotundifolium L.[35]	As *B. gerardii* All.	—
B. tenuissimum L.[233]	As *B. multinerve* DC.	—
Trinia glauca Dumort.[35]	$=CO\equiv_2\underset{R}{CH}=C_7H_{15}$ (R = H, OH)	—
T. kitaibelii Bieb.[35]	$=CO\equiv_2CH_2=C_7H_{15}$ $=CO\equiv=_3C_6H_{13}$	—
T. vulgaris DC.[35]	$=CO\equiv_2CH_2=C_7H_{15}$	—
Szovitsia callicarpa Fisch. *et* Mey.[35]	$=\underset{OH}{CH}\equiv_2CH_2=C_7H_{15}$	—
Cuminum cyminum L.[35]	$=CO\equiv_2\underset{R}{CH}=C_7H_{15}$ (R = H, OH) $=\underset{OH}{CH}\equiv_2\underset{OH}{CH}=C_7H_{15}$, $=CO\equiv_2CO=C_7H_{15}$	$=CO\equiv_2\underset{OH}{CH}=C_7H_{15}$
Apium graveolens L.[35,359]	$=CO\equiv_2\underset{R}{CH}=C_7H_{15}$ (R = H, OH) $=\underset{OH}{CH}\equiv_2\underset{OH}{CH}=C_7H_{15}$	—

Helosciadium inundatum (L.) Koch[35]	${\equiv}_2{=}$	—
H. nodiflorum Koch.[35]	${=}CO{\equiv}_2\underset{R}{C}H{=}C_7H_{15}$ (R = H, OH)	As roots
Sison amomun L.[35]	${=}CO{\equiv}_2\underset{R}{C}H{=}C_7H_{15}$ (R = H, OH)	${=}CO{\equiv}_2CH_2{=}C_7H_{15}$
Trachyspermum ammi (L.) Spr.[35]		$ROCH_2{=}{\equiv}_2{=}_2C_6H_{13}$ $ROCH_2CH_2CH_2{\equiv}_2{=}_2C_6H_{13}$ } R = H, Ac
Cicuta douglasii Coult. *et* Rose.[35]	$HOCH_2CH_2CH_2{\equiv}_2{=}_3\underset{OH}{C}HC_3H_7$	$HOCH_2(CH_2)_2{\equiv}_2{=}_3\underset{OH}{C}HR$ (R = C_3H_7, C_7H_{15})
	$HOCH_2CH_2CH_2{\equiv}_2{=}\underset{OH}{C}HC_7H_{15}$	$HOCH_2(CH_2)_2{\equiv}_2{=}_2\underset{OH}{C}HC_5H_{11}$
	$HOCH_2CH_2CH_2{\equiv}_2{=}_2\underset{R}{C}HC_5H_{11}$ (R = H, OH)	$ROCH_2(CH_2)_2{\equiv}_2{=}_3C_4H_9$ (R = H, Ac)
	$ROCH_2CH_2CH_2{\equiv}_2{=}_3C_4H_9$ (R = H, Ac)	$ROCH_2(CH_2)_2{\equiv}_2{=}_2C_6H_{13}$ (R = H, Ac)
	$ROCH_2CH_2CH_2{\equiv}_2{=}C_8H_{17}$ (R = H, Ac)	$ROCH_2(CH_2)_2{\equiv}_2{=}C_8H_{17}$ (R = H, Ac)
	$ACOCH_2CH_2CH_2{\equiv}_2{=}_2C_6H_{13}$	Cosmen
	$HOCH_2CH_2CH_2{\equiv}_2CH_2{=}C_7H_{15}$	
	Cosmen	
C. virosa L.[218]	$HOCH_2(CH_2)_2{\equiv}_2{=}_2C_6H_{13}$	$HOCH_2(CH_2)_2{\equiv}_2{=}_2C_6H_{13}$
	$HOCH_2(CH_2)_2{\equiv}_2{=}_3C_4H_9$	$HOCH_2(CH_2)_2{\equiv}_2{=}_3\underset{R}{C}HC_3H_7$ (R = H, OH)
	$HOCH_2(CH_2)_2{\equiv}_2{=}(CH_2)_4COC_3H_7$	
	$HOCH_2(CH_2)_2{\equiv}_2{=}_2COC_5H_{11}$	
	$HOCH_2(CH_2)_2{\equiv}_2{=}_2CH_2CH_2COC_3H_7$	$HOCH_2(CH_2)_2{\equiv}_2{=}(CH_2)_4COC_3H_7$
	$HOCH_2(CH_2)_2{\equiv}_3{=}_3COC_3H_7$	$HOCH_2(CH_2)_2{\equiv}_2{=}_2COC_5H_{11}$
	$HOCH_2(CH_2)_2{\equiv}_2{=}_3\underset{OH}{C}HC_3H_7$	$HOCH_2(CH_2)_2{\equiv}_2{=}_3COC_3H_7$
		$H_3C{=}{\equiv}_2{=}_3COC_3H_7$
	$HOCH_2(CH_2)_2{\equiv}_2{=}_2\underset{OH}{C}HC_5H_{11}$	
	$HOCH_2(CH_2)_2{\equiv}_2{=}\underset{OH}{C}HC_7H_{15}$	

TABLE XXXV—Continued

9a. Ammineae-Carinae	Constituents: Roots	Constituents: Leaves
Petroselinum crispum (Miller) A. W. Hill[35]	$=CO\equiv_2\underset{\vert \atop R}{CH}=C_7H_{15}$ (R = H, OH)	—
P. sativum Hoffm.[35]	$=CO\equiv_2CH_2=C_7H_{15}$ $=\underset{\vert \atop OH}{CH}\equiv_2CH_2=C_7H_{15}$, Coumarins	—
P. segetum Koch.[35]	$=CO\equiv_2CH_2=C_7H_{15}$	—
Ridolfia segetum (L.) Morris.[35]	$=CO\equiv_2CH_2=C_7H_{15}$	—
Zizia aurea Koch.[35]	$=CO\equiv_2\underset{\vert \atop R}{CH}=C_7H_{15}$ (R = H, OH) $=CH(OH)\equiv_2CH_2=C_7H_{15}$	—
Cryptotaenia canadensis DC.[35]	$=CO\equiv_2\underset{\vert \atop OH}{CH}=C_7H_{15}$	Trace $-C\equiv C-$
C. japonica Hassk.[35]	$=CO\equiv_2\underset{\vert \atop OH}{CH}=C_7H_{15}$	$=CO\equiv_2CH_2=C_7H_{15}$
Ammi visnaga Lam.[35]	$=CO\equiv_2\underset{\vert \atop OH}{CH}=C_7H_{15}$	—
Ptychotis amoides Koch.[35]	$HOCH_2=\equiv_2CH_2=R$ ($R = C_6H_{13}, C_7H_{15}, C_8H_{17}$) $OHC=\equiv_2CH_2=R$ ($R = C_6H_{13}, C_7H_{15}, C_8H_{17}$)	—
P. heterophylla Koch.[35]	$=CO\equiv_2\underset{\vert \atop R}{CH}=C_7H_{15}$ (R = H, OH)	$=CO\equiv_2\underset{\vert \atop OH}{CH}=C_7H_{15}$
P. saxifraga (L.) Loret *et* Bar.[35]	Trace $-C\equiv C-$	—
Falcaria vulgaris Bernh.[201,210]	$=CO\equiv_2CH_2=C_7H_{15}$ $=\underset{\vert \atop OH}{CH}\equiv_2\underset{\vert \atop R}{CH}=C_7H_{15}$ (R = H, OH) $=CO\equiv=_3C_6H_{13}$	—

Carum carvi L.[210]	$HOCH_2CH_2CO\equiv_2CH_2=C_7H_{15}$ $HOCH_2CH_2CO\equiv_2=C_8H_{17}$ $HOCH_2CH_2CH_2\equiv_2CH_2=C_7H_{15}$ $=CO\equiv_2\underset{\substack{\mid\\OH}}{C}H=C_7H_{15}$ $=CO\equiv_2CO=C_7H_{15}$	—
C. verticillatum Koch.[35]	As *Carum carvi* L.	—
Bunium alpinum W. *et* K.[35]	$=CO\equiv_2CH_2=C_7H_{15}$	$ROCH_2=\equiv_2=_2C_6H_{13}$ (R = H, Ac) $ROCH_2CH_2CH_2\equiv_2=_2C_6H_{13}$ (R = H, Ac)
B. bulbocastanum L.[35]	$=CO\equiv_2\underset{\substack{\mid\\R}}{C}H=C_7H_{15}$ (R = H, OH) $=\underset{\substack{\mid\\OH}}{C}H\equiv_2CH_2=C_7H_{15}$	
Muretia transitoria Korovin[35]	$=CO\equiv_2CH_2=C_7H_{15}$	$=CO\equiv_2\underset{\substack{\mid\\OH}}{C}H=C_7H_{15}$
Pimpinella affinis Ledeb.[35]	—	—
P. anisum L.[35]	—	—
P. koreana Nakai[358]	$-C\equiv C-$	—
P. major Huds.[35]	$=CO\equiv_2CH_2=C_7H_{15}$	—
P. peregrina L.[35]	—	—
P. saxifraga L.[35,358]	$-C\equiv C-$, Coumarins	—
Reutera aurea (DC.) Boiss.[35]	—	—
Aegopodium podograria L.[35]	$=CO\equiv_2\underset{\substack{\mid\\R}}{C}H=C_7H_{15}$ (R = H, OH) $=\underset{\substack{\mid\\OH}}{C}H\equiv_2CH_2=C_7H_{15}$ $HOCH_2CH_2CO\equiv_2\underset{\substack{\mid\\OH}}{C}H=C_7H_{15}$	$=\underset{\substack{\mid\\OH}}{C}H\equiv_2CH_2=C_7H_{15}$ $HOCH_2CH_2CO\equiv_2\underset{\substack{\mid\\OH}}{C}H=C_7H_{15}$
Sium erectum Huds.[35]	$=CO\equiv_2CH_2=C_7H_{15}$	—
S. lancifolium Schrank.[35]	$=CO\equiv_2\underset{\substack{\mid\\OH}}{C}H=C_7H_{15}$	—

TABLE XXXV—Continued

9a. Ammineae-Carinae	Constituents: Roots	Constituents: Leaves
S. latifolium L.[35]	$=CO\equiv_2\underset{OH}{CH}=C_7H_{15}$	—
S. nodifolium N.[35]	$=\underset{OH}{CH}\equiv_2CH_2=C_7H_{15}$	$=CO\equiv_2CH_2=C_7H_{15}$ $=\underset{OH}{CH}\equiv_2CH_2=C_7H_{15}$
S. sisarum L.[210]	$=CO\equiv_2\underset{OH}{CH}=C_7H_{15}$ $=CO\equiv_2CO=C_7H_{15}$	—
S. suave Walt.[35]	$-C\equiv C-$	—

9b. Ammineae-Seselinae	Roots	Leaves
Crithmum maritimum L.[35]	$=CO\equiv_2CH_2=C_7H_{15}$ $=\underset{OH}{CH}\equiv_2\underset{OH}{CH}=C_7H_{15}$	$=\underset{OH}{CH}\equiv_2\underset{OH}{CH}=C_7H_{15}$
Seseli hippomarathrum Jacq.[241]	$=CO\equiv_2CH_2=C_7H_{15}$ $=CO\equiv=_3C_6H_{13}$ $HC\equiv C-CO=$ 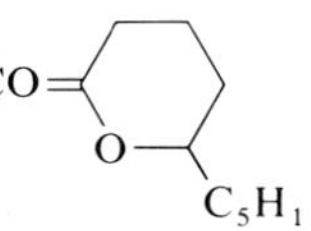 $HOCH_2=\equiv_2CH_2=C_7H_{15}$, 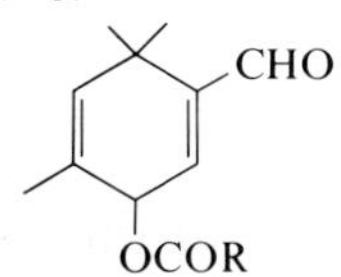	—

S. leucospermum Wldst. *et* Kit.[35]	$=CO\equiv_{2}\underset{\substack{\vert \\ R}}{C}H=C_7H_{15}$ (R = H, OH) $=\underset{\substack{\vert \\ OH}}{C}H\equiv_{2}CH_2=C_7H_{15}$	Coumarins
S. devenyense Simonk.[35]	$=CO\equiv_{2}\underset{\substack{\vert \\ R}}{C}H=C_7H_{15}$ (R = H, OH) $=CO\equiv=_{3}C_6H_{13}$, $=\underset{\substack{\vert \\ OH}}{C}H\equiv_{2}CH_2=C_7H_{15}$	—
S. glabratum Willd. *et* Schult.[35]	$=CO\equiv_{2}CH_2=C_7H_{15}$	$=CO\equiv_{2}\underset{\substack{\vert \\ OH}}{C}H=C_7H_{15}$
S. glaucum L.[35]	$=CO\equiv_{2}\underset{\substack{\vert \\ R}}{C}H=C_7H_{15}$ (R = H, OH) $=CO\equiv=_{3}C_6H_{11}$, $=\underset{\substack{\vert \\ OH}}{C}H\equiv_{2}CH_2=C_7H_{15}$	—
S. gracile Wldst. *et* Kit.	$=CO\equiv_{2}\underset{\substack{\vert \\ R}}{C}H=C_7H_{15}$ (R = H, OH) $=CO\equiv=_{3}C_6H_{13}$	Trace $-C\equiv C-$
S. libanotis (L.) Koch.[35,358]	Coumarins ($-C\equiv C-$)	—
S. montanum L.[35]	$=CO\equiv_{2}\underset{\substack{\vert \\ R}}{C}H=C_7H_{15}$ (R = H, OH) Coumarins	—
S. gummiferum Pall.[35,358]	$=CO\equiv_{2}\underset{\substack{\vert \\ R}}{C}H=C_7H_{15}$ (R = H, OH) $=\underset{\substack{\vert \\ OH}}{C}H\equiv_{2}CH_2=C_7H_{15}$ $=CO\equiv=_{3}C_6H_{13}$ $HOCH_2=\equiv_{2}\underset{\substack{\vert \\ R}}{C}H=C_7H_{15}$ (R = H, OAc) Coumarins	—

TABLE XXXV—Continued

9b. Ammineae-Seselinae	Constituents: Roots	Constituents: Leaves
S. osseum Crtz.[35]	$=CO\equiv_2CH(R)=C_7H_{15}$ (R = H, OH) $=CH(OH)\equiv_2CH_2=C_7H_{15}$, $=CO\equiv=_3C_6H_{13}$	—
S. rhodopeum Velen.[35]	Coumarins	Coumarins
S. varium Trev.[35]	$=CO\equiv_2CH(R)=C_7H_{15}$ (R = H, OH)	$-C\equiv C-$
Libanotis buchtormensis DC.[35]	$=CO\equiv_2CH_2=C_7H_{15}$, Coumarins	Coumarins
L. pyrenaica Bourg.[35]	$=CO\equiv_2CH_2=C_7H_{15}$, Coumarins	Coumarins
Discopleura nutallii DC.[35]	Trace $-C\equiv C-$	—
Oenanthe aquatica (L.) Poir.[217]	$=CO\equiv_2CH_2=C_7H_{15}$ $=CH(OH)\equiv_2CH_2=C_7H_{15}$ $H_3C=\equiv_2=_2C_4H_9$	As *O. crocata* + $H_3C=\equiv_2=CH_2CH_2COC_3H_7$ $HOCH_2=\equiv_2=_2C_4H_9$
O. crocata L.[216]	$ROCH_2=\equiv_2=_2CH_2CH_2CH(OH)C_3H_7$ (R = H, Ac) $ROCH_2(CH_2)_2\equiv_2=_2CH_2CH_2CH(OH)C_3H_7$ (R = H, Ac) $ROCH_2(CH_2)_2\equiv_2=CH(OH)C_7H_{15}$ (R = H, Ac) $HOCH_2=\equiv_2=CH(OH)C_7H_{15}$ $H_3C=\equiv_2=_2R$ (R = C_4H_9, C_6H_{13}) $H_3C=\equiv_2CH_2=C_7H_{15}$ $H_3C=\equiv_2=_2CH_2CH_2COC_3H_7$ $H_3C=\equiv_2=(CH_2)_4COC_3H_7$ $H_7C_3\equiv_2=_2CH_2CH_2COC_3H_7$	$RCH_2=\equiv_2=_2C_6H_{13}$ (R = H, OH, OAc) $ROCH_2(CH_2)_2\equiv_2=_2C_6H_{13}$ (R = H, Ac) $HOCH_2=\equiv_2=_2CH_2CH_2CH(OH)C_3H_7$ $HOCH_2(CH_2)_2\equiv_2=_2CH_2CH_2CH(OH)C_3H_7$ $H_3C=\equiv_2=CH(OH)C_5H_{11}$ $H_3C=\equiv_2=_2C_4H_9$

	$HOCH_2{=}{\equiv}_2{=}_2C_6H_{13}$ $HOCH_2(CH_2)_2{\equiv}_2{=}_2C_6H_{13}$ $={\underset{OH}{CH}}{\equiv}_2{=}_2C_6H_{13}$ $={\underset{OH}{CH}}{\equiv}_2{\underset{R}{CH}}{=}C_7H_{15}$ (R = H, OH) $H_3C{=}{\equiv}_2{=}{\underset{OH}{CH}}C_5H_{11}$ $HOCH_2{=}{\equiv}_2CH_2{=}C_7H_{15}$ $H_3C{=}{\equiv}_2{=}_2{\underset{OH}{CH}}C_3H_7$ $HOCH_2(CH_2)_2{\equiv}_2CH_2{=}C_7H_{15}$ $OHC{=}{\equiv}_2{=}{\underset{OH}{CH}}C_7H_{15}$ $HOCH_2{=}{\equiv}_2{=}_2CH_2CH_2COC_3H_7$ $HOCH_2(CH_2)_2{\equiv}_2{=}_2CH_2CH_2COC_3H_7$ $HOCH_2(CH_2)_2{\equiv}_2{=}(CH_2)_4COC_3H_7$	
O. fistulosa L.[35]	$={CO}{\equiv}_2CH_2{=}C_7H_{15}$	$RCH_2{=}{\equiv}_2{=}_2C_6H_{13}$ (R = H, OH, OAc) $={CO}{\equiv}_2CH_2{=}C_7H_{15}$
O. peucedanifolia Pol.[35]	$={CO}{\equiv}_2CH_2{=}C_7H_{15}$	—
O. phellandrium Lam.[35]	$={CO}{\equiv}_2CH_2{=}C_7H_{15}$	$AcOCH_2{=}{\equiv}_2{=}_2C_6H_{13}$
O. pimpinelloides L.[35,210]	$={CO}{\equiv}_2{\underset{R}{CH}}{=}C_7H_{15}$ (R = H, OH) $={\underset{OH}{CH}}{\equiv}_2CH_2{=}C_7H_{15}$ $={CO}{\equiv}_2CO{=}C_7H_{15}$	$ROCH_2{=}{\equiv}_2{=}_2C_6H_{13}$ (R = H, Ac) $ROCH_2(CH_2)_2{\equiv}_2{=}_2C_6H_{13}$ (R = H, Ac) $HOCH_2{=}{\equiv}_2{=}{\underset{OH}{CH}}C_7H_{15}$
O. silaifolia Bieb.[35]	As *O. crocata*	As *O. crocata*
O. stenoloba Schur.[35]	$={CO}{\equiv}_2{\underset{R}{CH}}{=}C_7H_{15}$ (R = H, OH) $={\underset{OH}{CH}}{\equiv}_2CH_2{=}C_7H_{15}$	$ROCH_2{=}{\equiv}_2{=}_2C_6H_{13}$ (R = H, Ac) $ROCH_2(CH_2)_2{\equiv}_2{=}_2C_6H_{13}$ (R = H, Ac) $HOCH_2{=}{\equiv}_2{=}{\underset{OH}{CH}}C_7H_{15}$

TABLE XXXV—Continued

9b. Ammineae-Seselinae	Constituents: Roots	Constituents: Leaves
O. sarmentosa Nutt.[35]	$=CO\equiv_2CH_2=C_7H_{15}$	$=CO\equiv_2CH_2=C_7H_{15}$, $H_3C=\equiv_2=_2C_6H_{13}$ $H_3C=\equiv_2=_2(CH_2)_2COC_3H_7$
Aethusa cynapium L.[257–259]	$H_3C=\equiv_2=_2C_2H_5$, $H_3C=\equiv_2=\underset{OH}{\underset{\vert}{C}}HC_3H_7$, $HOCH_2=\equiv_2=_2C_2H_5$	$H_3C=\equiv_2=_2C_2H_5$, $H_3C=\equiv_2CH_2=C_3H_7$ $RCH_2=\equiv_2=_2C_2H_5$ (R = OH, OAc), $H_3C=_2\equiv=_2C_2H_5$, $H_3C=\equiv_2=\underset{OR}{\underset{\vert}{C}}HC_3H_7$ (R = H, Ac), $H_3C=\equiv_2=COC_3H_7$ $HOCH_2(CH_2)_2\equiv_2=_2C_2H_5$ $RCH_2=\equiv_2=\underset{O}{\bigtriangledown}C_2H_5$ (R = H, OH) $H_3C=\equiv_2\underset{O}{\bigtriangledown}=C_2H_5$ $H_3C=\equiv_2CH_2\underset{O}{\bigtriangledown}=C_3H_7$ $H_3C=_2\equiv=R$ (R = CHO, CH_2OH) $H_3C=\equiv_2=_2CH_2CH_2CH_2OH$ $H_3C=\equiv_2CH_2CH_2\underset{OH}{\underset{\vert}{C}}HC_3H_7$ $H_3C=\equiv_2=CHO$, $H_3C=\equiv_2=CH_2COC_2H_5$ $H_3C=\equiv_2CH_2CH_2COC_3H_7$ $H_3C=\equiv_2COC_3H_7$, $H_3C\underset{OH\ OH}{\sqcap}=\equiv=_2C_2H_5$ $H_5C_2CO=\equiv=_2C_2H_5$, $HOCH_2=\equiv=_2C_2H_5$
Athamanta mattioli Jacq.[35]	—	—
Foeniculum capillaceum Gilib.[35]	$-C\equiv C-$	—
F. vulgare Miller[35]	—	—
Anethum graveolens L.[35]	—	$AcOCH_2=\equiv_2=_2C_6H_{13}$
Kundmannia sicula DC.[35]	$=CO\equiv_2\underset{R}{\underset{\vert}{C}}H=C_7H_{15}$ (R = H, OH)	—

Silaus alpestris Bess.[35]	Trace $-C\equiv C-$ Coumarins 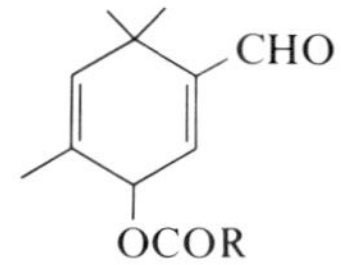	—	
S. besseri DC.[35]	$=CO\equiv_2CH_2=C_7H_{15}$	—	
S. flavescens Bernh.[35]	$=CO\equiv_2CH_2=C_7H_{15}$ 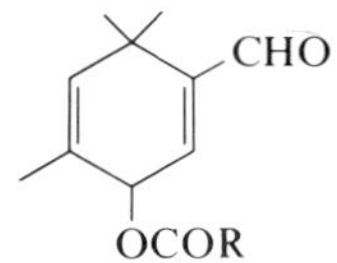	—	
S. pratensis Bess.[35]	$=CO\equiv_2CH_2=C_7H_{15}$ 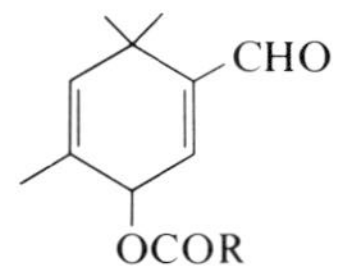	—	
S. tenuifolius DC.[35]	$=CO\equiv_2\underset{R}{CH}=C_7H_{15}$ (R = H, OH) $=CO\equiv_2CO=C_7H_{15}$ 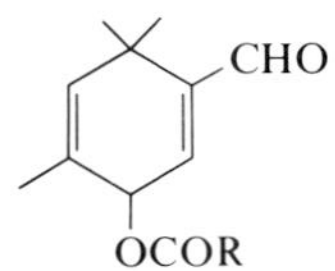	—	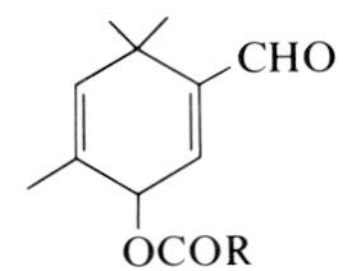
Silaum silaus Schinz. *et* Thell.[35]	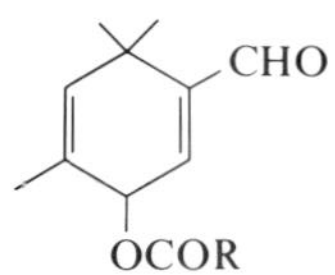 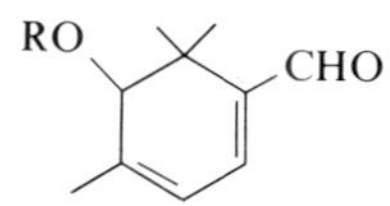	—	
Meum athamanticum Jacq.[35]	$=CO\equiv_2\underset{OH}{CH}=C_7H_{15}$	—	

TABLE XXXV—Continued

9b. Ammineae-Seselinae	Constituents: Roots	Constituents: Leaves	
Cnidium dahuricum Fisch. *et* Mey.[35]	Trace $-C\equiv C-$, 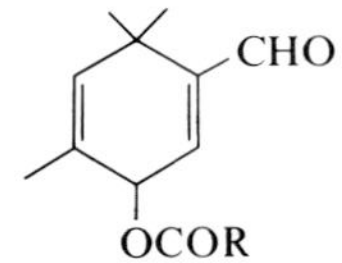	—	
Selinum carvifolia L.[35]	$=CO\equiv_2CH_2=C_7H_{15}$, 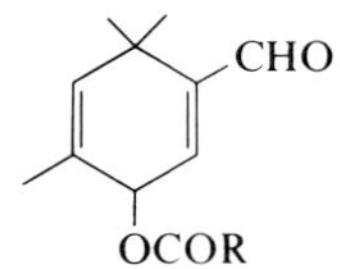	—	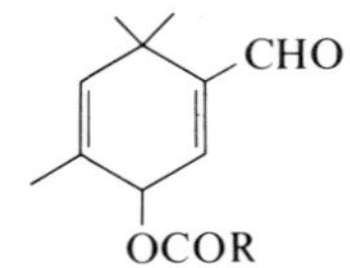
Cenolophium fischeri Koch.[35]	$=CO\equiv_2\underset{\displaystyle OH}{\underset{\mid}{C}}H=C_7H_{15}$, 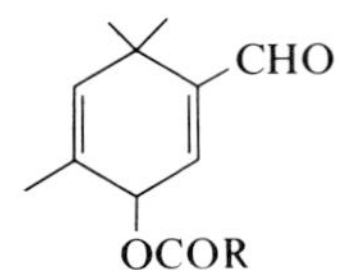	—	
Ligusticum lucidum Mill.[35]	— 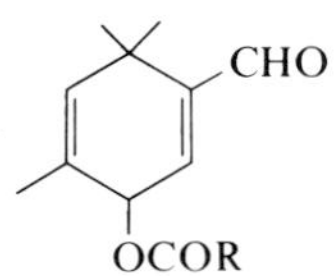 , Coumarins	—	
L. mucronatum Hort.[35]	$=CO\equiv_2\underset{\displaystyle OH}{\underset{\mid}{C}}H=C_7H_{15}$	—	
L. pyrenaicum Koch.[35]	Coumarins	—	
L. scoticum L.[35]	Trace $-C\equiv C-$	—	

	Roots	Leaves
Thaspium aureum Nutt.[35]	$=CO\equiv_2\underset{\mid \atop R}{C}H=C_7H_{15}$ (R = H, OH)	$ROCH_2=\equiv_2=_2C_6H_{13}$ (R = H, Ac) $ROCH_2(CH_2)_2\equiv_2=_2C_6H_{13}$ (R = H, Ac)
T. trifoliatum A. Gray.[35]	$-C\equiv C-$	—
10a. Peucedaneae-Angelicinae	Roots	Leaves
Conioselinum chinense[358]	$-C\equiv C-$	—
Angelica archangelica L.[35]	—	—
A. deculsiva[358]	$-C\equiv C-$	—
A. miqueliana Maxim.[358]	$-C\equiv C-$	—
A. polymorpha Maxim.[358]	$-C\equiv C-$	—
A. purpurascens Gilli.[35]	Coumarins	—
A. sylvestris L.[35]	$-C\equiv C-$	—
A. ursina Regel.[35]	Coumarins	—
Levisticum officinale Koch.[35]	$=CO\equiv_2\underset{\mid \atop OH}{C}H=C_7H_{15}$	—
10b. Peucedaneae-Ferulinae	Roots	Leaves
Capnophyllum dichotomum Lag.[35]	$=\underset{\mid \atop OH}{C}H\equiv_2\underset{\mid \atop OH}{C}H=C_7H_{15}$	Coumarins
C. peregrinum Lange.[35]	$=\underset{\mid \atop OH}{C}H\equiv_2\underset{\mid \atop OH}{C}H=C_7H_{15}$	Coumarins
Astydamia canariensis DC.[35]	$=CO\equiv_2CH_2=C_7H_{15}$	—
A. latifolia Baill.[35]	$=CO\equiv_2\underset{\mid \atop R}{C}H=C_7H_{15}$ (R = H, OH)	$=CO\equiv_2\underset{\mid \atop R}{C}H=C_7H_{15}$ (R = H, OH)
Dorema ammoniacum D. Don.[35]	$R=\equiv_2=_2C_2H_5$ (R = CH_3, $AcOCH_2$) $H_3C=\equiv_2=\underset{\mid \atop OH}{C}HC_3H_7$ Coumarins	$H_3C=\equiv_2=_2C_2H_5$ Coumarins
Ferula assa-foetida L.[35]	$=CO\equiv_2\underset{\mid \atop OH}{C}H=C_7H_{15}$	—

TABLE XXXV—Continued

10b. Peucedaneae-Ferulinae	Constituents: Roots		Constituents: Leaves	
F. communis L.[35]	$=CO\equiv_2\underset{OH}{CH}=C_7H_{15}$	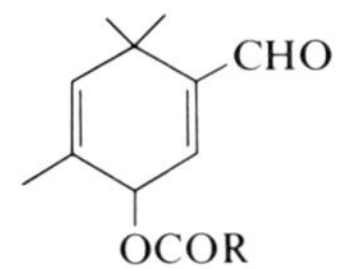	—	
	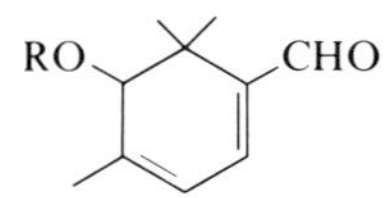			
F. galbaniflua Boiss. *et* Bhs.[35]	Trace $-C\equiv C-$	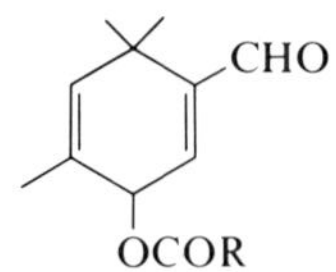		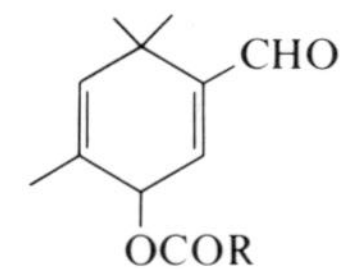
F. hispanica Rony.[35]	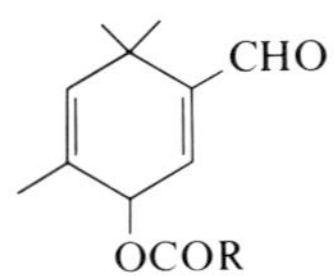		—	
Opopanax chironium Koch.[215]	$=CO\equiv_2CH_2=C_7H_{15}$ $=CO\equiv_2CO=C_7H_{15}$ $=\underset{OH}{CH}\equiv_2\underset{OH}{CH}=C_7H_{15}$		$H_3C=\equiv_2=_2CH_2CH_2COC_3H_7$ $ROCH_2=\equiv_2=_2C_6H_{13}$ (R = H, Ac) $ROCH_2(CH_2)_2\equiv_2=_2C_6H_{13}$ (R = H, Ac) $HOCH_2=\equiv_2=\underset{OH}{C}HC_7H_{15}$ $=\underset{OH}{CH}\equiv_2CH_2=C_7H_{15}$	

		$=\underset{\mid\atop \mathrm{OH}}{\mathrm{CH}}\equiv_2=_2\mathrm{C_6H_{13}}$ $\mathrm{HOCH_2}=\equiv_2\mathrm{CH_2}=\mathrm{C_7H_{15}}$ $\mathrm{H_3C}=\equiv_2=_2\mathrm{CH_2CH_2COC_3H_7}$ $\mathrm{ROCH_2}=\equiv_2=_2\mathrm{C_6H_{13}}$ (R = H, Ac) $\mathrm{ROCH_2(CH_2)_2}\equiv_2=_2\mathrm{C_6H_{13}}$ (R = H, Ac) $\mathrm{HOCH_2}=\equiv_2=\underset{\mid\atop \mathrm{OH}}{\mathrm{C}}\mathrm{HC_7H_{15}}$ $\mathrm{HOCH_2}=\equiv_2=_2\mathrm{CH_2CH_2}\underset{\mid\atop \mathrm{OH}}{\mathrm{C}}\mathrm{HC_3H_7}$
Peucedanum altissimum Desf.[35]	$\mathrm{H_3C}=\equiv_2=_2\mathrm{CH_3}$ $\mathrm{H_3C}=\equiv_2=\underset{\mid\atop \mathrm{OR}}{\mathrm{C}}\mathrm{HC_3H_7}$ (R = H, Ac) $\mathrm{AcOCH_2}=\equiv_2=_2\mathrm{C_2H_5}$	—
P. austriacum (Jacq.) Koch.[35]	as *P. altissimum*	
P. lancifolium Lange.[35]	Trace $-\mathrm{C}\equiv\mathrm{C}-$, 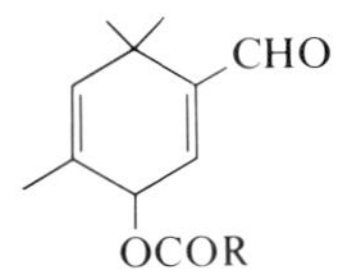	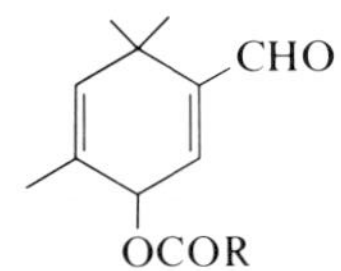
P. officinale L.[35]	$=\mathrm{CO}\equiv_2\underset{\mid\atop \mathrm{OH}}{\mathrm{C}}\mathrm{H}=\mathrm{C_7H_{15}}$	$\mathrm{ROCH_2}=\equiv_2=_2\mathrm{C_6H_{13}}$ (R = H, Ac) $\mathrm{ROCH_2(CH_2)_2}\equiv_2=_2\mathrm{C_6H_{13}}$ (R = H, Ac)
P. carvifolia Vill.[35]	$=\mathrm{CO}\equiv_2\underset{\mid\atop \mathrm{OH}}{\mathrm{C}}\mathrm{H}=\mathrm{C_7H_{15}}$	—
P. vittijugum Boiss.[35]	$=\mathrm{CO}\equiv_2\underset{\mid\atop \mathrm{OH}}{\mathrm{C}}\mathrm{H}=\mathrm{C_7H_{15}}$	—
P. oreoselinum (L.) Moench.[35]	$=\mathrm{CO}\equiv_2\underset{\mid\atop \mathrm{R}}{\mathrm{C}}\mathrm{H}=\mathrm{C_7H_{15}}$ (R = H, OH) Coumarins	Coumarins

TABLE XXXV—Continued

10b. Peucedaneae-Ferulinae	Constituents: Roots		Constituents: Leaves	
P. schottii Bess. *et* DC.[35]	$=CO\equiv_2CH_2=C_7H_{15}$ Coumarins	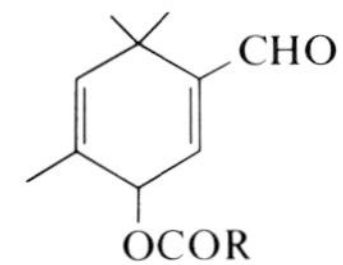	Trace $-C\equiv C-$	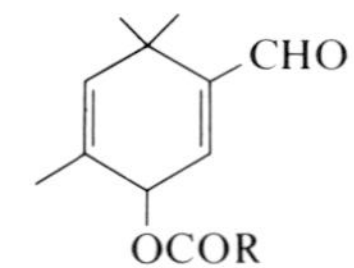
P. ruthenicum Bieb.[35]	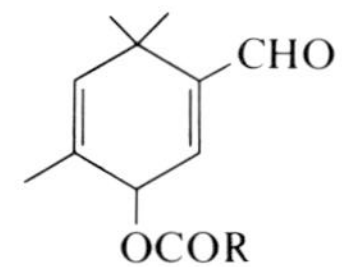		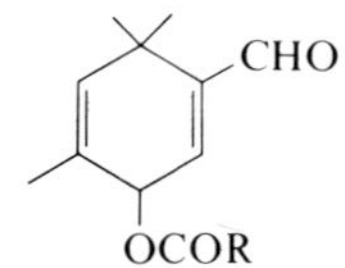	
P. tenuifolium Thunb.[35]	Trace $-C\equiv C-$	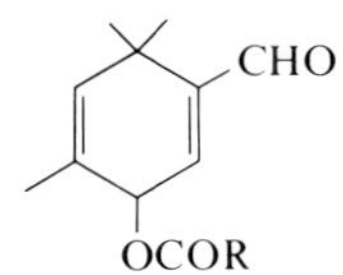	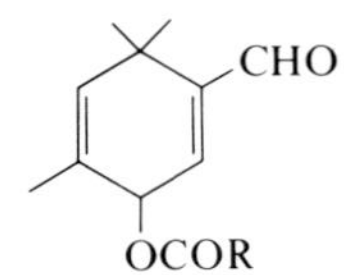	
	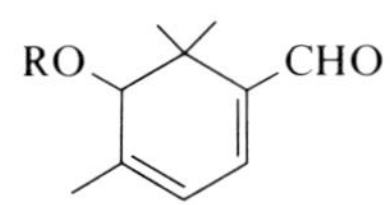			
P. venetum Koch.[35]	Trace $-C\equiv C-$,	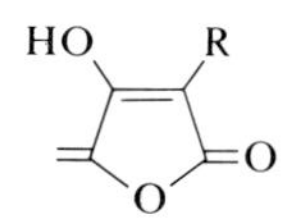	—	
P. verticillatum Rafin.[210]	$H_3C=\equiv_2=_2C_2H_5$		—	

Pastinaca sativa L.[35,190]	$=CO\equiv_2CH(R)=C_7H_{15}$ (R = H, OH) $=CH(OH)\equiv_2CH(OH)=C_7H_{15}$ Coumarins	Seeds: $=CO\equiv_2CH_2=(CH_2)_7CHO$
10c. Peucedaneae-Tordyliinae	Roots	Leaves
Heracleum sphondylium L.[217]	$H_3C=\equiv_2CH(OH)=C_7H_{15}$ $H_7C_3\equiv_2CH(OH)=C_7H_{15}$	—, Coumarins
Tordylium apulum L.[35]	Trace $-C\equiv C-$	Trace $-C\equiv C-$
T. maximum L.[35,358]	$=CO\equiv_2CH_2=C_7H_{15}$	—
Ainsworths trachycarpa Boiss.[35]	Trace $-C\equiv C-$	$=CO\equiv_2CH_2=(CH_2)_7CHO$
11a. Laserpitieae-Silerinae	Roots	Leaves
Siler trilobum Crtz.[35]	$=CH(OH)\equiv_2CH(OH)=C_7H_{15}$	—
11b. Laserpitieae-Elaeoselinae		
11c. Laserpitieae-Thapsiineae	Roots	Leaves
Laserpitium archangelica Wulf.[360]	$H_3C=\equiv_2=_2C_2H_5$ $H_3C=\equiv_2=CH(OR)C_3H_7$ (R = H, Ac) 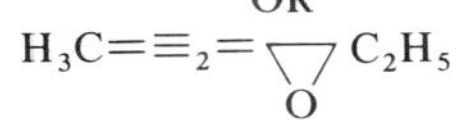 $AcOCH_2=\equiv_2=_2C_2H_5$ $H_3C=\equiv_2CH_2=C_3H_7$, Coumarins	$H_3C=\equiv_2=_2C_2H_5$ Coumarins

TABLE XXXV—Continued

11c. Laserpitieae-Thapsiineae	Constituents: Roots	Leaves
L. hispidum Bieb.[35]	Trace $-C\equiv C-$	—
L. prutenicum L.[361]	$=CO\equiv_2\underset{OH}{CH}=C_7H_{15}$ Sesquiterpenelactones	
Thapsia villosa L.[358]	$-C\equiv C-$	—
Melanoselinum decipiens Hoffm.[35]	—	—
12. Dauceae	**Roots**	**Leaves**
Daucus aureus Desf.[358]	$-C\equiv C-$	—
D. carota L.[214]	$=\underset{OH}{CH}\equiv_2\underset{R}{CH}=C_7H_{15}$ (R = H, OH) $=CO\equiv_2\underset{OH}{CH}=C_7H_{15}$ $=\underset{OAc}{CH}\equiv_2\underset{OH}{CH}=C_7H_{15}$	—
D. crinitus Desf.[358]	$-C\equiv C-$	—
D. guttatus Sibth.[358]	$-C\equiv C-$	—
D. maximus Desf.[35]	$=CO\equiv_2\underset{R}{CH}=C_7H_{15}$ (R = H, OH)	—
D. pusillus Michx.[358]	$-C\equiv C-$	—
Pseudorlaya pumila Grande[35,358]	$=CO\equiv_2CH_2=C_7H_{15}$	—

Surveying the distribution of acetylenes in the different investigated genera, falcarinone and its derivatives are the most widespread. However, in several genera other C_{17}-acetylenes are also predominant: *Cicuta*, *Oenanthe*, *Azorella* and *Heracleum*, while *Aethusa* has only a large variety of C_{13}-acetylenes.

Furthermore, the concentrations of acetylenes are very different. A few species are really rich in compounds of this type, whereas normally the concentrations are low. The relationship of the isolated acetylenes to those of other families seems to be a clear indication that the Umbelliferae are closely related to Araliaceae, to the Compositae and very probably to the Pittosporeae. All these families contain very similar acetylenes. A typical difference between Compositae and Umbelliferae, however, is the fact that in the former the C_{17}-acetylenes always have a vinyl endgroup saturated in the compounds existing in Umbelliferae.

3. Other Families

The occurrence of acetylenes in other families has not been very well investigated up to now, as shown below.

3.1. Araliaceae and Pittosporaceae

Only a few members have been investigated from the Araliaceae. As in the Umbelliferae falcarinone is again the most common constituent and sometimes occurs together with derivatives of this widespread ketone.

The family Pittosporaceae seems to be closely related to Araliaceae and Umbelliferae, although this family is not placed close to the latter two in

TABLE XXXVI
ARALIACEAE

	Constituents
Aralia californica S. Wats.[35]	$=CO\equiv_2\underset{OH}{CH}=C_7H_{15}$, $=CO\equiv_2CO=C_7H_{15}$
A. elata Seem.[35]	$=CO\equiv_2CH_2=C_7H_{15}$
A. mandschurica Seem.[35]	$=CO\equiv_2CH_2=C_7H_{15}$, $=CO\equiv_2\underset{OH}{CH}=C_7H_{15}$
A. racemosa L.[35]	$=CO\equiv_2CH_2=C_7H_{15}$, $=CO\equiv_2\underset{OH}{CH}=C_7H_{15}$
Hedera helix L.[210]	$=CO\equiv_2CH_2=C_7H_{15}$
Panax Ginseng C. A. Mey.[211]	$=CO\equiv_2CH_2=C_7H_{15}$
Polyscias fructicosa Harms.[35]	$=CO\equiv_2CH_2=C_7H_{15}$

TABLE XXXVII
PITTOSPORACEAE

	Constituents
Pittosporum buchanani Hook.[230]	$H_3C{=}{\equiv}_2{=}C_4H_9$, $H_3C{=}{\equiv}_2CH_2{=}C_5H_{11}$ $=CO{\equiv}_2{=}_2C_4H_9$, $=\underset{\mid\atop OH}{C}H{\equiv}_2{=}_2C_4H_9$ $=\underset{\mid\atop OH}{C}H{\equiv}_2CH_2{=}C_7H_{15}$

several plant systematics. Only *Pittosporum buchananii* has been investigated up to now and so a generalization about the acetylenic constituents is not possible. However, this species clearly indicates the close relationship between Araliaceae and Umbelliferae, as falcarinol is again present together with some C_{15}-acetylenes which are very similar to those of the families mentioned.

3.2. Campanulaceae

Although hitherto only primarily investigations of this family have been carried out, they indicate that acetylenes seem to be relatively common. The tetrahydropyranylethers, not previously isolated in the Compositae are the most widespread. The open chain precursors have been isolated together with these heterocyclic acetylenes. The whole number of acetylenes in this family appears to be more uniform as in the case of the Compositae to which the Campanulaceae are closely related botanically.

TABLE XXXVIII
CAMPANULACEAE

	Constituents
Adenophora lilifolia (L.) Fisch.[35]	$-C{\equiv}C-$
A. potaninii Korsh.[35]	$-C{\equiv}C-$
Campanula alliariaefolia Willd.[35]	$-C{\equiv}C-$
C. carpatica Jacq.[240]	$-C{\equiv}C-$
C. cochlearifolia Lam.[240]	$-C{\equiv}C-$
C. fenestrellata Feer.[240]	$-C{\equiv}C-$
C. garganica Ten.[240]	$-C{\equiv}C-$
C. glomerata L.[240]	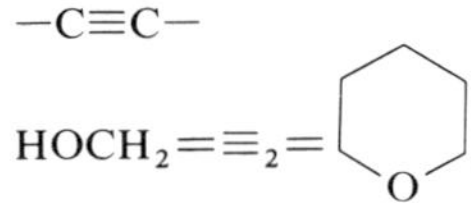, 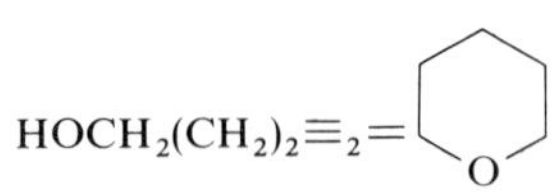, $H_3C{=}{\equiv}_2{=}CHO$

TABLE XXXVIII—Continued

	Constituents
C. isophylla Mor.[240]	—C≡C—
C. istriaca Feer.[240]	—C≡C—
C. lactiflora Bieb.[240]	—C≡C—
C. latiloba DC.[240]	—C≡C—
C. latifolia L.[240]	—C≡C—
C. medium L.[240]	—C≡C—
C. persicifolia L.[240]	—C≡C—
C. portenschlagiana Roem. *et* Schult.[240]	—C≡C—
C. porscharskyana Degen.[240]	—C≡C—
C. pyramidalis L.[240]	—C≡C—
C. rapuncoloides L.[240]	—C≡C—
C. rotundifolia L.[240]	—C≡C—
C. sarmatica Ker.[240]	—C≡C—
C. trachelium L.[240]	—C≡C—
C. vidalii H. C. Wats.[240]	—C≡C—
Codonopsis clematidea C. B. Clarke[240]	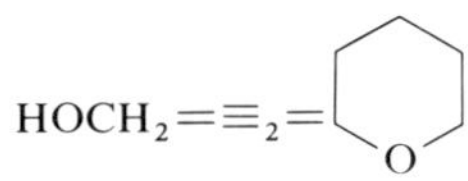, 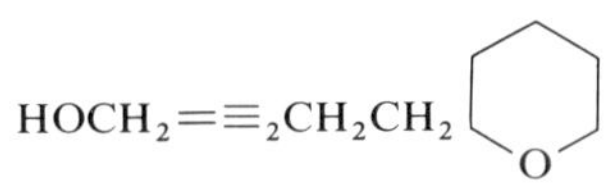
C. ovata Benth.[240]	—C≡C—
Jasione perennis Lam.[240]	—C≡C—
Lobelia cardinalis L.[240]	—C≡C—
L. erinus L.[240]	—C≡C—
Platycodon grandiflorum DC.[240]	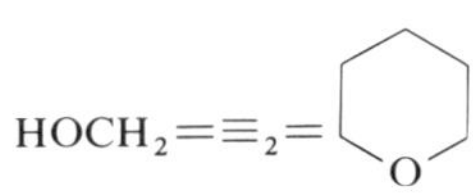, 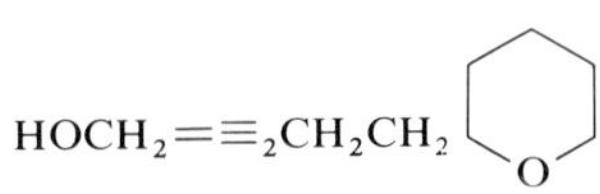
Symphyandra wanneri Heuff.[240]	—C≡C—
Trachelium caeruleum L.[35,240]	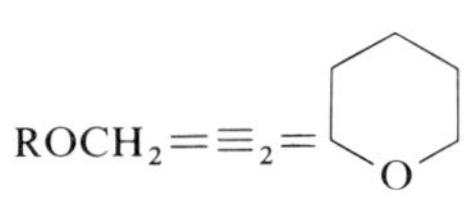 (R = H, Ac) 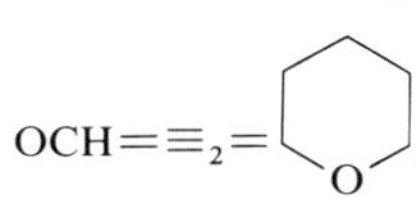, 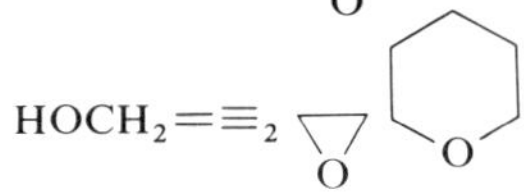

TABLE XXXVIII—Continued

	Constituents
Wahlenbergia gracilis DC.[35]	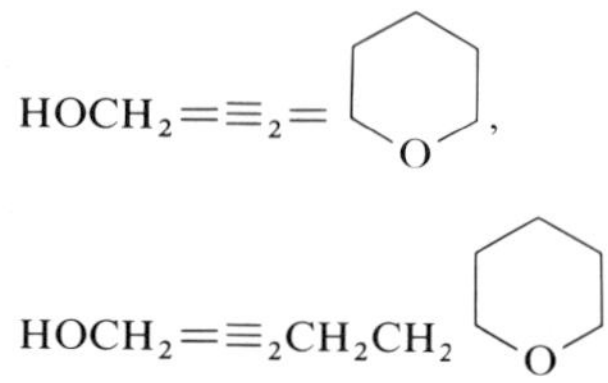

3.3. Further Families of the Higher Plants

In Table XXXIX families are listed which contain acetylenic fatty acids in their triglycerides. Up to now, compounds of this type have been isolated from six different families, more or less closely related botanically.

In the families listed next some very different acetylenes have been isolated, showing no relationship to other families.

The occurrence of acetylenes in the family Anonaceae, its structures being not fully established, is of special interest as this family is closely related to Magnoliaceae which is very old. Some botanists have made the assumption that the Compositae have developed from these families. Although the structures of the Anona-acetylenes are not closely related to those of Compositae, the isolation of acetylenes in this old family is perhaps of phylogenetic importance.

The occurrence of simple polyynes in Euphorbiaceae and Valerianaceae has to be confirmed, since only very small amounts of these acetylenes have been isolated from large quantities of plant material. Therefore, it cannot be excluded that perhaps some other plants may be the origin of the described acetylenes.

TABLE XXXIX

Simarubaceae	Constituents
Picramnia capinterae Polak[362]	$H_3C(CH_2)_{10}{\equiv}(CH_2)_4CO_2R$
P. camboita Engl.[362]	$H_3C(CH_2)_{10}{\equiv}(CH_2)_4CO_2R$
P. lindeniana Tul.[362]	$H_3C(CH_2)_{10}{\equiv}(CH_2)_4CO_2R$
P. tariri DC.[2,3]	$H_3C(CH_2)_{10}{\equiv}(CH_2)_4CO_2R$

TABLE XXXIX—Continued

Santalaceae	Constituents
Acanthosyris spinescens Griseb.[323]	$=(CH_2)_6\equiv(CH_2)_7CO_2R$ $C_6H_{13}=\equiv(CH_2)_nCO_2R$ (n = 6, 7) $=(CH_2)_4=\equiv(CH_2)_nCO_2R$ (n = 6, 7) $C_6H_{13}=\equiv CH(OH)(CH_2)_nCO_2R$ (n = 5, 6) $=(CH_2)_4=\equiv CH(OH)(CH_2)_nCO_2R$ (n = 5, 6)
Anthobolus leptemerioides R. Br.[363]	$=\equiv_2\cdots CO_2R$
Choretrum glomeratum R. Br.[363]	$C_4H_9=_2\equiv(CH_2)_7CO_2RC_6H_{13}=\equiv(CH_2)_7CO_2R$
C. pauciflorum A.D.C.[363]	$C_5H_9=_2\equiv(CH_2)_7CO_2RC_6H_{13}=\equiv(CH_2)_7CO_2R$
Comandra richardsiana Fern.[363]	$=\equiv_2\cdots CO_2R$
Eucarya acuminata Spr. *et* Summ.[364]	$=\equiv\cdots CO_2R$
E. murrayana Spr. *et* Summ.[364]	$=\equiv\cdots CO_2R$
E. spicata Spr. *et* Summ.[364]	$=\equiv\cdots CO_2R$
Exocarpus aphyllus R. Br.[363]	$C_4H_9=\equiv_2(CH_2)_7CO_2R$
E. cupressiformis Lab.[363,315]	$C_4H_9=\equiv_2(CH_2)_7CO_2R$ $H_3C(CH_2)_7\equiv(CH_2)_7CO_2R$
E. humifusus R. Br.[363]	$C_4H_9=\equiv_2(CH_2)_7CO_2R$
E. nanus Hook.[363]	$C_4H_9=\equiv_2(CH_2)_7CO_2R$
E. sparteus R. Br.[363]	$C_4H_9=\equiv_2(CH_2)_7CO_2R$
E. strictus R. Br.[363]	$C_4H_9=\equiv_2(CH_2)_7CO_2R$
Henslowia queenslandiae L. S. Smith[363]	$=_2\equiv_2$ and $=\equiv\cdots CO_2R$
Leptomeria acida R. Br.[363]	$=_2\equiv$, $=\equiv$, $C_4H_9=_2\equiv_2(CH_2)_7CO_2R$
L. aphylla R. Br.[363,364]	$C_4H_9=\equiv_2(CH_2)_7CO_2R$, $C_4H_9=_2\equiv_2(CH_2)_7CO_2R$
L. billandierii R. Br.[363]	$C_4H_9=_2\equiv(CH_2)_7CO_2R$ $C_2H_5=\equiv_3(CH_2)_7CO_2R$
Omphacomeria acerba A.D.C.[363]	$=\equiv_2\cdots CO_2R$
Pyrularia pubera Michx.[363,315]	$=(CH_2)_4=\equiv CH(CH_2)_6CO_2R$ (with R on CH) (R = H, OH) $H_3C(CH_2)_7\equiv(CH_2)_7CO_2R$
Santalum album L.[366]	$C_6H_{13}=\equiv(CH_2)_7CO_2R$
S. accuminatum DC.[363,315,319]	$C_6H_{13}=\equiv(CH_2)_7CO_2R$ $C_4H_9=_2\equiv(CH_2)_7CO_2R$ $C_4H_9=\equiv_2(CH_2)_7CO_2R$ $C_2H_5=_2\equiv_2(CH_2)_7CO_2R$ $C_2H_5=\equiv_3(CH_2)_7CO_2R$ $=_2\equiv_3(CH_2)_7CO_2R$
S. lanceolatum R. Br.[363]	$C_4H_9=\equiv_2(CH_2)_7CO_2R$
S. murrayana F.v.M.[365,363]	$C_4H_9=\equiv_2(CH_2)_7CO_2R$
S. freycinetianum Gaud.[363]	$C_4H_9=\equiv_2(CH_2)_7CO_2R$
Thesium australe R. Br.[363]	$==_2$ and $=_2=_2\cdots CO_2R$

Olacaceae	Constituents
Olax stricta R. Br.[363]	$=_2\equiv_2\cdots CO_2R$
Onguekoa klaineana Pierre[321]	$C_4H_9=\equiv_2(CH_2)_7CO_2R$, $=(CH_2)_2=\equiv_2(CH_2)_7CO_2R$, $C_6H_{13}\equiv_2(CH_2)_7CO_2R$, $=(CH_2)_4\equiv_2(CH_2)_7CO_2R$ $C_6H_{13}\equiv_2CH(CH_2)_6CO_2R$, (with OH on CH)

TABLE XXXIX—Continued

Olacaceae	Constituents
	$=(CH_2)_4\equiv_2\underset{\mid\atop OH}{CH}(CH_2)_6CO_2R$
	$=CH_2CH_2=\equiv_2\underset{\mid\atop OH}{CH}(CH_2)_6CO_2R$
Ximenia americana L.[363,316]	$C_6H_{13}=\equiv(CH_2)_7CO_2R$ $C_4H_9=_2\equiv(CH_2)_7CO_2R$ $C_4H_9=\equiv_2(CH_2)_7CO_2R$
X. americana microphylla[316]	$C_6H_{13}=\equiv(CH_2)_7CO_2R$
X. caffra Sond.[316]	$C_6H_{13}=\equiv\underset{\mid\atop R}{CH}(CH_2)_6CO_2R$ (R = H, OH) $C_4H_9=_2\equiv(CH_2)_7CO_2R$
X. caffra natalensis[316]	$C_6H_{13}=\equiv(CH_2)_6CO_2R$

Loranthaceae	Constituents
Nuytsia floribundus R. Br.[363,315]	$C_6H_{13}=\equiv(CH_2)_7CO_2R$, $C_4H_9=_2\equiv(CH_2)_7CO_2R$
Viscum album L.[363]	$=_2\equiv_2\cdots CO_2R$

Opiliceae	Constituents
Canjera leptostachya Benth.[363]	$=\equiv_2\cdots CO_2R$
Opilia amentaceae Roxb.[363]	—

Sterculiaceae	Constituents
Sterculia foetida L.[325]	$H_3C(CH_2)_7\equiv(CH_2)_7CO_2R$
S. alata Roxb.[325]	$\equiv(CH_2)_7$  $(CH_2)_6CO_2R$

Lauraceae	Constituents
Litsea odonifera Valet.[337]	$H_3C\overset{OCH_3\atop\mid}{C}H(CH_2)_nC\equiv CH$ (n = 7, 9)
Persea gratissima Gaertn.[338]	$\equiv(CH_2)_{11}$ , $\equiv(CH_2)_{11}COCH_2\underset{\mid\atop OH}{C}HCH_2OAc$, $\equiv(CH_2)_{11}\underset{\mid\atop OH}{C}HCH_2\underset{\mid\atop OH}{C}HCH_2OR$ (R = H, Ac)

TABLE XXXIX—Continued

Papilionaceae	Constituents
Vicia faba L.[242]	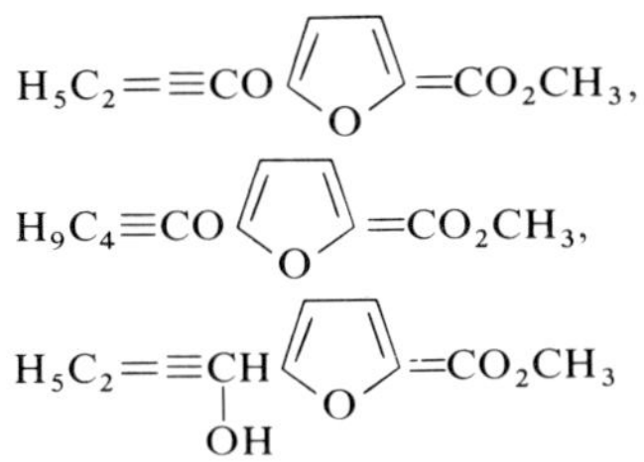

Myoporaceae	Constituents
Eromophilia freelingii[328]	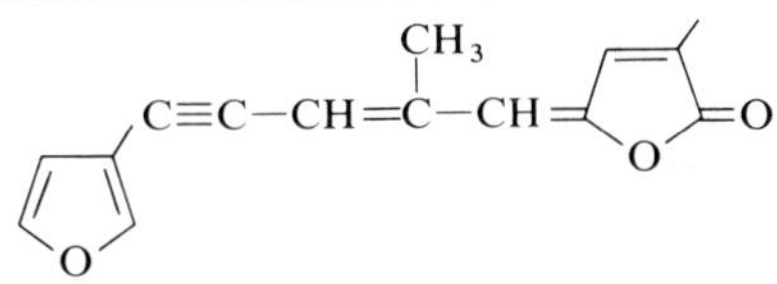

Sapindaceae	Constituents
Euphoria longan[343]	$HC{\equiv}CCH(CH_2R)CH_2CH(NH_2)CO_2H$ (R = H, OH) $HC{\equiv}CCH(OH)CH_2CH(NH_2)CO_2H$

Cupressaceae	Constituents
Chamaecyparis formosensis Matsum.[329]	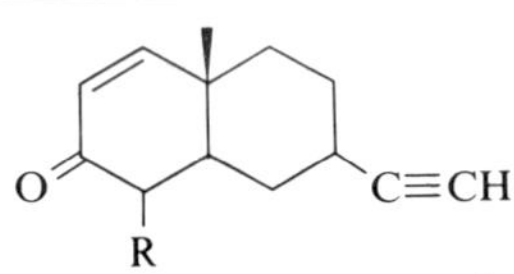 (R = α-CH_3, β-CH_3) 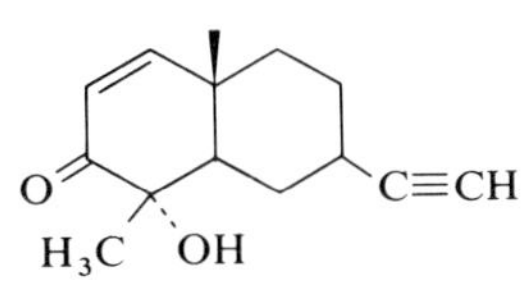 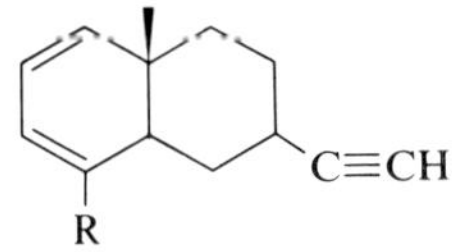 (R = CH_2OH, CHO) 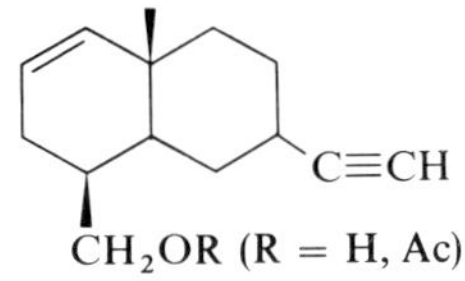

TABLE XXXIX—Continued

Annonaceae	Constituents
Alphonsea ventricosa[35]	$=(CH_2)_n=\equiv_2(CH_2)_n,\underset{OH}{CH}CH_2\underset{OH}{CH}CH_2OH$ (C_{21})

Euphorbiaceae	Constituents
Ricinus communis L.[367]	$H_3C\equiv_3=CO_2CH_3$ (?)

Valerianaceae	Constituents
Valeriana officinalis L.[368]	$H_3C\equiv_5=$ (?)

3.4. Algae and Marine Organisms

Up to now only a few organisms have been investigated for acetylenes. However already now several interesting acetylenic compounds are found, mainly carotenoids, but from *Laurentia* species also some interesting bromo-compounds with ethynyl endgroups.

TABLE XL
ALGAE

Rhodomelaceae	Constituents	
Laurentia nipponica Yamada[341]	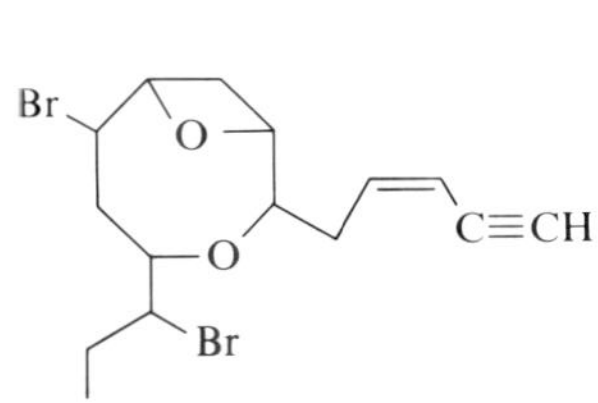	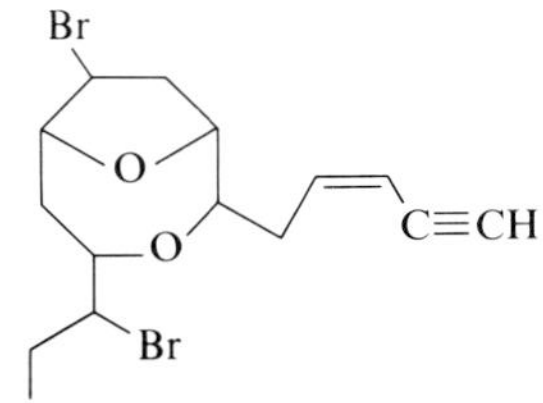
L. glandulifera Kutzing[341]	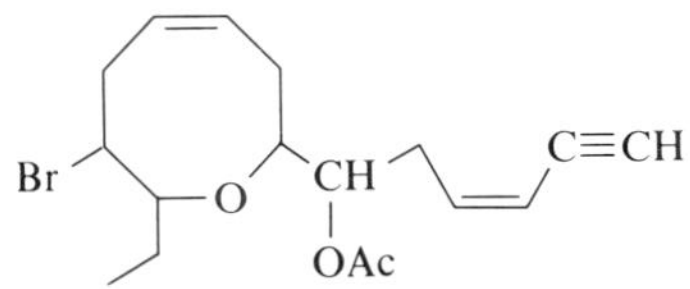	

Cryptophyceae	Constituents
Cryptomonas ovata var. palustris[332]	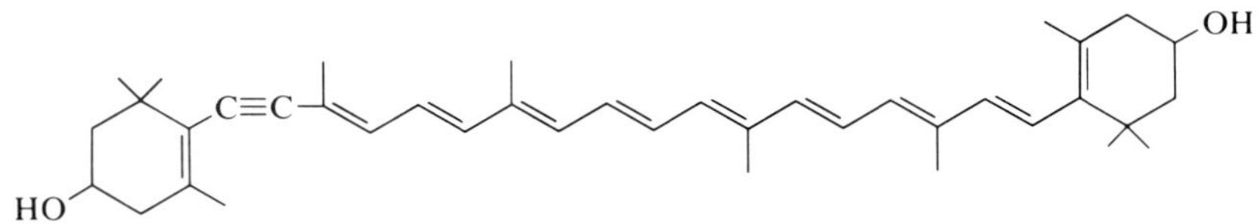
Rhodamonas Strain DC.[332]	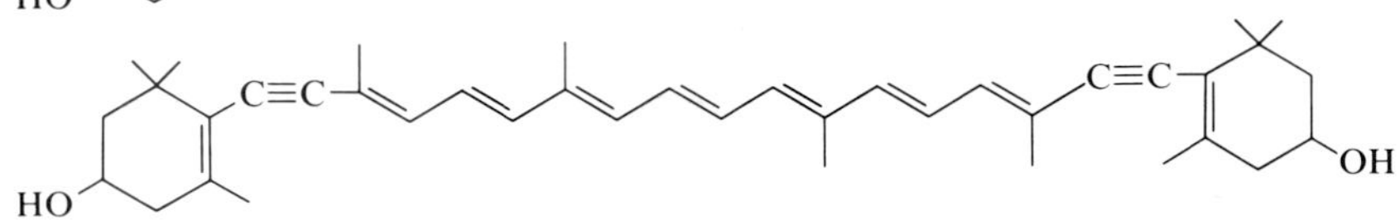
Hemiselmis virescens[332]	and others

TABLE XL—Continued

Marine organism	Constituents
Asterias rubeus[336]	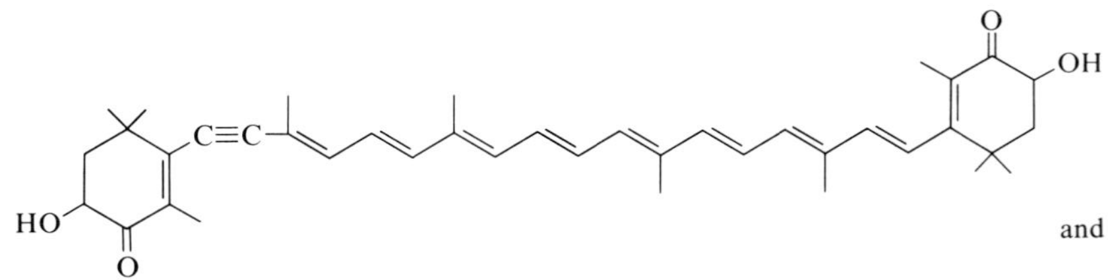 and others
Pecten maximus[333]	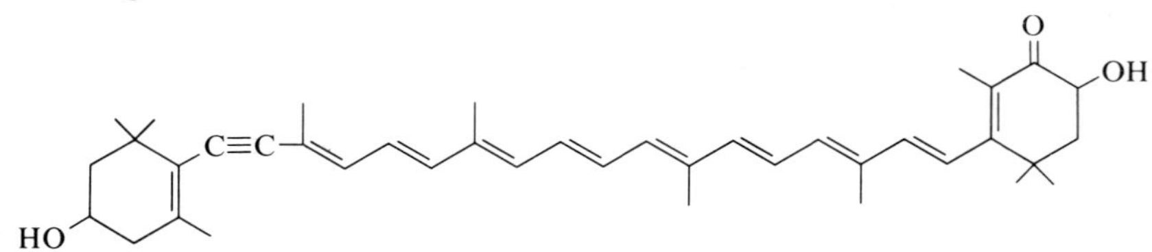
Halocynthia papillosa[333] *Mytilus edulis*[333]	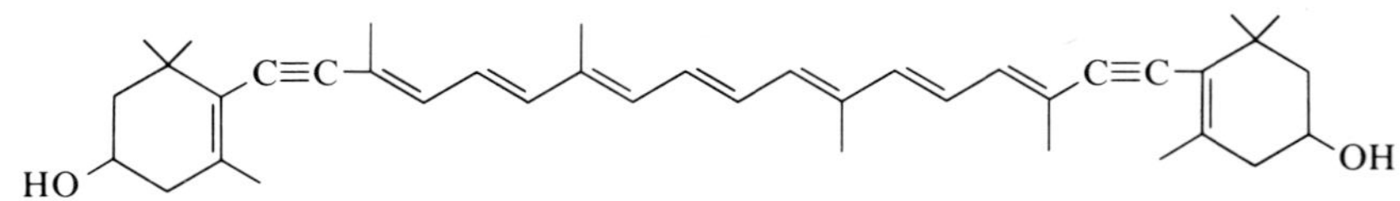
Dendrobates histrionicus[344]	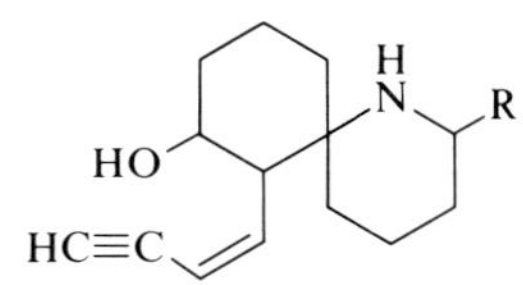 (R = $CH_2{=}C{\equiv}CH$, $CH_2CH_2{=}{=}$)

3.5. Micro-organisms

So far only one member of yeasts containing an acetylene is known and the same is true of bacteria. However, in the culture medium (and sometimes in the mycel of fungi) a large number of acetylenes have been isolated. Nearly all these fungi belong to the Basidiomycetes, especially to families of the orders Agaricales and Aphyllophorales. The only exceptions are Populaspora and Nocardia (Actinomytes) and Helminthosporium (*fungi imperfecti*). The latter contains an acetylene which is completely different from all others. Clear common botanical relationships have not yet been established but special acetylenes are nearly always present in some genera and are therefore typical. An example is the fully investigated genus *Clitocybe*.

TABLE XLI

MICRO-ORGANISMS

Yeast	Constituents
Rhodotorula glutinus var. lusitanica[326]	$HC{\equiv}C(CH_2)_8CO_2H$

Micro-organism	Constituents
Helminthosporium siccans g-207[327]	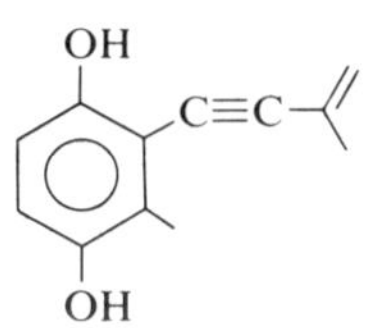
Bacterium coli[339]	$HC{\equiv}CCO_2H$
Streptomyces griseus[340]	$H_2NCOC{\equiv}CCONH_2$
S. chibaensis[340]	$H_2NCOC{\equiv}CCONH_2$
S. reticuli aquamyceticus[340]	$H_2NCOC{\equiv}CCONH_2$

Basidiomycetes	Constituents
Agrocybe dura[369]	$HOCH_2{\equiv}_3CONH_2$
Aleurodiscus roseus (Per. ex Fr.) v. Höhn. *et* Litsch.[175]	$H{\equiv}_3CH{-}CHCH_2OH$ (OH OH), $H{\equiv}_2{=}{\cdot}{=}CH_2CO_2R$ ($R = H, CH_3$) $H{\equiv}_2{=}{\cdot}{=}CH_2CH_2OH(+)$, $H{\equiv}_2{=}CH{-}CH{-}CH_2OH$ (OH OH), $H{\equiv}_2{=}(CH_2)_3OH$, $HOCH_2{=}{\equiv}_2CH_2CH_2OH$, $H_3C{\equiv}_3{=}R$ (R = CHO, COOH, CH_2OH) $H{\equiv}_2{=}{\cdot}{=}CH(CH_2)_2CO_2H$ (OH)

TABLE XLI—Continued

Basidiomycetes	Constituents
A 67[22]	$H{\equiv}_2{=}{=}CH(OH)(CH_2)_2CO_2H$, 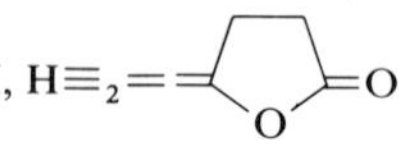
B 244[22]	$H_3CO_2C{=}{\equiv}_2(CH_2)_2CO_2CH_3$ $H_3CO_2C{=}{\equiv}_2(CH_2)_2CH_2OH$ $H_3C{=}{\equiv}_2CH_2CH_2CO_2CH_3$ $HOCH_2{=}{\equiv}_2(CH_2)_3OH$, $H_7C_3{\equiv}_2{=}CH_2OH$
B 285[146]	$H{\equiv}_2{=}{=}(CH_2)_3OH$, $H{\equiv}_2{=}{=}CH(OH)(CH_2)_3OH$
B481[139,370]	$H{\equiv}_2{=}{=}(CH_2)_4OH$, $H{\equiv}_2{=}{=}CH(OH)(CH_2)_2CO_2H$, 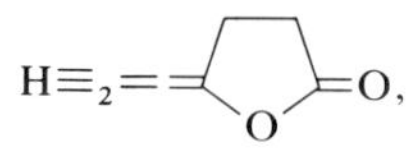 $H_3C{\equiv}_2{=}{=}CH(OR)(CH_2)_2CO_2H$ (R = H, xylose), 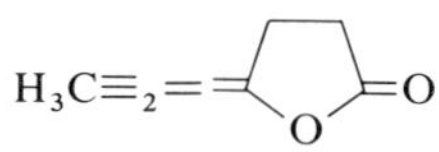
Clitocybe candida[172]	$H_3C{\equiv}_3{=}CH_2OR$ (R = H, CH_3)
C. cerussata (Fr.) Quel.[372]	$CC{\equiv}_2{=}CO_2H$, $H_2NCO{\equiv}_2{=}CH_2OH$, $HOCH_2{\equiv}_3{=}CO_2H$
C. cyathiformis[22]	$H_3C{\equiv}_3{=}CHO$
C. dealbata[153]	$HOCH_2{\equiv}_3{=}CO_2H$, $H_3C{\equiv}_3{=}CO_2H$
C. diatreta (Fr.) Quel.[371,372]	$R{\equiv}_2{=}R'$ (R = CN, R' = CO_2H, R = H_2NCO, R' = CH_2OH) $HOCH_2{\equiv}_3{=}CO_2H$
C. fragrans (Fr.) Quel.[22]	$NC{\equiv}_2{=}CO_2H$, $HOCH_2{\equiv}_3{=}CO_2H$
C. gallinacea (Fr.) Gillet.[373]	$NC{\equiv}_2{=}CO_2H$, $H_2NCO{\equiv}_2{=}CH_2OH$, $HOCH_2{\equiv}_3{=}CO_2H$
C. inversa (Fr.) Quel.[22]	$NC{\equiv}_2{=}CO_2H$, $HOCH_2{\equiv}_3{=}CO_2H$
C. nebularis (Fr.) Quel.[172]	$HOCH_2{\equiv}_3{=}CO_2H$
C. obbata[22]	$H_3C{\equiv}_3{=}R$ (R = CHO, CH_2OH)
C. rhizophora Velen.[374]	$H_5C_2CO{\equiv}_2CH(OH)CH_2OH$, $H_5C_2CH(OH){\equiv}_2CH(OH)CH_2OH$, 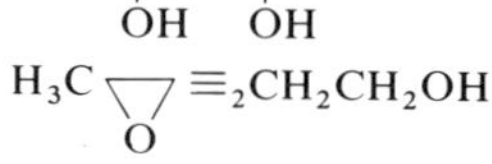
G. rivulosa (Fr.) Quel.[22]	$HOCH_2{\equiv}_3{=}CO_2H$
C. robusta Peck.[22]	$HOCH_2{\equiv}_3{=}CO_2H$, $H_3CO_2C{=}{\equiv}_2{=}CO_2CH_3$
C. truncicola (Fr.) Quel.[22]	$HOCH_2{\equiv}_3{=}CO_2H$, $H_3C{\equiv}_3{=}R$ (R = CO_2CH_3, CHO, CH_2OH)

TABLE XLI—Continued

Basidiomycetes	Constituents
Coprinus quadrifidus[174]	$H{\equiv}_3{=}R$ (R = CHO, CH_2OH), 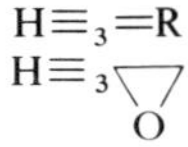 CH_2OH, $H{\equiv}_3CH(OH){-}CH(OH)CH_2OH$, $H{\equiv}_2{=}CH(OH){-}CH(OH)CH_2OH$, $OCH{\equiv}_3{=}CH_2OH$, $HOCH_2{\equiv}_3{=}CH_2OH$
Cortinellus berkeleyanus Ito and Imai[146]	$H{\equiv}_2{=}{=}CH_2OH$, (−) $H{\equiv}_2{=}{=}CH_2CH_2OH$, $H_3C{\equiv}_3{=}CH_2CO_2CH_3$
Daedalea juniperina (Murr.)[146,181,183]	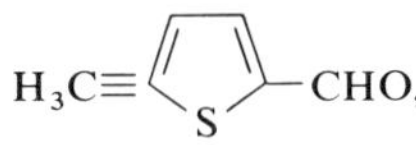, $H_3C{\equiv}_2{=}{=}CH_2CH_2OH$, 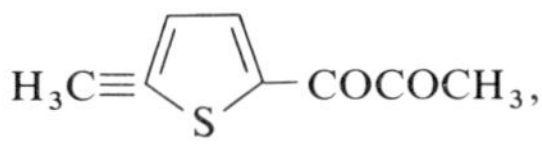, 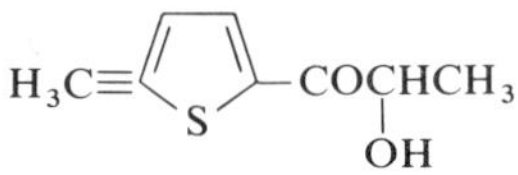
Drosophila sarcocephala (Fr.) Quel.[22]	$HOCH_2{\equiv}_3{=}CO_2H$
D. simivesta[22]	${=}{=}{\equiv}_2{=}CH_2CO_2H$
D. (*Psathyrella*) *subatrata*[147]	$H{\equiv}_2{=}CH_2CH_2CO_2H$, $HOCH_2{=}{\equiv}_2CH_2CO_2H$, $H{\equiv}CH_2{\equiv}_2{=}CH_2CO_2H$, ${=}{=}{\equiv}_2{=}CH_2CO_2H$, $HO_2C{=}{\equiv}_2(CH_2)_3CO_2H$, $H{\equiv}_2{=}{=}{=}_2CH_2CO_2H$
Fistulina hepatica (Huds.) Fr.[138]	$H_3C{\equiv}_3{=}CH_2OH$, $H_3C{\equiv}_4C_4H_9$, $H_3C{\equiv}_4CH(OH){-}CH(OH){-}CH(OH)CH_2OH$, $HOCH_2{-}{\equiv}_3C_4H_9$
F. pallida (Berk. *et* Rav.)[146]	$H_7C_3{\equiv}_3{=}CH(OH)CH_2OH$
Flammula sapinea Fr.[146]	$H{\equiv}_2{=}{=}(CH_2)_4OH$
Fomes annosus[180]	$H{\equiv}_3H$
Lepista diemii[22]	$NC{\equiv}_2{=}CO_2H$, $H{\equiv}_3{=}CO_2H$, $HOCH_2{\equiv}_3{=}CO_2H$
Leptoporus kymantodes[153]	$H_3C{\equiv}_2{=}CH_2OH$
Marasmius ramealis[176]	(−) $H{\equiv}_2{=}{=}CH_2CH_2OH$
Merulius lacrymans[153]	$HO_2C{\equiv}_3CH_2CH_2CO_2H$, $H{\equiv}_2{=}{=}CH(OH)CH_2CH_2OH$, $H{\equiv}_2{=}{=}(CH_2)_3OH$, $H{\equiv}_2{=}{=}CH(OH)(CH_2)_3OH$, $H{\equiv}_2{=}{=}(CH_2)_4OH$
Nocardia acidophilus[135]	$H{\equiv}_2{=}{=}{=}_2CH_2CO_2H$

TABLE XLI—Continued

Basidiomycetes	Constituents
Odontia bicolor[146]	$H{\equiv}_2{=}{=}\underset{OH}{C}HCH_2CH_2OH$, $H{\equiv}_2{=}{=}(CH_2)_3OH$, $H{\equiv}_2{=}{=}\underset{OH}{C}H(CH_2)_3OH$, $H{\equiv}_2{=}{=}(CH_2)_4OH$
Omphalia umbilicata (Fr.) Quel.[22]	$NC{\equiv}_2{=}CO_2H$, $H_2NCO{\equiv}_2{=}CH_2OH$
Papulospora polyspora[22]	$H_3C{\equiv}_3{=}CH_2OH$
Polyporus anthracophilus[307,373]	$HO_2C{=}{\equiv}{=}CO_2H$, $H_3CO_2C{=}{\equiv}_2{=}CO_2CH_3$, $R{=}{\equiv}_2{=}R$ ($R = CO_2H$, $R = CH_2OH$), $H_3C{=}{\equiv}_2{=}R$ ($R = CO_2H$, CO_2CH_3, CH_2OH, $CO_2CH_2{=}{\equiv}_2{=}CH_3$, $CO_2(CH_2)_3{\equiv}_2{=}CO_2CH_3$), $HO_2C{=}{\equiv}_2(CH_2)_3OH$, $H_3CO_2C{=}{\equiv}_2(CH_2)_3OH$, $HOCH_2{=}{\equiv}_2(CH_2)_3R$ ($R = H, OH$)
P. biformis[174,374]	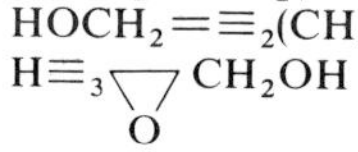
P. guttulatus[153]	$HO_2C{=}{\equiv}_2{=}CO_2H$, $H_3C{=}{\equiv}_2{=}CH_2OH$
Pleurotus ulmarius (Fr.) Quel.[153]	$H_3C{\equiv}_3{=}R$ ($R = CO_2H$, CHO, CH_2OH)
Poria colorea Overh. *et* Englerth.[375]	$H{\equiv}_2{=}{=}\underset{OH}{C}HCH_2CH_2CO_2H$, 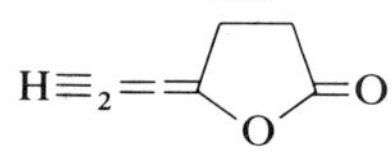
P. corticola[140,370]	$H{\equiv}_2{=}{=}\underset{OH}{C}HCH_2CH_2CO_2H$, 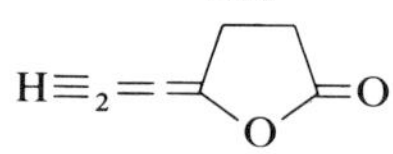
P. mutans Pk.[375]	$H{\equiv}_2{=}{=}\underset{OH}{C}HCH_2CH_2CO_2H$, 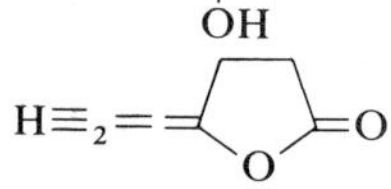
P. selecta Karst. *ex* Rom.[375]	$HOCH_2{=}{\equiv}_2CH_2CO_2H$
P. sinuosa Fr.[134]	$H{\equiv}_3{=}R$ [$R = CO_2H$, CH_2OH, $CONHCH(CO_2H)CH(CH_3)_2$] $HOCH_2{=}{\equiv}_2CH_2CH_2OH$, $HO_2C{\equiv}_3{=}CO_2H$, $HOCH_2{\equiv}_3{=}CH_2OH$, $H_3C{\equiv}_3{=}CH_2OH$, $HOCH_2{\equiv}_3CH_2\underset{OH}{C}HCO_2H$, $HOCH_2{\equiv}_2{=}CH_2\underset{OH}{C}HCO_2H$, $HO_2C{\equiv}_3CH_2{=}(CH_2)_3CO_2H$

TABLE XLI—Continued

Basidiomycetes	Constituents
P. subacida Pk.[375]	$H{\equiv}_2{=}{=}\underset{\mid}{C}H(CH_2)_2CO_2R$ (R = H, CH_3), OH 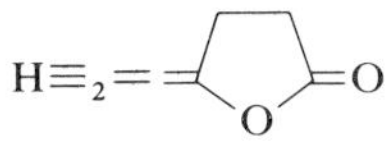
P. tenuis[140,370]	$H{\equiv}_2{=}{=}\underset{\mid}{C}H(CH_2)_2CO_2H$, 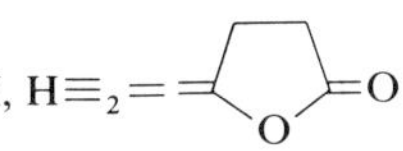 OH
Psilocybe sarcocephala[22]	$H{\equiv}_3{=}CO_2H$
Tricholoma grammopodium (Bull.) Fr.[153,188]	$HOCH_2{\equiv}_3{=}CH_2OH$, $H_3C{\equiv}_3{=}CH_2OH$, $HOCH_2{=}{\equiv}_2{=}CH_2OH$, $H_5C_2\underset{\mid}{C}H{\equiv}_2{=}CH_2OH$ OH
T. nudum (Bull.) Fr.[22]	$NC{\equiv}_2{=}CO_2H$, $HOCH_2{\equiv}_3{=}CO_2H$
T. panaeolum (Fr.) Quel.[22]	$HOCH_2{\equiv}_3{=}CO_2H$, $HO_2C{=}{\equiv}_2{=}CO_2H$
T. sordidum[22]	$NC{\equiv}_2{=}CO_2H$

4. General Chemotaxonomical Aspects

Surveying the results of the investigation of plants for acetylenes, the situation still appears to be complicated. Certainly, the occurrence of an acetylenic compound, in general, is not really important in plant taxonomy, but if we take a more detailed look at the results already available several conclusions become apparent; these might be important for plant systematics and phylogenetics. If we compare the accepted taxonomy with the distribution of a natural product or of a group of related products, we have the following question, stated quite simply: in the evolution of families and species, have certain chemical structures, which require a complex gene-determined enzyme system, descended in close association with the morphology on which the plant systematic is based, or have they descended independently? To answer such a question clearly is not very easy. However, it seems likely that some correlations between morphology and chemical constituents should indeed exist since we now know of many examples which clearly show such relationships in very different species and with completely different chemical constituents. However, as with the morphology, it is always possible that distinct chemical characters might have been developed at different stages in the evolution. If we look at the biogenesis of the natural acetylenes, it is obvious that the existence of these compounds in the plant kingdom is due to the development of an enzyme system which dehydrogenates the olefinic bonds. This capacity seems to be relatively widespread, though clearly not enough results are available. However, if we look for special groups of more complicated structures, many more enzyme systems have to cooperate in their biogenesis and the probability that these enzyme systems have been developed completely independently is much lower. Knowing the exact biogenesis of all natural acetylenes and knowing their distribution, it should be easy to establish the phylogenatic relationships of the families, tribus, genera and species. Unfortunately, we are not in such a position, although we already know a great deal about the biogenesis and the distribution of acetylenes.

If we look at the structures of the known natural acetylenes in connection with their biogenesis (as far as this is known), or if we can suggest probable pathways, the compounds must be divided into at least three groups.

1. Acetylenes formed by simple dehydrogenation of present olefinic compounds.
2. Compounds formed from stearolic acid.
3. Compounds formed from linolic and crepenynic acids.

Clearly, the preconditions for the occurrence of acetylenes of group 1 are much less than for those of group 2, while those for group 3 are the farthest going ones. Therefore, if we consider the chemotaxonomic importance of natural acetylenes those of group 3 are clearly the most useful ones. We can then define the special groups with more accuracy, increasing the number of necessary biochemical reaction steps, and leading to at least one special structure. However, it seems to be clear that special enzyme systems can sometimes operate with different substrates and also that different chemical structures (having common special features) are important in chemotaxonomy, e.g. the formation of enolthioethers in the genus *Anthemis*. Clearly, the capacity of the transformation of dehydromatricariaester and matricariaester to the corresponding thioethers is chemotaxonomically important. The occurrence of such compounds thus indicates a close botanical relationship. Since we know that the biogenesis of these thioethers is a two step reaction, the thiophene esters related to the dehydromatricariaester which are very probably formed from a common precursor and are also important in taxonomy.

The number of similar examples is already extremely large, especially in the family Compositae. Therefore, several genera are also well established by chemotaxonomic considerations, as already discussed in Sections 1.1–1.9.

However, as was also already mentioned, several genera do not appear to be very uniform—mostly those which are also complicated botanically. In some cases, separations seem to be indicated, especially in the cases of the large genera such as *Aster*, *Anthemis*, *Matricaria*, *Chrysanthemum*, *Artemisia* and *Centaurea*. Relationships between genera are also often very pronounced.

The resulting situation in the tribus Anthemideae should be pointed out. The following scheme shows the proposed connections which are generally in agreement with botanical aspects. Still questionable, however, are the positions of *Lasiospermum*, *Athanasia* and *Ursinia*, all of which contain no acetylenes but do contain similar sesquiterpene-furans.

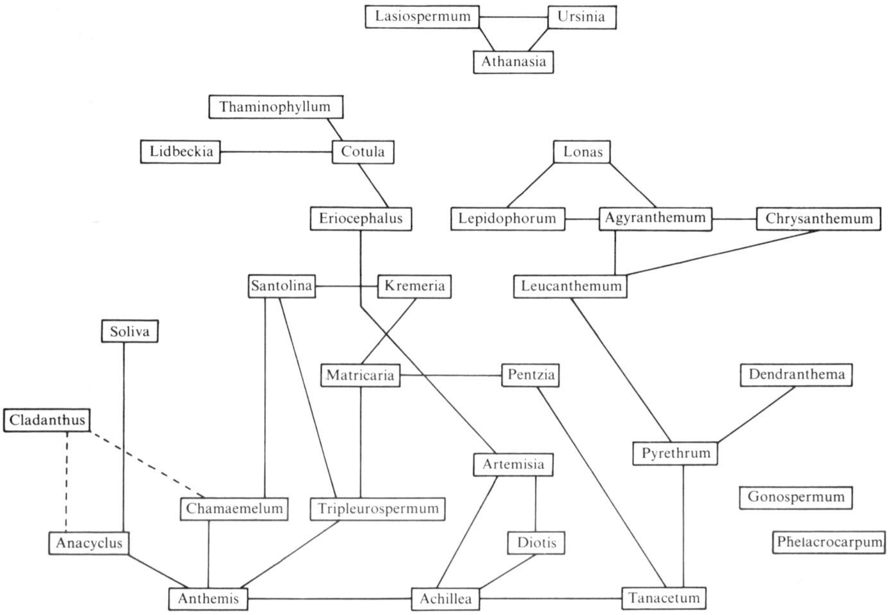

At the tribus level, the situation is more complicated. Only the tribus Anthemideae and Astereae are clearly separated from the others while both show close relationships, as C_{10}-acetylenes are only common in these two tribus. Some links between Anthemideae and Heliantheae are visible as dehydrofalcarinone is sometimes present in both tribus. The Heliantheae are related to Helenieae, Inuleae and Cynareae by the cooccurrence of several acetylenes. In the tribus Senecioneae, only some genera contain acetylenes (*Arnica*, *Doronicum* and *Gamolepis*); the other genera of this tribus also seem to be a more separated group in connection with their terpenoid compounds. The concentrations of acetylenes in the tribus Vernonieae, Eupatorieae, Calendulaceae, Arctotideae and Mutisieae are very low and the structure variation is very limited. Relationships with the tribus already discussed are thus not so clear. However, the occurrence of the widespread pentayn-ene with the ene-tetrayn-ene hydrocarbons indicates some general relations. The tribus Cichorieae has a special position. Special C_{10}-acetylenes have only been isolated in a few species so far. Perhaps the isolation of C_{10}-acetylenes would indicate some relation between Astereae Anthemideae; however, more results are necessary.

The results may be represented best in the following scheme*:

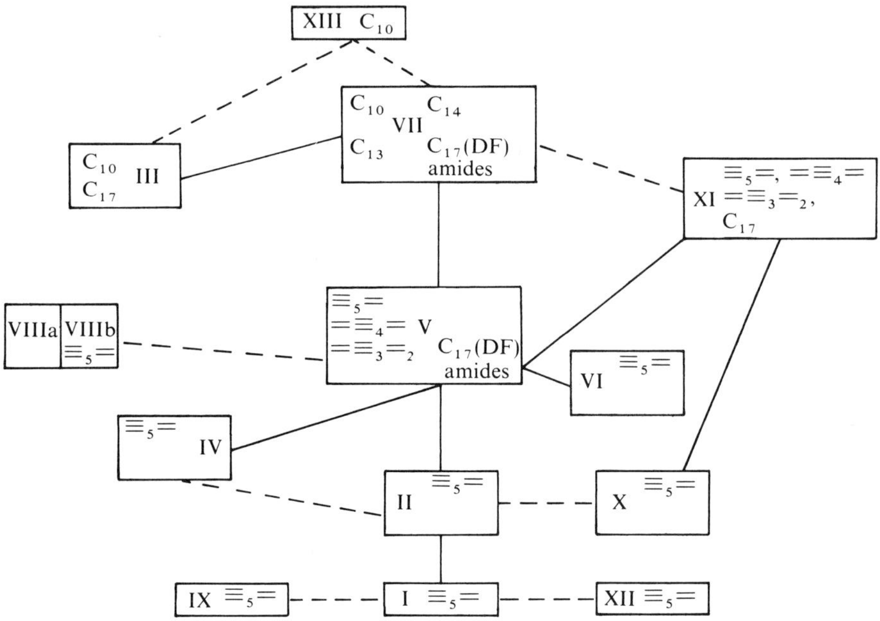

Some of these lines are also supported by the cooccurrence of other special natural products, such as sesquiterpenes, benzofuranes, benzopyranes and other aromatic compounds.

If we look at the acetylenes isolated from other families, it is obvious that the Umbelliferae, the Araliaceae, and the Campanulaceae are closely related to the Compositae. However, clear differences can be recognised. In the Umbelliferae C_{17}-acetylenes, falcarinone and related compounds are chiefly present, while in the Compositae we only have several species with dehydrofalcerinone—clearly differentiated by the vinyl endgroup. In the Araliaceae, only falcarinone and related acetylenes have so far been found, thus indicating the relationship with the Umbelliferae. The species of the Pittosporaceae

* Tribus numbering following Engler–Prantl, only typical chemical characters given.

investigated seems to establish the proposed relationship between this family and the Araliaceae.

The nature of the acetylenes from the Campanulaceae again shows the botanically accepted close relationship to the Compositae. However, the occurrence of tetrahydropyranes seems to be restricted to this family, although the general structural features are similar to those of the Compositae.

A further group of related families also possesses the common feature of the occurrence of stearolic acid and further dehydrogenation products in their triglycerides. The Santalaceae, Olacaceae, Loranthaceae and Opiliaceae, which have these acetylenes are also related botanically, while systematically the Simarubaceae are not normally placed so close to these families. Therefore, the occurrence of tariric acid, probably formed by dehydrogenation of petroselinic acid (also present in this family), might be an indication that there is no close relationship.

From higher plants the following further families are known to contain species with acetylenic compounds: Lauraceae, Papilionaceae, Cupressaceae, Myoparaceae, Sapindaceae, Anonaceae and, possibly, Euphorbiaceae and Valerianaceae.

The botanical phylogenetic systems are partially very different and it is therefore very difficult to discuss the occurrence of acetylenes in connection with the botanical classification. From the isolated chemical constituents and partly from the accepted botanical systematics, a scheme can be drawn up showing the proposed relationships.

5. Proposed Chemotaxonomic Relationship of the Families with Acetylenic Compounds

It is generally accepted that the Anonaceae are closely related to the oldest plants, the Magnoliaceae. However, the evolution of the Araliaceae, Umbelliferae and Compositae is treated very differently. The occurrence of coumarins and chromenes in both Rutaceae and Compositae as well as dienoic amides might be an indication that the suggested schemes are probable. These ideas are also suggested by R. Hegnauer[376] following further observations, e.g. also the cooccurrence of petroselinic acid in Rutaceae and Umbelliferae. The Araliaceae sometimes are placed far away from the Umbelliferae due to the woody nature of the species of this family. However, many botanical and other properties indicate a relationship with the Umbelliferae[376]. The same is true of Pittosporaceae[376], but the Campanulaceae are again always placed near Asterales. The connection between Anonaceae and Santalales is only indicated by the probable structures of the acetylenic triols isolated from an Anona species. Investigation of algae and

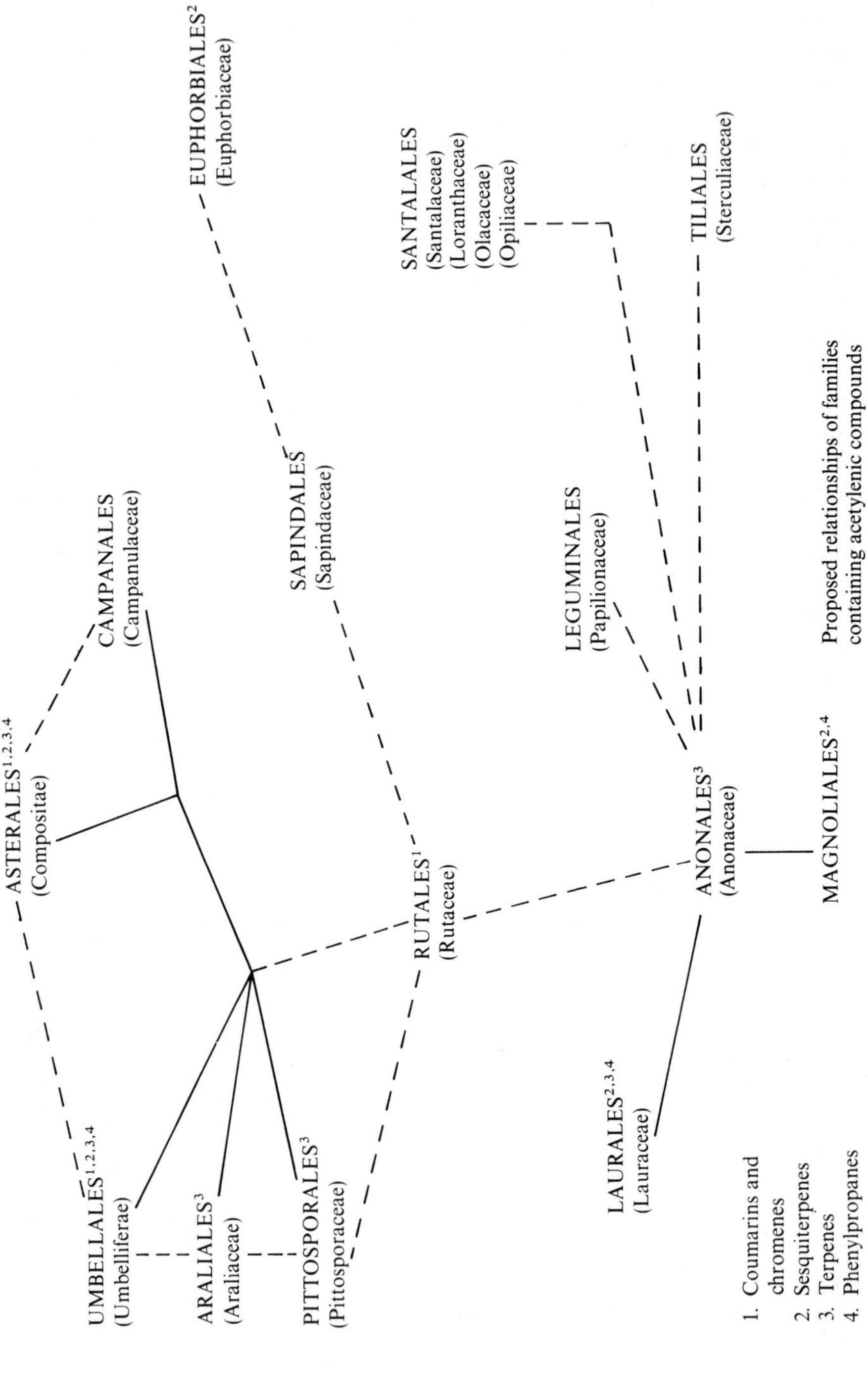

Proposed relationships of families containing acetylenic compounds

marine organism is still in its formative stages and so no detailed discussion is possible. All structures are very special however, and thus no biogenetical connections to the higher plants are visible.

Acetylenes from micro-organisms however show many relationships to those of Compositae. Crepinic acid is once again the common precursor. The nature of the isolated acetylenes, however, is normally quite different from those of the Compositae acetylenes, since most of the micro-organism compounds are highly polar with much greater oxygen functions. Furthermore, the chain lengths are normally C_{10} or less. Nevertheless, the general biogenetic pathway are very similar except for the lower degree of structure variations. Only in one case have thiophene acetylenes been found.

If we consider the distribution of the isolated acetylenes it appears that, with a few exceptions, all compounds have been isolated from higher fungi belonging to the Basidiomycetes, and mainly to the two families Agaricaceae and Polysporaceae. However, we should not forget that nearly all the compounds have been isolated from the culture medium of fungi and not from natural growing fungi. It is also well known that the production of fungal metabolites depends widely on the composition of the culture medium. It is therefore difficult to say whether in cases where no acetylenes have been found, perhaps under special culture conditions, such compounds may also be produced even more generally by other micro-organisms in which acetylenes have not yet been isolated.

7

Physiological and Pharmacological Aspects

Little is known about the plant physiology of acetylenes, but there are several indications that these compounds may be linked with special hormonal functions in the plant metabolism. In many cases, observations have been made showing that the concentrations of special acetylenes varies drastically during the vegetation period. This can be illustrated by the fact that the roots of *Anthemis tinctoria* L. in very young plants contain no acetylenes at all. Then the main compound is the dihydro-thio-enolether ester **399** (see Chapter 2, Section 5.2) and only traces of the aromatic cyclisation product **430** (see Chapter 2, Section 5.3). When the plants are in full blossom **430** is the main compound in a 100 times higher concentration.

The roots of *Helianthus annus* L. have the pentayn-ene **94** (see Chapter 2, Section 2.6), which disappears when the fruits begin to ripen.

Similar observations could be listed for many other plants, e.g. the composition of the acetylenes in *Aethusa cynapium* varies drastically during the vegetation.

In this case the acetylenes partially went into the fruits, while in many other cases no acetylenes are detectable in the seeds, as in the *Centaurea* species. However, if we take *Centaurea* species seeds and let them germinate, we can extract the acetylenes from the seedlings one day later even in etiolated ones. It is not very probable that at this stage the plants produce compounds useful in plant metabolism. Further observations may be of interest in connection with the usefulness of the acetylenes. If we follow the metabolism of an acetylene labelled with ^{14}C, we can estimate the half-life. Normally we find that this is only of the order of 6–48 hours, indicating a relatively rapid metabolism. But there are probably large differences and only about ten examples were really tested.

The absolute concentrations of acetylenes in different plants are extremely different, varying from $10^{-6}\%$ to 1% of the fresh plant material. Furthermore, the acetylene content of the different parts of the plants can be extremely different; sometimes nearly all the acetylenes occur in the roots and in other cases we find different compounds in the roots and in the aerial parts, or even the same compounds. Another point worth mentioning is the fact that in the seeds we sometimes have acetylenes not present in the other parts of the plant. An example of this is *Pastinaca sativa* L., where we have a C_{18}-aldehyde in the seeds and only falcarinone-type compounds are present in the roots. The seeds of *Crepis foetida* L. have large amounts of crepenynic acid in their triglycerides, while in the plants no acetylenes could be detected. This is perhaps very important, as we know that this acid is the common precursor of the acetylenes. It may be possible that physiologically important compounds are formed with acetylenes as a precursor. It is possible that these active products do not have a triple bond and therefore remain undetected. If there is an excess of the precursors, it is possible that large amounts of special acetylenes are formed, and these are usually detected. However, there are still other possibilities. Some of the acetylenes might only be storage products, other might be protecting compounds and a few might be real precursors of physiological active products. All investigations carried out to elucidate the last problem have failed so far. However, it always appears that the activity of labelled acetylenes introduced into the plants is rapidly transformed into highly polar compounds with structures as yet undetermined. Further studies are necessary in order to throw more light on to the question of whether polar compounds are biologically important.

As previously mentioned, some of the known natural acetylenes show powerful biological activity of various kinds, but so far only a very limited number of these compounds really could be investigated since they are only available in small amounts. As in the case of the fungal acetylenes, the acetylenes from higher plants also often show antibiotic activity, which is sometimes very high, but their toxicity and instability eliminate any practical use of the compounds.

One further point should be mentioned. So far it could be shown in two cases that special infections of plants cause accumulation of antifungal polyynes. The safflower—*Carthamus tinctorius* L.—is often infected by the pathogen *Phytophthora drechsleri* Tucker. In this case, the concentration of the highly antifungal ene-triyn-ene-diol (**413**) (see Chapter 3, Section 5.4) is enhanced twenty fold[267]. The second example is the broad bean *Vicia faba* L., which produces the acetylenic furan ketone wyerone (**308**) (see Chapter 3, Section 3.2) after infection with *Botrytis* in larger concentrations. These observations throw some light on the rôle of polyacetylenes as antifungals.

Several acetylenes are very toxic. For instance the compounds from *Cicuta virosa* L. and *Oenanthe crocata* L. are well known, but relatively small variations in structure lead to nearly non-toxic compounds. Some acetylenes have strong antifungal activities (capilline and related compounds), while others have strong nematodical activity (some of the thiophenes). Several acetylenes occur in plants which are used as food or for flavouring. In some cases these acetylenes are also toxic but the concentrations are too low to be really dangerous, e.g. carrots, where the toxic falcarinole exists in very small amounts. The same is true of parsley and celery roots. Several plants which have had a long history of use in medicine also contain acetylenes, but in no case could correlations really have been established of the observed activities with those of the acetylenes, though relatively little can be done in this field due to lack of availability and stability of these compounds.

References

1. Bretz and Eliesen. "Taschbuch für Chemiker und Apothoker" (1826), p. 61.
2. A. Arnaud. *Bull. Soc. Chim. France* (1892), [3] 233.
3. A. Arnaud. *Bull. Soc. Chim. France* (1902), [3] 489.
4. F. W. Semmler. *Chem. Zeitung* (1889), **13**, 1158.
5(a). H. Gilman, P. R. van Ess and R. R. Burtner. *J. Amer. Chem. Soc.* (1933), **55**, 3461.
5(b). A. S. Pfau, J. Pictet, P. Plattner and B. Susz. *Helv. Chim. Acta* (1955), **18**, 935.
6. W. W. Wiljams, W. S. Smirnow and W. P. Golmow. *Zhur. Obschei Khimij.* (1935), **5**, 1195. [*Chem. Zbl.* (1936), I, 3347.]
7. N. A. Sörensen and J. Stene. *Annalen* (1941), **549**, 80.
8. Sir E. R. H. Jones, *Proc. Chem. Soc.* (1960), 199.
9. N. A. Sörensen, *Proc. Chem. Soc.* (1961), 98.
10. F. Bohlmann, H. Bornowski and C. Arndt. *Fortschr. d. Chem. Forsch.* (1962), **4**, 138.
11. J. D. Bu'Lock. *Progr. in Org. Chem.* (1964), **6**, 86.
12. F. Bohlmann. *Fortschr. d. Chem. Forsch.* (1966), **6**, 65.
13. F. Bohlmann. *Fortschr. d. Chemie org. Naturstoffe* (1967), **XXV**, 1.
14. F. Bohlmann. *Chem. Ber.* (1953), **86**, 657.
15(a). Sir E. R. H. Jones, H. Lee and M. C. Whiting. *J. Chem. Soc.* (1960), 3483.
15(b). D. Walton. Unpublished work.
16. F. Bohlmann, C. Zdero, H. Bethke and D. Schumann. *Chem. Ber.* (1968), **101**, 1553.
17. F. Bohlmann, D. Schumann, H. Bethke and C. Zdero. *Chem. Ber.* (1967), **100**, 3706.
18. F. Bohlmann and H. G. Viehe. *Chem. Ber.* (1955), **88**, 1017.
19. J. D. Bu'Lock, D. C. Allport and W. B. Turner. *J. Chem. Soc.* (1961), 1654.
20. J. D. Bu'Lock and G. N. Smith. *J. Chem. Soc.* (1967), 332.
21. H. W. Sprecher, R. Maier, M. Barber and R. T. Holman. *Biochem.* (1965), **4**, 1856.
22. Sir E. R. H. Jones. Private communication.
23. F. Bohlmann and H. Schulz. *Tetrahedron Lett.* (1968), 4795.
24. J. Trénel. Dissertation, Technical University, Berlin (1971).
25. F. Bohlmann, R. Jente and R. Reinecke. *Chem. Ber.* (1969), **102**, 3283.
26. B. Hellström and N. Löfgren. *Acta Chem. Scand.* (1952), **6**, 1024.
27. N. Löfgren. *Acta Chem. Scand.* (1949), **3**, 82.
28. K. Stavholt and N. A. Sörensen. *Acta Chem. Scand.* (1950), **4**, 1567.
29. F. Bohlmann, E. Inhoffen and P. Herbst. *Chem. Ber.* (1957), **90**, 1661.
30. F. Bohlmann and T. Burkhardt. *Chem. Ber.* (1969), **102**, 1702.

31. R. Reinecke. Dissertation, Technical University, Berlin (1970).
32. F. Bohlmann and K.-M. Rode. *Chem. Ber.* (1967), **100**, 1940.
33. Sir E. R. H. Jones and J. Lee. *J. Chem. Soc.* (*C*). (1970), 314.
34. F. Bohlmann, E. Inhoffen and P. Herbst. *Chem. Ber.* (1957), **90**, 124.
35. F. Bohlmann. Unpublished work.
36. F. Bohlmann, K.-M. Rode and C. Zdero. *Chem. Ber.* (1966), **99**, 3544.
37. F. Bohlmann, L. Fanghänel, K.-M. Kleine, H.-D. Kramer, H. Mönch and J. Schuber. *Chem. Ber.* (1965), **98**, 2596.
38. F. Bohlmann, R. Jente, W. Lucas, J. Laser and H. Schulz. *Chem. Ber.* (1967), **100**, 3183.
39. W. Chodkiewicz. *Ann. Chimie* (XIII) (1957), **2**, 819.
40. F. Bohlmann and H.-G. Kapteyn. *Chem. Ber.* (1966), **99**, 1830.
41. F. Bohlmann, C. Zdero and H. Bethke. *Chem. Ber.* (1967), **100**, 2523.
42. F. Bohlmann and G. Florentz. *Chem. Ber.* (1966), **99**, 990.
43. F. Bohlmann, C. Arndt and H. Bornowski. *Chem. Ber.* (1960), **93**, 1937.
44. F. Bohlmann and H. Bornowski. *Chem. Ber.* (1961), **94**, 3189.
45. F. Bohlmann, W. v. Kap-herr, L. Fanghänel and C. Arndt. *Chem. Ber.* (1965), **98**, 1411.
46. F. Bohlmann, C. Arndt, K.-M. Kleine and H. Bornowski. *Chem. Ber.* (1965), **98**, 155.
47. F. Bohlmann and K.-M. Kleine. *Chem. Ber.* (1964), **97**, 1193.
48. F. Bohlmann, H. Mönch and P. Blaszkiewicz. *Chem. Ber.* (1967), **100**, 611.
49. F. Bohlmann, H.-D. Kramer and G. Ertingshausen. *Chem. Ber.* (1965), **98**, 2605.
50. F. Bohlmann, K.-M. Kleine and C. Arndt. *Chem. Ber.* (1964), **97**, 2125.
51. K. E. Schulte, J. Reisch and L. Hörner. *Chem. Ber.* (1962), **95**, 1943.
52. F. Bohlmann, H. Mannhardt and H. G. Viehe. *Chem. Ber.* (1955), **88**, 361.
53. F. Bohlmann and W. Thefeld. *Chem. Ber.* (1969), **102**, 1698.
54. F. Bohlmann, W. Karl and R. Zeisberg. *Chem. Ber.* (1970), **103**, 2860.
55. N. A. Sörensen. *Svensk Kemisk Tidskrift* (1960), **72**, 265.
56. F. Bohlmann, H. Jastrow, G. Ertingshausen and D. Kramer. *Chem. Ber.* (1964), **97**, 801.
57. F. Bohlmann, K.-M. Kleine and C. Arndt. *Chem. Ber.* (1966), **99**, 1642.
58. F. Bohlmann and K.-M. Kleine. *Chem. Ber.* (1963), **96**, 588.
59. S. C. Cascon, W. B. Mors, B. M. Tursch, R. T. Aplin and L. J. Durham. *J. Amer. Chem. Soc.* (1965), **87**, 5237.
60. C. Chin, M. C. Cutler, Sir E. R. H. Jones, J. Lee, S. Safe and V. Thaller. *J. Chem. Soc.* (*C*) (1970), 314.
61. F. Bohlmann and U. Niedballa. *Chem. Ber.* (1967), **100**, 3703.
62. F. Bohlmann, P. Herbst, C. Arndt, H. Schönowsky and H. Gleinig. *Chem. Ber.* (1961), **94**, 3193.
63. F. Bohlmann, P. Herbst and I. Dohrmann. *Chem. Ber.* (1963), **96**, 226.
64. F. Bohlmann, B. Diedrich, W. Gordon, L. Fanghänel and J. Schneider. *Tetrahedron Lett.* (1965), **19**, 1385.
65. F. Bohlmann, C. Arndt, H. Bornowski, K.-M. Kleine and P. Herbst. *Chem. Ber.* (1964), **97**, 1179.
66. F. Bohlmann and K.-M. Rode. *Chem. Ber.* (1966), **99**, 2416.
67. F. Bohlmann and H. Kapteyn. *Chem. Ber.* (1967), **100**, 1927.
68. F. Bohlmann, H. Bornowski and H. Schönowsky. *Chem. Ber.* (1962), **95**, 1733.
69. N. A. Sörensen and J. S. Sörensen. *Acta Chem. Scand.* (1958), **12**, 765.
70. F. Bohlmann, H. Bonnet and R. Jente. *Chem. Ber.* (1968), **101**, 855.

71. F. Bohlmann, H. Bornowski and K. M. Kleine. *Chem. Ber.* (1964), **97**, 2135.
72. J. S. Sörensen and N. A. Sörensen. *Acta Chem. Scand.* (1958), **12**, 771.
73. F. Bohlmann and C. Zdero. *Chem. Ber.* (1968), **101**, 3242.
74. F. Bohlmann, M. Grenz, M. Wotschokowsky and E. Berger. *Chem. Ber.* (1967), **100**, 2518.
75. J. S. Sörensen, D. Holme, E. T. Borlaug and N. A. Sörensen. *Acta Chem. Scand.* (1954), **8**, 1769.
76. E. R. H. Jones, L. Skatteböl and M. C. Whiting. *J. Chem. Soc.* (1958), 1054.
77. F. Bohlmann, C. Arndt, K.-M. Kleine and M. Wotschokowsky. *Chem. Ber.* (1965), **98**, 1228.
78. F. Bohlmann and E. Bresinsky. *Chem. Ber.* (1967), **100**, 1209.
79. F. Bohlmann and H. Mönch. *Chem. Ber.* (1967), **100**, 1944.
80. K. E. Schulte, J. Reisch and G. Rücker. *Arch. Pharm.* (1963), **296**, 273.
81. F. Bohlmann and K.-M. Kleine. *Chem. Ber.* (1963), **96**, 1229.
82. F. Bohlmann and A. Seyberlich. *Chem. Ber.* (1965), **98**, 3015.
83. F. Bohlmann, K.-M. Rode and C. Zdero. *Chem. Ber.* (1967), **100**, 537.
84. F. Bohlmann and U. Hinz. *Chem. Ber.* (1965), **98**, 876.
85. F. Bohlmann and K.-M. Kleine. *Chem. Ber.* (1966), **99**, 2096.
86. F. Bohlmann and C. Zdero. *Chem. Ber.* (1970), **103**, 834.
87. R. D. Stevens and C. E. Castro. *J. Org. Chem.* (1963), **28**, 3313.
88. F. Bohlmann, P. Blaszkiewicz and E. Bresinsky. *Chem. Ber.* (1968), **101**, 4163.
89. F. Bohlmann and K.-M. Kleine. *Chem. Ber.* (1965), **98**, 3081.
90. F. Bohlmann and E. Bresinsky. *Chem. Ber.* (1967), **100**, 107.
91. F. Bohlmann and E. Berger. *Chem. Ber.* (1965), **98**, 883.
92. J. H. Uhlenbroek and J. D. Bijloo. *Rec. trav. Chim.* (1959), **78**, 382.
93. F. Bohlmann and P. Herbst. *Chem. Ber.* (1962), **95**, 2945.
94. R. Atkinson, R. Curtis and G. Phillips. *J. Chem. Soc.* (1965), 1101.
95. F. Bohlmann, C. Zdero and W. Gordon. *Chem. Ber.* (1967), **100**, 1193.
96. L. Zechmeister and J. Sease. *J. Amer. Chem. Soc.* (1947), **69**, 270.
97. F. Bohlmann, M. Wotschokowsky, U. Hinz and W. Lucas. *Chem. Ber.* (1966), **99**, 984.
98. F. Bohlmann, M. Wotschokowsky, J. Laser, C. Zdero and K.-D. Bach. *Chem. Ber.* (1968), **101**, 2056.
99. F. Bohlmann, U. Hinz, A. Seyberlich and J. Repplinger. *Chem. Ber.* (1964), **97**, 809.
100. F. Bohlmann, W. Sucrow, H. Jastrow and H.-J. Koch. *Chem. Ber.* (1961), **94**, 3179.
101. F. Bohlmann and H. Jastrow. *Chem. Ber.* (1962), **95**, 2939.
102. F. Bohlmann and C. Arndt. *Chem. Ber.* (1965), **98**, 1416.
103. F. Bohlmann, C. Arndt and C. Zdero. *Chem. Ber.* (1966), **99**, 1648.
104. F. Bohlmann and W. Skuballa. *Chem. Ber.* (1971), **104**, 1962.
105. F. Bohlmann, C. Arndt, H. Bornowski, H. Jastrow and K.-M. Kleine. *Chem. Ber.* (1962), **95**, 1320.
106. F. Bohlmann and C. Zdero. *Chem. Ber.* (1970), **103**, 1879.
107. F. Bohlmann and H. Jastrow. *Chem. Ber.* (1962), **95**, 1742.
108. F. Bohlmann and C. Zdero. *Chem. Ber.* (1971), **104**, 958.
109. F. Bohlmann and C. Zdero. *Chem. Ber.* (1971), **104**, 954.
110. F. Bohlmann, W. Lucas, J. Laser and P.-H. Bonnet. *Chem. Ber.* (1968), **101**, 1176.
111. F. Bohlmann, H. Bornowski and P. Herbst. *Chem. Ber.* (1961), **93**, 1931.
112. P. K. Christensen. *Norges Tekniske Vitenskapsakademie* (1959), **2**, 7.

113. F. Bohlmann and T. Burkhardt. *Chem. Ber.* (1968), **101**, 861.
114. F. Bohlmann and H. Bethke. *Chem. Ber.* (1971), **104**, 11.
115. F. Bohlmann, H. Bornowski and C. Arndt. *Annalen* (1963), **668**, 51.
116. F. Bohlmann and K.-M. Kleine. *Chem. Ber.* (1962), **95**, 602.
117. F. Bohlmann, K.-D. Albrecht and G. Schmidt. *Chem. Ber.* (1966), **99**, 2822.
118. F. Bohlmann and K. Prezewowsky. *Chem. Ber.* (1964), **97**, 1176.
119. R. Harada, S. Noguchi and N. Sugiyana. *J. Chem. Soc. Japan* (1960), **81**, 654.
120. F. Bohlmann and K.-M. Kleine. *Chem. Ber.* (1962), **95**, 39.
121. F. Bohlmann and C. Zdero. *Chem. Ber.* (1970), **103**, 2856.
122. F. Bohlmann and R. Jente. *Chem. Ber.* (1966), **99**, 995.
123. F. Bohlmann and H. Mönch. *Chem. Ber.* (1969), **102**, 3298.
124. R. Harada. *Nippon Kagahu Zasshi* (1957), **78**, 415.
125. K. Imai. *J. Pharm. Soc.* (*Japan*) (1956), **76**, 405.
126. F. Bohlmann, H. Bornowski and D. Kramer. *Chem. Ber.* (1963), **96**, 584.
127. R. Harada. *Nippon Kagahu Zasshi* (1956), **77**, 990.
128. R. Harada. *Nippon Kagahu Zasshi* (1956), **77**, 1036.
129. F. Bohlmann and C. Zdero. *Chem. Ber.* (1966), **99**, 2413.
130. F. Bohlmann and C. Zdero. *Chem. Ber.* (1971), **104**, 1375.
131. F. Bohlmann and J. Schuber. *Chem. Ber.* (1969), **102**, 4209.
132. F. Bohlmann and C. Arndt. *Chem. Ber.* (1966), **99**, 135.
133. F. Bohlmann and C. Zdero. *Chem. Ber.* (1968), **101**, 2062.
134. R. C. Cambie, J. N. Gardner, E. R. H. Jones, G. Lowe and G. Read. *J. Chem. Soc.* (1963), 2056.
135. W. D. Celmer and I. A. Solomons. *J. Amer. Chem. Soc.* (1952), **74**, 1870.
136. F. Bohlmann and H. Viehe. *Chem. Ber.* (1954), **87**, 712.
137. F. Bohlmann and W. Sucrow. *Chem. Ber.* (1964), **97**, 1846.
138. Sir E. R. H. Jones, G. Lowe and P. V. R. Shannon. *J. Chem. Soc.* (*C*) (1966), 139.
139. J. D. Bu'Lock, E. R. H. Jones and P. R. Leeming. *J. Chem. Soc.* (1957), 1097.
140. J. D. Bu'Lock, E. R. H. Jones and P. R. Leeming. *J. Chem. Soc.* (1955), 4270.
141. J. N. Gardner, G. Lowe and G. Read. *J. Chem. Soc.* (1961), 1532.
142. J. D. Bu'Lock, E. R. H. Jones, P. R. Leeming and J. M. Thompson. *J. Chem. Soc.* (1956), 3767.
143. Sir E. R. H. Jones, J. B. Jones, L. Skatteböl and M. C. Whiting. *J. Chem. Soc.* (1960), 3489.
144. F. Bohlmann, P. Herbst and H. Gleinig. *Chem. Ber.* (1961), **94**, 948.
145. J. D. Bu'Lock and H. Gregory. *J. Chem. Soc.* (1960), 2280.
146. R. E. Bew, J. R. Chapman, Sir E. R. H. Jones, B. E. Lowe and G. Lowe. *J. Chem. Soc.* (*C*) (1966), 129.
147. Sir E. R. H. Jones, P. R. Leeming and W. A. Remers. *J. Chem. Soc.* (1960), 2257.
148. F. Bohlmann and H. Mannhardt. *Chem. Ber.* (1955), **88**, 429.
149. J. S. Sörensen, T. Brūun, D. Holme and N. A. Sörensen. *Acta Chem. Scand.* (1954), **8**, 26.
150. P. K. Christensen and N. A. Sörensen. *Acta Chem. Scand.* (1952), **6**, 602.
151. I. Bell, E. R. H. Jones and M. C. Whiting. *J. Chem. Soc.* (1958), 1313.
152. R. Bentley, Sir E. R. H. Jones and V. Thaller. *J. Chem. Soc.* (*C*) (1969), 1096.
153. J. N. Gardner, E. R. H. Jones, P. R. Leeming and J. S. Stephenson. *J. Chem. Soc.* (1960), 691.
154. F. Bohlmann, C. Zdero and G. Weickgenannt. *Annalen* (1970), **739**, 135.
155. F. Bohlmann, C. Arndt, H. Bornowski and K.-M. Kleine. *Chem. Ber.* (1963), **96**, 1485.

156. F. Bohlmann, K.-M. Kleine, C. Arndt and S. Köhn. *Chem. Ber.* (1965), **98**, 1616.
157. F. Bohlmann, W. v. Kap-herr, C. Rybak and J. Repplinger. *Chem. Ber.* (1965), **98**, 1736.
158. F. Bohlmann and A. Seyberlich. *Chem. Ber.* (1966), **99**, 138.
159. F. Bohlmann and W. Skuballa. *Chem. Ber.* (1970), **103**, 1886.
160. F. Bohlmann and C. Zdero. *Chem. Ber.* (1971), **104**, 1329.
161. F. Bohlmann and H.-C. Hummel. *Chem. Ber.* (1968), **101**, 2506.
162. F. Bohlmann, D. Bohm and C. Rybak. *Chem. Ber.* (1965), **98**, 3087.
163. F. Bohlmann and J. Laser. *Chem. Ber.* (1966), **99**, 1834.
164. F. Bohlmann and R. Reinecke. *Chem. Ber.* (1966), **99**, 3437.
165. F. Bohlmann, K.-M. Kleine and C. Arndt. *Annalen* (1966), **694**, 149.
166. F. Bohlmann, H.-C. Hummel and J. Laser. *Chem. Ber.* (1968), **101**, 3562.
167. F. Bohlmann and C. Zdero. *Chem. Ber.* (1966), **99**, 1226.
168. E. Guddal and N. A. Sörensen. *Acta Chem. Scand.* (1960), **13**, 1185.
169. L. Skatteböl. *Acta Chem. Scand.* (1959), **13**, 1460.
170. F. Bohlmann, K.-M. Kleine and H. Bornowski. *Chem. Ber.* (1962), **95**, 2934.
171. P. Hodge, Sir E. R. H. Jones and G. Lowe. *J. Chem. Soc.* (*C*) (1966), 1216.
172. E. J. McWhorter and M. Anchel. *J. Org. Chem.* (1965), **30**, 2359.
173. Sir E. R. H. Jones and J. S. Stephenson. *J. Chem. Soc.* (1959), 2197.
174. Sir E. R. H. Jones, J. S. Stephenson, W. B. Turner and M. C. Whiting. *J. Chem. Soc.* (1963), 2048.
175. R. C. Cambie, A. Hirschberg, E. R. H. Jones and G. Lowe. *J. Chem. Soc.* (1963), 4120.
176. G. Bendz. *Ark. Kemi* (1960), **15**, 131.
177. A. Landor. International Symposium of Acetylenes, Allenes and Cumulenes, Nottingham. 5–8 July, 1971.
178. M. Anchel. *Arch. Biochem. Biophys.* (1959), **85**, 569.
179. J. D. Bu'Lock, E. R. H. Jones, G. H. Mansfield, J. Thompson and M. C. Whiting. *Chem. and Ind.* (1954), 990.
180. A. T. Glen, S. A. Hutchinson and N. J. McCorkindale. *Tetrahedron Lett.* (1966), 4223.
181. J. H. Birkenshaw and P. Chaplen. *Biochem. J.* (1955), **60**, 255.
182. K. E. Schulte and N. Jantos. *Arch. Pharm.* (1959), **292**, 536.
183. R. F. Curtis and J. A. Tayler. *J. Chem. Soc.* (1969), 1813.
184. P. Hodge. *J. Chem. Soc.* (*C*) (1966), 1617.
185. K. L. Mikolajczek, C. R. Smith, Jr., M. O. Bagby and I. A. Wolff. *J. Org. Chem.* (1964), **29**, 318.
186. R. W. Bradshaw, A. C. Day, Sir E. R. H. Jones, C. B. Page, V. Thaller and (in part) R. A. Vere Hodge. *J. Chem. Soc.* (*C*) (1971), 1156.
187. R. G. Powell, C. R. Smith, Jr., C. A. Glass and I. A. Wolff. *J. Org. Chem.* (1965), **30**, 610.
188. J. D. Bu'Lock and H. M. Smalley. *J. Chem. Soc.* (1962), 4662.
189. F. Bohlmann, K.-M. Kleine and H. Bornowski. *Chem. Ber.* (1966), **99**, 142.
190. Sir E. R. H. Jones, S. Safe and V. Thaller. *J. Chem. Soc.* (*C*) (1966), 1220.
191. F. Bohlmann, U. Niedballa and J. Schneider. *Chem. Ber.* (1965), **98**, 3010.
192. F. Bohlmann and C. Zdero. *Chem. Ber.* (1971), **104**, 961.
193. F. Bohlmann and C. Zdero. *Chem. Ber.* (1967), **100**, 1910.
194. F. Bohlmann and C. Zdero. *Chem. Ber.* (1969), **102**, 1037.
195. F. Bohlmann and H.-C. Hummel. *Chem. Ber.* (1970), **103**, 3419.
196. F. Bohlmann, K.-M. Rode and E. Waldau. *Chem. Ber.* (1967), **100**, 1915.

197. F. Bohlmann and K.-M. Kleine. *Chem. Ber*. (1965), **98**, 872.
198. F. Bohlmann, M. Grenz and U. Niedballa. *Chem. Ber*. (1968), **101**, 532.
199. F. Bohlmann and C. Zdero. *Chem. Ber*. (1969), **102**, 1691.
200. F. Bohlmann, C. Zdero and U. Niedballa. *Chem. Ber*. (1968), **101**, 2987.
201. F. Bohlmann, U. Niedballa and K.-M. Rode. *Chem. Ber*. (1966), **99**, 3552.
202. F. Bohlmann, H. Mönch and U. Niedballa. *Chem. Ber*. (1966), **99**, 586.
203. J. Lam. *Phytochemistry* (1971), **10**, 647.
204. F. Bohlmann, S. Köhn and E. Waldau. *Chem. Ber*. (1966), **99**, 3201.
205. F. Bohlmann and K.-M. Kleine. *Chem. Ber*. (1966), **99**, 590.
206. F. Bohlmann, S. Postulka and J. Ruhnke. *Chem. Ber*. (1958), **91**, 1642.
207. F. Bohlmann and P. Herbst. *Chem. Ber*. (1959), **92**, 1319.
208. F. Bohlmann and C. Zdero. *Chem. Ber*. (1969), **102**, 1034.
209. F. Bohlmann, K.-M. Kleine and C. Arndt. *Chem. Ber*. (1965), **98**, 1225.
210. F. Bohlmann, C. Arndt, H. Bornowski and K.-M. Kleine. *Chem. Ber*. (1961), **94**, 958.
211. M. Takahashi and M. Yoshikura. *J. Pharmac. Soc. Japan* (1964), **84**, 752, 757.
212. D. G. Crosby and N. Aharonson. *Tetrahedron* (1967), **23**, 465.
213. R. K. Bentley and V. Thaller. *Chem. Comm*. (1967), 439.
214. R. K. Bentley, D. Bhattacharjee, Sir E. R. H. Jones and V. Thaller. *J. Chem. Soc*. (*C*) (1969), 685.
215. F. Bohlmann and K.-M. Rode. *Chem. Ber*. (1968), **101**, 525.
216. F. Bohlmann and K.-M. Rode. *Chem. Ber*. (1968), **101**, 1163.
217. F. Bohlmann, C. Zdero, J. Trénel, P. Hänel and M. Grenz. *Chem. Ber*. (1971), **104**, 1322.
218. F. Bohlmann and P. Hänel. *Chem. Ber*. (1969), **102**, 3293.
219. E. F. L. Anet, B. Lythgoe, M. H. Silk and S. Trippett. *J. Chem. Soc*. (1953), 309.
220. F. Bohlmann and H. Viehe. *Chem. Ber*. (1955), **88**, 1245.
221. J. S. Sörensen, B. Ve, T. Anthonsen and N. A. Sörensen. *Austr. J. Chem*. (1968), **21**, 2037.
222. B. E. Hill, B. Lythgoe, S. Mirvish and S. Trippett. *J. Chem. Soc*. (1955), 1770.
223. J. Pohl. *Arch. exper. Path. Pharmak*. (1894), **34**, 265.
224. H. H. Inhoffen, F. Bohlmann and G. Rummert. *Annalen* (1950), **569**, 226.
225. F. Bohlmann and H. Viehe. *Chem. Ber*. (1955), **88**, 1347.
226. F. Bohlmann and C. Zdero. *Chem. Ber*. (1970), **103**, 2327.
227. F. Bohlmann and J. Laser. *Chem. Ber*. (1970), **103**, 2100.
228. F. Bohlmann, H. Bornowski and C. Arndt. *Chem. Ber*. (1966), **99**, 2828.
229. F. Bohlmann and G. Haffer. *Chem. Ber*. (1967), **100**, 3450.
230. F. Bohlmann and K.-M. Rode. *Chem. Ber*. (1968), **101**, 1889.
231. F. Bohlmann and R. Miethe. *Chem. Ber*. (1971), **104**, 1362.
232. H. Oediger and F. Möller. *Angew. Chemie* (1967), **79**, 53.
233. F. Bohlmann, C. Zdero and W. Thefeld. *Chem. Ber*. (1971), **104**, 2030.
234. F. Bohlmann and C. Zdero. *Chem. Ber*. (1970), **103**, 2095.
235. J. Lam. Unpublished work.
236. F. Bohlmann, S. Köhn and C. Arndt. *Chem. Ber*. (1966), **99**, 3433.
237. F. Bohlmann, C. Zdero and P.-H. Bonnet. *Chem. Ber*. (1966), **99**, 3194.
238. F. Bohlmann and C. Zdero. *Chem. Ber*. (1969), **102**, 3765.
239. F. Bohlmann and G. Haffer. *Chem. Ber*. (1968), **101**, 2738.
240. R. K. Bentley, J. K. Jenkins, Sir E. R. H. Jones and V. Thaller. *J. Chem. Soc*. (*C*) (1969), 830.
241. F. Bohlmann and C. Zdero. *Chem. Ber*. (1971), **104**, 2354.

242. C. H. Fawcett, D. M. Spencer, R. L. Wain, A. G. Fallis, Sir E. R. H. Jones, M. M. Le Quen, G. B. Page, V. Thaller, D. C. Shubrock and P. M. Whitham. *J. Chem. Soc.* (*C*) (1968), 2455.
243. R. M. Letcher, D. A. Widdowson, B. J. Deverall and J. W. Mansfield. *Phytochemistry* (1970), **9**, 249.
244. J. S. Sörensen and N. A. Sörensen. *Acta Chem. Scand.* (1958), **12**, 756.
245. F. Bohlmann. *Chem. Ber.* (1955), **88**, 1755.
246. Sir E. R. H. Jones, S. Safe and V. Thaller. *J. Chem. Soc.* (*C*) (1967), 1038.
247. J. S. Sörensen and N. A. Sörensen. *Acta Chem. Scand.* (1954), **8**, 1741.
248. T. Brūun, L. Skatteböl and N. A. Sörensen. *Acta Chem. Scand.* (1954), **8**, 1757.
249. J. R. Fairbrother, Sir E. R. H. Jones and V. Thaller. *J. Chem. Soc.* (*C*) (1967), 1035.
250. F. Bohlmann and H. Bornowski. *Chem. Ber.* (1966), **99**, 1223.
251. F. Bohlmann and K.-M. Rode. *Chem. Ber.* (1967), **100**, 1507.
252(a). F. W. Semmler. *Ber.* (1906), **39**, 726.
252(b). F. W. Semmler. *Ber.* (1909), **42**, 2355.
253. R. Paul. *Compt. Rend.* (1936), **202**, 854.
254. I. Heilbronn, E. R. H. Jones, P. Smith and B. C. L. Weedon. *J. Chem. Soc.* (1946), 54.
255. F. Bohlmann and W. von Kap-herr. *Chem. Ber.* (1966), **99**, 148.
256. F. Bohlmann and H. J. Mannhardt. *Chem. Ber.* (1955), **88**, 1330.
257. F. Bohlmann, C. Arndt, H. Bornowski and P. Herbst. *Chem. Ber.* (1960), **93**, 981.
258. F. Bohlmann, H. J. Koch, S. Köhn and W. Herfurt. *Chem. Ber.* (1964), **97**, 2598.
259. F. Bohlmann, L. Fanghänel, M. Wotschokowsky and J. Laser. *Chem. Ber.* (1968), **101**, 2510.
260. F. Bohlmann and W. Karl. *Chem. Ber.* (1972), **105**, 335.
261. I. Yosioko, H. Hikino and Y. Sasaki. *Chem. and Pharm. Bull.* (*Japan*), **1960**, 949, 952, 957.
262. F. Bohlmann and P. Herbst. *Chem. Ber.* (1958), **91**, 1631.
263. F. Bohlmann, H. Bornowski and S. Köhn. *Chem. Ber.* (1964), **97**, 2583.
264. F. Bohlmann, C. Arndt and J. Starnick. *Tetrahedron Lett.* (1963), 1605.
265. J. S. Sörensen and N. S. Sörensen. *Acta Chem. Scand.* (1954), **8**, 1763.
266. F. Bohlmann and E. Inhoffen. *Chem. Ber.* (1956), **89**, 21.
267. E. H. Allen and C. A. Thomas. *Phytochemistry* (1971), **10**, 1579.
268. F. Bohlmann and W. Skuballa. *Tetrahedron Lett.* (1970), 2109.
269. F. Bohlmann, W. Skuballa, C. Zdero, T. Kühle and P. Steirl. *Annalen* (1971), **745**, 176.
270. E. R. H. Jones, J. M. Thompson and M. C. Whiting. *Acta Chem. Scand.* (1954), **8**, 1757.
271. F. Bohlmann, C. Arndt, H. Bornowski and K.-M. Kleine. *Chem. Ber.* (1962), **95**, 1315.
272. K. E. Schulte, G. Rücker and H. Stigler. *Archiv der Pharmazie* (1970), **303**, 7.
273. F. Pagani and G. Romussi. *Phytochemistry* (1971), 2233.
274. R. E. Atkinson and R. F. Curtis. *Tetrahedron Lett.* (1965), **10**, 297.
275. F. Bohlmann, P.-H. Bonnet and H. Hofmeister. *Chem. Ber.* (1967), **100**, 1200.
276. F. Bohlmann and E. Waldau. *Chem. Ber.* (1967), **100**, 1206.
277. L. Crombie. *J. Chem. Soc.* (1955), 999.
278. L. Crombie and M. Manzoor-I-Khuda. *J. Chem. Soc.* (1957), 2767.
279. F. Bohlmann and E. Inhoffen. *Chem. Ber.* (1956), **89**, 1276.
280. F. Bohlmann and C. Zdero. *Chem. Ber.* (1967), **100**, 104.
281. F. Bohlmann and R. Miethe. *Chem. Ber.* (1967), **100**, 3861.

282. R. S. Burden and L. Crombie. *J. Chem. Soc.* (*C*) (1969), 2477.
283(a). L. Horner, H. Hoffmann and G. Wippel. *Chem. Ber.* (1958), **91**, 61.
283(b). L. Horner, H. Hoffmann, G. Wippel and G. Klahre. *Chem. Ber.* (1959), **92**, 2499.
284. F. Bohlmann and M. Grenz. *Chem. Ber.* (1966), **99**, 3197.
285. L. Crombie. *J. Chem. Soc.* (1955), 995.
286. F. Acree, J. M. Jacobson and H. L. Haller. *J. Org. Chem.* (1947), **12**, 731.
287. K. S. Baalsrud, D. Holme, M. Nestoold, J. Pliva, J. S. Sörensen and N. A. Sörensen. *Acta Chem. Scand.* (1952), **6**, 883.
288. N. A. Sörensen. *Chem. and Ind.* (1953), 240.
289. D. Holme and N. A. Sörensen. *Acta Chem. Scand.* (1954), **8**, 34.
290. T. Brūun, P. K. Christensen, C. M. Haug, J. Stene and N. A. Sörensen. *Acta Chem. Scand.* (1951), **5**, 1244.
291. D. Holme and N. A. Sörensen. *Acta Chem. Scand.* (1954), **8**, 280.
292. F. Bohlmann, W. Thefeld and C. Zdero. *Chem. Ber.* (1970), **103**, 2245.
293. F. Bohlmann and C. Zdero. *Chem. Ber.* (1969), **102**, 1679.
294. F. Bohlmann and C. Zdero. *Chem. Ber.* (1970), **103**, 2853.
295. N. A. Sörensen and K. Stavholt. *Acta Chem. Scand.* (1950), **4**, 1080.
296. P. K. Christensen, N. A. Sörensen, J. Bell, E. R. H. Jones and M. C. Whiting. "Festschrift Arthur Stoll" (1957), p. 545, Birkhäuser Verlag, Basel.
297. F. Bohlmann, H. Bornowski and C. Arndt. *Chem. Ber.* (1965), **98**, 2236.
298. T. Brūun, C. M. Haug and N. A. Sörensen. *Acta Chem. Scand.* (1950), **4**, 850.
299. F. Bohlmann, C. Zdero and H. Kapteyn. *Chem. Ber.* (1969), **102**, 1682.
300. F. Bohlmann, K.-M. Kleine and C. Arndt. *Chem. Ber.* (1964), **97**, 3469.
301. F. Bohlmann and G. Grau. *Chem. Ber.* (1965), **98**, 2608.
302. F. Bohlmann and T. Burkhardt. *Chem. Ber.* (1972), **105**, 521.
303. F. Bohlmann and C. Zdero. *Tetrahedron Lett.* (1970), 2465.
304. F. Bohlmann, K.-M. Kleine and H. Bornowski. *Chem. Ber.* (1965), **98**, 369.
305. P. K. Christensen and N. A. Sörensen. *Acta Chem. Scand.* (1952), **6**, 893.
306. F. Bohlmann and B. M. Tietze. *Chem. Ber.* (1970), **103**, 561.
307. J. D. Bu'Lock, E. R. H. Jones and W. B. Turner. *J. Chem. Soc.* (1957), 1607.
308. R. E. Bew, R. C. Cambie, Sir E. R. H. Jones and G. Lowe. *J. Chem. Soc.* (C) (1966), 135.
309. Sir E. R. H. Jones, B. E. Lowe and G. Lowe. *J. Chem. Soc.* (1964), 1476.
310. F. Bohlmann and U. Hinz. *Chem. Ber.* (1964), **97**, 520.
311. G. C. Barlay, A. C. Day, U. Graf, Sir E. R. H. Jones, I. O'Neill, R. Tachikawa, V. Thaller and R. A. Vere Hodge. *J. Chem. Soc.* (C) (1971), 3308.
312. A. Arnoud. *Bull. Soc. Chim. France* (1902), **27**, 489.
313. P. B. Lumb and J. C. Smith. *J. Chem. Soc.* (1952), 5032.
314. A. Hebert. *Bull. Soc. Chim. France* (1896), **15**, 941.
315. C. Y. Hopkins and M. J. Chrisholm. *Tetrahedron Lett.* (1964), 3011.
316. S. P. Lighthelm, H. M. Schwartz and M. M. van Holdt. *J. Chem. Soc.* (1952), 1088.
317. S. P. Lighthelm. *Chem. and Ind.* (1954), 249.
318. H. H. Hatt, A. C. K. Triffett and P. C. Wailes. *Austr. J. Chem.* (1959), **12**, 190; (1960), **13**, 488.
319. J. D. Bu'Lock and G. N. Smith. *Bioch. J.* (1962), 35.
320. L. Crombie and J. C. Williams. *J. Chem. Soc.* (1962), 2449.
321(a). F. D. Gunstone and A. J. Sealy. *J. Chem. Soc.* (1963), 5772.
321(b). L. J. Morris. *J. Chem. Soc.* (1963), 5779.
322. A. Seher. *Annalen* (1954), **589**, 222.

323. R. G. Powell, C. R. Smith, Jr., C. A. Glass and I. A. Wolff. *J. Org. Chem.* (1966), **31**, 528.
324. J. D. Bu'Lock and G. N. Smith. *J. Chem. Soc.* (C) (1967), 332.
325. A. W. Jevans and C. Y. Hopkins. *Tetrahedron Lett.* (1968), 2167.
326. N. Prista. *Anais. Fac. Farm. Porto.* (1954), **14**, 19.
327. K. Ishibashi, K. Nose, T. Shindo, M. Arai and H. Mishina. *Ann. Report Sankyo Res. Lab.* (1968), **20**, 76.
328. R. A. Massy-Westropp, G. D. Reynolds and T. M. Spotswood. *Tetrahedron Lett.* (1966), 1939.
329(a). T. Nozoe, Y. S. Cheng and T. Toda. *Tetrahedron Lett.* (1966), 3663.
329(b). T. Asao, S. Ibe, K. Takase, Y. S. Cheng and T. Nozoe. *Tetrahedron Lett.* (1968), 3639.
329(c). T. Takase, S. Ibe, T. Asao, T. Nozoe, H. Shimanonchi and Y. Sasada. *Chem. and Ind.* (1968), 1638.
330. B. C. L. Weedon. *Rev. Pure and Appl. Chem.* (1970), 20.
331. S. Isbe, S. Katsumata, S. Be Hyeon and T. Sakan. *Tetrahedron Lett.* (1971), 1089.
332. A. K. Mallams, E. S. Waight, B. C. L. Weedon, D. T. Chapman, F. T. Haxo, T. W. Goodwin and D. M. Thomas. *Chem. Comm.* (1967), 301.
333. S. A. Campbell, A. K. Mallams, E. S. Waight, B. C. L. Weedon, M. Barbier, E. Lederer and A. Salaque. *Chem. Comm.* (1967), 941.
334. K. Aitzemüller, W. A. Svec, T. J. Katz and H. H. Strain. *Chem. Comm.* (1968), 32.
335. K. Egger, H. Nitsche and K. Kleinig. *Phytochemistry* (1969), **8**, 1583.
336. N. A. Sörensen, S. L. Jensen, B. Bordalen, A. Haug, C. Anzell and G. Francis. *Acta Chem. Scand.* (1968), **22**, 344.
337. W. S. Matthews, G. B. Pickering and A. T. Umoh. *Chem. and Ind.* (1963), 122.
338. Y. Kashman, I. Neeman and A. Lifshitz. *Tetrahedron* (1969), **25**, 4617.
339. W. F. Lange, *Proc. Soc. Exp. Biol. Med.* (1931–32), **29**, 1134.
340. S. Suzuki, G. Nakamara, K. O. Kuma and Y. Tomiyama. *J. Antibiotics* (*Japan*) (1958), **11A**, 81.
341(a). T. Irie, M. Izawa and E. Kurosawa, *Tetrahedron Lett.* (1968), 2091, 2735.
341(b). T. Irie, M. Izawa and E. Kurosawa, *Tetrahedron* (1970), **26**, 851.
341(c). T. Irie, M. Suzuki and T. Masume. *Tetrahedron* (1968), **24**, 4193.
342. E. Klenk and W. Knippraith. *Z. physiol. Chem.* (1959), **317**, 243.
343. M. L. Sung, L. Fowden, D. S. Millington and R. C. Sheppard. *Phytochemistry* (1969), **8**, 1277.
344. J. W. Daly, J. Karle, Ch. W. Myers, T. Tokuyama, J. A. Waters and B. Witkop. *Proc. Nat. Acad. Sci. U.S.A.* (1971), **68**, 1870.
345. R. E. Atkinson and R. F. Curtis. *Phytochemistry* (1971), **10**, 454.
346. G. M. Tronvold, M. Nestvold, D. Holme, J. S. Sörensen and N. A. Sörensen. *Acta Chem. Scand.* (1953), **7**, 1375.
347. K. E. Schulte, J. Reisch and J. Hopmann. *Arch. Pharm.* (1963), **296**, 353.
348. K. E. Schulte, G. Rücker and J. Perlick. *Arzneimittelforschung* (1967), **17**, 825.
349. S. L. Jensen and N. A. Sörensen. *Acta Chem. Scand.* (1961), **15**, 1885.
350. F. Bohlmann and M. Grenz. *Chem. Ber.* (1970), **103**, 90.
351. F. Bohlmann and M. Grenz. *Tetrahedron Letters* (1969), 1005.
352. J. T. Mortensen, J. S. Sörensen and N. A. Sörensen. *Acta Chem. Scand.* (1964), **18**, 2392.
353. J. S. Sörensen and N. A. Sörensen. *Austr. J. Chem.* (1969), **22**, 751.
354. J. S. Sörensen, J. T. Mortensen and N. A. Sörensen. *Acta Chem. Scand.* (1964), **18**, 2182.

355. E. Guddal and N. A. Sörensen. *Acta Chem. Scand.* (1959), **13**, 1185.
356. F. Bohlmann, K.-M. Rode and M. Grenz. *Chem. Ber.* (1967), **100**, 3201.
357. F. Bohlmann and C. Zdero. *Chem. Ber.* (1971), **104**, 1957.
358. R. K. Crown, J. B. Harborne and V. E. Heywood. *Phytochemistry* (1969), **8**, 1963.
359. F. Bohlmann. *Chem. Ber.* (1967), **100**, 3454.
360. F. Bohlmann and W. Thefeld. *Tetrahedron Letters* (1970), 3577.
361. F. Bohlmann and C. Zdero. *Chem. Ber.* (1971), **104**, 1611.
362. C. Grimme. *Chem. Rev. Fett. and Harz. Ind.* (1910), **17**, 158; (1912), **19**, 511.
363. H. H. Hatt, A. Meisters, A. C. K. Triffett and P. C. Wailes. *Austr.* (1967), **20**, 2285.
364. P. C. Wailes. Unpublished work.
365. H. H. Hatt and A. Z. Szumer. *Chem. and Ind.* (1954), 962.
366. F. D. Gunstone and M. A. McGee. *Chem. and Ind.* (1954), 1112.
367. K. E. Schulte, J. Reisch and H. Bornfleth. *Arch. Pharm.* (1964), **297**, 443.
368. K. E. Schulte, J. Reisch and P. Busch. *Arzneimittelforschung* (1964), **14**, 844.
369. P. J. Ashworth, E. R. H. Jones, G. H. Mansfield, K. Schlögl, J. M. Thompson and M. C. Whiting. *J. Chem. Soc.* (1958), 951.
370(a). M. Anchel, J. Polatnick and F. Kavanagh. *Arch. Biochem.* (1950), 208.
370(b). M. Anchel. *J. Am. Chem. Soc.* (1952), **74**, 1388.
371. M. Anchel. *Science* (1955), **121**, 607.
372. M. Anchel, W. B. Silverman, N. Valanju and C. T. Rogerson. *Mycologia* (1962), **54**, 240.
373. J. D. Bu'Lock, D. C. Allport and W. B. Turner. *J. Chem. Soc.* (1961), 1654.
374. M. Anchel and M. P. Cohen. *J. Bioch. Chem.* (1954), 319.
375. R. E. Bew, R. C. Cambie, Sir E. R. H. Jones and G. Lowe. *J. Chem. Soc.* (C) (1966), 135.
376. R. Hegnauer. "Chemotaxonomie der Pflanzen", Vol. 3 (1964), Vol. 4 (1966), Vol. 5 (1969). Birkhaüser Verlag, Basel and Stuttgart.

Author Index

The numbers in parentheses are the reference numbers

A

B

H

I

J

K

S

Y

Z

Botanical Index
(Including some marine organisms)

C

D

I

J

K

L

M

N

O

P

R

S

T